QK
524.4
T7.8
1982

D1074639

Ferns
and Allied Plants

Rolla M. Tryon
Alice F. Tryon
Harvard University

Ferns
and Allied Plants
*With Special Reference
to Tropical America*

Habitat Photography Principally
by Walter H. Hodge

Springer-Verlag
New York Heidelberg Berlin

Rolla M. Tryon
Alice F. Tryon
Harvard University Herbaria
22 Divinity Avenue
Cambridge, MA 02138, U.S.A.

With 2028 figures

Library of Congress Cataloging in Publication Data
Tryon, Rolla Milton, date
    Ferns and allied plants.

    Bibliography: p.
    Includes index.
    1. Ferns—America.   2. Pteridophyta—America.
3. Ferns—Tropics.   4. Pteridophyta—Tropics.   I. Tryon,
Alice F. (joint author) II. Hodge, W. H. (Walter Henricks), 1912-
III. Title.
QK524.4.T78   587′097   82-3248

© 1982 by Springer-Verlag New York Inc.
All rights reserved. No part of this book may be translated or repro-
duced in any form without written permission from Springer-Verlag,
175 Fifth Avenue, New York, New York 10010, U.S.A.

The use of general descriptive names, trade names, trademarks, etc. in
this publication, even if the former are not especially identified, is not to
be taken as a sign that such names, as understood by the Trade Marks
and Merchandise Marks Act, may accordingly be used freely by anyone.

Typesetting: Progressive Typographers, Inc., Emigsville, PA
Printing: Halliday Lithograph, West Hanover, MA
Binding: A. Horowitz & Sons, Fairfield, NJ

Printed in the United States of America

9 8 7 6 5 4 3 2 1

ISBN 0-387-90672-X   Springer-Verlag New York Heidelberg Berlin
ISBN 3-540-90672-X   Springer-Verlag Berlin Heidelberg New York

Preface

This systematic treatment of the ferns and allied plants provides a modern classification of the Pteridophyta based on an assessment of the wealth of new data published during the last few decades as well as on our own research. The accounts of the genera include systematics, ecology, geography, spores and cytology and often other aspects of their biology. The scope of the work has involved all genera and the Old World species of those genera represented in America. For a few American genera, that are especially complex in the Old World, it has been necessary to limit their treatment to the American representatives and this is specifically indicated in those cases. The eight American genera that do not occur in the tropics are treated in less detail. They are *Camptosorus, Gymnocarpium, Matteuccia* and *Onoclea* of North America, *Arthropteris* and *Thyrsopteris* of the Juan Fernandez Islands, *Pleurosorus* of southern South America, and the amphitropical *Cryptogramma.*

The complete generic nomenclature is included, except in a few specified cases where taxonomic complexities of the Old World elements have not been resolved. The synonyms and accepted names of subgeneric and sectional taxa are included only when pertinent to the generic nomenclature or to an infrageneric classification. Only the name of a publishing author is cited. There is no bibliographic purpose served by including the name of a person who was the source of, but did not validly publish, a name or epithet. Combinations published prior to January 1, 1953 (International Code of Botanical Nomenclature, Article 33) are accepted if there is an indication of the basionym even though it was not directly cited.

Geographic records are based on the collections of the Gray Herbarium and Arnold Arboretum, Harvard University, the New York Botanical Garden, the United States National Herbarium, Smithsonian Institution, and on monographic and floristic literature. The maps present the general distribution of the genera, based on countries and their major political subdivisions. The scale employed is shown in common and widely distributed genera such as *Elaphoglossum* (Fig. 96.4), *Asplenium* (Fig. 97.4), *Blechnum* (Fig. 104.4) and *Lycopodium* (Fig. 125.6). New names of some countries and islands have not been adopted because the previous name is prevalent in the literature and herbarium: Galápagos Islands is used for Archipiélago de Colón, British Honduras for Belize, British Guiana for Guyana, Staten Island for Isla del Estado, and Falkland Islands for Islas del Malvinas.

All of the spore figures are documented by a specimen citation, and these are in the collections at Harvard University unless another herbarium is cited.

Publications that have come to our attention prior to December, 1981 are included in the cited literature.

The text has been prepared jointly, with each author amending initial drafts prepared by the other. The book thus developed as a synthesis of our studies, although the nomenclature is

D. HIDEN RAMSEY LIBRARY
U.N.C. AT ASHEVILLE
ASHEVILLE. N. C. 28814

primarily the work of Rolla M. Tryon and the sections on spores and cytology of Alice F. Tryon.

Most of the materials in this work were developed with the financial support of National Science Foundation Grants G1064, G15949, G18906, G81693, GB4184, GB31170, DEB74-12319, and DEB78-03148. However, any opinions, findings, conclusions, recommendations expressed herein are those of the authors and do not necessarily reflect the views of the Foundation.

In addition to this essential support, we are indebted to the collaboration and assistance of many persons who have aided our work.

The treatment of the genus *Thelypteris* has been prepared with the collaboration of Alan R. Smith and *Lycopodium* and *Isoetes* with the collaboration of R. James Hickey. E. Hennipman has reviewed the treatment of *Platycerium*, K. U. Kramer the treatments of *Odontosoria, Lindsaea* and *Ormoloma*, B. Øllgaard, *Lycopodium,* M. Price, *Loxogramme,* and K. Iwatsuki the Hymenophyllaceae.

Stimulating discussions of fern classification with E. Hennipman, F. Jarrett, C. Jermy, and K. U. Kramer have aided in the resolution of many problems. We are indebted to the courtesies extended, during visits to major herbaria, by F. Badré, Paris, F. Jarrett, Kew, C. Jermy, British Museum (Natural History), D. B. Lellinger, Smithsonian Institution, J. T. Mickel, New York Botanical Garden, and R. G. Stolze, Field Museum of Natural History.

Walter H. Hodge has contributed the major portion of the field photographs, most of them obtained on several trips to the American tropics for this special purpose. His contribution is of major significance to our work. Other habitat photographs, acknowledged in the captions, have been contributed by D. S. Conant, L. D. Gómez, C. Haufler, K. U. Kramer, T. Lucansky, F. G. Meyer, B. Øllgaard, T. Reeves, C. S. Sperling, J. A. Steyermark, W. A. Weber, L. Wells, and P. G. Windisch. We are indebted for this collaboration, which has made it possible to illustrate all but a few genera as they grow in nature.

Other photography and art work has been by Lydia Vickers and principally by Emily Hoffman and Sarah Landry, whose talents have made it possible to provide extensive illustrations of the genera and many of the species. Edward Seling is responsible for the exceptionally fine quality of the SEM micrographs of the spores. We are also indebted to Carolyn Hesterberg, Helen Roca-Garcia, and Brooke Thompson-Mills for providing secretarial and editorial assistance.

Some of the figures of *Adiantopsis, Adiantum, Cheilanthes,* Cyatheaceae, *Doryopteris, Eriosorus, Jamesonia, Notholaena, Pellaea, Selaginella* and Vittariaceae have been previously published by the authors in Annals of the Missouri Botanical Garden 42. 1955 and 44. 1957; in Contributions from the Gray Herbarium of Harvard University 143. 1942, 179. 1956, 191. 1962, 194. 1964 and 200. 1970; and in Rhodora 73. 1971.

Fig. 24.3 (right) is from G. J. Gastony, Contributions from the Gray Herbarium of Harvard University 203. 1973; Figs. 35.3–

35.6 are from C. Haufler, Journal of the Arnold Arboretum 60. 1979; Fig. 45.6 is from E. Scamman, Contributions from the Gray Herbarium of Harvard University 187. 1960; Figs. Cyath. 8, 10–13 are from A. F. Tryon and L. Feldman, Canadian Journal of Botany 53. 1975; Figs. 93.11–93.17 are from E. Hennipman, Leiden Bot. Ser. 2. Leiden University Press. 1977; and Fig. 114.4 is from L. D. Gómez, Brenesia 4. 1974.

The base maps used are from the Goode Series, Copyright by the University of Chicago Geography Department, with permission of the copyright holder.

Cambridge, Massachusetts
December, 1981

Contents

Ferns
and Allied Plants

Introduction

Systematics

The classification of the pteridophytes adopted here includes about 9,000 species, 240 genera and 33 families. Other classifications have been proposed by Christensen (1938), Copeland (1947), Holttum (1949), Crabbe et al. (1975), and Pichi-Sermolli (1977). All of these differ in their systematic arrangement, especially among the ferns previously placed in the Polypodiaceae (families 14–22 of this treatment). Many of the differences are due to the recognition of a taxon at a different rank, as genus, subgenus or section. Others are due to problems of convergent evolution, which can lead to a different alignment of a group. As new data is directed towards solving these problems, a greater consistency in systems of classification can be expected. The recognition of small families, such as Danaeaceae, Mohriaceae and Onocleaceae, and of small genera such as *Osmundopteris, Cheiropleuria* and *Ormopteris* provides homogeneous groups, but does not express their relationships to allied taxa. Recognition of larger families and genera results in more heterogeneous taxa, which permits diversity to be recognized in infrafamilial and infrageneric ranks. The classification followed here generally recognizes larger taxa that allow evolutionary relationships to be most readily expressed.

The classification of pteridophytes has improved as a broader base of characters has been incorporated. Basic morphological and anatomical information is summarized by Bower (1923–1928), Bierhorst (1971) and Ogura (1972). Recent data of special importance has come from cytological work and from scanning electron microscope studies of spores. Current studies on stomatal patterns and phytochemistry supply additional data. New approaches to classification are discussed in Jermy et al. (1973) and data from experimental biology, reviewed by Dyer (1979), may also be incorporated into systematic work.

A general conclusion that has emerged from our studies is that many genera are neither clearly derived from, nor obviously ancestral to another genus. The major problem in depicting phyletic relationships in many groups is the lack of information for assessment of the primitive or advanced state of characters. Much of the direct evidence for the course of evolution has been lost through extinction and the fossil record (Andrews et al. 1970) is inadequate for phyletic analyses. Thus the classification presented here is based on character similarities rather than on presumed phyletic relationships.

Classification of Pteridophyta and American Genera

The following classification includes all families and tribes of pteridophytes within a framework of higher taxa. Those that do not occur in America are unnumbered. All of the genera, subgenera and sections in America are included. The genera that are confined to the Old World for which the classification is adequately known are noted in the systematic accounts of the families or tribes. The number of genera in a family or tribe is indicated, and also the number of species in a genus. The number of native American species is in parentheses, and the names of taxa only in America are in italic.

Division Pteridophyta
  Class Filicopsida
    Subclass Polypodiidae
      Order Ophioglossales
        1. Family Ophioglossaceae 3 genera
            1. Botrychium 25 species, (17 in America).
                Subgenus Sceptridium
                Subgenus Botrychium
                Subgenus Osmundopteris
            2. Ophioglossum 30,(12).
                Subgenus Ophioglossum
                Subgenus Cheiroglossa
      Order Marattiales
        2. Family Marattiaceae 7.
            3. Marattia 40,(8).
            4. *Danaea* (20).
    Order Polypodiales
      Suborder Polypodiineae
        3. Family Osmundaceae 3.
            5. Osmunda 6,(3).
        4. Family Schizaeaceae 4.
            6. Anemia 80,(65).
                Subgenus Coptophyllum
                Subgenus Anemia
                Subgenus *Anemirhiza*
            7. Lygodium 30,(6–8).
            8. Schizaea 30,(8–11).
                Subgenus Schizaea
                Subgenus Actinostachys
        5. Family Gleicheniaceae 2.
            9. Gleichenia 110,(40).
                Subgenus Diplopterygium
                Subgenus Mertensia
            10. Dicranopteris 10,(3).
                Subgenus Dicranopteris
                Subgenus *Acropterygium*
      Family Stromatopteridaceae 1.
      Family Matoniaceae 2.
      Family Dipteridaceae 1.
      Family Cheiropleuriaceae 1.
        6. Family Hymenophyllaceae 2.
            11. Hymenophyllum 300,(130).
                Subgenus Hymenophyllum
                Subgenus Mecodium
                Subgenus Leptocionium
                Subgenus *Hymenoglossum*
                Subgenus *Cycloglossum*
            12. Trichomanes 300,(115).
                Subgenus Trichomanes
                Subgenus Pachychaetum
                Subgenus Didymoglossum
                Subgenus Achomanes
        7. Family Loxomataceae 2.
            13. *Loxsomopsis* (1–2).
        8. Family *Hymenophyllopsidaceae* 1.
            14. *Hymenophyllopsis* (7).

9. Family Plagiogyriaceae 1.
    15. Plagiogyria 15,(1).
10. Family Dicksoniaceae 5.
    16. Culcita 7,(1).
    17. Dicksonia 20,(3).
    18. Cibotium 8,(2).
    19. *Thyrsopteris* (1).
11. Family *Lophosoriaceae* 1.
    20. *Lophosoria* (1).
12. Family *Metaxyaceae* 1.
    21. *Metaxya* (1).
13. Family Cyatheaceae 6.
    22. Sphaeropteris 120,(23).
        Subgenus *Sclephropteris*
        Subgenus Sphaeropteris
    23. Alsophila 230,(13).
    24. *Nephelea* (18).
    25. *Trichipteris* (50).
    26. *Cyathea* (40).
    27. *Cnemidaria* (25).
14. Family Pteridaceae 33–37.
    14a. Tribe Taenitideae 11.
        28. Pityrogramma 16–18,(12–13).
        29. Anogramma 5,(4).
        30. Eriosorus 25,(25).
        31. *Jamesonia* (19).
        32. *Pterozonium* (13).
        33. *Nephopteris* (1).
    Tribe Platyzomateae 1.
    14b. Tribe Cheilantheae 12–16.
        34. Cheilanthes 150,(85).
        35. *Bommeria* (4).
        36. *Adiantopsis* (7).
        37. *Notholaena* (39).
            Section *Argyrochosma*
            Section *Notholaena*
        38. *Hemionitis* (7).
        39. Pellaea 35,(20–22).
            Section Pellaea
            Section *Ormopteris*
        40. Doryopteris 25,(20).
            Section *Lytoneuron*
            Section Doryopteris
        41. Trachypteris 3,(2).
        42. Cryptogramma 2,(2).
        43. *Llavea* (1).
    14c. Tribe Ceratopterideae 1.
        44. Ceratopteris 3,(2).
    14d. Tribe Adianteae 1.
        45. Adiantum 150,(85).
    14e. Tribe Pterideae 7.
        46. Pteris 200,(55).
        47. *Anopteris* (1).
        48. *Neurocallis* (1).
        49. Acrostichum 3,(2).
15. Family Vittariaceae 6.
    50. *Hecistopteris* (1).
    51. Antrophyum 40,(10).
        Subgenus Antrophyum
        Subgenus *Polytaenium*
    52. Vittaria 50,(9).

84. *Adenoderris* (2).
85. Cystopteris 6,(3).
    Subgenus Cystopteris
86. Gymnocarpium 3–6,(2–3).
87. Woodsia 25,(11).
18c. Tribe Onocleeae 3.
88. *Onocleopsis* (1).
89. Matteuccia 2–3,(1).
90. Onoclea 1,(1).
18d. Tribe Oleandeae 2.
91. Arthropteris 15,(1).
92. Oleandra 40,(6–10).
18e. Tribe Bolbitideae 5.
93. Bolbitis 44,(14).
94. Lomariopsis 45,(15).
95. Lomagramma 20,(1).
96. Elaphoglossum 500,(350).
19. Family Aspleniaceae 7.
97. Asplenium 650,(150).
98. Pleurosorus 4,(1).
99. Camptosorus 2,(1).
100. *Holodictyum* (1).
101. *Schaffneria* (1).
20. Family Davalliaceae 9.
102. Nephrolepis 20,(6).
21. Family Blechnaceae 9.
103. Woodwardia 12,(4).
104. Blechnum 150,(50).
105. *Salpichlaena* (1).
22. Family Polypodiaceae 40.
22a. Tribe Polypodieae 36.
106. Polypodium 150,(135).
107. Pleopeltis 10,(10).
108. *Dicranoglossum* (3).
109. *Marginariopsis* (1).
110. Microgramma 13,(13).
111. *Campyloneurum* (20–25).
112. *Niphidium* (4–10).
113. *Neurodium* (1).
114. *Solanopteris* (3–4).
115. Platycerium 15,(1).
22b. Tribe Loxogrammeae 2.
116. Loxogramme 25,(1).
22c. Tribe Grammitideae 2.
117. Grammitis 400,(175).
    Section Cryptosorus
    Section *Glyphotaenium*
    Section Xiphopteris
    Section Grammitastrum
    Section Grammitis
    Section *Pleurogramme*
Suborder Marsileineae
23. Family Marsileaceae 3.
118. Marsilea 50,(12).
119. *Regnellidium* (1).
120. Pilularia 5,(1).
Suborder Salviniineae
24. Family Salviniaceae 2.
121. Salvinia 10,(7).
122. Azolla 6,(4).
    Section *Azolla*

Subclass Psilotidae
  Order Psilotales
    25. Family Psilotaceae 2.
        123. Psilotum 2,(2).
Class Equisetopsida
  Order Equisetales
    26. Family Equisetaceae 1.
        124. Equisetum 15,(13).
            Subgenus Hippochaete
            Subgenus Equisetum
Class Lycopodiopsida
  Order Lycopodiales
    27. Family Lycopodiaceae 2.
        125. Lycopodium 250,(125).
            Subgenus Cernuistachys
            Subgenus Lycopodium
            Subgenus Selago
  Order Selaginellales
    28. Family Selaginellaceae 1.
        126. Selaginella 700,(300).
            Subgenus Selaginella
            Subgenus Stachygynandrum
  Order Isoetales
    29. Family Isoetaceae 1.
        127. Isoetes 150,(50–60).

Biogeography

Pteridophytes are most common in wet and seasonally mild tropical and subtropical mountains. The greatest diversity of species is in two large regions where about 75% of the 9,000 species occur. One of these in America, including the Greater Antilles, southern Mexico and Central America, and the Andes of western Venezuela south to Bolivia, has about 2,250 species. The other center, southeastern Asia and Malesia, has about 4,500 species. The tropical and subtropical mountains of Africa have relatively few species, probably as a result of the decimation of a once richer flora by Pleistocene climatic changes.

There are about 3,250 species in America, with 3,000 of them in the tropics. The area of the American tropics is considered to include southern Florida and Mexico, south to northern Chile and Uruguay. Recent floras that are an essential source of biogeographic data are listed at the end of the Introduction.

The mountainous regions of tropical America, where most species occur, have high ecological diversity and a strongly developed mosaic of environments. Regions of low elevation usually have less ecological diversity and fewer but more extensive habitats. Although many species have a limited range, there are 300 or more that are widely distributed, as *Campyloneurum angustifolium*, *Gleichenia bifida*, *Pityrogramma calomelanos* and *Adiantum latifolium*. These species make up a basic element common to most of the tropical regions.

There are four regions with high species diversity and in each there is about 40% species endemism. The Greater Antilles have a pteridophyte flora of about 900 species, which are concentrated in eastern Cuba, Jamaica, and Hispaniola. These islands are especially rich in species of *Asplenium*, *Elaphoglossum*,

Thelypteris, Cyatheaceae, Hymenophyllaceae and Polypodiaceae. Examples of local endemism are the genera *Alsophila* and *Nephelea,* with a total of 16 species in the Greater Antilles, all endemic and 13 of them only on one island. The strongest relationships are with southern Mexico and Central America, and these include genera of limited distribution such as *Schaffneria, Adenoderris,* and *Maxonia.*

The region of southern Mexico and Central America also has about 900 species. The cloud forests support many epiphytic species of *Lycopodium, Asplenium, Elaphoglossum,* Hymenophyllaceae and Polypodiaceae, while the wet montane forests are rich in species of *Adiantum, Pteris, Selaginella* and Cyatheaceae. From Chiapas, Mexico species of this flora diminish in numbers northward in the Sierra Madre Oriental through Veracruz. There are strong relationships of this mesic flora to the Greater Antilles and especially to the Andes.

The Andean region includes a pteridophyte flora of about 1,500 species. It is especially well represented in the cloud forests by epiphytic species of *Lycopodium, Asplenium, Grammitis, Hymenophyllum* and *Elaphoglossum.* In the montane forests there are many terrestrial species of *Selaginella, Adiantum, Pteris, Thelypteris,* and Cyatheaceae. The lower slopes of the Andes, particularly to the east, often have additional species, and the semixeric inter-mountain regions have a small element of cheilanthoid ferns. Relationships are strong to adjacent Central America, as in some of the genera with limited ranges as *Loxsomopsis, Solanopteris,* and *Marginariopsis.*

Southeastern Brazil, from Minas Gerais and Espirito Santo south to Rio Grande do Sul, contains about 600 species. Most of these occur in the wet Serra do Mar, where there are many species of *Trichipteris, Lindsaea, Elaphoglossum, Asplenium,* and Polypodiaceae. Among the 20 American species of *Doryopteris,* 19 are in southeastern Brazil and 11 are endemic there. This flora has especially strong relationships with the Andes, with over 20 genera disjunct between the two regions, including *Danaea, Culcita, Lophosoria, Oleandra,* and *Niphidium.* At the species level, there are notable disjunctions from the Andes to the higher elevations of Serra do Itatiaia as in *Jamesonia brasiliensis* and *Eriosorus cheilanthoides.* Another element in southeastern Brazil is the campo limpo flora, especially developed on the sandy soils and sandstones of Minas Gerais. The flora is not rich but it is noted for species of *Pellaea* section *Ormopteris, Cheilanthes,* and *Anemia.* Nineteen of the 28 American species of *Anemia* subgenus *Coptophyllum* occur in southeastern Brazil and Goias, and 11 of them are endemic there.

Other regions of the American tropics have less species diversity. The Lesser Antilles have about 300 species that are mostly widely distributed. Guadeloupe, the most ecologically diverse of the islands, has the richest flora with 250 species. There is about 10% species endemism in the archipelago. Among species that are not widely distributed, the principal relationship is with northern South America.

The Guayana region is defined by the Roraima sandstone formation, characterized by isolated mesas surrounded by cliffs and basal talus slopes. It centers in Bolivar and Amazonas, Venezuela, with extensions to Surinam and eastern Colombia. The pteridophyte flora of about 450 species is particularly well developed in the genera *Adiantum, Lindsaea,* and *Trichomanes.* All 13 species of *Pterozonium* occur in the Guayana region and 11 of them are endemic there. The smaller genus *Hymenophyllopsis,* with about seven species, is confined to Guayana. In spite of the proximity of the Andes, there are few relationships among the pteridophytes except for widely distributed species. There are stronger alliances with the adjacent northern portion of the Amazon basin where sandstone and sandy soils also occur, and limited, disjunct ones with southeastern Brazil.

The Amazon basin has a poor pteridophyte flora, considering its large size. There are about 300 species in the Amazon of Brazil and these are mostly widely distributed. The flora is dominated by species of *Selaginella, Trichomanes, Lindsaea, Adiantum, Asplenium* and Polypodiaceae. A notable feature is the limited representation of large genera such as *Thelypteris* with only 12 species in the Brazilian Amazon, *Elaphoglossum* with nine, *Lycopodium* with five, and *Pteris* and *Hymenophyllum* with three each.

Two fern floras in the American tropics are of special interest, although they are not especially rich in species. One is the xeric and semixeric flora of northern and central Mexico. This is dominated by species of *Cheilanthes, Notholaena,* and *Pellaea,* all genera of the Pteridaceae tribe Cheilantheae. The center of diversity of these genera is in America, and especially in Mexico, where about 100 species form the richest xeric fern flora in the World. In *Notholaena,* 29 of the 39 species are in Mexico and 14 are endemic. Xeric species usually grow in rocky places, in crevices of cliffs or at the edge of rocks, where there may be local seepage, channeling of rainfall, or seasonal condensation of water. Many species have coriaceous leaves that are tolerant to short periods of desiccation. Other species have thin leaves that survive only the wet season, and often are deciduous by fracturing petioles when the leaf dies.

The World's richest alpine and subalpine pteridophyte flora is in the Andes. It is characterized by species of *Jamesonia* with linear, indeterminate leaves with many small pinnae, páramo species of *Lycopodium* subgenus *Selago,* and by *Gleichenia simplex,* with a small, unbranched, pinnatifid leaf. All of the 19 species of *Jamesonia* occur in the Andes and 18 of the 25 species of the closely allied *Eriosorus.* The páramo species of *Lycopodium* subgenus *Selago* are a group of perhaps 50 species with about two-thirds of them endemic. Other genera with terrestrial alpine and subalpine species are the commonly epiphytic *Grammitis, Elaphoglossum,* and *Asplenium.* The compact form of most of these ferns probably is related to the special environment of high tropical mountains, where there is high insolation and ultra-violet radiation and often hot days alternating with cold night.

The isolated islands of tropical America have relatively few species and apparently low species endemism, although the number of endemics is uncertain until modern monographic treatments critically compare them with related species on the mainlands. In the Pacific, Cocos Island has 60 species and six endemics, and the Galápagos Islands have 105 species and perhaps 10 endemics. In the Atlantic, Ilha Trindade has 13 species, of which a few are perhaps endemic.

Closest relationships between the tropical American flora and the paleotropical regions are to Africa. Several paleotropical genera such as *Platycerium, Blotiella,* and *Loxogramme* have a single American species most closely related to African elements of the genus; and American genera such as *Microgramma, Pleopeltis* and *Pellaea* subgenus *Pellaea* have a single species in Africa.

Relatively few American species are confined to the north or south of the tropics. Somewhat over half of the 350 species of the United States and Canada extend north from the tropics. The remaining temperate, boreal and arctic species have mostly circumboreal distributions or else have circumboreal relationships. A few, however, are endemic species more or less isolated in their genus, for example, *Lygodium palmatum, Woodwardia virginica, Woodwardia areolata, Diplazium pycnocarpon, Thelypteris hexagonoptera, Polystichum acrostichoides,* and *Dryopteris Goldiana.* About half of the some 100 species of austral South America are also tropical. Most of the others have relationships across the Pacific to New Zealand and Australia, or eastward to southern Africa. Among these are species of *Hypolepis, Asplenium, Blechnum, Grammitis, Gleichenia, Lycopodium,* and Hymenophyllaceae. The Juan Fernandez Islands have 54 species and among these perhaps 15 are endemic. These islands are noted for the endemic genus *Thyrsopteris* and for *Arthropteris,* which extends eastward across the Pacific.

The present distribution of species, the centers of species diversity, and centers of species endemism are the result of the processes of speciation operating on an ecological and geographic basis. They reflect previous climates that have promoted migration and extinction and have molded the ranges. The processes of ecogeographic speciation in ferns and the role of migration in the development of fern floras have been investigated by Tryon (1970, 1972) and Tryon & Gastony (1975).

Spores   The sequence of nuclear and cytoplasmic developments in spores leading to the formation of the mature wall ultimately results in three main strata. The *endospore,* innermost, is adjacent to the cell cytoplasm and consists of a relatively thin, cellulosic layer, formed after the deposition of exospore and perispore. The *exospore* is generally massive and usually formed of several layers. The major one is formed of plates (feuillettes) and additional infrastructural material between them, derived from the tapetum. In some families a clearly differentiated central layer is inserted between outer and inner strata especially in the primitive families, Ophioglossaceae and Osmundaceae. Canals penetrate the exospore, appearing as pits or pores at the surface, and are especially abundant at and adjacent to the apertural ridges. The *perispore* deposited by the tapetum forms the outer layer in all species, with few exceptions. Distinctions between genera, formerly made in regard to the presence or absence of perispore, need to be reexamined. The very thin perispore stratum in some groups is difficult to detect with light microscopy, but can be recognized by treating the spores with sodium hydroxide which causes the perispore to expand (Gastony 1974). The type of surface and stratification of the perispore is usually characteristic of genera or infrageneric groups.

Apertures (laesurae) in pteridophyte spores are either trilete or monolete. These basic types, and to some extent spore shape, are based on alignment of spores in the tetrad and the orientation of spindles. Trilete spores are generally tetrahedral-globose, often with three prominent angles, or spheroidal. Monolete spores are usually more or less elliptical but may be nearly spheroidal. Both aperture types have a central canal initially sealed at the apex by exospore and perispore deposit. A large amorphous body often present at the canal base appears to be involved with passage of material through the canal. The type of aperture is usually constant for a genus but both trilete and monolete forms occur in a few genera as *Gleichenia, Vittaria, Loxogramme,* and *Lindsaea.* Variation in aperture length or number often occurs in some species and is probably related to irregularities during meiosis or sporogenesis.

An evolutionary trend in spores of living ferns is evident in the type of aperture. Trilete apertures are characteristic of all of the primitive families as Ophioglossaceae, Dicksoniaceae, and Cyatheaceae. They also occur in more advanced families as Pteridaceae, but less commonly to rarely in the Polypodiaceae, Vittariaceae or Thelypteridaceae. Monolete spores are derived from trilete and apparently have been independently initiated in several groups as both trilete and monolete spores occur in distantly related families as the Schizaeaceae, Dennstaedtiaceae, and Thelypteridaceae. Advanced families with consistently monolete spores include the Dryopteridaeae, Aspleniaceae, and Blechnaceae.

A second evolutionary trend in these spores is evident in differences in the formation of the surface contours. Spores of primitive families have gross contours formed by the exospore and conforming to this a relatively thin perispore. This formation occurs in primitive families as the Ophioglossaceae, Marattiaceae, Schizaeaceae, and Gleicheniaceae. It also prevails in several genera of advanced families as in the tribe Taenitideae of the Pteridaceae where the thin perispore may be a persistent primitive character. Limited perispore may also be a derived character related to ecological differentiation as in epiphytic groups with relatively slight surface contours.

In most advanced families the surface contours of the spores are elaborated from perispore as in the Dryopteridaceae, Aspleniaceae, and Blechnaceae. Similarities in formation of the perispore of inflated folds in the Dryopteridaceae, Aspleniaceae, and Thelypteridaceae suggests that these families may be relatively closely related. Their spores are also monolete except for the small genus *Trigonospora* of the Thelypteridaceae. The widely diverse spore types in the Dennstaedtiaceae suggest it represents a heterogeneous family. Among 12 genera included in it from the American tropics, six have trilete spores, four are monolete, and two in the tribe Lindsaeae have monolete or trilete spores. In *Lindsaea* both types have a complex perispore with a smooth surface overlaying two lower strata.

Spores of the Polypodiaceae, tribe Polypodieae are consistently monolete with contours derived from the exospore. This type of exospore formation resembles that of spores in more primitive families and is clearly distinct from those with elaborated perispore in derived families as the Dryopteridaceae and Aspleniaceae.

Studies on the development and structure of *Psilotum* spores show the wall is similar to that of other genera in the Filicopsida. The compact exospore resembles the formation in monolete spores of *Gleichenia* and is unlike any of the diverse types in other classes of the Pteridophyta.

Scanning electron microscope studies of spores, at high magnification show the surface is composed of particulate material irregardless of the type of surface. In a biological context this surface is not ornamentation or sculpture but specific types of deposit. There are also commonly larger spherical bodies incorporated in and on the perispore in many genera, which are regarded as globules by Lugardon (1981) and the counterpart of the orbicules or Ubisch bodies in pollen. These and other resistant wall substances in pteridophyte spores and pollen are generally considered to be sporopollenin. Solubility tests of this in spores and pollen show that sporopollenin is in an extremely stable form, insoluble in 2–aminoethanol, in pteridophytes and differs from that of pollen walls (Southworth, 1974). Glycoproteins have been recognized in the surface coats of pteridophyte spores and are considered by Pettitt (1979) as having molecular structural capability by which diverse types of wall can be formed.

Spore wall development following meiosis through the formation of the mature wall has been examined in *Ophioglossum vulgatum, Osmunda regalis,* and *Blechnum Spicant,* by Lugardon (1971), and particulary the formation of exospore in *Botrychium Lunaria,* by Pettitt (1979). Two preliminary mucopolysaccharide layers, a "tetrad wall" which surrounds four members of the tetrad, and a "special wall" which isolates each spore in the tetrad are formed prior to the exospore deposition. Materials carried by vesicles generated from the golgi apparatus, are passed through the plasma membrane forming preliminary layers and later the initial wall deposits. The synchronous events and interactions between the cell and the tapetum forming the wall are not entirely clear. The pattern upon which the wall is constructed appears to be derived from the cell, although the structural materials and the execution of the major part of the exospore and perispore reside in the tapetum. The cytological changes during sporogenesis, reviewed by Bell (1979) indicate that there is an increase in vesicles prior to meiosis. These may relate to the elimination of ribonucleic acids, proteins, and other macromolecular material. This is regarded as a purge of sporophytic material prior to the formation of spores initiating the gametophytic phase of development. In *Pteridium* there is an increase in ribosome frequency at the end of meiosis that coincides with tubular extensions of the nucleus into the cytoplasm. These changes are noted by Bell as possibly related to transfer of information from the nucleus to the cytoplasm.

The interpretation of the role of microstructures forming the wall is based on transmission electron microscope studies. Data from scanning electron microscope studies supply an additional dimension on the nature and diversity of the major features of the wall. For each of the genera treated here descriptions provided for the spores are essential for comparison, but the terminology is inadequate to convey the complexity and detail which is evident in the SEM photographs.

Cytology

Chromosome numbers supply a source of data useful for assessing the affinities of pteridophyte genera and families. The numbers often indicate relationships as *Pleurosorus* with *Asplenium*, *Elaphoglossum* with the bolbitoid genera, and the Vittariaceae with the Pteridaceae. They also may reveal discrepancies as in *Hemidictyum* with 31 which is placed among dryopteroid genera with 40 and 41, and *Lonchitis* with 50 is exceptional among the genera of the Dennstaedtiaceae. The cytological studies on genera as *Asplenium*, *Dryopteris*, *Pteris* and *Ceratopteris* have also generated information relevant to polyploidy, hybridization, genetics and biogeography.

It is evident from the following list that the same numbers are known in several unrelated groups. Thirty-six is common to *Trichomanes*, *Asplenium*, and several genera of the Polypodiaceae. Forty is frequent in genera of Marattiaceae and eight genera of Dryopteridaceae, while 30 is known in *Lygodium*, *Dennstaedtia*, *Hypolepis*, and *Thelypteris* as well as genera of the cheilanthoids. Relationships must be based on the correlation of chromosome numbers with other characters rather than arithmetic schemes based on numbers alone. Additional cytological reports are essential as only about 20% of the species are known and records particularly from South America and China are lacking. Among the genera that are at present unknown are the following endemic to tropical America: *Adenoderris*, *Atalopteris*, *Hecistopteris*, *Holodictyum*, *Hymenophyllopsis*, *Nephopteris*, *Pterozonium* and *Schaffneria*. Many reports of approximate numbers need to be verified and some that are probably inaccurate, although these may be difficult to assess in genera where aneuploidy is well developed.

A remarkable feature of pteridophyte cytology is the high chromosome numbers derived mainly by an increase in polyploid levels especially among the homosporous ferns. Of 119 American genera that are cytologically known, nearly half have meiotic numbers of 75 or higher and the highest documented record, 631 in *Ophioglossum reticulatum* (Abraham & Ninan, 1954), is the greatest known in any organism. The heterosporous pteridophytes are characterized by low numbers as in *Selaginella* with $n = 10, 9, 8$, and $2n = 14$ in a species of New Guinea which is the lowest for the pteridophyta. Other heterosporous genera also have relatively low numbers—the highest, $n = 55$ in *Isoetes.* Pteridophyte chromosomes generally range from 1–8 μm and tend to be larger than those of angiosperms. The largest, in *Psilotum,* are comparable in size to the exceptionally large chromosomes of *Lilium,* and the smallest, about 1–1.5 μm, occur in *Selaginella.* Base numbers have been considered as the lowest number actually reported for a genus although such numbers as 30, 36 or 40 must represent derived levels. The many genera with either 40 or 41, with 36 or 37 or with 30 or 29 tend to support the proposals of ancestral base numbers of 10 or 9 and derivation of separate lineages upon these.

Original reports of chromosome numbers, widely scattered in the literature, have been assembled in several indices. They are included in the general indices of Ornduff (1967–1969), Moore (1970–1977), and Goldblatt (1981). Separate reports for the pteridophytes include the initial work by Chiarugi (1950) and supplements (Fabbri, 1953, 1960) that list n and $2n$ numbers as well as base numbers calculated on the lowest divisible number. The atlas of chromosome numbers for species of pteridophyta (Löve, Löve, & Pichi-Sermolli 1977) lists all reports as $2n$ including those originally made as n. The most extensive cytological work on tropical American pteridophyta is the study of Jamaican species by Walker (1966, 1973).

Chromosome Numbers of Genera of the Pteridophyta in America

The reports listed for the American genera include records known throughout their ranges. Numbers noted in the commentary are n numbers except as indicated.

Ophioglossaceae

Reports for *Botrychium* are mainly 45 or multiples except for 92 in *B. virginianum* which is treated in a distinct subgenus *Osmundopteris.*

The highest chromosome number, 631 in *Ophioglossum reticulatum,* is actually reported as 631 plus 10 fragments. A higher record of $n = 720$ for the species is undocumented. Polyploid changes appear to have occurred from 120 to 480 and, in addition, a series of higher numbers involve aneuploidy. The numbers included here represent a sample of the diverse reports.

A high chromosome number, 94, is also reported in *Helminostachys,* the third genus of the family.

   1. *Botrychium* $n = 45, 90, 92, 135$; $2n = 90, 180, 184$.

   2. *Ophioglossum* $n = 120, 240$, ca. $360, 436, 451, 480, 564, 566, 631$; $2n =$ ca. $300, 480$, ca. $720, 960, 1262$.

Marattiaceae

The cytology is known in four of the seven genera of the family. There are reports of 40 in each, including *Angiopteris* and $n = 80$ in *Christensenia.* Additional records of 39 and 78 for *Marattia* evidently represent aneuploid derivatives.

   3. *Marattia* $n = 39, 40, 78$; $2n = 156$.

   4. *Danaea* $n = 40$ (I + II), 80; $2n = 160$.

Osmundaceae

Cytologically the family is remarkably stable with 22 in each of the three genera, including several reports for *Osmunda* from a wide geographic range. Karotype analysis has indicated 11 as the base number of *Osmunda.*

   5. *Osmunda* $n = 22$; $2n = 44$.

Schizaeaceae

The three American genera have discrete chromosome numbers with *Lygodium* at 28, 29, 30, *Schizaea* at 77 or higher, and *Anemia* at 38. The African genus *Mohria* with 76 corresponds to reports for *Anemia* and similarities of the spores also indicate a close association between these genera. The numbers 29 and 30 in *Lygodium* are also characteristic of some genera of the Pteridaceae and suggest relations between these families.

6. *Anemia* $n = 38, 76; 2n = 76, 114, 152;$ apogamous 114.

7. *Lygodium* $n = 28, 29, 30, 56, 58,$ ca. 70, 87–90; $2n = 56, 58, 60, 112, 116.$

8. *Schizaea* $n = 77, 94, 96, 103,$ ca. 154, ca. 270, 350–370, 370.

Gleicheniaceae

The treatment of three subgenera in *Gleichenia* and two in *Dicranopteris* is supported by their distinct chromosome numbers. In *Gleichenia* reports of the Old World subgenus *Gleichenia* are either 20 or 22, of subgenus *Mertensia* 34, and of subgenus *Diplopterygium* 56. *Dicranopteris* subgenus *Dicranopteris* is definitely 39 and subgenus *Acropterygium* is 43 with a report of 44 that needs confirmation.

9. *Gleichenia* $n = 20, 22, 34, 56, 68; 2n = 40, 68, 136.$

10. *Dicranopteris* $n = 39, 43, 44, 78; 2n = 78, 156.$

Hymenophyllaceae

Meiotic chromosome numbers in *Hymenophyllum* tend to be somewhat lower than those in *Trichomanes,* although 36, the most frequent, occurs in both genera. The record of 9 in *Trichomanes* is the lowest reported for the ferns. A base number of 11 is proposed for *Hymenophyllum* from karyological studies but reports of 21 and 28 suggest a possible lower number.

11. *Hymenophyllum* $n = 11, 13, 18, 21, 22, 26, 27, 28, 36, 42, 56, 72; 2n = 56, 72, 108;$ apogamous 26.

12. *Trichomanes* $n = 9, 18, 32, 33, 34, 36, 64, 66, 68, 72, 128; 2n = 68, 72, 128;$ apogamous 26, 68, 108.

Loxomataceae

Record of 46 in the American *Loxsomopsis* emphasizes its isolated position in respect to the genus *Loxoma* reported as $n = 50$ from New Zealand.

13. *Loxsomopsis* $n = 46$

Hymenophyllopsidaceae

14. *Hymenophyllopsis*   not reported.

Plagiogyriaceae

The higher approximate numbers need confirmation although the chromosome number 66 is established for species in America and Asia. Relations of the family with either the Osmundaceae or Blechnaceae, as suggested by other characters, is supported by the record of 66, but not by the spores.

15. *Plagiogyria* $n = 66,$ ca. 75, ca. 100, ca. 125, ca. 132.

Dicksoniaceae

Reports of 65 in *Dicksonia* and 68 in *Cibotium* are based on Old World species. There appear to be two elements in *Culcita* based on 66 in the American *C. coniifolia* and 66–68 in the European *C. macrocarpa.* In contrast, the species of the Old World subgenus *Calochlaena* with $n =$ ca. 58 as in *C. dubia* appear to represent a distinct element. The report of 76–78 for *Thrysopteris,* although not precisely determined, is probably 76.

16. *Culcita* $n =$ ca. 58, 66, 66–68.

17. *Dicksonia* $n = 65; 2n = 130.$

18. *Cibotium* $n = 68.$

19. *Thyrsopteris* $n = 76–78.$

Lophosoriaceae

Relationships to the Dicksoniaceae may be shown in the chromosome number 65 also known in *Dicksonia* and by similarities of the spores to those of *Cibotium*.

    20. *Lophosoria* $n = 65$.

Metaxyaceae

Although exact numbers need to be determined, they appear to support recognition of the family as distinct from Dicksoniaceae and the Cyatheaceae.

    21. *Metaxya* $n = 94–96$; $2n = 190–192$.

Cyatheaceae

The reports of 69 for all six genera of the Cyatheaceae are unusually consistent and suggest that the record of $2n = 130–140$ is probably 138. The cytology of some hybrids recognized between *Alsophila* and *Nephelea* is exceptional as bivalents are formed during meiosis and viable spores produced.

    22. *Sphaeropteris* $n = 69$.
    23. *Alsophila* $n = 69$; $2n = 130–140$.
    24. *Nephelea* $n = 69$; $2n = 138$.
    25. *Trichipteris* $n = 69$.
    26. *Cyathea* $n = 69$.
    27. *Cnemidaria* $n = 69$.

Pteridaceae

Cytologically the genera of the Pteridaceae are relatively uniform. The 22 American genera included in the family are consistently 29, 30 or multiples, aside from *Ceratopteris*. The Old World genera for which reports are known, also based on 29 or 30, are *Actinopteris, Austrogramme, Coniogramme, Onychium,* and *Syngramma,* while *Idiopteris* is reported as 27 and *Taenitis* as 44 or 110.

    Taenitideae   28. *Pityrogramma* $n = 30, 58, 60, 116, 120$; $2n = $ ca. 90, ca. 232.
                   29. *Anogramma* $n = 29, 58$.
                   30. *Eriosorus* $n = 87, 174$.
                   31. *Jamesonia* $n = 87$.
                   32. *Pterozonium*   not reported
                   33. *Nephopteris*   not reported
    Cheilantheae   34. *Cheilanthes* $n = 29, 30, 56, 58, 60$; $2n = 60, 116, 120$; apogamous 87, 90.
                   35. *Bommeria* $n = 30$, $2n = 60$; apogamous 90.
                   36. *Adiantopsis* $2n = 60$.
                   37. *Notholaena* $n = 29$, ca. 30; $2n = 60$, ca. 116; ?apogamous 90
                   38. *Hemionitis* $n = 30$.
                   39. *Pellaea* $n = 29, 30, 58$; $2n = 58, 60$; apogamous 87, ca. 90, 116.
                   40. *Doryopteris* $n = 30, 60, 116$; $2n = 60, 120, 232$.
                   41. *Trachypteris* $n = 29$ or 30.
                   42. *Cryptogramma* $n = 30, 60$; $2n = 60, 120$.
                   43. *Llavea* $n = 29$.
    Ceratopterideae   44. *Ceratopteris* $n = 39, 77, 78$; $2n = 80, 154$.
    Adianteae   45. *Adiantum* $n = 29, 30, 57, 58, 60, 90, 114, 116, 150$; $2n = 58, 60, 114, 116, 120, 228$; apogamous 30, 60, 90, 171, ca. 175, ca. 180.
    Pterideae   46. *Pteris* $n = 29, 58, 116$; $2n = 58, 116$, ca. 232; apogamous 58, 87, ca. 90, 116, ca. 120.
                   47. *Anopteris* $n = 58$; $2n = 116$.
                   48. *Neurocallis* $n = 58$.
                   49. *Acrostichum* $n = 30$; $2n = 60$.

Vittariaceae

The chromosome number known in three of the American genera is consistently 60 or higher ploidy levels while 30 is known only in a Malesian species of *Monogramma*. Most records for the American genera are either 60 or 120, considered as tetraploid or octoploid based on 30. The chromosome number is one of the useful characteristics supporting the close relationships of the genera and position of the family near the Pteridaceae.

50. *Hecistopteris*	not reported
51. *Antrophyum*	$n = 60, 120, 180; 2n = $ ca. 180.
52. *Vittaria*	$n = 60, 120; 2n = 120, $ ca. 180, 240.
53. *Anetium*	$n = 60.$

Dennstaedtiaceae

Diversity in chromosome number of the genera is also complicated by many uncertain records for several groups. In *Dennstaedtia*, two series are proposed, 34 in the temperate American *D. punctilobula*, and 47 in American tropical species such as *D. circutaria*. Two series are also proposed for species of *Hypolepis* based on 29 or 26. The close relationship suggested by reports of $2n = 52$ in *Pteridium* and $n = 26$ in *Paesia* are not supported by characteristics of the spores nor are similarities of the spores in *Hypolepis* and *Blotiella* supported by the cytology. Several approximate numbers in *Lindsaea* are regarded as equal to or derived from 34, 44, or 47. The numbers in Old World species of *Odontosoria* probably also relate to 47, and 48, while 38 appears to represent a distinct element in the American tropics.

Dennstaedtieae

54. *Microlepia*	$n = 43, 44, 84–87, 86, 129; 2n = 84, 86, 160, $ ca. 170, 172.
55. *Dennstaedtia*	$n = 30, 33–34, 34, 46, 47, 60, 64, 65, 94; 2n = 60, $ ca. 94, 120, 128.
56. *Saccoloma*	$n = 188; 2n = 63 \pm 2$ (II), ca. 376.
57. *Pteridium*	$n = 52; 2n = 52, 208.$
58. *Paesia*	$n = 26, 104.$
59. *Hypolepis*	$n = 29, 39, 51–53, 52, $ ca. 92, 98, ca. 100, 104; $2n = $ ca. 92, 104, ca. 150, 208.
60. *Blotiella*	$n = 38, 76.$
61. *Lonchitis*	$n = 50, 100.$
62. *Histopteris*	$n = 48, 96.$

Lindseeae

63. *Odontosoria*	$n = 38, 39, 47, 48, $ ca. 88, 94, ca. 96, 100, 145–147; $2n = 76$; apogamous 47, 48, 94.
64. *Lindsaea*	$n = 34, $ ca. 40, 42, 44, 44 or 45, 47, ca. 50, ca. 84, ca. 87, 88, ca. 100, ca. 150, ca. 153, 155, ca. 220; $2n = $ ca. 100, ca. 176; ? apogamous 47, 80, 94, ca. 130.
65. *Ormoloma*	$n = 42.$

Thelypteridaceae

The numbers ranging from 27 to ca. 136 include an aneuploid series between 27 and 36, except for 28 and 33. Thirty-six is most frequent and occurs in five of the 11 American subgenera. On the basis of this number, the Thelypteridaceae have been regarded as derived from the same stock as the Aspleniaceae. Its origin has also been proposed from the Cyatheaceae with 36 derived from 69 through aneuploid reduction. However, cytological evidence has not been generated to substantiate this relationship.

66. *Thelypteris*	$n = 27, 29, 30, 31, 32, 34, 35, 36, 58, 60, 62, 64, 70, 72, 93, 116, $ ca. 136; $2n = 68, 70, 72, 124, 128, 144, 246;$ apogamous 90.

Dryopteridaceae

The chromosome numbers for this large family are remarkably consistent with 40 or 41 and multiples in most genera including some morphologically diverse forms. *Dryopteris* and *Polystichum* are perhaps cytologically best known among the large genera especially in temperate regions. There are a few divergent numbers in some genera as in the tribe Onocleeae with *Onoclea* consistently 37 and 39, as well as 40 in *Matteuccia*. In *Woodsia,* records of 33, 38, 39, and 41 suggest that there has been extensive aneuploid changes, while *Cystopteris* has differentiated several polyploid levels based on 42. Two reports of $n = 31$ for *Hemidictyum* raise a problem on the alignment of the genus with others in the family.

Dryopterideae

67. *Ctenitis*	$n = 41, 82.$
68. *Atalopteris*	not reported
69. *Tectaria*	$n = 40, 80, 120,$ ca. $160; 2n = 80, 160.$
70. *Hypoderris*	$n = 40; 2n = 80.$
71. *Cyclopeltis*	$n = 41.$
72. *Rumohra*	$n = 41.$
73. *Lastreopsis*	$n = 41, 82; 2n = 82.$
74. *Dryopteris*	$n = 41, 82, 123; 2n = 82, 164;$ apogamous 82, 123, 164.
75. *Cyrtomium*	$n = 41, 82;$ apogamous 123.
76. *Didymochlaena*	$n = 41.$
77. *Stigmatopteris*	$n = 41.$
78. *Polystichum*	$n = 41, 82, 164; 2n = 82,$ ca. $160, 164;$ apogamous 82, 123.
79. *Maxonia*	$n = 41.$
80. *Polybotrya*	$n = 41; 2n = 82.$

Physematieae

81. *Diplazium*	$n = 41, 82,$ ca. $121, 123, 164; 2n = 82, 164;$ apogamous 123, 164, ca. 200, 328.
82. *Athryium*	$n = 40, 80, 120,$ ca. $160; 2n = 80, 160, 240;$ apogamous 120.
83. *Hemidictyum*	$n = 31.$
84. *Adenoderris*	not reported
85. *Cystopteris*	$n = 42, 84, 126, 168; 2n = 126, 168, 252.$
86. *Gymnocarpium*	$n = 40, 80; 2n = 80, 160.$
87. *Woodsia*	$n = 33, 38, 39, 41, 76, 82; 2n = 66, 78, 82, 152, 156, 164.$

Onocleeae

88. *Onocleopsis*	$n = 40.$
89. *Matteuccia*	$n = 39, 40, 80; 2n = 78, 80.$
90. *Onoclea*	$n = 37; 2n = 74.$

Oleandreae

91. *Arthropteris*	$n = 41,$ ca. 42.
92. *Oleandra*	$n = 40, 41,$ ca. 80.

Bolbitideae

93. *Bolbitis*	$n = 41, 82; 2n = 82, 123.$
94. *Lomariopsis*	$n =$ ca. $39; 2n = 32, 62, 78, 164.$
95. *Lomagramma*	$2n = 82.$
96. *Elaphoglossum*	$n = 40, 41, 82,$ ca. $164; 2n = 82,$ ca. $164,$ ca. 250.

Aspleniaceae

Most cytological records for *Asplenium* are 36 or multiples to 288 aside from a species of Mexico with either 39 or 40 and one from Asia with 40. The extensive cytological work on *Asplenium* especially in Europe has provided general information on cytogenetic systems in the ferns. The studies are useful in depicting relations of species and alliances between the genera. Cytological data for the tropical American species will be helpful in the analyses of relationships especially involving hybrids.

97. *Asplenium*	$n = 36, 39$ or $40, 40, 72, 76, 108, 144, 216, 288; 2n = 72, 80, 116, 120, 144, 152, 216;$ apogamous 108, 210 or 211, ca. 212, 216, ca. 270, 277–280, 288, ca. 330–357, ca. 346.

98. *Pleurosorus*	$n = 36, 72; 2n = 72$.
99. *Camptosorus*	$n = 36; 2n = 72$.
100. *Holodictyum*	not reported.
101. *Schaffneria*	not reported.

Davalliaceae

Chromosome numbers of either 40, 41 or multiples in all of the cytologically known genera support the general position of the Davalliaceae among the derived families. It mainly consists of Old World genera, largely with reports of 40 except for *Leucostegia* and *Nephrolepis* with 41.

102. *Nephrolepis*	$n = 41, 82; 2n = 82$.

Blechnaceae

Records of chromosome numbers of *Blechnum,* the largest genus of the family, range between 28 and 99 with aneuploid series between 28 and 36 except for 30 and 35. The report of 40 for the American genus *Salpichlaena* is not readily aligned with those of other genera included in the family. In addition to the extensive aneuploid numbers in *Blechnum,* polyploids reaching a hexaploid level based on 33, as well as evidence of hybridization, indicate cytological complexities in the genus. Correlation of chromosome numbers and geography in some species suggest that cytological data are useful in the systematics of the genus.

103. *Woodwardia*	$n = 34, 35, 68; 2n = 68$.
104. *Blechnum*	$n = 28, 29, 31, 32, 33, 34, 36,$ ca. $37, 56, 62, 64, 68, 99; 2n = 66, 68, 124$.
105. *Salpichlaena*	$n = 40$.

Polypodiaceae

Chromosome numbers in 12 American genera range from 32–111, but reports are mostly of 37 or 74. Lower numbers such as 35 and 34 in *Pleopeltis,* 35 in *Marginariopsis,* and 36, 33, and 32 in *Grammitis* evidently involve aneuploid changes. Ploidy levels in the whole family are exceptionally low, mainly tetraploid with few hexaploids in genera as *Pleopeltis, Polypodium,* and *Pyrrosia.*

Polypodieae

106. *Polypodium*	$n = 37, 74, 111$; apogamous 111.
107. *Pleopeltis*	$n = 34, 35, 37,$ ca. $70, 74; 2n =$ ca. 210.
108. *Dicranoglossum*	$n = 36$.
109. *Marginariopsis*	$n = 35$.
110. *Microgramma*	$n = 37, 74$.
111. *Campyloneurum*	$n = 37, 74$.
112. *Niphidium*	$n = 74$.
113. *Neurodium*	$n = 37$.
114. *Solanopteris*	$n = 37$.
115. *Platycerium*	$n = 37; 2n = 74$.

Loxogrammeae

116. *Loxogramme*	$n = 35, 36, 70; 2n = 72$.

Grammitideae

117. *Grammitis*	$n = 32, 33, 36, 37, 74, 132–138$ (I); $2n = 72, 74$.

Marsileaceae

Cytological counts are 10 or multiples except for *Regnellidium* in which B chromosomes are reported and the numbers vary between 19 and 20 or $2n = 38$ and 40.

118. *Marsilea*	$n = 20; 2n = 40, 60$.
119. *Regnellidium*	$n = 19, 20; 2n = 38, 40$.
120. *Pilularia*	$n = 10; 2n = 20$.

Salviniaceae

The chromosomes of *Salvinia* are exceptional for pteridophytes in their relatively low number and large size, one of the largest among ferns. The number for *Azolla* is also low and possibly based on 11.

121. *Salvinia*	$n = 9$, chromosomes 45, 63: $2n = 18, 54$.
122. *Azolla*	$n = 22; 2n = 44$.

Psilotaceae

Reports for *Psilotum*, based on collections from the Old World are mostly tetraploid $n = 104$, or diploid 52. Records for the related genus *Tmesipteris,* confined to the Old World, are higher with mostly tetraploid $n = 104$ and also octoploid $= 208$. These records correspond to those of *Psilotum* and support the close relationship of these genera. Chromosomes of *Psilotum* up to 18 μm are the largest reported for Pteridophyta.

123. *Psilotum* $n = 52, 104, \text{ca. } 210; 2n = 104, 208.$

Equisetaceae

Cytological records for *Equisetum* are consistently $n = 108$ which is considered to be a 24-ploid level derived on a base number of 9. Reports have been made for nearly all species although few of these are based on tropical American material.

124. *Equisetum* $n = 108; 2n = 216.$

Lycopodiaceae

The chromosome numbers reported for *Lycopodium* range between 23–264 with records of $2n = 502$–510 for the related genus *Phylloglossum* of Australia and New Zealand. The chromosome numbers appear to correlate with other features of the plants in characterizing the subgenera. In subgenus *Lycopodium* species reported as 23 or 24 indicate two discrete elements. Those in subgenus *Cernuistachys* range between 35 and 208 for which two base numbers, 35 or 78 have also been proposed. The highest numbers, ranging between 128 and 264, are known in subgenus *Selago* with 44 and 45 proposed as base numbers.

125. *Lycopodium* $n = 23, 34, 35, 48, 70, 78, 104, 110, \text{ca. } 128,$
          $\text{ca. } 132, 136, 156, \text{ca. } 165, 208, 264;$
          $2n = 46, 48, 68, 264, 272, \text{ca. } 340, 528.$

Selaginellaceae

The chromosome reports for *Selaginella* range from 8–36 and 6, 7, 8, 9, 10 have been proposed as base numbers. The lowest three are uncommon as most species seem to be based on either 9 or 10. The most frequent report, $2n = 18$, is characteristic of temperate species, while records of neotropical species are frequently $2n = 20$.

126. *Selaginella* $n = 8, 9, 10, 12, 18, 36; 2n = 14, 16, 18, 20,$
          $24, 27, 36, 48–50, 60.$

Isoetaceae

The chromosome numbers reported for *Isoetes* range between 10–55 + 1, the lowest known from collections of Europe and North Africa. New reports from tropical America are either $2n = 22$ or 44. Species of temperate America are also mostly $2n = 22$ or 44, with a high report of 110.

The chromosomes are readily fragmented and this may account for the frequent reports of an extra chromosome. The base number 6 has been proposed for *Isoetes* from analyses of the karyotype.

127. *Isoetes* $n = 10, 11, \text{ca. } 29, \text{ca. } 48, 54–56, 55 + 1;$
          $2n = \text{ca. } 20, 22, 24–26, 44, 44 + 1, \text{ca. }$
          $48, \text{ca. } 50, \text{ca. } 58, 66, \text{ca. } 100, 110;$
          apogamous $22 + 1, 33 + 1.$

Literature

1. References cited in the Introduction.

Andrews, H. N., C. A. Arnold, E. Boureau, J. Doubinger, and S. Leclerq. 1970. Filicophyta, Traité de paleobotanique (E. Boureau, Ed.) 4(1). 519 pp. Masson, Paris.

Bell, P. R. 1979. The contribution of the ferns to an understanding of the life cycles of vascular plants. pp. 58–85. *in*: A. F. Dyer (Ed.), The experimental biology of ferns. Internat. Series Monogr. 14. Univ. Sussex, England.

Bierhorst, D. W. 1971. Morphology of vascular plants. 560 pp. Macmillan. New York.

Bower, F. O. 1923–1928. The ferns 1, 359 pp. 1923; 2, 344 pp. 1926; 3, 306 pp. 1928. Cambridge Univ. Press.

Chiarugi, A. 1950. Tavole cromosomiche delle Pteridophyta, Caryologia 13: 27–150.

Christensen, C. 1938. Filicinae, in Manual of Pteridology (F. Verdoorn, Ed.) 522–550. Nijhoff. The Hague.

Copeland, E. B. 1947. Genera filicum. 247 pp. Chronica Botanica, Waltham, Mass.

Crabbe, J. A., A. C. Jermy, and J. M. Mickel. 1975. A new arrangement for the pteridophyte herbarium. Fern Gaz. 11: 141–162.

Dyer, A. F. (Ed.). 1979. The experimental biology of ferns. 657 pp. Academic Press.

Fabbri, F. 1963. Primo supplemento alle tavole cromosomiche delle Pteridophyta di Alberto Chiarugi. Caryologia 16: 237–335.

Fabbri, F. 1965. Secundo supplemento alle tavole chromosomiche della Pteridophyta di Alberto Chiarugi. Caryologia 18: 675–731.

Gastony, G. J. 1974. Spore morphology in the Cyatheaceae. I. The perine and sporangial capacity: General considerations. Amer. Jour. Bot. 61: 672–680.

Goldblatt, P. 1981. Index to plant chromosome numbers for 1975–1978. Mo. Bot. Gard. Monogr. Syst. Bot. 5: 1–553.

Holttum, R. E. 1949. The classification of ferns. Biol. Reviews 24: 267–296.

Jermy, A. C., J. A. Crabbe, and B. A. Thomas (Eds.). 1973. The phylogeny and classification of the ferns. Bot. Jour. Linn. Soc. 67, Suppl. 1: 1–283.

Löve, A., D. Löve, & R. E. G. Pichi-Sermolli. 1977. Cytotaxonomical atlas of the Pteridophyta. 398 pp. Cramer, Vaduz.

Lugardon, B. 1971. Contribution á la connaissance de la morphogénèse et la structure des parois sporales chez les Filicinées isosporées. 257 pp. These Univ. Paul Sabatier, Toulouse.

Lugardon, B. 1981. Sur la formation du sporoderme chez *Psilotum triquetrum* Sw. (Psilotaceae). Grana 18: 145–165.

Moore, R. J. 1970–1977. Index to plant chromosome numbers for 1968–1974. Regnum Veget. 68: 1–115; 77: 1–112; 84: 1–134; 90: 1–539; 91: 1–108; 96: 1–257.

Ogura, Y. 1972. Comparative anatomy of vegetative organs of the pteridophytes, ed. 2. 502 pp. Borntraeger. Berlin.

Ornduff, R. 1967–1969. Index to plant chromosome numbers for 1965–1967. Regnum Veget. 50: 1–128; 55: 1–126; 59: 1–129.

Pettitt, J. M. 1979. Ultrastructure and cytochemistry of spore wall morphogenesis. pp. 213–252. *in*: A. F. Dyer (Ed.), The experimental biology of ferns. Internat. Series Monogr. 14. Univ. Sussex, England.

Pichi-Sermolli, R. E. G. 1977. Tentamen pteridophytorum in taxonomicum ordinem redigendi. Webbia 31: 313–512.

Southworth, D. 1974. Solubility of pollen exines. Amer. Jour. Bot. 61: 36–44.

Tryon, R. 1970. Development and evolution of fern floras of oceanic islands. Biotropica 2: 76–84.

Tryon, R. 1972. Endemic areas and geographic speciation in tropical American ferns. Biotropica 4: 121–131.

Tryon, R. and G. J. Gastony. 1975. The biogeography of endemism in the Cyatheaceae. Fern Gaz. 11: 73–79.

Walker, T. G. 1966. A cytotaxonomic survey of the pteridophytes of Jamaica. Trans. Roy. Soc. Edinburgh 66: 169–237.

Walker, T. G. 1973. Additional cytotaxonomic notes on the pteridophytes of Jamaica. *ibidem* 69: 109–135.

2. Floras of tropical and austral America.

Southern Florida

Long, R. W. and O. Lakela. 1976. A flora of tropical Florida. 962 pp. Banyan Books, Miami.

West Indies

Christensen, C. 1937. The collection of Pteridophyta made in Hispaniola by E. L. Ekman 1917 and 1924–1930. Kungl. Svenska Vetenskapsakad. Handl., III, 16: 1–93.

Duek, J. J. 1971. Lista de las especies cubanas de Lycopodiophyta, Psilotophyta, Equisetophyta y Polypodiophyta (Pteridophyta). Adansonia 11: 559–578, 717–731.

Hodge, W. H. 1954. Flora of Dominica, B. W. I. part I. Lloydia 17: 1–238.

Kramer, K. U. 1962. Pteridophyta, Flora of the Netherlands Antilles, 1, A. L. Stoffers (Ed.) Naturwetschap. Stud. Suriname Nederl. Antillen 25: 1–84.

Maxon, W. R. 1926. Pteridophyta, Sci. Survey Porto Rico Virgin Islands 6: 373–521. New York Acad. Sciences.

Proctor, G. R. 1953. A preliminary checklist of the Jamaican pteridophytes. Bull. Inst. Jam. Sci. Ser. 5: 1–89.

Proctor, G. R. 1977. Pteridophyta, flora of the Lesser Antilles 2, R. A. Howard (Ed.), 414 pp. Arnold Arboretum, Jamaica Plain.

Mexico and Central America

Gómez, L. D. 1975. Contribuciones a la pteridologia costarricense, VII. Pteridofitos de la Isla de Cocos. Brenesia 6: 33–48.

Knobloch, I. W. and D. S. Correll. 1962. Ferns and fern allies of Chihuahua, Mexico. Contrib. Texas Res. Found. 3. 198 pp. Renner, Texas.

Matuda, E. 1956. Los helechos del Valle de Mexico y alrededores. Anal. Instit. Biol. Mex. 27: 49–168.

Seiler, R. 1980. Una guia taxonomica para helechos de El Salvador. 58 pp. San Salvador.

Smith, A. R. 1981. Pteridophytes, Flora of Chiapas, part 2, D. E. Breedlove (Ed.), 370 pp. California Academy of Sciences. San Francisco.

Stolze, R. G. 1976. Ferns and fern allies of Guatemala, part I. Fieldiana: Bot. 39: 1–130.

Stolze, R. G. 1981. Ferns and fern allies of Guatemala, part II. Fieldiana: Bot. n.s. 6: 1–522.

Tropical South America

Brade, A. C. 1969. Algumas espécias novae Pteridophyta de Ilha brasileira Trindade coletadas por J. Becker em 1965/1966. Bradea 1: 3–10.

Capurro, R. H. 1938. Catalogo de las pteridofitas Argentinas. Anais Prim. Reun. Sul–Amer. Bot. 2: 69–210.

Foster, R. C. 1958. A catalogue of the ferns and flowering plants of Bolivia. Contrib. Gray Herb. 184: 1–223.

Kramer, K. U. 1978. The pteridophytes of Suriname. Naturwetschap. Stud. Suriname Nederl. Antillen 93: 1–198.

Murillo, M. T. Pteridophyta, Catalogo illustrada de las plantas de Cundinamarca 2, L. D. Mora (Ed.), 153 pp. Bogotá.

Sehnem, A. 1967–1979. (18 families of Pteridophyta), Flora illustrada catarinense, P. R. Reitz (Ed.). Itajai, Santa Catarina.

Sota, E. R. de la. 1972–1977. Sinopsis de las pteridofitas del noroeste de Argentina, I. Darwiniana 17: 11–103. 1972; II, *ibidem* 18: 173–263. 1973; III, *ibidem* 20: 225–232. 1976; IV, *ibidem* 21: 120–138. 1977.

Sota, E. R. de la. 1977. Pteridophyta, Flora de la Provincia de Jujuy Republica Argentina 2, A. L. Cabrera (Ed.), 275 pp. Col. Cientifica INTA, Buenos Aires.

Tryon, R. 1964. The ferns of Peru, Polypodiaceae (Dennstaedtieae to Oleandreae). Contrib. Gray Herb. 194: 1–253.

Tryon, R. M. and D. S. Conant. 1975. The ferns of Brazilian Amazonia. Acta Amazonica 5: 23–34.

Vareschi, V. 1969. Helechos, flora de Venezuela 1. 1033 pp. Instituto Botanico, Caracas.

Werff, H. H. van. 1978. The vegetation of the Galápagos Islands. 102 pp. Lakenman & Ochtman. Zierikzee.

Wiggins, I. L. and D. M. Porter. 1971. Flora of the Galápagos Islands. 898 pp. Stanford University Press. Stanford.

Austral South America

Diem, J. 1943. Flora de parque nacional de Nauhel-Huapí, 1. Helechos y los demas cryptógamas vasculares. 118 pp. J. Diem. Buenos Aires.

Duek, J. J. and R. Rodriguez. 1972. Lista preliminar de las especies de Pteridophyta en Chile continental y insular. Bol. Soc. Biol. Concepcion 45: 129–174.

Kunkel, G. 1965. Catalogue of the pteriodophytes of the Juan Fernandez Islands (Chile). Nova Hedwigia 9: 245–284.

Looser, G. 1955. The ferns of southern Chile. Amer. Fern Jour. 38: 33–44, 71–87.

Looser, G. 1955. Los helechos (pteridofitos) de Chile central. Moliniana 1: 5–95.

Looser, G. 1961–1968. Los pteridofitos o helechos de Chile, I. Rev. Univ. Catolica de Chile 46: 213–262. 1961; II. *ibidem* 47: 17–31. 1962; III. *ibidem* 50–51: 75–93. 1965–1966; IV. *ibidem* 53: 27–39. 1968.

Key to Families of Pteridophyta in America

a. Two or more sporangia joined in a synangium.   b.
   b. Synangia many on the abaxial surface of a leaf, stem stout to massive, subterranean.   2. Marattiaceae, p. 39
   b. Synangium single in the axial of a bifid, bract-like leaf, aerial stem slender, green, dichotomously forked.   25. Psilotaceae, p. 782
   b. Two elongate synangia at the apex of a spike, stem small, subterranean.   1. Ophioglossaceae, p. 25
a. Sporangia separate.   c.
   c. Sporangia borne on the inner side of a peltate sporangiophore in an apical strobilus, leaves much reduced, forming a sheath at the apex of each elongate, ridged internode (joint) of the stem.   26. Equisetaceae, p. 788
   c. Sporangia single in or near the axil of a leaf, or on one lobe of a bilobed leaf, or several to many sporangia borne on a leaf: at the margin, on the abaxial surface, on a specialized portion or on a specialized leaf. d.
      d. A single sporangium borne in or near the axil of a leaf.   e.
         e. Homosporous, leaves without a ligule, borne along an elongate stem.   27. Lycopodiaceae, p. 796
         e. Heterosporous, with megasporangia and microsporangia, leaves with a ligule.   f.
            f. Leaves less than 1 cm long, borne along an elongate stem, fertile leaves in an apical strobilus.   28. Selaginellaceae, p. 812
            f. Leaves 2 cm long or usually longer, clustered at the apex of a compact to slightly elongate stem, all of them usually fertile.   29. Isoetaceae, p. 826
      d. Several to many sporangia borne on a leaf, or a single sporangium borne on one lobe of a bilobed leaf.   g.
         g. Heterosporous, the leaf bearing megasporangia and (or) microsporangia enclosed in small, specialized structures.   h.
            h. Stem rooted in wet soil or under water, leaves filiform, or with 2 or 4 leaflets at the apex of the petiole.   23. Marsileaceae, p. 759
            h. Floating aquatics, the floating leaves entire, oblong to suborbicular, or unequally bilobed with one lobe submerged.   24. Salviniaceae, p. 770

g. Homosporous, the isomorphic sporangia exposed at the margin or on the abaxial surface of a leaf, or on a specialized portion, or on a specialized leaf, sometimes enclosed, prior to maturity, within an indusium or an enrolled pinna or segment. i.

  i. Sporangia with a 1–3 rowed stalk, the annulus vertical or nearly so, interrupted by the stalk. j.

    j. Petiole articulate, or continuous with the stem and then the spores spheroidal-trilete and green, or if ellipsoidal-monolete then the leaves sessile and spaced on the stem.
        22. Polypodiaceae, p. 684

    j. Petiole continuous with the stem, the spores spheroidal-trilete and not green, or if ellipsoidal-monolete then the leaves petioled or if sessile than clustered. k.

      k. Stem scales clathrate, if the petiole with two curved xylem strands toward the base then these facing outward. l.

        l. Indusium absent, lamina entire to furcate.
          15. Vittariaceae, p. 354

        l. Indusium present, or if absent then the lamina pinnate. 19. Aspleniaceae, p. 627

      k. Stem scales not calthrate, or if so then the petiole with two curved xylem strands toward the base and these facing inward. m.

        m. Spores trilete, or if monolete then the sori marginal. n.

          n. Stem with trichomes only and the sori exindusiate, or the stem with scales and an abaxial indusium absent. 14. Pteridaceae, p. 213

          n. Stem with trichomes only and the sori indusiate, or with scales and an abaxial indusium present.
            16. Dennstaedtiaceae, p. 370

        m. Spores monolete and the sori abaxial. o.

          o. Sori elongate, adjacent and parallel to the segment axis. 21. Blechnaceae, p. 662

          o. Sori roundish, or if elongate then most of all of them neither adjacent nor parallel to the segment axis, or the sori marginal. p.

            p. Petiole with two vascular bundles, leaves with unicellular, acicular or variously branched trichomes.
              17. Thelypteridaceae, p. 432

            p. Petiole with three or more vascular bundles, the only trichomes on the leaf minute, simple and blunt. q.

              q. Pinna-stalks continuous with the rachis, or if articulate then either the sori exindusiate or the pinnae with a large basal auricle on the basiscopic side.
                18. Dryopteridaceae, p. 454

              q. Pinnae articulate, sori indusiate, pinnae cordate at the base or less developed on the basiscopic side, lamina 1-pinnate.
                20. Davallliaceae, p. 654

  i. Sporangia sessile or subsessile, or with a stalk of 4–8 or more rows of cells, annulus absent, or if present then lateral. r.

    r. Sporangia lacking an annulus or with a usually poorly differentiated annulus. s.

      s. Sporangia lacking an annulus, borne on a specialized fertile branch of a leaf, spores lacking chlorophyll.
        1. Ophioglossaceae, p. 25

      s. Sporangia with a poorly differentiated lateral annulus, borne on partly or wholly fertile pinnae, spores with chlorophyll (green). 3. Osmundaceae, p. 50

    r. Sporangia with a well differentiated, oblique to apical annulus. t.

      t. Sporangia on the abaxial surface of a fertile portion of a leaf, remotely attached in a cluster or single on a vein, or in wholly fertile panicles. 4. Schizaeaceae, p. 58

      t. Sporangia contiguous on the receptacle of marginal or abaxial sori. u.

u. Sporangia in marginal sori. v.
  v. Stem with scales, siphonostelic.
        8. Hymenophyllopsidaceae, p. 129
  v. Stem with trichomes. w.
    w. Leaves very thin, one to a few cells thick, translucent, lacking stomata, stem protostelic.   6. Hymenophyllaceae, p. 97
    w. Leaves thickened, with stomata, stem siphonostelic or dictyostelic. x.
      x. Stem long-creeping, with scattered, short, stiff trichomes, receptacle elongate.   7. Loxomataceae, p. 125
      x. Stem massive, arborescent to decumbent, with a dense mass of very long, soft trichomes, receptacle short or globose.
        10. Dicksoniaceae, p. 138
u. Sporangia in abaxial sori. y.
  y. Stem and leaves lacking evident indument, a mucilaginous secretion which becomes flaky when dry sometimes present, leaves strongly dimorphic.   9. Plagiogyriaceae, p. 134
  y. Stem and usually the leaves with evident trichomes and (or) scales, leaves monomorphic to somewhat dimorphic. z.
    z. Stem slender, long-creeping, subterranean, freely branched, leaves usually partly pseudodichotomously branched with axillary arrested or dormant buds.
        5. Gleicheniaceae, p. 83
    z. Stem stout to massive, more or less epigeous or arborescent, sparingly if at all branched, leaves wholly pinnately branched. aa.
      aa. Stem and petiole bearing scales, trichomes may be present or absent.
        13. Cyatheaceae, p. 166
      aa. Stem and petiole bearing only trichomes. bb.
        bb. Lamina 2-pinnate-pinnatifid to 3-pinnate-pinnatisect.
          11. Lophosoriaceae, p. 156
        bb. Lamina 1-pinnate.
          12. Metaxyaceae, p. 162

Family 1. Ophioglossaceae

Ophioglossaceae Agardh, Aphor. Bot. 113. 1822. Type: *Ophioglossum* L.
Botrychiaceae Nakai, Jour. Jap. Bot. 24: 9. 1949. Type: *Botrychium* Sw.
Helminthostachyaceae Ching, Bull. Fan Mem. Instit. Biol. Bot. 10: 235.
     1941. Type: *Helminthostachys* Kaulf.

Description

Stem erect, unbranched, or prostrate and rarely branched, small, siphonostelic or rarely dictyostelic, fleshy, without indument, or pubescent-scaly at the apex; leaves small (usually several cm to ca. 50 cm long), entire, pinnate or broadly and deeply more or less palmate or dichotomous, glabrous or thinly pubescent erect or folded, or rarely somewhat circinate in bud, the petiole basally expanded into a membranous, stipular sheath, the sterile leaf with an expanded lamina, fertile leaf with sporangia borne on a special branch (or branches) arising from the base or below the sterile lamina, or (in *Ophioglossum simplex*) the sterile lamina reduced or absent; sporangia sessile or subsessile, separate or laterally joined in a synangium, lacking an annulus; homosporous, spores without chlorophyll. Gametophyte subterranean, without chlorophyll, mycorrhizic, fleshy, irregularly elongate, sometimes branched or nearly spherical, radially symmetrical with antheridia and archeogonia rather uniformly distributed or dorsiventral with the gametangia on the upper surface, the antheridium of many cells.

Comments on the Family

The family Ophioglossaceae is composed of three genera, two occurring worldwide and the third, *Helminthostachys* Kaulf., only in the wet tropics of Ceylon, southeast Asia and the western Pacific. The Ophioglossaceae are an isolated family, not closely allied to other living Pteridophyta. It is certainly much older than the earliest fossils, which are spores ascribed to *Ophioglossum* from the Jurassic. The family has primitive characters such as the large, massive sporangia with a high spore capacity. The leaves have fertile branches that are difficult to interpret in relation to the fertile leaves of most ferns. There are also specialized characters such as the mycorrhizal relation and usually simple leaf structure. Some species are ecologically specialized, as they occur in arenicolous habitats or are epiphytes.

The species tend to be very widely distributed, with several having worldwide or pantropic ranges. There are a few local endemics but these may actually be geographic variations of other species. The subterranean gametophytes evidently inhibit hybridization and outcrossing. Thus a minor variation of a species may, through inbreeding, form a colony; or different variations may grow together without intergrading through crossing. The resulting complex patterns of local morphological variation combined with the relatively few characters of the plants have made the taxonomy of the species difficult. The widespread occurrence of Ophioglossaceae in recently glaciated or climatically altered regions indicates a high colonizing capacity.

Transmission electron microscope studies of spores of Lugardon (1971) include a special series on the development and elaboration of the complex wall in *Ophioglossum vulgatum*. The spores of Ophioglossaceae are considered to have the least specialized form of exospore and represent one of the five main types in the Filicopsida.

Key to Genera of Ophioglossaceae in America

a. Sporangia separate on the paniculate, racemose, rarely spicate fertile branch, veins free.   1. *Botrychium*
a. Sporangia joined laterally in a synangial spike, veins anastomosing.
                                                                                        2. *Ophioglossum*

Literature

Clausen, R. T. 1938. A monograph of the Ophioglossaceae. Mem. Torrey Bot. Cl. 19 (2): 1–177.

de Lichtenstein, J. S. 1944. Las Ofioglosáceas de la Argentina, Chile y Uruguay. Darwiniana 6: 380–441.

Lugardon, B. 1971. Contribution à la connaissance de la morphogénèse et de la structure des parois sporales chez les Filicinées isoporées. 257 pp. Thèse, Univ. Paul Sabatier, Toulouse.

Nishida, M. 1952. A new system of *Ophioglossales*. Jour. Jap. Bot. 27: 271–278.

Walker, T. G. 1966. A cytotaxonomic survey of the pteridophytes of Jamaica. Trans. Roy. Soc. Edinburgh 66: 169–237.

1. *Botrychium*
Figs. 1.1–1.15

Botrychium Sw., Jour. Bot. (Schrad.) 1800 (2): 8, 110, 1802. Type: *Botrychium Lunaria* (L.) Sw. (*Osmunda Lunaria* L.).

Botrypus Michx., Fl. Bor. Amer. 2: 274. 1803. Type: *Botrypus virginianus* (L.). Michx. (as *virginicus*) (*Osmunda virginiana* L.) = *Botrychium virginianum* (L.) Sw. [Subgenus *Osmundopteris*].

Botrychium section *Osmundopteris* Milde, Verh. Zool. Bot. Ges. Wien 19: 96. 1896. Type: *Botrychium virginianum* (L.) Sw. *Botrychium* subgenus *Osmundopteris* (Milde) Clausen, Mem. Torrey Bot. Cl. 19: 93. 1938. *Osmundopteris* (Milde) Small, Ferns SE. United States, 377, 482. 1938.

Botrychium section *Phyllotrichium* Prantl, Ber. Deutsch Bot. Ges. 1: 349. 1883. Type: Has not been selected from the eight species listed. [Subgenus *Sceptridium*].

Sceptridium Lyon, Bot. Gaz. 40: 457. 1905. Type: *Sceptridium obliquum* (Willd.) Lyon (*Botrychium obliquum* Willd.) = *Botrychium dissectum* Spreng. *Botrychium* subgenus *Sceptridium* (Lyon) Clausen, Mem. Torrey Bot. Cl. 19: 24. 1938.

Japanobotrychium Masam., Jour. Soc. Trop. Agric. Formosa 3: 246. 1931. Type: *Japanobotrychium arisanense* Masam. = *Botrychium lanuginosum* Hook. & Grev. *Botrychium* subgenus *Japanobotrychium* (Masam.) Kato & Sahashi, Acta Phytotax. Geobot. 28: 150. 1977.

Description   Terrestrial, or (in *B. lanuginosum*) rarely epiphytic; stem erect, elongate, small, lacking indument, bearing thickened roots; leaf usually single, rarely several, usually a few cm to ca. 25 cm long (rarely to 75 cm), glabrous or especially the leaf bud thinly to densely pilose-pubescent, the green, expanded lamina pinnatifid, 1-pinnate to decompound, rarely simple and entire, bud of the next leaf wholly or partly enclosed by the expanded stipular

Fig. 1.1. *Botrychium dissectum* ssp. *decompositum,* Volcán Barba, Costa Rica, older leaf, lower left, and younger leaf with fertile branch not fully expanded. (Photo W. H. Hodge.)

base of the current leaf, veins free, the fertile branch arising at or well below the base of the sessile or long-stalked lamina, or (in *B. lanuginosum*) on the lamina, sporangia borne separately on the paniculate, racemose or rarely spicate fertile branch; spores tetrahedral-globose or globose, trilete, the laesurae $\frac{1}{2}$ the radius of the proximal face, the sporoderm granulate, shallowly to prominently tuberculate, or verrucate, usually more prominent on distal face. Chromosome number: $n = 45, 90, 92, 135$; $2n = 90, 180, 184$.

Botrychium roots are thickened, especially in species of subgenus *Sceptridium* (Fig. 3). Leaf buds are preformed several years ahead, each bud within the stipular base of the previous leaf. The leaf parts are differentiated prior to the final expansion of the bud. The late season bud of *B. virginianum* (Fig. 8) shows the well-formed pinnae and pinnules. Some species, especially in subgenus *Sceptridium,* have a ternate lamina (Figs. 3, 5) and thickened segments. In other species, the lamina is thin (Fig. 6) and the free veins are clearly apparent (Fig. 7). The sporangia are usually large, and although crowded on the fertile branch, each is discrete (Fig. 9).

Systematics

Botrychium is a widely distributed genus of about 25 species. Four groups, too closely related for generic recognition, are treated as subgenera. The infrageneric treatment follows Clausen (1938) except that subgenus *Osmundopteris* section *Lanuginosae* is treated as subgenus *Japanobotrychium* (Kato and Sahashi, 1977).

Fig. 1.2. American distribution of *Bo-trychium,* south of lat. 35° N.

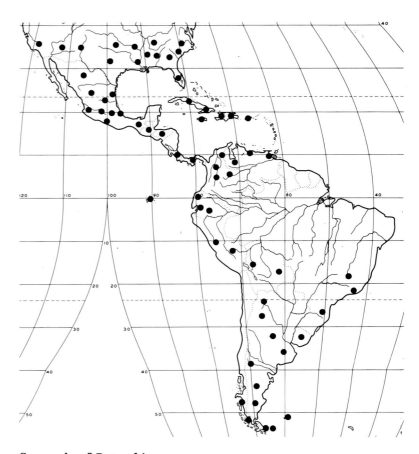

Synopsis of *Botrychium*

Subgenus *Sceptridium*

Leaf buds usually pilose with long, slender, whitish trichomes, completely enclosed by the sheath of the current leaf; lamina deltoid, ternate, usually decompound, thickened; the fertile branch arising well below the lamina which is then long-stalked.

About 12 species, eight in America and four in tropical America.

Subgenus *Japanobotrychium*

Leaf buds densely pilose with long, slender, whitish trichomes, completely enclosed by the sheath of the current leaf; lamina deltoid to elongate-deltoid, decompound, usually thickened; the fertile branch arising above the basal pinnae of the lamina, rarely at its base.

A single species, *Botrychium lanuginosum* Hook. & Grev. of southeastern Asia and Malesia.

Subgenus *Botrychium*

Leaf buds glabrous, completely enclosed by the sheath of the current leaf; usually elongate, rarely deltoid, lamina entire to 2-pinnate (rarely more), thin to rather thick; the fertile branch arising at or somewhat below the lamina which is then sessile to short-stalked.

Six or more species, all in America, but not tropical. The taxonomic status of four new species described by Wagner and

Figs. 1.3–1.7. *Botrychium*. 3. Plant of ***Botrychium Underwoodianum,*** the fertile branch, left, not fully expanded, × 0.25. **4.** Portion of lamina, ***B. dissectum*** ssp. ***decompositum,*** × 0.75. **5.** Ternate lamina, ***B. Schaffneri,*** × 0.75. **6.** Basal pinna of lamina, ***B. virginianum,*** × 0.75. **7.** Segment of lamina with free veins, ***B. virginianum,*** × 3.

Wagner (1981) must await assessment within a treatment of the whole subgenus.

Subgenus *Osmundopteris*

Leaf buds pilose, with long, slender, whitish trichomes, partially exposed by the sheath of the current leaf, lamina deltoid, decompound, thin; the fertile branch arising at the base of the lamina which is then sessile.

Three species, one in tropical and temperate America.

8   9

Figs. 1.8, 1.9. *Botrychium.* **8.** Leaf bud of *Botrychium virginianum* (petiole base of the leaf of the current year, right, has been displaced to expose the bud), ×1.5. **9.** Sporangia on paniculate fertile branch, *B. dissectum,* ×3.

Tropical American Species

Botrychium is not well represented in the American tropics. The four species of subgenus *Sceptridium* are: *B. dissectum* Spreng. ssp. *decompositum* (Mart. & Gal.) Clausen (Fig. 4), *B. Jenmanii* Underw., *B. Schaffneri* Underw. (Fig. 5), and *B. Underwoodianum* Maxon (Fig. 3). Subgenus *Osmundopteris* is represented by one species: *B. virginianum* (L.) Sw. (Fig. 6), including *B. cicutarium* (Sav.) Sw.

Ecology (Fig. 1)

Botrychium is primarily a genus of open habitats but also occurs in forests. In the American tropics the genus grows in a wide variety of habitats such as cloud forests, mossy forests, oak or pine woods, grasslands, alpine meadows, and sandy lake shores, from 1000 to 4000 m. It frequently colonizes partially vegetated, disturbed areas and persists through successional stages of vegetation.

Geography (Fig. 2)

The distribution of *Botrychium* is nearly worldwide. In tropical America it occurs in Mexico and Central America, all of the Greater Antilles, and in Andean Venezuela and Colombia, south to Bolivia, Argentina and southeastern Brazil. It also occurs southward to Staten Island, Tierra del Fuego, and Chile; also on the Galápagos Islands and Falkland Islands. In North America, it occurs northward to Labrador and Alaska.

Species in each of the three subgenera are widely distributed and there are some notable geographic disjunctions in American *Botrychium*. In subgenus *Sceptridium*, *B. australe* R. Br. is represented in Patagonia by ssp. *Negri* (Christ) Clausen, and in Australia and New Zealand by ssp. *australe*. Two species of subgenus *Botrychium* are widely disjunct north and south of the American tropics. *Botrychium Lunaria* (L.) Sw. and *B. matricariifolium* A. Br. are both in boreal America and in southern South America.

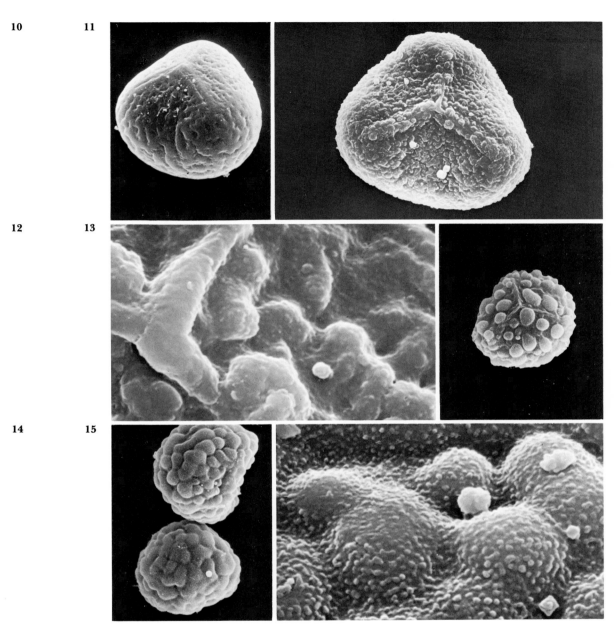

Figs. 1.10–1.15. *Botrychium* spores, × 10,000, surface detail, × 10,000. **10.** *B. Jenmanii,* proximal face, Hispaniola, *Ekman 13814.* **11, 12.** *B. lunaria* var. *Dusenii,* Patagonia, *Furlong 10.* **11.** Proximal face, irregularly papillate. **12.** Surface detail, portion of laesura, left. **13–15.** *B. virginianum.* **13.** Proximal face, verrucate, Panama, *Pittier 3026.* **14.** Distal face, below, lateral view above, verrucate, Panama, *Davidson 778.* **15.** Surface detail of distal face, as in Fig. 14.

Spores

The three main spore types in *Botrychium* generally correspond to the subdivisions of the genus. Those of subgenus *Sceptridium* have the surface shallow to prominently rugose (Fig. 10). Spores of subgenus *Botrychium* have a course and irregularly papillate surface (Figs. 11, 12). Those of *B. Lunaria* (Fig. 11) are notably larger than other species and may reflect a higher polyploid level. In subgenus *Osmundopteris* the spores are strongly verrucate (Figs. 13–15). Studies on the ultrastructure of the wall, especially the exospore, show the deposition of four strata overlaid by the thin perispore (Pettitt, 1979).

Cytology

The reports of chromosome numbers of tropical American species of *Botrychium* by Walker (1966, 1973) are consistently $n = 45$. This represents the lowest number for the genus and is considered the diploid level. In *B. virginianum* there are reports of $n = 92$ from North America (Fabbri, 1963) and Japan (Nishida et al. 1964) but $n = 90$ is also reported in the related *B. lanuginosum* of southeast Asia.

Observations

Gametophytes of all genera of the Ophioglossaceae are subterranean and have endophytic fungi. Gametophyte structure and the young sporophyte development in these genera have been reviewed by Bierhorst (1971). In experiments by Whittier (1972) spores of *Botrychium dissectum* were germinated and gametophytes grown to maturity without fungi. The larger gametophytes in the culture were of the same form as those collected in their native habitat. This work raises questions regarding the obligate nature of mycorrhizae and their role in growth and development of these plants.

Literature

Bierhorst, D. W. 1971. Morphology of vascular plants. 560 pp. Macmillan, New York.

Clausen, R. T. 1938. Reference under the family.

Fabbri, F. 1963. Tavole cromosomiche delle Pteridophyta, Prim. Suppl. Caryologia 16: 237–335.

Kato, M., and N. Sahashi. 1977. The systematic position of *Botrychium lanuginosum*. Acta Phytotax. Geobot. 28: 143–151.

Nishida, M., S. Kurita, and S. Nizzeki, 1964. Cytotaxonomy of *Ophioglossales,* II. Chromosome number in *Sceptridium.* Jour. Jap. Bot. 39: 140–144.

Pettitt, J. M. 1979. Ultrastructure and cytochemistry of spore wall morphogenesis, *in*: A. F. Dyer, The Experimental Biology of Ferns, pp. 213–252. Academic Press, London.

Wagner, W. H., Jr., and F. S. Wagner. 1981. New species of moonworts, *Botrychium* subgenus *Botrychium* (Ophioglossaceae) from North America. Amer. Fern Jour. 71: 20–30.

Walker, T. G. 1966. Reference under the family.

Walker, T. G. 1973. Additional cytotaxonomic notes on the pteridophytes of Jamaica. Trans. Roy. Soc. Edinburgh 69: 109–135.

Whittier, D. P. 1972. Gametophytes of *Botrychium dissectum* as grown in sterile culture. Bot. Gaz. 133: 336–339.

2. *Ophioglossum*
Figs. 2.1–2.19

Ophioglossum L., Sp. Pl. 1062. 1753; Gen. Pl. ed. 5, 484. 1754. Type: *Ophioglossum vulgatum* L.

Ophioderma (Blume) Endlicher, Gen. Pl. 66. 1836. *Ophioglossum* section *Ophioderma* Blume, Enum. Pl. Jav. 259. 1828. Type: *Ophioglossum pendulum* L. *Ophioglossum* subgenus *Ophioderma* (Blume) Clausen, Mem. Torrey Bot. Cl. 19: 114. 1938.

Cheiroglossa Presl, Suppl. Tent. Pterid. 56. 1845. Type: *Cheiroglossa palmata* (L.) Presl = *Ophioglossum palmatum* L. *Ophioglossum* subgenus *Cheiroglossa* (Presl) Clausen, Mem. Torrey Bot. Cl. 19: 112. 1938.

Rhizoglossum Presl, *ibidem* 47. 1845. Type: *Rhizoglossum Bergianum* (Schlect.) Presl = *Ophioglossum Bergianum* Schlect. *Ophioglossum* subgenus *Rhizoglossum* (Presl) Clausen, Mem. Torrey Bot. Cl. 19: 163. 1938.

Description

Terrestrial or epiphytic; stem erect, globose or short-prostrate, small, without indument or pubescent-scaly at the apex, roots thickened, sometimes with proliferous buds; leaf usually single, sometimes several, small, ca. 1.6–25 cm long, or rarely (in *O. pendulum*) over 1 m long, the sterile lamina simple, entire, and narrowly elliptic to nearly circular or elongate, or deeply, more or less palmately or dichotomously branched, glabrous, the bud of the next leaf free from the eroded stipular base of the current leaf, veins anastomosing, frequently with included free veinlets, the single fertile branch arising at or below the base of the sterile lamina, or (in subgenus *Ophioderma*) borne centrally on the sterile lamina, or (in subgenus *Cheiroglossa*) usually a few to several on the margin of the base of the sterile lamina and (or) at the apex of the petiole, very rarely (in *O. Bergianum* and *O. simplex* Ridley) the fertile leaf lacking a sterile lamina, sporangia joined laterally in a synangial spike; spores globose, trilete, the laesurae $\frac{1}{2}$ to $\frac{2}{3}$ the radius of the proximal face, the sporoderm surface with fused cones and pits that appear more or less reticulate. Chromosome number: among numerous reports are $n = 120$, 240, ca. 360, 436, 451, 480, 564, 566, 631; $2n =$ ca. 300, 480, 720, 960, 1262.

Plants of *Ophioglossum* are usually at least 15 cm tall, but there are some minute species as in Fig. 4. The stem is usually elongate or rarely globose (Fig. 5), bearing somewhat thickened roots (Fig. 6). The stipular leaf bases usually wither and deteriorate along with the rest of the leaf, but may persist (Fig. 6). Venation is areolate with free veinlets within the areolae (Figs. 7, 8, 11). The sporangia are laterally fused into a synangium (Fig. 10) and the spores disperse through a transverse slit.

Systematics

Ophioglossum is a widely distributed genus of about 30 species grouped into four subgenera. These are sometimes recognized as genera but are quite similar, especially in the fertile parts, and are best maintained in a single genus. The infrageneric classification follows Clausen (1938).

Fig. 2.1. *Ophioglossum reticulatum* on an Inca terrace, Macchu Picchu, Peru. (Photo Alice F. Tryon.)

Synopsis of *Ophioglossum*

Subgenus *Ophioglossum*

Terrestrial, stem glabrous, fertile leaf with a single spike arising at or below the base of the entire, small sterile lamina.

About 22 species; 10 of the 11 American species are in the tropics.

Subgenus *Rhizoglossum*

Terrestrial, stem glabrous, fertile leaf with a single spike and no sterile lamina, sterile leaf with an entire, elongate, small lamina.

One species, *O. Bergianum* Schlect., of South Africa.

Subgenus *Ophioderma*

Terrestrial or epiphytic, stem somewhat scaly at the apex, fertile leaf with a single spike arising more or less centrally on the entire, elongate, sometimes very long, sterile lamina, or the fertile leaf lacking a sterile lamina.

Three species of Malesia and the Pacific.

Subgenus *Cheiroglossa*

Epiphytic, stem densely pubescent-scaly at the apex, fertile leaf with a few to several (rarely one) spikes arising from the margin

Fig. 2.2. *Ophioglossum palmatum,* epiphyte in montane forest, Mera, Prov. Pastaza, Ecuador. (Photo B. Øllgaard.)

at, or somewhat below, the base of the usually deeply and irregularly palmately lobed sterile lamina.

One species, *O. palmatum,* of the American tropics and rare in the paleotropics.

Tropical American Species

The following species of subgenus *Ophioglossum* occur in the American tropics: *Ophioglossum crotalophoroides* Walt., *O. ellipticum* Hook. & Grev., *O. Engelmannii* Prantl, *O. Harrisii* Underw., *O. lusitanicum* L., *O. nudicaule* L. f., (Fig. 4), *O. petiolatum* Hook., *O. reticulatum* L. (Fig. 1), *O. scariosum* Clausen, and *O. vulgatum* L. In addition, *O. fernandezianum* C. Chr. is on Juan Fernandez and

Fig. 2.3. American distribution of *Ophioglossum*, south of lat. 35° N.

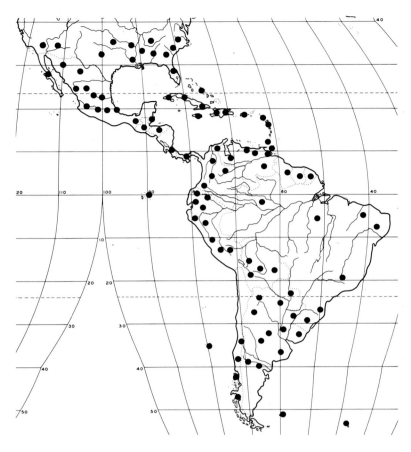

Easter Islands. *Ophioglossum palmatum* L. (Figs. 2, 9), of subgenus *Cheiroglossa,* is a variable species, especially in the degree of lobing of the sterile lamina and in the number and placement of the fertile spikes. Gómez (1976) has discussed the variation and the reasons for recognizing a single species. In a study of the epidermis, Maroti (1965) shows different cell types and patterns that may provide useful taxonomic characters.

Ecology (Figs. 1, 2)

Species of subgenus *Ophioglossum* grow in a wide range of habitats, generally in open soil or among grasses. In tropical America, plants are most frequent in open woods, savannahs and marshy areas, but also grow in soil pockets in lava beds, in forests, in *Larrea* scrub (Mexico), in rocky places, on coastal lomas (Peru), and on stream banks. Species are frequent pioneers on roadside borders and banks, in pastures and early successional stages of cleared and disturbed lands. *Ophioglossum Engelmannii* grows in nonacidic soils, especially in thin soils over limestone. The species usually occur from 500 to 3000 m, although there are a few records beyond this range. *Ophioglossum palmatum* is an epiphyte in wet forests usually from 500 to 2000 m.

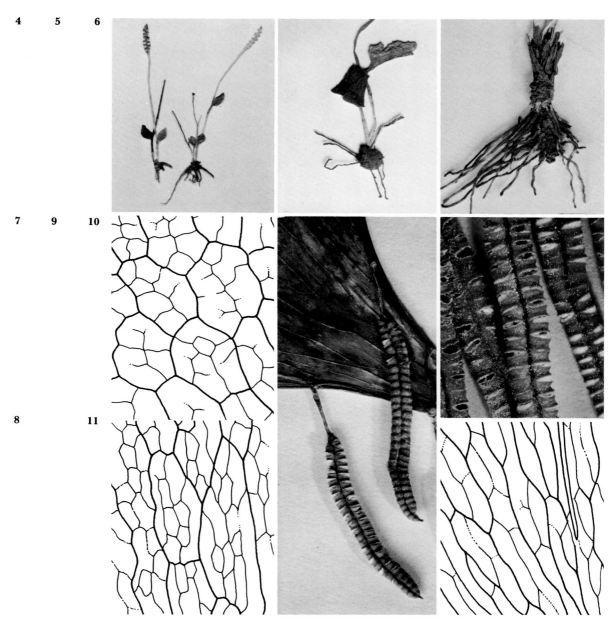

Figs. 2.4–2.11. *Ophioglossum.* **4.** Plants of *Ophioglossum nudicale,* ×1.5. **5.** Globose stem of *O. crotalophoroides,* ×1.5. **6.** Root system and unusually persistent leaf bases, *O. Engelmannii,* ×1.5. **7, 8.** Anastomosing veins of the sterile lamina, ×5. **7.** *O. reticulatum.* **8.** *O. Engelmannii.* **9.** Fertile branches of *O. palmatum,* on the margins at the base of the sterile lamina, ×2.5. **10.** Portions of synangia, each sporangium with a transverse opening, *O. palmatum,* ×5. **11.** Anastomosing veins, *O. palmatum,* ×5.

Geography (Fig. 3)

The distribution of *Ophioglossum* is essentially worldwide; in tropical America it grows from Mexico and Central America, through the Antilles, and generally throughout South America. It also occurs in southern Chile and Argentina, the Galápagos, Juan Fernandez, Falkland Islands, South Georgia, and in North America north to Quebec in Canada, and southern Alaska.

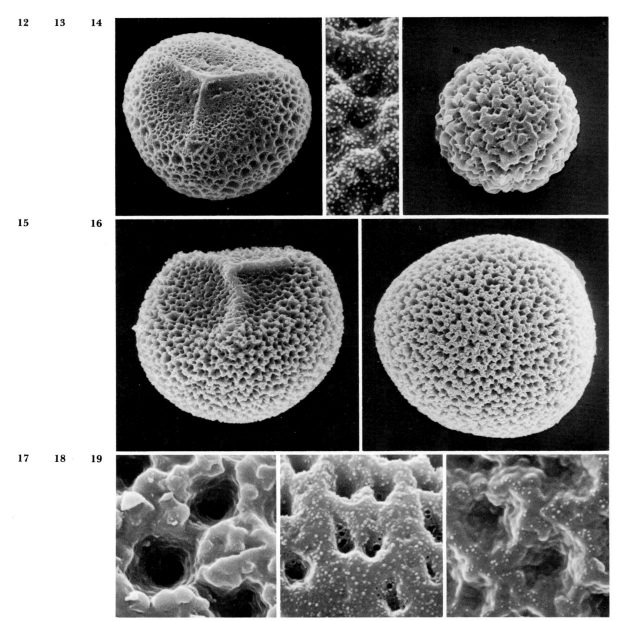

Figs. 2.12–2.19. *Ophioglossum* spores, × 1000, surface detail, × 10,000. **12–13.** *O. Engelmannii,* Mexico, *Pringle 8649.* **12.** Proximal face somethat reticulate in equatorial area. **13.** Detail, particulate surface. **14.** *O. crotalophoroides,* distal face, Brazil, *Leite 1604.* **15–17.** *O. palmatum,* Dominican Republic, *Howard 12214.* **15.** Proximal face, tilted. **16.** Distal face. **17.** Detail of pitted, distal face. **18.** *O. reticulatum,* detail of small pits and particulate deposit, proximal face, Peru, *Tryon & Tryon 5400.* **19.** *O. ellipticum,* surface detail, distal face, British Honduras, *Gentle 2997.*

Spores

The tropical American species have spores with a basically uniform cone-pitted surface that appears more or less reticulate (Figs. 12–16). Variations are mainly in the size and degree of coalescence of the cones and the density of the pits (Figs. 17–19). Apertures at the base of pits (Fig. 18) and the particulate surface deposit are evident at higher magnification (Fig. 13). Studies of wall development in *O. vulgatum* show a thin outer perispore formed over a thick, stratified exospore (Lugardon, 1972).

Cytology

Cytologically the genus is most remarkable in the record of $n = 631$, (indicated as $631 + 10$ fragments) in *Ophioglossum reticulatum* (Abraham and Ninan, 1954), which is the highest chromosome number known for any living organism. A higher report $n = 720$ has not been photographically documented. The report of a chromosome number, $n = 480$ for this species from Jamaica, by Walker (1966) is regarded as an octoploid. The same number is also known for populations from India (Bir, 1973). Several polyploids reported for this species range from $n = 120-631$ and there also are aneuploid records as $n = 436, 451, 564, 566$.

Literature

Abraham, A., and C. A. Ninan. 1954. The chromosomes of *Ophioglossum reticulatum*. Curr. Science 23: 213–214.

Bir, S. S. 1973. Cytology of Indian pteridophytes. Glimpses in plant research 1: 28–119.

Clausen, R. T. 1938. Reference under the family.

Gómez, L. D. 1976. Variation in Costa Rican *Ophioglossum palmatum* and nomenclature of the species. Amer. Fern Jour. 66: 89–92.

Lugardon, B. 1972. La structure fine de l'exospore et de la périspore des filicinées isosporées. Pollen et Spores 14: 227–261

Maroti, I. 1965. Vergleichende Anatomische Untersuchungen an den blattern der Ophioglossaceae. Acta Biologica (Acta Univ. Szeged.), n.s. 11: 55–71.

Verma, S. C. 1956. Cytology of *Ophioglossum*. Curr. Science 25: 398–399.

Walker, T. G. 1966. Reference under the family.

Family 2.   Marattiaceae.

Marattiaceae Bercht. & J. S. Presl, Přirozen. Rostl. 1: 272. 1820. Type: *Marattia* Sw.

Danaeaceae Agardh, Aphor. Bot. 117. 1822. Type: *Danaea* Sm.

Angiopteridaceae Bommer, Bull. Soc. Roy. Bot. Belg. 5: 359. 1866. Type: *Angiopteris* Hoffm.

Christenseniaceae Ching, Bull. Fan Mem. Instit. Biol. Bot. 10: 227. 1940. Type: *Christensenia* Maxon (*Kaulfussia* Blume, not Dennst., nor Nees).

Stem erect, decumbent, or prostrate, unbranched, moderately stout to massive, dissected dictyostelic or nearly polystelic, fleshy, with sometimes persistent scales; leaves ca. 50 cm to 3 m or more long, pinnate, rarely entire, or (in *Christensenia*) palmate, very sparingly to moderately scaly, especially on the axes, circinate in the bud, petiole enlarged at the base with fleshy, rather thin-margined, sometimes proliferous stipules, rachis often with a swollen node (pulvinus) at the base of each pair of pinnae, similar nodes sometimes also on the petiole; sporangia several to many in a compact sorus on the abaxial surface of the lamina, sessile or joined in a massive, indurated synangium, annulus lacking; homosporous, spores without chlorophyll. Gametophyte epigeal, with chlorophyll, somewhat to definitely elongate, sometimes branched, with a thickened central region and thinner lateral margins, the archegonia borne on the lower surface of the thickened portion, the many celled antheridia borne mostly on the lower, less often on the upper surface, or on the margins.

Comments on the Family

The Marattiaceae are represented by two genera in America: *Danaea* which is endemic, and *Marattia* which is also paleotropical. The other five genera, *Angiopteris* Hoffm., *Macroglossum* Copel., *Archangiopteris* Christ & Giesenh., *Christensenia* Maxon, and the recently reinstated *Protomarattia* Hayata (Pichi-Sermolli, 1968), all center in the tropics of southeastern Asia. The genera have been placed in two, or sometimes in four, families. However, the characters common to all of them indicate that they represent one major evolutionary line and justify their treatment as a single family. Cytological conformity based on $n = 40$, reported in several genera, support this relationship. Biochemical studies of flavonoids by Swain and Cooper-Driver (1973) show the occurrence of leucocyanidins in both *Marattia* and *Angiopteris*.

The family is an ancient one with a fossil record back to the Carboniferous (Mamay, 1950; Stidd, 1974). The living members form a distinctive family and alliances to other living ferns have not been clearly established. The unusual forms of synangia are particularly unique among living ferns and appear to be similar to those of fossil genera such as *Scolecopteris* Zenker or *Ptychocarpus* Weiss.

The whole family is ecologically specialized in being well adapted for growth on the shaded floor of wet, tropical forests. There is little detailed ecological information and no recent systematic study on either *Marattia* or *Danaea*.

Key to Genera of Marattiaceae in America

a. Lamina 2- to 4-pinnate, gradually reduced at the apex, synangia superficial on the unmodified fertile segments.   3. *Marattia*
a. Lamina rarely entire, usually 1-pinnate and imparipinnate, or the rachis tip proliferous, synangia deeply sunken between vertical partitions formed by the leaf tissue between the veins.   4. *Danaea*

Literature

Mamay, S. H. 1950. Some American Carboniferous fern fructifications. Ann. Mo. Bot. Gard. 37: 409–476.

Pichi-Sermolli, R. E. G. 1968. The fern genus *Protomarattia* Hayata. Webbia 23: 153–158.

Stidd, B. M. 1974. Evolutionary trends in the Marattiales. Ann. Mo. Bot. Gard. 61: 388–407.

Swain, T., and G. Cooper-Driver. 1973. Biochemical systematics in the Filicopsida, *in*: The phylogeny and classification of the ferns. Bot. Jour. Linn. Soc. Suppl. 1, 67: 111–134.

Walker, T. G. 1966. A cytological survey of the pteridophytes of Jamaica. Trans. Roy. Soc. Edinburgh 66: 169–237.

3. *Marattia*
Figs. 3.1–3.13

Marattia Sw., Prod. Fl. Ind. Occ. 8, 128. 1788. Type: *Marattia alata* Sw.
Myriotheca Juss., Gen. Pl. 15. 1789. Type: *Myriotheca fraxinea* (Sm.) Poir.
  (Lam. Encycl. 403. 1797) = *Marattia fraxinea* Sm.
Eupodium Hook., Gen. Fil. *t. 118*. 1842. Type: *Eupodium Kaulfussii*
  (Hook.) Hook. (*Marattia Kaulfussii* Hook.) = *Marattia laevis* Sm.
Discostegia Presl, Suppl. Tent. Pterid. 11. 1845. *Discostegia alata* (Sw.)
  Presl = *Marattia alata* Sw.
Gymnotheca Presl, *ibidem* 12. 1845. Type: *Gymnotheca cicutifolia* (Kaulf.)
  Presl = *Marattia cicutifolia* Kaulf.
Stibasia Presl, *ibidem* 15. 1845. Type: *Stibasia Douglasii* Presl = *Marattia
  Douglasii* (Presl) Hook. & Baker.

Description

Terrestrial, stem erect, globose and often large, including the persistent petiole bases, bearing sometimes persistent scales and thickened roots; leaves sometimes solitary, monomorphic, to ca. 2–3.5 m long, lamina 2- to 4-pinnate, gradually reduced at the apex, pinnae and pinnules often opposite glabrous or (especially beneath) somewhat scaly and with a few trichomes, veins free; synangial sori on the unmodified fertile segments, supramedial to nearly marginal or less often medial, sessile to definitely stalked, apically bivalvate, opening lengthwise with the valves spreading to expose the oblique, elongate sporangial pores; spores more or less ellipsoidal, monolete, the laesurae short, mostly obscure, the sporoderm a coarse, low, rugate lower layer superimposed by a thinner, more or less pelleted outer sheath, or echinate. Chromosome number: $n = 39, 40, 78; 2n = 156$.

The leaves of *Marattia* have a basically ternate lamina architecture with the basal pinnae elaborated as in Fig. 4. The figure represents a small leaf but they often reach a meter or more in length and have stout petioles. The petioles usually have a moderate cover of broad, tan or bicolorous scales (Fig. 3) which are darker at the point of attachment. The ultimate axes are alate (Fig. 5) especially in decompound leaves. The synangia may be stalked; and the spores are released through a grid of elongate apertures (Fig. 6). These unusual spore producing structures have been interpreted as a single synangium or two synnangia fused into a sporocarp.

Systematics

Marattia is a pantropical genus of perhaps 40 species, although these are not adequately known. Species with a stalked synangium (*Eupodium*) may merit recognition as a subgenus but a modern revision is needed before an infrageneric classification can be presented. Underwood (1909) treated six North American species.

Tropical American Species

Examples of the several species are: *Marattia alata* Sw., *M. cicutifolia* Kaulf., *M. excavata* Underw., *M. interposita* Christ, *M. laevis* Sm. (*M. Kaulfussii* Hook.), and *M. weimanniifolia* Liebm. These and others such as *M. chiricana* Maxon, *M. laxa* Kze. and *M. Raddii* Desv. require the interpretation of a modern monograph based on field studies and complete collections.

Fig. 3.1. *Marattia laevis,* cloud forest, Monte Jayuya, Puerto Rico. (Photo D. S. Conant.)

Fig. 3.2. Distribution of *Marattia* in America.

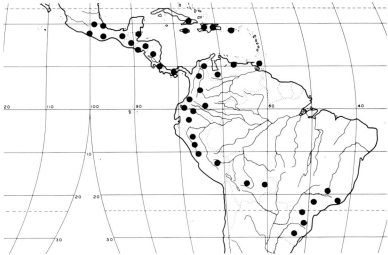

Ecology (Fig. 1)

Marattia is a genus of the wet tropics. In America it grows on stream banks, in ravines, in wet mountain forests and cloud forests, from 700 to 3000 m. It is more restricted to the higher and cooler areas than *Danaea*.

Geography (Fig. 2)

The genus ranges in the paleotropics from Africa to the Pacific, including New Zealand and the Hawaiian Islands. In America it occurs in southern Mexico and Central America, the Greater Antilles and the Andes from Venezuela to Bolivia; it is somewhat isolated in south-eastern Brazil.

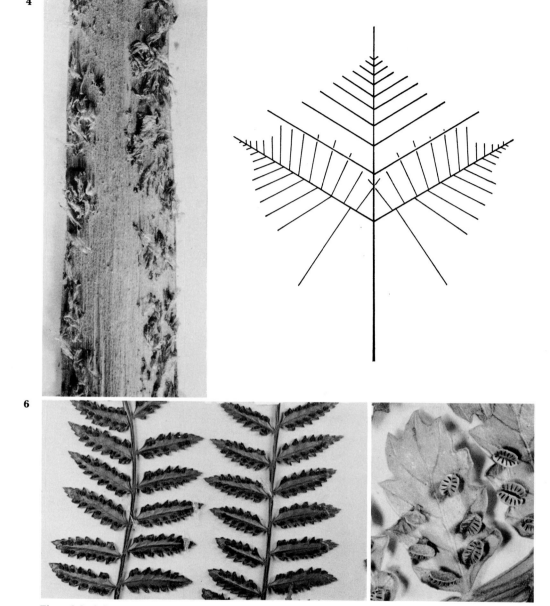

Figs. 3.3–3.6. *Maratia*. 3. Scales on petiole, ***Marattia interposita***, ×1.5. **4.** Diagram of architecture of small leaf of **M. laevis**, ×0.25. **5.** Portions of two fertile pinnules from a central pinna of **M. laevis**, ×1. **6.** Fertile segment and synangia, with two bent (lower left), showing the stalk, **M. laevis**, ×5.

Spores

The spores of *Marattia* (Figs. 7–13) are among the smallest ranging from 15 to 32 μm at the longest axis. The perispore in the Marattiaceae is regarded to be of a unique, cleaved form by Lugardon (1974) and discontinuities were shown in a wall section of *Marattia fraxinea* Sm. This separation is not evident in the abraded surface of the wall of *Marattia* shown in Fig. 11 with the exospore (E) below the perispore (P). Most tropical American species of *Marattia* have spores with a low, rugose surface as in

7   8   9

10   11

12   13

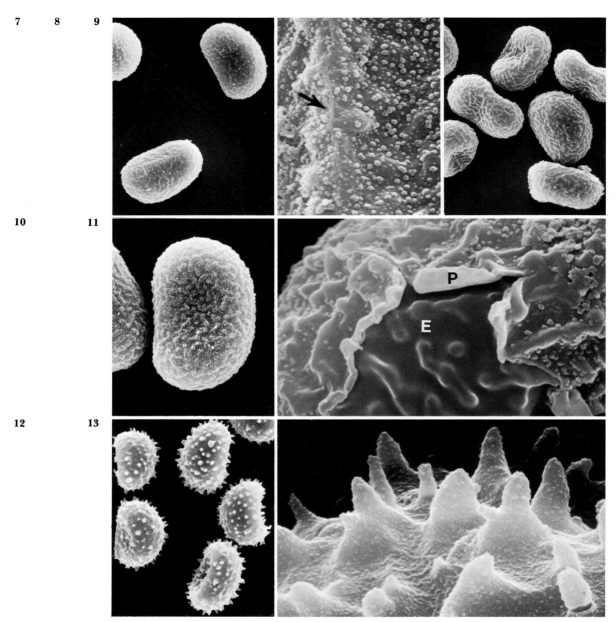

Figs. 3.7–3.13. *Marattia* spores, × 1000, except Fig. 10; surface detail, × 10,000. **7, 8. *M. excavata,*** Honduras, *Yunker et al. 6163.* **7.** Low, rugate, monolete spores, proximal face below, lateral view above. **8.** Surface detail, laesura left. **9. *M. laxa,*** Mexico, *Bourgeau 2108.* **10, 11. *M. weinmanniifolia,*** Guatemala, *Tuerckheim 1006.* **10.** Rugose spores, × 2000. **11.** Wall structure, exospore (E), perispore (P). **12.** Echinate spores, ***M. alata.*** Brazil, *Dusén 2751.* **13.** Detail of echinate surface, ***M. laevis,*** Dominican Republic, *Gastony et al. 385.*

M. weimanniifolia (Figs. 10, 11). In contrast to these, *M. alata* (Fig. 12) and *M. laevis* (Fig. 13) have distinctive, echinate spores resembling those of *Danaea.*

Cytology

Marattia alata is reported as *n* = 40, a diploid from Jamaica (Walker, 1966). The number is also known for *M. fraxinea* from west Africa. Collections of this species from southern India are reported as $n = 78$, $2n = 156$, and *M. salicina* Sm. from New Zealand as $n = 39$. There clearly appear to be two diploid numbers in the genus 78 and 40 with 39 possibly derived from 40.

Literature

Lugardon, B. 1974. La structure fine de l'exospore et de la périspore des filicinées isosporées. 2. Filicales. Commentaires. Pollen et Spores 16: 161–226.
Underwood, L. M. 1909. *Marattia,* in No. Amer. Fl. 16: 21–23.
Walker, T. G. 1966. Reference under the family.

4. *Danaea*
Figs. 4.1–4.20

Danaea Sm., Mém. Acad. Turin 5: 420. 1793, *nom. conserv.,* not Allioni, 1785. Type: *Danaea nodosa* (L) Sm. (*Acrostichum nodosum* L.).
Heterodanaea Presl, Suppl. Tent. Pterid. 38. 1845. Type: *Heterodanaea stenophylla* (Kze.) Presl = *Danaea stenophylla* Kze.

Description

Terrestrial; stem elongate, prostrate and creeping to decumbent or erect, rather slender to moderately stout including the persistent petiole bases, bearing sometimes persistent scales and thickened roots; leaves monomorphic to slightly dimorphic (the fertile ones more erect and with smaller pinnae), few to several on a plant, often 30–50 cm long, to ca. 2 m long, lamina rarely entire, or usually 1-pinnate and imparipinnate or with the rachis tip proliferous, pinnae entire, opposite, the rachis nodose at the base, glabrous or usually slightly scaly, especially beneath, petiole with one to few nodes, veins free; synangial sori deeply sunken between vertical partitions formed by the leaf tissue between the veins, extending nearly from the costa to the margin, sessile, not valvate, with apical, more or less circular sporangial pores; spores more or less ellipsoidal, monolete with short, obscure laesurae, the surface prominently echinate. Chromosome number: $n = 40$ I + II, 80; $2n = 160$.

Open, dichotomous venation (Figs. 12, 13) is characteristic of the genus. The petiole is somewhat scaly to glabrate. Many scales of the leaf are peltate and appressed. The synangia are densely packed and imperforate when young (Fig. 14), and later a series of pores develop through which the spores are dispersed (Fig. 15).

Systematics

Danaea is a tropical American genus of perhaps 20 or more species. There is no recent study of the genus and it is in need of a systematic revision. The genus *Danaeopsis* Presl (1845, p. 39), described on the basis of sterile material, is sometimes cited as a synonym of *Danaea,* with which Presl associated it. However, it has anastomosing veins and is probably a *Bolbitis.*

Tropical American Species

Characteristic species of *Danaea* are: *D. alata* Sm., *D. elliptica* Sm. (Fig. 1), *D. grandifolia* Underw. (Fig. 3), *D. jamaicensis* Underw., *D. longifolia* Desv., *D. Moritziana* Presl (Figs. 8, 9), *D. nodosa* (L.)

Fig. 4.1. *Danaea elliptica,* Monte del Estado, near Maricao, Puerto Rico. (Photo L. Wells.)

Sm., *D. simplicifolia* Rudge (Fig. 7), *D. trifoliata* Kze. (Fig. 6), *D. Urbanii* Maxon (Fig. 5), *D. Wendlandii* Reichb. f. (Figs. 4, 8, 9), and *D. Wrightii* Underw.

The leaf form in *Danaea* is illustrated in Figs. 3–7. *Danaea simplicifolia* (Fig. 7) and *D. carrillensis* Christ are the only species with an entire lamina. *Danaea trifoliata* has a large terminal segment (Fig. 6) more or less equivalent in size to the lamina of *D. simplicifolia,* but with a pair of small basal pinnae. This may represent a stage in lamina reduction leading to the entire form as in *D. simplicifolia.*

Fig. 4.2. Distribution of *Danaea.*

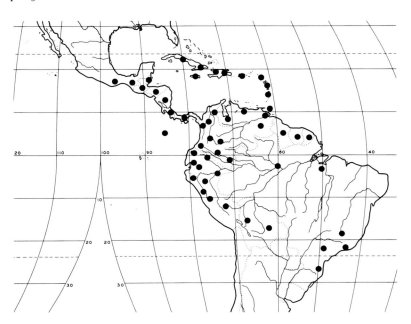

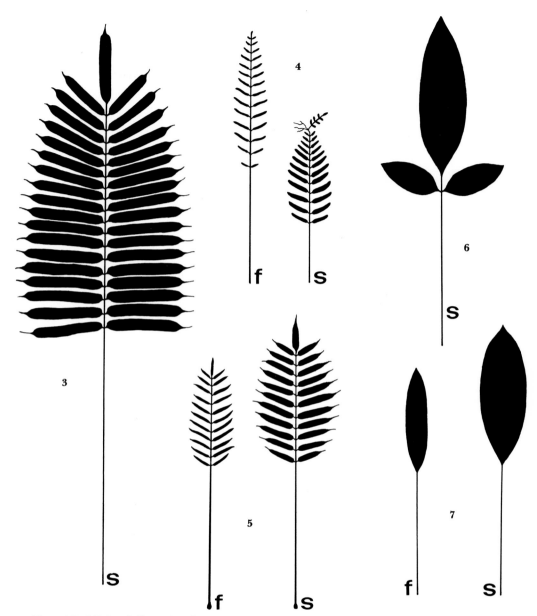

Figs. 4.3–4.7. Leaf diagrams of Danaea species (s, sterile leaf; f, fertile leaf). **3.** ***D. grandifolia,*** ×0.1. **4.** ***D. Wendlandii,*** ×0.2. **5.** ***D. Urbanii,*** ×0.2. **6.** ***D. trifoliata,*** ×0.2. **7.** ***D. simplicifolia,*** ×0.2.

Ecology (Fig. 1)

Danaea grows in ravines, on stream banks, sometimes in rocky places, in wet forests and in rain forest from near sea level to ca. 2300 m. In the Lesser Antilles it occurs mostly between 600 and 1000 m and is a good indicator of rain forest.

Geography (Fig. 2)

Danaea ranges from southern Mexico to Panama, through the Antilles, Guianas and Venezuela to Bolivia and adjacent Amazonian Brazil. It is disjunct in southeastern Brazil and also occurs on Cocos Island.

8

9   11

10

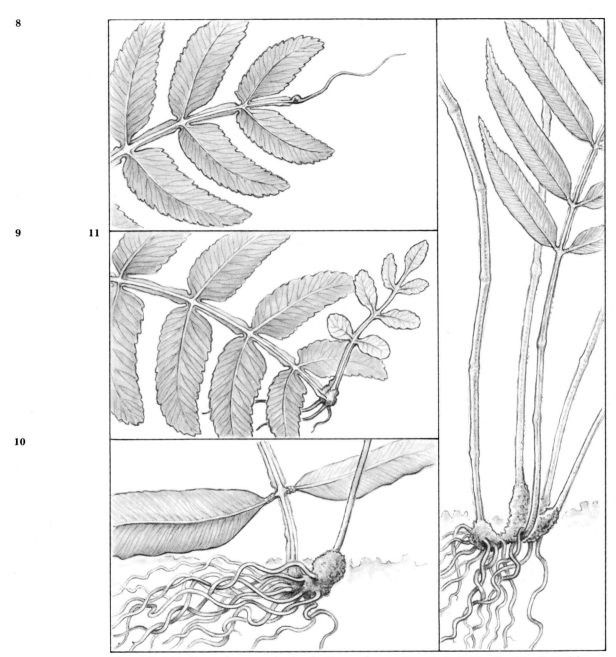

Figs. 4.8–4.11. Vegetative reproduction in *Danaea* leaves, ×1. **8, 9.** *D. Wendlandii.* **8.** Proliferous rachis tip, with one root. **9.** Small plant at rachis tip. **10, 11.** *D. Moritziana.* **10.** Apex of leaf and established plant, with roots and small stem bearing one petiole. **11.** An older plant with four leaves (right), the persistent rachis of the old leaf (far left) without pinnae.

Spores

Danaea spores (Figs. 16–20) are generally larger than those of *Marattia* but vary considerably in size and shape. The sporoderm is prominently echinate but varies in density and size of the projections which may be coarse to somewhat rugose as in *D. simplicifolia* (Fig. 20). The wall section in *D. longifolia* (Fig. 17) shows the echinate projections formed by the compact exospore.

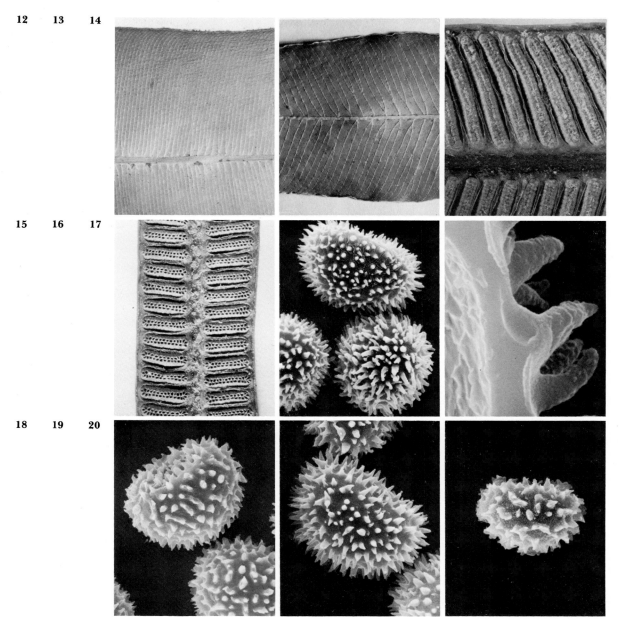

Figs. 4.12–4.20. *Danaea.* **12.** Venation of sterile pinna of *Danaea grandifolia,* ×1.5. **13.** Venation of sterile pinna of *D. elliptica,* ×1.5. **14.** Nearly mature synangia of *D. longifolia,* the sporangial pores not open, ×5. **15.** Mature synangia of *D. alata,* with open sporangial pores, ×5. **16–20.** Echinate spores of *Danaea,* ×1000. **16, 17.** *D. longifolia,* Brazil, *Dusén 14686.* **16.** Lateral view. **17.** Wall section, the compact exospore forms the echinate structure, ×10,000. **18.** *D. Wrightii,* Cuba, *Clement 976.* **19.** *D. grandifolia,* Panama, *Stern & Chambers 185.* **20.** *D. simplicifolia,* French Guiana, *Leprieur 168.*

Cytology

Several species of *Danaea* from Jamaica, Trinidad and Surinam are reported as $n = 80$ and regarded as tetraploids by Walker (1966). A population of hybrids between one of the tetraploids, *D. Jenmanii* Underw. and *D. jamaicensis,* in Jamaica was reported by Walker as triploid with 40 pairs and 40 univalents. Thus *D.*

jamaicensis represents the diploid element in this hybrid. A report of *D. nodosa* from Puerto Rico as 116, by Sorsa (1970), apparently is an approximate count of another triploid.

Observations

New plants are often vegetatively formed from leaves. In *D. Wendlandii* a bud is initiated at the apex of the rachis (Fig. 8) and a young plant develops from this (Fig. 9). In *D. Moritziana* the stem becomes thickened and many adventitious roots develop while the young plant is attached to the rachis (Fig. 10). The old rachis often persists after the young plants are well developed (Fig. 11). Nodes are frequent on the petioles (Fig. 11) and have been interpreted as sites of former pairs of pinnae (Brebner, 1902; Underwood, 1909). However, they are spaced differently than the pinnae and their possible function and origin need to be investigated.

Literature

Brebner, G. 1902. On the anatomy of *Danaea* and other Marattiaceae. Ann. Bot. 16: 517–552.

Presl, C. B. 1845. Supplementum Tentamen Pteridographiae. 119 pp. Pragae.

Sorsa, V. 1970. Fern cytology and the radiation field, *in:* H. T. Odum, ed., A Tropical Rain Forest, pp. G39–50. Off. Tech. Inform., U.S. Atomic Energy Comm., U.S. Dept. Commerce, Springfield, Virginia.

Underwood, L. M. 1909. *Danaea,* in No. Amer. Fl. 16: 17–21.

Walker, T. G. 1966. Reference under the family.

Family 3.   Osmundaceae

Osmundaceae Bercht. & J. S. Presl, Přirozen. Rostl. 1: 272. 1820. Type: *Osmunda* L.

Description   Stem erect to decumbent, massive, usually dichotomously branched, or with a single arborescent trunk, medullated protostelic, becoming dictyostelic, indurated, lacking indument; leaves usually ca. 1–2 m long, pinnate, bearing trichomes, at least when young, circinate in the bud, petiole with an expanded, stipular base; sporangia separate or in loose clusters, borne on wholly fertile parts of the lamina or on the abaxial surface of relatively unmodified segments, with a short, many-rowed stalk, and a poorly differentiated annulus; homosporous, spores with chlorophyll (green). Gametophyte epigeal, with chlorophyll, obcordate to elongate, with a thickened center and thin margins, the archegonia borne on the lower surface in rows along both sides of the thickened portion, the several- to many-celled antheridia mostly on the lower surface of the margins, less often at their edge.

Comments on the Family

The family Osmundaceae has three genera: *Osmunda,* nearly worldwide, *Todea* Bernh. of South Africa, New Guinea, Australia and New Zealand, and *Leptopteris* Presl, an arborescent genus, with filmy leaves lacking stomates, in western Polynesia, New Guinea, Australia and New Zealand.

The Osmundaceae are an old, distinctive family with a fossil record from the Carboniferous. The study by Miller (1971) reviews characters of stems, leaf bases and origin of the roots with the purpose of showing phyletic relationships. The living groups of *Osmunda* are related to Paleocene species, but some reservation is expressed on allying the older fossils from the Paleozoic to the Cretaceous to the living osmundas. Although there is a relatively rich fossil record, there are still significant gaps that preclude a more complete synthesis of the evolution of the family. A detailed study of the living species of the family by Hewitson (1962) includes morphological aspects of the stem and leaves and the anatomy of the leaf base, sporangia and epidermal structure. The disposition of sclerenchyma in the leaf base was regarded as the most reliable anatomical feature for characterizing species and genera.

The Osmundaceae are more advanced than the Ophioglossaceae or Marattiaceae on the basis of the sporangium having an annulus and with walls only one cell thick. However, the relatively large size of the sporangium, the rudimentary type of annulus, and the stipular leaf bases are more primitive characters as compared to those in the families that follow.

The spores of *Osmunda* and *Leptopteris* shown in the work of Lugardon (1971, 1972) have a similar massive, rugose exospore, and above this a thin, echinate perispore. The exospore completely envelops the spore as in the Ophioglossaceae and Marattiaceae, but does not overlie the scar apex as in those families.

The chromosome numbers are uniformily $n = 22$ in the family and they are exceptionally large. The base number is $x = 11$.

Literature

Hewitson, W. 1962. Comparative morphology of the Osmundaceae. Ann. Mo. Bot. Gard. 49: 57–93.

Lugardon, B. 1971. Contribution à la connaissance de la morphogénese et la structure des parois sporales chez les Filicinées isosporées. 257 pp. Thèse, Univ. Paul Sabatier, Toulouse.

Lugardon, B. 1972. La structure fine de l'exospore et de la périspore des filicinées isosporées. 1. Généralitiés. Eusporangiées et Osmundales. Pollen et Spores 16: 227–261.

Miller, C. N., Jr. 1971. Evolution of the fern family Osmundaceae based on anatomical studies. Contrib. Mus. Paleont. Univ. Mich. 23: 105–169.

5. *Osmunda*
Figs. 5.1–5.8

Osmunda L., Sp. Pl. 1063. 1753; Gen. Pl. ed. 5, 484, 1754. Type: *Osmunda regalis* L.

Struthopteris Bernh., Jour. Bot. (Schrad.) 1800 (2): 126. 1802, not *Struthiopteris* Scop., 1760 (= *Blechnum*). Type: *Osmunda regalis* L (No combinations were made).

Aphyllocalpa Lagasca, Garcia & Clemente, Anal. Cien. 5: 164. 1802. Type: *Aphyllocalpa regalis* (L.) Lagasca et al. = *Osmunda regalis* L.

Plenasium Presl, Tent. Pterid. 109. 1836. Type: *Plenasium banksiifolium* (Presl) Presl (*Nephrodium banksiifolium* Presl) = *Osmunda banksiifolia* (Presl) Kuhn. *Osmunda* subgenus *Plenasium* (Presl) Presl, Suppl. Tent. Pterid. 66. 1845.

Osmundastrum (Presl) Presl, Gefässb. Stipes Farrn. 18. 1847. *Osmunda* subgenus *Osmundastrum* Presl, Suppl. Tent. Pterid. 68. 1845. Type: *Osmunda cinnamomea* L.

Description

Terrestrial; stem erect to decumbent, often massive, including the persistent petiole bases, lacking indument, covered by long-persistent petiole bases, bearing wiry roots; leaves wholly or partially dimorphic, ca. 1–2 m long, with the croziers densely covered by matted trichomes, nearly glabrous at maturity, sterile lamina (or sterile portions) 1-pinnate, 1-pinnate-pinnatified or 2-pinnate, veins free; sporangia separate, sometimes in loose clusters, borne on wholly to partially fertile, usually 2-pinnate pinnae; spores tetrahedral-globose, trilete, the laesurae ca. $\frac{3}{4}$ the radius, the surface coarsely rugose or crested with slender echinate processes. Chromosome number: $n = 22$; $2n = 44$.

The stipular leaf bases enclose the stem apex, protecting several sets of leaf primordia and young croziers. In mature leaves, these bases are laterally expanded and flattened, becoming thinner toward the margin (Fig. 4). The fertile pinnae in *Osmunda* are more complex than the sterile, for example, in *O. regalis* the sterile pinnae are 1-pinnate (Fig. 6), while the fertile pinnae are 2-pinnate (Fig. 3).

Systematics

Osmunda is a widely distributed genus of six or perhaps a few more species, three American and two of them tropical. There are three species-groups that have often been recognized as genera (Bobrov, 1967) or as subgenera: *Osmunda*—*O. regalis* and *O. lancea* Thunb.; *Osmundastrum*—*O. cinnamomea*; *Plenasium*—*O. banksiifolia* (Presl) Kuhn and *O. javanica* Blume. *Osmunda Claytoniana* L., sometimes placed with *O. regalis* and sometimes with *O. cinnamomea*, is better considered as representing a fourth group.

A long history of intensive morphological and anatomical work on *Osmunda* has placed undue emphasis on differences within this relatively homogeneous group. The genus is small and while there are distinctive elements within it, they do not appear to represent evolutionary lines as strong as those in other genera of primitive families. Thus an infrageneric classification is not adopted.

Fig. 5.1. *Osmunda cinnamomea* (center) growing with **Blechnum serrulatum, Lindsaea portoricensis** (foreground) and other pteridophytes, Mason river savannah, Clarendon, Jamaica. (Photo Alice F. Tryon.)

Tropical American Species

Tropical specimens of both *O. cinnamomea* (Fig. 5) and *O. regalis* (Fig. 6) are usually smaller than those from temperate areas, and tend to have more coriaceous segments, with prominent veins. These and other differences have led to the recognition of several species that seem to be poorly defined variations of the basic species.

Key to Species of *Osmunda* in Tropical America

a. Fertile leaf with all pinnae fertile, sterile lamina pinnate-pinnatifid.
O. cinnamomea L.
a. Fertile leaf with apical fertile pinnae, sterile lamina 2-pinnate.
O. regalis L.

Ecology (Fig. 1)

Osmunda is a genus of wet habitats, rarely well-drained situations, most often growing in open habitats but also in shaded woods. In tropical America, plants usually occur in open or brushy habitats which are nearly constantly wet as swamps, bogs,

Fig. 5.2. American distribution of *Osmunda,* south of lat. 35° N.

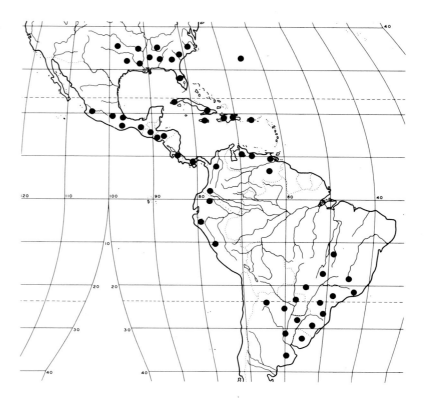

marshes, wet savannahs and lake margins. It may invade wet areas in pasture and meadows, and also occur in shaded habitats in or along streams, on wet brushy banks or wet areas in woods. Plants that develop in suitable habitats form large colonies and may be a few hundred years in age. *Osmunda* occurs mostly at altitudes from 1000 to 1500 m but there are records as low as 700 m or up to 2000 m and in Uruguay it grows nearly at sea level.

Geography (Fig. 2)

Osmunda is locally but widely distributed throughout most of the world except in cold and arid climates, and in the islands of the Pacific. In tropical America, it is local, often rare and widely scattered in central Mexico to Central America, the Greater Antilles to northern Peru, northwest Argentina to southeastern Brazil and south to Buenos Aires in Argentina; also on Bermuda; *Osmunda cinnamomea* and *O. regalis* extend northward to eastern boreal North America where *O. Claytoniana* also occurs.

Spores

Studies of spore wall development in *Osmunda regalis* by Lugardon (1971) show formation of a massive, rugose exospore. The spore surface consists of a relatively thin perispore with delicate echinate projections overlaying the rugose exospore (Figs. 7, 8). The echinate structures are composed of compacted rods and microbacules that readily become dissociated causing erosion of the perispore. The chemical stability of the spore wall of *O. cinnamomea* has been shown in the solubility tests of pollen and

3 4

5 6

Figs. 5.3–5.6. *Osmunda.*   **3.** Sporangia on fertile pinnules, *Osmunda regalis,* × 10.   **4.** Stipular petiole bases, *O. regalis,* × 2.   **5.** Portion of sterile lamina, *O. cinnamomea,* Brazil, × 1.5.   **6.** Portion of sterile pinna, *O. regalis,* Brazil, × 1.5.

spores by Southworth (1974). *Osmunda* spore walls were not disintegrated by hydrolytic acids or inorganic bases, such as 2-aminoethanol, which dissolved sporopollenin in most pollen.

Cytology

Cytological records of *Osmunda* from the American tropics are uniformly $n = 22$ and correspond to the numerous reports of the genus from temperate and paleotropical areas. Autotetraploid and triploid plants have been experimentally produced by Manton (1950). The tetraploid gametophyte was generally smaller, while cell size of the antheridia, antherzoids and rhizoids was larger than in specimens at lower ploidy levels. Karyotype studies of the Japanese species of *Osmunda* by Tatuno and Yoshida (1966, 1967) show that the chromosomes are relatively similar except for one pair of submedian B-type chromosomes. On the basis of arm length and satellites the 44 chromosomes can be arranged into five sets of eight and one set of four chromosomes.

Observations

Biologically *Osmunda* is one of the most familiar genera of ferns because the species have been utilized in experiments on fundamental processes. The coiling and uncoiling mechanisms of the

7   8

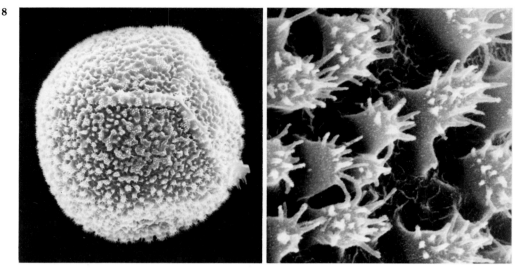

Figs. 5.7, 5.8. *Osmunda regalis* spores. Honduras, *Molina 2436.* **7.** Proximal face, tilted, × 1000. **8.** Detail of echinate perispore over the coarsely verrucate exospore surface, × 10,000.

leaves of *O. cinnamomea,* studied by Briggs and Steeves (1958, 1959) and Steeves (1963), are based on differential cell division and elongation between the abaxial and adaxial regions of the leaf. Auxin produced by the expanding pinnae is an important regulator of the process. The developmental potential of leaf primordia has been studied by Haight and Kuehnert (1969), Caponetti and Steeves (1970) and others. Young excised leaves and primordia isolated from shoot systems in *O. cinnamomea* show autonomous control of morphogenetic processes.

Osmunda regalis has been utilized in studies of cytological and genetic systems by Klekowski (1973). A large percentage of gametophytes of this species were unable to produce embryos. Crossing experiments showed that the failure to form sporophytes may be attributed to the presence of recessive lethals. *Osmunda regalis* was also used for bioassay of stream pollution (Klekowski and Berger, 1976); there was a higher mutation rate (chromosome aberrancy) among plants in polluted than nonpolluted habitats. The plants form large, dichotomously branched stem systems developed from a single zygote. This allows the calculation of rhizome growth rate and an estimate of the time of pollution can be made.

A study of the degree of DNA base sequence homology by Stein et al. (1979) utilized three species of *Osmunda.* The results indicated that all three species were equally divergent. Spore protein analysis by Petersen and Fairbrothers (1971) showed more similarity between *Osmunda cinnamomea* and *O. Claytoniana* than either of them to *O. regalis.* Miller (1967) related *O. Claytoniana* closer to *O. regalis,* and this agrees with the implied relationship of those species as the parents of *Osmunda × Ruggii* Tryon, the only known hybrid in the family (Tryon, 1940; Wagner et al., 1977). Considering these different lines of evidence, it is probable that the three species are somewhat distantly related to each other.

Literature

Bobrov, A. E. 1967. The family Osmundaceae (R. Br.) Kaulf, its taxonomy and geography. Bot. Žurn. (Acad. Nauk SSSR) 52: 1600–1610.

Briggs, W. R., and T. A. Steeves. 1958. Morphogenetic studies on *Osmunda cinnamomea* L.—The expansion and maturation of vegetative fronds. Phytomorph. 8: 234–248.

Briggs, W. R., and T. A. Steeves. 1959. Morphogenetic studies on *Osmunda cinnamomea* L.—The mechanism of crozier uncoiling. Phytomorph. 9: 134–147.

Caponetti, J. D., and T. A. Steeves. 1970. Morphogenetic studies on excised leaves of *Osmunda cinnamomea*: histological studies of leaf development in sterile nutrient culture. Canad. Jour. Bot. 48: 1005–1016.

Haight, T. H., and C. C. Kuehnert. 1969. Developmental potentialities of leaf primordia of *Osmunda cinnamomea*. V. Toward greater understanding of the final morphogenetic expression of isolated set I cinnamon fern leaf primordia. Canad. Jour. Bot. 47: 481–488.

Klekowski, E. J., Jr. 1973. Genetic load in *Osmunda regalis* populations Amer. Jour. Bot. 60: 146–154.

Klekowski, E. J., Jr., and B. B. Berger. 1976. Chromosome mutations in a fern population growing in a polluted environment: A bioassay for mutagens in aquatic environments. Amer. Jour. Bot. 63: 239–246.

Lugardon, B. 1971. Reference under the family.

Manton, I. 1950. Problems of Cytology and Evolution in the Pteridophyta. Cambridge Univ. 316 pp. Cambridge, England.

Miller, C. N., Jr. 1967. Evolution of the fern genus *Osmunda*. Contrib. Mus. Paleont. Univ. Mich. 21: 139–203.

Petersen, R. L., and D. E. Fairbrothers. 1971. North American *Osmunda* species: A serologic and disc electrophoretic analysis of spore proteins. Amer. Midl. Nat. 85: 437–457.

Southworth, D. 1974. Solubility of pollen exines. Amer. Jour. Bot. 61: 36–44.

Steeves, T. A. 1963. Morphogenetic studies of fern leaves. Jour. Linn. Soc. (Bot.) 58: 401–415.

Stein, D. B., W. F. Thompson, and H. S. Belford. 1979. Studies on DNA sequences in the Osmundaceae. Jour. Molec. Evol. 13: 215–232.

Tatuno, S., and H. Yoshida. 1966. Karyologische utersuchungen über Osmundaceae. I. Chromosomes der gattung *Osmunda* aus Japan. Bot. Mag. (Tokyo) 79: 244–252.

Tatuno, S., and H. Yoshida. 1967. Karyological studies of Osmundaceae. II. Chromosome of the Genus *Osmundastrum* and *Plenasium* in Japan. Bot. Mag. (Tokyo) 80: 130–138.

Tryon, R. 1940. An *Osmunda* hybrid. Amer. Fern Jour. 30: 65–68.

Wagner, W. H., Jr., F. S. Wagner, C. N. Miller, Jr., and D. H. Wagner, 1978. New observations on the royal fern hybrid *Osmunda* × *Ruggii* Tryon. Rhodora 80: 92–106.

Family 4.   Schizaeaceae

Schizaeaceae Kaulf., Wesen Farrenkr. 119. 1827. Type: *Schizaea* Sm.
Anemiaceae Link, Handb. Erken. Gew. 3: 8. 1833. Type: *Anemia* Sw.
Lygodiaceae Presl, Suppl. Tent. Pterid. 98. 1845. Type: *Lygodium* Sw.
Mohriaceae (Presl) Reed, Bol. Soc. Broter. II, 21: 168. 1948. Type:
   Mohria Sw.

Description   Stem erect to decumbent; usually small and sometimes
branched, or long-creeping, slender and freely branched, protostelic, siphonostelic or dictyostelic, indurated, bearing trichomes, or (in *Mohria*) scales; leaves ca. 2–3 cm, mostly ca. 20–
50 cm long, or (in *Lygodium*) to many meters long, entire and
filiform, or dichotomous, or pinnate, glabrous, pubescent, or (in
Mohria) with scales, circinate in the bud, petiole without stipules;
sporangia borne abaxially on slightly to strongly modified portions of the leaf, separate, or crowded on each side of a vein, or
in loose clusters on wholly fertile panicles, sessile, or with a short,
many-rowed stalk, and an apical annulus; homosporous, spores
without chlorophyll. Gametophyte with a 3- to rarely 5-celled
antheridium, epigeal, with chlorophyll, usually asymmetrically
obcordate, thickened, often with small trichomes, the archegonia borne on the lower surface of the central thickened portion, the antheridia borne in the same areas as the archegonia,
or sometimes on the upper surface; or (in *Schizaea* subgenus
Schizaea) more or less epigeal and the exposed areas with chlorophyll, filamentous, branched, partially mycorrhizic, the archegonia borne mostly on tufts of upright filaments, the antheridia
on short filaments; or (*Schizaea* subgenus *Actinostachys*) subterranean, without chlorophyll, elongate, usually branched, becoming irregularly tuberous with age, mycorrhizic, with the archegonia and antheridia distributed over the surface.

Comments   The Schizaeaceae are a family of four very distinctive genera, all
on the   in America except *Mohria* Sw., a genus of Africa, Madagascar
Family   and adjacent islands, distinguished from the others by scales,
rather than trichomes on the stem and leaves. The family is an
old one, with records from the Jurassic or earlier, and the living
groups are undoubtedly the remnants of a long history of divergence and evolution. Because of considerable differences, the
genera have sometimes been segregated as families (Reed,
1947). *Schizaea* is the most distinctive, in morphology of the gametophytes as well as the sporophytes, while *Anemia* and *Mohria*
are relatively closely allied. The genera form a diverse but natural group that is best regarded as one family.

Taxonomic treatments of the family in the American tropics
are those by Maxon (1909) for North America and by Lellinger
(1969) for the Guayana region.

The Schizaeaceae are frequently considered as a basic family
in the evolution of ferns with phyletic relations to the Pteridaceae, Vittariaceae, Marsileaeeceae, Salviniaceae, Hymenophyllaceae and the Dicksoniaceae.

Experimental studies on *Anemia* and *Lygodium* gametophytes
(Näf, 1959, 1960) showed that they produce a substance capable

of promoting the development of antheridia in young gameto-phytes. The effect of the antheridogen is similar to that produced in *Pteridium* gametophytes. However, the *Pteridium* antheridogen is also effective in a wide range of other genera (Miller, 1968; Näf et al., 1975; Voeller 1964, 1971) while the antheridogen of *Anemia* is effective only in that genus and in *Lygodium*, and the *Lygodium* antheridogen is not effective in other genera.

Most fern spores require light for germination, but investigations show that germination may occur in the dark if antherodogen of the same species is supplied to the medium. The role of an antheridogen in the dark reaction closely parallels its specificity as a stimulus to antheridia production on young gametophytes (Näf et al., 1975, Voeller, 1971). The evolutionary significance of antheridogen is not clear, but it seems to increase the opportunities of outcrossing in natural populations of a species (Tryon and Vitale, 1977).

Key to Genera of *Schizaeaceae* in America

a. Each sporangium covered by a laminar flange, leaf with widely spaced alternate pinnae, the pinnae short-stalked with an arrested bud in the axil of the first, pseudodichotomous, branch. 7. *Lygodium*
a. Sporangia naked. b.
   b. Leaf rarely pinnatifid, usually with opposite to closely alternate pinnae which are entire to pinnately divided. 6. *Anemia*
   b. Leaf filiform or with dichotomous veins or branches. 8. *Schizaea*

Literature

Lellinger, D. B. 1969. Schizaeaceae (Filicales), *in*: The Botany of the Guayana Highland—Part VIII, Mem. New York Bot. Gard. 18: 2–11.

Maxon, W. R. 1909. Schizaeaceae. North America Flora 16: 31–52.

Miller, J. H. 1968. Fern gametophytes as experimental materials. Bot. Review 34: 361–440.

Näf, U. 1959. Control of antheridium formation in the fern species *Anemia phyllitidis*. Nature 184: 798–800.

Näf, U. 1960. On the control of antheridium formation in the fern species *Lygodium japonicum*. Proc. Soc. Exper. Biol. Med. 105: 82–86.

Näf, U, K. Nakanishi, and M. Endo. 1975. On the physiology and chemistry of fern antheridogens. Bot. Review 41: 315–359.

Reed, C. F. 1947. The phylogeny and ontogeny of the Pteropsida, I. Schizaeales. Bol. Soc. Broter. II, 21: 71–197.

Tryon, R. M., and G. Vitale. 1977. Evidence for antheridogen production and its mediation of a mating system in natural populations of fern gametophytes. Jour. Linn. Soc. Bot. 74: 243–249.

Voeller, B. 1964. Antheridogens in ferns, *in*: Regulateurs Naturels de la croissance végétale. Colloques Centre Nat. Rech. Sci. 123: 665–684.

Voeller, B. 1971. Developmental physiology of fern gametophytes: Relevance for biology. BioScience 21: 266–270.

6. *Anemia*
Figs. 6.1–6.29.

Anemia Sw., Syn. Fil. 6, 155. 1806, *nom. conserv.,* sometimes as *Aneimia;* not *Anemia* Nutt. 1838. (Saururaceae). Type: *Anemia phyllitidis* (L.) Sw. (*Osmunda phyllitidis* L.).

Ornithopteris Bernh., Neues Jour. Bot. (Schrad.) 1 (2): 40. 1806. Type: *Ornithopteris adiantifolia* (L.) Bernh. (*Osmunda adiantifolia* L.) = *Anemia adiantifolia* (L.) Sw. [Subgenus *Anemirhiza*].

Anemidictyon Hook., Gen. Fil. *t. 103.* 1842. Type: *Anemidictyon phyllitidis* (L.) Hook. (*Osmunda phyllitidis* L.) = *Anemia phyllitidis* (L.) Sw. [Subgenus *Anemia*].

Trochopteris Gardn., Lond. Jour. Bot. 1: 74. 1842. Type: *Trochopteris elegans* Gardn. = *Anemia elegans* (Gardn.) Presl. *Anemia* subgenus *Trochopteris* (Gardn.) Sturm, Mart. Fl. Bras. 1 (2): 187. 1859. [Subgenus *Coptophyllum*].

Coptophyllum Gardn., *ibidem* 133. 1842, not Korthals 1851 (*nom. conserv.,* Rubiaceae). Type: *Coptophyllum buniifolium* Gardn. = *Anemia buniifolia* (Gardn.) Moore. *Anemia* subgenus *Coptophyllum* (Gardn.) Presl, Suppl. Tent. Pterid. 79. 1845. *Hemianemia* subgenus *Coptophyllum* (Gardn.) Reed, Bol. Soc. Broter. II, 21: 157. 1947.

Anemirhiza J. Sm., Seemann, Bot. Voy. Herald 243. 1854. Type: *Anemirhiza adiantifolia* (L.) J. Sm. (*Osmunda adiantifolia* L.) = *Anemia adiantifolia* (L.) Sw. *Anemia* subgenus *Anemirhiza* (J. Sm.) Prantl, Morph. Gefässkrypt. II, Schizaeaceen. 88. 1881, sometimes as *Anemiorrhiza,* or as *Anemirrhiza.*

Anemiaebotrys Fée, Crypt. Vasc. Brésil 1: 267. 1869. Type: *Anemiaebotrys aspera* Fée = *Anemia aspera* (Fée) Baker. *Hemianemia* subgenus *Anemiaebotrys* (Fée) Reed, *ibidem* 156. 1947. [Subgenus *Coptophyllum*].

Hemianemia (Prantl) Reed, *ibidem* 154. 1947. *Anemia* subgenus *Hemianemia* Prantl, *ibidem* 86. 1881. Type: *Anemia tomentosa* (Sav.) Sw. (*Osmunda tomentosa* Sav.). [Subgenus *Coptophyllum*].

Description

Terrestrial; stem decumbent to short- or rather long-creeping, rarely erect, rather small, bearing short to long trichomes and several to many fibrous roots; leaves usually partially or wholly dimorphic, with a pair of fertile pinnae at the base of, or below, the sterile lamina, or with similar leaves longer and more erect than the wholly sterile leaves, or the whole leaf either sterile or fertile, ca. 1 cm (in *A. elegans*) to mostly 30–50 cm, to 75 cm long, clustered to spaced, usually 1-pinnate to 2-pinnate-pinnatifid, the veins free to rarely anastomosing without included free veinlets; sporangia borne on the fertile segments which are reduced to axes or have narrow borders of laminar tissue along the axes, rarely monomorphic (in *A. elegans*) with the sporangia on lobes of a pinnately lobed lamina, or subdimorphic (in *A. colimensis* and *A. aspera*) with sporangia on the slightly contracted basal pinnae; spores tetrahedral-globose, trilete with prominent parallel ridges especially in the equatorial region, or the ridges often echinate with coarse, peglike projections or with short spicules, or spores coarsely rugose-reticulate. Chromosome number: $n = 38, 76$; $2n = 76, 114, 152$; apogamous 114.

A compact, short-creeping dictyostelic rhizome is illustrated in Fig. 4, and a long-creeping siphonostelic rhizome of subgenus *Anemirhiza* in Fig. 7. Anastomosing veins (Fig. 5) are rare in the genus; a few species have coriaceous segments with the veins raised on the upper surface (Fig. 8). Fertile branches may have a

Fig. 6.1. *Anemia adiantifolia,* Puente Nacional, Veracruz, Mexico. (Photo W. H. Hodge.)

narrow border of laminar tissue (Fig. 3), or they may be reduced to axes (Fig. 6).

Systematics

Anemia is a genus of about 80 species, ranging from America east to southern India. The three subgenera are too closely allied to merit generic rank. Mickel has revised subgenus *Coptophyllum* (1962) and subgenus *Anemirhiza* (1981).

Synopsis of *Anemia*

Subgenus *Coptophyllum*

Stem compact, with dark red to orange trichomes; leaves polystichous, the sterile lamina with gland-tipped trichomes, fertile segments with narrow borders of laminar tissue along the axes; spores usually with compact ridges, the grooves less than half as broad as the ridges, the distal face often with three strong angles, or in a few species with grooves as wide as the ridges.

Thirty-nine species, 29 of them in the American tropics.

Fig. 6.2. Distribution of *Anemia* in America.

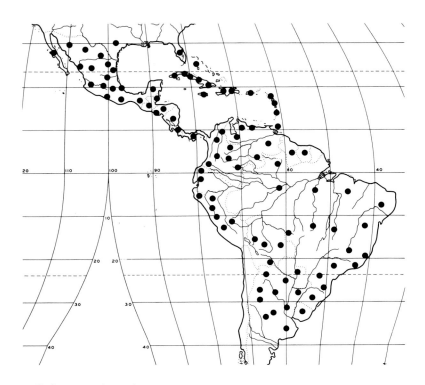

Subgenus *Anemia*

Stem compact, with dark red to orange trichomes; leaves polystichous, the sterile lamina with gland-tipped trichomes, fertile segments reduced to the axes, spores with grooves as broad as, or broader than, the ridges, beset with coarse, peglike projections or echinate with short, dense spicules.

About 25 species, all but one in the American tropics.

Subgenus *Anemirhiza*

Stem more or less long-creeping, with brown to dark brown trichomes, leaves distichous, the sterile lamina with acute- to acuminate-tipped trichomes, fertile segments with narrow borders of laminar tissue along the axes, spores with grooves as broad as, or broader than, the ridges which are mostly parallel in the equatorial area and often more or less interrupted on the distal and proximal faces, or in a few species coarsely rugose-reticulate, especially on the distal face.

About 12 tropical American species.

Tropical American Species

Some of the variation in leaf dimorphism, as well as in the architecture of the sterile lamina, is illustrated in Figs. 9–17. In *A. elegans* (Fig. 10), the basal lobes of a leaf are fertile and slightly lobed. The usual condition of wholly fertile and erect basal pinnae is illustrated in Figs. 9, 11, and 14–16. In *A. humilis* (Fig. 14), the fertile pinnae are especially long in relation to the sterile lamina; in *A. glareosa* (Fig. 12), the fertile leaves have a rather small, sterile lamina and are taller and more erect than the wholly sterile leaves. Complete leaf dimorphism is seen in *A. mil-*

Figs. 6.3–6.8 *Anemia.* **3.** Fertile segments of *Anemia Karwinskyana,* with narrow laminar tissue along the axes, ×5. **4.** Compact stem of *A. affinis,* ×1.5. **5.** Anastomosing veins of *A. phyllitidis,* ×3. **6.** Fertile segments of *A. phyllitidis,* with nonlaminar axes, ×10. **7.** Long creeping stem of *A. adiantifolia,* ×1.5. **8.** Coriaceous segments of *A. aurita,* with free veins, ×3.

lefolia Fig. 13, with the sterile lamina highly dissected, and in *A. portoricensis* (Fig. 17).

Examples of the numerous American species are the following.

Subgenus *Coptophyllum:* Mexico, Central America: *Anemia guatemalensis* Maxon, *A. Karwinskyana* (Presl) Prantl. South America: *A. aspera* (Fée) Baker, *A. elegans* (Gardn.) Presl (Fig. 10), *A. flexuosa* (Sav.) Sw. (Fig. 11), *A. glareosa* Gardn. (Fig. 12), *A. millefolia* (Gardn.) Presl (Fig. 13), *A. myriophylla* Christ, *A. organensis*

Figs. 6.9–6.17. Leaves of *Anemia*. **9.** *A. organensis*, × 0.25. **10.** *A. elegans*, × 1. **11.** *A. flexuosa*, × 0.25. **12.** *A. glareosa*, × 0.5. **13.** *A. millefolia*, × 0.5. **14.** *A. humilis*, × 0.5. **15.** *A. ouropretana*, × 0.25. **16.** *A. tenella*, × 0.5. **17.** *A. portoricensis*, × 0.5.

Rosenst. (Fig. 9), *A. simplicior* (Christ) Mickel, *A. tomentosa* (Sav.) Sw., *A. villosa* Willd.

Subgenus *Anemia:* Wide ranging: *Anemia hirsuta* (L.) Sw., *A. hirta* (L.) Sw., *A. humilis* (Cav.) Sw. (Fig. 14), *A. oblongifolia* (Cav.)

Sw., *A. phyllitidis* (L.) Sw. Mexico: *A. affinis* Baker, Brazil: *A. man-dioccana* Raddi, *A. ouropretana* Christ (Fig. 15), *A. tenella* (Cav.) Sw. (Fig. 16).

Subgenus *Anemirhiza:* Northern tropics: *Anemia adiantifolia* (L.) Sw. Mexico: *A. colimensis* Mickel, *A. mexicana* Kl. Greater Antilles: *A. Abbottii* Maxon, *A. aurita* Sw., *A. portoricensis* Maxon (Fig. 17), *A. Wrightii* Baker.

About 20 interspecific hybrids have been proposed (Mickel, 1962) involving some 25 species. This apparently widespread hybridization is probably a recent phenomenon in disturbed habitats.

Ecology (Fig. 1)

Anemia is primarily a genus of open habitats and well-drained sites. In America it usually grows on ravine banks, stream borders, on shrubby hillsides, among rock outcrops and on small cliffs, and also on road banks. It sometimes grows in savannahs, less often in open forests or in rain forests. Species of subgenus *Anemirhiza* grow especially on ledges or in eroded pockets of limestone, or in calcareous soils. *Anemia* is often associated with other ferns as *Pellaea* and *Doryopteris* and also angiosperm familes as the Eriocaulaceae and Velloziaceae in the campo limpo of southeastern Brazil. In this region plants were observed growing in sandy soil with stems slightly below ground level in an area that had been burned. Although old leaf bases were charred, there were clusters of vigorous, new leaves on the plants. *Anemia* grows from sea level to usually 500 m in the West Indies, mostly up to 1000 m in Central America, and to 2000 m or rarely up to 3200 m in the Andes.

Geography (Fig. 2)

Anemia is primarily a genus of the American tropics, where 80% of the species grow. Nine species of subgenus *Coptophyllum* occur in Africa-Madagascar and one in southern India. A single species of subgenus *Anemia* occurs in South Africa. Subgenus *Anemirhiza* is Caribbean and circum-Caribbean in America.

Anemia ranges throughout tropical America, from Texas and southern Florida, in the United States, and the Bahamas, south to Cordoba, San Luis and Buenos Aires in Argentina.

Species of *Anemia* tend to have small ranges, as compared to *Lygodium* and *Schizaea* which mostly have extensive species ranges. There are three centers of species diversity, that are also centers of endemism: the Greater Antilles have 16 species, about half endemic; Mexico and Central America have 21 species, about half endemic; and central and southeastern Brazil have 33 species, about three-fourths endemic.

Spores

The spore wall sections of *Anemia phyllitidis* by Lugardon (1974) consist largely of an homogeneous exospore forming the basic ridged structure. The perispore is relatively thin, somewhat denser adjacent to the exospore and granular toward the outer surface. Three main spore types generally correlate with the subgenera. In subgenus *Anemirhiza*, the spores usually have few parallel ridges and broad grooves as in *A. Abbottii* (Fig. 18). An unusual spore type in this subgenus is coarsely rugose-reticulate as *A. Wrightii* (Fig. 19). Spores of subgenus *Anemia* usually have prominently pegged ridges as in *A. hirta*, *A. phyllitidis* and *A. mandioccana* (Figs. 20–22) and a somewhat papillate surface more or less fused into low ridges (Fig. 23). A few species as *A. hirsuta* and *A. oblongifolia* (Figs. 24, 25) have ridged spores with short, echinate projections and may possibly represent a separate element in the subgenus. Spores of subgenus *Coptophyllum* have prominent, compact ridges as in *A. guatemalensis* (Figs. 26, 27). The surface has sharp, echinate processes as in Fig. 29 and the three angles may strongly project as in *A. elegans* (Fig. 28). *Anemia myriophylla* and *A. simplicior* in this subgenus have spores with few ridges similar to those in subgenus *Anemirhiza*. The SEM studies by Hill, of spores in subgenus *Coptophyllum* (1977), show variation in the echinate surface elements, and varied pegged ridges in those of subgenus *Anemia* (1979). However, the spores are generally characteristic of the subgenera illustrated here.

The study of recent and fossil spores of the Schizaeaceae by Bolkhovitina (1961) includes fossils as *A. Cooksonii* (Balme) Bolk. with broad grooves resembling those in subgenus *Anemirhiza* and several with compact ridges as in subgenus *Coptophyllum*. A number of genera of the Pteridaceae, such as *Anogramma, Ceratopteris, Eriosorus, Jamesonia* and *Pityrogramma,* have spores with prominent parallel equatorial ridges similar to forms in *Anemia*.

Cytology

Cytological records of *Anemia* from the neotropics include reports of mostly $n = 38$ or 76 from Mexico, Jamaica, Brazil and Argentina. Complexes involving different ploidy levels have been recognized in a few species. Cytological reports of several populations of *A. adiantifolia* in Jamaica by Walker (1962) showed a predominance of diploids with $n = 38$. Triploids with 38 pairs and 38 singles are known where diploids and tetraploids are associated, and the triploid condition is evident from the intermediate morphology of the leaves. A hybrid between *A. adiantifolia* and *A. mexicana,* from San Luis Potosí, Mexico, with 38 irregularly pairing chromosomes was reported by Mickel (1962). Records for several other American species in that work include diploids with $n = 38$, tetraploids with $n = 76$, and hexaploid sexual plants with $n = 114$, as well as an apparently apogamous form with a somatic number of 114.

18 19

20 21

22 23

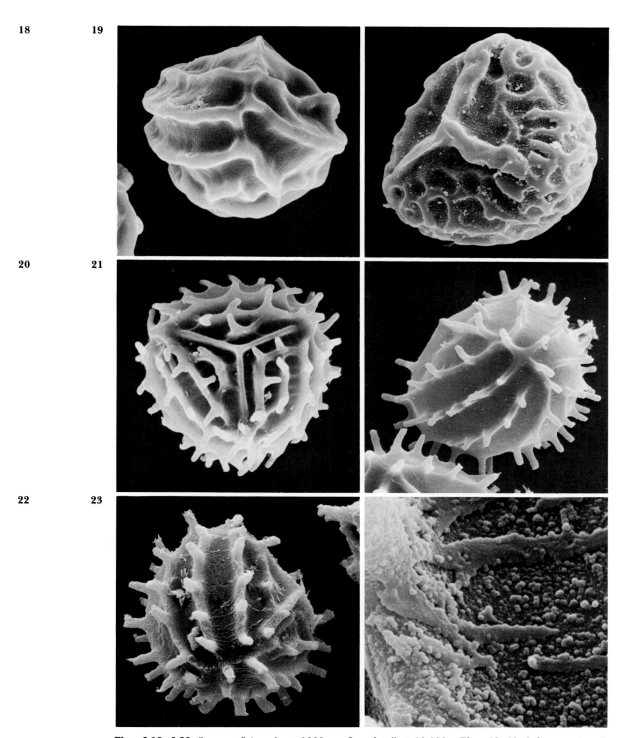

Figs. 6.18–6.23. Spores of *Anemia*, × 1000; surface detail, × 10,000. **Figs. 18, 19.** Subgenus *Anemirhiza.* **Figs. 20–23.** Subgenus *Anemia.* **18.** *A. Abbottii,* ridged, equatorial area, proximal face above, Dominican Republic, *Abbott 1309.* **19.** *A. Wrightii,* rugose-reticulate proximal face, Bahama, *Small & Carter 8806.* **20.** *A. hirta,* pegged-ridges proximal face, Brazil, *Silva 58425.* **21.** *A. phyllitidis,* equatorial area, proximal face above, left, Brazil, *Irwin et al. 19219.* **22, 23.** *A. mandioccana,* Brazil, *Dusén 711a.* **22.** Equatorial area, proximal face right. **23.** Detail of papillate surface, laesura, left.

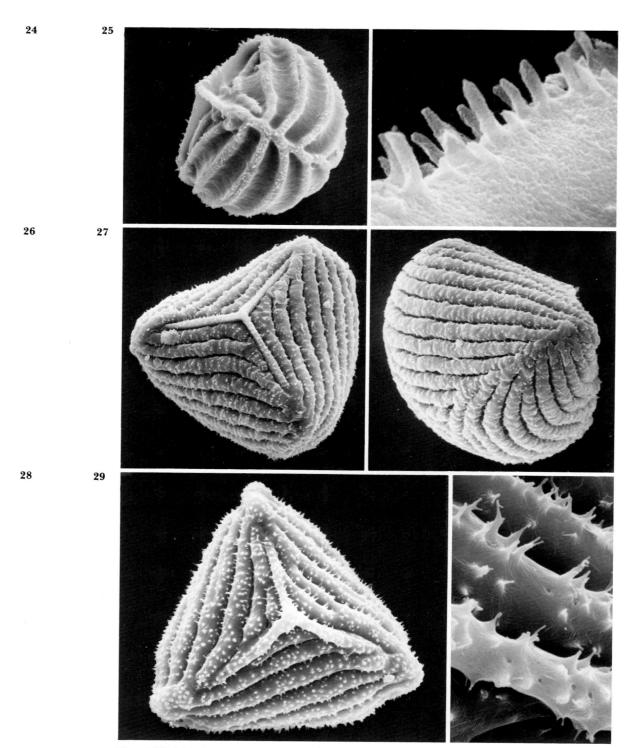

Figs. 6.24–6.29. Spores of *Anemia.* **Figs. 24, 25.** Subgenus *Anemia.* **Figs. 26–29.** Subgenus *Coptophyllum.* **24.** *A. hirsuta,* echinate ridges, equatorial area, proximal face above, left, × 1000. Mexico, *Kenoyer 1662,* × 1000. **25.** *A. oblongifolia,* detail of echinate ridges, Mexico, *Hinton 9472,* × 10,000. **26, 27.** *A. guatemalensis,* Honduras, *Molina 18621,* × 660. **26.** Proximal face. **27.** Distal face tilted, with convergent ridges below one of the angles. **28, 29.** *A. elegans,* Brazil, *Heringer* in 1966. **28.** Proximal face with three projecting angles, × 1000. **29.** Detail of echinate ridges, × 5000.

Observations

The genus *Anemia* is noted for its two unusual types of stomata, known to occur only in ferns. One type has the guard cells attached by one (or a double) wall to the otherwise completely surrounding subsidiary cell. This type, termed desmo-mesogenous by Fryns-Claessens and Van Cotthem (1973), is known in *Anemia* subgenus *Coptophyllum* and subgenus *Anemirhiza,* in *Mohria, Pyrrosia* and *Monogramma.* The second type has the guard cells unattached to the surrounding subsidiary cell. It is termed euperimesogenous (also free or floating) and occurs in all subgenera of *Anemia,* and in *Monogramma, Platycerium,* and various polypodioid genera such as *Pyrrosia* and *Drymoglossum*

Literature

Bolkhovitina, N. A. 1961. Fossil and recent spores of the Schizaeaceae. Moscow, Trud. Geol. Inst. Nauk SSSR 40: 1–176.

Fryns-Claessens, E., and W. Van Cotthem. 1973. A new classification of the ontogenetic types of stomata. Bot. Review 39: 71–138.

Hill, S. R. 1977. Spore morphology of *Anemia* subgenus *Coptophyllum.* Amer. Fern Jour. 67: 11–17.

Hill, S. R. 1979. Spore morphology of *Anemia* subgenus *Anemia.* Amer. Fern Jour. 69: 71–79.

Lugardon, B. 1974. La structure fine de l'exospore et de la périspore des filicinées isosporées. 2. Filicales. Pollen et Spores 16: 161–226.

Mickel, J. T. 1962. A monographic study of the fern genus *Anemia,* subgenus *Coptophyllum.* Iowa State Jour. Sci. 36: 349–482.

Mickel, J. T. 1981. Revision of *Anemia* subgenus *Anemiorrhiza* (Schizaeaceae). Brittonia 33: 413–429.

Walker, T. G. 1962. The *Anemia adiantifolia* complex in Jamaica. New Phytol. 61: 291–298.

7. *Lygodium*
Figs. 7.1–7.14

Lygodium Sw., Jour. Bot. (Schrad.) 1800 (2): 106. 1802, *nom. conserv.* Type: *Lygodium scandens* (L.) Sw. (*Ophioglossum scandens* L.).

Ugena Cav., Icon. Descr. Pl. 6: 73. 1801. Type: *Ugena semihastata* Cav. = *Lygodium semihastatum* (Cav.) Desv.

Ramondia Mirbel, Bull. Sci. Soc. Philom. Paris 2: 179. 1801. Type: *Ramondia flexuosa* (L.) Mirbel (*Ophioglossum flexuosum* L.) = *Lygodium flexuosum* (L.) Sw.

Odontopteris Bernh., Jour. Bot. (Schrad.) 1800 (2): 127. 1802. Type: *Odontopteris scandens* (L.) Bernh. (*Ophioglossum scandens* L.) = *Lygodium scandens* (L.) Sw. *Lygodium* subgenus *Odontopteris* (Bernh.) Reed, Bol. Soc. Broter. II, 21: 144, 1947.

Gissopteris Bernh., *ibidem* 129. 1802. Type: *Gissopteris palmata* Bernh. = *Lygodium palmatum* (Bernh.) Sw. *Lygodium* subgenus *Gissopteris* (Bernh.) Reed, *ibidem* 141. 1947.

Hydroglossum Willd., Akad. Wiss. Erfurt 2 (6th paper): 20. 1802. Type: *Hydroglossum longifolium* Willd. = *Lygodium circinnatum* (Burm. f.) Sw.

Cteisium Michx., Fl. Bor. Amer. 2: 275. 1803. Type: *Cteisium paniculatum* Michx. = *Lygodium palmatum* (Bernh.) Sw.

Vallifilix Thouars, Gen. Nov. Madagas. 1. 1808; also in Romer, Collect. Bot. 195. 1809. Type: *Ophioglossum scandens* L. = *Lygodium scandens* (L.) Sw.

Lygodictyon Hook., Gen. Fil. *t. 111b.* 1842. Type: *Lygodictyon Forsteri* J. Sm. = *Lygodium reticulatum* Schkuhr.

Description

Terrestrial; stem short- to long-creeping, slender, frequently branched, protostelic, bearing short trichomes and a few to many fibrous roots; leaves partially dimorphic, with the fertile portions with marginal fertile lobes, or somewhat contracted and wholly fertile, sometimes with a different architecture than the sterile, close or rather widely spaced, ca. 1 to often more than 10 m long, climbing, widely alternately pinnate, glabrous to somewhat pubescent, pinnae short-stalked, pseudodichotomously branched with an arrested bud in the axil, each primary pinna-branch palmately or radiately lobed or branched or pinnate, veins free or rarely anastomosing without included free veinlets; sporangia borne separately on marginal lobes of a pinna-segment or on a wholly fertile segment, each covered by a laminar outgrowth (flange); spores tetrahedral-globose, trilete, the laesurae ca. $\frac{3}{4}$ the radius, the surface with spherical deposition, the distal face often verrucate and the proximal one with a prominent ridge connecting the ends of the laesurae, or rugose to reticulate. Chromosome number: $n = 28, 29, 30, 56, 58,$ ca. $70, 87-90; 2n = 56, 58, 60, 112, 116.$

The unique pinna structure of *Lygodium* (Fig. 4) shows the arrested bud terminating the pinna-rachis and subtended by the primary pinna-branches. Marginal lobes bearing sporangia shown in Fig. 7 are characteristic of many species. The laminar flange that covers each sporangium (Fig. 5) serves as an indusium and is unique in the family. The slender stem is protostelic.

Systematics

Lygodium is a widely distributed genus of about 30 species, six to eight in the American tropics and two species, one of which is introduced, in the eastern United States. A modern study to determine evolutionary groups among the species is needed before a reliable infrageneric classification can be established.

The American species of *Lygodium* were treated in a brief revision by Duek (1978).

Tropical American Species

It is expected that further study will reduce the number of American species of *Lygodium*, as *L. mexicanum* Presl is evidently a synonym of *L. venustum*, and *L. micans* Sturm and *L. volubile* are not clearly distinct. *Lygodium pedicellatum* is known from limited material and additional collections are needed for an assessment of the species. *Lygodium polymorphum* (Cav.) HBK, has often been incorrectly used as an earlier name for *L. venustum* but it is a synonym for the paleotropical *L. flexuosum* (L.) Sw. Among the American species, anastomosing veins occur only in *L. heterodoxum* (Fig. 6). *Lygodium oligostachyum*, a small leaved species of Hispaniola, has lamina segments with unusual cuneate bases (Fig. 8). A hybrid between *L. venustum* and *L. volubile* (as *L. micans*) has been proposed from Trinidad (Fay, 1973).

Key to Species of *Lygodium* in America

a. Veins anastomosing.   *L. heterodoxum* Kze.
a. Veins free.   b.
   b. Primary pinna-branch palmate or radiate, moderate to deeply lobed, the segments joined at the base.   c.
     c. Pinnae petiolulate; fertile pinnae more complex than the sterile, branched into fertile segments; eastern United States.   *L. palmatum* (Bernh.) Sw.
     c. Pinnae subsessile; the pinna-stalk nearly absent, fertile pinnae similar to the sterile, with many small, fertile lobes along the segment margins.   *L. radiatum* Prantl
   b. Primary pinna-branch pinnate, the pinnules stalked, except sometimes near the apex of the primary branch.   d.
     d. Sterile pinnules entire, rarely some basal ones lobed.   *L. volubile* Sw.
     d. Sterile pinnules lobed or pinnate.   e.
       e. Sterile pinnules palmately lobed at base.   *L. venustum* Sw.
       e. Sterile pinnules mostly with stalked segments.   f.
         f. Sterile stalked and sessile segments cuneate.   *L. oligostachyum* (Willd.) Desv.
         f. Sterile stalked and sessile segments subcordate to semicuneate or semicordate at base.   g.
           g. Fertile lobes sessile on the segment margins; Cuba.   *L. cubense* HBK.
           g. Fertile lobes pedicellate on the segment margins; Hispaniola.   *L. pedicellatum* C. Chr. & Maxon

Ecology (Fig. 1)

Lygodium characteristically occurs in open forest especially along the borders where the climbing leaves can reach well-lighted situations. In tropical America, it sometimes grows in rain forests, but more commonly in gallery forest, shrubby savannahs or along the borders of streams or river banks. It frequently becomes established in disturbed vegetation as secondary forests or brushy clearings in which the leaves form a tangled growth. The plants frequently invade old coconut palm and banana plantations. It most commonly occurs from sea level to about 350 m, as a characteristic element of the low, humid tropics, and less often grows up to 1000 m. One of the small species, *L. cubense* has rather xeromorphic leaves and sometimes occurs on serpentine barrens in Cuba.

Geography (Fig. 2)

Lygodium is a pantropical genus with extensions in temperate South Africa, New Zealand, Japan, and the eastern United States. In tropical America it occurs in Mexico, Central America, the Greater Antilles, Grenada, Tobago and Trinidad, and throughout tropical South America south to Bolivia, Chaco and Missiones in northern Argentina, and southern Brazil. It is absent from Puerto Rico and most of the Lesser Antilles. In the eastern United States *Lygodium palmatum* is distributed from Georgia to Vermont and Michigan, and is disjunct in the northern part of the range.

Two species of *Lygodium* of the Old World, *L. japonicum* (Thunb.) Sw., and *L. microphyllum* (Cav.) R. Br. are adventive in the southeastern United States.

Fig. 7.1. *Lygodium venustum,* Puente
Nacional, Veracruz, Mexico. (Photo
W. H. Hodge.)

Fig. 7.2. Distribution of *Lygodium* in
tropical America.

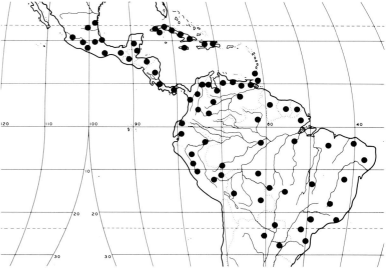

Fig. 7.3. A complete leaf of *Lygodium venustum,* 3.5 m long (diagrammatic), ×0.15.

Spores

Lygodium japonicum spores have an extensive, stratified perispore, shown in the studies of Lugardon (1974). The lower part is formed largely of porous material with a thicker, dense stratum above. Silica was detected on the inner surface of the outer wall in this species by Edman (1932). *Lygodium* spores are exceptionally large thus the micrographs are about half the magnification of those of most other genera. The coarse spherical deposition is characteristic of the spores of American species (Fig. 12).

Spores of the tropical American species are of two main types. One has a relatively smooth base with more or less dense, spherical deposit as in *L. venustum* and *L. heterodoxum* (Figs. 9, 10). Another form is coarsely verrucate especially on the distal face with a prominent ridge connecting the laesurae as in *L. volubile* and *L.*

4   5

6   7   8

Figs. 7.4–7.8. *Lygodium*.   **4.** Base of pinna of *Lygodium venustum* the rachis vertical, left, showing pseudo-dichotomous branching and arrested bud in the axil of the branches, × 1.   **5.** Fertile lobes of *L. venustum*, with laminar flanges on each side of the midvein of the lobe, each covering a sporangium, one is exposed (arrow), × 10.   **6.** Anastomosing veins of *L. heterodoxum*, × 1.   **7.** Free veins of *L. radiatum*, fertile marginal lobes on segment, × 1.   **8.** Cuneate segments of *L. oligostachyum*, × 1.

cubense (Figs. 11, 13) and *L. oligostachyum*. Some paleotropical species as *L. reticulatum* Schkuhr (Fig. 14) and *L. scandens* (L.) Sw. have distinctive rugulose-reticulate spores.

Cytology

A collection of *Lygodium volubile* from Jamaica was reported as $n = 87-90$ and considered as hexaploid by Walker (1966) on the basis of a record of $n = 29$ for a cultivated plant of the species. The occurence of three base numbers in *Lygodium*, $n = 28, 29, 30$, and higher ploidy levels as $n = 56$, and $n = 58$ indicate a relatively complex cytology. The problem of establishing a common base number by arithmetic calculation is especially evident in this group and is discussed in the cytological summary of the genus by Roy and Manton (1965). They suggest that the gametic number of $n = 29$ may be a relatively late innovation in *Lygodium* and possibly derived from an ancient base number of seven. The chromosome numbers in *Lygodium* contrast with other members of the family as 38 in *Anemia*, 76 in *Mohria*, and 77 in *Schizaea*, and indicate relatively distant relationships among these genera.

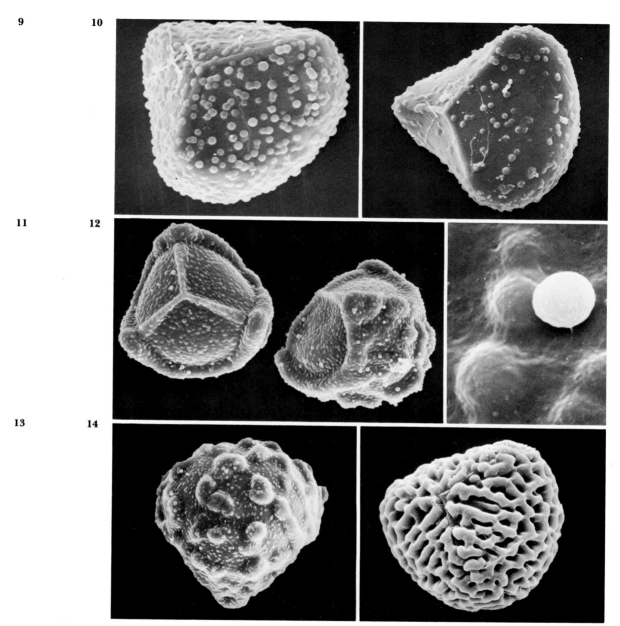

Figs. 7.9–7.14. Spores of *Lygodium*. **9.** *L. venustum,* proximal face, fungal hyphae at apex, Guatemala, *Deam 447,* ×600. **10.** *L. heterodoxum,* proximal face, fungal hyphae at apex, Guatemala, *Türckheim* in 1913, ×500. **11, 12.** *L. volubile,* British Guiana, *de la Cruz 4025.* **11.** Proximal face, left, distal tilted, right, ×500. **12.** Detail of proximal face with spheres below and on the surface, ×10,000. **13.** *L. cubense,* verrucate, distal face, Cuba, *Bües* in 1927, ×350. **14.** *L. reticulatum,* reticulate-rugose, proximal face, Australia, *Goy 424,* ×500.

Observations

The elongate leaf of *Lygodium* (Fig. 3) is adapted to scrambling or climbing. The apex is indeterminate and, as it grows, the stem twines around small stems and branches of trees and shrubs. The long, weak rachis is also supported by the divergent, forked pinnae that become entangled with other vegetation.

Climbing leaves that reach considerable heights are unusual in ferns and are best developed in *Lygodium* and in *Salpichlaena volubilis*. The scandent leaves of many species of Gleicheniaceae usually form dense thickets and only rarely are high climbers.

Literature

Duek, J. J. 1978. A taxonomical revision of *Lygodium* (Filicinae) in America. Fedde Repert. 89: 411–423.

Edman, G. 1932. Verkieselung und Verholzung der Sporenmembran der *Lygodium japonicum*. Svensk Bot. Tidskr. 26: 313–325.

Fay, A. 1973. A natural *Lygodium* hybrid found in Trinidad. Amer. Fern Jour. 63: 165.

Lugardon, B. 1974. La structure fine de l'exospore et de la périspore des filicinées isosporées. 2. Filicales. Commentaires. Pollen et Spores 14: 161–226.

Roy, S. K., and I. Manton, 1965. A new base number in the genus *Lygodium*. New Phytol. 64: 286–292.

Walker, T. G. 1966. A cytotaxonomic survey of the pteridophytes of Jamaica. Trans. Roy. Soc. Edinburgh 66: 169–237.

8. *Schizaea*
Figs. 8.1–8.15

Schizaea Sm., Mém. Acad. Turin 5: 419. 1793, *nom. conserv.* Type: *Schizaea dichotoma* (L.) Sm. (*Acrostichum dichotomum* L.)

Lophidium Rich., Act. Soc. Hist. Nat. Paris 1: 114. 1792., Type: *Lophidium latifolium* Rich. = *Schizaea elegans* (Vahl) Sw. *Schizaea* subgenus *Lophidium* (Rich.) Reed, Bol., Soc. Broter. II, 21: 119. 1947 [Subgenus *Schizaea*].

Ripidium Bernh., Jour. Bot. (Schrad.) 1800 (2): 127. 1802. Type: *Ripidium dichotomum* (L.) Bernh. (*Acrostichum dichotomum* L.) = *Schizaea dichotoma* (L.) Sm. [Subgenus *Schizaea*].

Actinostachys Wall., List, 1. 1829, description from R. Br., Prod. Fl. Nov. Holl. 162. 1810. Type: *Actinostachys digitata* (L.) Wall. (*Acrostichum digitatum* L.) = *Schizaea digitata* (L.) Sw. *Schizaea* subgenus *Actinostachys* (Wall.) Sturm, Mart. Fl. Brasil. 1 (2): 179. 1859.

Microschizaea Reed, *ibidem* 133. 1947. Type: *Microschizaea fistulosa* (Labill.) Reed = *Schizaea fistulosa* Labill. [Subgenus *Schizaea*].

Schizaea subgenus *Paraschizaea* Reed, *ibidem* 120. 1947. Type: *Schizaea pectinata* (L.) Sw. [Subgenus *Schizaea*].

Description Terrestrial; stem erect, or ascending, rather small, bearing trichomes and few to many slender, fibrous roots; leaves ca. 5–50 cm long, glabrous or with more or less scattered, small trichomes, partially to wholly dimorphic, the fertile with elongate fertile segments borne pinnately or subdigitately at the apex of the laminar axes which are simple or dichotomously branched and then not joined to fully joined by laminar tissue to form 2- to many-veined segments with free veins; sporangia in one or more (then sometimes indistinct) rows on each side of the vein, intermixed or not with long trichomes; spores ellipsoidal, monolete, with sparse to dense, spherical deposition on the otherwise smooth, slightly tuberculate, striate or pitted surface. Chromosome number: $n = 77, 94, 96, 103$, ca. 154, ca. 270, 350–370, 370.

Pinnate fertile branches (Figs. 5, 7) are characteristic of spe-

Fig. 8.1. *Schizaea pennula* (left) and *S. incurvata* (right), sand savannah, Surinam. (Photo K. U. Kramer.)

cies of subgenus *Schizaea,* while, in subgenus *Actinostachys,* they are subdigitate (dichotomous in their development). *Schizaea Poeppigiana* (Figs. 2, 4) is one of a few species with strongly dimorphic leaves. The stem of *Schizaea* is nearly siphonostelic.

Systematics *Schizaea* is a widely distributed genus of about 30 species. The two subgenera, *Schizaea* and *Actinostachys,* are rather closely allied, in spite of differences, especially in the gametophytes. Within the context of the family their relationship seems better expressed by recognition as a single genus.

Synopsis of *Schizaea*

Subgenus *Schizaea*
Fertile branch pinnate, its segments with narrow margins, bearing sporangia in a single row on each side of the vein.

About 18 species, with four to six in tropical America, one in eastern North America and one in Chile and the Falkland Islands.

Subgenus *Actinostachys*
Fertile branch subdigitate, its segments with broad margins, bearing sporangia usually in two or more, sometimes indistinct rows on each side of the vein.

Ten or 11 species, with two or three in tropical America.

Tropical American Species

The systematics of tropical American *Schizaea* is not adequately known largely due to the lack of field studies necessary to establish the extent of variation within the species. Variation in the number of axes and the extent of webbing between them, as in

S. elegans (Figs. 8, 9) suggests other species may also include a greater morphological diversity than currently recognized. Three species are not included in the key as they are considered possible variants of other species: *Schizaea fluminensis* Sturm has a lamina architecture consisting of very few, narrow segments, but its basic structure is similar to *S. elegans* and it is probably a form of that species. *Schizaea stricta* Lell., with dimorphic leaves, forked several times, appears to be a geographic variant of *S. incurvata* (Fig. 6). In *S. Germanii* (Fée) Prantl, similarities of the leaves to those of *S. pennula* suggest that it may represent a small ecological form of that species.

Schizaea penicillata Willd. has no systematic status since the name is a nomenclatural synonym of *S. pennula;* it has been applied to small specimens of that species and of *S. subtrijuga.*

Key to Species of *Schizaea* in America

a. Fertile branch pinnate (subgenus *Schizaea*).   b.
  b. Axis of leaf branched, at least in part, with two to many axes laterally joined by laminar tissue, the segment then with two to many veins.   c.
    c. Leaf expanded toward the apex into one to few (then flabellate) to many laminar segments, fertile branch with ca. 5–15 pairs of segments.   *S. elegans* (Vahl) Sw.
    c. Leaf expanded centrally into a single broadened segment that branches apically, fertile branch with ca. 25 pairs of segments.   *S. Sprucei* Baker
  b. Axis of leaf unbranched, or branched and the axes not or very slightly expanded, joined only at the base.   d.
    d. Axis of the fertile branch strongly recurved at maturity.   *S. Poeppigiana* Sturm
    d. Axis of the fertile branch straight to slightly curved at maturity.   e.
      e. Trichomes lacking among the sporangia; Chile and Falkland Islands.   *S. fistulosa* Labill.
      e. Numerous trichomes among the sporangia.   f.
        f. Sterile and fertile leaf branched, or unbranched and nearly straight.   *S. incurvata* Schkuhr
        f. Sterile and fertile leaf unbranched, the sterile curved to tortuous, the fertile nearly straight; eastern North America.   *S. pusilla* Pursh
a. Fertile branch subdigitate (subgenus *Actinostachys*).   g.
  g. Leaf axis with a moderate to well developed abaxial ridge, the axis margins straight or nearly so, spores pitted.   *S. pennula* Sw.
  g. Leaf axis with a broad, low, abaxial thickening, the axis margins recurved, spores striate.   *S. subtrijuga* Mart.

Ecology (Figs. 1, 2)

Schizaea is a genus of diverse habitats but most often grows in soils deficient in moisture or nutrients; very rarely it is epiphytic. The species appear to be adapted to environments in which there is a lack of competition with other vigorously growing plants. In tropical America most species occur in open habitats in sandy soil, or among rocks, or they grow in wet savannahs or boggy sites. A few, as *S. elegans,* are forest species sometimes growing in densely shaded, humus soil or on rotten wood. The species mostly grow below 500 m, occasionally to 1500 m and rarely to 2000 m.

Schizaea incurvata and *S. pennula* are characteristic of the savannahs and campinas of the Guayana region. These two species are the only ferns growing in this habitat characterized by open

Fig. 8.2. *Schizaea Poeppigiana,* Mason River savannah, Jamaica. (Photo W. H. Hodge.)

areas of white sand where there is high light intensity, temperatures up to 42°C near the ground, and seasonally dry soil. The erect leaves with very slender laminar axes are evidently well adapted to these extreme conditions. It has been observed that *S. incurvata* tends to grow in the open sand or at borders of shrubs, while *S. pennula* is more often under the shade of shrubby vegetation, although sometimes, as in Fig. 1, they do grow together. Takeuchi (1960) made an ecological study of the *Schizaea* species of Amazonian Brazil.

Geography (Fig. 3)

Schizaea is primarily a pantropical genus with a few extratropical species. In tropical America the genus occurs in southern Florida, Mexico and Central America, some of the Greater and

Fig. 8.3. Distribution of *Schizaea* in America.

Lesser Antilles, and northern South America southward to Bolivia and southern Brazil. There are two extratropical American species. *Schizaea fistulosa* occurs in Chile and the Falkland Islands and is also in New Zealand and islands of the southwest Pacific. A coastal plain species of eastern North America, *S. pusilla* is disjunct in New Jersey, Long Island (New York), Nova Scotia and Newfoundland.

Spores

Spores of the tropical American species of *Schizaea* are uniformly monolete and the laesurae, often prominently project (Fig. 13). There are two main spore types in tropical American species which generally correspond to the groups based on leaf form. Species with filiform leaf axes in subgenus *Schizaea* as *S. incurvata* (Fig. 11) have shallowly tuberculate spores. Those with broader lamina segments, as *S. Sprucei* and *S. Poeppigiana*, have smoother spores usually with dense spherical deposition as in *S. elegans* (Fig. 12) which has exceptionally small spores. In subgenus *Actinostachys* the leaves have subdigitate fertile branches and pitted spores as in *S. pennula* (Fig. 13). In this group, *Schi-*

Figs. 8.4–8.10. *Schizaea*. **4.** Plant of ***Schizaea Poeppigiana,*** with dimorphic leaves, ×0.2. **5.** Fertile segments of ***S. Poeppigiana,*** ×2.25. **6.** Plants of ***S. incurvata,*** ×0.2. **7.** Fertile segments of ***S. incurvata,*** ×3.5. **8. *S. elegans,*** with leaf cleft into few, narrow lobes, ×0.3. **9. *S. elegans,*** with leaf cleft into two very broad lobes, ×0.3. **10.** Fertile segments of ***S. pennula,*** ×1.5.

zaea subtrijuga has exceptional spores with low, parallel ridges and dense spherical surface deposit. (Figs. 14, 15).

There are other distinctive types of spores in the extratropical American species. Those of *Schizaea pusilla* are densely pitted, and *S. fistulosa* spores are monolete but nearly spherical. Some paleotropical species as *S. digitata* (L.) Sw., *S. laevigata* Mett., and *S. melanesica* Sell. have spores with prominent spiral ridges similar to those in *Anemia* subgenus *Coptophyllum.* The utility of the spores in distinguishing and assessing relationships of species

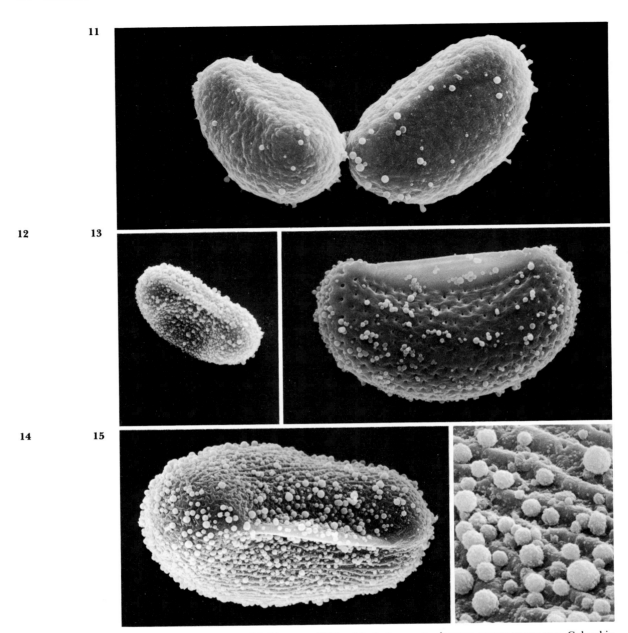

11

12   13

14   15

Figs. 8.11–8.15. Spores of *Schizaea*, ×1000.   **11. *S. incurvata*,** laesura across spore top, Colombia, *Schultes & Cabrera 16920*.   **12. *S. elegans*,** dense spherical deposit, Peru, *Tryon & Tryon 5292*.   **13. *S. pennula*,** raised laesura at top, Brazil, *Conant 1116*.   **14, 15. *S. subtrijuga*,** Columbia, *Schultes et al. 18291*.   **14.** Proximal face with ridges, laesura at center.   **15.** Detail of ridges and spherical deposit, ×5000.

was recognized by Selling (1944) in his study of the recent and fossil spores of *Schizaea*.

Cytology

Report of n = ca. 154 for *Schizaea incurvata* occurring near Manaus, Brazil (Tryon et al., 1975) supplies an insight into cytology of the genus in tropical America. The lowest number known for the genus, n = 77 in *S. asperula* Wakef. from New Zealand, also known in *S. dichotoma* (L.) Sm. from Ceylon, is considered as the diploid level in *Schizaea*. On the basis of this, the Brazilian report can be regarded as tetraploid. A report of n = ca. 540 for a New Zealand specimen of *S. dichotoma* and reports of other species between 94 and 370 indicate cytological complexities in the genus involving both polyploidy and aneuploidy. The cytogenetic mechanisms appear to be similar to those of *Ophioglossum* in which a chromosome number of n = 631 has been reported.

Literature

Selling, O. H. 1944. Studies in the recent and fossil spores of *Schizaea*, with particular reference to the spore characters. Meddel. Goteb. Bot. Tradg. 16: 1–112.

Takeuchi, M. 1960. O genero *Schizaea* no Amazonia. Bol. Mus. Goeldi, n.s. Bot. 5: 1–26.

Tryon, A. F., H. P. Bautista, and I. da Silva Araújo. 1975. Chromosome studies of Brazilian ferns. Acta Amazonica 5: 35–43.

Family 5. Gleicheniaceae

Gleicheniaceae (R. Br.) Presl, Rel. Haenk. 1: 70. 1825, as Order Gleicheneae. *Gleicheneae* R. Br., Prod. Fl. Nov. Holl. 160. 1810, as Tribe of Filices. Type: *Gleichenia* Sm.

Dicranopteridaceae Ching, Acta Phytotax. Sinica 3: 94. 1954. Type: *Dicranopteris* Bernh.

Description Stem very long-creeping, slender, freely branched, protostelic or siphonostelic, indurated, bearing trichomes and (or) scales; leaves ca. 30 cm to 10 m or more long and scrambling or climbing, lamina and its branches (except the ultimate ones) usually pseudodichotomous, indeterminate, with a permanently arrested bud at the fork of the branches, or the lamina partially pinnately branched, the rachis and sometimes the pinna-rachis with a periodically dormant bud between the last developed branches, a pair of accessory segments sometimes present at the base of a pair of otherwise usually naked axes, small stipular segments sometimes associated with the laminar buds, or rarely the lamina (more often the pinnae), wholly pinnately branched and determinate, ultimate segments confined to the ultimate branches or also present on the axes of a lower order, leaf with trichomes and (or) scales, circinate in the bud, petiole without stipules; sporangia borne in exindusiate sori on the abaxial surface of the ultimate segments, with a short, many-rowed stalk of cells, with a central to oblique or nearly apical annulus not interrupted by the stalk; homosporous, spores lacking chlorophyll. Gametophyte epigeal, with chlorophyll, obcordate to somewhat elongate, thickened centrally, with thin margins, and often with small trichomes, archegonia borne on the lower surface of the thickened portion, antheridia ca. 6- to 12-celled, borne on the lower surface, mostly initiated before the archegonia.

Comments on the Family

The classification of Holttum (1957a) is adopted, but *Stromatopteris* of New Caledonia is placed in its own family following the treatment of Bierhorst (1968). There are two pantropical (sometimes extratropical) genera: *Gleichenia* with three subgenera and *Dicranopteris* with two subgenera. This classification recognizes the major species groups and their relationships. Most previous classifications have maintained the same five taxa (our subgenera) at some rank although Nakai (1950) recognized eight genera. Pichi-Sermolli (1970) has discussed the reasons for recognizing Presl as the author of the family name.

The Gleicheniaceae are undoubtedly an ancient family, with the earliest fossils from the Carboniferous. Cretaceous fossils show characteristic branching of the leaves with some specimens closely resembling living species.

The family is moderately advanced, especially in the characters of the sporangium and the highly specialized lamina architecture of most of the subgenera. The Gleicheniaceae are isolated except for a relationship with Stromatopteridaceae. While the family is often considered to be ancestral to other ferns, especially to the polypodioid and grammitid ferns, these relations have not been established with certainty.

The leaves of Gleicheniaceae usually have a complex lamina architecture based on pseudodichotomous branching and periodically dormant or permanently arrested laminar buds at the fork of a pair of branches. While the forked axis is referred to as pseudodichotomous, anatomically it is a compact sympodial dichotomy (Chrysler, 1943).

The primitive leaf was probably (Holttum, 1957b) fully expanded and pinnate, perhaps with a 2-pinnate-pinnatifid lamina. Leaves with a similar arrangement as in *Gleichenia* subgenus *Diplopterygium* and subgenus *Gleichenia* could possibly be derived by dormancy in the apex of the primary or secondary axes. However, complex leaves with segments of the fith order or higher and accessory and stipular segments, are difficult to homologize with the parts of a complex pinnate lamina. Considerable elaboration from the primitive condition seems to have occurred in such leaves. Some species have determinate leaves and a rather simple lamina architecture, as in the 1-pinnate *Gleichenia simplex* and forms of *G. vulcanica* which are 1-pinnate-pinnatifid. The relatively simple leaves of these species resemble juvenile leaves of species with more complex adult leaves, and they probably evolved through neotony. The evolution of the different kinds of leaves in the family is evidently based on an unusually labile growth pattern.

Spores of the Gleicheniaceae are unusual among the Filicopsida in the trilete and monolete forms in both genera and unique wall stratification. Spores from herbarium specimens mostly appear incompletely formed with collapsed walls and sparse surface deposits, as in Fig. 13. The photographs in Erdtman and Sorsa (1971), and Selling (1946), and the drawing of *Dicranopteris linearis* in Harris (1955) also illustrate this condition. The unusual, thin exposure formation, in the distal part of the spore, discussed by Lugardon (1974), may account for the collapsed walls of dry specimens.

Key to Genera of Gleicheniaceae

a. Stem with scales, the leaves, especially the laminar buds, with scales and sometimes also stellate trichomes, veins of an ultimate segment simple or 1-forked (rarely to 3-forked), sori of usually 2–4 (rarely more) sporangia, usually paraphysate.   9. *Gleichenia*

a. Stem with trichomes, the leaves, especially the laminar buds with usually branched, rarely stellate, trichomes, veins of an ultimate segment 2- to 4-forked, sori of usually ca. 8–15 sporangia, without paraphyses.   10. *Dicranopteris*

Literature

Bierhorst, D. W. 1968. On the Stromatopteridaceae (Fam. nov.) and on the Psilotaceae. Phytomorph. 7: 168–268.

Chrysler, M. A. 1943. The vascular structure of the leaf of *Gleichenia,* I. The anatomy of the branching regions. Amer. Jour. Bot. 30: 735–743.

Erdtman, G., and P. Sorsa. 1971. Pollen and Spore Morphology/Plant Taxonomy, Pteridophyta. 302 pp. Almqvist & Wiksell, Stockholm.

Harris, W. F. 1955. A manual of the spores of New Zealand Pteridophytes. 186 pp. Bull. 116. Dept. Sci. & Ind. Research, Wellington.

Holttum, R. E. 1957a. Florae Malesianae Praecursores XVI. On the taxonomic subdivision of the Gleicheniaceae, with description of new Malaysian species and varieties. Reinwardtia 4: 257–280.

Holttum, R. E. 1957b. Morphology, growth-habit and classification in the family Gleicheniaceae. Phytomorph. 7: 168–184.

Lugardon, B. 1974. La structure fine de l'exospore de la perispore des Filicinées. 2. Filicales. Commentaires. Pollen et Spores 16: 161–226.

Maxon, W. R. 1909. Gleicheniaceae, in No. Amer. Fl. 16 (1): 53–63.

Nakai, T. 1950. A new classification of the Gleicheniales. Bull. Nat. Sci. Mus. Tokyo 29: 1–71.

Pichi-Sermolli, R. E. G. 1970. A provisional catalogue of the family names of living Pteridophytes. Webbia 25: 219–297.

Selling, O. H. 1946. Studies in Hawaiian Pollen Statistics. Pt. 1. The spores of the Hawaiian Pteridophytes. 87 pp. Bishop Museum Special Publ. 37, Honolulu.

Underwood, L. M. 1907. American ferns VIII. A preliminary review of the North American Gleicheniaceae. Bull. Torrey Bot. Cl. 34: 243–262.

Walker, T. G. 1966. A cytotaxonomic survey of the pteridophytes of Jamaica. Trans. Roy. Soc. Edinburgh 66: 169–237.

Walker, T. G. 1973. Additional cytotaxonomic notes on the Pteridophytes of Jamaica. Trans. Roy. Soc. Edinburgh 69: 109–135.

9.   *Gleichenia*
Figs. 9.1–9.15

Gleichenia Sm., Mém. Acad. Turin 5: 419. 1793. Type: *Gleichenia polypodioides* (L.) Sm. (*Onoclea polypodioides* L.).

Calymella Presl, Tent. Pterid. 48. 1836. Type: *Calymella alpina* (R. Br.) Presl = *Gleichenia alpina* R. Br. [Subgenus *Gleichenia*].

Sticherus Presl, *ibidem* 51. 1836. Type: *Sticherus laevigatus* (Willd.) Presl (*Mertensia laevigata* Willd.) = *Gleichenia truncata* (Willd.) Spreng. [Subgenus *Mertensia*].

Gleichenia subgenus *Mertensia* Hook., Sp. Fil. 1: 4. 1844, not *Mertensia* Willd. Type: *Gleichenia laevigata* (Willd.) Hook. = *Gleichenia truncata* (Willd.) Spreng.

Gleicheniastrum Presl, Gefässb. Stipes Farrn. 30. 1847. Type: *Gleicheniastrum microphyllum* (R. Br.) Presl = *Gleichenia microphylla* R. Br. [Subgenus *Gleichenia*].

Gleichenia section *Holopterygium* Diels, Nat. Pflanz. 1 (4): 353. 1900. Type: *Gleichenia pubescens* (Willd.) HBK. = *Gleichenia furcata* (L.) Spreng. [Subgenus *Mertensia*].

Diplopterygium (Diels) Nakai, Bull. Nat. Sci. Mus. Tokyo 29: 47. 1950. *Gleichenia* section *Diplopterygium* Diels, Nat. Pflanz. 1 (4): 353. 1900. Type. *Gleichenia glauca* (Houtt.) Hook. (*Polypodium glaucum* Houtt. earlier than Thunb. usually cited). *Gleichenia subgenus Diplopterygium* (Diels) Holtt., Reinwardtia 4: 261. 1957.

Description

Terrestrial; stem very widely creeping, slender, bearing often fimbriate scales and sometimes stellate trichomes, and slender, fibrous roots at irregular intervals; leaves monomorphic, widely spaced, ca. 25 cm to often 1–3 m to very rarely 15 m long, not branched (in *G. simplex* and high altitude plants of *G. vulcanica* Blume) to several times pseudodichotomously branched, with pinnatisect penultimate segments, laminar buds usually with fimbriate scales, other parts of the lamina, especially when young, with often fimbriate scales and stellate trichomes, veins free, simple or 1-forked, rarely to 3-forked; sori abaxial, usually paraphysate, exindusiate, with 2–4 (rarely more) sporangia; spores more or less ellipsoidal monolete, or spheroidal, trilete with slightly papillate surface, sometimes with sparse spherical deposit. Chromosome number: $n = 20, 22, 34, 56, 68$; $2n = 40, 68, 136$.

The characters of the genus are the simple or 1-forked (Fig. 3) veins, the few sporangia in a sorus (Fig. 4), and the scales which are especially evident on the laminar buds (Fig. 5). The stem is protostelic.

A sample of the simple to complex types of leaf architecture is shown in Figs. 6–11. The different positions of the periodically dormant and permanently arrested laminar buds are also illustrated.

Systematics

Gleichenia is a pantropical and sometimes extratropical genus of about 110 species. The three subgenera, *Gleichenia, Diplopterygium* and *Mertensia,* are treated as genera by some authors.

Gleichenia subgenus *Mertensia* is sometimes considered as subgenus *Mertensia* (Willd.) Hooker; however, it must be considered as a new subgenus because *Mertensia* Willd. is illegitimate and cannot be transfered to another rank (see *Dicranopteris*).

Synopsis of *Gleichenia*

Subgenus *Gleichenia*

Lamina architecture wholly to predominately pinnate, the lamina with pinnately arranged ultimate and penultimate segments, and usually with several pairs of pinnae, some portions of the lamina may be pseudodichotomously branched, veins simple, each ultimate segment bearing a single sorus, spores trilete.

A paleotropical and austral group of about 10 species, none in America.

Fig. 9.1. *Gleichenia bifida,* Cerro Azul, San Martin, Peru, on road cut through forest. (Photo Alice F. Tryon.)

Subgenus *Diplopterygium*

Lamina architecture wholly pinnate, the lamina with pinnately arranged ultimate and penultimate segments and when fully expanded with many pairs of pinnae. Veins 1-forked, several to many sori on each ultimate segment, spores trilete.

About 10 species, Indo-Malesia to Japan, with one species, *G. Bancroftii* in the American tropics.

Subgenus *Mertensia*

Lamina architecture predominantly pseudodichotomous, only the ultimate segments pinnately arranged (these are primary segments in *G. simplex*), or sometimes with a few pairs of pinnae, veins usually 1- to rarely 3-forked, several to many sori on an ultimate segment, spores monolete.

A pantropical and austral subgenus of about 90 or more species with about 40 in the American tropics and in southern South America.

Fig. 9.2. Distribution of *Gleichenia* in America.

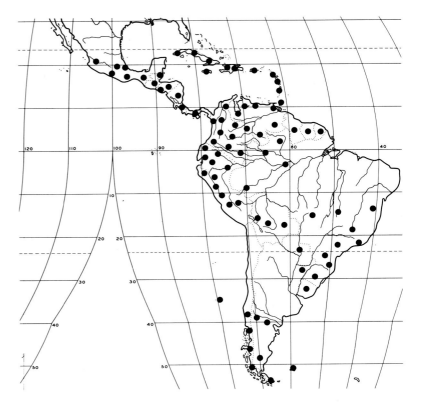

Tropical American Species

Gleichenia Bancroftii Hook. (Fig. 10) is the only American species of subgenus *Diplopterygium*. All others are members of subgenus *Mertensia,* for example, *Gleichenia bifida* (Willd.) Spreng. (Fig. 1), *G. jamaicensis* (Underw.) C. Chr., *G. longipes* (Fée) Christ (Fig. 7), *G. nuda* (Reichardt) Moore (Fig. 11), *G. palmata* (Fée) Moore, *G. pteridella* Christ (Fig. 8), *G. retroflexa* Christ, *G. rubiginosa* Mett., *G. simplex* (Desv.) Hook. (Fig. 9), *G. squamosa* (Fée) Brade, *G. tomentosa* (Sw.) Spreng., and *G. Underwoodiana* (Maxon) C. Chr. (Fig. 6).

The species of the Andes and of southeastern Brazil are poorly known and a modern revision is needed. Details of the ultimate segments and of the lamina indument, and the position of buds in relation to leaf development need to be investigated.

American species of *Gleichenia* have often been referred to *Dicranopteris,* following the treatments of Underwood (1907) and Maxon (1909) who restricted *Gleichenia* to the subgenus *Gleichenia* (above), while referring all other species to *Dicranopteris.*

Gleichenia × *Leonis* (Maxon) C. Chr. (*pro species*), a hybrid of *G. bifida* and *G. palmata* (Duek, 1974), is infrequent in Cuba. It has abortive spores and lamina characters intermediate between the parental species.

Ecology (Fig. 1)

Most of the species are plants of open habitats, often growing in sterile soil. Some, especially of subgenus *Diplopterygium,* form dense thickets or climb at the edge of forest.

3   4   5

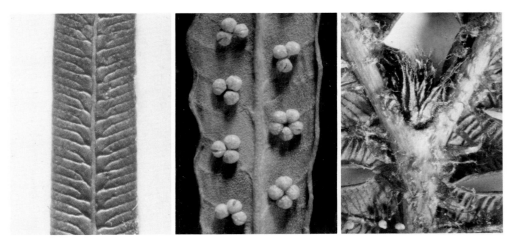

Figs. 9.3–9.5.   3. *Gleichenia squamosa*, venation, × 6.5.   **4. *G. retroflexa*,** sori, × 10.   **5. *G. rubiginosa*,** pseudodichotomous branching of leaf, and permanently arrested, scaly bud between the branches, × 5.

In America *Gleichenia bifida* is the common thicket species and sometimes also a climber. Most other species are colonial but do not form dense thickets. They grow on river banks, ravines, rarely in rain forests, in oak and pine woods (Mexico), in damp thickets in open rocky or grassy places. Species often pioneer on landslides, road cuts and fills, especially in forest clearings and old pastures. *Gleichenia simplex* is a species of the páramo and puna from Colombia to Peru, which grows, in wet, grassy sod, on rocky banks and shrubby slopes. The species are found usually above 500 m, most commonly between 1000 and 2500 m, rarely to 3000 m, except *G. simplex* which grows from 3000 to 3800 m.

Gleichenia species are colonial because of their long-creeping and freely branching stem. They usually require high light intensity for vigorous growth and the size and architecture of the leaves, and the orientation of their branches show adaptations to their habitat. Some species such as *G. bifida* and *G. Brancroftii* produce several pairs of pinnae, each pair becoming nearly fully developed before the periodically dormant rachis bud continues growth to form another pair of pinnae. The lower pinnae provide support for the growth by leaning on other *Gleichenia* leaves or the adjacent vegetation. These species often produce dense thickets, or when growing at a forest border, they can climb to many meters. The pinnae and their branches are oriented toward the light, and shade the supporting vegetation. Other species, for example, *Gleichenia nuda* (Fig. 11) and *G. longipes* (Fig. 7), produce only one or a few pairs of pinnae, the leaf is more erect and self-supporting than in the thicket species. The pinna branches spread horizontally and the species grow in light shade or in more open places. *Gleichenia simplex* with small, 1-pinnate leaves is the most reduced species in the genus (Fig. 9). Its leaf architecture is evidently adapted to the high altitude open habitats in which it grows.

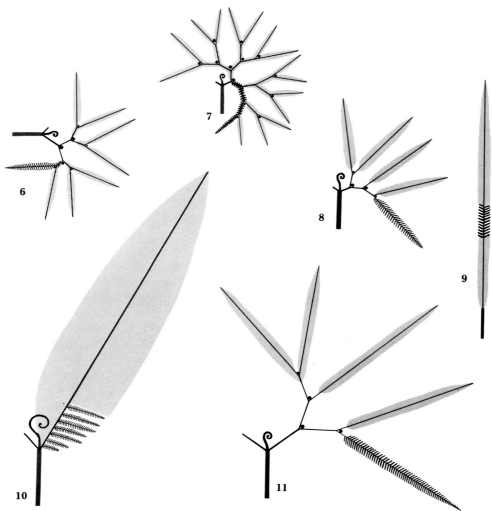

Figs. 9.6–9.11. Architecture of *Gleichenia* leaves, the coil indicating a periodically dormant bud, the black dots indicating position of permanently arrested buds, all ×0.25, except Fig. 8, ×0.5. **6.** *G. Underwoodiana,* ½ of one pinna. **7.** *G. longipes,* ½ of lamina. **8.** *G. pteridella,* ½ of lamina. **9.** *G. simplex,* lamina. **10.** *G. Bancroftii,* one pinna of several pairs, the rachis bud periodically dormant. **11.** *G. nuda,* one pinna of two pairs.

Geography (Fig. 2)

Gleichenia is a widely distributed pantropical and sometimes extratropical genus. In America it is distributed from central Mexico to Panama, through the Antilles, and in South America south to Bolivia, northeastern Argentina and Uruguay; it also is found in southern Chile and Argentina, the Juan Fernandez Islands and the Falkland Islands.

Subgenus *Mertensia* has a single wide-ranging species, *G. bifida,* which occurs nearly throughout the tropical range of the genus in America. Distributions of other species of the subgenus are much less extensive. There are three centers of species diversity: Costa Rica with ca. 15 species, the Andes from Colombia to Bolivia with ca. 20 species, and southeastern Brazil with ca. 10 species.

Gleichenia Bancroftii, the only American species of the sub-

12 14

13 15

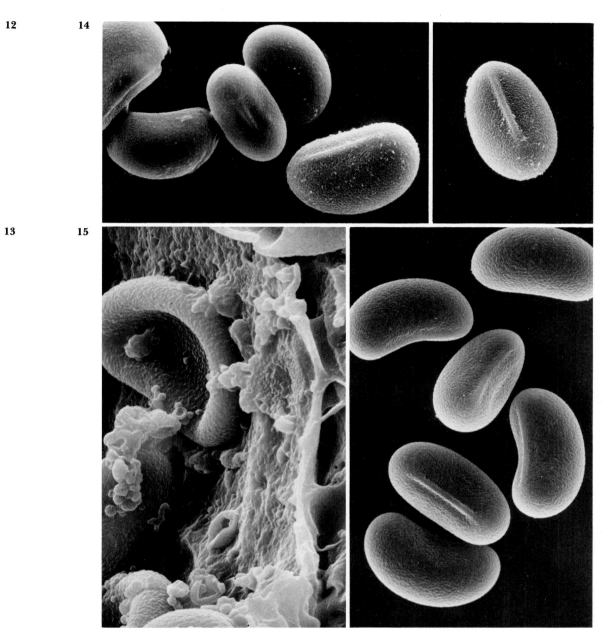

Figs. 9.12–9.15. Spores of *Gleichenia,* × 1000.   **12.** *G. tomentosa,* with spherical deposit, Paraguay, *Rojas 12726.*   **13.** *G. tomentosa,* portion of the inner wall of the sporangium with young spores and irregular tapetal material, × 2000.   **14.** *G. pennigera,* Brazil, *Mexia 4856.*   **15.** *G. bifida,* St. Lucia, *Cooley 8682.*

genus *Diplopterygium,* occurs in Mexico and Central America, the Greater Antilles, Martinique and Guadeloupe, and in the mountains from Venezuela to Bolivia.

Spores

American species of subgenus *Mertensia* have monolete spores as in *G. tomentosa* (Fig. 12), *G. pennigera* (Mart.) Moore (Fig. 14) and *G. bifida* (Fig. 15). The scar is prominently raised and the slightly irregular surface may be overlayed with a sparse, spherical deposit, as in Fig. 12. A portion of the inner sporangium wall of *G. tomentosa* (Fig. 13) shows incompletely formed spores and tapetal particles. Spores of *G. Bancroftii* (subgenus *Diplopterygium*) are uniformly trilete. Sections of the wall of this species in Lugardon's work (1974) show unusual development of an interrupted middle exospore layer that characterizes the family. Spores of the paleotropical subgenus *Gleichenia* are also trilete.

Cytology

The cytological reports of American species of *Gleichenia* are based on the studies of Walker in Jamaica (1966, 1973) and Sorsa (1968) in Puerto Rico. There are distinctive chromosome numbers for each of the three subgenera. Subgenus *Mertensia* is $n = 34$, based on *G. bifida* and *G. palmata,* while *G. jamaicensis* with $n = 68$ is regarded as a tetraploid. There are also reports of several paleotropical species in this subgenus with $n = 34$ or 68, from Australia and New Caledonia. In subgenus *Diploptergyium,* *G. Bancroftii* is reported as $n = 56$ from Jamaica (Walker, 1966) and Oaxaca, Mexico (Mickel et al., 1966). Subgenus *Gleichenia,* which occurs exclusively in the Old World, has been reported as $n = 20$ or 22, from New Zealand and New Caledonia.

Literature

Dueck, J. J. 1974. A newly recognized *Gleichenia* hybrid from Cuba. Amer. Fern Jour. 64: 74–76.
Lugardon, B. 1974. Reference under the family.
Mickel, J. T., W. H. Wagner, and K. L. Chen. 1966. Chromosome observations on the ferns of Mexico. Caryologia 19: 95–102.
Underwood, L. M. 1907. Reference under the family.
Walker, T. G. 1966. Reference under the family.
Walker, T. G. 1973. Reference under the family.

10. *Dicranopteris*
Figs. 10.1–10.10

Dicranopteris Bernh., Neues Jour. Bot. (Schrad.) 1 (2): 38. 1806, *nom. nov.* for *Mertensia* Willd. (not Roth) and with the same type.
Mertensia Willd., Kongl. Vetensk. Acad. Nya Handl. 25: 165. 1804, *illegit.*, not Roth 1790 (Boraginaceae), not *Gleichenia* subgenus *Mertensia* Hook. Type: *Mertensia dichotoma* (Murray) Willd. (*Polypodium dichotomum* Murray, earlier than Thunb. usually cited), *Dicranopteris dichotoma* (Murray) Bernh. = *Dicranopteris linearis* (Burm. f.) Underw. [Subgenus *Dicranopteris*].
Hicriopteris Presl, Epim. Bot. 26. 1851. Type: *Hicriopteris speciosa* Presl = *Dicranopteris speciosa* (Presl) Holtt. [Subgenus *Dicranopteris*].
Mesosorus Hassk., Acta Soc. Sci. Indo-Nêerl. 1 (7): 2. 1856, a later *nom. nov.* for *Mertensia* Willd. and with the same type.
Gleichenia section *Heteropterygium* Diels, Nat. Pflanz. 1 (4): 355. 1900. Type: *Gleichenia linearis* (Burm. f.) Bedd. = *Dicranopteris linearis* (Burm. f.) Underw. [Subgenus *Dicranopteris*].
Gleichenella Ching, Sunyatsenia 5: 276. 1940. Type: *Gleichenella pectinata* (Willd.) Ching = *Dicranopteris pectinata* (Willd.) Underw. [Subgenus *Acropterygium*].

Fig. 10.1. *Dicranopteris nervosa,* Mt. Itatiaia, Rio de Janeiro, Brazil. (Photo Alice F. Tryon.)

Acropterygium (Diels) Nakai, Bull. Nat. Sci. Mus. Tokyo 29: 5. 1950, *nom. superfl.* for *Gleichenella* Ching and with same type. *Gleichenia* section *Acropterygium* Diels, Nat. Pflanz. 1 (4): 353. 1900. Type: *Gleichenia pectinata* (Willd.) Presl = *Dicranopteris pectinata* (Willd.) Underw. *Dicranopteris* subgenus *Acropterygium* (Diels) Holtt., Reinwardtia 4: 261. 1957.

Description

Terrestrial; stem very widely creeping, slender, bearing often irregularly branched trichomes and slender, fibrous roots at irregular intervals; leaves monomorphic, widely spaced, less than 1 m to usually 2–4 m or very rarely to 30 m long, 1- or 2-forked, or to several times forked, with pinnatisect, rarely 1-pinnate penultimate segments, laminar buds with trichomes, other parts of the lamina, especially when young, with usually irregularly branched or (in *D. pectinata*) sometimes stellate trichomes, veins free 2- to 4-forked; sori abaxial, not paraphysate, exindusiate, with usually 8–15 sporangia; spores tetrahedral-globose, trilete or more or less ellipsoidal, monolete, the surface smooth or granulate. Chromosome numbers: $n = 39, 43, 44, 78$; $2n = 78, 156$.

The principal characters distinguishing *Dicranopteris* from *Gleichenia* are the twice or more forked veins (Fig. 3), the greater number of sporangia in a sorus (Fig. 4), and the laminar bud in the axil of a fork bearing trichomes (Fig. 5) rather than scales. The complex lamina architecture of *D. flexuosa* and *D. pectinata,* and the simpler *D. nervosa* are shown in Figs. 6–8. The stem is protostelic in subgenus *Dicranopteris* and siphonostelic in subgenus *Acropterygium.*

Systematics

Dicranopteris is a pantropical, sometimes subtropical genus of about 10 species. Two subgenera are recognized: subgenus *Dicranopteris* and subgenus *Acropterygium. Dicranopteris* is adopted here in a restricted sense, rather than the broad circumscription of Underwood (1907) which included species of *Gleichenia* in the genus.

The correct nomenclatural status of the name *Dicranopteris* is difficult to determine. Pichi-Sermolli (1972) regards it as a re-

Fig. 10.2. Distribution of *Dicranopteris* in America. In the southeast United States doubtless as waifs.

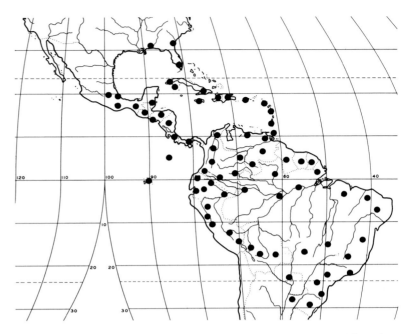

naming of *Mertensia* Willd., while Holttum (1973) considers it a new genus. We believe that the evidence is equivocal, but adopt the former interpretation. Since Bernhardi cited only one species under *Dicranopteris,* this serves to typify the synonymous generic name *Mertensia.*

Synopsis of *Dicranopteris*

Subgenus *Dicranopteris*

Stem protostelic, accessory segments usually present, spores trilete or monolete.

A pantropical or subtropical subgenus of about 10 species, or fewer if *D. linearis* is treated in a broad sense. *Dicranopteris flexuosa, D. nervosa* and possibly others are in the American tropics.

Subgenus *Acropterygium*

Stem siphonostelic, acessory segments lacking, spores monolete.
One species, *D. pectinata,* of the American tropics.

Tropical American Species

The species of subgenus *Dicranopteris, D. flexuosa* (Schrad.) Underw., with the lamina several times forked, and *D. nervosa* (Kaulf.) Ching, with the lamina usually 1-forked, are both polymorphic. Each may represent a variable species, or a small complex of species or varieties. The lamina of *D. nervosa* is rarely 2-forked and the somewhat more complex *D. Schomburgkiana* (Sturm) Morton of Mount Roraima is perhaps a geographic variant of it.

Incomplete specimens of *Dicranopteris pectinata* (Willd.) Underw. (subgenus *Acropterygium*) and of *D. flexuosa* can usually be distinguished by the unequal lamina branches (except for the ultimate pairs) of the former species and the nearly equal branches of the latter.

Figs. 10.3–10.5. **3.** *Dicranopteris flexuosa,* venation, ×6.5. **4.** *D. pectinata,* sori, ×10. **5.** *D. flexuosa,* pseudodichotomous branching of leaf, arrested bud with trichomes in fork, ×4.

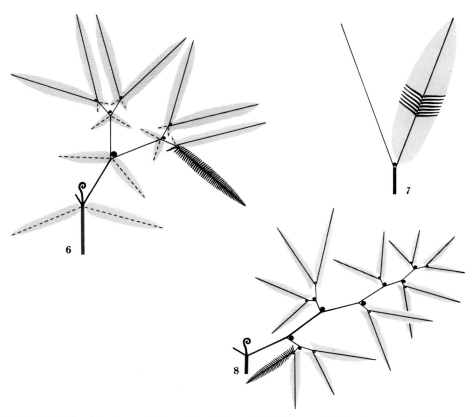

Figs. 10.6–10.8. Leaf architecture of ***Dicranopteris,*** the coil indicats a periodically dormant leaf bud, the black dot a permanently arrested bud, the basal black bar indicates petiole, ×2.5. **6.** *Dicranopteris flexuosa,* basal primary branch of lamina with accessory segments at base of the axes. **7.** *D. nervosa* lamina, 1-forked. **8.** *D. pectinata,* basal primary branch of lamina.

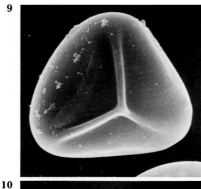

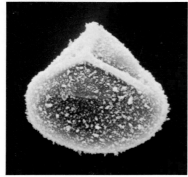

Figs. 10.9, 10.10. Spores of *Dicranopteris*, × 1000. **9.** *Dicranopteris nervosa* nearly smooth surface, Brazil, *Reitz & Klein 4829* (US). **10.** *D. flexuosa* unusually dense deposit, Brazil, *R. Windisch & Ghillany 351.*

Ecology (Fig. 1)

Dicranopteris is primarily a genus of scrambling and sometimes climbing ferns which form dense thickets in forest openings, on open banks, in grassy areas and in disturbed sites.

In America it grows in grassland, along forest borders where the leaves may climb to 10 m, and on brushy banks or in shaded ravines and forest slopes. It is a frequent pioneer in old pastures and forest clearings, on landslides, river bank slumps and road cuts where it is an important soil binder. It usually grows in well drained sterile soils, but sometimes in marshy or boggy places. *Dicranopteris* grows from sea level to 3000 m but most often between 500 and 1500 m. In contrast to *Gleichenia* it is predominently a genus of lower elevations.

Geography (Fig. 2)

Dicranopteris is a widely distributed tropical and subtropical genus. In America it occurs from central Mexico to Central America, in the Antilles, and in South America south to Bolivia, northeastern Argentina and southeastern Brazil. *Dicranopteris pectinata* is on Cocos Island and *D. flexuosa* on the Galápagos Islands. *Dicranopteris nervosa* occurs from Bolivia to southeastern Brazil; the other two species are widely distributed. *Dicranopteris flexuosa* is reported from the southeastern United States where it occurs as a waif from the south rather than a permanent member of the flora.

Spores

Spores of subgenus *Dicranopteris* are trilete with smooth to granulate surface as in *D. flexuosa* and *D. nervosa* (Figs. 9, 10). Those of the paleotropical *D. linearis* (Burm. f.) Underw. in the same subgenus are similar. Wall sections of this species, studied by Lugardon (1974), show a middle exospore layer extends nearly to the ends of the laesurae on the proximal face. The spores of *D. pectinata* (subgenus *Acroptergyium*) are monolete with a smooth sporoderm.

Cytology

Cytological data on the American species are based on the studies of Walker (1966, 1973) in Jamaica and Sorsa (1968) in Puerto Rico. The subgenera, as in *Gleichenia,* have distinctive chromosome numbers. In subgenus *Dicranopteris, D. flexuosa* is reported as n = ca. 78 from Jamaica and is considered a tetraploid based on 39. A collection of the same species from Puerto Rico (as *D. rigida* (Kze.) Nakai) was also reported as n = 78, but with irregular pairing. Several collections of *D. linearis,* a paleotropical species with n = 39 or 78, are consistent with records from the neotropics. In subgenus *Acropterygium, D. pectinata* is reported as n = 43 from Jamaica while several populations from Puerto Rico are reported as n = 44 or ca. 44. The numbers need to be confirmed to determine whether there are aneuploid differences.

Literature

Holttum, R. E. 1973. On the typification of *Mertensia* Willd. non Roth (Glecheniaceae) with notes on *Sticherus* Presl and *Hicriopteris* Presl. Taxon 22: 447–450.

Lugardon, B. 1974. Reference under the family.

Pichi-Sermolli, R. E. G. 1972. Names and types of fern genera, 3. Webbia 26: 491–536.

Sorsa, V. 1968. Chromosome studies on Puerto Rican ferns (Gleicheniaceae). Caryologia 21: 97–103.

Underwood, L. M. 1907. Reference under the family.

Walker, T. G. 1966. Reference under the family.

Walker, T. G. 1973. Reference under the family.

Family 6.   Hymenophyllaceae

Hymenophyllaceae Link, Handb. Erken. Gew. 3: 36. 1833. Type: *Hymenophyllum* Sm.

Trichomanaceae Kunkel, Fedde Report. 70: 155. 1965. Type: *Trichomanes* L.

Description

Stem erect to decumbent, small, or usually long-creeping and very slender, sometimes branched, protostelic, more or less indurated, bearing usually short trichomes; leaves small, sometimes minute (2–5 mm), or rarely to 2 m long, entire to pinnate, glabrous to pubescent, thin and lacking stomates, circinate in bud unless very small, petiole without stipules; sporangia borne in marginal sori on a short to elongate receptacle, enclosed by a bivalved to tubular indusium, with a short, ca. 6-rowed stalk, and an oblique annulus not interrupted by the stalk; homosporous, spores with chlorophyll (green). Gametophyte epigeal, with chlorophyll, narrowly thalloid or filamentous, freely branched, sometimes with gemmae, the archegonia borne on the lower surface or in clusters on a filamentous branch, the antheridia with ca. 5 to many cells, mostly on the lower surface or on filamentous branches.

Comments on the Family

The Hymenophyllaceae are a nearly worldwide family of tropical and wet temperate regions, consisting of two genera, *Hymenophyllum* with 7 subgenera, and *Trichomanes* with 5 subgenera.

There are wide differences in the number of genera recognized in the family, for example, Christensen (1938) accepted four genera, and Morton (1968) six, while Copeland (1938) included 33 and later (1947) 34 genera. The most recent treatment by Pichi-Sermolli (1977) has 42 genera. The two main groups are recognized here as genera and the more distinctive elements within each as subgenera. This classification generally follows that of Morton but does not include his seven sections of *Hymenophyllum* and 26 of *Trichomanes* except for those listed in the account of the cytology. The names of all genera and subgenera applicable to American species are included in the generic nomenclature. A reassessment of the family classification is being made by K. Iwatsuki (1975, 1977a, 1977b, 1978). Which will probably result in the recognition of about eight genera.

The fossil record of the family is meager and undoubtedly does not reflect its age, the earliest record being *Trichomanides laxus* Ten.-Woods from the Jurassic in Queensland. Later records of *Hymenophyllum* are from the Lower Cretaceous and Tertiary. Spores from the Upper Cretaceous (Oldman formation) in Alberta (Rouse, 1957) have been identified as *Hymenophyllumsporites deltoidea* Rouse. The weakly granulate to laevigate surface and strongly trilete ridges do not conform with the recent spores of the family.

The Hymenophyllaceae are a specialized, hygrophilous family and its relation to others is uncertain. It clearly occupies an intermediate evolutionary position with respect to the structure of the sporangium and antheridium, but it is primitive in its retention of low chromosome numbers as $n = 9$, and 11 which are among the lowest reports for ferns. Relationships have been suggested with the Loxomataceae and Hymenophyllopsidaceae based on similarities of the sorus and indusium, but these are not supported by the spores or other characters. The structure of the gametophyte shows features found in the Schizaeaceae and Pichi-Sermolli (1977) notes other resemblances to that family, as the indusium and flange associated with the sporangium in *Lygodium*.

The **ecological** tolerances to desiccation, as determined by Shreve (1911) in a study of Jamaican Hymenophyllaceae, correlate with the occurrence of species at different levels of the environmental gradient from the floor to the canopy of the rain forest. *Trichomanes rigidum* shows little tolerance to desiccation and is terrestrial while *Hymenophyllum polyanthos* and *H. sericeum* (Sw.) Sw., both with considerable tolerance, occur as epiphytes well above the ground and even on higher branches. Although the leaves readily lose water, they also function in water absorption, especially in the species with few or no roots that occur in very wet sites.

Unusual **anatomical** features of the Hymenophyllaceae that relate to specialization to very humid environments are the thin lamina (one or few cells thick), absence of stomates, and lack of roots (particularly in *Trichomanes* subgenus *Didymoglossum*). The small protostele, illustrated in species of the Congo, by Mauri (1969), may be reduced to a single line of tracheids as in *Trichomanes latealatum* (v. d. Bosch) Christ and *T. proliferum* Bl., or the stele may be wholly lacking as in *T. Motleyi* v. d. Bosch (Bierhorst, 1971). A study of the stem apex of *Trichomanes* by Hébant-Mauri (1973) shows cell lineages derived from the single, tetrahedral apical cell. Bierhorst (1974) reviews the plasticity of stem-leaf determination in the family. In some species the leaf is a direct continuation of the stem, while in others the leaves are distinct, dorsiventral structures from their inception.

The **position of the sori** on the ultimate vein system has been considered to be of systematic importance and the special terms applied to these conditions by Prantl (1875) were revived by Morton (1968). Epitactic applies to sori on the apical veins of an ultimate branch system of a lamina with catadromic branching (the first branch of a segment is given off toward the base of the lamina, pinna or pinnule). Paratactic applies to sori on the basal veins of an ultimate branch system of a lamina with anadromic branching (the first branch of a segment, is given off toward the apex of the lamina, pinna or pinnule). Pantotactic applies to cases in which the sori are on all or nearly all veins of an ultimate branch system. These terms are not used here because of difficulty in their application to leaves with an entire to 1-pinnate lamina, and they are redundant to terms for the kinds of branching patterns.

Gametophytes may germinate in the sporangium before dehiscence. Very young gametophytes have a triangular form as the result of an unusual division into three cells following spore germination. These and later developmental stages of the gametophyte are well illustrated in Stokey's work (1940) on germination and vegetative stages of gametophytes of *Hymenophyllum* and *Trichomanes*. In that study gametophytes which were maintained in culture for more than 10 years, showed that while both filamentous and thalloid forms may occur in both genera, the filamentous type is more characteristic of *Trichomanes,* and the thalloid of *Hymenophyllum.*

Spores of the Hymenophyllaceae are usually bright green as they contain abundant chloroplasts and often germinate within the sporangium. They are generally small; thus the SEM magnification is twice that of most genera. However, they are relatively large in some species as *Trichomanes Martiusii* (Fig. 12.29) and *Hymenophyllum lobatoalatum* (Fig. 11.25) possibly reflecting higher ploidy levels or differences in number of spores per sporangium. The spores are trilete and the proximal face more or less concave in dry specimens. The surface is usually papillate, varying in density and prominence of the papillae, or may be low and coarsely echinate. The walls consist of four distinct layers according to Lugardon (1980). He recognized two thin inner layers, the endospore and pseudoendospore that surround the cell, the exospore which comprises the major part of the wall, and an outer perispore that is thin and sometimes lacking. The absence of perispore is regarded by Lugardon as the most unusual feature.

A cytological assessment of chromosome numbers is presented in Tables 1 and 2 arranged by subgenera and with sections following Morton (1968). Thirty-six is the most frequent number reported in both genera. In *Hymenophyllum* the numbers are generally lower and more diverse than in *Trichomanes*. Two low numbers for the family, 11 and 13, are reported in *Hymenophyllum* and there are five polyploid series in the genus based on these numbers and 18, 21, and 28. Aneuploid differences are also evident, especially in the subgenus *Mecodium,* where there is an array of numbers at 21, 27, 28, and 36.

In *Trichomanes* the numbers range from 9 to 128 and the genus seems to be more homogeneous, with four polyploid series based on 9, 32, 33, and 34. The lowest number for the family $n = 9$ is reported by Tilquin (1978a, 1978b) in *Trichomanes*

Table 1. Synopsis of chromosome numbers for sections and subgenera in *Hymenophyllum*. Chromosome numbers (*n*) of *Hymenophyllum*. Underlined names include reports of American species. Subgenus *Hymenophyllum* (H), sections: *Hymenophyllum*, *Myriodon*, and *Ptychophyllum*; subgenus *Leptocionium* (L), section *Leptocionium*; and subgenus *Mecodium* (M).

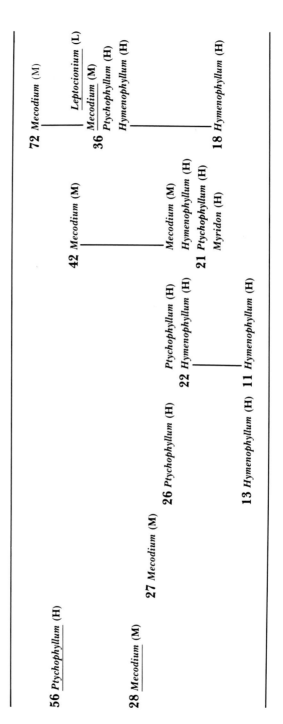

Table 2. Synopsis of chromosome numbers for sections and subgenera in *Trichomanes*.
Chromosome numbers (*n*) of *Trichomanes*. Underlined names include reports for American species. Subgenus *Trichomanes* (T), sections: *Crepidium, Crepidomanes, Gonocormus, Lacosteopsis, Microtrichomanes, Phlebiophyllum, Pleuromanes, Trichomanes;* subgenus *Pachychaetum* (P), sections: *Callistopteris, Cephalomanes, Davalliopsis, Nesopteris, Pachychaetum;* subgenus *Didymoglossum* (D), sections: *Didymoglossum, Lecanium, Microgonium;* subgenus *Achomanes* (A), sections: *Acarpacrium, Achomanes, Feea, Lacostea, Neurophyllum, Ragatelus, Trigonophyllum.*

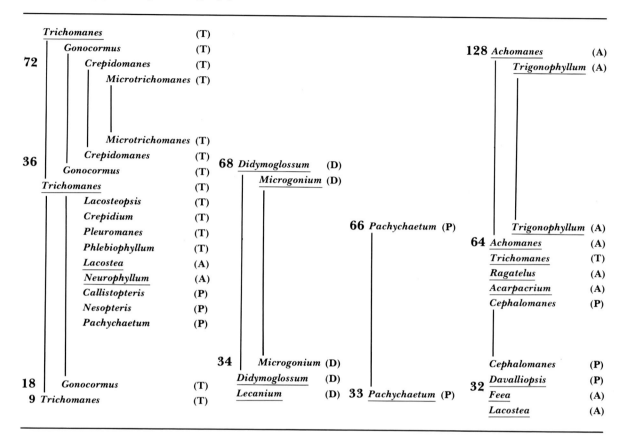

Chevalieri (C. Chr.) Kunkel (as *Vandenboschia*) from Nigeria and the highest $n = 479$ in apogamous plants of *T. chamaedrys* Tanton (as *Gonocormus*) from Zäire.

In subgenus *Didymoglossum*, the close relationship, proposed by Boer (1962), between the mainly neotropical section *Didymoglossum* and the largely paleotropical one *Microgonium*, is supported by reports of $n = 34$ and 68 in both sections, and section *Lecanium* (*T. membranaceum*) also has $n = 34$. Subgenus *Achomanes* appears to represent a homogeneous line based largely on 32, except for sections *Lacostea* and *Neurophyllum* reported as $n = 36$, although the report for the latter section, based on *T. pinnatum*, is uncertain. Subgenus *Pachychaetum* appears to be based on 33 although species in sections *Callistopteris* and *Nesopteris* are reported with 36.

Chromosome numbers are useful in depicting relationships but they also raise problems in regard to the taxonomic treatments. Assessment of groups with several aneuploid records, as in *Pachychaetum* and *Mecodium*, are especially difficult. In developing a phylogenetic system for the family, Vessey and Barlow (1963) derived two base chromosome numbers $x = 7$ and 11, with 18 regarded as a central number arising from hybridization. An alternate scheme was proposed by Walker (1966) based on known or ancestral numbers 6, 7, 8, 9, 11, 13, and a derived one of 17. This system includes six polyploid series and one at 17, as well as gain or loss of single chromosomes. These schemes show the utility of chromosome numbers in evaluating classifications and in depicting relationships.

The relatively few chromosome numbers for the family from the American tropics are largely from Walker's studies in Jamaica (1966). Many more are known from the paleotropics particularly from the work of Braithwaite (1969, 1975) which includes reports for 56 species from islands of the South Pacific and of Tilquin (1978a, 1978b) on 14 species of west Africa. About one-third of the species occur in the American tropics and additional reports of these are essential for further cytological assessments of relationships in this large family.

Key to Genera of Hymenophyllaceae

a. Indusium bivalved, the valves ½ or more the length of the indusium (rarely less), receptacle included to somewhat exerted; stem long, creeping, slender; lamina architecture anadromic, segments sometimes toothed, without false veins.   11. *Hymenophyllum*
a. Indusium entire, or with two flaring lips, rarely bivalved, the valves not more than ⅓ the length of the indusium, receptacle exerted, usually strongly so; stem sometimes thickened, creeping or erect; lamina architecture anadromic or catadromic, segments not toothed, false veins sometimes present.
                                  12. *Trichomanes*

Literature

Bierhorst, D. W. 1971. Morphology of Vascular Plants. 560 pp. Macmillan, New York.

Bierhorst, D. W. 1974. Variable expression of the appendicular status of the megaphyll in extant ferns with particular reference to the Hymenophyllaceae. Ann. Mo. Bot. Gard. 61: 408–426.

Boer, J. G. W. 1962. Reference under *Trichomanes*.

Braithwaite, A. F. 1969. The cytology of some Hymenophyllaceae from the Solomon Islands. Brit. Fern Gaz. 10: 81–91.

Braithwaite, A. F. 1975. Cytotaxonomic observations on some Hymenophyllaceae from the New Hebrides, Fiji and New Caledonia. Jour. Linn. Soc. (Bot.) 71: 167–190.

Christensen, C. 1938. Filicinae, *in:* F. Verdoorn, ed., Manual of Pteridology, pp. 522–550. Martinus Nijhoff, The Hague.

Copeland, E. B. 1938. Genera Hymenophyllacearum. Phil. Jour. Sci. 67: 1–110, pl. 1–11.

Copeland, E. B. 1947. Genera Filicum. 247 pp. Chronica Botanica, Waltham, Mass.

Diem, J., and J. S. Lichtenstein. 1959. Las Himenofiláceas del area argentino-chilena del sur. Darwiniana 11: 611–760.

Hébant-Mauri, R. 1973. Fonctionnement apical et ramification chez quelques fougéres du genre *Trichomanes* L. (Hyménophyllacées). Adansonia 13: 495–526.

Iwatsuki, K. 1975. Studies in the systematics of filmy ferns, I. A note on the identity of *Microtrichomanes*. Brit. Fern Gaz. 11: 115–124.

Iwatsuki, K. 1977a. Studies in the systematics of filmy ferns, II. A note on *Meringium* and the taxa allied to this. Gard. Bull. Singapore 30: 63–74.

Iwatsuki, K. 1977b. Studies in the systematics of filmy ferns, III. An observation on the involucres. Bot. Mag. (Tokyo) 90: 259–267.

Iwatsuki, K. 1978. Studies in the systematics of filmy ferns, IV. Notes on the species with false veinlets. Mem. Coll. Sci. Kyoto Series B 7: 31–43.

Lugardon, B. 1980. TEM observations on the sporoderm of Hymenophyllaceae and Hymenophyllopsidaceae. Intern. Palynol. Conf., Cambridge, 5: 236.

Mauri, R. 1969. Premiéres observations sur Hymenophyllacées du Congo-Brazzaville. Nat. Monspeliensia 20: 197–207.

Morton, C. V. 1968. The genera, subgenera, and sections of the Hymenophyllaceae. Contrib. U.S. Nat. Herb. 38: 153–214.

Pichi-Sermolli, R. E. G. 1977. Tentamen Pteridophytorum genera in taxonomicum ordinem redigendi. Webbia 31: 313–512.

Prantl, K. B. 1875. Untersuchungen zur Morphologie der Gefässkryptogamen, I. Die Hymenophyllaceen, die niedrigste Entwicklungsreihe der Farne. pp. 1–73. Leipzig.

Rouse, G. E. 1957. The application of a nomenclatural approach to upper Cretaceous plant microfossils from western Canada. Canad. Jour. Bot. 35: 349–375.

Shreve, F. 1911. Studies on Jamaican Hymenophyllaceae. Bot. Gaz. 51: 184–209.

Stokey, A. G. 1940. Spore germination and vegetative stages of the gametophytes of *Hymenophyllum* and *Trichomanes*. Bot. Gaz. 101: 759–790.

Tilquin, J. P. 1978a. Observations cytotaxonomiques sur des Hyménophyllacées Africaines. I. Genre *Didymoglossum* Desv. et *Microgonium* Presl. Caryologia 31: 23–42.

Tilquin, J. P. 1978b. Observations cytotaxonomiques sur des Hyménophyllacées Africaines. II. d'autres genres. Caryologia 31: 191–209.

Vessey, J., and B. A. Barlow. 1963. Chromosome numbers and phylogeny in the Hymenophyllaceae. Proc. Linn. Soc. New South Wales 88: 301–306.

Walker, T. G. 1966. A cytological survey of the pteridophytes of Jamaica. Trans. Roy. Soc. Edinburgh 66: 169–237.

11. *Hymenophyllum*
Figs. 11.1–11.28

Hymenophyllum Sm., Mém. Acad. Turin 5: 418. 1793. Type: *Hymenophyllum tunbridgense* (L.) Sm.

Meringium Presl, *Hymenophyllaceae*. 24. 1843, (Abhandl. Böhm. Gesell. V, 3). Type: *Meringium Meyenianum* Presl = *Hymenophyllum Meyenianum* (Presl) Copel. *Hymenophyllum* subgenus *Meringium* (Presl) Copel., Phil. Jour. Sci. 64: 14. 1937. [Subgenus *Hymenophyllum*].

Leptocionium Presl, *ibidem* 26. 1843. Type: *Leptocionium dicranotrichum* Presl = *Hymenophyllum dicranotrichum* (Presl) Sadeb. *Hymenophyllum* subgenus *Leptocionium* (Presl) Christ, Farnkr. Erde 20. 1897.

Myrmecostylum Presl, *ibidem* 27. 1843. Type: *Myrmecostylum tortuosum* (Hook. & Grev.) Presl = *Hymenophyllum tortuosum* Hook. & Grev. [Subgenus *Hymenophyllum*].

Ptychophyllum Presl, *ibidem* 28. 1843. Type: *Ptychophyllum plicatum* (Kaulf.) Presl = *Hymenophyllum plicatum* Kaulf. [Subgenus *Hymenophyllum*].

Hymenophyllum subgenus *Sphaerodium* Presl, *ibidem* 31. 1843. Type: *Hymenophyllum Wilsonii* Hook. = *Hymenophyllum peltatum* (Poir.) Desv. [Subgenus *Hymenophyllum*].

Hymenophyllum subgenus *Cycloglossum* Presl, *ibidem* 32. 1843. Type: *Hymenophyllum caespitosum* Gaud.

Sphaerocionium Presl, *ibidem* 33. 1843. Type: *Sphaerocionium hirsutum* (L.) Presl (*Trichomanes hirsutum* L.) = *Hymenophyllum hirsutum* (L.) Sw. *Hymenophyllum* subgenus *Sphaerocionium* (Presl) C. Chr., Ind. Fil. Suppl. 3: 5. 1934. [*Subgenus Leptocionium*].

Hymenoglossum Presl, *ibidem* 35. 1843. Type: *Hymenoglossum cruentum* (Cav.) Presl = *Hymenophyllum cruentum* Cav. *Hymenophyllum* subgenus *Hymenoglossum* (Presl) R. & A. Tryon, Rhodora 83: 134. 1981.

Serpyllopsis v. d. Bosch, Versl. Meded. Konink. Akad. Wetens. Afd. Natuurk. 11: 318. 1861. Type: *Trichomanes caespitosum* (Gaud.) Hook. = *Hymenophyllum caespitosum* Gaud. (*Serpyllopsis caespitosa* (Gaud.) C. Chr.). *Trichomanes* subgenus *Serpyllopsis* (v. d. Bosch) Christ, Farnkr. Erde 23. 1897. [*Subgenus Cycloglossum*].

Diplophyllum v. d. Bosch, *ibidem* 322. 1861, not Lehm. 1814. Type: *Trichomanes dilatatum* Forst. = *Hymenophyllum dilatatum* (Forst.) Sw. [Subgenus *Mecodium*].

Mecodium (Copel.) Copel., Phil. Jour. Sci. 67: 17, 1938. *Hymenophyllum* subgenus *Mecodium* Copel., Phil. Jour. Sci. 64: 93. 1937. Type: *Hymenophyllum polyanthos* (Sw.) Sw.

Buesia (Morton) Copel., Phil. Jour. Sci. 67: 47. 1938. *Hymenophyllum* subgenus *Buesia* Morton, Bot. Gaz. 93: 336. 1932. Type: *Hymenophyllum mirificum* Morton. [Subgenus *Hymenophyllum*].

Description   Epiphytic, or sometimes rupestral, rarely terrestrial; stem filamentous, long-creeping, bearing usually scattered trichomes and small, delicate roots; leaves monomorphic mostly 1–20 cm long, sometimes to 50 cm or (in *H. speciosum*) to nearly 2 m, borne singly, lamina entire to ca. 4-pinnate, glabrous or sparingly to densely pubescent with simple or branched trichomes, veins free or (in subgenus *Rosenstockia*) anastomosing without included free veinlets, ultimate segments usually with a single vein, 1 to a few cells thick; sori terminal on the veins, with a more or less elongate, sometimes expanded, receptacle included within or infrequently somewhat exerted beyond the indusium, not paraphysate, indusium more or less deeply bivalvate; spores tetrahedral-globose, trilete, the laesurae usually more than $\frac{3}{4}$ the radius of the proximal face, the sporoderm surface is papillate with usually dense, more or less prominent papillae. Chromosome numbers: $n = 11, 13, 18, 21, 22, 26, 27, 28, 36, 42, 56, 72$; $2n = 56, 72, 108$; apogamous 26.

The indusium of *Hymenophyllum* is usually strongly bivalved and encloses the receptacle as shown in Fig. 8. Some important characters of the subgenera are the toothed segments in *Hymenophyllum* (Fig. 10), pubescent segments in *Leptocionium* (Figs. 11, 12), the simple lamina of *Hymenoglossum* (Fig. 13), and the small 1-pinnate lamina of *Cycloglossum* (Fig. 5). Some patterns of soral arrangement are shown in Figs. 6, 7, 9, and 13, and a selection of the diversity of leaf form among the American species is illustrated in Figs. 15–22. The characteristic anadromic architecture is shown in Figs. 7, 9, 17, and 19.

The subgenera *Craspedophyllum, Rosenstockia* and *Hymenoglossum* (Fig. 14) have dark, usually black, borders along the segment margins. Accessory wings which are somewhat irregular outgrowths usually perpendicular to the plane of the lamina occur in some species. *Hymenophyllum* leaves are generally small but *H. speciosum* v. d. Bosch of southern Peru and Bolivia has pendent leaves nearly 2 m long.

Fig. 11.1. *Hymenophyllum myriocarpum* on tree trunk, Cerro de Zurqui, Costa Rica. (Photo W. H. Hodge.)

Systematics

Hymenophyllum is a genus of about 300 species, included in seven subgenera. Three of these, *Hymenophyllum*, *Mecodium* and *Leptocionium*, have many species and are widely distributed. The monotypic subgenera *Hymenoglossum* and *Cycloglossum* are American. Subgenus *Craspedophyllum* Presl (with two species, *H. marginatum* Hook. & Grev. and *H. Armstrongii* Baker), occurs in southeastern Australia, Tasmania and New Zealand, and the monotypic subgenus *Rosenstockia* (Copel.) R. & A. Tryon, Rhodora 83: 134. 1981 (*H. Rolandiprincipis* Ros.) is endemic to New Caledonia.

Iwatsuki (1977) has placed *Hymenophyllum Baileyanum* Domin, with the denticulate margined species, which makes the monotypic subgenus *Hemicyatheon* a synonym of subgenus *Hymenophyllum*.

Hymenophyllum subgenus *Mecodium* Copel. must be considered to be based on *Diplophyllum* v. d. Bosch since the lack of the required Latin description precludes the interpretation that it is a name for a new taxon. The publication of the genus *Mecodium* Copel. presents a number of difficult problems which seem to be best solved by considering it as a new name for *Diplophyllum* v. d. Bosch.

The American species of *Leptocionium*, with the exception of two austral ones, were revised by Morton (1947) under the name *Sphaerocionium*. The austral South American species have been treated very thoroughly by Diem and Lichtenstein (1959).

Synopsis of American *Hymenophyllum*

Subgenus *Hymenophyllum*

Lamina 1-pinnate-lobed to 4-pinnate, glabrous, or rarely with each marginal tooth extended into a short trichome, the margins

Fig. 11.2. *Hymenophyllum asplenioides* on tree trunk at 1800 m, Cerro de Zurqui, Costa Rica. (Photo W. H. Hodge.)

without a black border, toothed, receptacle included, or slightly exerted.

About 20 species in tropical America and perhaps 80 in the Old World. In southern South America there is a concentration of 14 species. One species occurs in the southeastern United States.

Subgenus *Mecodium*

Lamina pinnatifid to 3-pinnate-lacinate, glabrous, the margins without a black border, entire, receptacle included.

About 35 species in the American tropics and about twice as many in the Old World. Four species occur in austral South America, and one in northwestern North America.

Subgenus *Leptocionium*

Lamina pinnatifid-laciniate to 2-pinnate-pinnatifid, pubescent, often densely so, or only on the margins and veins, the margins without a black border, entire or very slightly toothed, receptacle included, or somewhat exerted in *H. dicranotrichium* (Presl) Sadeb.

About 50 species in the American tropics, two in austral South America, and about 16 in the Old World.

Subgenus *Hymenoglossum*

Lamina simple and entire, the margin sometimes undulate, glabrous, margins with a narrow blackish border, receptacle slightly exerted.

Fig. 11.3. Distribution of *Hymenophyllum* in America, triangles indicate records based on gametophytes.

A monotypic subgenus, the single species *Hymenophyllum cruentum* Cav. (Fig. 22) native to Chile, and the Juan Fernandez Islands.

Subgenus *Cycloglossum*

Lamina 1-pinnate, with a few to many large, reddish-brown trichomes, especially beneath, the margins without a black border, entire, receptacle somewhat exerted.

A monotypic subgenus, *Hymenophyllum caespitosum* Gaud. (Fig. 15), growing in southern South America, the Juan Fernandez Islands, Staten Island and the Falkland Islands.

Tropical American Species

Several species groups of *Hymenophyllum* in America require taxonomic revision. They have been little studied in the field, and a number of species are known from only few collections. Some currently recognized species appear to be ecological or geographic variants while other widely distributed complexes, which presently are treated as a single species, may require segregation at least at infraspecific ranks. The treatment of *Hymenophyllum* by Stolze (1976) discusses some of these problems.

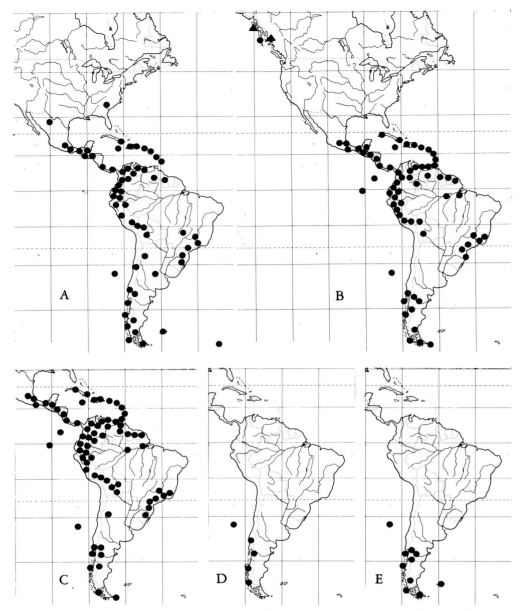

Fig. 11.4. Distribution of subgenera of *Hymenophyllum* in America: A, *Hymenophyllum;* B, *Mecodium,* triangles as in Fig. 11.3; C, *Leptocionium;* D, *Hymenoglossum;* E, *Cycloglossum.*

Some of the numerous species included in the three tropical subgenera are the following:

Subgenus *Hymenophyllum: H. ectocarpon* Fée, *H. fucoides* (Sw.) Sw. (Fig. 19), *H. pedicellatum* Kl., and *H. tunbridgense* (L.) Sm.

Subgenus *Mecodium: H. asplenioides* Sw. (Fig. 18), *H. axillare* Sw., *H. brevifrons* Kze., *H. caudiculatum* Mart. (Fig. 16), *H. endivii-folium* Desv., *H. myriocarpum* Hook., *H. polyanthos* (Sw.) Sw. (Fig. 17) and *H. undulatum* Sw.

Subgenus *Leptocionium: H. crispum* HBK., *H. fragile* (Hedw.) Morton, *H. hirsutum* (L.) Sw., *H. lanatum* Fée, *H. lobatoalatum* Kl. (Fig. 21), *H. microcarpum* Desv., *H. plumosum* Kaulf., and *H. tomentosum* Kze. (Fig. 20).

Figs. 11.5–11.13. *Hymenophyllum.* **5.** *H. caespitosum,* leaf, × 2.25. **6.** *H. asplenioides,* central part of lamina, × 1.25. **7.** *H. polyanthos,* central part of lamina, × 1.25. **8.** *H. polyanthos,* terminal lobes with bivalvate indusia, × 10. **9.** *H. fucoides,* portion of lamina with sori, × 1.25. **10.** *H. pedicellatum,* segments with toothed margins, × 6.5. **11.** *H. tomentosum,* pubescent segments, × 18.0. **12.** *H. hirsutum,* segment with forked trichomes, × 20. **13.** *H. cruentum,* fertile lamina, sori terminal on veins, × 1.25.

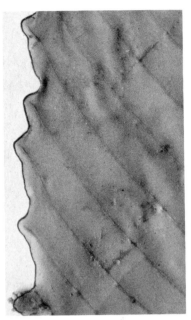

Fig. 11.14. *Hymenophyllum cruentum,* black lamina border, × 5.0.

Ecology (Figs. 1, 2)

Hymenophyllum is restricted to areas or local habitats that are nearly constantly moist. The genus is best developed in wet north and south temperate regions and in tropical mountains.

In tropical America the genus occurs especially in cloud forests, mossy (elfin) forests, and wet montane forests. It is less frequent in rain forests at lower altitudes. Many species are pendent epiphytes on trunks and branches of trees. Others grow at the base of cliffs, on mossy ledges, on wet rocks along streams, or on rotten trunks or stumps. Plants are only rarely truly terrestrial in humus-rich soil. The genus most commonly occurs at 1500–2500 m, although some species occur below 500 m, and in the Andes several occur to 3000 m or higher. Some widely distributed species also have broad altitudinal ranges, as *H. tunbridgense* which occurs from 250 to 3000 m, and *H. polyanthos* from 50 to 4400 m.

Geography (Figs. 3, 4)

Hymenophyllum is a widely distributed pantropical genus, with some extensions into north and south temperate regions, especially southern Chile, New Zealand and Japan; there is one species in Norway, and one in British Columbia.

In tropical America, the genus is distributed from Chihuahua in Mexico through Central America and the West Indies, and from French Guiana west to Colombia and south to Bolivia, in southeastern Brazil, and sparingly in the Amazon basin; also on Cocos Island and the Galápagos Islands. In austral South America, it extends south to Tierra del Fuego, Staten Island, the Falkland Islands and South Georgia. The subgenera *Mecodium* and *Leptocionium* have nearly identical distributions. In the tropics they both range from central Mexico to Panama, through the Greater and Lesser Antilles, and with few exceptions, in a broad arc around the Amazon basin and adjacent dry uplands of Brazil, from French Guiana to Bolivia and are disjunct in southeastern Brazil. Subgenus *Hymenophyllum* has a similar distribution but is somewhat less represented at lower altitudes.

The center of species diversity and of endemism is in the Andes from Venezuela to Bolivia, and especially from Colombia to Peru. Secondary centers are in Central America, especially Costa Rica, in the Greater Antilles and in southeastern Brazil.

Hymenophyllum tunbridgense has an isolated occurrence in South Carolina in the United States. *Hymenophyllum Wrightii* v. d. Bosch occurs in the Queen Charlotte Islands in British Columbia and in Japan. It is also represented by gametophyte colonies on the adjacent coast of British Columbia and in southern Alaska (Taylor, 1967). There is a notable concentration of 22 species of *Hymenophyllum* in austral South America. All but two species are endemic to the region and 14 are species of subgenus *Hymenophyllum.*

Figs. 11.15–11.22. Leaves of *Hymenophyllum.* **15.** *H. caespitosum,* subgenus *Cycloglossum,* ×1.0. **16.** *H. caudiculatum,* subgenus *Mecodium,* ×0.5. **17.** *H. polyanthos,* subgenus *Mecodium,* ×0.5. **18.** *H. asplenioides,* subgenus *Mecodium,* ×0.5. **19.** *H. fucoides,* subgenus *Hymenophyllum,* ×0.5. **20.** *H. tomentosum,* subgenus *Leptocionium,* ×0.5. **21.** *H. lobatoalatum,* subgenus *Leptocionium,* ×0.5. **22.** *H. cruentum,* subgenus *Hymenoglossum,* ×0.5.

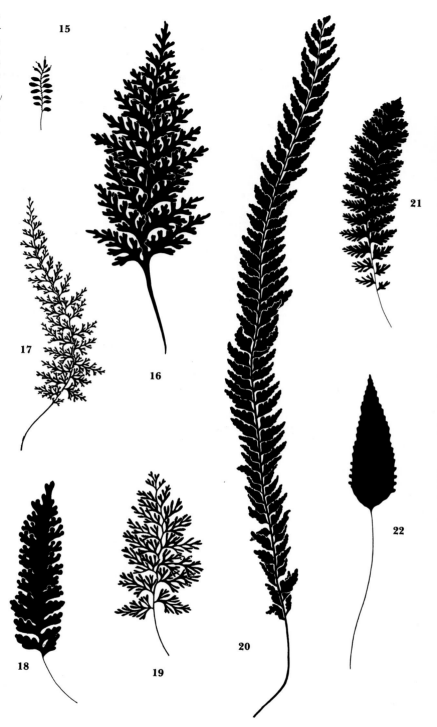

Spores

The surface is usually papillate as in *H. fucoides* (Figs. 23, 24) and in *H. lobatoalatum* (Figs. 25, 26) or the papillae may be diffuse and less prominent than in the latter figures. The laesurae are often more densely papillate than the spore surface as in *H. loba-*

23   24

25   26

27   28

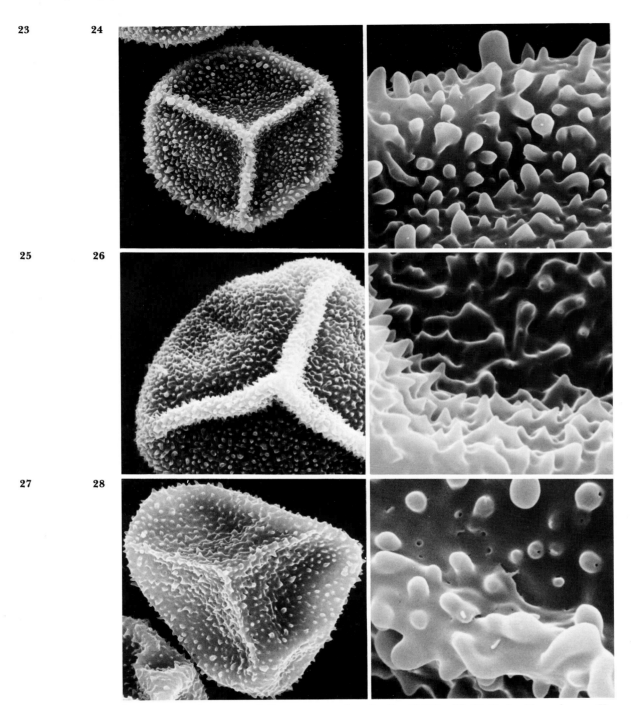

Figs. 11.23–11.28. *Hymenophyllum* spores, surface detail × 10,000.   **23, 24.** *H. fucoides,* subgenus *Hymenophyllum.*   **23.** Papillate proximal face, Ecuador, *Prieto P-135,* × 1000.   **24.** Surface detail, proximal face.   **25, 26.** *H. lobatoalatum,* subgenus *Leptocionium,* Peru, *Klug 3248.*   **25.** Portion of proximal face with dense, fairly prominent papillate surface, × 2000.   **26.** Surface detail, the laesura below, somewhat echinate.   **27.** *H. tunbridgense,* subgenus *Hymenophyllum,* with triradiate laesurae, Jamaica, *Maxon & Killip 4258,* × 2000.   **28.** *H. cruentum,* subgenus *Hymenoglossum,* papillate surface, laesura below, Chile, *Cuming 1866–69.*

toalatum. The spores of *H. cruentum* (subgenus *Hymenoglossum*) are small and papillate (Fig. 28). The similarity of these spores to those of other species in *Hymenophyllum* does not support the treatment of *Hymenoglossum* as a monotypic genus. There has been some question whether the spores of *H. tunbridgense* are trilete; however, those of a Jamaican collection (Fig. 27) clearly show the laesurae.

Cytology

The few chromosome numbers reported for American species of *Hymenophyllum* are based on Walker's studies in Jamaica (1966). More numerous extra-American records range between $n = 11$ and 72, representing both polyploid and aneuploid series. The most frequent numbers are 21 and 36. In the subgenus *Leptocionium* section *Leptocionium* species in both the neo- and paleotropics are consistently $n = 36$. In subgenus *Mecodium* chromosome counts for tropical American species are most often 28 or 36 with other reports of 21, 42, 27, and 72, which indicates the possibility of more than one evolutionary lineage. There are no reports for subgenus *Hymenophyllum* section *Hymenophyllum* from the American tropics but counts of 11, 13, 18, 21, 22, and 36 from other areas also indicate divergent elements within this group. The chromosome numbers of *Hymenophyllum* are reviewed in Table 1 (p. 100).

Among the pteridophytes *Hymenophyllum* chromosomes are some of the largest and have conspicuous terminal structures and euchromatic areas. Six sets of mitotic chromosomes were recognized by Tatuno and Takei (1969) in species of subgenera *Hymenophyllum* and *Mecodium*. They indicate a primary base number of 11 for these groups and suggest that the present series from 11 to 72 were formed by the elimination of members of the complement following polyploidy.

Literature

Diem, J., and J. S. Lichtenstein. 1959. Reference under the family.

Iwatsuki, K. 1977. Reference under the family, 1977a.

Morton, C. V. 1947. The American Species of *Hymenophyllum* section *Sphaerocionium*. Contrib. U.S. Nat. Herb. 29: 139–201.

Stolze, R. G. 1976. *Hymenophyllum,* in Ferns and fern allies of Guatemala I. Ophioglossaceae through Cyatheaceae. Fieldiana 39: 1–130.

Tatuno, S. and M. Takei, 1969. Karyological studies in the Hymenophyllaceae I. Chromosomes of the genus *Hymenophyllum* and *Mecodium* in Japan. Bot. Mag. (Tokyo) 82: 121–129.

Taylor, T. M. C. 1967. *Mecodium Wrightii* in British Columbia and Alaska. Amer. Fern Jour. 57: 1–6.

Walker, T. G. 1966. Reference under the family.

12. *Trichomanes*

Figs. 12.1–12.31

Trichomanes L., Sp. Pl. 1097. 1753; Gen. Pl., ed. 5. 485. 1754. Type: *Trichomanes scandens* L.

Feea Bory, Dict. Class d'Hist. Nat. 6: 446. 1824. Type: *Feea polypodina* Bory = *Trichomanes osmundoides* DC. *Trichomanes* subgenus *Feea* (Bory) Hook., Sp. Fil. 1: 114. 1844. [*Subgenus Achomanes*].

Hymenostachys Bory, *ibidem* 6: 588. 1824; 8: 462. 1825 Type: *Hymenostachys diversifrons* Bory (*Trichomanes elegans* Rudge, not L. C. Rich.) = *Trichomanes diversifrons* (Bory) Mett. *Trichomanes* subgenus *Hymenostachys* (Bory) Hook., Sp. Fil. 1: 114. 1844 [Subgenus *Achomanes*].

Didymoglossum Desv., Mém. Soc. Linn. Paris 6: 330. 1827. Type: *Didymoglossum muscoides* (Sw.) Desv. (*Trichomanes muscoides* Sw.) = *Trichomanes hymenoides* Hedw. *Trichomanes* subgenus *Didymoglossum* (Desv.) C. Chr., Ind. Fil. xiv. 1906.

Lecanium Presl, Hymenophyllaceae. 11. 1843 (Abhandl. Böhm. Gesell. V, 3), not Reinw. 1825. Type: *Lecanium membranaceum* (L.) Presl = *Trichomanes membranaceum* L. [Subgenus *Didymoglossum*].

Microgonium Presl, *ibidem* 19. 1843. Type: *Microgonium cuspidatum* (Willd.) Presl = *Trichomanes cuspidatum* Willd. [Subgenus *Didymoglossum*].

Neurophyllum Presl, *ibidem* 18. 1843, not Torrey & Gray 1840. Type: *Neurophylum pinnatum* (Hedw.) Presl = *Trichomanes pinnatum* Hedw. [Subgenus *Achomanes*].

Ragatelus Presl, *ibidem* 14. 1843. Type: *Ragatelus crinitus* (Sw.) Presl = *Trichomanes crinitum* Sw. [Subgenus *Achomanes*].

Trichomanes subgenus *Achomanes* Presl, *ibidem* 15. 1843. Type: *Trichomanes crispum* L.

Trichomanes subgenus *Pachychaetum* Presl, *ibidem* 16. 1843. Type: *Trichomanes rigidum* Sw.

Homoeotes Presl, Abhandl. Böhm. Gesell. V, 5: 331. 1848. Type: *Homoeotes heterophylla* (Willd.) Presl = *Trichomanes heterophyllum* Willd. [Subgenus *Achomanes*].

Odontomanes Presl, Epim. Bot. 20. 1849 (1851). Type: *Odontomanes Hostmannianum* (Kl.) Presl = *Trichomanes Hostmannianum* (Kl.) Kze. (*Neurophyllum Hostmannianum* Kl.). [Subgenus *Achomanes*].

Trichomanes subgenus *Pseudachomanes* Presl, *ibidem* 16. 1851. Type: *Trichomanes sinuosum* L. C. Rich. = *Trichomanes polypodioides* L. [Subgenus *Achomanes*].

Crepidomanes (Presl) Presl, *ibidem* 258. 1851. *Trichomanes* subgenus *Crepidomanes* Presl, *ibidem* 17. 1851. Type: *Trichomanes intramarginale* Hook. & Grev. [Subgenus *Trichomanes*].

Neuromanes v. d. Bosch, Ned. Kruid. Arch. 4: 347. 1859, *nom. nov.* for *Neurophyllum* Presl (not Torrey & Gray) and with the same type.

Lacostea v.d. Bosch, Versl. Akad. Wetens. Afd. Natuurk. 11: 320. 1861. Type: *Trichomanes brachypus* Kze. = *Trichomanes pedicellatum* Desv. *Trichomanes* subgenus *Lacostea* (v.d. Bosch) C. Chr., Ind. Fil. 634. 1906. [Subgenus *Achomanes*].

Maschalosorus v. d. Bosch, *ibidem* 320. 1861. Type: *Trichomanes Mougeotii* v. d. Bosch = *Trichomanes osmundoides* DC. [Subgenus *Achomanes*].

Ptilophyllum v. d. Bosch, *ibidem* 11: 321. 1861 (not Reichenb. 1841), *nom. superfl.* for *Ragatelus* Presl and with the same type. [Subgenus *Achomanes*].

Davalliopsis v. d. Bosch, *ibidem* 11: 323. 1861. Type: *Trichomanes Prieurii* Kze. = *Trichomanes elegans* L. C. Rich. [Subgenus *Pachychaetum*].

Trichomanes subgenus *Holophlebium* Christ, Farnkr. Erde 27. 1897, including subgenus *Trichomanes*.

Trichomanes subgenus *Ptilophyllum* C. Chr., Ind. Fil. 634. 1906. Type: *Trichomanes crinitum* Sw. (Morton, 1968, p. 198). [Subgenus *Achomanes*].

Macroglena (Presl) Copel., Phil. Jour. Sci. 67: 82. 1938. *Trichomanes* subgenus *Macroglena* Presl, Abhandl. Böhm. Gesell. V, 5: 333. 1848.
Type: *Trichomanes meifolium* Willd. [Subgenus *Pachychaetum*].

Selenodesmium (Prantl) Copel., Phil. Jour. Sci. 67: 80. 1938. *Trichomanes*
section *Selenodesmium* Prantl, Untersuch. Morph. Gefässkrypt. 1: 53.
1875. Type: *Trichomanes rigidum* Sw. [Subgenus *Pachychaetum*].

Vandenboschia Copel., Phil. Jour. Sci. 67: 51. 1938. Type: *Vandenboschia
radicans* (Sw.) Copel = *Trichomanes radicans* Sw. *Trichomanes* subgenus
Vandenboschia (Copel.) Allan, Fl. New Zealand 34. 1961. [Subgenus
Trichomanes].

Lecanolepis Pic-Ser., Webbia 28: 449. 1973, *nom. nov.* for *Lecanium* Presl,
not Reinw. and with the same type. [Subgenus *Didymoglossum*].

Mortoniopteris Pic.-Ser., Webbia 31: 243. 1977. Type: *Mortoniopteris scandens* (L.) Pic.-Ser. = *Trichomanes scandens* L. [The name is evidently not
superfluous since Pichi-Sermolli considered *T. crispum* as the type of
Trichomanes.] [Subgenus *Trichomanes*].

Pteromanes Pic.-Ser., Webbia 31: 244. 1977. Type: *Pteromanes Martiusii*
(Presl) Pic.-Ser. = *Trichomanes Martiusii* Presl. [Subgenus *Achomanes*].

Description Terrestrial, rupestral, or epiphytic; stem erect or decumbent
and short-creeping, moderately stout to slender, to very long-
creeping and filamentous, bearing scattered to dense, often stiff
trichomes and small, delicate to large, fibrous roots, or (especially in subgenus *Didymoglossum*) lacking roots; leaves monomorphic or dimorphic (the fertile more erect than the sterile
and with a reduced lamina) minute to mostly 1–40 cm or longer,
borne singly, or in a cluster or crown, the lamina entire to ca.
5-pinnate, veins free, or (in *T. diversifrons*) anastomosing without
included free veinlets, ultimate segments with a single or many
veins, 1 to a few cells thick; sori terminal on the veins, with an
elongate receptacle exerted beyond the indusium, not paraphysate, indusium mostly tubular, its margin entire or with two flaring lips, or rarely nearly bivalved; spores tetrahedral-globose, or
spheroidal, trilete, the laesurae $\frac{3}{4}$ to nearly equal the radius, the
surface usually dense, more or less prominently papillate, or
echinate, the papillae sometimes fused. Chromosome number
$n = 9, 18, 32, 33, 34, 36, 64, 66, 68, 72, 128$; $2n = 68, 72, 128$;
apogamous 26, 68, 108.

The elongate indusium and long, exerted receptacle (Fig. 10)
are characteristic of *Trichomanes*. The indusia are usually
oriented in the plane of the segment but may be directed downward as in *T. elegans* (Fig. 9). Diversity of leaf form ranging from
highly dissected to entire is shown in Figs. 14–25. Dimorphic
leaves have developed in several species particularly in subgenus
Achomanes, as in *T. diversifrons* (Figs. 5, 6). Trichomes may occur
especially along the leaf margin in members of subgenus *Achomanes*, as in *T. crinitum* (Fig. 7). In subgenus *Didymoglossum* the
leaves may also have marginal trichomes and are usually small
and entire as in *T. hymenoides* and *T. punctatum* (Figs. 12, 13).
False veins formed of dark, narrow cells may be at right angles
to true veins as in *T. pinnatum* (Fig. 8), or parallel to them, as in
T. pyxidiferum (Fig. 11). The lamina may have accessory wings as
in *Hymenophyllum*, but lacks the dark margins that rarely occur in
that genus.

Fig. 12.1. *Trichomanes radicans,* ephiphytic in Liquidambar cloud forest, Veracruz, Mexico. (Photo W. H. Hodge.)

Systematics

Trichomanes is a genus of about 300 species included in five subgenera. Four of these—*Trichomanes, Pachychaetum, Didymoglossum,* and *Achomanes*—have many species, while subgenus *Cardiomanes* (Presl) Christ is monotypic with *T. reniforme* Forst. of New Zealand.

The American species of subgenus *Didymoglossum* have been revised by Boer (1962), and the species of subgenus *Achomanes* section *Achomanes,* by Windisch (1977).

Synopsis of American *Trichomanes*

Subgenus *Trichomanes*

Stem long-creeping, lamina mostly bipinnatifid, to 4-pinnate-pinnatifid, anadromic, segments lacking trichomes on the margin (except *T. scandens*), false veins absent or (in *T. pyxidiferum*) present and nearly parallel to the veins.

About 120 species, some 18 in the American tropics, three in southern South America, and one in the southeastern to east-central United States.

Fig. 12.2. *Trichomanes elegans,* in humus on forest floor, Laudat, Dominica. (Photo W. H. Hodge.)

Subgenus *Pachychaetum*

Stem short, erect to ascending, the leaves clustered, lamina 3-pinnate to 3-pinnate-laciniate, anadromic, segments lacking trichomes on the margin, false veins absent.

About 60 species, only four in the American tropics.

Subgenus *Didymoglossum*

Stem long-creeping, lamina simple, entire, to 1-pinnate-pinnatifid, catadromic when more than 1-pinnate, with trichomes on the margin, false veins present, nearly parallel to the veins.

About 45 species, 24 in the American tropics, and one also in the southeastern United States.

Subgenus *Achomanes*

Stem erect to long-creeping, lamina entire or 1-pinnate to 2-pinnate-pinnatified, catadromic when more than 1-pinnate, segments often with trichomes on the margins, false veins absent or (in *T. pinnatum* and *T. Vittaria*) present and more or less at right angles to the veins.

About 65 species, all occurring in the American tropics except *T. crenatum* v. d. Bosch (*T. crispiforme* Alston) of west Africa.

Fig. 12.3. Distribution of *Trichomanes* in America, triangles indicate records based on gametophytes.

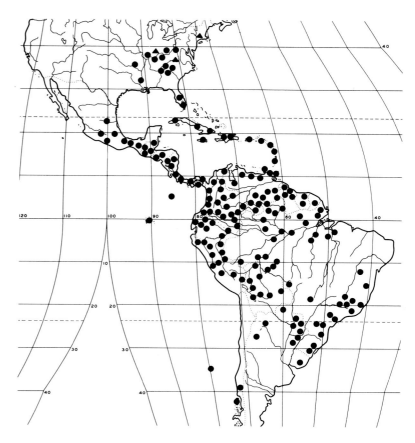

Tropical American Species

The species of *Trichomanes* are usually better defined than those of *Hymenophyllum*, but further taxonomic study is needed in a number of species groups. All American species of subgenus *Pachychaetum* are included in the following list, and some representatives of the other subgenera.

Subgenus *Trichomanes*: *T. angustatum* Carm., *T. axillare* Sod., *T. capillaceum* L., *T. hymenophylloides* v. d. Bosch, *T. pyxidiferum* L., *T. radicans* Sw. and *T. scandens* L.

Subgenus *Pachychaetum*: *T. cellulosum* Kl., *T. elegans* L. C. Rich., *T. rigidum* Sw. and *T. Sprucei* Baker.

Subgenus *Didymoglossum*: *T. angustifrons* (Fée) Boer, *T. hymenoides* Hedw., *T. membranaceum* L., *T. ovale* (Fourn.) Boer, *T. punctatum* Poir. and *T. reptans* Sw.

Subgenus *Achomanes*: *T. alatum* Sw., *T. arbuscula* Desv., *T. crinitum* Sw., *T. crispum* L., *T. diversifrons* (Bory) Mett., *T. lucens* Sw., *T. pedicellatum* Desv., *T. pellucens* Kze., *T. pinnatum* Hedw., *T. polypodioides* L. and *T. Vittaria* DC.

Ecology (Figs. 1, 2)

Trichomanes usually grows in wet, shaded places and is especially well developed in wet, tropical forests. Some species, particularly the terrestrial ones, may occur in a variety of other habitats.

In tropical America the genus grows in cloud forests, in wet ravines and along streams in dense mountain forests, in lower

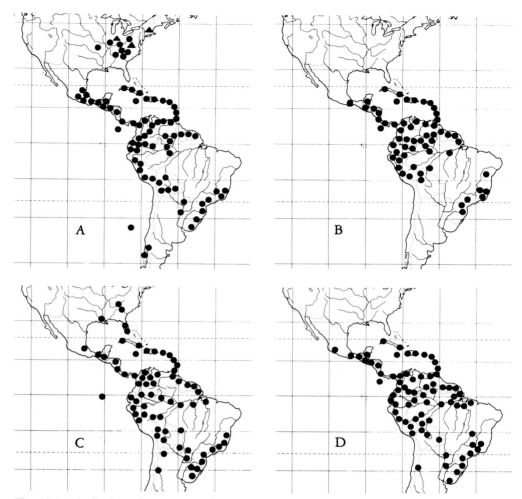

Fig. 12.4. Distribution of subgenera of *Trichomanes* in America: A, *Trichomanes,* triangles as in Fig. 12.3; B, *Pachychaetum;* C, *Didymoglossum;* D, *Achomanes.*

altitude rain forests, in gallery forests, and in moist thickets in savannahs, and among rocks in moist places. Many species are epiphytic on tree trunks and branches, sometimes twining on small twigs; others grow on wet rocks along streams or on ledges, at the base of cliffs, or on mossy boulders. The terrestrial species grow in humus-rich forest soil, on clay banks, rotten wood, or in lightly shaded, sandy soil. Some species may be both terrestrial or epiphytic, especially members of subgenus *Achomanes. Trichomanes crispum* has a broad ecological range occurring on rocks as well as on soil, or as an epiphyte. Some rupestral species grow on rocks that are periodically inundated. *Trichomanes Hostmannianum* (Kl.) Kze., a terrestrial forest species, grows in low areas near rivers that seasonally flood the land for a considerable period. Epiphytes such as *Trichomanes Tuerckheimii* Christ, *T. Ankersii* Hook. & Grev., and *T. pedicellatum* are originally established in soil at the base of a tree. Their long stems grow upward with both stems and leaves tightly appressed to the tree trunk. The genus generally grows at lower altitudes than *Hymenophyllum.* Most species occur below 1500 m but some, especially members

Figs. 12.5–12.13. *Trichomanes.* **5.** *T. diversifrons,* venation in central part of sterile lamina, × 2. **6.** *T. diversifrons,* central part of fertile lamina with elongate receptacles in marginal sori, × 2. **7.** *T. crinitum* sterile portion of lamina with trichomes along margin and veins, × 10. **8.** *T. pinnatum,* sterile lamina with false veins perpendicular to larger, true veins, × 20. **9.** *T. elegans,* central part of pinna with conical indusia, × 4.0. **10.** *T. hymenophylloides,* tubular indusia with flared margins, one with central, elongate receptacle, × 8.0. **11.** *T. pyxidiferum,* segments with false veins parallel to midvein, × 10. **12.** *T. hymenoides,* leaf with dichotomous venation, × 3. **13.** *T. punctatum,* circular, sterile leaf with stellate marginal trichomes, × 6.0.

of subgenera *Trichomanes* and *Achomanes,* grow to 3000 m or rarely higher, as *Trichomanes lucens* at 3800 m.

Geography (Figs. 3, 4)

Trichomanes is a pantropical genus but extends into austral regions and in temperate areas in North America and Japan, also in southwestern Europe and the British Isles.

In tropical America it occurs from central Mexico through Central America, south Florida and the West Indies, southward to Bolivia, northwest Argentina, Paraguay and Uruguay; also Cocos Island, and the Galápagos and Juan Fernandez Islands. The subgenera have rather similar distributions in tropical America. In the Amazon basin, *Trichomanes* is one of the best represented genera of ferns. In Brazilian Amazonia there are 27 species of *Trichomanes* (Tryon and Conant, 1975), while there are only three species of *Hymenophyllum* in the region.

Three species of subgenus *Trichomanes* occur in austral South America (Diem and Lichtenstein, 1959): *T. exsectum* Kze. of Chile, and *T. Ingae* C. Chr. & Skottsb. and *T. Phillipianum* Sturm both of the Juan Fernandez Islands. *Trichomanes Boschianum* v. d. Bosch occurs in the southeastern United States as far north as West Virginia and southern Illinois (Evers, 1961). Colonies of asexually reproducing gametophytes, probably of this species, extend the range of the genus north to western Massachusetts. *Trichomanes Petersii* A. Gray occurs from northern Florida and Louisiana north to North Carolina and Tennessee; it is disjunct in southern Mexico and Guatemala.

Spores

Many species have relatively small spores with thin walls that are often collapsed when dry, or some are larger as *Trichomanes lucens* (Fig. 31), *T. Martiusii* Presl (Fig. 29), and *T. radicans* (Fig. 28). In the latter species both diploid and tetraploid elements are known, and the larger one from Brazil, figured here, may represent a tetraploid. Papillate surfaces as in *T. radicans* (Fig. 28) and *T. elegans* (Fig. 30) are characteristic of most species. Some species as *T. arbuscula* (Fig. 26) have prominately echinate spores, or the surface may consist of compact, more or less fused elements as in *T. lucens* (Fig. 31). Such differences in surface detail do not appear to correlate with other characters distinguishing subgeneric groups. This generalization is based on SEM studies of species in each of the main sections as treated by Morton, including a large sample of species in subgenus *Achomanes.*

Cytology

There are relatively few records of chromosome numbers from the American tropics except for those from Jamaica by Walker (1966) and several from Brazil (Tryon et al. 1975). The chromosome numbers reviewed in Table 2 (p. 101) are based largely on records from the paleotropics. These are reported for most of the sections listed by Morton (1968). These data are useful for

Figs. 12.14–12.25. Leaves of *Trichomanes*. **14.** *T. capillaceum*, subgenus *Trichomanes*, ×1.0. **15.** *T. diversifrons*, subgenus *Achomanes*, ×0.5. **16.** *T. elegans*, subgenus *Pachychaetum*, ×0.5. **17.** *T. radicans*, subgenus *Trichomanes*, ×0.25. **18.** *T. pellucens*, subgenus *Achomanes*, ×0.25. **19.** *T. membranaceum*, subgenus *Didymoglossum*, ×0.5. **20.** *T. alatum*, subgenus *Achomanes*, ×0.5. **21.** *T. Vittaria*, subgenus *Achomanes*, ×0.5. **22.** *T. reptans*, subgenus *Didymoglossum*, ×1.0. **23.** *T. polypodioides*, subgenus *Achomanes*, ×0.5. **24.** *T. pinnatum*, subgenus *Achomanes*, ×0.25. **25.** *T. punctatum*, subgenus *Didymoglossum*, ×1.0.

26   27

28   29

30   31

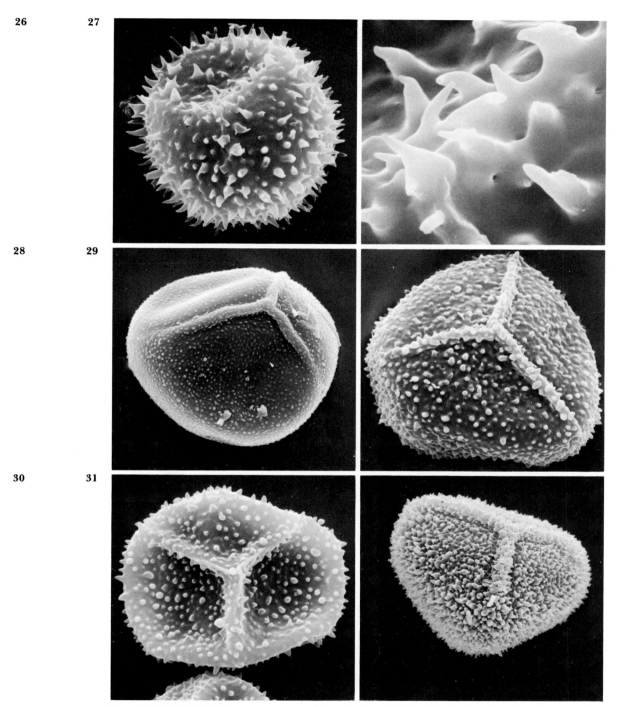

Figs. 12.26–12.31. *Trichomanes* spores, surface detail, × 10,000.   **26, 27. *T. arbuscula,*** subgenus ***Acho-manes.*** Surinam, *Kramer et al. 3242.*   **26.** Proximal face tilted, echinate, × 2000.   **27.** Detail, equatorial area.   **28. *T. radicans,*** subgenus ***Trichomanes,*** proximal face tilted, with low, finely papillate, surface Brazil, *Mexia 4236,* × 1000.   **29. *T. Martiusii,*** subgenus ***Achomanes,*** coarsely papillate, Surinam, *Hass & Tawjeon 10704,* × 1000.   **30. *T. elegans,*** subgenus ***Pachychaetum,*** the surfaces of proximal face concave, Venezuela, *Steyermark 94811,* × 2000.   **31. *T. lucens,*** subgenus ***Achomanes*** lateral view, proximal face with prominent, compact papillate to echinate surface, Ecuador, *Espinosa 1009,* × 1000.

geographic analyses and also assessment of the relationships of groups. The lowest number $n = 9$ in the genus, and indeed in the family, is reported in *Trichomanes Chevalieri* (as *Vandenboschia*), from Nigeria, by Tilquin (1978a, 1978b). In this paper *T. Mannii* Hook. (as *Gonocormus*) is reported as $n = 18$, along with some unusual numbers as 27 and 54, and reports up to $n = 479$ in apogamous plants from Zäire. The meiotic number 32 or its multiples predominate in nine sections including largely neotropical species, while 36 occurs in 13 sections comprised largely of paleotropical species. Section *Didymoglossum* differs with reports of either $n = 34$ or 68. Studies of this group by Boer (1962) indicated a close relationship of this mainly neotropical section with the largely paleotropical one, *Microgonium*. Reports of $n = 68$ for paleotropical species of that section support this alliance. Sections with pantropical ranges as *Lacosteopsis* (subgenus *Trichomanes*) and *Pachychaetum* have mostly $n = 36$. However, in *Pachychaetum* there are a few reports of $n = 33$ or 66 for both neo- and paleotropical species. These numbers suggest a closer relationship among these species than to others in the section with $n = 36$. Reports of $n = 32$ and 64, in the exclusively paleotropical section *Cephalomanes* (subgenus *Pachychaetum*), represent a divergent element almong sections confined to the Old World. This number is characteristic of several sections in the American tropics and suggests relationships between neo- and paleotropical groups.

Literature

Boer, J. G. W. 1962. The new world species of *Trichomanes* sect. *Didymoglossum* and *Microgonium*. Acta Bot. Neerland. 11: 277–330.

Diem, J., and J. S. Lichstenstein. 1959. Reference under the family.

Evers, R. A. 1961. The filmy fern in Illinois. Biological notes 44. 15 pp. Nat. Hist. Sur. Div. Urbana, Illinois.

Morton, C. V. 1968. Reference under the family.

Tilquin, J. P. 1978a. Reference under the family.

Tilquin, J. P. 1978b. Reference under the family.

Tryon, A. F., Bautista, H. P., and I. da Silva Araújo. 1975. Chromosome studies of Brazilian ferns. Acta Amazonica 5: 35–43.

Tryon, A. F., H. P. Bautista, and I. da Silva Araújo. 1975. Chromosome Acta Amazonica 5: 23–34.

Walker, T. G. 1966. Reference under the family.

Windisch, P. G. 1977. Systematic studies in tropical American ferns. Ph.D. Thesis, Harvard University, Cambridge, Mass. (Includes a revision of *Trichomanes* subgenus *Achomanes* section *Achomanes*.)

Family 7. Loxomataceae

Loxomataceae Presl, Gefassb. Stipes Farrn 31. 1847, as Loxsomaceae. Type: *Loxoma* Cunn., often altered to *Loxsoma*.

Description

Stem long-creeping, slender to rather stout, branched, with a siphonostele indurated, densely pubescent with stiff trichomes enlarged at the base; leaves ca. 0.5–5 m long, pinnate, glabrous or pubescent, circinate in the bud, petiole without stipules; sporangia borne in marginal sori on an elongate receptacle, within a more or less urceolate indusium, with a short, ca. 6-rowed stalk, with an oblique annulus not interrupted by the stalk; homosporous, spores lacking chlorophyll. Gametophyte epigeal, with chlorophyll, obcordate or somewhat elongate, centrally thickened, with thin margins, bearing long, multicellular trichomes, the archegonia borne on the lower surface on the thickened portion, especially apically, the antheridia several- to many-celled, on the lower surface, less often on the upper surface.

Comments on the Family

A small family of two genera, *Loxoma* Cunn. of North Island, New Zealand, with one species, and *Loxsomopsis* of Costa Rica and the Andes, with one or perhaps two species. The genera are distinguished principally by the characters of the sporangium. The family is an isolated one, with strongest affinities to the Dennstaediaceae as indicated by the similarities in the spores, sorus, indusium, and chromosome number. Early work noting affinity with the Hymenophyllaceae or Cyatheaceae is not supported by recent studies of those groups. The fossil genus *Stachypteris* Pomel of the Jurassic has a similar sorus and possibly belongs in the family.

Key to Genera of Loxomataceae

a. Central portion of the rachis with a lateral ridge on each side, these ridges continuous upward with the decurrent pinna base; sporangia with only the apical annulus cells thickened, opening longitudinally.   *Loxoma*
a. Central portion of the rachis terete or subterete; sporangia with most of the annulus cells thickened, opening transversely.   13. *Loxsomopsis*

13. *Loxsomopsis*
Figs. 13.1–13.9

Loxsomopsis Christ, Bull. Herb. Boiss. s. II, 4: 399. 1904. Type: *Loxsomopsis costaricensis* Christ.

Description

Terrestrial; stem long-creeping, slender to rather stout, bearing numerous stiff, dark trichomes that are enlarged at the base, and sparse fibrous roots; leaves monomorphic, spaced, ca. 0.5–5 m long, lamina 2-pinnate-pinnatifid to 2-pinnate-pinnatisect, glabrous or pubescent beneath, veins free; sorus marginal, paraphysate, the indusium narrowly cyathiform to urceolate, with an entire rim, the elongate receptacle exerted; spores tetrahedral-globose, trilete, the laesurae $\frac{1}{2}$ to nearly equal the radius, the

Fig. 13.1. *Loxsomopsis costaricensis,* Volcán Poas, Costa Rica. (Photo W. H. Hodge.)

surface coarsely tuberculate especially on the distal face, the tubercules overlaid usually by dense, particulate deposit. Chromosome number: $n = 46$.

The characteristic lamina architecture, the branched stem, and a sorus are shown in Figs. 3–5. The sori are marginal but directed downward from the lamina surface, and the receptacle is barren of sporangia at its base.

Systematics

Loxsomopsis costaricensis Christ is rather uniform and appears to be a distinct species. However, specimens from the Andes are similar to the Costa Rican species in some characters. There are few collections of this rare fern and more ample material will probably show that *Loxsomopsis Pearcei* (Baker) Maxon, *L. notabilis* Slosson and *L. Lehmannii* Hieron. are all Andean elements belonging to *L. costaricensis.*

Fig. 13.2. Distribution of *Loxsomopsis*.

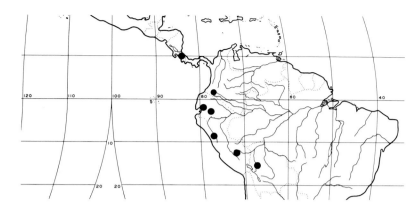

3

4   **5**

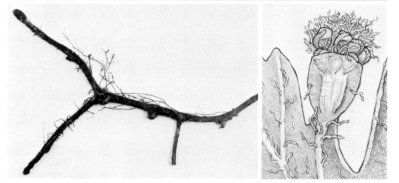

Figs. 13.3–13.5. *Loxsomopsis costaricensis.* **3.** Basal pinna of small leaf, ×1. **4.** Dichotomously branched stem (a petiole base is at lower right), ×0.6. **5.** Sorus and indusium, ×15.

Ecology (Fig. 1)

Loxsomopsis is a genus of the high montane forest and elfin cloud forest. It grows in partially open habitats in ravines, on brushy slopes and open woodland. The leaves are sometimes long and lean on adjacent shrubby plants with the apical portion often pendent.

It grows from 1600 to 2900 m.

6   7

8   9

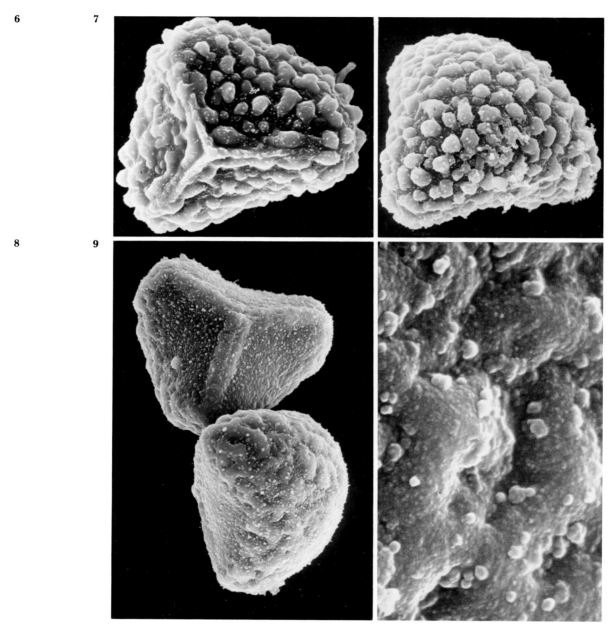

Figs. 13.6–13.9. *Loxsomopsis costaricensis* spores, × 1000, surface detail × 10,000.   **6, 7.** Costa Rica, *Lent 512*.   **6.** Proximal face with short laesurae, coarsely tuberculate, with particulate deposit.   **7.** Distal face.   **8, 9.** Peru, *McBride 4521*.   **8.** The proximal face above with dense particulate deposit and sparse, low tubercules, below, shallowly tuberculate, distal face.   **9.** Detail of irregular particulate deposit.

Geography (Fig. 2)

Loxsomopsis is a tropical American genus in the mountains of Costa Rica, Colombia, Ecuador, Peru, and Bolivia. It has been rarely collected, even in areas that are botanically rather well known.

Spores

Specimens from Costa Rica and Peru have basically similar, coarsely tuberculate spores with irregularly granulate deposit (Figs. 6–9). Those from Costa Rica have more prominent tubercules on the proximal face than spores from Peru. These resemble spores of some species of *Dennstaedtia* and along with other characters noted in comments on the family, suggest an affinity with that genus.

Cytology

Loxsomopsis costariensis has been reported as $n = 46$ from Cerro Vueltas, Costa Rica by Wagner (1980). *Loxoma* was first reported by Brownlie (1965) as $n = $ ca. 47 and later as definitely $n = 50$ by Brownsey (1975) who included a figure showing several small bivalents. Both 46 and 50 are unusual chromosome numbers in the ferns although they are reported in some genera of the Dennstaedtiaceae. There are records of 47 and 48 especially in the lindsaeoid genera and 46 is known in *Dennstaedtia dissecta* and 50 in *Lonchitis*. These chromosome numbers suggest an association of the Loxomataceae with the Dennstaedtiaceae rather than the Hymenophyllaceae or Cyatheaceae.

Literature

Brownlie, G. 1965. Chromosome numbers in some Pacific Pteridophyta. Pacific Sci. 19: 493–497.
Brownsey, P. J. 1975. A chromosome count in *Loxoma*. New Zealand Jour. Bot. 13: 355–360.
Wagner, F. S. 1980. New basic chromosome numbers for genera of neotropical ferns. Amer. Jour. Bot. 67: 733–738.

Family 8.   Hymenophyllopsidaceae

Hymenophyllopsidaceae Pic.-Ser., Webbia 24: 712. 1970. Type: *Hymenophyllopsis* Goebel.

Description   Stem erect or ascending, small, seldom to frequently branched, siphonostelic, or nearly so, rather indurated, bearing scales; leaves ca. 10–40 cm long, pinnate, thin and lacking stomates, circinate in the bud, petiole base without stipules; sporangia borne in marginal sori on a short receptacle, enclosed by a bivalved indusium, subsessile, with an oblique annulus not interrupted by the stalk; homosporous, spores lacking chlorophyll. Gametophyte not known.

Comments on the Family

The Hymenophyllopsidaceae are a small American family of one genus with several species. It is usually considered to have a unique combination of characters and uncertain affinity with other ferns. Relationships with Hymenophyllaceae, Cyatheaceae, *Asplenium, Lindsaea,* and dennstaedtioid ferns have been variously proposed. The spore surface structure and similarities of the indusium suggest relationships to the dennstaedtioid ferns, especially to *Leptolepia.* However, the stem scales are more advanced and the sporangium is more primitive than in the dennstaedtioid ferns. The thin leaves lacking stomates are probably adaptive to the habitat and not indicative of a relationship with Hymenophyllaceae.

14. *Hymenophyllopsis*
Figs. 14.1–14.11

Hymenophyllopsis Goebel, Flora 124 (n.f. 24): 3, 21. 1929. Type: *Hymenophyllopsis dejecta* Goebel.

Description

Rupestral, or rarely terrestrial; stem erect or ascending, small, rather short to long and slender, bearing abundant scales and fibrous roots; leaves monomorphic, ca. 10–40 cm long, lamina 1-pinnate-pinnatifid to 4-pinnate, glabrous or slightly scaly or with a few trichomes to thinly pubescent, veins free; sori marginal, not paraphysate, the sporangia borne on a short receptacle, enclosed by a bivalved, entire to deeply lacerate indusium; spores tetrahedral-globose, trilete, the laesurae $\frac{3}{4}$ the radius, usually depressed and obscured by the surface structure, the surface verrucate, overlaid by strands forming a somewhat open, reticulate to cristate structure. Chromosome number: not reported.

The characteristic bivalved indusium and enclosed sorus are shown in Fig. 3. Some species have a short stem bearing long scales (Fig. 4), while others have an elongate stem and short scales (Fig. 5). Variations in the lamina architecture are shown by a portion of a leaf with a highly dissected lamina (Fig. 6) and one with a coarsely dissected lamina (Fig. 7).

Systematics

Hymenophyllopsis is a little known American genus of about seven species, including some which are undescribed. It is probably related to ancestors of *Dennstaedtia.*

Tropical American Species

The following five species have been described: *Hymenophyllopsis asplenioides* A. C. Sm., *H. dejecta* Goebel (*Hymenophyllum dejectum* Baker, is a heterotypic synonym), *H. hymenophylloides* Gómez, *H. Steyermarkii* Vareschi, and *H. universitatis* Vareschi. As more ample collections become available, a better assessment of the taxonomy of the genus, and especially of the variation within species, may be made. Goebel (1929) gives a morphological account of *Hymenophyllopsis dejecta* and Vareschi (1958) distinguished three species, partly on anatomical characters.

Fig. 14.1. *Hymenophyllopsis asplenioides,* Cerro Jaua, Bolivar, Venezuela. (Photo J. A. Steyermark.)

Fig. 14.2. Distribution of *Hymenophyllopsis* (the States of Amazonas and Bolivar, Venezuela).

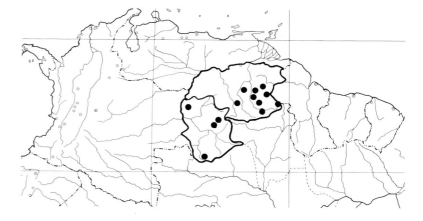

Ecology (Fig. 1)

Hymenophyllopsis is restricted to the Roraima sandstone formation, where it grows on the mesas (tepuis) on wet shaded sandstone ledges, cliffs or rock faces, or in crevices or recesses of rocks, usually in very humid sites, but sometimes on rather dry rocks or in sandy or clay soil among rocks. It grows from 1000 to 2700 m, usually from 1600 to 2200 m.

Geography (Fig. 2)

Hymenophyllopsis is confined to the Roraima formation of the states of Amazonas and Bolivar in Venezuela. *Pterozonium* is centered in the same region, but extends also to Surinam, Colombia and Peru. The pattern of geography of *Hymenophyllopsis* is correlated with the size of the mesas, up to four species occurring on

Figs. 14.3–14.7. *Hymenophyllopsis*. 3. Sorus and bivalved indusium of *H. universitatis,* × 10.   **4.** Short stem of *H. universitatis,* with abundant long scales, × 2.   **5.** Long, erect stem of *H. Steyermarkii,* bearing short scales, × 5.   **6.** Highly dissected lamina of *H. Steyermarkii,* × 2.   **7.** Coarsely dissected lamina of *H. universitatis,* × 1.25.

large ones such as Auyan-tepui and the Chimanta massif, while smaller ones have mostly a single species.

Spores

Spores are coarsely verrucate (Figs, 8, 9, 11) with a surface of reticulate strands incorporating spherical deposits (Fig. 10). The relatively elaborate surface of these spores is clearly distinct from the simple, papillate spores of the Hymenophyllaceae and suggests that this genus is more derived. Although there is no report of the chromosome number, the size differences shown in Figs. 8 and 11 may reflect different ploidy levels.

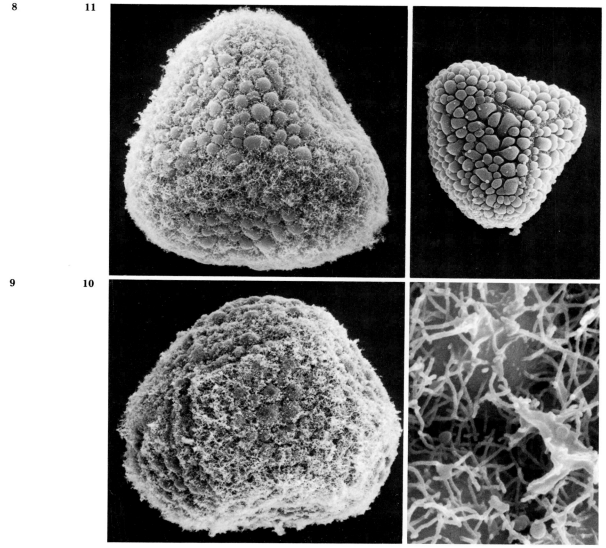

Figs. 14.8–14.11. *Hymenophyllopsis* spores, Venezuela, × 1000, surface detail, × 10,000. **8, 9.** *H. asplenioides, Maguire et al. 29889.* **8.** Proximal face, the three laesurae arms slightly depressed, reticulate to cristate surface over verrucate exospore. **9.** Distal face with denser deposit. **10.** Cristate sheath formed between strands, *Steyermark & Dunsterville 92581.* **11.** *H. hymenophylloides,* verrucate proximal face tilted, the laesurae arms right *Steyermark 93817.*

Literature

Goebel, K. 1929. Archegoniatenstudien, XVIII. Roraimafarne (1. *Hymenophyllopsis dejecta* n.g.). Flora 124 (n.f. 24): 3–21.

Vareschi, V. 1958. *Hymenophyllopsis universitatis.* Acta Biol. Venez. 2: 151–162.

Family 9. Plagiogyriaceae

Plagiogyriaceae Bower, Ann. Bot. 40: 484. 1926. Type: *Plagiogyria*
(Kze.) Mett.

Description

Stem erect to decumbent, rather stout, often branched, with a slightly dissected siphonostele, very indurated, lacking indument; leaves to 2 m long, pinnate, with minute mucilage-secreting trichomes when young, otherwise glabrous, circinate in the bud, petiole expanded basally into indurated stipules, usually with aerophores at the base abaxially; sporangia borne in sori on the veins on the abaxial surface of the pinnae, exindusiate, with a long 4- to 6-rowed stalk, with an oblique annulus not interrupted by the stalk; homosporous, spores lacking chlorophyll. Gametophyte epigeal, with chlorophyll, obcordate or somewhat elongate, thickened centrally, especially with age, with thin margins, the archegonia borne especially on the lower surface near the apex, also on the upper surface, the antheridia of several to many cells, generally distributed on both surfaces.

Comments on the Family

The Plagiogyriaceae are a small family of one genus in Asia, Australasia and America. It is an isolated family, but one that has been central to many considerations of fern phylogeny. Similarities to the Osmundaceae may be seen especially in the stem anatomy, the petiole base with stipules, and the mucilage. However, these characters are not sufficient to establish a strong affinity to that family. Diverse relationships have been proposed with groups of more advanced ferns, for example, with *Blechnum* and with the taenitoid ferns, but these are supported by few characters.

15. *Plagiogyria*
Figs. 15.1–15.8

Plagiogyria (Kze.) Mett., Abhandl. Senken. Naturf. Ges. 2: 265 (Ueber ein. Farngatt. II. *Plagiogyria*). 1858. *Lomaria* section *Plagiogyria* Kze., Farrnkr. 2: 61, 63. 1850. Type: *Lomaria euphlebia* Kze. = *Plagiogyria euphlebia* (Kze.) Mett.

Description

Terrestrial; stem short, decumbent to erect, rather stout including the persistent petiole bases, sometimes stoloniferous, lacking indument, bearing numerous wiry roots; leaves dimorphic (fertile leaves erect and with contracted pinnae), up to ca. 2 m long, borne in a spreading crown, lamina pinnatisect to 1-pinnate, the pinnae entire, lacking indument except when young with mucilage-secreting trichomes, the secretion becomes flaky when dry, veins free; sori elongate on somewhat thickened veins, sometimes extending below the fork of a vein, not paraphysate, exindusiate; spores tetrahedral-globose, trilete, the laesurae about $\frac{3}{4}$ the radius, areas between laesura depressed, the surface irregularly tuberculate with numerous scattered spheres and smaller, fused particles covering both proximal and distal faces. Chromosome number: n = 66, ca. 75, ca. 100, ca. 125, ca. 132.

Fig. 15.1. *Plagiogyria semicordata,* Cerro de la Muerte, Costa Rica. (Photo W. H. Hodge.)

The stem is very hard and compact, with persistent petiole bases (Fig. 3). The fertile and sterile leaves (Figs. 1, 4, 5) are quite dimorphic. The sori (Fig. 6) have sporangia confluent at maturity. When young, they are covered by the margin which becomes reflexed as the sporangia mature.

Systematics

Monographic studies have recognized about 50 species grouped into two sections, one with four subsections (Copeland, 1929; Ching, 1958; Lellinger, 1971). The genus seems to be much more homogeneous than these classifications would suggest, and perhaps consists of a few species-groups and 15 species. The American element of the genus is closely related to *P. Matsumuraeana* Makino of Japan and a few other species of eastern Asia.

Tropical American Species

The genus is represented in America by a single variable species, *Plagiogyria semicordata* (Presl) Christ. Although Copeland (1929) recognized nine species and Lellinger (1971) six, these are not adequately characterized by stable qualitative characters. The earliest name for the American species would be *Plagiogyria serrulata* (Willd.) Lell., if the type (Plum. Fil. Amer. t. 81) belongs to this genus. This is questionable since the plate shows a fern with a long-creeping stem quite different than that of *Plagiogyria*.

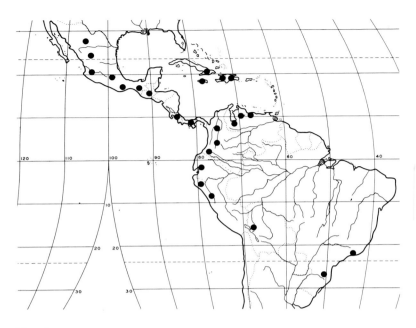

Fig. 15.2. Distribution of *Plagiogyria* in America.

Ecology (Fig. 1)

Plagiogyria is a genus of mountain forests, or of brushy habitats at higher elevations. In America it grows on wet forested mountain slopes, sometimes in coniferous forest (Mexico, Hispaniola), or in ravines. At higher elevations it occurs in subpáramo scrub or in protected sites on the páramo, occasionally in poorly drained places. It occurs from 1200 to 3500 m, usually from 2000 to 3000 m.

Geography (Fig. 2)

Plagiogyria occurs from the Himalayas to Japan, east and southward to New Guinea and northeastern Australia, and in tropical America. The center of species diversity is in China. In America it grows from Chihuahua in Mexico to Panama, Cuba, Jamaica and Hispaniola, the Andes of Venezuela south to Bolivia, and in southeastern Brazil.

Spores

Spores of *Plagiogyria semicordata* from Colombia have scattered spherical deposition (Figs. 7, 8) similar to that of *P. Henryi* Christ of China (Erdtman, 1972). The surface of some spores of *P. pycnophylla* (Kunze) Mett. are described as having irregular, coarse, flattened tubercule-like areas (Erdtman and Sorsa, 1971) but this is not evident in their figures. The spores of Formosan species vary from nearly smooth to densely, and irregularly tuberculate (Seong, 1976). The relatively simple tetrahedral-globose shape and spherical surface deposit in *Plagiogyria* are generally similar to spores of *Lygodium venustum.* However, differences in respect to other characters, suggest that similarity of spores may be superficial.

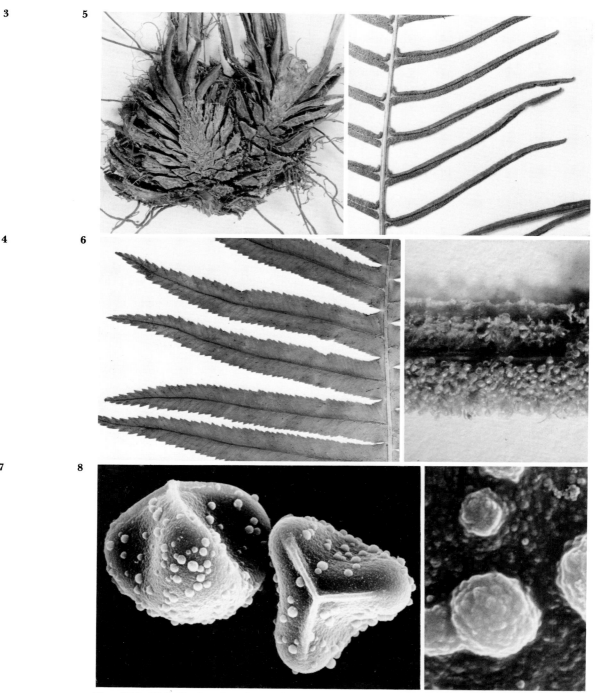

Figs. 15.3–15.8. *Plagiogyria.* **3.** Longisection of erect stem with branch (right) and compact petiole bases, × 1. **4.** Sterile pinnae with dichotomous veins near costa, × 1. **5.** Fertile pinnae with confluent sori and strongly reflexed margins, × 1. **6.** Portion of fertile pinnae with sporangia removed above showing the modified marginal indusium, × 15. **7, 8.** *Plagiogyria semicordata* spores, Colombia, *Little & Little 8189.* **7.** Spherical deposit somewhat denser on distal face, lateral view left, proximal, right, × 1000. **8.** Detail of particulate surface, × 10,000.

Cytology

The report of $n = 66$ for *Plagiogyria semicordata* from Blue Mountain Peak, Jamaica is clearly documented by Walker (1966) and the same number is reported for a species of Asia along with another ca. 123 (Walker, 1973). These are not readily aligned with the reports of ca. $n = 75$, 100, 125 for species of Japan (Kurita, 1963). Walker reports that great difference in chromosome size within a cell is a common feature of the genus. This along with the high number suggests a compound derivation of the complement. The position of the genus is in some doubt; however, consideration of the chromosome number along with the spores indicates that *Plagiogyria* is allied among the less derived families rather than with the Blechnaceae.

Literature

Ching, R. C. 1958. The fern genus *Plagiogyria* on the mainland of Asia. Acta Phytotax. Sinica 7: 105–154.

Copeland, E. B. 1929. The fern genus *Plagiogyria*. Phil. Jour. Sci. 38: 377–415.

Erdtman, G. 1972. Pollen and Spore Morphology/Plant Taxonomy. 127 pp. Hafner Publishing, New York.

Erdtman, G., and P. Sorsa. 1971. Pollen and Spore Morphology. 302 pp. Almqvist & Wiksell, Stockholm.

Lellinger, D. B. 1971. The American species of *Plagiogyria* section *Carinatae*. Amer. Fern Jour. 61: 110–118.

Kurita, 1963. Cytotaxonomical studies on some leptosporangiate ferns. Coll. Arts Sci. Chiba Univ., Nat. Sci. Ser., 4: 43–52.

Seong, L. F. 1976. Scanning electron microscopical studies on the spores of Pteridophytes. VII. The Family Plagiogyriaceae. Taiwania 21: 37–49.

Walker, T. G. 1966. A cytotaxonomic survey of the pteridophytes of Jamaica. Trans. Roy. Soc. Edinburgh 66: 169–237.

Walker, T. G. 1973. Evidence from cytology in the classification of ferns. *in*: A. C. Jermy, et al. The phylogeny and classification of the ferns. Jour. Linn. Soc. Bot. 67, Suppl. 1: 91–110.

Family 10.   Dicksoniaceae

Dicksoniaceae Bower, Origin Land Flora 591. 1908, as Dicksonieae. Type: *Dicksonia* L'Hérit.

Thyrsopteridaceae Presl, Gefässb. Stipes Farrn 22 (in footnote), 38. 1847, as Ordo Thyrsopterideae. Type: *Thyrsopteris* Kze.

Culcitaceae Pic.-Ser., Webbia 24: 702. 1970. Type: *Culcita* Presl.

Description   Stem erect, sometimes a single arborescent trunk, to decumbent or prostrate-creeping, stout to massive, usually unbranched, siphonostelic or dictyostelic, indurated, bearing a dense covering of long trichomes; leaves usually large, ca. 1–3 m long, pinnate, glabrous to pubescent, circinate in the bud, petiole without stipules; sporangia borne at the margin of the segments in sori with the receptacle usually slightly elevated, or (in *Thyrsopteris*) clavate, and more or less laterally extended and joined to the base of, or the adaxial portion of, the indusium, the indusia either joined at the base with the adaxial indusium usually firm, partly green, tan or brown at margin, and covering (when young) part of the thinner, tan or light brown abaxial indusium, or (in *Culcita dubia*) the indusia separate, or (in *Cystodium* and *Thyrsopteris*) completely joined, sporangial stalk long, to 4- to 6-rowed, with a slightly oblique annulus not interrupted by the stalk; homosporous, spores without chlorophyll. Gametophyte epigeal, with chlorophyll, obcordate or cordate-elongate, somewhat thickened centrally, the archegonia and antheridia borne on the lower surface, the antheridium with usually five or a few more cells.

Comments on the Family

The family Dicksoniaceae consists of five distinctive genera. Three of these, *Cibotium, Culcita* and *Dicksonia,* are in both the neo- and paleotropics; *Thyrsopteris* is on the Juan Fernandez Islands; and *Cystodium* Hook. is Malesian. The last two genera are the most distinctive and probably are not closely related to the others. The later name Dicksoniaceae is used pending its conservation.

The separate indusia of *Culcita dubia* appear to exemplify the primitive condition. In this species the adaxial indusium is formed from a marginal lobe of the segment which is somewhat thinner than the leaf-tissue and whitish at the margin. The abaxial indusium is separate, membranous, tan to brownish and slightly joined to the base of the receptacle. In other species of the Dicksoniaceae partial to complete fusion of indusia is regarded as derived.

The Dicksoniaceae have a rather extensive fossil record extending from the Jurassic with at least six genera including *Dicksonia, Coniopteris* Brongn. and *Eboracia* H. H. Thom. represented in that period. This early diversity and the distinctiveness of the living genera suggest a long history of evolution and also extinction within the family as well in other allied groups.

The treatment of the Dicksoniaceae has differed in several earlier studies. It has been maintained as a separate family (Christensen, 1938); united with a large family Pteridaceae (Copeland, 1947); united with the Cyatheaceae, including *Lophosoria* and *Metaxya* (Holttum and Sen, 1961); and recognized as three families, Dicksoniaceae, Culcitaceae and Thyrsopteridaceae (Pichi-Sermolli, 1977).

The relations of the family seem most close to the Lophosoriaceae on the basis of the long, matted trichomes on the stem and at the base of the petiole, the chromosome number of $n = 65$ in *Dicksonia* and *Lophosoria,* and the flanged spores and glaucous leaves of *Cibotium* and *Lophosoria.* However, the two families can be distinguished by the marginal sorus enclosed by distinctive marginal indusia in the Dicksoniaceae and the superficial, exindusiate sorus of the Lophosoriaceae. The Dicksoniaceae is less closely related to the Cyatheaceae in which the sorus is superficial with a single indusium (when present), and the stem and petiole have scales.

An association between the Dicksoniaceae and the Dennstaedtiaceae has also been proposed. However, the few similarities may be convergent and not reflect a close evolutionary relation of the two groups.

Key to Genera of Dicksoniaceae

a. Lamina 2-pinnate.   *Cystodium*
a. Lamina 2-pinnate-pinnatifid or more complex.   b.
   b. Axes of lamina adaxially grooved.   c.
      c. Fertile portions of the lamina similar to the sterile; receptacle slightly elevated, paraphysate; indusia separate or joined at the base.
                           16. *Culcita*
      c. Fertile segments strongly modified; receptacle clavate, not paraphysate; indusium asymmetrically cyathiform.   19. *Thyrsopteris*
   b. Axes of the secondary and tertiary segments of the lamina adaxially ridged.   d.
      d. Adaxial indusium gradually differentiated from the segment tissue; lamina reduced at base, green beneath.   17. *Dicksonia*
      d. Adaxial indusium rather sharply differentiated from the segment tissue; lamina broadly ovate, usually glaucous beneath.   18. *Cibotium*

Literature

Christensen, C. 1938. Filicinae, *in*: F. Verdoorn, ed., Manual of Pteridology, pp. 522–550. Martinus Nijhoff, The Hague.

Copeland, E. B. 1947. Genera Filicum. 247 pp. Chronica Botanica. Waltham, Mass.

Holttum, R. E., and S. U. Sen. 1961. Morphology and classification of the tree ferns. Phytomorphology 11: 406–420.

Pichi-Sermolli, R. E. G. 1977. Tentamen Pteridophytorum genera in taxonomicum ordinem redigendi. Webbia 31: 313–512.

16.   *Culcita*
Figs. 16.1–16.8

Culcita Presl, Tent. Pterid. 135. 1836. Type: *Culcita macrocarpa* Presl (*Dicksonia Culcita* L'Hérit.).
Culcita subgenus *Calochlaena* Maxon, Jour. Wash. Acad. Sci. 12: 459. 1922. Type: *Culcita dubia* (R. Br.) Maxon (*Davallia dubia* R. Br.).

Description   Terrestrial; stem prostrate-creeping to decumbent, rarely to 3 m tall, stout, bearing many long trichomes, or fewer shorter ones and fibrous roots; leaves monomorphic, ca. 1–3 m long, borne in a loose cluster, lamina 4- to 5-pinnate, deltoid, with the axes grooved adaxially, glabrate to somewhat pubescent, the petiole with long trichomes matted at the base, veins free; sori marginal, receptacle slightly elevated, laterally extended, paraphysate, the adaxial indusium scarcely differentiated from the segment tissue, basally joined to the usually thinner, often dentate to ciliate abaxial indusium of nearly the same shape, or (in *C. dubia*) indusia separate, distinct in shape and texture; spores tetrahedral-globose, with prominent angles, trilete, the laesurae ca. $\frac{3}{4}$ the radius, the areas between them concave with slightly papillate sporoderm, the distal face irregularly pitted, or spores spheroidal without prominent angles, both faces with coarse tubercules more or less fused into ridges especially parallel to the laesurae. Chromosome number: n = ca. 58, 66, 66–68.

Characteristic features of the genus are the highly dissected lamina (Fig. 4) and the grooves on the adaxial surface of the axes

Fig. 16.1. *Culcita coniifolia,* Cordillera de Talamanca, Costa Rica. (Photo W. H. Hodge.)

which distinguish this genus from *Dicksonia* and *Cibotium* (Fig. 5). A portion of a fertile segment in Fig. 6 shows the position of the sori as well as the sporangia enclosed within the adaxial and abaxial indusia. The croziers (Fig. 2) are densely covered with trichomes, as in other genera of the family. The diagram of the sorus in section (Fig. 8) shows the union of the receptacle on the abaxial indusium. *Culcita dubia* (R. Br.) Maxon, of the paleotropics has the indusia separate at the base, a condition considered to represent the primitive type of indusia in the family. Indusia fused only at the base are regarded as an intermediate type while the strongly fused indusia as in *Thyrsopteris* are the most specialized, derived form in the family.

Systematics *Culcita* is a genus of about seven species. A subgeneric classification, proposed by Maxon (1922) was enlarged by Holttum (1963) to include all austral Malesian-Pacific species in subgenus *Calochlaena,* and *C. macrocarpa* Presl and *C. coniifolia* in subgenus *Culcita.* A review of the critical characters of all species and especially additional chromosome records are essential to assess this classification. Differences in surface structure of the spores support recognition of two subgenera or perhaps genera.

There is only a single variable species, *C. coniifolia* (Hook.) Maxon in the American tropics.

Ecology (Fig. 1)

Culcita is a genus of tropical or temperate mountains. It is primarily a forest genus but also grows in thickets or more open habitats and will persist in clearings. In Spain, *C. macrocarpa* occurs along streams, and in oak woods at 350 m.

Fig. 16.2. *Culcita coniifolia,* Cordillera de Talamanca, Costa Rica. Young crozier densely invested with long trichomes. (Photo W. H. Hodge.)

Culcita coniifolia grows especially on steep mountain slopes and in cloud and elfin forest, also in sheltered ravines and less often in wet pine land or at higher altitudes in páramo-thicket or in sphagnum bogs. Its usual range is between 2000 and 3000 m, but sometimes occurs lower to 1500 m or up to 3500 m.

Geography (Fig. 3)

Culcita is widely distributed through Malesia and in eastern Australia east to Samoa. *Culcita macrocarpa* occurs in the Azores, Madeira, Tenerife in the Canary Islands, Portugal and southern Spain.

Culcita coniifolia grows in southern Mexico, Central America, the Greater Antilles (except Puerto Rico), and in South America from Venezuela to Colombia, south to Peru, and with a disjunct occurrence on Mt. Itatiaia in Brazil.

Spores

Spores of *Culcita* are of diverse forms. The American species *C. coniifolia* has strongly 3-lobed spores and relatively smooth areas adjacent to the laesurae (Fig. 7). The somewhat rugulose proximal and pitted distal faces may represent the exospore surface. Paleotropical species as *C. dubia, C. straminea* (Labill.) Maxon and *C. villosa* C. Chr. have quite spheroidal spores with coarse tuberculate projections that may be more or less fused into ridges especially near the laesurae. Young spores of *Culcita macrocarpa,* from Spain have delicate, strandlike echinate structures and are strongly 3-lobed. These and the rugulose-pitted form differ from the strongly tuberculate type in the paleotropics and suggest that the genus may include more that one element. A detailed analysis of the diverse spore types in *Culcita* and related genera by Gastony (1981) reviews the problems of recognition of perispore.

Fig. 16.3. American distribution of *Culcita* (*C. coniifolia*).

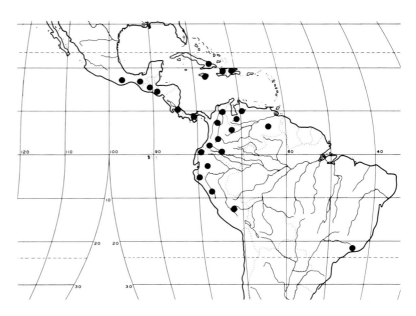

Figs. 16.4–16.7. *Culcita coniifolia*.   4. Pinnules from central part of pinna, ×0.5.   **5.** Adaxial surface of central portion of a pinnule, the axes grooved, ×5.   **6.** Sori and indusia, ×5.   **7.** Spores, lateral view, a portion of the pitted distal face, right, Colombia, *Killip & Smith 20960,* ×1000.

Cytology

The chromosome number $n = 66$ listed for *Culcita coniifolia* from Costa Rica (Gómez Pignataro, 1971) correlates with the approximate numbers of 66–68 for *C. macrocarpa* from the Azores (Manton, 1958). These records are within the same range as other dicksonioids including *Cibotium* with 68 and *Dicksonia* with 65. Neither reports of *n* ca. 55 nor of 58 for a cultivated specimen of *Culcita dubia,* by Manton (in Holttum, 1963), or 55–58 (Roy and Holttum, 1965) are in alignment with other records known for *Culcita*. However, the spores of the paleotropical species suggest that the species may represent a distinct element in the genus.

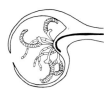

Fig. 16.8. Longitudinal section of marginal sorus and indusia of **Culcita coniifolia,** diagrammatic, enlarged; with the adaxial indusium at top, the vein (thick line) extended in the slightly elevated receptacle with sporangia and paraphyses.

Literature

Gastony, G. J. 1981. Spore morphology in the Dicksoniaceae 1. The genera *Cystodium, Thrysopteris* and *Culcita.* Amer. Jour. Bot. 68: 808–819.

Gómez Pignataro, L. D. 1971. Ricerche citoloiche sulle Pteridofite della Costa Rica, 1. Atti Ist. Bot. Univ. Pavia 7: 29–31.

Holttum, R. E. 1963. Cyatheaceae, Flora Malesiana, Ser. II, 1: 65–176.

Manton, I. 1958. Chromosomes and fern phylogeny with special reference to "Pteridaceae." Jour. Linn. Soc. Bot. 56: 73–91.

Maxon, W. R. 1922. The genus *Culcita.* Jour. Wash. Acad. Sci. 12: 454–460.

Roy, S. K., and R. E. Holttum. 1965. New cytological records for *Cystodium* and *Dicksonia.* Amer. Fern. Jour. 55: 35–37.

17. *Dicksonia*
Figs. 17.1–17.14

Dicksonia L'Hérit., Sert. Anglicum 30. 1788. Type: *Dicksonia arborescens* L'Hérit.

Balantium Kaulf., Enum. Fil. 228. 1824. Type: *Balantium auricomum* Kaulf. = *Dicksonia arborescens* L'Hérit.

Description   Terrestrial; stem usually erect, arborescent to 10 m tall, or basally decumbent, usually massive, bearing long, dense trichomes and many fibrous roots which may occur from the base nearly to the apex; leaves monomorphic or partially to wholly dimorphic (the fertile pinnae sometimes strongly modified to narrow laminar tissue), ca. 1–3.5 m long, borne in a spreading crown, lamina 2-pinnate-pinnatifid to 4-pinnate, reduced at the base, usually pubescent, with long, dense matted trichomes or sometimes shorter ones at the base of the petiole, axes of the secondary and tertiary segments ridged adaxially, veins free; sori marginal, rarely terminal on narrowly laminar axes, receptacle elevated, laterally extended, usually paraphysate, the adaxial indusium slightly differentiated from the segment tissue, basally joined to the thinner, sometimes dentate abaxial indusium; spores tetrahedral-globose, trilete, the laesurae $\frac{3}{4}$ the radius, with prominent angles and areas between often depressed, and the surface granulate, or reticulate with strands of fused spheres surrounding somewhat depressed areolae. Chromosome number: $n = 65$; $2n = 130$.

A portion of a fertile pinna of *Dicksonia Sellowiana* (Fig. 3) and of a sterile one (Fig. 5) contrast with the less complex pinnae of *D. Stuebelii* (Fig. 4). The adaxially ridged axes (Fig. 6) and the corrugated dictyostele (Fig. 8) are important generic characters. The marginal sori with associated indusia that envelop the sporangia are shown in Fig. 7. In a sectional view of the sorus (Fig. 9), the fusion of the receptacle with the base of the indusia is evident.

Systematics   *Dicksonia* is a relatively uniform genus of perhaps 20 species, two of them in tropical America and one in the Juan Fernandez Islands. It is rather closely related to *Cibotium* and more distantly

Fig. 17.1. *Dicksonia Sellowiana,* growing with *Trichipteris bicrenata* (background), cloud forest north of Jalapa, Veracruz, Mexico. (Photo W. H. Hodge.)

Fig. 17.2. American distribution of *Dicksonia.*

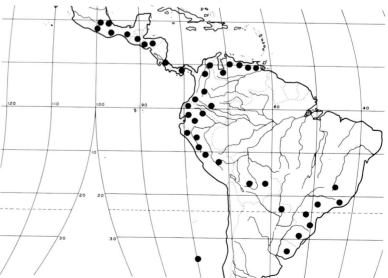

3

6

4

7

5

8

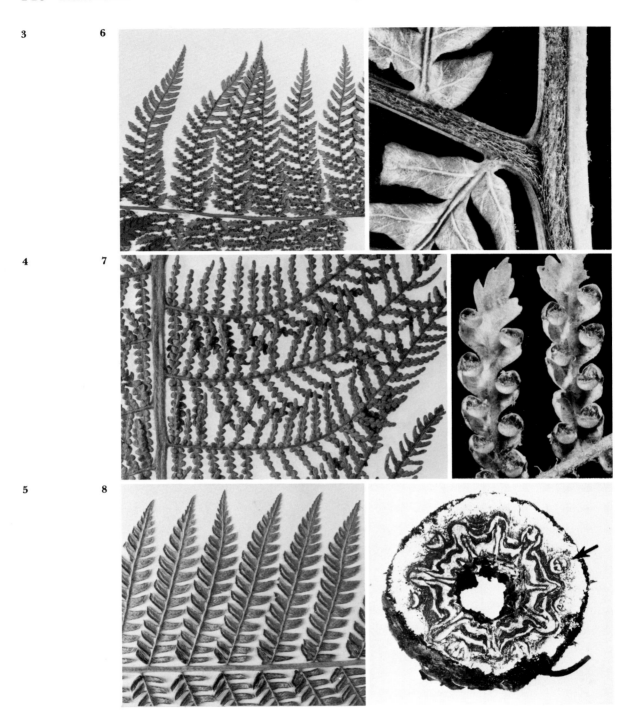

Figs. 17.3–17.8. *Dicksonia.* **3.** Portion of a fertile pinna of *D. Sellowiana,* ×0.5. **4.** Portion of the lamina of *D. Stuebelii* with fertile pinnae, ×0.75. **5.** Portion of a sterile pinna of *D. Sellowiana,* ×0.75. **6.** Pubescent adaxial surface of rachis and pinna-rachis, and ridged axes of secondary and tertiary segments of *D. Sellowiana,* ×0.5. **7.** Marginal sori with adaxial and abaxial indusia enclosing sporangia, *D. Sellowiana,* ×8.0. **8.** Transverse stem section of *D. Sellowiana,* the dark, sclerotic tissue outlines the corregated dictyostele, a leaf trace at arrow, ×0.25.

Fig. 17.9. Longitudinal section of sorus and indusia of ***Dicksonia Sellowiana,*** diagrammatic, enlarged; the adaxial indusium at top, the vein (thick line) extended in the elevated receptacle with sporangia and paraphyses.

to the other genera of the family. Maxon (1913) revised the North American species; however, the variability of characters was not recognized since the work was based on limited collections.

Tropical American Species

Dicksonia Sellowiana Hook. is treated here as a variable species which includes the often recognized *D. Ghiesbreghtii* Maxon, *D. gigantea* Karst., *D. Karsteniana* (K1.) Karst. and *D. lobulata* Christ. These species have been based on unstable characters such as the size of the fertile segments, the number of veins and extent to which they branch in a segment, and size of the indusia. *Dicksonia Sellowiana* has counterparts in other montane species that are variable and widely distributed, such as *Culcita coniifolia* and *Lophosoria quadripinnata*. *Dicksonia Stuebelii* is a distinctive species related to *D. Sellowiana*. Neither of these is closely related to *D. Berteriana* which is allied to species of the western Pacific.

Key to American Species of *Dicksonia*

a. Secondary segments of the fertile pinnae simple, or nearly so, each with a single sorus.   *D. Stuebelii* Hieron.
a. Secondary segments of the fertile pinnae (except apical ones) pinnatifid, each with few to several sori.   b.
   b. Fertile segments lacking a leafy apex or with a small one.   *D. Berteriana* (Colla) C. Chr.
   b. Fertile segments with a definitive leafy apex.   *D. Sellowiana* Hook.

Ecology (Fig. 1)

Dicksonia is primarily a genus of wet mountain forests especially in the tropics. In the temperate parts of its range it also grows in wet habitats but more open ones such as gallery forest, thicket-scrub and exposed wet slopes.

In tropical America the genus usually grows in wet montane forests and cloud forests on steep slopes or in ravines or along stream banks. Sometimes it grows in oak woods (Mexico) or in páramo-thicket (Andes). It usually occurs at ca. 1500–2500 m, sometimes up to 3500 m, or especially in Brazil, at lower elevations. Plants frequently persist in cutover lands.

Geography (Fig. 2)

Most species of *Dicksonia* occur in Malesia south to Tasmania and east to Samoa. *Dicksonia arborescens* L'Hért. is endemic to St. Helena.

In America, *Dicksonia Sellowiana* is wide ranging from southern Mexico through Central America, and in South America from Venezuela to Colombia, south to Bolivia, Paraguay, Uruguay and southeastern Brazil. In contrast, *Dicksonia Stuebelii* is known only from the Andes of northern Peru, and *D. Berteriana* is endemic to the Juan Fernandez Islands.

10   **11**

12

13   **14**

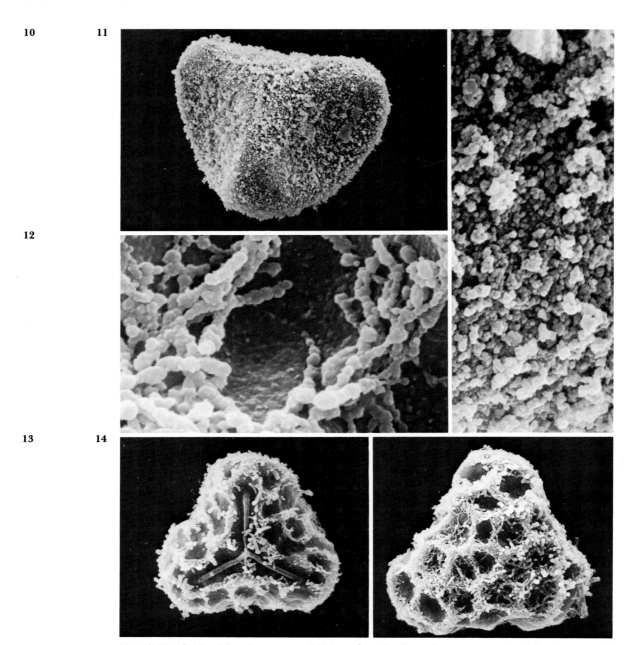

Fig. 17.10–17.14. *Dicksonia* spores, × 1000, surface detail, × 10,000. **10, 11. *D. Sellowiana,*** Brazil, *Reiss 61.* **10.** Proximal face 3-lobed with concave adjacent areas, granulate surface. **11.** Detail of granulate deposit. **12–14.** Reticulate spores of ***D. Berteriana,*** Juan Fernandez, Chile. *Reed,* Oct. 1872. **12.** Detail of strands of fused spheres, and central, depressed areola. **13.** Proximal face, the areas adjacent to the laesurae depressed. **14.** Distal face, strands partly obscure areolae.

Spores

There are two types of spores among American species of *Dicksonia.* Those of *D. Sellowiana* (Fig. 10) and *D. Stuebelii* are strongly 3-lobed, the areas adjacent to the laesurae are depressed, and the surface is densely, granulate (Fig. 10). Spores of *D. Berteriana* are reticulate or coarsely ridged and less prominently 3-lobed (Figs. 13, 14). The ridges are formed of strands of fused spheres that surround low areolae (Fig. 12). These spores are similar to those of Old World species as *D. squarrosa* (Forst.) Sw. of New Zealand, *D. Hieronymi* Brause and *D. sciurus* C. Chr. of New Guinea and *D. arborescens* of St. Helena.

Cytology

Reports for several tropical or temperate species of the Old World are mostly $n = 65$, but a tetraploid, $n = 130$, is reported from Java (Lovis, 1977). *Lophosoria* is also n 65 which suggests a possible alliance between these genera.

Literature

Lovis, J. D. 1977. Evolutionary patterns and processes in ferns. Adv. Bot. Research 4: 229–415.
Maxon, W. R. 1913. The North American ferns of the genus *Dicksonia*. Contrib. U.S. Nat. Herb. 17: 153–156.

18. *Cibotium*
Figs. 18.1–18.11

Cibotium Kaulf., Berl. Jahrb. Pharm. 21: 53. 1820. Type: *Cibotium Chamissoi* Kaulf.
Pinonia Gaud., Ann. Sci. Nat. 3: 507. 1824. Type: *Pinonia splendens* Gaud. = *Cibotium splendens* (Gaud.) Skottsb.

Description Terrestrial; stem decumbent to erect and arborescent, to 8 m tall, stout to massive, bearing very long, matted trichomes and many fibrous roots, which may occur from the base nearly to the apex; leaves monomorphic, ca. 2–4 m long, borne in a loose cluster or a spreading crown, lamina 2-pinnate-pinnatifid to 3-pinnate-pinnatifid, broadly ovate, glabrate to slightly pubescent, axes of secondary and tertiary segments ridged adaxially, the base of the petiole with long, matted trichomes, veins free; sori marginal, receptacle slightly elevated, laterally extended, paraphysate or not, the adaxial indusium rigid, rather sharply differentiated from the segment tissue, basally joined to the usually narrower and longer, entire abaxial indusium of similar texture; spores tetrahedral-globose, trilete with prominent angles and the adjacent areas often concave and more or less coarsely ridged, often with a strong equatorial flange and usually more or less ridged especially parallel to the laesurae, and on the distal face, the ridges sometimes anastomosed or somewhat tuberculate. Chromosome number: $n = 68$.

Some important characters of the genus are the long, matted trichomes at the base of the petiole (Fig. 4) and the adaxially ridged secondary and tertiary axes (Fig. 5). The form and division of fertile pinnules from the central part of the pinna of *C. Schiedei* are shown in Fig. 3 and attenuate sterile pinnules near the pinna apex of *C. regale* are shown in Fig. 6. Marginal sori with rigid, recurved adaxial indusia and somewhat tongue-shaped, longer abaxial indusia and included sporangia are shown in Fig. 7. The section of the sorus and indusium in Fig. 8

Fig. 18.1. *Cibotium Schiedei* on a hillside bordering Pedregal Esquilón, near Jalapa, Veracruz, Mexico. (Photo W. H. Hodge.)

Fig. 18.2. American distribution of *Cibotium*.

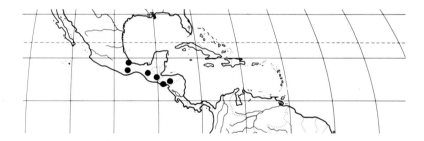

shows the union of the receptacle to the base of the abaxial indusium and a vein near the sporangia and paraphyses.

Systematics

Cibotium is a genus of eight species, *C. Schiedei* Schlect. & Cham. and *C. regale* Versch. & Lem. in America, four species in the Hawaiian Islands, and two in southeastern Asia and Malesia. The American species were revised by Maxon (1912) but because of limited specimens and variation in characters of the material the study was not definitive. In discussing the variability of the Central American plants, Stolze (1976) concludes that *C. Wendlandii* Kuhn and *C. guatemalense* Kuhn are elements within the variation of *C. regale*.

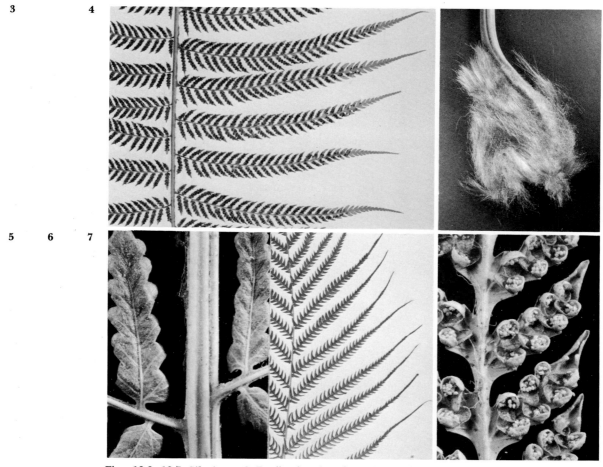

Figs. 18.3–18.7. *Cibotium.* **3.** Fertile pinnules of *C. Schiedei,* from central part of a pinna, ×0.5. **4.** Petiole base of *C. Schiedei* with long, matted trichomes, ×0.25. **5.** Adaxial surface of pinna-rachis and ridged axes of the secondary and tertiary segments, *C. Schiedei,* ×5. **6.** Sterile pinnules of *C. regale,* near the pinna apex, ×0.5. **7.** Marginal sori of *C. Schiedei,* ×5.

Fig. 18.8. Longitudinal section of sorus and indusia of *Cibotium Schiedei,* diagrammatic, enlarged; the adaxial indusium at top, the vein (thick line) extended into the scarcely enlarged receptacle with sporangia and paraphyses.

Ecology (Fig. 1)

Cibotium is a thicket and forest genus of tropical mountains. In America it grows on wooded slopes on river banks, in wet ravines, in cloud and montane forests, and rarely in pine forests or at the base of cliffs. Its usual altitudinal range is 1500–2500 m but it may occur up to 3000 m, or as low as 750 m.

Geography (Fig. 2)

Cibotium grows in southeastern Asia, from Assam to southern China, and south to Sumatra and Java, also in the Philippine Islands, New Guinea, the Hawaiian Islands, and in Mexico and Central America.

In America, *Cibotium Schiedei* grows in Veracruz and Oaxaca, in Mexico, and *C. regale* in southern Chiapas, in Mexico, Guatemala, Honduras and El Salvador.

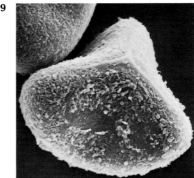

Spores

Spores of the American species of *Cibotium* are generally of two forms. Those of *C. Schiedei* tend to have a less prominently ridged surface or may essentially lack ridges (Fig. 9). The outer perispore deposit is more or less fused in short strands and below this is a more compact papillate formation (Fig. 10). Spores of *Cibotium regale* have similar surface deposition usually over stronger ridges (Fig. 11). Variation in *Cibotium* spores, reviewed by Gastony (1982), indicated similarities of these species. However, the spores of *C. Schiedei* appear to represent a less prominently ridged type in the northern part of the American range of the genus.

Cytology

Reports of the chromosome number $n = 68$ in *Cibotium* are based on Manton's study (1958) of two species in cultivation, believed to have originated in Hawaii.

Literature

Gastony, G. J. 1982. Spore morphology in the Dicksoniaceae II. The genus *Cibotium*. Canad. Jour. Bot. 60(6).

Maxon, W. R. 1912. The American species of *Cibotium*. Contrib. U.S. Nat. Herb. 16: 54–58.

Manton, I. 1958. Chromosomes and fern phylogeny with special reference to Pteridaceae. Jour. Linn. Soc. Bot. 56: 73–92.

Stolze, R. G. 1976. Ferns and fern allies of Guatemala, I. Ophioglossaceae through Cyatheaceae. Fieldiana Bot. 39: 1–130.

Fig. 18.9–18.11. *Cibotium* spores, × 1000.   **9, 10.** *C. Schiedei*, Veracruz, Mexico, *Adams*, in March 1976.   **9.** Proximal face the areas between laesura with sparse deposit.   **10.** Detail of abraded surface the papillate formation (top right) overlaid by fused, irregular spherical deposit in lower part of figure, × 10,000.   **11.** *C. regale*, proximal face above, distal below, with coarse ridges and particulate deposit, Honduras, *Molina & Molina 14017*.

19. *Thyrsopteris*
Figs. 19.1–19.8

Thyrsopteris Kze., Linnaea 9: 507. 1834. Type: *Thyrsopteris elegans* Kze.
Panicularia Colla, Mem. Acad. Torino 39: 33. 1836, not Fabricus, 1763.
Type: *Panicularia Berterii* Colla = *Thyrsopteris elegans* Kze.

Description

Terrestrial; stem decumbent to erect and arborescent, to ca. 2 m tall, massive, bearing long, matted trichomes and fibrous roots, sometimes producing slender stolons; leaves partially dimorphic (the fertile ones with strongly modified fertile portions borne at the base of the lower pinnae or also beyond), to 2–3.5 m long, borne in a cluster, lamina ca. 5-pinnate, the fertile portions more complex (to 7-pinnate), broadly ovate, axes grooved adaxially, glabrate, the base of the petiole with long, matted trichomes, veins free; sori terminal on very narrow laminar axes, receptacle clavate, usually somewhat laterally expanded, not paraphysate, indusium firm, entire, asymmetrically cyathiform, bilateral and apically bilobed when immature; spores tetrahedral-globose, trilete, with prominent angles and adjacent depressed areas, the laesurae $\frac{3}{4}$ the radius, the surface irregularly granulate. Chromosome number: n = ca. 76–78.

Comments on the Genus

The single species, *Thyrsopteris elegans* Kze., is endemic to the Juan Fernandez Islands (Fig. 2). The dissection of sterile portions of the lamina (Fig. 4) and the adaxially grooved axes (Fig. 5) are characteristic of the genus. A portion of an old fertile segment (Fig. 6) shows the sori and indusia borne on slender laminar axes, each sorus with an enlarged receptacle within the indusium. A longitudinal section of the sorus and indusium (Fig. 3) shows the receptacle strongly fused to the abaxial portion of the indusium and served by a broadly expanded vein. Young in-

Fig. 19.1. *Thyrsopteris elegans,* Mas Afuera, Juan Fernandez Islands. (Photo F. G. Meyer.)

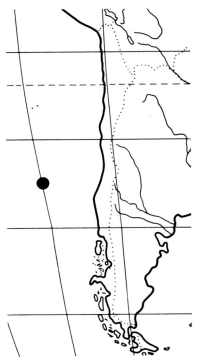

Fig. 19.2. Distribution of *Thyrsopteris*, Juan Fernandez Islands.

Fig. 19.3. Longitudinal section of the sorus and indusium of *Thyrsopteris elegans,* diagrammatic, enlarged; the adaxial side at top, the vein thickened in the elongate receptacle with sporangia.

dusia have a slightly greater development on the adaxial side, which partly encloses the abaxial portion.

Bower (1926) briefly discusses the stolons (runners) of *Thyrsopteris,* but Looser (1965–66) specifically says they are not present. His observations were evidently made on material cultivated in Valparaiso. It is possible that plants produce stolons or not, depending on the growing conditions. Material collected by Tod F. Stuessy and grown at Duke University by R. A. White produced short stolons at the base of the stem and longer ones above the base that grew downward.

The surface of *Thrysopteris* spores is formed by perispore as shown by expansion of the outer papillate material when spores are treated with sodium hydroxide (Gastony, 1981). The prominent angles and fused granulate deposit of *Thyrsopteris* spores (Figs. 7, 8) resemble the spores of *Dicksonia Sellowiana* and *Cibotium Schiedei.*

The cytological record is based on the material at Duke University, reported as $n = 76$–78, by M. D. Turner and R. A. White. This does not appear to be aligned with the series of numbers $n = 58$ or 66–68 reported for *Culcita,* which is considered to be closely related by Holttum (1963).

Thrysopteris elegans grows in the upper woodland and higher heaths, often with *Dicksonia Berteriana,* and usually at 500–1000 m.

Literature

Bower, F. O. 1926. The Ferns, 2: 260. Cambridge Univ. Press.

Gastony, J. G. 1981. Spore morphology in the Dicksoniaceae. 1. The genera *Cystodium, Thyrsopteris* and *Culcita.* Amer. Jour. Bot. 68: 808–819.

Holttum, R. E. 1973. Cyatheaceae, Flora Malesiana, Ser. II, 1: 65–176.

Looser, G. 1965–66. Los pteridofitos o helechos de Chile, III. Revis. Univers. Católica Chile 50, 51: 77–93.

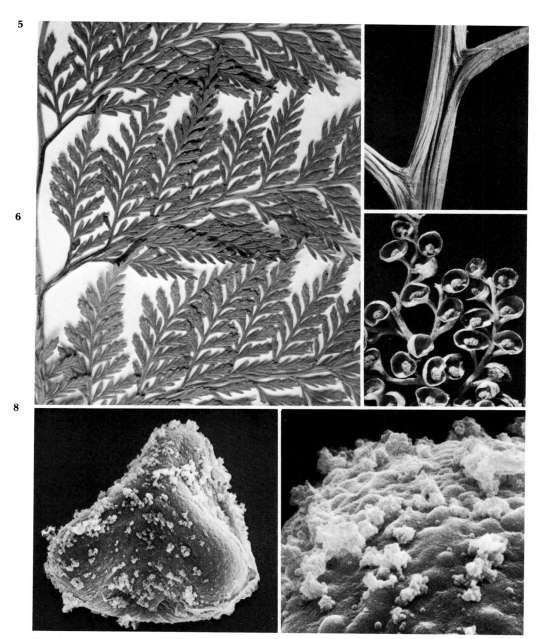

Figs. 19.4–19.8. *Thyrsopteris elegans.* **4.** Portions of sterile pinnae from the apical portion of a lamina, ×1.0. **5.** Adaxial side of a pinna-rachis, the groove continuous into the tertiary axes, ×5. **6.** Portion of an old fertile segment showing indusia and clavate receptacles on which the sporangia were attached, ×5. **7, 8.** Spores, Mas Afuera, *Meyer 9393.* **7.** Proximal face, granulate deposit above irregularly tuberculate perispore, ×1000. **8.** Detail of perispore surface on distal face, ×5000.

Family 11.   Lophosoriaceae

Lophosoriaceae Pic.-Ser., Webbia 24: 700. 1970. Type: *Lophosoria* Presl

Description

Stem decumbent to erect, to 5 m tall, massive, sometimes branching, with a siphonostele or sometimes a dictyostele, indurated, bearing a dense covering of long trichomes; leaves large, ca. 1–5 m long, pinnate, slightly to densely pubescent, circinate in the bud, the petiole without stipules; sporangia borne on the abaxial surface of the pinnae, in exindusiate sori, with a short, 6-rowed stalk and an oblique annulus not interrupted by the stalk; homosporous, spores without chlorophyll. Gametophyte epigeal, with chlorophyll, obcordate-elongate, slightly thickened centrally, rarely with multicellular trichomes, the archegonia and antheridia on the lower surface, the antheridium of five cells.

Comments on the Family

The Lophosoriaceae include the single American genus *Lophosoria.* This has usually been variously placed in other families as the "Proto-Cyatheaceae" (Bower, 1926), among the scaly Cyatheaceae (Christensen, 1938, Copeland, 1947), or in a large family including the Dicksoniaceae and the Cyatheaceae (Holttum and Sen, 1961). It is treated as a separate family by Pichi-Sermolli (1970, 1977). Classification of the Lophosoriaceae, Metaxyaceae, Dicksoniaceae and Cyatheaceae is uncertain because of reticulate relations expressed by some of the characters of the genera. On the basis of similarities of *Lophosoria* with *Cibotium* and *Dicksonia,* discussed in the treatment of the Dicksoniaceae, it is evidently closer to the Dicksoniaceae than to the Cyatheaceae. However, in soral characters there are marked similarities between the Lophosoriaceae and the Cyatheaceae, especially *Sphaeropteris* subgenus *Sclephropteris.*

Recent studies of the stem and petiole anatomy of *Metaxya* and *Lophosoria* by Lucansky (1974) and Lucansky and White (1974) show that these genera are quite distinct from the scaly Cyatheaeae and that they have some anatomical characters in common such as the short leaf-gaps and corrugated vascular tissue in the petiole with the ends incurved on the adaxial side. These features, and the simple sorus, the similar stem indument and soral position support a relationship of the two genera. However, the different number of bundles in the petiole, the spore characters and the discrete chromosome numbers suggest that the relation is not a close one.

Literature

Bower, F. O. 1926. The Ferns, 2: chapt. xxxii. Cambridge Univ. Press.

Christensen, C. 1938. Fillicinae, *in:* F. Verdoorn, ed., Manual of Pteridology, pp. 522–550. Martinus Nijhoff, The Hague.

Copeland, E. B. 1947. Genera Filicum. 247 pp. Chronica Botanica, Waltham, Mass.

Holttum, R. E., and U. Sen. 1961. Morphology and classification of the tree ferns. Phytomorphology 11: 406–420.

Lucansky, T. W. 1974. Comparative studies of the nodal and vascular anatomy in the tropical Cyatheaceae. I. *Metaxya* and *Lophosoria.* Amer. Jour. Bot. 61: 464–471.

Lucansky, T. W., and R. A. White. 1974. Comparative studies of the nodal and vascular anatomy in the neotropical Cyatheaceae. III. Nodal and petiole patterns; summary and conclusions. Amer. Jour. Bot. 61: 818–828.

Pichi-Sermolli, R. E. G. 1970. Fragmenta Pteriodologiae, II. Webbia 24: 699–722.

Pichi-Sermolli, R. E. G. 1977. Tentamen Pteridophytorum genera in taxonomicum ordinem redigendi. Webbia 31: 313–512.

20. *Lophosoria*
Figs. 20.1–20.10

Lophosoria Presl, Gefässb. Stipes Farrn, 36. 1847. Type: *Lophosoria pruinata* (Sw.) Presl [*Polypodium pruinatum* Sw. (*Polypodium glaucum* Sw. 1788, not Houtt. 1783)] = *Lophosoria quadripinnata* (Gmel.). C. Chr. *Trichosorus* Liebm., Vid. Selsk. Skr. V, 1: 281. 1849, *nom. superfl.* for *Lophosoria* and with the same type.

Description Terrestrial; stem decumbent to erect, to 5 m tall, massive, bearing a dense mat of long trichomes and many fibrous roots; leaves monomorphic, 0.3 to usually 2–3 to 5 m long, borne in a usually spreading cluster, 2-pinnate-pinnatifid to 3-pinnate-pinnatisect, the segments densely to very sparsely pubescent beneath and often glaucous, veins free; sori abaxial, round, borne singly on the fertile veins, the receptacle scarcely elevated, paraphysate, exindusiate; spores tetrahedral-globose, trilete, the faces unequal with a large, rounded flange closer to the proximal than the distal pole, the proximal face coarsely tuberculate, the distal pitted, more or less coarsely rugose, the surface granulate. Chromosome number: $n = 65$.

The petiole of *Lophosoria* has three convoluted vascular bundles, and the stomates, which are sunken near the veins, have two subsidiary cells. The lamina architecture of a portion of a typical large leaf is shown in Figs. 4 and 5, and the leaf of the less complex juvenile plant in Fig. 7. The paraphysate sori are illustrated in Fig. 9. *Lophosoria* has a prostrate-creeping, occasionally branched, stem that bears scale-leaves; the apex eventually turns upward and produces a stout erect stem with foliage leaves.

Systematics The single species of the genus, *Lophosoria quadripinnata* (Gmel.) C. Chr., is widely distributed in the American tropics and also occurs in wet, south temperate regions. Commonly used synonyms of the species are: *Alsophila pruinata* (Sw.) Kze., *Alsophila quadripinnata* (Gmel.) C. Chr. and *Lophosoria pruinata* (Sw.) Presl.

Ecology (Figs. 1,2)

Lophosoria is primarily a cloud forest genus in the tropics; it also occurs in similar cool, wet regions of southern South America. It grows in a variety of habitats such as oak woods (Mexico), mon-

Fig. 20.1. *Lophosoria quadripinnata,* near El Empalme, Costa Rica. (Photo W. H. Hodge.)

Fig. 20.2. *Lophosoria quadripinnata* var. *contracta* in scrubby forest, at 2900 m between Ona and Saraguro, Loja, Ecuador. (Photo B. Øllgaard.)

Fig. 20.3. Distribution of *Lophosoria*.

tane rain forests, elfin forests, páramo-thickets, high altitude grasslands and páramo. It readily colonizes disturbed sites such as road cuts and fills and landslips. It may persist in cutover forest, in pastures and burned areas.

Very large leaves seem to develop on plants in sites with constant high moisture and partial shade, as often seen in the Blue Mountains of Jamaica. Old plants, especially in pastures where they are periodically injured by cutting or burning, may form large multicipital stem mounds up to 3 m across. *Lophosoria quadripinnata* var. *contracta* (Hieron.) R. & A. Tryon, with very small leaves and short, imbricate, ascending pinnae (Figs. 2, 8), grows at high elevations in Ecuador.

In the tropics *Lophosoria* usually grows between 1000 and 3000 m, sometimes to 3800 m, and especially in southeastern Brazil, as low as to 500 m. In its south temperate range it grows from sea level to 1000 m.

Geography (Fig. 3)

Lophosoria quadripinnata occurs in tropical America from Veracruz and Hidalgo in Mexico, through Central America, in all of the Greater Antilles, and from Trinidad and Venezuela to Colombia, south to Bolivia and disjunct in southeastern Brazil. It is also in southern Chile and adjacent Argentina and in the Juan Fernandez Islands.

Fig. 20.4–20.8. *Lophosoria quadripinnata.* **4.** Leaf apex, ×0.5. **5.** Pinnules on central part of pinna, ×0.5. **6.** Soft, matted trichomes on the petiole base and portion of the stem, ×0.25. **7.** Juvenile leaf, the lamina once pinnate-pinnatifid, ×0.5. **8.** Leaf of var. *contracta* attached to part of the rhizome, ×0.25.

Fig. 20.9. *Lophosoria* sori with paraphyses and sporangia, × 10.

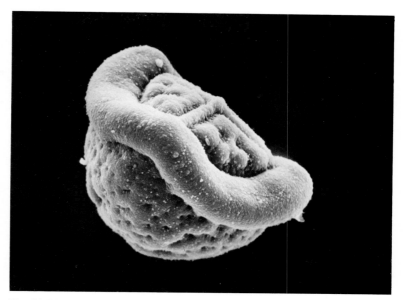

Fig. 20.10. *Lophosoria* spore, lateral view with broad equatorial flange, coarsely tuberculate proximal face above, distal face rugose, pitted, Colombia, *Barkley & Gutierrez 1433*, × 1000.

Spores

Lophosoria spores have a large, rounded equatorial flange which unequally divides the irregularly tuberculate proximal face from the pitted distal one (Fig. 10). Dense granulate deposit, especially on older spores often obscures the contours. The account of the Cyatheaceae spores by Gastony and Tryon (1976) indicates that perispore is present but does not form an inflated layer when *Lophosoria* spores are treated with sodium hydroxide. The spores are generally similar to those of some species of *Cibotium* which have prominent equatorial ridges, as in *C. Barometz*. In *Lophosoria,* as in *Cibotium,* the surface of compact granulate deposit may be more or less eroded.

Cytology

The record of $n = 65$ in *Lophosoria quadripinnata* from Jamaica is clearly documented by a figure in the work of Walker (1966). Additional cytological reports from other areas over the wide geographic range of the species may help in the assessment of morphological variation in populations. The same chromosome number is known in several species of *Dicksonia* and suggests a relationship with the Dicksoniaceae.

Literature

Gastony, G. J., and R. Tryon. 1976. Spore morphology in the Cyatheaceae. II. The genera *Lophosoria, Metaxya, Sphaeropteris, Alsophila,* and *Nephelea*. Amer. Jour. Bot. 63: 738–758.
Walker, T. G. 1966. A cytological survey of the Pteridophytes of Jamaica. Trans. Roy. Soc. Edinburgh 66: 169–237.

Family 12.   Metaxyaceae

Metaxyaceae Pic.-Ser., Webbia 24: 701. 1970. Type: *Metaxya* Presl

Description

Stem prostrate, decumbent, to nearly erect, rather stout, sometimes branching, with a siphonostele, indurated, bearing rather long and dense trichomes, especially toward the apex; leaves large, ca. 1–2 m long, pinnate, glabrate, circinate in the bud, the petiole without stipules; sporangia borne on the abaxial surface of the pinnae, in exindusiate sori, with a short, 4-rowed stalk and slightly oblique annulus that is nearly interrupted by the stalk; homosporous, spores without chlorophyll. Gametophyte epigeal, with chlorophyll, cordate, somewhat thickened centrally, antheridia and archegonia borne on the lower surface, the antheridium with usually five or a few more cells.

Comments on the Family

The family Metaxyaceae includes the single American genus *Metaxya.* This has been variously placed, along with *Lophosoria,* in other families such as the "Proto-Cyatheaceae," the scaly Cyatheaceae, and a large family including the Dicksoniaceae and Cyatheaceae (see Lophosoriaceae for references). Although *Metaxya* is, in a general way, allied to these groups, it is not closely related to any of them and is best recognized as a separate family. The description of the gametophyte is based on sections prepared by M. D. Turner, of material collected in Costa Rica.

21.   *Metaxya*
Figs. 21.1–21.10

Metaxya Presl, Tent. Pterid. 59. 1836. Type: *Metaxya rostrata* (HBK.) Presl [*Aspidium rostratum* HBK. (*Polypodium rostratum* Willd., 1810, not Burm., 1768)].
Amphidesmium J. Sm., Ferns Brit. For. 167. 1866. Type: *Amphidesmium blechnoides* (Hook.) J. Sm. (*Alsophila blechnoides* Hook.) = *Metaxya rostrata* (HBK.) Presl.

Description

Terrestrial, or sometimes on the base of trees; stem prostrate, decumbent to nearly erect, rather stout, bearing rather long trichomes, especially at the apex and many fibrous roots; leaves monomorphic, ca. 1–2 m long, borne in a loose cluster, 1-pinnate, the pinnae entire, axes grooved on the adaxial surface, glabrous or very sparsely pubescent, veins free; sori abaxial roundish to elongated, 1–3 borne on each fertile vein, irregularly disposed, receptacle nearly flat, paraphysate, exindusiate; spores globose, trilete, the laesurae $\frac{3}{4}$ to nearly equal to the radius, the perispore of fused granulate material forming irregular rods or strands that overlay a fairly smooth exospore. Chromosome number: $n = 94–96$; $2n = 190–192$.

The petiole of *Metaxya* has a single convoluted vascular bundle, and the stomates have three subsidiary cells. A nearly erect stem is shown in Fig. 4, and the numerous paraphyses in the sori

Fig. 21.1. *Metaxya rostrata,* lowland rain forest, Osa Peninsula, Costa Rica. (Photo T. W. Lucansky.)

Fig. 21.2. Distribution of *Metaxya*.

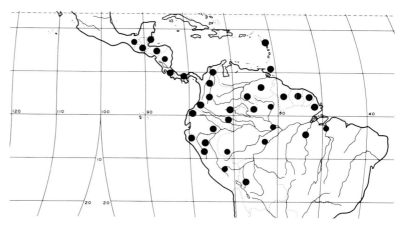

in Fig. 5. An unusual feature of the genus is that the very young plants (Fig. 7) have the pinnae lobed or pinnatifid, more complex than those of somewhat older (Fig. 6) or adult (Fig. 3) plants.

Systematics

The single species, *Metaxya rostrata* (HBK.) Presl, is relatively uniform throughout its range in tropical America. It was commonly referred to as *Alsophila blechnoides* Hook. or *Amphidesmium blechnoides* Kl. in older literature.

Ecology (Fig. 1)

Metaxya usually grows in densely shaded lowland forests, in ravines, on slopes, or the banks of ravines. It occurs less often at the edge of wet thickets or in wet savannahs, more rarely among

3   4

5   6   7

Figs. 21.3–21.7. *Metaxya rostrata.*   **3.** Apex of mature leaf, ×0.25.   **4.** Stem with three long petiole bases at top, the apex with matted trichomes, ×0.25.   **5.** Sori with dense paraphyses enmeshing the sporangia, the costa below, ×8.0.   **6.** Leaf of juvenile plant, the pinna margins strongly dentate, ×0.5.   **7.** Young plant, the first leaves more complex than older ones (Fig. 6), ×0.8.

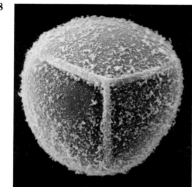

8

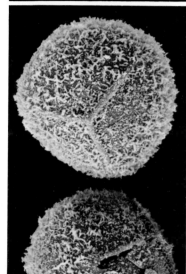

9

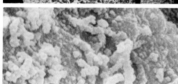

10

Figs. 21.8–21.10. *Metaxya* spores, Bolivia, *Krukoff 10959,* × 1000. **8.** Proximal face with initial granulate perispore deposit. **9.** Proximal face with additional granulate deposit fused into rods, the perispore ruptured in the lower spore. **10.** Detail of, granulate surface, × 10,000.

rocks, but is frequent in sandy soils. In the varzea creek-forests of Amazonian Brazil, it may become established on tree bases, especially palm trunks. It is a genus of low altitudes, growing from near sea level to 750 m.

Geography (Fig. 2)

Metaxya grows from Chiapas in Mexico to Panama, in the Antilles on Guadeloupe, where it is perhaps extinct, and on Trinidad, in French Guiana west to Colombia, and south to Bolivia, and in Amazonian Brazil.

Spores

The spheroidal shape of *Metaxya* spores and the granulate surface differs from those of the Lophosoriaceae or Dicksoniaceae that usually have prominent angles on the proximal face. The wall is thin and often broken in dry specimens (Fig. 9). The surface deposit of short rods or strands that form the perispore is readily separated from the lower, smooth exospore layer.

Cytology

The chromosome reports for *Metaxya rostrata* from Surinam, in cultivation, of 94–96 in meiotic cells and not less than 190 nor more than 192 in root tips by Roy and Holttum (1965) help in the assessment of the genus. Although these records are not precise it is evident that they represent a cytological level discrete from those of the Lophosoriaceae, Dicksoniaceae, and Cyatheaceae and support treatment of the genus as a separate family.

Literature

Roy, S. K., and R. E. Holttum. 1965. Cytological and morphological observations on *Metaxya rostrata* (HBK.) Presl. Amer. Fern Jour. 55: 158–164.

Family 13.   Cyatheaceae

Figs. Cyath. 1–15

Cyatheaceae Kaulf., Wesen Farrenkr. 119. 1827. Type: *Cyathea* Sm.
Alsophilaceae Presl, Gefässb. Stipes Farrn 32. 1847. Type: *Alsophila* R. Br.

Description

Stem usually erect, a single arborescent trunk to ca. 20 m tall, sometimes decumbent and short-creeping, usually massive, rarely small, usually unbranched, dictyostelic with medullary bundles, indurated, bearing scales and sometimes spines; leaves usually large, ca. 2–3 m, sometimes to 5 m long, rarely small, 50 cm or less long, pinnate, or (in *Alsophila sinuata*) entire, glabrous, pubescent and (or) scaly, circinate in the bud, petiole without stipules; sporangia borne in sori on the abaxial surface of segments, with a low to usually elevated receptacle, exindusiate or the indusium small and scalelike to completely enclosing the sorus, sporangia with a short, 4-rowed stalk and an oblique annulus that is not interrupted by the stalk; homosporous, spores lacking chlorophyll. Gametophyte epigeal, with chlorophyll, obcordate to slightly elongate, somewhat thickened centrally, with multicellular trichomes, archegonia and antheridia borne on the lower surface, the antheridium 5-celled.

Comments on the Family

The Cyatheaceae is a family of six genera: *Sphaeropteris, Alsophila, Nephelea, Trichipteris, Cyathea* and *Cnemidaria*. All of these occur in the American tropics with *Alsophila* and *Sphaeropteris* also in the paleotropics. The classification of the family has recently been intensively studied by Holttum (1963, 1964, 1965), and Tryon (1970). Major alliances, some more distinctive than others, are recognized here as genera. The six that are adopted place cohesive lineages at the most useful taxonomic level. Further studies may provide evidence for some realignment of the groups or the recognition of additional ones.

Evolutionary trends in the family and relationships of the genera are derived from a variety of data on the petiole scales, petiole spines, indusium, spore capacity of the sporangium, spore ornamentation, and stem anatomy. *Sphaeropteris,* with undifferentiated petiole scales, contains elements allied to the other two major evolutionary groups in the family. Subgenus *Sphaeropteris* is related to *Alsophila* by the dark seta at the apex of the petiole scales, and *Nephelea* is a specialized development from a group in *Alsophila*. The relationship of *Sphaeropteris* subgenus *Sclephropteris* is with *Trichipteris* and *Cyathea* as shown by the undifferentiated apex of the petiole scales. *Cnemidaria,* a specialized development from *Cyathea,* is also a member of this alliance.

Fossil Cyatheaceae, such as *Ciboticaulis* Ogura, the earliest record based on a stem from the Jurrasic (Ogura, 1927b), *Cyathea tyrmensis* (Seward) Krassilov, fertile foliage from the late Jurrasic (Krassilov, 1978), and *Cyathea colombiensis* Pons, fertile foliage from the Upper Tertiary (Pons, 1965, 1975), are probably correctly placed in the family although it is difficult to relate these

Fig. Cyath. 1. Colony of *Cyathea fulva*, south of Misantla, Veracruz, Mexico. (Photo A. Gómez-Pompa.)

with certainty to living genera. The family is a rather close morphological group that seems to have developed through relatively recent evolution of species and genera. However, it probably represents a long and diverse evolutionary series that may well extend farther back in the geological record than the earliest fossils. Morphological evidence indicates affinities of the Cyatheaceae with the Lophosoriaceae or perhaps with the Dicksoniaceae. The relationships of the Cyatheaceae to more advanced ferns must also rest on morphological evidence from living groups. Bower (1928) considered *Woodsia* as a derivative, while Holttum (1971) regards *Thelypteris* as arising from the Cyatheaceae. Basic differences in the sporangia, indusia, and

Fig. Cyath. 2. Croziers of *Sphaeropteris horrida* within a crown of leaves, Guatemala, ×0.25. (Photo W. H. Hodge.)

chromosome numbers between the Cyatheaceae and Thelypteridaceae suggest that certain similarities may be independently derived in the two families.

The **arborescent habit** of the Cyatheaceae is one of the most handsome plant forms. Although the stem has only primary tissues, it may grow to a height of 20 m and produce a crown of leaves with a spread of 6 m. The species often form groves on mountain sides (Fig. 1), and are a prominent part of the vegetation. Little is known of the growth rates and age of the plants, but Conant (1976) has studied two species in Puerto Rico. The stem of *Cyathea arborea,* a pioneer species in open sites, grows rapidly at a rate of about 30 cm a year and plants live for about 30–35 years. *Alsophila bryophila,* a species of the cloud forest, grows slowly at the rate of about 5 cm a year and has a life-span of up to 130 years.

Reproduction of the Cyatheaceae is almost wholly dependent on reproduction by spores; vegetative reproduction is rare and

Fig. Cyath. 3. Portion of stem section of *Cyathea pallescens*, with dark medullary bundles in the center of the stem, three meristeles (and parts of two others) each with blackish sclerotic tissue at the periphery and central, dark vascular tissue; at top, scales attached to two old petiole bases and to the stem surface, ×1.5.

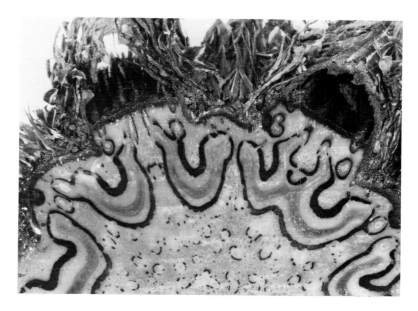

not effective for dispersal. Martius (1834) estimated the spore production of a leaf of *Cyathea Delgadii* (as *C. vestita* Mart.) to be nearly six billion. Studies on *Cyathea arborea* by Conant (1976) give evidence for spore production of 1,250,000,000,000 during the life of one plant. The enormous spore production ensures the establishment of new plants and is of special interest in relation to energy allocation.

The **biogeography** of some species of Cyatheaceae reflect the high reproductive and dispersal capacity of the family, for example, the species with wide distributions and those that occur on isolated oceanic islands 600–1500 km from the nearest source. However, among some 500 species in the family, not more than 10% have wide ranges, and on the isolated islands of the Pacific, east of Fiji, only 16 species occur. Most species grow in the forests of tropical mountains, in wet montane forests or in cloud forests, and many of them have restricted ranges within these vegetational zones. For example, about 60 endemics are found in the Andes, many of which are of local occurrence. It is evident that most species have narrow ecological adaptations and their high reproductive and dispersal capacities are not realized. The complex mosaic of environments in mountain forests provide excellent opportunities for the evolution of ecologically specialized species (Tryon and Gastony, 1975).

The **stem apex** bears prominent leaf buds, commonly called croziers (Fig. 2). Observations on *Trichipteris bicrenata* (Feldman and Nardi, 1973) show that the croziers in that species are borne in sets of six. Although they appear as a whorl, each set actually is in a compact spiral. Within these croziers there are eight sets of leaf primordia that are progressively smaller toward the stem apex. Although the phyllotaxy is usually a spiral, in some species it is a true whorl. The single apical cell of the stem is situated at the top of a moundlike apical meristem, as in *Cyathea fulva* (Fig. 14). The apex and leaf primordia are closely invested by a dense cover of scales. These scales and the tightly packed young cro-

4

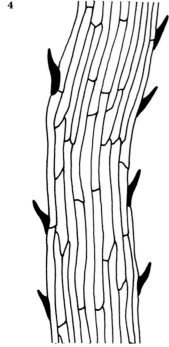

5

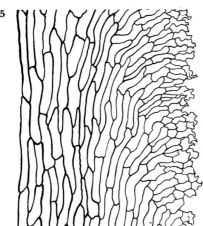

Figs. Cyath. 4, 5. Petiole scales. **4.** Portion of conform scale of ***Sphaerop-teris Brunei,*** with dark teeth on the edges, × 60. **5.** Portion of marginate scale of ***Alsophila Salvinii*** showing differences in size and orientation between cells of the central portion (left) and those of the margin, × 60.

ziers afford protection for the apical meristem. The mass of scales may also absorb water from mist and rain to supplement the water requirements of the crown of large leaves.

The uncoiling of the **croziers** is accomplished by a remarkable system of controlled growth (Voeller, 1966). Uncoiling proceeds by the elongation of the cells on the inner (adaxial) side of the petiole and rachis until these are equal in length to the cells of the outer (abaxial) side which also elongate but to a lesser extent. The process is under the influence of an auxin produced by the developing pinnae. The differential elongation begins at the base of the petiole and proceeds upwards, finally resulting in a nearly straight leaf axis as the cells on each side become equivalent in length.

Stem and petiole anatomy of the American species of Cyatheaceae has been studied especially by Lucansky (1974) and Lucansky and White (1974) and similar studies of paleotropical species were undertaken by Ogura (1927a). The stem (Fig. 3) is a dictyostele, with individual meristeles, each nearly surrounded by dark sclerotic tissue. Within the sclerotic tissue is a zone of light-colored parenchyma on either side of the darker meristele. Many medullary bundles occur in the central region of the stem and sometimes bundles also develop in the cortex. Variations of these basic patterns occur in some genera. *Cnemidaria* has a reduced, more simplified anatomy, and *Nephelea* exhibits an elaboration of the nodal and vascular anatomy. The complexly organized pathways of the medullary bundles as they anastomose through the stem has been investigated by Adams (1977). The petiole (Fig. 6) has numerous vascular bundles, arranged on the basic plan of an abaxial arc and an adaxial arc, but may have additional bundles. The leaf scars form symmetrical patterns on the trunk (Fig. 26.13).

Petiole scales are of special utility in the classification of the family and special terms are applied to their parts. The *body* is the whole scale except for a row of cells, sometimes including cilia, teeth, or setae, which form the *edge*. Scales of *Sphaeropteris* have the body cells all more or less alike and with the same orientation; these are *conform* scales (Fig. 4). Scales of the other genera have the cells of the body differentiated. The *central portion* is formed of elongate cells oriented with their long axis parallel to that of the scale. A zone of cells, between the edge and the central portion, forms the *margin*. These cells differ in size, orientation and usually in shape and color, from those of the central area and may form a broad or narrow zone; these are *marginate* scales (Fig. 5).

The **petiole spines** of *Nephelea* are of a special type unknown in other ferns. These are blackish, indurated, with a sharp apex, and are fully developed on the petiole of the crozier (Fig. 7). They are *squaminate* spines that have originated from petiole scales. Petiole spines of other genera, originating from the outer tissues of the cortex, are *corticinate* spines. These are soft and immature on the petiole of the crozier, bear a scale at their apex, and become indurated as the lamina expands.

The **indusium** was formerly considered to be an important

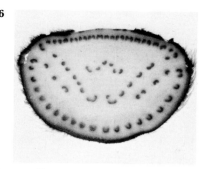

Figs. Cyath. 6, 7. **6.** Petiole section of ***Sphaeropteris Brunei*** with arcs of vascular bundles, × 0.75. **7.** Petiole base of a crozier of ***Nephelea polystichoides*** with scales and large, black, squaminate spines, × 1.0.

character for the classification of the family. Indusia are variable in form and similar types occur in different genera, as the hemitelioid indusia in *Cnemidaria mutica* (Fig. 8) and in *Sphaeropteris dissimilis* (Fig. 9). A study of the development and diversity of indusia by Tryon and Feldman (1975) included species of both the paleo- and neotropics. Indusial development, examined in *Cyathea fulva* from Veracruz, Mexico, showed initiation from protoderm cells on the abaxial leaf surface closer to the midvein than to the margin of the segment. A section of a young sorus shows its dorsal position and the unequal development of indusial lobes prior to their fusion into a sphaeropteroid indusium (Fig. 13). This evidence of a dorsal origin of the indusium differs from earlier interpretations which considered it as originating from the margin.

The diverse form of the indusia varies from small scalelike structures to large globose ones that wholly envelope the sorus. The three major forms are hemitelioid, cyatheoid and sphaeropteroid. *Hemitelioid* indusia do not completely surround the base of the receptacle. They may be mostly appressed to the leaf tissue beneath the sorus (Fig. 8) or they may partly arch over the sorus (Fig. 9). The smallest scalelike hemitelioid indusia are *minute*. *Cyatheoid* indusia completely surround the base of the receptacle and usually have an entire edge. Three types can be distinguished on the basis of their development: *meniscoid* indusia are disc- or saucer-shaped with an upturned edge, *cyathiform* indusia are cup-shaped (Fig. 10), and *urceolate* indusia are longer, with a constricted opening (Fig. 11). *Sphaeropteroid* indusia (Fig. 12) are globose, completely enclose the sorus and usually have an apical umbo. When these indusia split into segments at maturity, the original form can often be identified by the umbo which is retained on one of the segments.

The **sporangial capacity and spore characters** correlate with other characters establishing generic relations. The spore capacity of the sporangium of most genera is 64, with exceptional records of 32 or 16. However, reduction in the number of spores to 16 per sporangium is well established in the *Alsophila-Nephelea* alliance (Gastony, 1974). The sporangium of all species of *Nephelea* consistently has 16 spores, and reports for *Alsophila* are mostly of 16, but some species have 64. *Alsophila* and *Nephelea* spores (Gastony and Tryon, 1976) have a unique ridged sporoderm distinct from other genera. *Cnemidaria* spores lack an outer perispore layer, and have a porate surface with three large pores situated in the equatorial area equidistant from the ends of the laesurae. *Cyathea* and *Sphaeropteris* spores also may lack perispore or this may be deposited late in sporogenesis. The exospore in these genera and in *Trichipteris* may be strongly verrucate with the pores usually uniform in size and symmetrically placed. Perispore in most species of *Cyathea*, *Sphaeropteris* and *Trichipteris* consists of strands forming a delicate echinate surface, or are more or less fused to somewhat reticulate.

Cytological reports for species of Cyatheaceae, uniformly $n = 69$, are summarized by Löve et al. (1977). The original reports from America are from studies in Jamaica (Walker, 1966),

8　　9　　10

11　　12

13　　14

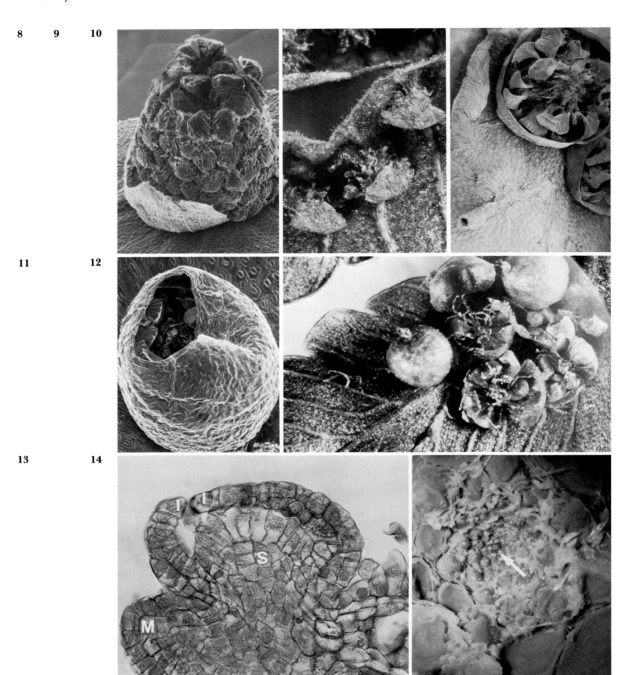

Figs. Cyath. 8–14. Cyatheaceae.　**8.** Sorus of *Cnemidaria mutica* with the hemitelioid indusium mostly obscured beneath clusters of nearly mature sporangia, a portion exposed at lower left, × 35.　**9.** Sori of *Sphaeropteris dissimilis* with hemitelioid indusium arching over the sorus, × 20.　**10.** Sorus of *Cyathea arborea,* with cyanthiform indusia (an undeveloped indusium, lower left), × 35.　**11.** Urceolate indusium of *Nephelea Grevilleana,* × 35.　**12.** Sphaeropteroid indusia of *Cyathea Lechleri,* unopened indusium with apical umbo, left, split indusia with sporangia and paraphyses, right, × 20.　**13.** Young sorus of *Cyathea fulva* (s) developing in a dorsal position, the segment margin (m) left, the two unequal lobes of the indusium (i) nearly fused, × 400.　**14.** Dissected stem apex of *Cyathea fulva,* the central apical meristem (arrow) surrounded by leaf primordia, × 7.0.

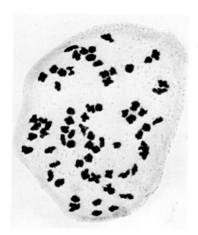

Fig. Cyath. 15. Meiotic chromosomes of **Trichipteris microdonta,** Veracruz, Mexico, *Gastony & Gastony 1020,* $n = 69$, × 1400.

Puerto Rico (Sorsa, 1970) and Costa Rica (Gómez, 1971). The photograph of meiotic chromosomes ($n = 69$) of *Trichipteris microdonta* (Fig. 15) is an original record from Veracruz, Mexico. Chromosome numbers for the family have not been reported from South America. The uniform number that occurs throughout the family is exceptional among ferns and indicates an absence of polyploidy. A base number $x = 23$ proposed for the family is also unusual and does not indicate evolutionary affinities with other groups. Cytological studies on *Alsophila* by Conant (1976) show bivalent associations in intergeneric hybrids indicating a marked cytogenetic uniformity within the family.

Hybrids have been proposed involving five of the genera in America (Conant, 1975, 1976; Tryon, 1976). Nineteen hybrids have been recognized, nine in the *Alsophila-Nephelea* alliance and ten in the *Cnemidaria-Cyathea-Trichipteris* alliance. Most of these are between species of different genera and only three are intrageneric hybrids. Hybrids between *Cnemidaria horrida* and *Cyathea arborea,* crosses between very distinctive species, demonstrate that genetic barriers to crossing have not evolved concomitantly with morphological divergence.

Conant and Cooper-Driver (1980) studied a colony of fertile diploid hybrids in Puerto Rico, involving two species of *Alsophila* and one species of *Nephelea.* They proposed a new mode of speciation, called autogamous allohomoploidy, based on the dispersal of second generation hybrids and their morphological stabilization through inbreeding.

Apogamous sporophytes have been reported by Stokey (1918, 1930) in studies on the gametophytes of the Cyatheaceae. It appears that these apogamously developed sporophytes were the result of cultural conditions, for apogamy has not been recognized in natural populations.

Key to Genera of Cyatheaceae

a. Petiole scales conform (Fig. 4), the cells of the body similar in orientation, shape, and (usually) in size and color; the edge may bear cilia, teeth or setae.

22. *Sphaeropteris*

a. Petiole scales marginate (Fig. 5), with a narrow or broad margin of cells different in orientation, size, and (usually) in shape and color from those of the central portion. b.

  b. Petiole scales with a dark, opaque apical seta. c.

    c. Petiole lacking spines, or with corticinate spines, which bear, at least when young, a scale at the apex, croziers lacking spines.

23. *Alsophila*

    c. Petiole with squaminate spines, many of these large, black, with a slender apex (Fig. 7), and present on the croziers, petiole scales borne on the petiole surface. 24. *Nephelea*

  b. Petiole scales lacking a differentiated apical seta, the apex rounded to filamentous. d.

    d. Indusium absent. 25. *Trichipteris*

    d. Indusium present. e.

      e. Spores lacking large pores, or with variously distributed small pits or pores; costae and costules adaxially with trichomes; veins free, the basal ones of adjacent segments extending to the margin above the sinus (in two species some costal areolae present).

26. *Cyathea*

      e. Spores with three large equatorial pores, often also with smaller pits or pores; costae and costules adaxially glabrous; veins usually anastomosing to form costal areolae, or if free, then the basal veins of adjacent segments connivent to the sinus (in one species veins extending above the sinus). 27. *Cnemidaria*

Literature

Adams, D. C. 1977. Ciné analysis of the medullary bundle system in *Cyathea fulva*. Amer. Fern Jour, 67: 73–80.

Bower, F. O. 1928. The Ferns. 3: Dryopteroid ferns (I. Woodsieae) 99–103. Cambridge Univ. Press.

Conant, D. S. 1975. Hybrids in American Cyatheaceae. Rhodora 77: 441–455.

Conant, D. S. 1976. Ecogeographic and Systematic Studies in American Cyatheaceae. Ph.D. Thesis, Harvard University, Cambridge, Mass.

Conant, D. S., and G. Cooper-Driver. 1980. Autogamous allohomoploidy in *Alsophila* and *Nephelea* (Cyatheaceae): A new hypothesis for speciation in homoploid homosporous ferns. Amer. Jour. Bot. 67: 1269–1288.

Feldman, L., and J. Nardi. 1973. Tree fern apices and starch storage. *in:* Tryon et al., eds. Fern Biology in Mexico. BioScience 23: 31, 32.

Gastony, G. J. 1974. Spore morphology in the Cyatheaceae, 1. The perine and sporangial capacity: General considerations. Amer. Jour. Bot. 61: 672–680.

Gastony, G. J., and R. Tryon. 1976. Spore morphology in the Cyatheaceae 2. The genera *Lophosoria, Metaxya, Sphaeropteris, Alsophila,* and *Nephelea*. Amer. Jour. Bot. 63: 738–758.

Gómez Pignataro, L. D. 1971. Ricerche citologiche sulle pteridofite della Costa Rica. 1. Atti Ist. Bot. Univ. Pavia 7: 29–31.

Holttum, R. E. 1963. Cyatheaceae, Fl. Males. Ser. II, 1: 65–176.

Holttum, R. E. 1964. The tree ferns of the genus *Cyathea* in Australasia and the Pacific. Blumea 12: 241–274.

Holttum, R. E. 1965. Tree ferns of the genus *Cyathea* Sm. in Asia (excluding Malaysia). Kew Bull. 19: 463–487.

Holttum, R. E. 1971. Studies in the family Thelypteridaceae, III. A new system of genera in the Old World. Blumea 19: 17–52.

Krassilov, V. 1978. Mesozoic lycopods and ferns from the Bureja Basin. Palaeontographica 166, Abt. B: 16–29.

Löve, A., D. Löve, and R. Pichi-Sermolli. 1977. Cytotaxonomical Atlas of the Pteridophyta. 398 pp. Cramer, Vaduz.

Lucansky, T. W. 1974. Comparative studies of the nodal and vascular anatomy in the neotropical Cyatheaceae, II. Squamate genera. Amer. Jour. Bot. 61: 472–480.

Lucansky, T. W., and R. A. White. 1974. Comparative studies of the nodal and vascular anatomy in the neotropical Cyatheaceae, III. Nodal and petiole patterns: Summary and conclusions. Amer. Jour. Bot. 61: 818–828.

Martius, C. F. P. 1834. Icon. Plantarum Cryptogam. Brazil. 138 pp. Munich. C. Wolf.

Ogura, Y. 1927a. Comparative anatomy of Japanese Cyatheaceae. Jour. Fac. Sci. Univ. Tokyo (Bot.) 1: 141–350.

Ogura, Y. 1927b. On the structure and affinities of some fossil tree-ferns from Japan. Jour. Fac. Sci. Univ. Tokyo (Bot.) 1: 351–380.

Pons, D. 1965. Sur des empreintes foliares de Cyathéacées fossiles de Colombie. Bol. Geol. Univ. Industr. Santander. 20: 5–26.

Pons, D. 1975. A propos d'une fougére tertiare de Colombie: *Cyathea colombiensis* D. Pons. Argumenta Paleobot. 4: 39–44.

Sorsa, V. 1970. Fern Cytology and the radiation field. Chap. G3: pp. 39–50, *in:* H. T. Odum and R. F. Pigeon, A Tropical Rain Forest. A Study of Irradiation and Ecology at El Verde. Puerto Rico. Division of Technical Information. U.S. Atomic Energy Commission.

Stokey, A. G. 1918. Apogamy in the Cyatheaceae. Bot. Gaz. 65: 97–102.

Stokey, A. G. 1930. Prothallia of the Cyatheaceae. Bot. Gaz. 90: 1–45.

Tryon, A. F., and L. J. Feldman. 1975. Tree fern indusia: studies of development and diversity. Canad. Jour. Bot. 53: 2260–2273.

Tryon, R. 1970. The classification of the Cyatheaceae. Contr. Gray Herb. 200: 3–53.

Tryon, R. 1976. Reference under *Cyathea*.

Tryon, R., and G. J. Gastony. 1975. The biogeography of endemism in the Cyatheaceae. Brit. Fern Gaz. 11: 73–79.

Voeller, B. R. 1966. Crozier uncoiling of ferns. Rockefeller Univ. Rev. 4: 4–19.

Walker, T. G. 1966. A cytotaxonomic survey of the Pteridophytes of Jamaica. Trans. Roy. Soc. Edinburgh 66: 169–237.

22. *Sphaeropteris*
Figs. 22.1–22.21

Sphaeropteris Bernh., Jour. Bot. (Schrad.) 1800 (2): 122. 1802, not Wall., 1830 (= *Peranema*). Type: *Sphaeropteris medullaris* (Forst.) Bernh. (*Polypodium medullare* Forst.).

Cyathea subgenus *Sphaeropteris* (Bernh.) Holtt., Fl. Malesiana II, 1: 124. 1963.

Schizocaena Hook., Gen. Fil. *t. 2.* 1838. Type: *Schizocaena brunonis* Hook. = *Sphaeropteris moluccana* (Desv.) Tryon. *Cyathea* section *Schizocaena* (Hook.) Holtt., Fl. Malesiana, II, 1: 141. 1963. *Sphaeropteris* section *Schizocaena* (Hook.) Windisch, Bot. Jahrb. Syst. 98: 196. 1977. [Subgenus *Sphaeropteris*]

Eatoniopteris Bomm., Bull. Soc. Bot. France 20: xix. 1873. Lectotype: *Cyathea insignis* D. C. Eaton [Bommer made no combinations for the names of the 19 species he placed in his new genus] = *Sphaeropteris insignis* (D. C. Eaton) Tryon [Subgenus *Sphaeropteris*].

Fourniera Bomm., *ibidem* 20: xix. 1873. Type: *Fourniera novaecaledoniae* (Mett.) Bomm. (*Alsophila novae-caledoniae* Mett.) = *Sphaeropteris novaecaledoniae* (Mett.) Tyron [Subgenus *Sphaeropteris*].

Sphaeropteris subgenus *Sclephropteris* Windisch, Bot. Jahrb. Syst. 98: 181. 1977. Type: *Sphaeropteris aterrima* (Hook.) Tryon (*Alsophila aterrima* Hook.).

Description Terrestrial; stem erect, stout, usually tall, rarely short-decumbent, unbranched or very rarely branched, with scales, a dense mat of fibrous roots developed especially at the base; leaves monomorphic to slightly dimorphic, usually 1–3 m, rarely shorter or to 5 m long, borne in a spreading crown, lamina 1-pinnate to usually 2-pinnate-pinnatifid, to rarely 3-pinnate-pinnatifid, monomorphic to slightly dimorphic, with conform scales (especially on the crozier and usually persistent at the base of the petiole), these with the apical cell unmodified or with a dark apical seta, often with cilia, teeth or dark setae on the edges, petiole smooth to tuberculate, or with corticinate spines, veins free, or (in *S. Bradei*) partially anastomosing without included free veinlets; sori round, borne on the veins, often at a fork, receptacle slightly elevated to elongate, with short or long (and then sometimes branched) paraphyses, exindusiate (several closely investing scales may surround the sorus in species of the Pacific), or with a hemitelioid (minute or larger) to sphaeropteroid indusium; spores tetrahedral-globose, with prominent angles, trilete, the laesurae $\frac{3}{4}$ the radius, the sporoderm surface usually of more or less fused, delicately echinate strands, or with coarse, diffuse, echinate projections, a lower stratum may be porate, smooth, or verrucate. Chromosome number: $n = 69$.

Fig. 22.1. American distribution of *Sphaeropteris:* subgenus *Sclephropteris* (stars), subgenus *Sphaeropteris* (dots).

Petiole scales of *Sphaeropteris* (Figs. 2–6) show the conform cellular structure. Portions of broad scales are shown in Figs. 4–6, from the central portion to the edge, which may be plain (Fig. 4), ciliate (Fig. 5), or toothed (Fig. 6). Subgenus *Sclephropteris* has an undifferentiated apical cell (Fig. 2), which contrasts to the dark seta in subgenus *Sphaeropteris* (Fig. 3). The edge of the scale in subgenus *Sphaeropteris* also usually has dark setae (Fig. 3) or dark teeth (Fig. 6). The petiole is sometimes densely and persistently clothed with scales as in *S. Atahuallpa* (Fig. 12). The typical 2-pinnate-pinnatifid lamina architecture is shown in *S. elongata* (Fig. 13), and the less common 2-pinnate architecture in *S. dissimilis* (Fig. 10). Sori and indusia shown in Figs. 7–9 include exindusiate sori with prominent paraphyses (Fig. 7), hemitelioid indusia (Fig. 8) and sphaeropteroid indusia (Fig. 9). A single species, *S. Bradei*, has the veins joining to form costal areolae (Fig. 11).

Systematics

Sphaeropteris is a genus of 120 species with a nearly pantropical distribution except absent from Africa and Madagascar. Two subgenera are based on characters of the petiole scales. Subgenus *Sphaeropteris,* which is related to *Alsophila,* has petiole scales with the dark apical setae characteristic of that genus. Subgenus *Sclephropteris* is related to *Trichipteris.* Some of the exindusiate species of this subgenus are nearly intermediate between the two genera as shown by petiole scales that are ciliate (Fig. 4), or otherwise have slightly modified edges or outer margins.

The paleotropical species have been revised (under *Cyathea* subgenus *Sphaeropteris*) by Holttum (1963, 1964, 1965), and the American species by Windisch (1977, 1978) and Tryon (1971, 1972).

Synopsis of *Sphaeropteris*

Subgenus *Sclephropteris*
Petiole scales with unmodified cells at the apex and along the edges or the edges ciliate (dark setae and teeth lacking).

Fifteen species of the American tropics, forming three species groups.

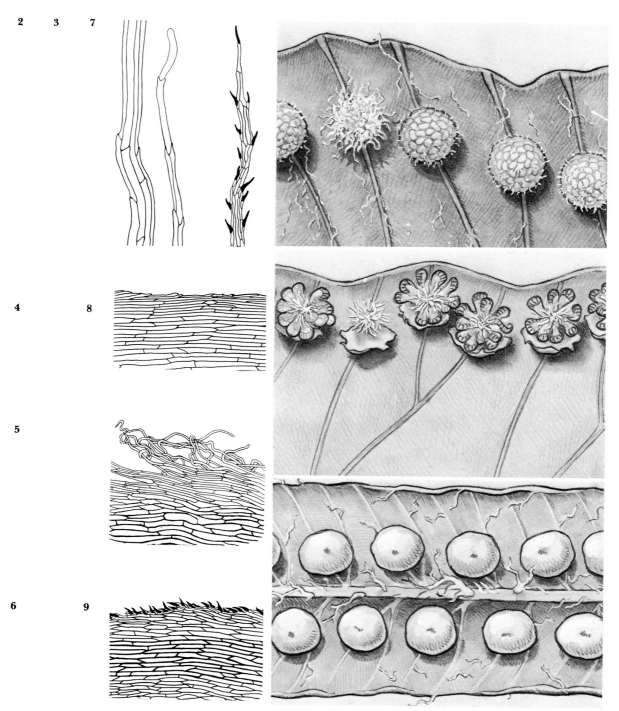

Figs. 22.2–22.9. *Sphaeropteris.* **2.** Petiole scales of *S. aterrima,* central portion and apex of a narrow scale. ×60. **3.** Apex of petiole scale of *S. horrida,* ×30. **4.** Central portion to edge of a broad petiole scale of *S. aterrima,* ×30. **5.** Central portion to ciliate edge of a petiole scale of *S. marginalis,* ×30. **6.** Central portion to edge of a petiole scale of *S. elongata,* ×30. **7.** Exindusiate sori of *S. marginalis* with sporangia, these removed in one showing paraphyses, ×15. **8.** Sori and hemitelioid indusia of *S. dissimilis* with sporangia and paraphyses, sporangia removed from one sorus, ×15. **9.** Sori of *S. quindiuensis* enclosed by a sphaeropteroid indusia with apical umbo, dissected scales on costa, ×12.

Figs. 22.10–22.13. *Sphaeropteris.*   **10.** Central portion of pinna of **S. *dissimilis*** with sori and hemitelioid indusia near margin, × 1.0.   **11.** Central portion of a pinnule of **S. *Bradei,*** showing costal areolae, × 3.   **12.** Dense, persistent scales on a portion of petiole of **S. *Atahuallpa,*** × 0.3.   **13.** Central portion of a pinna of **S. *elongata,*** × 0.75.

Subgenus *Sphaeropteris*

Petiole scales with a rigid, usually dark apical seta and usually with dark teeth or setae on the edges.

About 100 species, with eight in the American tropics. This subgenus may be further classified into two sections, each with two subsections (Holttum, 1963; Windisch, 1978). The American species all belong to section *Sphaeropteris* subsection *Sphaeropteris* and form two species groups.

Tropical American Species

In subgenus *Sclephropteris* the exindusiate species compose the group of *S. aterrima* (Hook.) Tryon, consisting of six species, including *S. Lockwoodiana* Windisch and *S. marginalis* (Kl.) Tryon. The group of *S. hirsuta* (Desv.) Tryon includes eight species with hemitelioid indusia, as *S. Bradei* Windisch (Fig. 11), *S. cyatheoides* (Desv.) Windisch, *S. dissimilis* (Morton) Windisch (Fig. 10), *S. macrocarpa* (Presl) Tryon, and *S. macrosora* (Baker) Windisch. One species, *S. Atahuallpa* Tryon has a sphaeropteroid indusium; in this and other characters it forms a discrete group.

Subgenus *Sphaeropteris* includes the *S. horrida* group of indusiate species: *S. Brunei* (Christ) Tryon, *S. Cuatrecasasii* Tryon, *S. Gardneri* (Hook.) Tryon, *S. horrida* (Liebm.) Tryon, *S. insignis* (D. C. Eaton) Tryon, and *S. quindiuensis* (Karst.) Tryon. The two exindusiate species, *S. elongata* (Hook.) Tryon (Fig. 13) and *S. myosuroides* (Liebm.) Tryon, are closely related to each other.

Ecology

Sphaeropteris grows in montane forests and low elevation rain forests on mountain slopes, in ravines, on forest borders, and sometimes in rocky places or in swamp forest. Some species pioneer in open habitats as landslides, stream-cuts and road banks.

In America the two subgenera have distinct ecologies. Subgenus *Sclephropteris* is mostly in the Roraima sandstone formation and its derived sandy soils. *Sphaeropteris hirsuta* and *S. cyatheoides* are usually small understory species. The subgenus is most common up to 1000 m. Subgenus *Sphaeropteris* is primarily pan-Andean and usually grows in mountainous areas between 750 and 2000 m. Species of this subgenus are often in cloud forest areas where plants may form a part of the canopy.

Geography (Fig. 1)

Sphaeropteris occurs in the American tropics with subgenus *Sphaeropteris* also in the paleotropics from India and southeastern Asia to New Zealand, and in the Pacific islands east to Pitcairn Island. In America the genus ranges from the Greater Antilles and Veracruz in Mexico, south to La Paz in Bolivia and São Paulo in Brazil.

The two subgenera have nearly distinct distributions (Fig. 1) with different centers of diversity. Subgenus *Sclephropteris* is confined to Panama and northern South America, south to Peru and Amazonian Brazil. All but one of the species grows in the Guayana Highlands or in the low Amazon basin and adjacent regions. Species such as *S. sipapoensis* Tryon, *S. dissimilis* and *S. intramarginalis* Windisch are local endemics in Guayana. Subgenus *Sphaeropteris* is basically montane, in the Greater Antilles, Mexico and Central America, Colombia to Bolivia and southeastern Brazil. Five of the eight species occur in Central America and Colombia. The indusiate species form a closely related allopatric group. *Sphaeropteris insignis* of the Greater Antilles and *S. horrida* of Mexico and northern Central America, are the most primitive. More advanced species such as *S. Brunei* and *S. quindiuensis* occur progressively southward, with the most advanced, *S. Gardneri,* in southeastern Brazil. This group illustrates, in a diagrammatic way, geographic speciation in the family.

Spores

Sphaeropteris spores, especially those of American species, are of three main types. They usually have the perispore formed of more or less connected strands (Figs. 14, 15) of particulate material (Fig. 16) as in *S. macrocarpa*. The exospore may be quite smooth as in *S. macrocarpa,* somewhat rugose and perforate as in *S. hirsuta* (Figs. 20, 21), or prominently verrucate in *S. marginalis*. A second distinct type has the sporoderm foraminate with large pits more or less covered with granulate perispore as in *S. myosuroides* (Fig. 17). A third type of spore has a coarsely echinate perispore with short somewhat irregular projections as in *S. quindiuensis* (Figs. 18, 19). This type is characteristic of the paleotropical species and the American group of *S. horrida*. The diversity in both neo- and paleotropical species of *Sphaeropteris* is well illustrated in the survey of Cyatheaceae spores by Gastony and Tryon (1976) and correlated with the systematic treatment.

Cytology

The report of $n = 69$ in *S. Brunei* (as *Cyathea*) from Costa Rica by Gómez Pignataro (1971) is consistent with paleotropical records from Ceylon, India, New Zealand and Japan. There is relatively little cytological work on these plants, especially from the Malesian and Australian-Pacific areas, considering the richness and diversity of species in those regions.

Literature

Gastony, G. J., and R. Tryon. 1976. Reference under the family.

Gómez Pignataro, L. D. 1971. Reference under the family.

Holttum, R. E. 1963, 1964, 1965. References under the family.

Tryon, R. 1971. The American tree ferns allied to *Sphaeropteris horrida*. Rhodora 73. 1–19.

Tryon, R. 1972. Taxonomic fern notes, VI. New species of American Cyatheaceae. Rhodora 74: 441–450.

Windisch, P. G. 1977. Synopsis of the genus *Sphaeropteris* (Cyatheaceae), with a revision of the neotropical exindusiate species. Bot. Jahrb. Syst. 98: 176–198.

Windisch, P. G. 1978. *Sphaeropteris* (Cyatheaceae). The systematics of the group of *Sphaeropteris hirsuta*. Mem. New York Bot. Gard. 29: 2–22.

Figs. 22.14–22.21. *Sphaeropteris* spores, ×1000, surface detail, ×10,000. **14, 15.** *S. macrocarpa,* Surinam, *Maguire 24359*. **14.** Proximal face with dense strands. **15.** Detail of connected strands. **16.** *S. Lockwoodiana,* detail of particulate surface of strands, Colombia, *Schultes 6535*. **17.** *S. myosuroides,* granulate perispore deposit on foraminate exospore, Mexico, *Schultes & Reko 839*. **18, 19.** *S. quindiuensis,* Peru, *Soejarto 1427*. **18.** Coarsely echinate surface. **19.** Detail of coarse echinate structures. **20, 21.** *S. hirsuta,* British Guiana, *Hitchcock 17254*. **20.** Shallowly rugose spores, proximal face, above, lateral, below. **21.** Detail of somewhat rugose, perforate exospore.

14 15

16

17 19

21

18 20

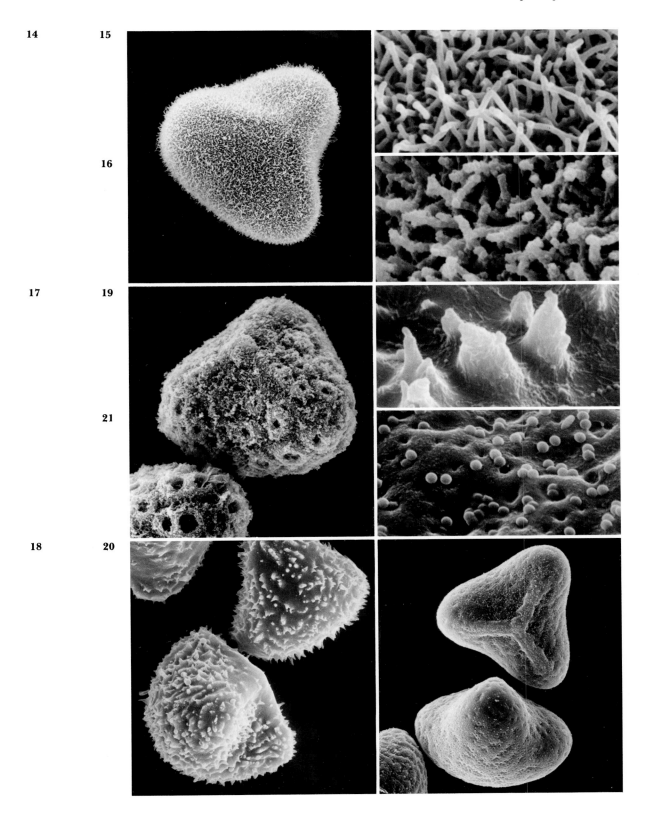

23. *Alsophila*

Figs. 23.1–23.14

Alsophila R. Br., Prod. Fl. Nov. Holl. 158. 1810. Type: *Alsophila australis* R. Br.

Gymnosphaera Bl., Enum. Pl. Jav. 242. 1828. Type: *Gymnosphaera glabra* Bl. = *Alsophila glabra* (Bl.) Hook. *Cyathea* subgenus *Gymnosphaera* (Bl.) Tindale, Contrib. N. S. Wales Natl. Herb. 2: 331. 1956. *Cyathea* section *Gymnosphaera* (Bl.) Holtt., Fl. Malesiana, II, 1: 115. 1963.

Amphicosmia Gardn., Lond. Jour. Bot. 1: 441. 1842. Lectotype: *Amphicosmia riparia* (Willd.) Gardn. (*Cyathea riparia* Willd.) = *Alsophila capensis* (L. f.) J. Sm.

Dichorexia Presl, Gefässb. Stipes Farrn, 36. 1847 (preprint from Abhandl. Böhm. Ges. V, 5: 344. 1848). Type: *Dichorexia latebrosa* (Hook.) Presl = *Alsophila latebrosa* Hook.

Thysanobotrya vAvR., Bull. Jard. Bot. Buitenz. II, 28: 66. 1918. Type: *Thysanobotrya arfakensis* (Gepp) vAvR. (*Polybotrya arfakensis* Gepp) = *Alsophila biformis* Rosenst.

Description Terrestrial; stem erect, stout, usually tall, very rarely short-decumbent or (in *A. biformis* Rosenst.) climbing to 3 m, rarely branching by adventitious buds, with scales, and very rarely spines, a dense mat of fibrous roots developed especially at the base; leaves monomorphic to sometimes dimorphic (the fertile with reduced and sometimes more complex pinnules), rarely 0.3 m long, usually 1–3 m, or up to 5 m long, usually borne in a spreading crown, lamina simple and entire [in *A. sinuata* (Hook. & Grev.) Tryon] to usually 2-pinnate-pinnatifid, to rarely nearly 4-pinnate, with marginate scales, especially on the crozier and usually persistent at the base on the petiole, these with a usually dark apical seta and often with setae on the edges and (or) body, petiole smooth to tuberculate or with corticinate spines, veins free; sori round, on the veins, often at a fork, receptacle globose to elongate, with short to long paraphyses, exindusiate or with a hemitelioid to sphaeropteroid indusium; spores tetrahedral-globose, with prominent angles, trilete, the laesurae $\frac{3}{4}$ to nearly equal to the radius, with prominent forked or branched ridges usually extending from the ends of the laesurae and parallel to the equator. Chromosome number: $n = 69$; $2n = 130–140$.

Typical petiole scales of *Alsophila* have a dark apical seta and a well-differentiated margin (Figs. 3, 4). One American species, *A. Salvinii* (Fig. 6), is exindusiate, and another, *A. capensis,* (Fig. 7), has hemitelioid indusia; most others, as *A. Urbanii* (Fig. 8) have cyatheoid indusia.

Systematics *Alsophila* is a large, pantropical genus of perhaps 230 species; only 13 of them American. Copeland (1947) recognized *Gymnosphaera* as a genus, while Tindale (1956) and Holttum (1963) recognized it as a subgenus and section, respectively, of *Cyathea*. These different circumscriptions of the taxon indicate that further study is needed to assess the characters and to clarify the status of the group. *Gymnosphaera* includes species with very dark lamina axes, pinna-rachises that are not pubescent beneath, dimorphic fertile and sterile leaves (or pinnae) and exindusiate sori. Among the American species, *Alsophila Salvinii* may belong

Fig. 23.1. *Alsophila bryophila,* elfin forest, El Yunque, Luquillo National Forest, Puerto Rico. (Photo D. S. Conant.)

here and represent a disjunction of the group between Central America and Madagascar.

Alsophila is related to *Sphaeropteris* subgenus *Sphaeropteris* by the dark apical seta on the petiole scales.

The American species have been revised by Conant (1976) and nearly all species of the paleotropics by Holttum (1963, 1964, 1965) and Tardieu-Blot (1951, 1953).

Fig. 23.2. Distribution of *Alsophila* in America (**A. Salvinii,** Mexico and Central America; **A. Engelii,** Venezuela, southern Colombia; **A. paucifolia,** Ecuador; **A. capensis,** southeastern Brazil).

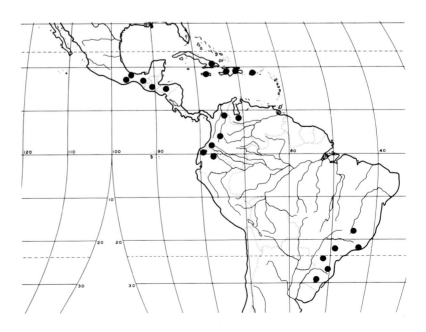

Tropical American Species

Most American species have the lamina 1-pinnate-pinnatifid (Fig. 9) to 2-pinnate. These are evidently related to species of Africa-Madagascar with similar lamina architecture. They form a compact natural group of distinctive species. The three other species with leaves having a more complex lamina, *A. capensis, A. Engelii* and *A. Salvinii* (Fig. 10), are also distinctive. Each is probably related to a different source among the Africa-Madagascar species of *Alsophila*.

There are reports of several hybrids involving *Alsophila* (Conant, 1975, 1976; Conant and Cooper-Driver, 1980). One is an intrageneric hybrid, between *A. bryophila* and *A. Aminta* (as *A. dryopteroides*), while the others are crosses with *Nephelea*. There is evidence, in one area in Puerto Rico, that hybrids involving *A. bryophila, A. Aminta* and *Nephelea portoricensis* are backcrossing with all of the parental species.

Morphological-Geographic Key to American *Alsophila*

a. Lamina 1-pinnate-pinnatifid to 2-pinnate. b.
  b. Indusium meniscoid. c.
    c. Hispaniola.   *A. hotteana* (C. Chr. & Ekman) Tryon
    c. Puerto Rico.   *A. bryophila* Tryon
  b. Indusium cyathiform. d.
    d. Colombia.   *A. rupestris* (Maxon) Tryon
    d. Jamaica.   *A. Nockii* (Jenm.) Tryon
    d. Puerto Rico, Hispaniola and (or) Cuba. e.
      e. Base of lamina gradually reduced. f.
        f. Indusium glabrous. g.
          g. Hispaniola.   *A. Abbottii* (Maxon) Tryon
          g. Puerto Rico.   *A. Aminta* Conant
                              (*A. dryopteroides* (Maxon) Tryon)
        f. Indusium pubescent; Hispaniola, Cuba.
                            A. minor (D. C. Eaton) Tryon
      e. Base of lamina abrupt. h.
        h. Pinnules 3–4 mm broad; Cuba, Hispaniola, Puerto Rico.
                            A. Brooksii (Maxon) Tryon
        h. Pinnules 5–8 mm broad; Hispaniola.
                            A. Urbanii (Brause) Tryon
  b. Indusium sphaeropteroid; Ecuador.   *A. paucifolia* Baker
a. Lamina 2-pinnate-pinnatifid to nearly 4-pinnate. i.
  i. Indusium absent; Mexico, Guatemala, Honduras.   *A. Salvinii* Hook.
  i. Indusium hemitelioid; southeastern Brazil.   *A. capensis* (L. f.) J. Sm.
  i. Indusium sphaeropteroid; Venezuela, Colombia.   *A. Engelii* Tryon

Ecology (Fig. 1)

Alsophila usually grows in cloud forests, wet montane or elfin forests, in both America and the paleotropics. It is principally a genus of the understory, or small species, such as *A. rupestris* and *A. Nockii,* may form part of the ground cover. At higher altitudes the crown of leaves may be in the low canopy, or plants may grow in more open sites.

In America, *Alsophila* grows on mountain slopes, in ravines, along streams and sometimes in rocky places. *Alsophila Brooksii,* in Puerto Rico and Hispaniola, sometimes grows on serpentine soils. *Alsophila bryophila* of Puerto Rico usually is rather low-growing, but rarely has stems up to 7 m tall. The age of large plants of this species has been analyzed and estimated to be

Figs. 23.3–23.8. *Alsophila.*   **3.** Apex of marginate petiole scale of *A. Salvinii* with dark seta, ×30.   **4.** Apex and central portion of marginate petiole scale of *A. Nockii,* the central part especially with distinctive marginal cells, × 30.   **5.** Aphlebium of *A. capensis,* × 0.5.   **6.** Exindusiate sori of *A. Salvinii,* globose receptacle with sporangia removed, bullate scales on costa, × 15.   **7.** Hemitelioid indusium and sori of *A. capensis,* elongate receptacle with sporangia removed, × 15.   **8.** Cyathiform indusia and sori of *A. Urbanii,* somewhat elongate receptacle with sporangia removed, one indusium and sorus removed, small scales on costa, × 15.

about 130 years. Most species grow from 1000 to 2000 m, but rarely occur as low as 250 m, or as high as 3000 m.

Geography (Fig. 2)

Alsophila occurs in the American tropics, tropical and southern Africa, eastward to Rapa and the Marquesas in the Pacific, north to southern Japan and south to the Auckland Islands. In America it occurs in four distinct regions, unlike the other genera which have few disjunctions in range. There are eight species, characterized by a 1-pinnate-pinnatifid lamina in the Greater Antilles, all of them endemic mostly to single islands. They represent a compact species group evolved by insular radiation. *Alsophila Salvinii* occurs from southern Mexico to Honduras. Three species, *A. rupestris, A. Engelii* and *A. paucifolia,* are restricted to the Andes from western Venezuela to Ecuador. *Alsophila capensis,* isolated in southeastern Brazil, has been distinguished as a separate subspecies (Conant, 1976) from the element in southern Africa. The species is the only one in the family that occurs in both neo- and paleotropical regions.

Spores

The neotropical species of *Alsophila* consistently have 16 spores per sporangium. The perispore is elaborated into irregular ridges as in *A. minor* (Figs. 11–13) often fewer near the laesurae (Figs. 13, 14) or somewhat fused (Fig. 12). The survey of the number and morphology of spores of both neo- and paleotropical species of *Alsophila* by Gastony and Tryon (1976) shows a greater diversity of ridged forms in the paleotropical species. A few species of *Alsophila* of Africa and Australia have spores with strands similar to those of *Sphaeropteris.* The porate exospore in *Alsophila decurrens* is similar to that characteristic of *Cnemidaria* spores.

Cytology

The chromosome number $n = 69$ has been documented for *Alsophila bryophila* and *A. Aminta* (as *A. dryopteroides*) from Puerto Rico (Conant, 1976). This is the same as reported for several species over a wide geographic range in the paleotropics in Ceylon, India, Africa, China and also in New Zealand. Meiotic figures of the hybrids *A. bryophila* × *Nephelea portoricensis* and *A. Aminta* (as *A. dryopteroides*) × *Nephelea portoricensis, with* $n = $ ca. 69, show a high degree of pairing.

Observations

An unusual morphological modification occurring especially in *Alsophila* is the skeletonized or reduced basal pinnae. These are usually near the base of the petiole and distant from the normal pinnae (Tardieu-Blot, 1941; Tindale, 1956). Aphlebia (Fig.5) represent the extreme of this development. They may form a

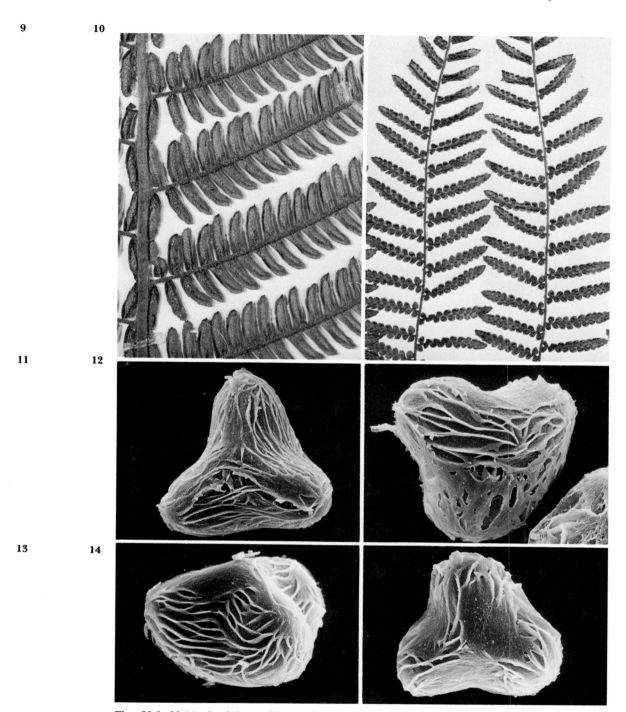

Figs. 23.9–23.14. *Alsophila.* **9.** Pinnae of *A. minor,* central portion of lamina, × 1.0. **10.** Pinnules of *A. Salvinii,* deeply lobed, × 1.0. **11–14.** Spores of *Alsophila,* × 1000. **11–13.** *A. minor,* Dominican Republic, *Türckheim 3115.* **11.** Coarse perispore ridges of young spore, proximal face. **12.** Ridges and fused perispore, distal face. **13.** Lateral view. **14.** *A. bryophila,* proximal face, Puerto Rico, *Conant 1592.*

tuft near the apex of the stem and early botanists sometimes mistook them for epiphytes. The aphlebia of *Alsophila capensis* were twice described as species of *Trichomanes* (Hymenophyllaceae).

Literature

Conant, D. S. 1975. Reference under the family.

Conant, D. S. 1976. Ecogeographic and Systematic Studies in American Cyatheaceae. Ph.D. Thesis. Harvard University, Cambridge, Mass. (Includes a revision of American *Alsophila*.)

Conant, D. S., and G. Cooper-Driver. 1980. Reference under the family.

Copeland, E. B. 1947. Genera Filicum, pp. 1–247. Waltham, Mass.

Gastony, G. J., and R. Tryon. 1976. Reference under the family.

Holttum, R. E. 1963, 1964, 1965. References under the family.

Tardieu-Blot, M. 1941. Sur les aphlebia des cyathéacées Malgaches. Bull. Soc. Bot. France 88: 522–531.

Tardieu-Blot, M. 1951. 4 fam. Cyathéacées, in Humbert, Fl. Madagas. Comores, pp. 1–45/ Firmin-Didot, Paris.

Tardieu-Blot, M. 1953. Les ptéridophytes de l'Afrique intertropicale Française. Mém. Instit. Franc. Afr. Noir 28. pp. 1–241

Tindale, M. D. 1956. The Cyatheaceae of Australia. Contrib. N. S. Wales Natl. Herb. 2: 327–361.

24. *Nephelea*
Figs. 24.1–24.17

Nephelea Tryon, Contrib. Gray Herb. 200: 37. 1970. Type: *Nephelea polystichoides* (Christ) Tryon. (*Cyathea polystichoides* Christ).
Cyathea subgenus *Nephelea* (Tryon) Proctor, Fl. Lesser Antilles 2 (Pteridophyta): 102, 1977.

Description Terrestrial; stem erect, stout, usually tall, sometimes branching by adventitious buds, with scales and spines, a dense mat of fibrous roots developed especially at the base; leaves monomorphic, 2–4 m long, borne in a spreading crown, lamina 1-pinnate-pinnatifid to 3-pinnate-pinnatifid, with marginate scales (especially on the croziers and usually persistent at the base of the petiole), these with a dark apical seta and sometimes with dark setae on the edges and (or) body, petiole with black squaminate spines, veins free; sori round, borne on the veins, usually at a fork, receptacle globose to elongate, with short paraphyses, exindusiate or usually with a hemitelioid to sphaeropteroid indusium; spores with three prominent lobes, trilete, tetrahedral-globose, the laesurae $\frac{3}{4}$ to nearly equal to the radius, often obscure, the surface strongly ridged with long branched ridges mostly parallel to the laesurae or the equator, these sometimes shorter and more or less fused. Chromosome number: $n = 69$; $2n = 138$

Fig. 24.1. *Nephelea mexicana,* in Liquidambar cloud forest north of Jalapa, Veracruz, Mexico. (Photo W. H. Hodge).

The petiole scales are similar to those of *Alsophila,* with dark apical setae and well-differentiated margins (Fig. 3). The form of the indusium varies, but it is usually cyathiform or urceolate (Fig. 12). A stem section (Fig. 8) illustrates the pattern of meristeles, sclerotic tissue and medullary bundles. The most characteristic feature of the genus is the large black spines on the stem (Fig. 9) and petiole (Fig. 10). The petiole spines have their origin from a scale rather than from outer tissues of the petiole as in the other genera. Transitional forms (Fig. 4) between scales and spines are frequently present. The epidermis often has a stomatal arrangement common in the Cyatheaceae. This is a polocytic type with the guard cells largely connected to a single subsidiary cell (Fig. 13).

Systematics

Nephelea is a tropical American genus of 18 species. It is closely related to *Alsophila* species of the Greater Antilles and perhaps to some of Madagascar that tend toward a development of squaminate spines. There are two main species groups. The first, is characterized by an acutely tapered lamina apex and petiole scales with the apical seta usually up to 0.25 mm long. This includes all species of the Greater Antilles as well as *N. Tryoniana*

Fig. 24.2. Distribution of *Nephelea.*

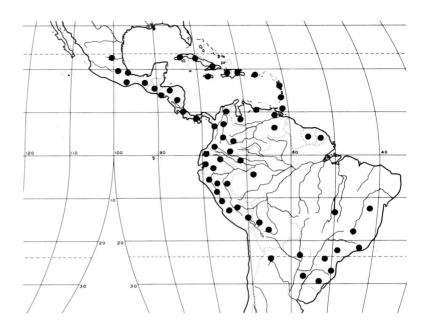

Fig. 24.3. Apex of petiole scale of *Nephelea Sternbergii* with apical seta, ×30 (left), and of *N. cuspidata* with setae at apex and edge, ×40 (right).

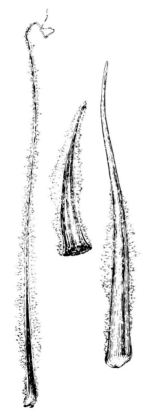

Fig. 24.4. Structures transitional between petiole scales and spines, *Nephelea erinacea,* ×8.

Gastony of Central America. The other group, characterized by the lamina being abruptly reduced to a pinna-like apex and the petiole scales with an apical seta up to 1.0 mm long (or more), includes continental species, one of which, *N. Imrayana,* extends into the Lesser Antilles.

Nephelea has been revised by Gastony (1973).

Tropical American Species

Species of the Greater Antilles include *N. balanocarpa* (D. C. Eaton) Tryon, *N. portoricensis* (Kuhn) Tryon (Fig. 5), *N. pubescens* (Kuhn) Tryon, and *N. woodwardioides* (kaulf.) Gastony. The only species in the Lesser Antilles is *Nephelea Imrayana* (Hook.) Tryon. Species of Central America, which may extend to Mexico, or to the Andes are *N. cuspidata* (Kze.) Tryon, *N. erinacea* (Karst.) Tryon (Fig. 7), *N. mexicana* (Schlect. & Cham.) Tryon, and *N. polystichoides* (Christ) Tryon, which has the most complex lamina in the genus (Fig. 6). There are two species in Brazil, *N. setosa* (Kaulf.) Tryon, which has aphlebia and resembles *Alsophila capensis* of the same area, and *N. Sternbergii* (Sternb.) Tryon.

Several hybrids have been recognized by Conant (1975, 1976). These include an intrageneric cross between *N. balanocarpa* and *N. woodwardioides,* and other hybrids with species of *Alsophila.*

Ecology (Fig. 1)

Nephelea is a genus primarily of cloud and wet montane forests where it grows on mountain slopes, in ravines and along stream banks. It rarely grows in pine woods, or in gallery forest. In the Greater Antilles, species such as *N. woodwardioides* and *N. fulgens* (C. Chr.) Gastony may grow on calcareous soils.

Most species grow between 500 and 2000 m, rarely up to 2500 m. There is an exceptional record for *N. erinacea* var. *purpurescens* (Sod.) Gastony occurring up to 2800 m on volcanos in Ecuador. Young plants of *Nephelea* have a prostrate stem,

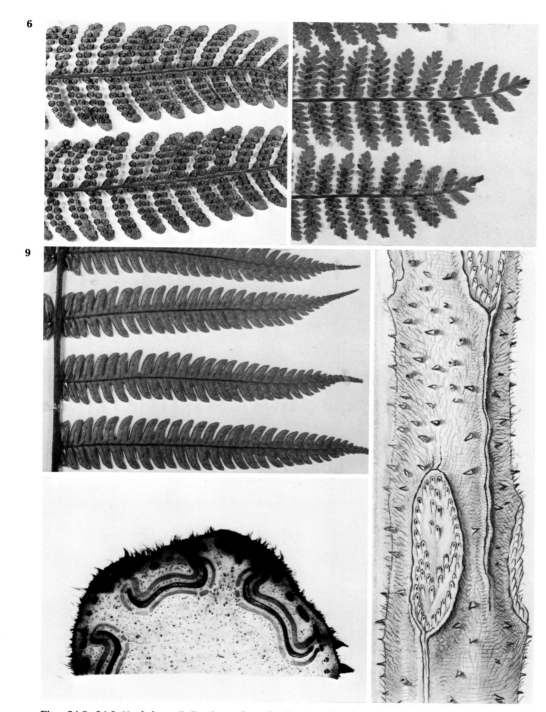

Figs. 24.5–24.9. *Nephelea*.   **5.** Portions of two fertile pinnules of *N. portoricensis*, × 1.   **6.** Apical portion of fertile pinnules of *N. polystichoides*, × 1.   **7.** Sterile central pinnules of *N. erinacea*, × 0.75.   **8.** Section of stem of *N. erinacea* with two meristeles and part of a third, and numerous small medullary bundles, × 1.   **9.** Stem of *N. erinacea* with spines and petiole scars, with protruding vascular bundles × 0.5.

Fig. 24.10. Croziers at apex of stem of *Nephelea Imrayana,* with dense scales and large spines, Morne Plat Pays, Dominica. (Photo W. H. Hodge.)

Fig. 24.11. Prostrate stems of young plants of *Nephelea mexicana* with the apices ascending, ×0.5.

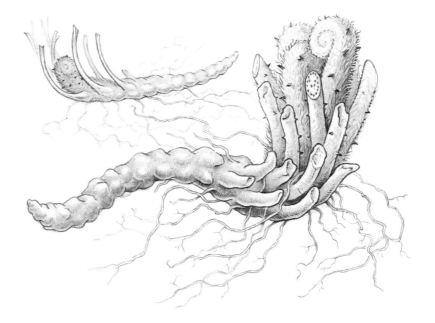

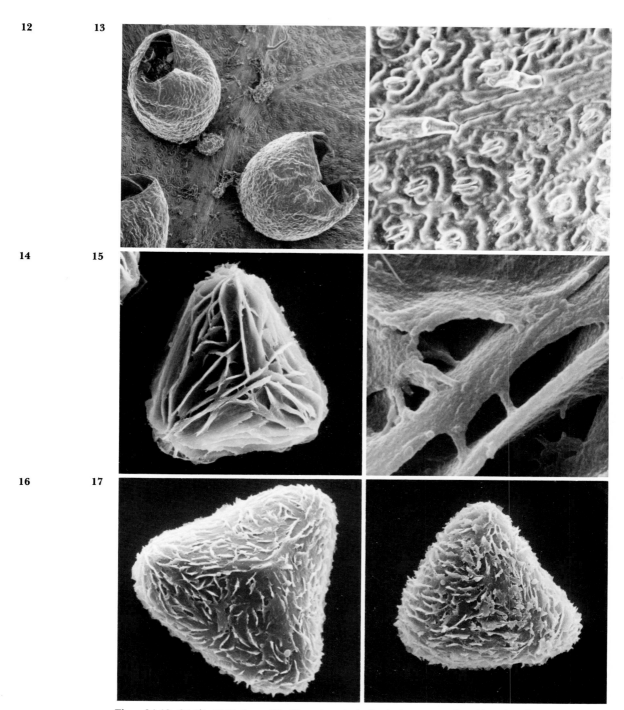

Figs. 24.12–24.17. *Nephelea*. 12. Urceolate indusia, indument and epidermis of *N. Grevilleana*, × 20. **13.** Abaxial epidermis of leaf with stomata, *N. Grevilleana* × 80. **14–17.** *Nephelea* spores, surface detail, × 10,000. **14, 15.** *N. Imrayana*, Dominica, *Lellinger 401*. **14.** Coarse ridges of proximal face. × 1000. **15.** Detail of particulate ridge surface. **16, 17.** *N. cuspidata*, Costa Rica, *Burger & Stolze 5857*. **16.** Short ridges, proximal face, × 1200. **17.** Short, compact ridges, distal face, × 1000.

slightly ascending at the apex (Fig. 11). The stems become somewhat larger in older plants and the apex becomes erect initiating the arborescent habit of the mature plant. The growth of the stem is often slow, leaving closely placed petiole scars on the stem, or it may be rapid and leave widely spaced scars (Fig. 9).

Geography (Fig. 2)

Nephelea ranges from the Greater Antilles and Hidalgo in Mexico, south to Missiones in Argentina and Rio Grande do Sul in Brazil. It is notably absent from the Amazon basin. The center of diversity and of endemism is in the Greater Antilles where nine of the 18 species occur, all but one of them endemic on single islands. Six species occur in Mexico and Central America, five in the Andes from Colombia to Bolivia and two in southeastern Brazil.

Spores

All species of *Nephelea* have sporangia consistently bearing 16 spores which are rather uniformly strongly ridged. The ridges are usually long as in *Nephelea Imrayana* (Fig. 14) but may be somewhat shorter and cristate in *N. cuspidata* (Figs. 16, 17). The perispore surface is formed of particulate material (Fig. 15) and overlays a relatively even exospore. The ridged formation in *Nephelea* has been identified as perispore by Gastony (1974) based on treatment of the spores with sodium hydroxide.

Cytology

The chromosome numbers are uniformly $n = 69$ for *Nephelea Grevilleana* (Mart.) Tryon, *N. pubescens,* and *N. Tussacii* (Desv.) Tryon from Jamaica, (Walker, 1966), and *N. mexicana* from Costa Rica (Gómez Pignataro, 1971) (all reported as *Cyathea*) and consistent with others in the Cyatheaceae.

Observations

In contrast to other genera of the Cyatheaceae, plants of *Nephelea* more commonly bear short branch buds on the stem. They may also have starch storage throughout the stem, while in other genera starch is usually stored only near the apex. These two features are probably related developments in *Nephelea,* since the growth of branch buds may depend on the proximity of a source of energy.

Literature

Conant, D. S. 1975, 1976. References under the family.
Gastony, G. J. 1973. A revision of the fern genus *Nephelea*. Contrib. Gray Herb. 203: 81–148.
Gastony, G. J. 1974. Reference under the family.
Gastony, G. J., and R. Tryon. 1976. Reference under the family.
Gómez Pignataro, L. D. 1971. Reference under the family.
Walker, T. G. 1966. Reference under the family.

25. *Trichipteris*

Figs. 25.1–25.18

Trichipteris Presl, Delic. Prag. 1: 172. 1822, sometimes but incorrectly as *Trichopteris*. Type: *Trichipteris excelsa* Presl = *Trichipteris corcovadensis* (Raddi) Copel.

Chnoophora Kaulf., Enum. fil. 250. 1824. Type: *Chnoophora Humboldtii* (Kaulf., *nom. superfl.* for *Cyathea villosa* Willd. = *Trichipteris villosa* (Willd.) Tryon.

Description Terrestrial; stem erect, stout, tall, or rarely short-decumbent, rarely branching by adventitious buds, with scales, a dense mat of fibrous roots developed especially at the base; leaves monomorphic, 0.5 m to usually 2–3 m, to 4 m long, usually borne in a spreading crown, lamina 1-pinnate with entire pinnae to usually 2-pinnate-pinnatifid, to rarely 3-pinnate-pinnatifid, with marginate scales (especially on the crozier and usually persistent at least at the base of the petiole) these lacking dark setae, the apex rounded to filiform, petiole smooth to tuberculate or with corticinate spines, veins free; sori round, borne on the veins, often at a fork, receptacle globose to elongate, with short to long (and then sometimes branched) paraphyses, exindusiate (one or sometimes more scales may be closely associated with the sorus); spores with three prominent angles, trilete, tetrahedral-globose, the laesurae $\frac{1}{2}$ to $\frac{3}{4}$ the radius, perispore usually of prominently projecting mostly interconnected strands or rods overlaying a more or less uniformly porate exospore, or a granulate deposit over a large-porate, verrucate exospore. Chromosome number: $n = 69$.

The marginate petiole scales (Fig. 3) may bear processes along the edges, as a row of dark teeth or long cilia (Fig. 4); the apex (Fig. 5) is not differentiated into a seta. The sorus is exindusiate and in most species has either short (Fig. 7) or long paraphyses (Fig. 8). Species as *T. aspera* (Fig. 6), *T. costaricensis* and *T. sagittifolia* regularly have one or more laminar scales closely associated with the sorus. These scales are clearly cellular, narrowed basally, attached at one point, partly erect, with an attenuate apex. These characters distinguish them from small indusia, as in *Cyathea,* which are somewhat thickened with obscure cellular construction, broadly attached at the base, usually appressed to the leaf tissue beneath the sorus, and rounded at the apex. Many species have the sori on simple veins (Fig. 9), while others have them borne at the fork of a vein (Fig. 10).

Systematics *Trichipteris* is a tropical American genus of about 50 species. One group of 10 species has sori borne on simple veins, or above the fork of a vein, short paraphyses, and minute, scaly indument on the petiole. Another group of 14 species has dark teeth on the edges of the petiole scales, sori borne at the fork of a vein, and a glabrous or pubescent petiole. The other species are difficult to place in well-defined groups. *Trichipteris* is closely related to *Sphaeropteris* subgenus *Sclephropteris,* from which it has evidently been derived. The principle difference between the groups is in the petiole scales which are marginate in *Trichipteris* and con-

Fig. 25.1. *Trichipteris bicrenata,* Liquidambar cloud forest, north of Jalapa, Veracruz, Mexico. (Photo Alice F. Tryon.)

Fig. 25.2. Distribution of *Trichipteris.*

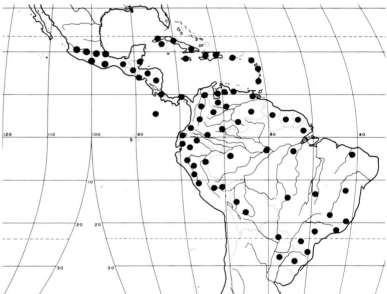

form in *Sphaeropteris.* Proctor (1977) has treated *Trichipteris* as a subgenus of an inclusive genus *Cyathea,* but without providing a valid name. He also follows Morton (1971) in spelling the name *Trichopteris.* However, the correct spelling *Trichipteris,* as originally used by Presl, is clearly established by the Code of Nomenclature revised at Leningrad in 1975.

Trichipteris has been revised by Barrington (1978), except for the *T. armata* group revised by Riba (1969).

Tropical American Species

Examples of the species group with the sori borne on simple veins, mentioned above, are *T. decomposita* (Karst.) Tryon, *T. phalaenolepis* (C.Chr.) Tryon, *T. procera* (Willd.) Tryon, and *T. sagittifolia* (Hook.) Tryon. Species included in the group with dark

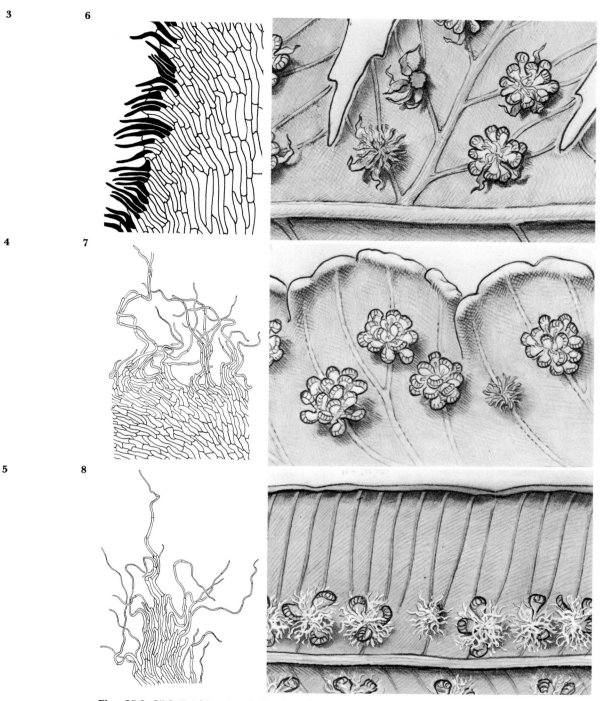

Figs. 25.3–25.8. *Trichipteris.* **3.** Portion of marginate petiole scale of *T. stipularis,* ×120. **4.** Central portion of petiole scale of *T. mexicana* with fimbriate edge, ×30. **5.** Apex of petiole scale of *T. mexicana,* ×30. **6.** Lamina scales closely associated with sori, sporangia removed showing paraphyses and scales (left, below), the sorus removed showing scales (left, above), *T. aspera,* ×15. **7.** Sori, one with sporangia removed showing paraphyses, *T. demissa,* ×15. **8.** Sori with paraphyses and sporangia, one with sporangia removed, *T. corcovadensis,* ×15.

9 **11**

10 **12**

Figs. 25.9–25.12. Trichipteris. 9. Central portion of a pinnule with free veins, *T. procera,* × 3. **10.** Pinnule segments with forked veins, *T. Wendlandii,* × 3. **11.** Pinnules of central portion of pinnae with gradually attenuate apices, *T. stipularis,* × 1.0. **12.** Pinnae of central portion of lamina with acute apices, *T. pubescens,* × 0.75.

teeth on the petiole scales are *T. armata* (Sw.) Tryon, *T. bicrenata* (Liebm.) Tryon, *T. conjugata* (Hook.) Tryon, *T. mexicana* (Mart.) Tryon, and *T. stipularis* (Christ) Tryon (Fig. 11).

Among the numerous other species are: *T. aspera* (L.) Tryon, *T. corcovadensis* (Raddi) Copel., *T. costaricensis* (Kuhn) Barr., *T. demissa* (Morton) Tryon, *T. microdonta* (Desv.) Tryon, *T. pubescens* (Baker) Tryon, a 1-pinnate-pinnatifid species (Fig. 12), and *T. Wendlandii* (Kuhn) Tryon.

Several hybrids involving *Trichipteris* species are proposed by Tryon (1976), most with *Cyathea* and one with *Cnemidaria grandifolia.*

Ecology (Fig. 1)

Trichipteris is ecologically the most diverse genus of the family in America. Species usually grow in wet montane or in cloud forests, but also occur in low elevation rain forests and sometimes in savannahs, shrubby sites in grasslands or swamp forests. In the Amazon basin of Peru, *T. microdonta* and *T. procera* invade aban-

13 14

15 16

17 18

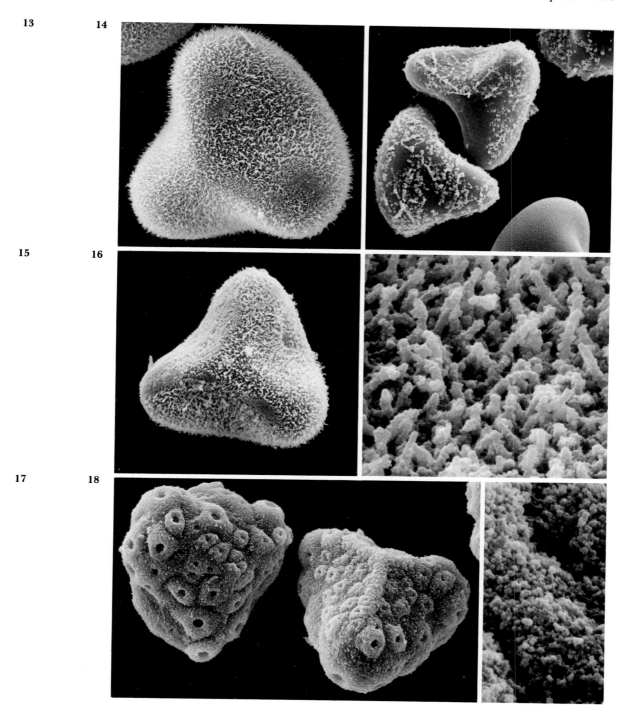

Figs. 25.13–25.18. *Trichipteris* spores, × 1000, surface detail, × 10,000. **13.** *T. aspera,* proximal face with compact perispore strands, Dominica, *Wilbur et al. 8314.* **14.** *T. costaricensis,* young spores with initial perispore deposit, proximal face above, the distal below, part of distal face of spore with perforate exospore (right, below), Guatemala, *Bequaert 28.* **15, 16.** *T. corcorvadensis,* Brazil, *Leite 3575.* **15.** Proximal face, the perispore of more or less connected strands. **16.** Detail of particulate surface of strands. **17, 18.** *T. mexicana,* Guatemala, *Türckheim 1007.* **17.** Spores with verrucate, foraminate exospore, and granulate perispore surface, proximal face, right, distal face, left. **18.** Detail of granulate perispore deposit.

doned fields and other secondary vegetation. *Trichipteris micro-donta* may also grow in swamps, or rarely in brackish water. *Trichipteris costaricensis* may grow in arroyas in oak forests (Mexico) and *T. villosa* (Willd.) Tryon in open grasslands (Venezuela). Near the crest of the Serra do Mar in São Paulo, Brazil, *T. atrovirens* (Langsd. & Fisch.) Tryon has been observed in burned areas, surviving fires that destroy most of the ground cover. The species also survives depredation by man in removing leaves for commercial, decorative purposes.

The genus most commonly grows up to 2000 m. Some species grow as high as 3000–3500 m, as *T. mexicana, T. pauciflora* (Kuhn) Tryon, and *T. frigida* (Karst.) Tryon.

Geography (Fig. 2)

Trichipteris is distributed from the Greater Antilles and Veracruz west to Nayarit in Mexico, south to Corrientes in Argentina, and Rio Grande do Sul in Brazil. The genus has a broader distribution than the others in America primarily due to wide-ranging species as *T. microdonta, T. procera* and *T. nigra* (Mart.) Tryon in the Amazon basin. *Trichipteris nesiotica* (Maxon) Tryon is an endemic of Cocos Island.

The center of species diversity is in the Andes from Colombia to Peru, where 20 species, including many endemics, are found. Secondary centers are in Mexico and Central America with 14 species and southeastern Brazil with 12 species.

Spores

Spores of *Trichipteris* are of two main forms based on the type of exospore and perispore. Most species have a perispore of more or less connected strands (Figs. 13, 15) of particulate material (Fig. 16), as in *T. aspera* and *T. corcovadensis*. The perispore overlays an even, perforate exospore as in *T. costaricensis* (Fig. 14) or a strongly verrucate to rugose formation. A second type of spore is prominently verrucate, with foraminate pits as in *T. mexicana* (Fig. 17), and with a granulate deposit rather than strands forming the perispore (Fig. 18). The spore morphology of nearly all species of *Trichipteris* was surveyed and profusely illustrated by Gastony (1979). Details of the exospore and perispore formation noted in that review reinforce the close relationship between *Trichipteris* and *Sphaeropteris* subgenus *Sclephropteris*.

Cytology

Our record of *Trichipteris microdonta,* from Veracruz, Mexico, included under the family treatment (Fig. Cyath. 15), is a meiotic figure with $n = 69$. This number is also listed by Walker (1966) for *T. armata* and *T. microdonta* (as *Cyathea*) from Jamaica, for *T. borinquena* (Maxon) Tryon (as *Alsophila*) from Puerto Rico by Sorsa (1970) and for *T. stipularis* (as *Alsophila*) from Costa Rica by Gómez (1971).

Literature

Barrington, D. S. 1978. A revision of *Trichipteris* (Cyatheaceae). Contrib. Gray Herb. 208: 3–93.

Gastony, G. J. 1979. Spore morphology in the Cyatheaceae, III. The genus *Trichipteris*. Amer. Jour. Bot. 66: 1238–1260.

Gómez Pignataro, L. D. 1971. Reference under the family.

Morton, C. V. 1971. Review of R. Tryon, "Classification of the Cyatheaceae." Amer. Fern Jour. 61: 142–143.

Proctor, G. R. 1977. Pteridophyta, *in*: R. A. Howard, Flora of the Lesser Antilles, 2: 103. Arnold Arboretum, Harvard Univ., Cambridge, Mass.

Riba, R. 1969. Revisión monográfica del complejo *Alsophila Swartziana* Martius (Cyatheaceae). Am. Instit. Biol. Univ. Mex. 38, ser. Bot. 1: 61–100 (1967).

Sorsa, V. 1970. Reference under the family.

Tryon, R. 1976. A revision of the genus *Cyathea*. Contrib. Gray Herb. 206: 19–98.

Walker, T. G. 1966. Reference under the family.

26. *Cyathea*
Figs. 26.1–26.20

Cyathea Sm., Mém. Acad. Turin, 5: 416. 1793. Type: *Cyathea arborea* (L.) Sm. (*Polypodium arboreum* L.).

Hemitelia R. Br., Prod. Fl. Nov. Holl. 158. 1810. Type: *Hemitelia multiflora* (Sm.) Spreng. = *Cyathea multiflora* Sm. [Brown did not make any combinations for the names of the species of his new genus].

Disphenia Presl, Tent. Pterid. 55. 1836, *nom. superfl.* for *Cyathea* and with the same type. [Of the species originally included in *Cyathea*, all but *C. arborea* had been removed to other genera prior to Presl's publication: two species of the original six to *Cystopteris* and three species to *Hemitelia*].

Cormophyllum Newm., Phytol. 5: 237. 1856, *nom superfl.* for *Cyathea* and with the same type.

Description Terrestrial; stem erect, stout, tall, or rarely very short, rarely branching by adventitious buds, with scales, with a dense mat of fibrous roots especially at the base; leaves monomorphic to somewhat dimorphic, rarely 0.25 to usually 2–4, to 6 m long, borne in a spreading crown, lamina 1-pinnate, with entire pinnae, to usually 2-pinnate-pinnatifid, to 4-pinnate, with marginate scales (especially on the croziers and often persistent at the base of the petiole), these lacking dark setae, the apex rounded to filiform, petiole smooth to tuberculate, or with corticinate spines, veins free or rarely anastomosing without included free veinlets to form some costal areolae; sori round, borne on the veins, often at a fork, receptacle globose to elongate, with short to moderately long paraphyses, indusium hemitelioid (minute or larger) to sphaeropteroid; spores with three prominent angles, tetrahedral-globose, trilete, the laesurae usually $\frac{3}{4}$ the radius, the surface relatively even or verrucate, porate and usually covered with a deposit of compact, more or less branched strands. Chromosome number: $n = 69$.

The marginate petiole scale of *Cyathea* (Fig. 3) has differentiated cells extending to the apical portion (Fig. 4). The indusium varies from hemitelioid (Fig. 6) to cyathiform (Fig. 7) to

Fig. 26.1. *Cyathea arborea*, between Roseau and Buena Vista, Dominica. (Photo W. H. Hodge.)

Fig. 26.2. Distribution of *Cyathea*.

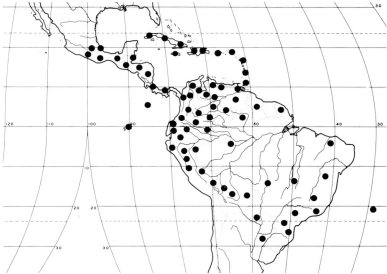

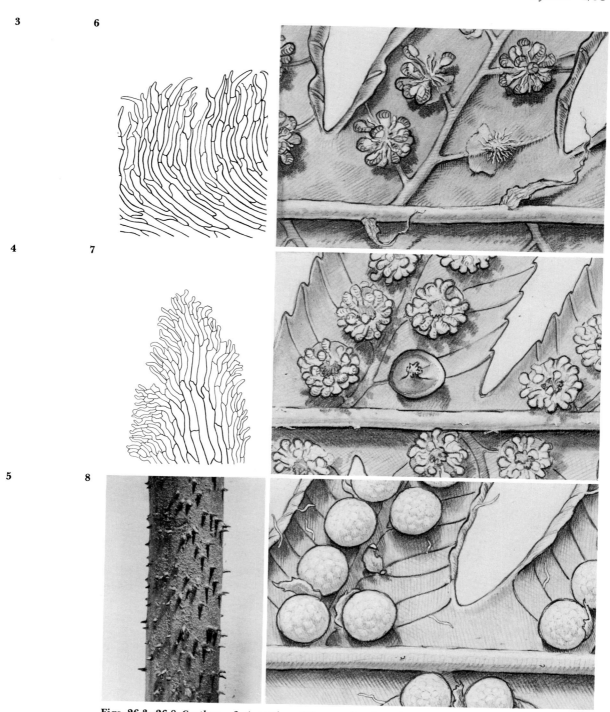

Figs. 26.3–26.8. *Cyathea*. **3.** A portion of the margin of a petiole scale of *C. platylepis,* ×90. **4.** Apex of petiole scale of *C. parvula,* ×90. **5.** Corticinate spines on petiole of *C. Delgadii,* ×1.5. **6.** Sori of *C. multiflora,* one with sporangia removed to show the hemitelioid indusium and paraphyses, ×15. **7.** Sori of *C. arborea,* one with the sporangia removed to show the cyathiform indusium and central receptacle, ×15. **8.** Sphaeropteroid indusia of *C. divergens* enclosing sori, one sorus removed to show the subtending scale, ×15.

9 11

10 12

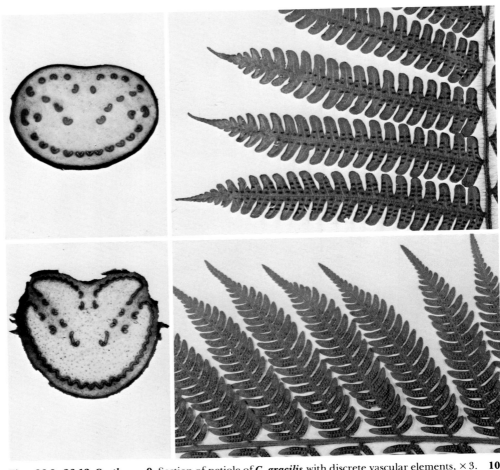

Figs. 26.9–26.12. *Cyathea.* **9.** Section of petiole of *C. gracilis* with discrete vascular elements, × 3. **10.** Section of petiole of *C. fulva* with peripheral vascular elements fused, × 2. **11.** Fertile pinnae of *C. decorata* with coarse, blunt lobes, × 0.75. **12.** Fertile pinnules of *C. Delgadii* with gradually attenuate apices, × 1.0.

sphaeropteroid (Fig. 8). If the petiole has spines, as *C. Delgadii,* they are of a corticinate type (Fig. 5). Petiole sections (Figs. 9, 10) show some of the variation in the arcs formed by the vascular strands.

Systematics

The name *Cyathea* has been variously applied to genera of widely different scope and definition, sometimes including nearly all members of the family. It is restricted here to a genus of 40 species of the American tropics. Species of *Cyathea* with small indusia are most closely allied to *Trichipteris.* Except for these, the two genera appear to represent distinct evolutionary groups. There are ten species groups in *Cyathea* but relationships among them are not established with certainty.

The genus has been revised by Tryon (1976).

Tropical American Species

There are 16 species with a hemitelioid indusium, including *Cyathea conformis* (Tryon) Stolze, *C. decorata* (Maxon) Tryon, (Fig. 11), *C. multiflora* Sm., *C. parvula* (Jenm.) Domin, *C. petiolata*

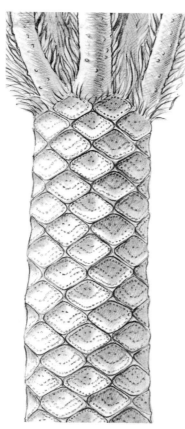

Fig. 26.13. Upper portion of ***Cyathea caracasana*** stem with compact petiole scars, × 0.25.

(Hook.) Tryon, *C. platylepis* (Hook.) Domin, and *C. speciosa* Willd. *Cyathea Haughtii* (Maxon) Tryon of Colombia, with leaves ca. 25 cm long and a 1-pinnate lamina, is one of the smallest species in the family. Only three species have a cyatheoid indusium, *C. Alstonii* Tryon and *C. peladensis* (Hieron.) Domin of Colombia, and the well-known *C. arborea* (L.) Sm. of the West Indies. Among the 21 species with a sphaeropteroid indusium, typical ones with the usual 2-pinnate-pinnatifid lamina are *C. caracasana* (Kl.) Domin, *C. Delgadii* Sternb. (Fig. 12), *C. divergens* Kze., *C. fulva* (Mart. & Gal.) Fée, and *C. gracilis* Griesb. There are difficult taxonomic problems in this group, which contains several critical and highly variable species.

Several hybrids have been proposed by Tryon (1976), one between *Cyathea divergens* and *C. fulva,* and the others with species of *Trichipteris* and *Cnemidaria.* The hybrid status of one of the latter, *Cnemidaria horrida* × *Cyathea arborea,* formerly recognized as *Cyathea Wilsonii* (Hook.) Domin, has been clarified by the field studies of Conant (1975) in Puerto Rico.

Ecology (Fig. 1)

Cyathea is primarily a genus of montane forests and cloud forests, although it may occur in low rain forests, especially in Central America. It grows on steep slopes, in ravines, along streams, and less often in wet savannahs (*C. parvula*) or in wet pinelands (*C. Harrisii* Maxon) in the Greater Antilles; in gallery forests (*C. Delgadii*) in Brazil, in rocky places (*C. platylepis*) in the Guayana region, or in Chusquea thickets or subpáramo scrub (*C. caracasana* var. *boliviensis* (Rosenst.) Tryon and *C. fulva*) in the Andes.

Species mostly are tall and their crowns form part of the irregular canopy of cloud or montane forest; small ones such as *C. decorata* and *C. parva* (Maxon) Tryon are understory species. A few, such as *Cyathea arborea,* are pioneer species in landslide areas, road cuts, and slippage sites along stream banks. Species are usually rather slow growing as *C. caracasana* with compact petiole scars on the stem (Fig. 13), but others, such as *C. arborea,* grow at a rate of ca. 30 cm a year.

Most species occur at 1500–2000 m. Some, especially those with hemitelioid indusia, grow below 500 m, and others particularly those with sphaeropteroid indusia grow to 3000 m in the Andes. Plants of *C. fulva* and *C. caracasana* rarely grow up to 4200 m, the highest elevation known for the family.

Geography (Fig. 2)

Cyathea ranges from the Greater Antilles and southern Mexico, south to Corrientes in Argentina. It also occurs on Cocos Island, the Galápagos Islands and on Ilha Trindade.

The diversity in the genus is concentrated in an area extending from Venezuela to Peru which includes 28 species. This is also a center of endemism, with 18 species confined to the region and 15 of these are restricted to one country. The genus is represented on Cocos Island by two endemics, *C. Alphonsiana* Gómez and C. *notabilis* Domin, and on the Galápagos Islands by

Fig. 26.14. Positively geotrophic branches on stem of ***Cyathea parvula,*** petioles and part of lamina erect (left), the stem surface covered with roots and bases of several erect petioles, × 0.5.

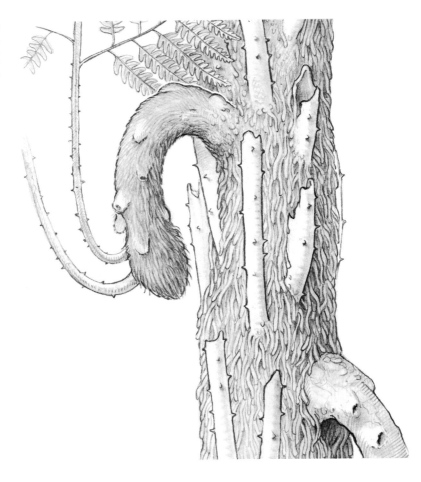

the endemic *C. Weatherbyana* (Morton) Morton. The widely distributed *C. Delgadii* occurs on Ihla Trindade.

Spores

Cyathea spores usually are verrucate with a well developed perispore of more or less connected strands (Figs. 15, 18) of particulate material (Fig. 16), as in *C. Delgadii* and *C. pallescens* (Sod.) Domin. The exospore surface may be perforate and even as in *C. speciosa* (Fig. 20) rather than verrucate as in *C. caracasana* (Fig. 17). The perispore is often sparse or may be lacking in spores with perforate exospore as in *C. speciosa* (Fig. 20) and *C. arborea.* A few species have spores with porate, somewhat verrucate exospore and sparse perispore as in *C. Weatherbyana* (Fig. 19).

Figs. 26.15–26.20. *Cyathea* spores, × 1000. **15, 16.** *C. Delgadii,* Brazil, *Tryon & Tryon 6658.* **15.** Perispore strands covering verrucate exospore, proximal face, below, part of distal above. **16.** Detail of particulate structure of strands, × 10,000. **17.** *C. caracasana* var. **boliviensis,** verrucate exospore, proximal face (center), part of distal face with perispore strands (right), Venezuela, *Lindig 242.* **18.** *C. pallescens,* detail of connected strands, Colombia, *Madison 861,* × 5000. **19.** *C. Weatherbyana,* porate, somewhat verrucate exospore with sparse perispore strands, proximal face, left, distal face, right, above, Chatham Island, Ecuador, *Stewart 895.* **20.** *C. speciosa,* perforate exospore, proximal face, below, distal face, above, Venezuela, *Steyermark 92132.*

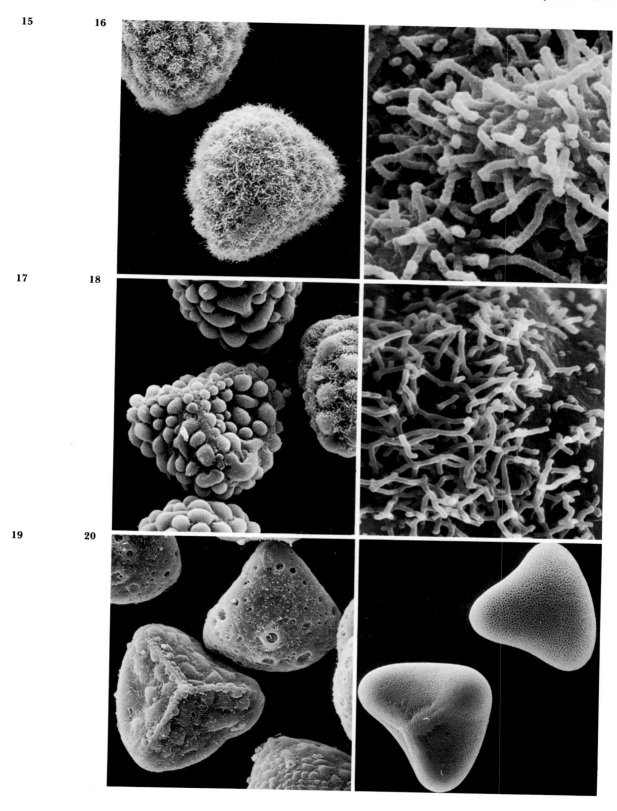

Cyathea spores are similar to those of several other genera. Spores with verrucate or even exospore surface and prominent perispore strands also occur in *Sphaeropteris* and *Trichipteris,* and some species of *Alsophila* have similar porate spores.

Cytology

The chromosome number is consistently $n = 69$ for *Cyathea arborea, C. furfuracea* Bak., *C. Harrisii,* and *C. parvula,* from Jamaica (Walker, 1966), and *C. multiflora* from Costa Rica (Gómez Pignataro, 1971).

Observations

In Puerto Rico, *Cyathea parvula* sometimes bears unusual positively geotrophic branches (Fig. 14). These arise at irregular intervals from near the base of the stem to at least 1 m up the trunk. These branches are rigid and grow toward the ground bearing a tuft of small leaves that are nearly upright. The branches root when they reach the ground. A similar type of branching is reported for *Alsophila Manniana* (Hook.) Tryon (as *Cyathea*) of Africa by Hallé (1966).

Literature

Conant, D. S. 1975. Reference under the family.
Gómez Pignataro, L. D. 1971. Reference under the family.
Hallé, F. 1966. Etude de la ramification du tronc chez quelques fougères arborescentes. Adansonia 6: 405–424.
Tryon, R. 1976. A revision of the genus *Cyathea.* Contrib. Gray Herb. 206: 19–98.
Walker, T. G. 1966. Reference under the family.

27. *Cnemidaria*
Figs. 27.1–27.10

Cnemidaria Presl, Tent. Pterid. 56. 1836. Type: *Cnemidaria speciosa* Presl.

Hemitelia subgenus *Cnemidaria* (Presl) C. Chr., Ind. Fil. xvii. 1906.

Cnemidopteris Reichenb., Deutsche Botaniker 1 (Repert, Herb. Nomencl. Gen. Pl.), Abteil. 2: 148, 235. 1841. [An illegitimate correction of the name *Cnemidaria* Presl].

Microstegnus Presl, Gefässb. Stipes Farrn, 45. 1847 (preprint from Abhandl. Böhm. Ges. V, 5: 353. 1848). Type: *Microstegnus grandifolius* (Willd.) Presl (*Cyathea grandifolia* Willd.) = *Cnemidaria grandifolia* (Willd.) Proctor.

Hemistegia Presl, *ibidem* 46. 1847. Lectotype: *Hemistegia Kohautiana* (Presl) Presl (*Cnemidaria Kohautiana* Presl) = *Cnemidaria grandifolia* (Willd.) Proctor.

Actinophlebia Presl, *ibidem* 47. 1847 Type: *Actinophlebia horrida* (L.) Presl (*Polypodium horridum* L.) = *Cnemidaria horrida* (L.) Presl.

Description   Terrestrial; stem erect, short to moderately tall, or often short-decumbent, sometimes branched, with scales, and usually many fibrous roots; leaves monomorphic, 1–3.5 m long, borne in a spreading crown or cluster, lamina 1-pinnate with entire pinnae,

Fig. 27.1. *Cnemidaria horrida,* growing with one taller plant of *Cyathea arborea* (upper center), Luquillo Mountains, Puerto Rico. (Photo L. Wells.)

to 1-pinnate-pinnatisect, with marginate scales (especially on the crozier and sometimes persistent at the base of the petiole), these lacking dark setae, the apex rounded to filamentous, petiole smooth to tuberculate or with corticinate spines, veins free or regularly anastomosing without included free veinlets to form costal areolae; sori round, borne on the veins, rarely at a fork, receptacle subglobose, with very short paraphyses, indusium hemitelioid to rarely cyatheoid; spores with three prominent angles, tetrahedral-globose, trilete, the laesurae usually $\frac{3}{4}$ of the radius, the surface relatively even, always with three large equatorial pores equidistant from the laesurae, and smaller pores irregularly distributed. Chromosome number: $n = 69$.

Petiole scales of *Cnemidaria* are marginate (Fig. 3) and the indusium usually hemitelioid (Fig. 4). Although the petioles are smaller than in species of other genera with larger leaves, they have similar vascular patterns (Fig. 5). Variations in pinna architecture, venation, and soral arrangement are shown in Figs. 6–8. The genus is characterized by reduced laminar architecture, sporoderm with large pores, and the absence of trichomes on the adaxial side of the costae. The free veined species of *Cnemidaria,* with one exception, differ from those of other genera in having the basal veins of adjacent segments connivent to the sinus. The exception, *Cn. amabilis,* has the basal veins extending to the margin above the sinus, as in other genera.

Systematics

Cnemidaria is a tropical American genus of 25 species. Its closest relations are with *Cyathea,* especially with a group of species having hemitelioid indusia and a 2-pinnate lamina, as *Cyathea petiolata* and *C. conformis* rather than to species of *Cyathea* with a 1-pinnate lamina, such as *C. Haughtii* and *C. speciosa.* These 1-pinnate species probably represent a separate line of lamina reduction within *Cyathea.*

The genus has been revised by Stolze (1974).

Fig. 27.2. Distribution of *Cnemidaria*.

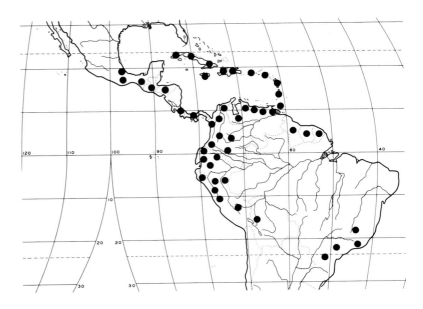

Tropical American Species

Most of the species are quite distinctive. The widely distributed *Cn. horrida* shows little variation throughout its range. In contrast, *Cnemdaria mutica* of Costa Rica and Panama has four major sympatric varieties. Species that are consistently free veined include *Cn. amabilis* (Morton) Tryon, *Cn. apiculata* (Hook.) Stolze, *Cn. mutica* (Christ) Tryon, and *Cn. Tryoniana* Stolze. *Cnemidaria Uleana* (Samp.) Tryon has both free veins and costal areolae. Species that have regular costal areolae include *Cn. grandifolia* (Willd.) Proctor, *Cn. horrida* (L.) Presl, (Fig. 8) *Cn. nervosa* (Maxon) Tryon (Fig. 7), *Cn. quitensis* (Domin) Tryon, *Cn. speciosa* Presl (Fig. 6), and *Cn. spectabilis* (Kze.) Tryon.

Tryon (1976) lists four hybrids of *Cnemidaria*, three with species of *Cyathea* and one with *Trichipteris*. The hybrid *Cnemidaria horrida* × *Cyathea arborea* was studied in detail by Conant (1975) in Puerto Rico.

Ecology (Fig. 1)

Cnemidaria is a genus of wet, usually montane, forests, growing in deep shade on forested slopes, on stream banks or near waterfalls. Some species may grow in more open habitats such as road banks or in shrubby places. Most species are small and occur as ground cover, although *Cnemidaria horrida* may have a stem to 4 m tall and be an understory species. *Cnemidaria Uleana* may occur among rocks, an unusual substrate for these plants. *Cnemidaria* grows most commonly up to 1500 m, rarely up to 2200 m.

Geography (Fig. 2)

Cnemidaria is distributed from the Greater Antilles and southern Mexico, south to La Paz in Bolivia, and is disjunct in southeastern Brazil. The distribution of most species is limited. However,

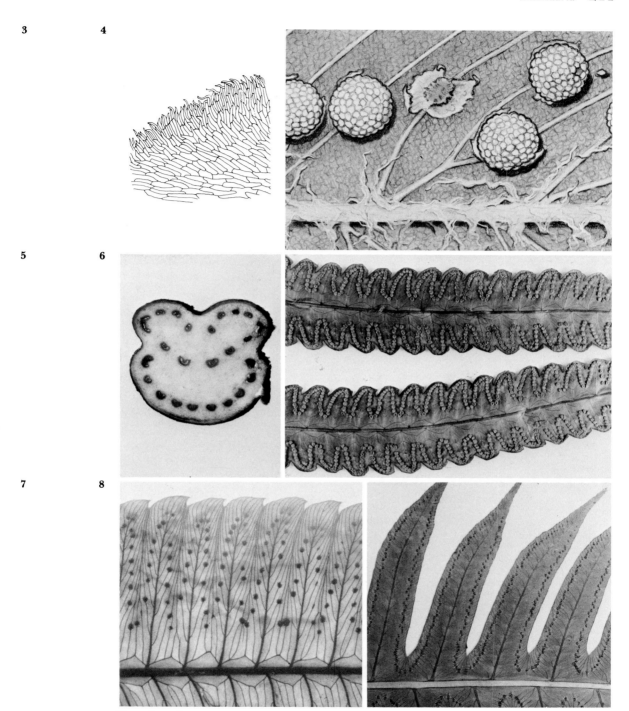

Figs. 27.3–27.8. *Cnemidaria.* **3.** A portion of marginate petiole scale of *Cn. spectabilis* with fimbriate edge, × 30. **4.** Sori and hemitelioid indusium of *Cn. quitensis,* one with compact sporangia removed showing indusium and central receptacle, scales on costa, × 15. **5.** Section of petiole of *Cn. mutica,* the vascular bundles arranged in arcs, × 3. **6.** Portion of pinnae of *Cn. speciosa,* the sori symmetrically aligned, × 1.0. **7.** Portion of pinna of *Cn. nervosa,* with large areolae adjacent to costa, the sori somewhat irregularly aligned, × 1.0. **8.** Portion of pinna of *Cn. horrida,* the sori aligned parallel to the margins, × 1.0.

9

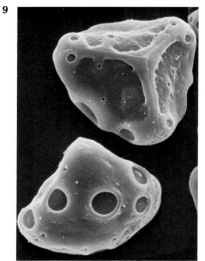

10

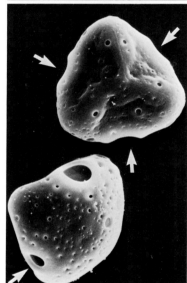

Cn. horrida has a considerable range from the Greater Antilles and Costa Rica to Venezuela and Peru. *Cnemidaria Uleana* has a disjunct range with var. *Uleana* in southeastern Brazil and Peru, and var. *abitaguensis* (Domin) Stolze in Colombia. *Cnemidaria spectabilis* is also disjunct with var. *spectabilis* occurring from French Guiana to central northern Venezuela, and var. *colombiensis* Stolze in east-central Colombia. There appears to be no primary center of diversity or endemism for the genus.

Spores

The perispore is absent in *Cnemidaria* and the spores usually are characterized by three pores in the equatorial area, equidistant between the laesurae and often with smaller pores (Figs. 9, 10). Spores of other genera of the Cyatheaceae have a similar porate exospore as in *Alsophila decurrens* Hook. and *Cyathea Weatherbyana* but in these the perispore usually overlays the exospore surface.

Cytology

The few records of chromosome numbers for *Cnemidaria* are consistently $n = 69$. *Cnemidaria horrida* has been cytologically examined from two localities in Jamaica by Walker (1966) as well as in Puerto Rico by Sorsa (1970), and *Cn. choricarpa* (Maxon) Tryon (as *Cyathea*) is reported from Costa Rica by Gómez Pignataro (1971).

Literature

Conant, D. S. 1975. Reference under the family.
Gómez Pignataro, L. D. 1971. Reference under the family.
Sorsa, V. 1970. Reference under the family.
Stolze, R. G. 1974. A taxonomic revision of the genus *Cnemidaria* (Cyatheaceae). Fieldiana (Bot.) 37: 1–98.
Tryon, R. 1976. Reference under *Cyathea*.
Walker, T. G. 1966. Reference under the family.

Figs. 27.9, 27.10. *Cnemidaria mutica* spores, Costa Rica, × 1000. **9.** Porate exospore, proximal face, right, distal face, left. *Burger & Stolze 5685.* **10.** Spores with three large equatorial pores (at arrows), proximal face, right, lateral view, left, *Lellinger & White 1494* (US).

Family 14. Pteridaceae

Pteridaceae Reichenb., Hand. Nat. Pflanz. 138, 1837, as Pteroideae.
Type: *Pteris* L.

Family synonyms of the American Pteridaceae are placed under the tribes, where they correspond more closely to the taxonomy.

Description

Stem erect, decumbent to long-creeping, very small and poorly developed (in *Anogramma*) to massive, with a siphonostele or dictyostele, usually indurated or rarely (in *Pteris*) fleshy, or (in *Cryptogramma*) succulent-brittle, bearing trichomes or scales or both; leaves ca. 3 cm to 4 m long, entire, radiate, pedate, palmate, helicoid, or usually pinnate, circinate in the bud or less often not circinate or only partly so, petiole without stipules; sporangia borne in sori on the abaxial surface of the segments, along the veins or at the margin, or on a usually marginal vascular commissure, or the sporangia distant or contiguous along anastomosing veins, sometimes also between them, exindusiate or the recurved margin modified as a marginal indusium; sporangia short- to usually long-stalked, the stalk 2- to 3-rowed or (in *Cryptogramma*) to 5-rowed below its apex, the annulus vertical or rarely oblique, interrupted by the stalk; homosporous, spores lacking chlorophyll. Gametophyte epigeal, with chlorophyll, obcordate to reniform, sometimes asymmetrical, not or slightly to strongly thickened centrally, with thin, sometimes raised margins, glabrous or (in *Notholaena*) sometimes with glandular trichomes, archegonia borne on the lower surface, usually on the central cushion, antheridia 3-celled, borne on the lower surface, mostly apart from the archegonia, or (in *Ceratopteris*) sometimes at or near the margins.

Comments on the Family

The Pteridaceae are a large and diverse family of nearly worldwide distribution with some 35 genera, 22 of them American. It is characterized by sporangia borne abaxially on unmodified veins or borne marginally and then often covered by a marginal indusium, by trilete spores, and by a chromosome number of $n = 29$ or 30, or multiples of these. There are a few departures from these characters such as an acrostichoid arrangement of sporangia in some genera, and chromosome numbers of $n = 38$ in *Platyzoma,* 44 or 110 in *Taenitis,* and 39 or 40 in *Ceratopteris.*

The family is scarcely known from the fossil record, being represented with certainty only in the Eocene by *Acrostichum preaureum* Arnold & Daugherty. Jurassic fossils referred to *Adiantites* Goepp are of uncertain affinity. Pteridaceae Reichenb. is a superfluous name, but has been consistently used for a segregate group including *Pteris* and is adopted here.

The family is represented by six tribes which probably are independent evolutionary lines or relics from sources among now extinct relatives of the Schizaeaceae. While some of the tribes, especially Ceratopterideae, and Platyzomateae are very distinctive, their affinities are with this major evolutionary line of advanced leptosporangiate ferns. The tribes may be characterized as follows.

The tribe **Taenitideae** has stems with trichomes (in most genera), sori along usually free veins, sometimes paraphysate, fertile segment without the margin modified as an indusium, and spores usually with one or more equatorial flanges and other prominent ridges or tubercles. An unusual element in this tribe is the genus *Jamesonia* which represents a specialized evolutionary development of alpine ferns.

The tribe **Platyzomateae** (Nakai) R. & A. Tryon Rhodora 83: 135. 1981. has trimorphic leaves and is nearly heterosporous with large and small spores and dioecious gametophytes (A. Tryon, 1964). It is represented by the monotypic genus *Platyzoma* R. Br. of Queensland.

The tribe **Cheilantheae** has stems with scales rarely intermixed with trichomes, sori mostly marginal, either at the apex of veins or on a marginal vascular commissure, sometimes extending along much of the vein, not paraphysate, sporangia rarely borne on anastomosing veins, and spores diversely ornamented, without an equatorial flange. The Cheilantheae are notable as being the largest group of xeric and semixeric ferns.

The tribe **Ceratopterideae,** represented by the single genus *Ceratopteris,* is often placed in its own family, but many of its unusual features undoubtedly relate to its adaptation to the aquatic habit. The sporangia are distant on the veins, often with few or no indurated annulus cells, and the spores are exceptionally large and strongly ridged.

The tribe **Adianteae,** represented by the single genus *Adiantum,* is a large and distinctive group. It is unique in having the fertile veins enter the indusium where they bear the sori. The genus is notable for its diversity of leaf architecture.

The tribe **Pterideae** is a small tribe of seven genera, with one large genus *Pteris* and other small ones. The paraphysate sori are on a marginal commissure, or sporangia are on anastomosing veins and between them. In *Pteris* there is unusual diversity in leaf architecture and some species have exceptionally large leaves to 6 m long.

Key to American Genera of Pteridaceae

a. Margin of the fertile segments modified as an indusium.   b.
  b. Sporangia distant on the veins of the 1-pinnate or more complex fertile lamina.   44. *Ceratopteris,* p. 312
  b. Sporangia contiguous in compact groups, or sometimes 1 to few on a vein tip, or scattered and the lamina entire.   c.
    c. Sori paraphysate.   d.
      d. Veins anastomosing, or if free, then the ultimate segments joined, adnate or rounded at the base, or if rarely cuneate, the sterile veins running to, or nearly to, the margin, spores usually with a prominent equatorial flange.   46. *Pteris,* p. 332
      d. Veins free, ultimate segments broadly to very narrowly cuneate, sterile veins ending well back of the margin, spores without an equatorial flange.   47. *Anopteris,* p. 341
    c. Sori not paraphysate.   e.
      e. Leaves strongly dimorphic.   43. *Llavea,* p. 309
      e. Leaves monomorphic to somewhat dimorphic.   f.
        f. Stem succulent-brittle, greenish-yellow.
                                   42. *Cryptogramma,* p. 306
        f. Stem indurated, dark in color.   g.

g. Veins extending into the indusium where they bear the sori. 45. *Adiantum*, p. 319
g. Indusium without veins. h.
  h. Lamina pinnate, imparipinnate, most or all of the ultimate segments sessile to stalked. 39. *Pellaea*, p. 284
  h. Lamina entire, pedate, palmate, or radiate, or usually pinnate and the lamina 1-pinnate or more complex and the pinnae gradually to abruptly reduced at the apex, ultimate segments often joined or adnate. i.
    i. Lamina entire, 3-lobed, or palmate, or usually pedate and the lamina not farinaceous and the petiole terete or rarely flattened. 40. *Doryopteris*, p. 293
    i. Lamina pinnate or radiate, or if pedate then the lamina farinaceous or the petiole ridged. j.
      j. Lamina axes ridged, many to all ultimate segments subsessile to stalked, asymmetrical, spores echinate. 36. *Adiantopsis*, p. 266
      j. Lamina axes terete or flattened, or if ridged, then the ultimate segments adnate or symmetrical, spores cristate, reticulate-cristate, rugose, or verrucate. 34. *Cheilanthes*, p. 249
a. Margin of the fertile segments not or rarely modified, but not or hardly covering the sporangia. k.
  k. Sporangia borne on the anastomosing veins and between them. l.
    l. Sterile lamina pedate, densely scaly beneath. 41. *Trachypteris*, p. 302
    l. Sterile lamina 1-pinnate, glabrous to slightly pubescent beneath. m.
      m. Margin of fertile pinnae with a thin membranous border. 48. *Neurocallis*, p. 345
      m. Margin of fertile pinnae coriaceous. 49. *Acrostichum*, p. 348
  k. Sporangia borne only on veins, a vascular receptacle or vascular commissure. n.
    n. Stem scarcely developed, not bearing leaf bases from previous years. 29. *Anogramma*, p. 224
    n. Stem well developed, bearing leaf bases from previous years. o.
      o. Stem with trichomes, sometimes also with scales. p.
        p. Stem with trichomes only. q.
          q. Leaves strongly dimorphic. 33. *Nephopteris*, p. 246
          q. Leaves monomorphic or nearly so. r.
            r. Lamina entire, or 1-pinnate (or mostly so) and the pinnae long-stalked and the lamina imparipinnate. 32. *Pterozonium*, p. 241
            r. Lamina pinnatisect, or more than 1-pinnate, or 1-pinnate and the pinnae adnate to short-stalked and the lamina gradually reduced at the apex. s.
              s. Pinnae pinnatifid to decompound, or entire and then the lamina determinate and the pinnae stalked. 30. *Eriosorus*, p. 228
              s. Pinnae entire or shallowly lobed, the lamina indeterminate, or determinate and the pinnae adnate. 31. *Jamesonia*, p. 235
        p. Stem with scales in addition to trichomes. t.
          t. Lamina pedate, bipinnatifid to tripinnatifid. 35. *Bommeria*, p. 262
          t. Lamina entire, pedately or palmately lobed, or pinnate. u.
            u. Lamina partly 3-pinnate, or less complex. 38. *Hemionitis*, p. 278
            u. Lamina 4- to 5-pinnate. 30. *Eriosorus*, p. 228
      o. Stem with scales only. v.
        v. Leaves strongly dimorphic. 42. *Cryptogramma*, p. 306
        v. Leaves monomorphic or nearly so. w.
          w. Lamina entire, or 1-pinnate and the pinnae lacking a costa. 32. *Pterozonium*, p. 241
          w. Lamina pinnatifid or more complex, when 1-pinnate the pinnae with a costa. x.
            x. Lamina farinaceous beneath (this sometimes concealed by other indument), or if not, then the lamina pinnate and the sporangia extending along $\frac{1}{4}$ or more of the vein and the lamina glabrous or pubescent. y.

    y. Petiole with two vascular bundles near the base (at least in larger leaves), sporangia along most to nearly all of the vein, spores usually tan with dark brown ridges.   28. *Pityrogramma,* p. 216

    y. Petiole with one vascular bundle near the base, sporangia at the vein tip, or if extending along ¼ or more of the vein then the spores coarsely cristate.   37. *Notholaena,* p. 270

x. Lamina glabrous or variously indumented, not farinaceous beneath, lamina pinnate and the sporangia confined to the vein tip or nearly so, or if extending along the vein then the lamina pedate or densely scaly. z.

    z. Lamina pedate, sporangia borne on ¼ or more of the vein.   35. *Bommeria,* p. 262

    z. Lamina pinnate, or if pedate the sporangia borne at the vein tip.   34. *Cheilanthes,* p. 249

Literature

Tryon, A. 1964. *Platyzoma*—A Queensland fern with incipient heterospory. Amer. Jour. Bot. 51: 939–942.

14a. Tribe Taenitideae

Taenitideae Presl, Tent. Pterid. 222. 1836. Type: *Taenitis* Schkuhr
Taenitidaceae (Presl) Pic.-Ser., Webbia 29: 1. 1975.

This tribe contains 11 genera, six of them are in America. The Old World genera are: *Syngramma* J. Sm., *Austrogramme* Fourn. (see Hennipman, Brit. Fern Gaz. 11: 61–72. 1975), *Coniogramme* Fée, *Taenitis* Schkuhr (see Holttum, Blumea 16: 87–95. 1975), and *Cerosora* (Baker) Domin.

28. *Pityrogramma*
Figs. 28.1–28.19

Pityrogramma Link, Handb. Gewächse 3: 19. 1833. Type: *Pityrogramma chrysophylla* (Sw.) Link (*Acrostichum chrysophyllum* Sw.).

Ceropteris Link, Fil. Sp. 141. 1841, *nom. superfl.* for *Pityrogramma* and with the same type.

Trismeria Fée, Mém. Fam. Foug. 5 (Gen. Fil.): 164. 1852. Type: *Trismeria aurea* Fée = *Trismeria trifoliata* (L.) Diels = *Pityrogramma trifoliata* (L.) Tryon.

Gymnogramma section *Cerogramma* Diels, Nat. Pflanz. 1 (4): 260. 1899. Lectotype: *Gymnogramma sulphurea* (Sw.) Desv. (*Acrostichum sulphureum* Sw.) = *Pityrogramma sulphurea* (Sw.) Maxon.

Gymnogramma section *Isgnogramma* Hieron., Engl. Bot. Jahrb. 34: 474. 1904. Type: *Gymnogramma Lehmannii* Hieron. = *Pityrogramma Lehmannii* (Hieron.) Tryon.

Pityrogramma section *Ceropteris* Domin, Publ. Fac. Sci. Univ. Charles 88: 4. 1928. Lectotype: *Pityrogramma chrysophylla* (Sw.) Link (*Acrostichum chrysophyllum* Sw.).

Piryrogramma section *Oligolepis* Domin, *ibidem* 88: 4. 1928. Type: *Pityrogramma sulphurea* (Sw.) Maxon (*Acrostichum sulphureum* Sw.).

Pityrogramma section *Trichophylla* Domin, *ibidem* 88: 4. 1928. Type: *Pityrogramma ferruginea* (Kze.) Maxon (*Gymnogramma ferruginea* Kze.).

Description   Terrestrial; stem erect to decumbent or rarely short-creeping, small to somewhat stout, usually short, bearing scales and, espe-

Fig. 28.1. *Pityrogramma chrysophylla,* Laudat, Dominica. (Photo W. H. Hodge.)

cially at the base, many fibrous roots; leaves monomorphic to slightly dimorphic (the fertile longer and more erect), ca. 0.15–1 m long, borne in a cluster or rarely spaced, lamina pinnatisect to 4-pinnate-pinnatifid, usually with capitate glands beneath which produce copious white or yellow, rarely roseate, farinaceous indument, or sometimes pubescent, or rarely glabrous, veins free; sori elongate on the unmodified veins, sometimes extending almost to the vein-tips and below the fork of a vein, not paraphysate, exindusiate; spores tetrahedral, somewhat globose, trilete, the laesurae $\frac{3}{4}$ the radius, coarsely tuberculate or more or less reticulate, usually with a prominent equatorial flange and 1–4 parallel ridges, or with coarse granulate deposit and without ridges. Chromosome number: $n = 30, 58, 60, 116, 120, 2n = 90,$ ca. 232.

The broad segments of *Pityrogramma Lehmannii* (Figs. 3, 11) illustrate especially well the lines of sporangia along the veins and the dense farinaceous indument characteristic of the genus. The

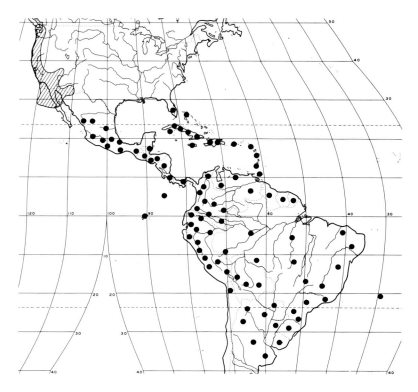

segments of this species are unique in lacking a costa, the veins arising from the primary leaf axis. The extreme forms of lamina architecture are shown in the pinnatisect lamina of *P. Lehmannii* (Fig. 3) and the highly complex one of *P. Pearcei* (Fig. 7). The leaf of *P. ferruginea* (Fig. 10) is exceptional in being densely lanate to the very base of the petiole. The stem is usually short and erect, as in Figs. 9 and 10.

Systematics

Pityrogramma is a genus of about 16 species, most of them in tropical America, one centered in the western United States, and four or five in Africa-Madagascar. It is most closely related to *Anogramma*. Three species, *P. calomelanos, P. tartarea* and *P. triangularis,* each have three distinctive varieties. Several species that often have been recognized are poorly known and their status requires confirmation. Sectional classifications for *Pityrogramma* are unsatisfactory, as most of the species form a rather homogeneous group. Among other species, *Pityrogramma triangularis* is the most distinctive, with a petiole that is terete rather than grooved, stem scales that are bicolorous rather than essentially concolorous, and spores that lack an equatorial ridge. This species might be better placed among the cheilanthoid ferns but its affinity to that group is not clear. Other distinctive species are *P. Lehmannii* with a short creeping stem and a pinnatisect lamina, and *P. ferruginea* with a densely lanate lamina.

A synopsis of the American species of *Pityrogramma* has been published by Tryon (1962).

Tropical American Species

A number of the American species of *Pityrogramma* are adequately known and relatively distinctive, for example, *P. calomel-*

3   4   5

6   7

8   9   10

Figs. 28.3–28.10. *Pityrogramma.*   **3.** Central portion of lamina of *P. Lehmannii*, × 1.   **4.** Central portion of lamina of *P. dealbata*, × 0.75.   **5.** Portion of a lamina of *P. tartarea*, × 1.   **6.** Central portion of a lamina of *P. calomelanos*, × 0.75.   **7.** Apical portion of a lamina of *P. Pearcei*, × 0.75.   **8.** Central portion of lamina of *P. sulphurea*, × 0.75.   **9.** Base of plant of *P. tartarea*, × 1.   **10.** Base of plant of *P. ferruginea*, × 1.

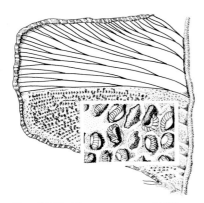

Fig. 28.11. Pinna-segment of *Pityrogramma Lehmannii*, showing venation and sporangia along the veins embedded in the farinaceous indument, × 2.

anos (L.) Link, with three varieties, var. *calomelanos* (Fig. 6), var. *austroamericana* (Domin) Farw. (var. *aureoflava* of Tryon, 1962), var. *ochracea* (Presl) Tryon, *P. chrysoconia* (Desv.) Maxon, *P. chrysophylla* (Sw.) Link (Fig. 1), *P. dealbata* (Presl) Tryon (Fig. 4), *P. ferruginea* (Kze.) Maxon (Fig. 10), *P. Lehmannii* (Hieron.) Tryon (Fig. 3), *P. Pearcei* (Moore) Domin (Fig. 7), *P. sulphurea* (Sw.) Maxon (Fig. 8), *P. tartarea* (Cav.) Maxon, with three varieties, var. *tartarea* (Fig. 5), var. *aurata* (Moore) Tryon, and var. *Jamesonii* (Baker) Tryon, *P. triangularis* (Kaulf.) Maxon, with three varieties, var. *pallida* Weath., var. *triangularis* and var. *viscosa* D. C. Eaton, and *P. trifoliata* (L.) Tryon.

Most of these species are quite variable, especially in leaf architecture. Plants that are more or less intermediate between two species are probably of hybrid origin. Many species of *Pityrogramma* were among the most popular cultivated ferns in the nineteenth century, and there is considerable evidence from horticulture that species can cross and produce fertile hybrids (Domin, 1929). Since most of the species are colonizers, the construction of roads and other disturbances by man have undoubtedly created opportunities for the species to grow together and hybridize. Tryon (1962) mentions hybrids of *Pityrogramma trifoliata* with *P. ferruginea* and with three varieties of *P. calomelanos*.

Other recognized species, such as *P. Dukei* Lell., *P. pulchella* (Moore) Domin, *P. schizophylla* (Jenm.) Maxon, *P. triangulata* (Jenm.) Maxon, and *P. Williamsii* Proctor, are poorly understood. For the most part, records of these species are few; they may represent extreme variants of a species, or segregated hybrids, or local endemics.

Pityrogramma ebenea (L.) Proctor has erroneously been used for *P. tartarea* (Proctor, 1965); it is a synonym of *P. calomelanos* (Anon., 1975).

Ecology (Fig. 1)

Pityrogramma is predominantly a genus of moist, open habitats although some species are tolerant to limited, rarely long, dry periods. A number of species, as *P. sulphurea,* are short-lived and have small stems. Others, such as *P. tartarea* and *P. calomelanos,* have larger stems, are long-lived and usually occur in more stable habitats. Species grow on landslides, on gravel bars in streams, on rocky banks, cliffs, or among volcanic cinders, also in open woods, swamp forests, or brushy savannahs. *Pityro-*

Figs. 28.12–28.19. *Pityrogramma* spores, × 1000, surface detail, × 10,000. **12.** *P. Pearcei* lateral view, center, proximal face, left, portion of distal face, lower right, Costa Rica, *Tryon & Tryon 7127.* **13.** *P. ferruginea,* the proximal face, with several parallel ridges, left, distal face tilted, right, Peru, *Tryon & Tryon 5451.* **14, 15.** *P. calomelanos* var. *calomelanos,* Surinam, *Maguire 23748.* **14.** Reticulate distal face tilted, right, proximal face, left. **15.** Detail of proximal face, laesura between two coarser, parallel ridges with particulate deposit. **16.** *P. Lehmannii,* distal face with a broad equatorial flange at periphery and ridges forming two triangles, a central aerola with tubercles, Colombia, *Madison 883.* **17.** *P. triangularis* proximal face irregularly tuberculate, California, *Alexander & Kellogg 2857.* **18.** *P. trifoliata* proximal face, granulate, Paraguay *Fiebrig 1010.* **19.** *P. trifoliata* detail of granulate deposit, part of laesura at top, Costa Rica, *Burger 4037.*

12 **16**

13 **17**

14 **18**

15 **19**

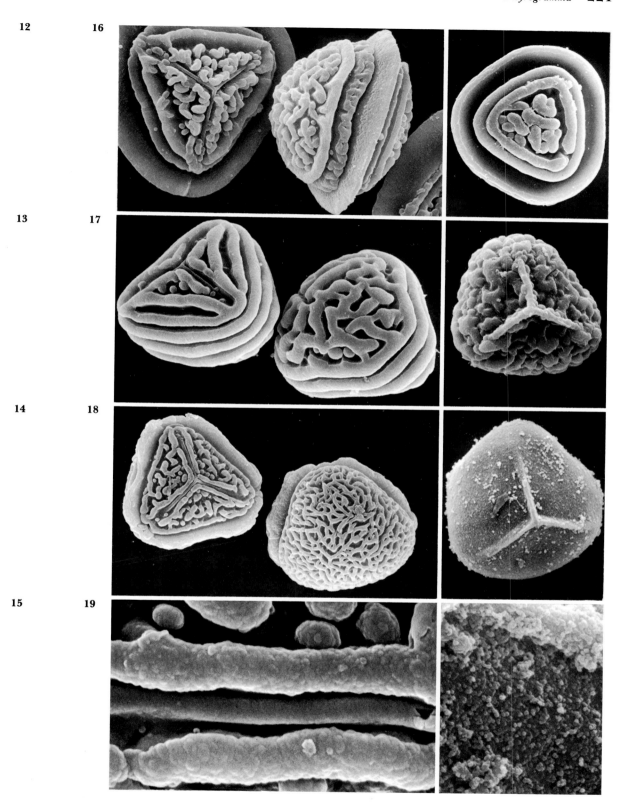

gramma calomelanos, especially in America, may invade palm and banana plantations and rough pastures. This species is also widely adventive in the paleotropics where it is sometimes weedy (Wardlaw, 1962). In America, *Pityrogramma* grows from sea level to usually 1000–2000 m and sometimes as high as 3500 m.

Geography (Fig. 2)

Pityrogramma is native to America and Africa-Madagascar, and adventive through most of the paleotropics (Panigrahi, 1976). In tropical America it ranges from Sinaloa in Mexico through Central America, in Florida and the Bahamas, nearly all of the West Indies, and through South America to Arica in northern Chile and Buenos Aires in Argentina; also on Cocos Island, the Galápagos Islands, and Ilha Trindade. *Pityrogramma calomelanos* is the only species in the Amazon Basin and northeastern Brazil. *Pityrogramma triangularis* is centered in the Pacific United States (Fig. 2). In America, species are primarily centered in the Andes of Colombia to Peru where there are at least 10 species. Secondary centers with about eight species are in Costa Rica and the Greater Antilles. It is difficult to assess the effect of plants escaped from cultivation on the present geography of the American species. For example, the Central American *P. dealbata* also occurs in Guadeloupe, the Andean *P. calomelanos* var. *austroamericana* is in southeastern Brazil, and *P. calomelanos* var. *calomelanos* occurs in the Bahamas. Reports from these disjunct areas may all represent adventive, rather than natural, occurrences.

Spores

The spores of *Pityrogramma* are bicolorous, tan with dark brown ridges, uniformly dark brown, or rarely tan. An equatorial flange and coarse, parallel ridges, as in *P. Pearcei* (Fig. 12), are characteristic of most species, but the position and number of ridges vary. In *P. calomelanos* (Fig. 14) there are fewer ridges, with a well-developed reticulum on the distal face. The fused surface particles are evident especially on two ridges parallel to a central portion of a laesura (Fig. 15). Ridges on the distal face of *P. Lehmannii* spores form concentric triangles (Fig. 16) and spores of *P. ferruginea* have a series of ridges on the proximal face (Fig. 13). Two species have spores that lack ridges: those of *P. triangularis* are coarsely tuberculate with more or less fused tubercules (Fig. 17), and spores of *P. trifoliata* are granulate (Figs. 18, 19).

Cytology

Pityrogramma sulphurea was reported from Jamaica as tetraploid, $n = 60$, and *P. calomelanos, P. tartarea* and *P. trifoliata* as octoploid with $n = $ ca. 120 (Walker 1966). There are reports for *P. calomelanos* from Manaus, Brazil and also of adventive plants from India as $n = 116$. In the *Pityrogramma triangularis* complex there are records of diploids, triploids and tetraploids based on

$n = 30$ (Alt and Grant, 1960; Smith et al., 1971). These reports, although few, suggest there possibly are base numbers of both 29 and 30 in the genus.

Observations

The color of the farinaceous indument of *Pityrogramma* leaves is determined by particular compounds or mixtures of them, primarily chalcones or dihydrochalcones (Wollenweber, 1978; Wollenweber and Dietz, 1980). Sometimes a compound is present only in one taxon, as ceroptin in some of the races of *Pityrogramma triangularis*. In other cases, the same compound may be present in quite different species as dihydrochalcone-5 in *P. Lehmannii* and *P. calomelanos* var. *calomelanos*. A detailed study of the flavonoids of *Pityrogramma triangularis* by Smith (1980) identifies several compounds and relates their occurrence to the cytotypes of the species.

Literature

Alt, K. S., and V. Grant. 1960. Cytotaxonomic observations on the goldback fern. Brittonia 12: 153–170.

Anon. 1975. Report of the standing committee on stabilization of specific names. Taxon 24: 171–200.

Domin, K. 1929. The hybrids and garden forms of the genus *Pityrogramma* (Link). Rozpr. II Tr. Ceské Akad. 38 (4): 1–80.

Panigrahi, G. 1976. The genus *Pityrogramma* (Hemionitidaceae) in Asia. Kew Bull. 30(4): 657–667.

Proctor, G. R. 1965. Taxonomic notes on Jamaican ferns. Brit. Fern Gaz. 9: 213–221.

Smith, D. M. 1980. Flavonoid analysis of the *Pityrogramma triangularis* complex. Bull. Torrey Bot. Cl. 107: 134–145.

Smith, D. M., S. P. Craig, and J. Santarosa. 1971. Cytological and chemical variation in *Pityrogramma triangularis*. Amer. Jour. Bot. 58: 292–299.

Tryon, R. 1962. Taxonomic Fern Notes 2. *Pityrogramma* (including *Trismeria*) and *Anogramma*. Contrib. Gray Herb. 189: 52–76.

Wardlaw, C. W. 1962. A note on *Pityrogramma calomelanos* (L.) Link, a fern nuisance in Cameroon plantations. Jour. Ecology 50: 129–131.

Walker, T. G. 1966. A cytotaxonomic survey of the pteridophytes of Jamaica. Trans. Roy. Soc. Edinburgh 66: 169–237.

Wollenweber, E. 1978. The distribution and chemical constituents of the farinose exudates in gymnogrammoid ferns. Amer. Fern Jour. 68: 13–28.

Wollenweber E., and V. H. Dietz. 1980. Flavonoid patterns in the farina of goldenback and silverback ferns. Biochem. Syst. Ecol. 8: 21–33.

29. *Anogramma*

Figs. 29.1–29.13

Anogramma Link, Sp. Fil. 137. 1841. Type: *Anogramma leptophylla* (L.) Link. (*Polypodium leptophyllum* L.).

Pityrogramma subgenus *Anogramma* (Link) Domin, Publ. Fac. Sci. Univ. Charles 88: 9. 1928.

Description

Terrestrial; stem erect, small to minute, bearing trichomes or scales and many roots; leaves monomorphic, usually short, 1–10 cm or sometimes to ca. 40 cm long, clustered, lamina 1- to 4-pinnate, elongate-triangular, the basal pinnae usually broader than the apical, glabrous or rarely pubescent, veins free; sori more or less elongate along the unmodified veins, usually extending almost to the vein-tips and sometimes below the fork of a vein, not paraphysate, exindusiate; spores tetrahedral-globose, trilete, laesurae $\frac{3}{4}$ or equal to the radius, coarsely ridged or tuberculate with a prominent equatorial flange and one or more parallel ridges, the distal face with more or less fused, irregular tubercules, or irregularly verrucate. Chromosome number: $n = 29, 58$.

Systematics

Anogramma is a genus of about five species of tropical and temperate regions. It is distinguished especially by the annual habit of the sporophyte and the persistence of the gametophyte by means of tubercules. The genus is probably a specialized derivative of *Pityrogramma*, except for *A. Osteniana*. That species lacks the gametophyte tubercules (Baroutis, 1976) and also differs in

Fig. 29.1. *Anogramma leptophylla,* Cartago Province, Costa Rica. (Photo L. D. Gómez.)

Fig. 29.2. Distribution of *Anogramma* in America.

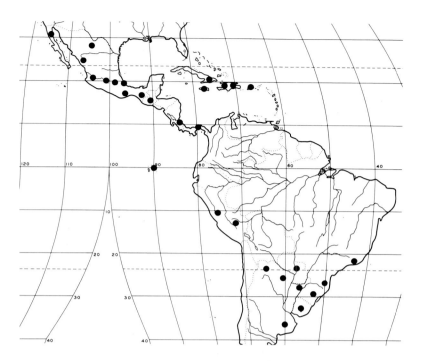

having glandular trichomes on the leaf. In these characters, as well as its spores it shows similarities to *Eriosorus,* a possible relationship that needs further investigation.

Tropical American Species

Anogramma leptophylla (Figs. 6, 7), the most widely distributed species, is also the most polymorphic. Robust elements of this species have been distinguished as *A. guatemalensis* (Domin) C. Chr. of Guatemala and Mexico, and as *A. caespitosa* Pic.-Ser. of Mt. Kilimanjaro in east Africa. *Anogramma chaerophylla* (Figs. 4, 5) and *A. ascensionis* are clearly related to each other on the basis of their rhizome scales and spores. Both species also have ultimate segments with acute lobes (Fig. 4). *Anogramma Lorentzii* (Fig. 3) is one of the small, variable species and *Gymnogramma Regnelliana* Lindm. is included within the range of its variation. *Anogramma Osteniana* (Figs. 8, 9) is distinctive; it has, as mentioned above, similarities to *Eriosorus.*

Key to Species of Anogramma

a. Spores dark brown, the stem with trichomes, leaves with the ultimate lobes obtuse, mostly with forked veins ending short of the margin.   b.
  b. Leaves, especially the petioles, with many glandular trichomes.
                                               A. Osteniana Dutra
  b. Leaves glabrous or rarely with trichomes at the base of the petiole.   c.
    c. Plants small, with fertile leaves 1- to 2-pinnate, less than 3 cm long.
                                    A. Lorentzii (Hieron.) Diels
    c. Plants moderately large, fertile leaves 3-pinnate, 5–20 cm long.
                                     A. leptophylla (L.) Link
a. Spores tan, the stem with scales, leaves with ultimate lobes acute, mostly with a single vein extending nearly to the margin.   d.
  d. Leaves 4-pinnate, usually moderately large, up to 40 cm long, the segment margins entire or slightly dentate.   *A. chaerophylla* (Desv.) Link
  d. Leaves 1- or 2-pinnate, usually small, less than 10 cm, the segment margins erose or dentate.   *A. ascensionis* (Hook.) Diels

Figs. 29.3–29.9. *Anogramma.* **3.** *A. Lorentzii,* plant, ×1. **4, 5.** *A. chaerophylla.* **4.** Pinnule, with sporangia on veins of one segment and their stalks on other veins, ×4. **5.** Small plant, ×0.5. **6, 7.** *A. leptophylla.* **6.** Plant, with juvenile and older leaf, ×0.4. **7.** Plant, with leaf not fully expanded, ×0.5. **8, 9.** *A. Osteniana.* **8.** Plant, with one leaf, ×1.25. **9.** Pinnule, with glandular trichomes, and sporangia on veins of one segment ×12.

Ecology (Fig. 1)

Anogramma grows in moist areas such as wet ravines, the edge of pools, among rocks, under ledges and frequently along road banks or trail sides. The unusual annual habit of the plants requires an open habitat, usually a niche of bare soil, or at least freedom from competition of other plants.

In the American tropics, *Anogramma* often grows in seasonally varied environments, as at Loma Luchay, Lima, Peru, which is moist during the winter months and dry in the summer. It commonly grows from 500 to 2400 m and sometimes to 3200 m.

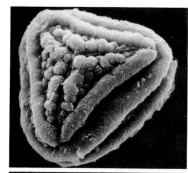

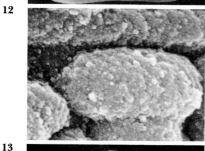

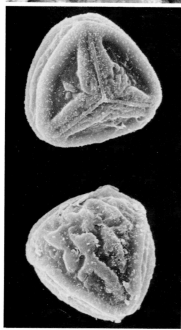

Figs. 29.10–29.13. *Anogramma* spores, × 1000.   **10–12.** *A. leptophylla.*   **10, 11.** Peru, *Tryon & Tryon 5416.*   **10.** Proximal face tilted.   **11.** Irregularly verrucate *distal face.*   **12.** Detail, granulate surface, Cameroon Mountain, west Africa, *Tryon & Tryon 6561,* × 10,000.   **13.** *A. ascensionis,* proximal face above, distal below, Ascension Island, *Loomis,* in 1887.

Geography (Fig. 2)

Anogramma occurs in the American tropics, on Ascension Island, and from southern Europe and Africa east to New Zealand. In America it grows from Baja California in Mexico, south to Panama, in the Greater Antilles, on the Galápagos Islands, and from Peru to southeastern Brazil and south to Buenos Aires in Argentina.

Anogramma leptophylla has an exceptionally wide range occurring through most of the range of the genus. *Anogramma chaerophylla* is widely distributed in America while the closely related *A. ascensionis* is endemic to Ascension Island in the South Atlantic. The range of *A. Osteniana* centers in southern Brazil and Uruguay while that of *A. Lorentizii* is similar with an extension southward into Argentina.

Spores

The strongly ridged and tuberculate spores of *Anogramma* are basically similar to those of related genera, particularly *Pityrogramma.* They are usually dark brown but tan in *A. chaerophylla* and *A. ascensionis.* Three prominent ridges in the equatorial area (Fig. 10) are characteristic for the genus. In *A. leptophylla* the distal face is verrucate to coarsely tuberculate (Fig. 11) with a granulate surface (Fig. 12). In *A. ascensionis* (Fig. 13) and *A. chaerophylla* the distal face is irregularly tuberculate. The variation in size and many undeveloped spores in *A. Osteniana,* as well as incomplete or exceptionally broad surface ridges, indicate some irregularity in spore development.

The study of spore size in *Anogramma* by Baroutis (1976) shows a considerable range in equatorial diameter between ca. 35–98 μm. The maximum spore size of 68 μm in populations from South Africa and New Zealand, contrast with specimens from Greece with a maximum of 54.5 μm. These may reflect different polyploid levels, although a tetraploid is known only from India. Exceptionally large spores of *A. Lorentzii* have a diameter exceeding 98 μm that seems to correlate with the tetraploid level reported for the species.

Cytology

Chromosome records are reported for all of the species except *A. ascensionis* (Baroutis, 1976; Baroutis and Gastony, 1978). The widely distributed species *A. leptophylla* and *A. chaerophylla* have both diploid and tetraploid populations. Several reports of *A. leptophylla* are consistently $n = 29$ from Mexico, Guatemala, South Africa, the Canary Islands, and New Zealand, but two records from India, $n =$ ca. 58 and $n =$ ca. 56, 57, probably represent tetraploids. Plants of *A. chaerophylla* from Jamaica are diploid and a tetraploid is reported from Rio Grande do Sul, Brazil. *Anogramma Osteniana* is diploid with $n = 29$ and *A. Lorentzii* is tetraploid with $n = 58$ (Gastony and Baroutis, 1975). The diploid and tetraploid records of *Anogramma* represent less derived cytological levels, in comparison to the hexaploids and dodecaploids known in *Eriosorus* and *Jamesonia*.

Observations

Anogramma is exceptional among ferns in the persistence of gametophytes bearing tubercules from which new sporophytes can be produced. Studies of Baroutis (1976) show tubercules are formed in all of the American species except *A. Osteniana*. Experimental studies on the effect of temperature on tubercule formation in *A. chaerophylla* showed that an increase in temperature after fertilization promotes formation of a tubercule in association with an embryo. The embryo remains dormant until the temperature is lowered, after which it will grow into a mature sporophyte. Mehra and Sandhu (1976) give a detailed account of the sporophyte and gametophyte of *Anogramma leptophylla*.

Literature

Baroutis, J. G. 1976. Cytology, Morphology, and Developmental Biology of the fern genus *Anogramma*. Ph. D. Thesis. Indiana University, Bloomington, Indiana.

Baroutis, J. G., and G. J. Gastony. 1978. Chromosome numbers in the fern genus *Anogramma*, 2. Amer. Fern Jour. 68: 3–6.

Gastony, G. J., and J. G. Baroutis. 1975. Chromosome numbers in the genus *Anogramma*. Amer. Fern Jour. 65: 71–75.

Mehra, P. N., and R. S. Sandhu. 1976. Morphology of the fern *Anogramma leptophylla*. Phytomorphology 26: 60–76.

30. *Eriosorus*
Figs. 30.1–30.30

Eriosorus Fée, Mém. Fam. Foug. (Gen. Fil.) 5: 152. 1852. Type: *Eriosorus scandens* Fée = *Eriosorus aureonitens* (Hook.) Copel.

Description Terrestrial; stem creeping or decumbent, slender, often bearing glandular trichomes or bristles (very rarely scales), and many slender roots; leaves monomorphic, usually 8–60 cm long, or up to 4.5 m if scandent, usually closely placed on the rhizome, lamina 1-pinnate to 4-pinnate-pinnatifid, rarely to 6-pinnate, elongate-triangular, rhomboid, trullate, or linear, usually pubescent, sometimes densely so, or glandular, veins free; sori more or less elongate along the unmodified veins, usually not to the vein-tip, sometimes extending below the fork of a vein, at maturity sometimes nearly confluent and forming a broad band, not paraphysate, exindusiate; spores tetrahedral-globose, trilete, the laesurae $\frac{1}{2}$ the radius, coarsely ridged or tuburculate, the equatorial flange usually prominent, the distal face with ridges forming a triangle. Chromosome number: $n = 87, 174$.

Fig. 30.1. *E. myriophyllus* along roadside, near Palmares, Rio de Janeiro, Brazil. (Photo Alice F. Tryon.)

Fig. 30.2. *E. cheilanthoides* at base of boulder, planalto of Mt. Itatiaia, Rio de Janeiro, Brazil. (Photo Alice F. Tryon.)

There are usually trichomes or glands on the leaves (Fig. 4) or trichomes which may form a tomentum as in *Eriosorus Stuebelii* (Fig. 7). The shape of the segments of the leaf varies as in Figs. 4–10. The simple, pinnate leaf of *Eriosorus longipetiolatus* has coriaceous pinnae (Fig. 6) with strongly incurved margins. The pinnules of *E. Sellowianus* (Fig. 4) are strongly enrolled and beadlike. The most complex, scandent leaves of *E. flexuosus* and *E. glaberrimus* have bifid ultimate segments (Fig. 8).

The sporangia are usually along the veins as in *E. insignis* (Fig. 10), but may be on the distal portions of the ultimate veins as in *E. hispidulus* (Fig. 9). In *E. Orbignyanus* the sori form a submarginal band (Fig. 5).

The stem usually bears pluricellular trichomes as in *E. Warscewiczii* and *E. congestus* (Figs. 11, 12) or rigid bristles, a few cells broad, as in *E. glaberrimus* (Fig. 13). There are rarely scales as in *E. flexuosus* var. *galeanus* A. Tryon (Fig. 14)

Fig. 30.3. Distribution of *Eriosorus*. The inset shows the Tristan da Cunha Islands and Gough Island in the South Atlantic (distances in km).

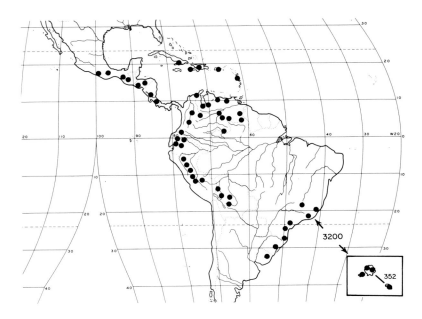

Systematics

Eriosorus is a genus of 25 species of the tropical American highlands. It is most closely related to *Jamesonia* among the genera of the tribe Taenitideae. Species were formerly sometimes placed in the genus *Psilogramme*, a synonym of *Jamesonia*.

Hybridization among species of *Eriosorus* and species of *Jamesonia* (× *Eriosonia* Pic.-Ser.) has made systematic definition and relationships complex. Hybrids with *Jamesonia* have been established on the basis of field and cytological study in Costa Rica. Several others are recognized in South America where ranges are sympatric, on the basis of irregular spores and intermediate morphology of the leaves.

An account of the species, variability and relationships is included in a monograph of *Eriosorus* by Tryon (1970).

Tropical American Species

The remarkable diversity of leaf architecture seems to have developed through two main modifications. Plants with a scandent habit produce large, often flexuous leaves, while those from higher altitudes have developed smaller leaves with a linear lamina that may be only 1-pinnate. The elongate triangular lamina of *E. hirtus* (HBK.) Copel. (Fig. 15) represents a basic form that is characteristic of most species such as *E. congestus* (Christ) Copel., *E. myriophyllus* (Sw.) Copel., *E. hispidulus* (Kze.) Vareschi, *E. rufescens* (Fée) A. Tryon, *E. insignis* (Kuhn) A. Tryon, and *E. Wurdackii* A. Tryon. Large scandent leaves represent a derived type characteristic of *E. flexuosus* (HBK.) Copel., *E. glaberrimus* (Maxon) Scamman (Fig. 18), and *E. Ewanii* A. Tryon, which may form a closely related alliance. However, other scandent forms, such as *E. Orbignyanus* (Kuhn) A. Tryon and *E. aureonitens* (Hook.) Copel., appear to be independently derived from other species. Several species have reduced basal pinnae and the lamina has a rhomboid or trullate form as in *E. Stuebelii* (Hieron.) A. Tryon (Fig. 19), *E. paucifolius* (A. C. Sm.) Vareschi, *E. velleus*

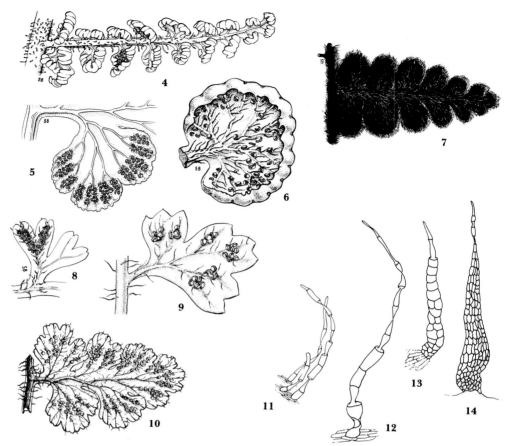

Fig. 30.4–30.14. Segments of *Eriosorus* leaves, indument and sporangia. **4. *E. Sellowianus,*** pinna, with pinnules enrolled, × 6.0 **5. *E. Orbignyanus,*** segments of 4-pinnate leaf, the sporangia in submarginal bands. × 4.0. **6. *E. longipetiolatus,*** pinna, the veins with rigid trichomes, sporangia, × 5.0. **7. *E. Stuebelii,*** tomentose adaxial surface of pinna, × 1.5. **8. *E. glaberrimus,*** bifid ultimate segments, × 4.0. **9. *E. hispidulus,*** pinna, × 3.0. **10. *E. insignis,*** pinnule, × 2.0 **11. *E. Warscewiczii,*** stem trichomes × 30. **12. *E. congestus,*** stem trichome, × 30. **13. *E. glaberrimus,*** stem bristle, × 30. **14. *E. flexuosus*** var. ***galeanus,*** stem scale, × 15.

(Baker) A. Tryon, *E. Warscewiczii* (Mett.) Copel, *E. novogranatensis* A. Tryon, and *E. Biardii* (Fée) A. Tryon. Reduction of the leaf to a linear lamina has occurred in several species lineages, the most extreme forms are in *E. longipetiolatus* (Hieron.) A. Tryon (Fig. 17) and in *E. cheilanthoides* (Sw.) A. Tryon (Fig. 16) a highly derived, polyploid species. Others with linear leaves, such as *E. hirsutulus* (Mett.) A. Tryon, *E. Lindigii* (Mett.) Vareschi, *E. setulosus* (Hieron.) A. Tryon and *E. Sellowianus* (Kuhn) Copel., are independently derived from broader leaved forms.

Ecology (Figs. 1, 2)

Species grow between 600 and 4200 m, more than half of them occurring above 2200 m, and only three wholly below 1800 m. They are largely in páramo regions, or cloud forests or pine-oak forests, often along the forest border, among rocks and at the edge of boulders. The scandent forms usually produce large leaves climbing and entangled with other vegetation. *Eriosorus*

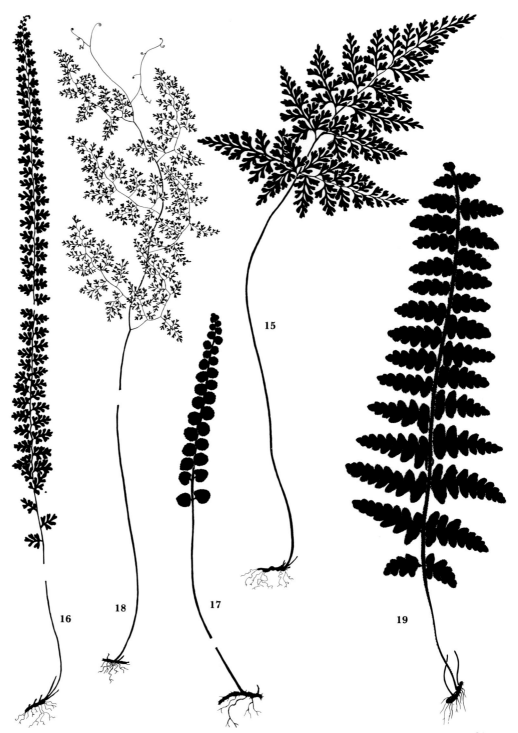

Figs. 30.15–30.19. Leaves of *Eriosorus*, deleted portion indicated by gap in the petiole.   **15.** *E. hirtus*, ×0.5.   **16.** *E. cheilanthoides*, ×0.5.   **17.** *E. longipetiolatus*, ×0.5.   **18.** *E. glaberrimus*, ×0.33.   **19.** *E. Stuebelii*, ×0.5.

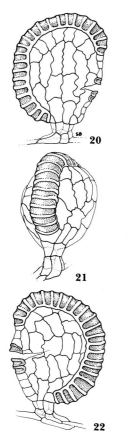

Figs. 30.20–30.22. Sporangia of *E. congestus,* × 90. **20.** Capsule face, stomium right. **21.** Lateral view, annulus excentric. **22.** Capsule face, stomium left.

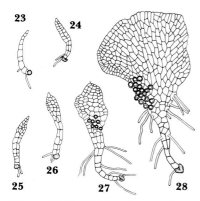

Figs. 30.23–30.28. Gametophytes with attached spore and rhizoids. *E. insignis,* × 40. **23.** Single filament, 7 weeks. **24.** Two rowed filament, 7 weeks. **25.** Early development of thallus plate, 7 weeks. **26.** Early thallus plate, 7 weeks. **27.** Thallus with antheridia, 11 weeks. **28.** Thallus with lateral meristem (left) and antheridia, 5 months.

myriophyllus (Fig. 1) occurs at the lowest altitude of 600 m. *Eriosorus cheilanthoides* (Fig. 2) grows among boulders at high altitudes. Several species on Volcán Poas in Costa Rica appear to have ecological differences but in places where the vegetation has been burned or disturbed, hybrids have been established. *Eriosorus Warscewiczii* grows among boulders in grassy areas on Cerro de la Muerte, Costa Rica, where *Jamesonia* also occurs and hybrids are frequent.

Geography (Fig. 3)

Eriosorus ranges from Guerrero in Mexico, to Costa Rica, eastern Cuba, Hispaniola, Puerto Rico and Dominica, and in South America from British Guiana (Roriama) and Venezuela to Colombia and south to Santa Cruz in Bolivia; it is disjunct in eastern Brazil (ranging from Minas Gerais to Rio Grande do Sul) and Uruguay (Cerro Largo); it is also found on Tristan da Cunha and Gough Islands in the mid-Atlantic some 3200 km east of Brazil.

The species are concentrated in the Andes with 19 occurring from Colombia to Ecuador at altitudes above 3000 m and five are endemic in Colombia. There are secondary centers with four species in Costa Rica and six in Brazil, four of them endemic. A few species occur in the Andean regions of Venezuela, and *E. paucifolius* is largely on the tepuis of the Roraima sandstone formation.

Spores

Eriosorus spores are tetrahedral with a prominent equatorial flange (Fig. 29), and three coarse ridges form a triangular base on the distal face (Fig. 30). Coarse tubercules are formed within the triangle and also fuse into ridges parallel to the laesurae on the proximal face. Wall sections, examined in *E. congestus* and *E. Lindigii,* show two distinct perispore strata overlay the exospore which is formed of thick outer and thin inner layers.

Fossil spores of *Eriosorus* have been used by Hafsten (1960) to analyze the former vegetation on Tristan da Cunha and Gough Island in the South Atlantic. They occur in the lowest part of the peat sample from Tristan and at two higher levels. A more continuous record is found in the sample from Gough, with the lowest level there dated about 5000 B.P. Study of the pleistocene vegetation near Bogotá, Colombia, by van der Hammen and Gonzales (1960) shows an abundance of these spores, identified as *Jamesonia,* throughout the core with a base dated at 21,000 (± 600) B.C. The material may well also include *Eriosorus* since both genera presently occur in the area, and their spores are not readily distinguished. These were the most abundant pteridophyte spores in the sample except for "Cyatheaceae," and suggest that these genera flourished in the region from the Würm stage of the Pleistocene to the present.

29   30

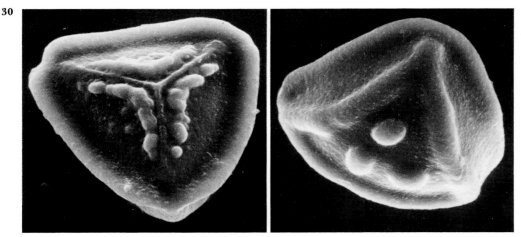

Figs. 30.29, 30.30. Spores of *Eriosorus congestus,* Costa Rica, *Tryon & Tryon 6995,* × 1000. **29.** Proximal face, equatorial flange at perimeter. **30.** Distal face tilted with triangular base.

Cytology

The report of $2n = 174$ for *Eriosorus hirtus* from Chiapas, Mexico (Smith and Mickel, 1977) and of four species of Costa Rica with $n = 87$ (Tryon, 1970) indicate a relatively derived hexaploid level for the genus. The report of $n = 174$ in *E. cheilanthoides* from Tristan da Cunha (Manton and Vida, 1968), which is the same for the species from Mt. Itatiaia, Brazil (Tryon, 1970), indicates a still higher dodecaploid level in the genus. Records for the Andean species are necessary to determine if the chromosome number is uniformly high in this group.

Cytology is especially useful for the recognition of hybrids between species as *E. flexuosus* × *E. Warscewiczii* and also the intergeneric hybrid *E. Warscewiczii* × *Jamesonia Scammanae,* both with 174 mostly unpaired chromosomes at meiosis.

Observations

Sporangia (Figs. 20–22) have usually short stalks that may form a thickened cushion of cells by intercalary division. The capsule is pyriform or orbicular with marked asymmetry as in the lateral aspect and faces of the sporangia of *E. congestus.*

Gametophytes of *Eriosorus insignis* cultured on soil are spathulate. Several developmental stages (Figs. 23–28) include the initial filament of cells (Fig. 23), which becomes two rowed (Fig. 24), and after seven weeks forms a thallus plate (Figs. 25, 26). Antheridia, produced near the base of the thallus, and a marginal meristem were evident at 11 weeks (Fig. 27). After five months the gametophytes were spathulate, and bore autheridia but no archegonia (Fig. 28). Asymmetrical gametophytes with a persistent lateral meristem are also reported in *Anogramma* (Baroutis, 1976) and in *Pityrogramma,* where they become asymmetrical-cordate with age (Momose, 1964). Thus there are similarities among these genera in the gametophyte as well as the sporophyte.

Literature

Baroutis, J. G. 1976. Cytology, Morphology and Developmental Biology of the Fern Genus *Anogramma*. Ph.D. Thesis. Indiana University, Bloomington, Indiana.

Hafsten, U. 1960. Pleistocene development of vegetation and climate in Tristan da Cunha and Gough Island. Univ. Bergen Arb. Naturv. 20: 1–45.

Hammen, T. van der, and E. Gonzales. 1960. Upper Pleistocene and Holocene climate and vegetation of the "Sabana de Bogotá," Colombia, South America. Leidse Geol. Meded. 25: 261–315.

Manton, I., and G. Vida. 1968. Cytology of the fern flora of Tristan da Cunha. Proc. Royal Soc. (B). 170: 361–379.

Momose, S. 1968. The prothallium of *Ceropteris*. Jour. Jap. Bot. 39: 225–235.

Smith, A. R., and J. Mickel. 1977. Chromosome counts for Mexican Ferns. Brittonia 29: 391–398.

Tryon, A. F. 1970. A monograph of the fern genus *Eriosorus*. Contrib. Gray Herb. 200: 54–174.

31. *Jamesonia*
Figs. 31.1–31.17

Jamesonia Hook. & Grev., Icon. Fil. 2: *t. 178.* 1830. Type: *Jamesonia pulchra* Hook. & Grev.

Gymnogramma section *Jamesonia* (Hook. & Grev.) Kl., Linnaea 20: 407. 1847.

Psilogramme Kuhn, Festsch. 50 Jub. Reals. Berlin 332 (reprint, Chaetopterides, 12) 1882, *nom. superfl.* for *Jamesonia* and with the same type.

Psilogramme section *Jamesonia* (Hook. & Grev.) Kuhn, *ibidem* 332, (12). 1882.

Description Terrestrial; stem long-creeping or decumbent, slender, bearing sometimes glandular trichomes, or bristles, and many rather coarse roots; leaves monomorphic, 15–60 cm long or rarely exceeding 90 cm, usually well spaced, 1-pinnate, linear, usually pubescent or tomentose especially on the abaxial surface, or glandular, veins free; sori elongate along the unmodified veins, extending almost to the vein-tips, and sometimes below the fork of a vein, not paraphysate, exindusiate; spores tetrahedral-globose, trilete, the laesurae $\frac{1}{2}$ to $\frac{3}{4}$ the radius, coarsely ridged or tuberculate, with a prominent equatorial flange, the distal face with ridges forming a triangle sometimes with additional parallel ridges. Chromosome number: $n = 87$.

The genus is readily recognized by the unusual linear leaves (as in *Jamesonia Scammanae,* Fig. 4) which are usually indeterminate (Figs. 1, 2, 4, 15) and may be sparsely pubescent, glandular, or tomentose (Fig. 1). The stem indument varies from delicate and thin, to coarse and rigid trichomes with apical glands as in *J. Alstonii* (Fig. 13) or thickened bristles in *J. Cuatrecasasii* (Fig. 14).

Systematics *Jamesonia* is a tropical American genus of 19 species. It is most closely related to *Eriosorus,* as evident in many characters, and appears to represent elements derived in more than one lineage

Fig. 31.1. Part of a large colony of *Jamesonia canescens* on Páramo de la Negra, Táchira, Venezuela. (Photo Alice F. Tryon.)

Fig. 31.2. Plants of *Jamesonia brasiliensis* at base of boulder, Mt. Itatiaia, Rio de Janeiro, Brazil. (Photo Alice F. Tryon.)

from that genus. The linear, 1-pinnate lamina in *Jamesonia* clearly represents a derived form of leaf architecture, and the genus is one of the specialized elements in the *Pityrogramma* alliance.

Detailed discussion of the relationships, the morphology and geography of the species are included in a monographic treatment by Tryon (1962).

Tropical American Species

The species differ in size and form of the pinnae and in less conspicuous characters of their margins, stem indument and spores. The margins may be strongly modified and somewhat indusiate as *J. pulchra* (Fig. 9), coarsely dentate as in *J. rotundifolia* (Fig. 12), strongly ciliate as in *J. Alstonii* (Fig. 11). Sporangia are often obscured by dense tomentum as in *J. Scammanae* (Fig. 5). The pin-

Fig. 31.3. Distribution of *Jamesonia*.

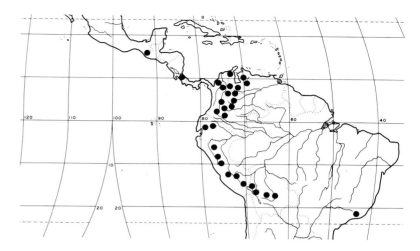

nae may be adnate to the rachis as in *J. verticalis* (Fig. 6), with a stalk that is straight, or bent as in *J. brasiliensis* (Fig. 7) or strongly twisted as in *J. imbricata* (Fig. 8).

The species can be aligned in six related groups and a few isolated elements. *Jamesonia rotundifolia* Fée is least specialized and three others *J. laxa* (Kuhn) Diels, *J. auriculata* A. Tryon and *J. Cuatrecasasii* A. Tryon are similar in the light colored lamina indument and general shape of the pinnae and margins. Several species with densely tomentose pinnae, *J. robusta* Karst., *J. pulchra* Hook. & Grev., *J. bogotensis* Karst., and *J. canescens* Kze., are allied but variation among the species suggests that hybridization has complicated relationships. The most difficult complex is *J. imbricata* (Sw.) Hook. & Grev., in which hybrids are involved and varieties are recognized to distinguish the main forms. *Jamesonia brasiliensis* Christ and *J. scalaris* Kze. also belong to this complex. *Jamesonia Alstonii* A. Tryon and *J. Goudotii* (Hieron.) C. Chr. form the closest species pair, characterized by coriaceous pinnae with erose margins. Three other coriaceous species, *J. verticalis* Kze., *J. blepharum* A. Tryon, and *J. cinnamomea* Kze., have densely ciliate pinna margins and represent another discrete element. Three species with small pinnae, *J. peruviana* A. Tryon, *J. boliviensis* A. Tryon and *J. Scammanae* A. Tryon each have different kinds of lamina indument and represent distinctive species difficult to relate to others.

Ecology (Figs. 1, 2)

Jamesonia is a genus characteristic of the páramo, although not restricted to it, usually growing between 3000 and 4000 m. The term páramo is usually included in ecological data on collections and applies to the Andean highlands south to Ecuador, between 3200 and 5000 m, above the forest and below permanent snow. Critical ecological features of the páramo to which the vegetation is adapted are wind, strong insolation, high soil moisture, clouds or fog, and temperatures ranging between 20° and -2°C. The highest parts of the páramo have marked diurnal fluctuations in temperature with hot sun during the day and freezing at night. Special accounts of páramo vegetation are in-

Figs. 31.4–31.15. *Jamesonia.* **4.** *J. Scammanae,* plant with many petiole bases along stem, the lamina indeterminate, × 0.5. **5.** *J. Scammanae,* abaxial surface of pinna with dense tomentum, × 10. **6.** *J. verticalis,* the pinna adnate to the rachis, the margins enrolled, ciliate at vein-ends, × 8. **7.** *J. brasiliensis,* the pinna stalk bent, pinna margin ciliate, × 10. **8.** *J. imbricata,* the pinna stalk bent, pinna margin broad, indusioid, × 10. **9.** *J. pulchra,* the pinna stalk bent, pinna with broad, indusioid entire or shallowly toothed margin, × 10. **10.** *J peruviana,* pinna stalk straight, the pinnae inequilateral, the margins long cilate or papillate, × 10. **11.** *J. Alstonii,* the pinnae coriaceous with erose margin, × 10. **12.** *J. rotundifolia,* pinnae inequilateral, the margin slightly toothed, × 5. **13.** *J. Alstonii,* stem trichomes with one or two cells at base, the apical cell glandular, × 30. **14.** *J. Cuatrecasasii,* stem bristle with several basal cells, × 30. **15.** *Jamesonia Scammanae* × *Eriosorus Warscewiczii,* plant, part of stem, leaves and petiole bases, × 0.5.

cluded in the geobotanical observations by J. Cuatrecasas (1934), which includes excellent photographs of páramo vegetation and of *Jamesonia,* and in the description of the páramo vegetation at Sumapáz, south of Bogotá, by Fosberg (1944).

Plants of *Jamesonia* occur at the base of boulders (Fig. 2) and sometimes in open grassy meadows where the stems are deeply embedded in turf. Some of these grasslands are known to be burned and the cover of turf affords the stems of *Jamesonia* protection from fire. Several species become established in open soil of new road cuts. Stems of *Jamesonia* are freely branched and plants often form large colonies as *Jamesonia canescens* (Fig. 1) on the highest part of Páramo de la Negra, Táchira, Venezuela. The white-tomentose leaves of this colony appeared at a distance to form a white cover over the field.

Geography (Fig. 3)

Jamesonia ranges from southern Mexico and Costa Rica to the highlands of eastern Venezuela, south to Bolivia; it is disjunct on Mt. Itatiaia, in eastern Brazil. The species are mainly concentrated in the Andes, all of them occurring in the area from Mérida, Venezuela, to La Paz, Bolivia. The region of greatest diversity, where 11 of 21 taxa occur, is between Bogotá in Colombia and Cuenca in Ecuador. There are a few wide-ranging species, as *J. Scammanae,* which is distributed from Costa Rica to Bolivia, but several are localized especially in western Venezuela near Mérida. The wide ranges and also disjunctions of several species which occur at high altitudes have probably been influenced by Pleistocene glaciation. During that time, the present páramo vegetation may have descended 400 m lower than at present, allowing for a greater continuity of habitat.

Spores

The spores are characterized by an equatorial flange and coarse tubercules (Figs. 16, 17). Wall sections of *J. brasiliensis* show two perispore strata above a thicker exospore, as in spores of *Eriosorus. Jamesonia* spores are not readily distinguished from those of *Eriosorus* thus palynological samples of Pleistocene age near Bogotá examined by Van der Hammen and Gonzales (1960) may include some of both genera which presently occur near Bogotá. A single spore from a late Oligocene deposit in Puerto

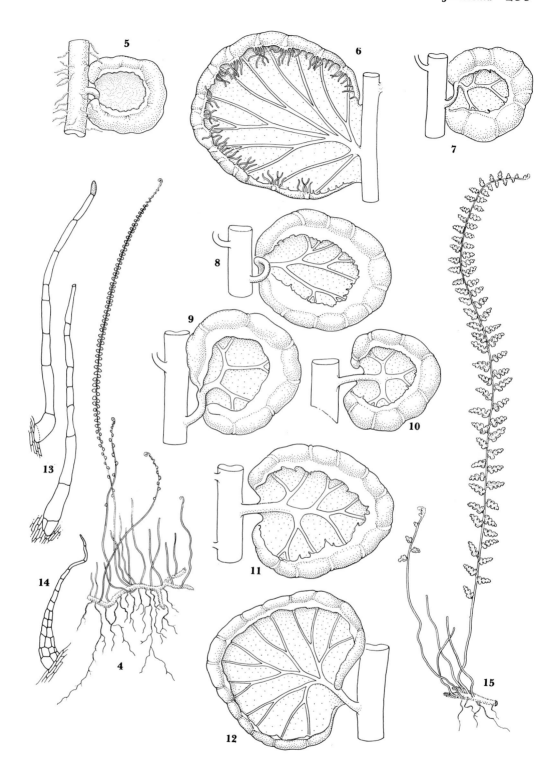

16. **17.**

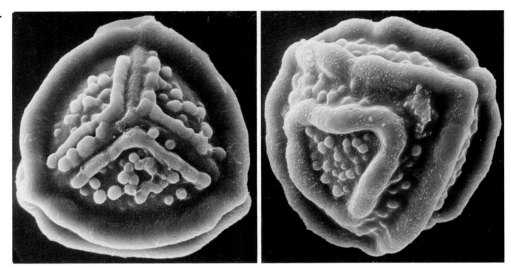

Fig. 31.16, 31.17. *Jamesonia Goudotii* spores, Colombia, *Cleef 8587.* **16.** Proximal face with equatorial flange at perimeter, × 1000. **17.** Distal face, portions of equatorial flange at top and right, and two sets of concentric basal ridges, × 900.

Rico was identified as *Jamesonia* by Graham and Garzen (1969). This may represent *Eriosorus hispidulus,* which occurs in Puerto Rico, as there is no record of *Jamesonia* in the Caribbean area.

Cytology

Meiotic cells with 87 bivalents have been observed in *Jamesonia bogotensis* from Colombia and *J. Scammanae* from Costa Rica, and are considered hexaploids based on 29. Hybrids between *Eriosorus Warscewiczii* and *Jamesonia Scammanae* grow with the parental species in Costa Rica and are morphologically intermediate. The parents have 87 bivalents at meiosis and the hybrids have 174 largely unpaired chromosomes. The base number may be 29, although it is not known whether this occurs in *Jamesonia* or is derived from a lower level in an ancestral genus.

Literature

Cuatrecasas, J. 1934. Observations geobotánicas en Colombia. Trab. Mus. Nac. Cien. Nat. Madrid. Ser. Bot. 27: 1–144.

Fosberg, F. R. 1944. El páramo de Sumapáz. Jour. N.Y. Bot. Gard. 45: 226–234.

Graham, A., and D. M. Jarzen. 1969. Studies in neotropical paleobotany. I. The Oligocene communities of Puerto Rico. Ann. Mo. Bot. Gard. 56: 308–357.

Hammen, T. van der, and E. Gonzales. 1960. Upper Pleistocene and Holocene climate and vegetation of the "Sabana de Bogotá," Colombia, South America. Leidse Geol. Meded. 25: 261–315.

Tryon, A. F. 1962. A monograph of the fern genus *Jamesonia.* Contrib. Gray Herb. 191: 109–197.

32. *Pterozonium*
Figs. 32.1–32.23

Pterozonium Fée, Mém. Soc. Mus. Hist. Nat. Strasbourg 4: 202. 1850, and Gen. Fil. (Mém. Fam. Foug. 5): 37, 178. 1852. Type: *Pterozonium reniforme* (Mart.) Fée (*Gymnogramma reniformis* Mart.).
Syngrammatopsis Alston, Mutisia 7: 7. 1952. Type: *Syngrammatopsis elaphoglossoides* (Baker) Alston (*Gymnogramma elaphoglossoides* Baker) = *Pterozonium elaphoglossoides* (Baker) Lell.

Description

Rupestral or sometimes terrestrial; stem prostrate-creeping, decumbent or erect, slender to rather stout, bearing dense trichomes, bristles or scales, and few to many long, fibrous roots; leaves monomorphic, ca. 0.1–1.0 m long, clustered or sometimes widely spaced, lamina simple, entire or 1-pinnate (rarely partially 2-pinnate), glabrous or minutely pubescent along the costa, or rarely definitely pubescent, usually very coriaceous, veins free; sori somewhat sunken, sometimes short and at the somewhat modified vein-tip, or usually long, extending nearly all along the unmodified veins, or a portion of them, and at maturity sometimes confluent and forming a broad band, sometimes with yellow farinaceous indument among the sporangia, not paraphysate, exindusiate; spores tetrahedral with prominent angles and equatorial flange, trilete, the laesurae about $\frac{1}{2}$ the radius, the distal face with ridges extending from the angles to a central area, the surface with fairly dense spherical deposit, smooth, or especially the proximal face, irregularly rugose or somewhat tuberculate. Chromosome number: not reported.

The stem indument is unusually variable in this genus, and Figs. 7–11 illustrate the scales, trichomes and bristles that occur in different species. The sporangia are usually borne along a considerable portion of the vein, (Figs. 4, 6) but in some species the sori are short and near the margin (Fig. 3). Although the sori have been described as paraphysate, the trichomes among the sporangia are similar to those borne, usually sparingly, elsewhere on the lamina.

Systematics

Pterozonium is a distinctive genus of 13 species strongly centered on the Roraima sandstone in Venezuela. The species with scales were originally placed in the paleotropical genus *Syngramma* and later segregated by Alston as *Syngrammatopsis*. Pichi-Sermolli (1977) recognizes Alston's segregate and also *Pterozonium;* however, we follow Lellinger (1967) in treating all species in *Pterozonium* without an infrageneric classification. This treatment is adopted since the relationships of the species appear complex. Although the stem indument can be referred to as scales or trichomes there are varied forms as in Figs. 7–11. Species with a 1-pinnate lamina, *P. Maguirei* and *P. spectabile,* are not necessarily closely related, and the small species *P. Steyermarkii* is quite distinct from others with small leaves such as *P. cyclosorum.*

Pterozonium has been revised by Lellinger (1967) on the basis of ample recent collections from the Guayana region of Venezuela.

Fig. 32.1. *Pterozonium cyclophyllum*, Mt. Roraima, Venezuela. (Photo J. A. Steyermark.)

Fig. 32.2. Distribution of *Pterozonium:* The dot nearest the letter is the location of Chimanta (C), Sipapo (S), Duida (D), Roraima (R), Auyan-tepui (A). See text for explanation.

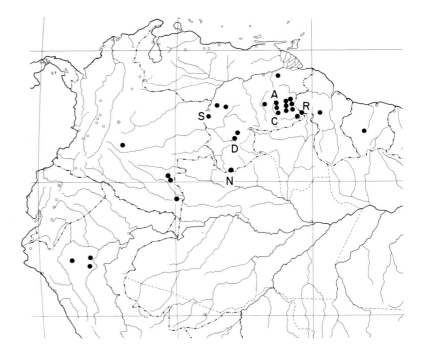

Tropical American Species

A group of six species have a more or less elongate simple lamina with a well-developed costa, pinnate venation, and scales on the stem although these may be narrow in *P. scopulinum* Fig. 10). These are *Pterozonium terrestre* Lell., *P. lineare* Lell., *P. brevifrons* (A. C. Sm.) Lell., *P. paraphysatum* (A. C. Sm.) Lell., *P. elaphoglossoides* (Baker) Lell., and *P. scopulinum* Lell. (Fig. 13). Another

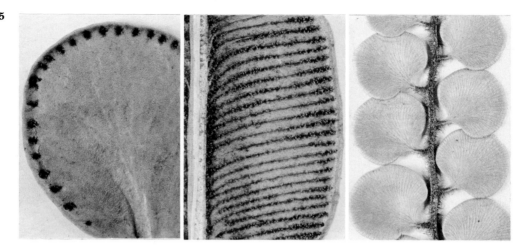

Figs. 32.3–32.5. Detail of *Pterozonium* leaves. **3.** *P. cyclosorum,* sori at vein ends, ×2. **4.** *P. terrestre,* sori along parallel veins, ×2. **5.** *P. spectabile,* pinnae on central portion of lamina, ×1.5.

group of four species has a broad to reniform simple lamina without a costa, with flabellate, venation, and lax or rigid trichomes on the stem. These are *Pterozonium reniforme* (Mart.) Fée (Fig. 15), *P. Tatei* A. C. Sm. (Fig. 14), *P. cyclophyllum* (Baker) Diels, and *P. cyclosorum* A. C. Sm. (Fig. 16). *Pterozonium Steyermarkii* Vareschi (Fig. 17) is similar to the previous group but has an elongate lamina with a partly developed costa and ascending veins. The l-pinnate species are *P. Maguirei* Lell. (Fig. 12) which has definite scales on the stem and *P. spectabile* Maxon & Smith (Fig. 5) which has mostly rigid uniseriate trichomes on the stem and some bristles.

Ecology (Fig. 1)

Most species of *Pterozonium* are rupestral on sandstone, growing in wet to rather dry, rocky places such as crevices and ledges of cliffs or among rocks. On the Guayana tepuis, species grow near the base of the escarpment to the summit. They may occur in somewhat open forest, in shrubby places or open rocky sites. *Pterozonium terrestre* and *P. Maguirei,* especially, are terrestrial in sandy soil, the latter sometimes forming a dense growth. The genus usually grows at 1400–2000 m.; in southeastern Colombia it grows at 700 m.

Geography (Fig. 2)

Pterozonium strongly centers in the states of Bolivar and Amazonas of Venezuela, where all of the species occur. Four extend beyond this region: *P. scopulinum* at Kaieteur Falls in British Guiana, *P. paraphysatum* on Tafelberg in Surinam, and *P. reniforme* and *P. brevifrons* in Colombia and Peru. The largest massifs (tepuis) have the most species. Chimanta has seven, Sipapo six, Neblina five, and Duida, Roraima and Auyan-tepui each have four. All others have one or two species. Eleven of the 13 species occur on Neblina and (or) Chimanta, and all 13 occur in the arc from Neblina to Duida, Chimanta and Roraima.

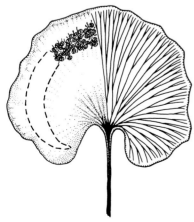

Fig. 32.6. Fertile leaf of *P. reniforme:* venation and position of sori, ×1.

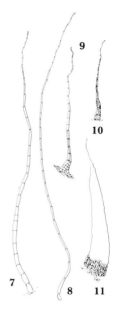

Figs. 32.7–32.11. Stem indument of *Pterozonium*, × 7. **7.** *P. reniforme.* **8.** *P. cyclophyllum.* **9.** *P. Spectabile.* **10.** *P. scopulinum.* **11.** *P. paraphysatum.*

Spores

Pterozonium spores have a more or less pominent equatorial flange (Figs. 18, 21) but lack the parallel ridges as in most related genera. On the distal face ridges extend from the three angles in the equatorial region and fuse in the central area as in *P. cyclosorum* and *P. Maguirei* (Figs. 19, 20). Spherical particles are more or less densely deposited on a relatively smooth surface (Figs. 22, 23). In *P. paraphysatum* the surface may be shallowly rugose or somewhat tuberculate, and similar on the proximal face in *P. cyclosorum* (Fig. 18). The relatively smooth spores with central ridge and basal triangle are generally like but somewhat simpler than those of *Eriosorus* and *Jamesonia*. They appear less elaborate than the more prominently ridged and strongly tuberculate spores of *Pityrogramma* and *Anogramma*.

Literature

Lellinger, D. B. 1967. *Pterozonium. in:* B. Maguire and collaborators, The Botany of the Guayana Highland—Part VII. Mem. New York Bot. Gard. 17: 2–23.

Pichi-Sermolli, R. E. G. 1977. Tentamen pteridophytorum genera in taxonomicum ordinem redigendi. Webbia 31: 313–512.

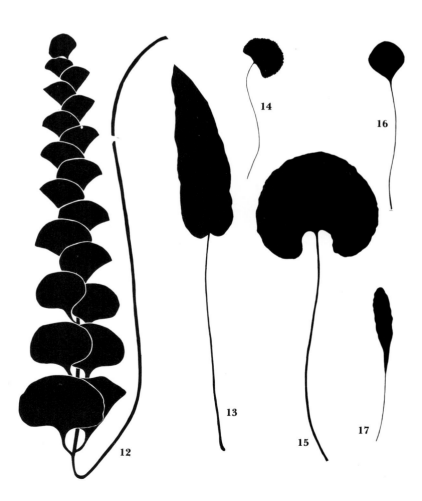

Figs. 32.12–32.17. Leaves of *Pterozonium*, ×0.5. **12.** *P. Maguirei.* **13.** *P. scopulinum.* **14.** *P. Tatei.* **15.** *P. reniforme.* **16.** *P. cyclosorum.* **17.** *P. Steyermarkii.*

18   19

20   21

22   23

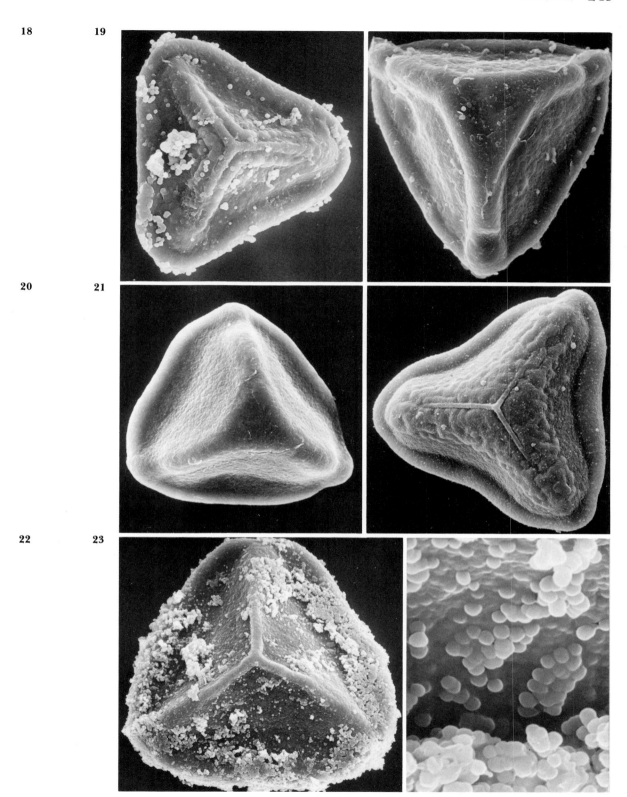

Figs. 32.18–32.23. *Pterozonium* spores, × 1000, surface detail, × 10,000.   **18, 19.** *P. cyclosorum,* Venezuela, *Maguire 33005.*   **18.** Proximal face, peripheral equatorial ridge.   **19.** Distal face, with projecting basal triangle.   **20.** *P. Maguirei,* distal face with projecting basal triangle, Venezuela, *Maguire 42448.*   **21.** *P. paraphysatum,* proximal face shallowly rugulose, Venezuela, *Wurdack 34251.*   **22, 23.** *P. Steyermarkii,* Venezuela, *Wurdack 34163.*   **22.** Proximal face spherical deposit somewhat eroded.   **23.** detail of spherical deposit, proximal face.

33. *Nephopteris*
Figs. 33.1–33.7

Nephopteris Lell., Amer. Fern Jour. 56: 180. 1966. Type: *Nephopteris Maxonii* Lell.

Description

Terrestrial; stem prostrate, short, slender, bearing trichomes and a few coarse roots; leaves dimorphic (the sterile ca. 2 cm long, spreading, 1-pinnate, slightly pubescent to glabrate, the fertile ca. 5–10 cm long, erect, with a long petiole, 2-pinnate) loosely clustered, the abaxial surface with dense, matted trichomes, veins free; sporangia borne in elongate sori along the veins, not paraphysate, exindusiate; spores tetrahedral, with the distal face more or less flat, trilete, the laesurae equal to the radius or nearly so, with a prominent equatorial flange, the proximal face rugose, the distal with coarse ridges forming a triangle with the angles connected to the flange. Chromosome number: not reported.

Systematics

Nephopteris is a rare, monotypic genus of the Colombian Andes, known from only three collections. The exindusiate sporangia borne along unmodified veins (Fig. 5) and ridged spores (Fig. 7) show relation of the genus to the group of pityrogrammoid ferns. However, the unusual strongly dimorphic leaves (Fig. 2), surcurrent segments, and definitely stalked sporangia make it difficult to align *Nephopteris* with the other genera. The architecture of the sterile lamina of *Nephopteris Maxonii* Lell. resembles the small leaves of *Anogramma Lorentzii*. The dense leaf tomentum resembles that in some species of *Pityrogramma, Eriosorus* and *Jamesonia*. The coriaceous segments with flabellate veins (Fig. 4) are also characteristic of *Pterozonium* and rigid stem trichomes occur in most of the allied genera. The trichomes along the veins among sporangia were considered paraphyses in the original description, but are not regarded as paraphyses here since they are undifferentiated from other lamina indument.

Ecology

Nephopteris Maxonii is known from a grassy páramo and from peaty banks beside a mule trail on a steep hillside, at altitudes of 2900–3200 m.

Geography (Fig. 1)

The plants were first collected in 1918 on Páramo de Chaquiro in the Cordillera Occidental of Colombia, and later at two sites in the Cordillera Oriental, at Sierra Nevada de Cocuy and southward at Veho Hondo. The occurrence of these plants in two Cordilleras of northern Colombia, some 400 km miles apart, suggests that the genus may be more widely distributed.

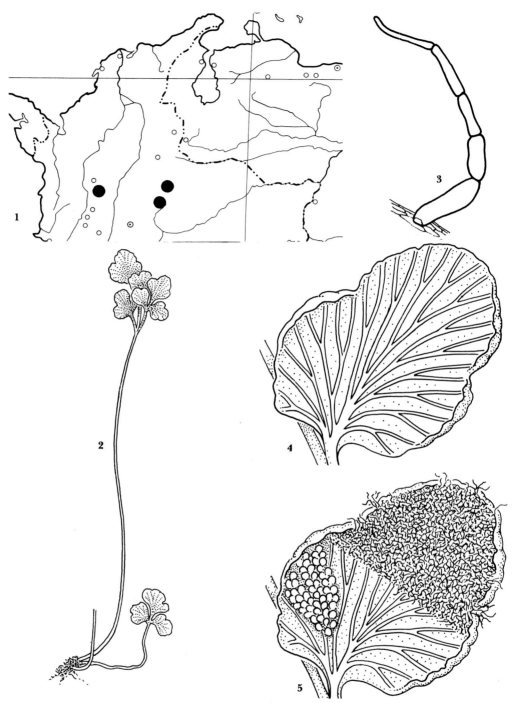

Fig. 33.1–33.5. *Nephopteris Maxonii.* **1.** Distribution (northern Colombia). **2.** Habit of plant, × 1. **3.** Stem trichome, × 20. **4.** Abaxial surface of fertile pinna, sporangia and indument removed, × 4. **5.** Abaxial surface of fertile pinna, the tomentum and sporangia, partly removed, × 4.

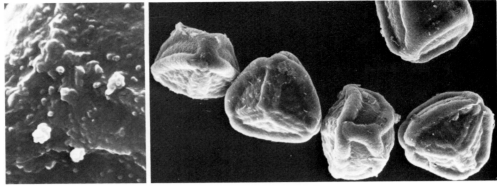

Figs. 33.6, 33.7. *Nephopteris* spores, Columbia, *Grubb & Guymer P68.* **6.** Surface detail, × 2600. **7.** Spores, the proximal face second from left, part of the distal face, right, above, × 500.

Spores

Nephopteris spores are generally smaller than those of *Jamesonia* and *Eriosorus.* The variously oriented spores in Fig. 7 are at half the magnification of figures in those genera. The equatorial flange, basal triangle, and granulate surface (Fig. 6) are similar to those of *Jamesonia* and *Eriosorus* spores.

14b. Tribe Cheilantheae

Cheilantheae J. Sm., Hist. Fil. 277. 1875. Type: *Cheilanthes* Sw.
Sinopteridaceae Koidz., Acta Phytotax. Geobot. 3: 50. 1934. Type: *Sinopteris* C. Chr. & Ching.
Cryptogrammaceae Pic.-Ser., Webbia 17: 299. 1963. Type: *Cryptogramma* R. Br.
Hemionitidaceae Pic.-Ser., Webbia 21: 487. 1966. Type: *Hemionitis* L.
Cheilanthaceae Nayar, Taxon 19: 233. 1970. Type: *Cheilanthes* Sw.

This tribe consists of 12-16 genera, ten of them in America. Among the Old World genera are *Actinopteris* Link and *Onychium* Kaulf.; and also *Negripteris* Pic.-Ser., *Neurosoria* Kuhn, *Paraceterach* Copel., and *Sinopteris* C. Chr. & Ching, which are doubtfully distinct from *Cheilanthes.*

The cheilanthoid ferns have been the most contentious group of ferns with respect to a practical and natural generic classification. For this reason, considerable time has been devoted to their study and as a result a number of changes in the current classifications have been made. These changes are based on the evidence that convergence of single characters is especially common in this group and that evolutionary groups have a strong tendency to be confined to the three large continental/archipelago land areas of America, Africa and Asia-Malesia.

Cheilanthes has been enlarged by adding to it elements from *Notholaena, Pellaea* and *Doryopteris. Adiantopsis,* although sometimes merged into *Cheilanthes,* is recognized primarily on the basis of spore characters. *Notholaena* is restricted to the predominently farinaceous species of America, the other species are placed in *Cheilanthes. Hemionitis* is united with *Gymnopteris* and confined to American species; those of the Old World are excluded. *Pellaea* is restricted to four sections and various species usually referred to it are placed in *Cheilanthes*. *Doryopteris* is made more homogeneous by the transfer of the *concolor* group to *Cheilanthes*. Finally, the monotypic genus *Saffordia* is combined with *Trachypteris*.

These changes clarify the taxonomy and relationships of the genera except for *Cheilanthes* which becomes a larger and more heterogeneous group. Further study is needed to adequately characterize the evolutionary groups within it or to propose clearly merited generic segregates.

Literature

Knobloch, I. W. 1967. Chromosome numbers in *Cheilanthes, Notholaena, Llavea* and *Polypodium*. Amer. Jour. Bot. 54: 461–464.

Manton, I., and W. A. Sledge. 1954. Observations on the cytology and taxonomy of the pteridophyte flora of Ceylon. Phil. Trans. Roy. Soc. London 238: 127–185.

Tryon, R. M., and A. F. Tryon. 1973. Geography, spores and evolutionary relations in the cheilanthoid ferns, *in:* A. C. Jermy et al., eds. The phylogeny and classification of the ferns, pp. 145–153. Bot. Jour. Linn. Soc. 67, Suppl. 1.

Wagner, W. H., Jr. 1963. A biosystematic survey of United States ferns. Amer. Fern Jour. 53: 1–16.

Walker, T. G. 1966. A cytotaxonomic survey of the pteridophytes of Jamaica. Trans. Roy. Soc. Edinburgh 66: 169–237.

34. *Cheilanthes*
Figs. 34.1–34.40

Cheilanthes Sw., Syn. Fil. 5, 126. 1806, *nom. conserv.* Type: *Cheilanthes micropteris* Sw. [Group 8, see Systematics].

Allosorus Bernh., Neues Jour. Bot. (Schrad.) 1(2): 36. 1805. Type: *Allosorus pusillus* Bernh. (*Pteris acrostica* Balb.) = *Cheilanthes fragrans* (L. f.) Sw.

Myriopteris Fée, Gen. Fil. (Mém. Fam. Foug. 5): 148. 1852. Type: *Myriopteris marsupianthes* Fée = *Cheilanthes lendigera* (Cav.) Sw. [Group 9].

Aleuritopteris Fée, ibidem 153. 1852. Type: *Aleuritopteris farinosa* (Forsk.) Fée (*Pteris farinosa* Forsk.) = *Cheilanthes farinosa* (Forsk.) Kaulf. [Group 6].

Cheiloplecton Fée, Mém. Fam. Foug. 7: 33, *t. 20.* 1857. Type: *Cheiloplecton rigidum* (Sw.) Fée (*Pteris rigida* Sw.) = *Cheilanthes rigida* (Sw.) Domin. [Group 11].

Cosentinia Todaro, Syn. Pl. Acot. Vasc. Sicil. 14. 1866 and Giorn. Sci. Nat. Palmero 1: 219. 1866. Type: *Cosentinia vellea* (Ait.) Todaro (*Acrostichum velleum* Ait.) = *Cheilanthes vellea* (Ait.) F. Muell.

Mildella Trev., Rend. Ist. Lombardo II, 9: 810. 1876. Type: *Mildella intramarginalis* (Link) Trev. (*Pteris intramarginalis* Link) = *Cheilanthes intramarginalis* (Link) Hook. [Group 5].

Cheilosoria Trev., Atti Ist. Veneto V, 3: 579. 1877. Type: *Cheilosoria allosuroides* (Mett.) Trev. = *Cheilanthes allosuroides* Mett. [Group 1].

Choristosoria Kuhn, von Decken's Reisen Ost-Afrika III (3) Bot.: 13. 1879. Type: *Choristosoria pteroides* (Sw.) Kuhn = *Cheilanthes pteroides* Sw.

Pomatophytum Jones, Contrib. West. Bot. 16: 12. 1930. Type: *Pomatophytum pocillatum* Jones = *Cheilanthes lendigera* (Cav.) Sw. [Group 9].

Aspidotis Copel., Gen. Fil. 68. 1947. Type: *Aspidotis californica* (Hook.) Copel. (*Hypolepis californica* Hook.) = *Cheilanthes californica* (Hook.) Mett. [Group 2].

Description

Terrestrial or rupestral; stem erect or decumbent, small and compact to long-creeping and slender, bearing scales and fibrous roots; leaves monomorphic to very rarely dimorphic, ca. 5–75 cm long, borne in a cluster or somewhat distant, lamina 1- to 5-pinnate, or if pedate to 4-pinnatifid, glabrous, glandular, glandular-pubescent, pubescent, scaly, and (or) farinaceous, veins free or (in *C. decora*) anastomosing without included free veinlets; sori usually with few sporangia, rarely one, on unmodified to slightly modified vein-tips, or extending along the apical portion of the veins, or rarely on a discontinuous to continuous marginal commissure, not paraphysate, exindusiate, the margin flat to somewhat recurved and unmodified, or with an indusium gradually and slightly to abruptly and strongly differentiated from the recurved margin, covering one to many sori; spores globose to tetrahedral-globose, the laesurae usually $\frac{3}{4}$ the radius, the surface reticulate-cristate, cristate, or more or less rugose, verrucate, or granulate. Chromosome number: n = 29, 30, 56, 58, 60; $2n$ = 60, 116, 120; apogamous 87, 90.

The exceptional morphological diversity of *Cheilanthes*, especially of the leaves, indusium and indument, makes it difficult to characterize the genus. Fertile segments may be exindusiate or may have a slightly (Fig. 6) or strongly (Figs. 7, 8) differentiated indusium. Most species have a pinnate lamina architecture (Figs. 17–23) but some are pedate. The leaves are usually conspicuously pubescent or scaly (Figs. 9–16) but they may be farinose, glandular or glabrous.

Systematics

Cheilanthes is a large and diverse genus with 150 or more species and nearly worldwide distribution. There appears to be a close relationship with *Adiantopsis, Bommeria, Doryopteris, Notholaena* and *Pellaea*. Within *Cheilanthes*, both major and smaller species-groups can be recognized, but there are also a considerable number of morphologically isolated species. Several species-groups are proposed here to provide a general arrangement for most of the American species. These ferns have evidently undergone strong divergent and convergent evolution as well as considerable extinction. Most species occur in xeric environments that have a mosaic distribution and such conditions would readily influence evolutionary change. The complexity of morphological patterns that have evolved is not amenable to classification into discrete subgenera and sections.

The Old World species of *Cheilanthes*, including all those sometimes placed in *Notholaena*, have not been adequately studied to enable us to place most of them into groups. With a few exceptions, they seem to represent evolutionary lines independent from those in America. Groups such as *Aleuritopteris* and *Choristosoria* are not sufficiently distinct to merit recognition as genera, and considering the heterogeneity of *Cheilanthes*, perhaps the genera *Negripteris* Pic.-Ser., *Sinopteris* C. Chr. & Ching,

Fig. 34.1. *Cheilanthes intramarginalis,* Pedregal Esquilón, north of Jalapa, Veracruz, Mexico. (Photo W. H. Hodge.)

Paraceterach Copel., and *Neurosoria* Kuhn should also be placed in *Cheilanthes.* The relations of *Cheilanthes vellea* (Ait.) F. v. Muell. of Macronesia, the Mediterranean and near East, need further study. Its spores, among other characters are similar to those of *Pityrogramma ferruginea.*

There is no modern revisionary study of *Cheilanthes.* Tryon (1966) contains a revision of the American species then referred to *Notholaena.*

Groups of American *Cheilanthes*

The following groups are recognized, with all or representative species listed, to show the main lines of diversity in America and to provide a framework for further studies. Characterization of the groups is not parallel but includes the salient features of each. A suitable formal classification can perhaps be developed as new data are acquired.

1. *Cheilanthes microphylla* Group

Lamina elongate, with dark axes, glabrous, or short and thinly pubescent. Species: *C. aemula* Maxon, *C. alabamensis* (Buckl.) Kze., *C. allosuroides* Mett. (Fig. 6), *C. microphylla* (Sw.) Sw., and *C. notholaenoides* (Desv.) Weath. (Fig. 21). *Cheilanthes regularis* Mett. (*Adiantopsis regularis* (Mett.) Moore probably belongs here.

Fig. 34.2. *Cheilanthes bonariensis,* Pedregal las Vigas, above Jalapa, Veracruz, Mexico. (Photo W. H. Hodge.)

Fig. 34.3. *Cheilanthes myriophylla,* Pedregal, near San Angel, Mexico City. (Photo W. H. Hodge.)

Fig. 34.4. *Cheilanthes scariosa,* exposed calcareous rocks above Tarma, Junín, Peru. (Photo Alice F. Tryon.)

2. *Cheilanthes californica* Group

Lamina deltoid, finely dissected, glabrous, indusium strongly differentiated. Species: *C. californica* (Hook.) Mett. and *C. meifolia* D. C. Eaton (Fig. 7). *Cheilanthes Schimperi* Kze., not included in this group, is considered to represent a convergent species.

3. *Cheilanthes concolor* Group

Lamina pedate or subpedate, glabrous or rarely pubescent, petiole and the rachis, when present, sulcate or ridged, veins rarely anastomosing, sporangia often borne on an incomplete to complete vascular commissure. One pantropic species, *Cheilanthes concolor* (Langsd. & Fisch.) R. & A. Tryon is in America. *Cheilanthes decora* (Brack.) R & A. Tryon in the Hawaiian Islands, and about five other species of Madagascar-Africa belong to this group.

4. *Cheilanthes Brandegei* Group

Lamina yellow to nearly orange-farinaceous, or sparsely pubescent, petioles fractiferous, with large, thin scales at the base. Species: *C. aurantiaca* Moore, *C. aurea* Baker (Fig. 27), *C. Brandegei* D. C. Eaton, *C. fractifera* Tryon, and *C. Palmeri* D. C. Eaton.

5. *Cheilanthes marginata* Group

Lamina glabrous or sometimes glandular, indusium covering a few to usually many sori, usually either confined to cuneate fertile segments or extending along the stalk. Among the species are: *C. angustifolia* HBK. (Fig. 17), *C. intramarginalis* (Link) Hook. (Figs. 1, 24), *C. Kaulfussii* Kze., *C. marginata* HBK. (Fig. 8), *C. membranacea* (Davenp.) Maxon, *C. Poeppigiana* Kuhn, and perhaps *C. siliquosa* Maxon. Other primarily Chinese species referred to *Mildella* by Hall and Lellinger (1967) differ in their spores from *C. intramarginalis* and probably represent a separate evolutionary line.

Fig. 34.5. Distribution of *Cheilanthes* in America.

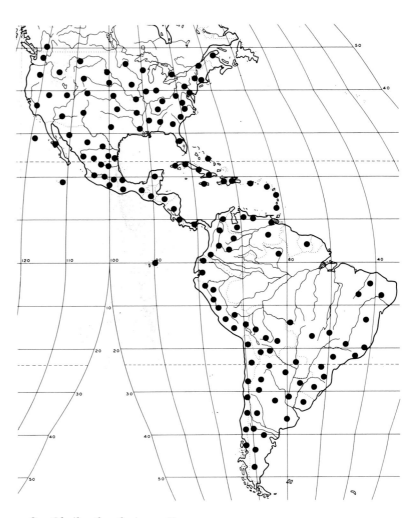

6. *Cheilanthes farinosa* Group

Lamina white to rarely yellow-farinaceous, or the farina sparse and nearly colorless, indusium well differentiated. *Cheilanthes farinosa* (Forsk.) Kaulf. (Fig. 26) is the only species in America, where its scattered distribution suggests that it may have been introduced, perhaps originally in Mexico. About 10 other species, mostly Sino-Himalayan, also belong to this group.

7. *Cheilanthes Fraseri* Group

Lamina pubescent, not scaly, fertile segments with unmodified margins, or the margins gradually and only slightly modified. Among the species are: *C. bonariensis* (Willd.) Proctor (Figs. 2, 9) (*Notholaena aurea* (Poir.) Desv., not *Cheilanthes aurea* Baker), *C. Feei* Moore, *C. Fraseri* Kuhn (Fig. 19), *C. geraniifolia* (Weath.) R. & A. Tryon, *C. goyazensis* (Taubert) Domin (Fig. 13), *C. hypoleuca* (Kze.) Mett. (*Notholaena tomentosa* Desv., not *Cheilanthes tomentosa* Link), *C. lanosa* (Michx.) D. C. Eaton, *C. mollis* (Kze.) Presl (Figs. 11, 23), *C. Newberryi* (D. C. Eaton) Domin (Fig. 16), *C. Parryi* (D. C. Eaton) Domin (Fig. 10), *C. Pohliana* (Kze.) Mett. (Fig. 18) and *C. venusta* (Brade) R. & A. Tryon (Fig. 25). Most of these species have recently been treated in *Notholaena* (Tryon, 1956, species nos. 8–21).

Fig. 34.6. Portion of a fertile segment of *Cheilanthes allosuroides* with slightly differentiated indusium, × 8.

Fig. 34.7. Fertile segments of *Cheilanthes meifolia*, × 7.

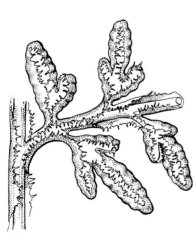

Fig. 34.8. Fertile segments of *Cheilanthes marginata,* the modified margin extending to the axes of the segments, × 5.

8. *Cheilanthes micropteris* Group

Lamina glandular-viscid or glandular-pubescent. Species: *C. micropteris* Sw., *C. pilosa* Goldm., *C. pruinata* Kaulf. and *C. Thellungii* Hert.

9. *Cheilanthes myriophylla* Group

Lamina ca. 3-pinnate, with scales and usually also with trichomes, ultimate segments mostly small to very small, indusium rather poorly differentiated, or rarely strongly so. Among the species are: *C. castanea* Maxon, *C. Eatonii* Baker, *C. horridula* Maxon, *C. lendigera* (Cav.) Sw., *C. Lindheimeri* Hook., *C. myriophylla* Desv. (Figs. 3, 12) and *C. tomentosa* Link.

10. *Cheilanthes squamosa* Group

Lamina with scales, without trichomes, spores verrucate. Species: *C. arequipensis* (Maxon) R. & A. Tryon, *C. incarum* Maxon (Fig. 15), *C. lonchophylla* (Tryon) R. & A. Tryon (Fig. 20), *C. peruviana* (Desv.) Moore, *C. scariosa* (Sw.) Presl and *C. squamosa* Hook. & Grev. Most of these species were recently treated in *Notholaena* (Tryon, 1956, species nos. 1–4.). The group is best characterized by the unusual verrucate spores.

11. Morphologically Isolated Species

The following species are among those that cannot be placed in one of the preceding groups. Additional data may clarify their relations to the other American species. Species: *Cheilanthes brachypus* Kze. (Fig. 14), *C. dichotoma* Sw. (*Adiantopsis dichotoma* (Sw.) Moore), *C. glauca* (Cav.) Mett., *C. leucopoda* Link, *C. Lozanii* (Maxon) R. & A. Tryon (Fig. 22), *C. rigida* (Sw.) Domin, *C. sinuata* (Sw.) Domin and *C. Skinneri* (Hook.) R. & A. Tryon (Fig. 28).

Ecology (Figs. 1–4)

Cheilanthes is primarily a genus of open habitats, especially in semiarid regions. It also grows in more mesic regions in habitats that are seasonally dry.

In tropical America plants occur on cliffs and ledges, on rocky slopes, on stone walls and less often on road banks or steep slopes. The genus is most frequent in regions of xeric scrub and open parkland and also occurs on cliffs or in rocky places in forested areas. Species grow on a great variety of rock types; some are apparently restricted to either acidic or basic substrates. For example, *Cheilanthes leucopoda* of Texas and Mexico and *C. scariosa* of Peru and Bolivia are evidently confined to calcareous rocks. *Cheilanthes* grows mostly at 1000–3000 m, sometimes as low as 100 m, and in the Andes species such as *C. pruinata* can reach 4400 m.

Many species with creeping stems form large clumps at the base of boulders or in ledges of cliffs. Species are usually heliophiles with leaves exposed in sun while the stems are in cooler, moist microniches under rocks. Temperatures taken on lava rocks in Mexico, near Jalapa, indicated that most *Cheilanthes* species growing there had stems well below the surface of the rocks where the temperature was at least 15°F less than at the surface.

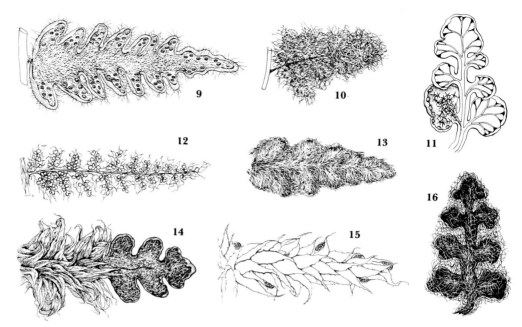

Figs. 34.9–34.16. Leaf segments of *Cheilanthes* species.   **9.** *C. bonariensis,* lower surface, × 4.5.   **10.** *C. Parryi,* upper surface, × 2.5.   **11.** *C. mollis,* lower surface, stellate trichomes removed except in one area, enlarged vein-tips indicate sori, × 4.5.   **12.** *C. myriophylla,* upper surface, × 2.   **13.** *C. goyazensis,* upper surface, × 2.5.   **14.** *C. brachypus,* lower surface, scales partly removed, × 4.5.   **15.** *C. incarum,* lower surface, × 2.   **16.** *C. Newberryi,* upper surface, × 9.

A detailed account of the adaptive physiology of *Cheilanthes Parryi* (as *Notholaena*) by Nobel (1978), emphasizes the importance of rainfall channeling by rocks, which provides a relatively mesic niche for the stems and roots. In species that grow in areas having recurrent dry periods, the plants have stems and leaves adapted to survive periodic dessication.

Geography (Fig. 5)

Cheilanthes has nearly worldwide distribution, although relatively few species grow outside of the semiarid tropics and subtropics.

In America the genus occurs from Quebec, Ontario, and British Columbia in southern Canada south to Llanquihue in Chile and Santa Cruz in Argentina. It is absent from the Amazon basin and adjacent areas, and is uncommon in most of Brazil. The center of diversity of the genus is in Mexico and the adjacent United States (Texas to California), where about 50 species occur. A secondary center with about 30 species is in the Andes of Colombia to Bolivia. Since most species have limited or very local ranges, these centers of diversity are also centers of endemism. Few species, such as *Cheilanthes sinuata, C. bonariensis* and *C. myriophylla,* have very broad ranges, from the southwestern United States and (or) Mexico to Argentina.

Spores

There are several distinctive types of spores in *Cheilanthes* but the predominate form is reticulate-cristate as reported in a review of the spores of the cheilanthoids (Tryon and Tryon,

Figs. 34.17–34.23. *Cheilanthes* leaves, all ×0.5. **17.** *C. angustifolia.* **18.** *C. Pohliana.* **19.** *C. Fraseri.* **20.** *C. lonchophylla.* **21.** *C. notholaenoides.* **22.** *C. Lozanii.* **23.** *C. mollis.*

1973). The characteristic surface in *C. membranacea* (Fig. 29) is formed of intricate strands which fuse into cristae (Fig. 30). The cristate structure is elaborated in *C. Skinneri* (Fig. 37) and especially in *C. leucopoda* (Fig. 39). A compact, rugose surface is frequent in several groups, and quite consistently in the *C. Fraseri* group (*C. geraniifolia,* Fig. 31) and the *C. myriophylla* group (*C. myriophylla,* Fig. 35). The surface is compact and granulate in a few species as *C. bonariensis* and *C. rigida* (Figs. 33, 34). The unique verrucate spores in *C. peruviana* (Fig. 40) are characteristic of a group of six Andean species.

Figs. 34.24–34.28. *Cheilanthes* leaves, all ×0.5.   **24.** *C. intramarginalis.*   **25.** *C. venusta.*   **26.** *C. farinosa.*   **27.** *C. aurea.*   **28.** *C. Skinneri.*

Abraded spores show the laminated structure of the wall (Figs. 32, 36, 38). The rugulose spores of *C. geraniifolia* have a relatively smooth basal stratum overlaid by a more or less compact reticulum or stranded formation and the outer rugulose surface is formed of fused strands (Fig. 32). There are marked differences in size, as in *C. rigida* (Fig. 33), which has large spores ca. 90 μm in diameter in contrast to those of *C. Skinneri* (Fig. 37), which are ca. 30 μm. The limited cytological records for species of *Cheilanthes* restrict correlations with spore size, however report of a triploid in *C. myriophylla* from Peru does correlate with large spores (Fig. 35). To some extent, the surface contours are useful in depicting group relations, as in the species with unique, verrucate spores, but affinities are not as readily shown among species with reticulate and cristate spores.

29   30

31   32

33   34

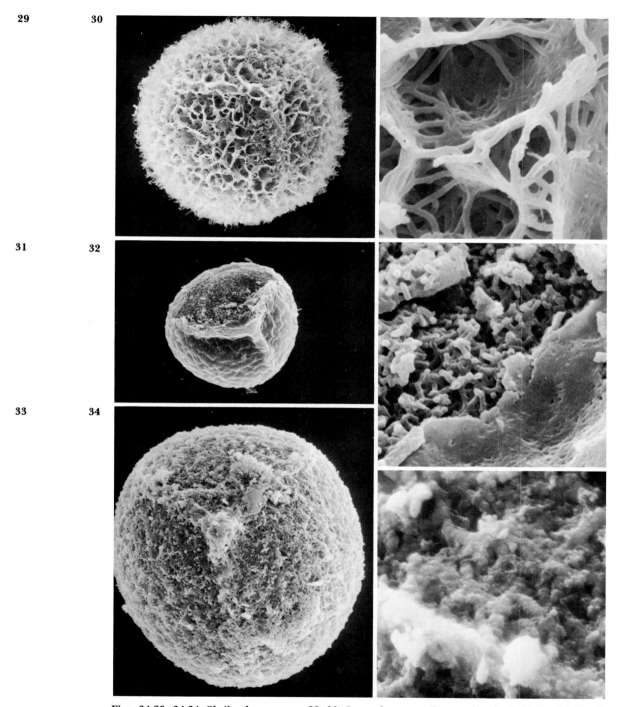

Figs. 34.29–34.34. *Cheilanthes* spores. **29, 30.** *C. membranacea,* Guatemala, *Skutch 1059.* **29.** Proximal face, reticulate-cristate, × 1000. **30.** Detail of stranded reticulate structure, × 10,000. **31, 32** *C. geraniifolia,* Brazil, *Ule 7239.* **31.** Proximal face, rugose, partially abraded, × 1000. **32.** Detail of abraded area with outer rugulose surface, below, and inner reticulate formation, above, × 5000. **33, 34.** *C. rigida.* **33.** Proximal face, Mexico, *Hinton 13374,* × 1000. **34.** Detail of granulate surface, part of laesura, lower left, Mexico, *Mexia 8826,* × 5000.

Cytology

Two polyploid levels based on either 29 or 30 are known from a relatively broad sample in *Cheilanthes* species. Reports of 13 species from northern Mexico by Knobloch and collaborators (1966, 1967, 1975) include one tetraploid and seven triploids based on either 29 or 30. The survey of nine species of *Cheilanthes* of Europe and the Canary Islands (Vida et al., 1970) includes eight tetraploids based on either 29 or 30 and two species with both diploid and tetraploid populations. Apogamy is known in several species; in *Cheilanthes alabamensis* Whittier (1965) reports a triploid with 87 that produced apogamous gametophytes in culture. In *C. farinosa* diploids with $n = 30$ and tetraploids with $n = 60$ as well as apogamous plants with 90 were reported from Ceylon by Manton and Sledge (1954). The limited cytological records of *Cheilanthes* from the American tropics do not supply sufficient data on the affinities of groups or the relationships between species based on either 29 or 30. The present record of only diploid, triploid and tetraploid levels in *Cheilanthes* indicate that polyploidy has not reached levels as high as in many genera.

Literature

Hall, C. C., and D. B. Lellinger. 1967. A revision of the fern genus *Mildella*. Amer. Fern Jour. 57: 113–134.

Knobloch, I. W. 1966. A preliminary review of spore numbers and apogamy within the genus *Cheilanthes*. Amer. Fern Jour. 56: 163–167.

Knobloch, I. W. 1967. Reference under the tribe Cheilantheae.

Knobloch, I. W., W. Tai, and T. Adangappuram. 1975. Chromosome counts in *Cheilanthes* and *Aspidotis* with a conspectus of the cytology of the Sinopteridaceae. Amer. Jour. Bot. 62: 649–654.

Manton, I., and W. A. Sledge. 1954. Reference under the tribe Cheilantheae.

Nobel, P. S. 1978. Microhabitat, water relations, and photosynthesis in a desert fern, *Notholaena Parryi*. Oecologia 31: 293–309.

Tryon, R. 1956. A revision of the American species of *Notholaena*. Contrib. Gray Herb. 179: 1–106.

Tryon, R. and A. F. Tryon. 1973. Reference under the tribe Cheilantheae.

Vida, G., C. N. Page, T. G. Walker, and T. Reichstein. 1970. Cytology der Fern-Gattung *Cheilanthes* in Europa und auf den Canarischen Inseln. Bauhinia 4: 223–252.

Whittier, D. P. 1965. Obligate apogamy in *Cheilanthes tomentosa* and *C. alabamensis*. Bot. Gaz. 126: 275–281.

35   36

37   38

39   40

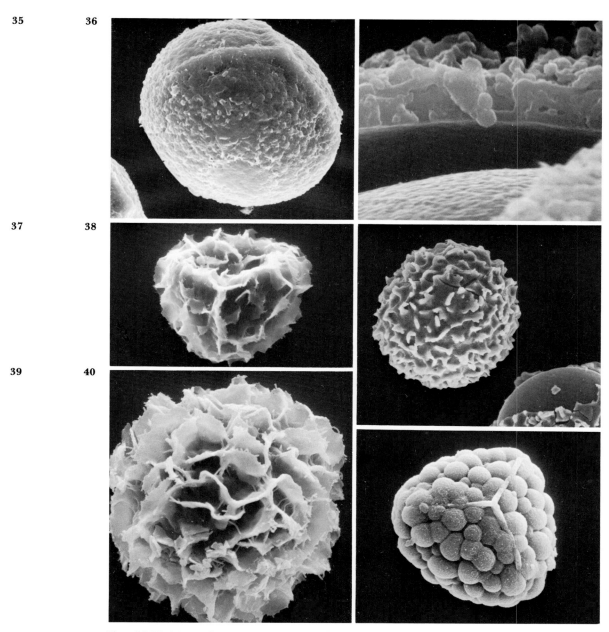

Figs. 34.35–34.40. *Cheilanthes* spores, × 1000, surface detail, × 10,000.   **35.** *C. myriophylla,* rugose proximal face tilted, Venezuela, *Tryon & Tryon 5759.*   **36.** *C. horridula,* wall section of compact reticulate formation, above, separated by a gap from the rugulose stratum, below, Texas, *Correll & Correll 12788.*   **37.** *C. Skinneri,* cristate proximal face, Mexico, *Goldsmith 102.*   **38.** *C. farinosa,* distal face coarsely cristate, portion of abraded spore with smooth lower stratum, below, Mexico, *Purpus 1599.*   **39.** *C. leucopoda,* cristate with slender connecting strands, Texas, *Cory 8608.*   **40.** *C. peruviana,* verrucate proximal face, Peru, *López et al. 3688.*

35. *Bommeria*
Figs. 35.1–35.13

Bommeria Fourn., Dict. Bot. (Baillon) 1: 448. 1876. Type: *Gymnogramma Ehrenbergiana* Kl. = *Bommeria Ehrenbergiana* (Kl.) Underw.

Gymnogramma section *Stenogramme* Kl., Linnaea 20: 411. 1847. Type: *Gymnogramma Ehrenbergiana* Kl. = *Bommeria Ehrenbergiana* (Kl.) Underw.

Gymnopteris subgenus *Bommeria* (Fourn.) Christ, Farnkr. Erde 67. 1897.

Description Terrestrial or rupestral; stem small, very short-creeping to long-creeping and slender, bearing scales which (in *B. hispida*) may intergrade to trichomes, and many slender, fibrous roots, these often branching at right angles; leaves monomorphic, ca. 5–35 cm long, borne in a loose cluster, or rather distant, lamina bipinnatifid to tripinnatifid, pedate, pubescent and somewhat scaly, veins free or partially anastomosing without included free veinlets; sori on unmodified veins, sometimes extending below a fork, forming a narrow to broad marginal band, not paraphysate, exindusiate; spores globose, trilete, the laesurae $\frac{3}{4}$ the radius or nearly so, the surface prominently reticulate to cristate, formed of more or less fused strands. Chromosome number: $n = 30$; $2n = 60$; apogamous = 90.

Systematics *Bommeria* is an American genus of four species, characterized by pedate leaves (Figs. 3–6) and sori in marginal bands (Figs. 7, 8). It belongs to the complex of cheilanthoid ferns, although its exact relationships are not certain. There are similarities to *Hemionitis,* especially in soral characters, and also to elements in *Cheilanthes.* The reticulate-cristate spores are similar to those of *Cheilanthes Palmeri* and *C. aurantiaca* of the *C. Brandegei* group, and the pedate, pubescent leaves to species of the *Cheilanthes Fraseri* group.

  Bommeria was revised by Maxon (1913) and by Haufler (1979). The key is adapted from the latter paper.

Key to Species of *Bommeria*

a. Veins free.   b.
  b. Stem long-creeping, the petioles distant, the abaxial surface of the lamina with some long, nearly straight trichomes and many shorter, tortuous ones, sporangium with 64 spores; Texas to California, Mexico, Nicaragua. (Fig. 6)   *B. hispida* (Kuhn) Underw.
  b. Stem short-creeping, the petioles adjacent, abaxial surface of the lamina with long, nearly straight trichomes, sporangium with 32 spores; Mexico to Costa Rica. (Figs. 4, 7)   *B. pedata* (Sw.) Fourn.
a. Veins partially anastomosing.   c.
  c. The dark color of the petiole not extending into the primary costa on the abaxial side of the lamina; Mexico. (Fig. 3)   *B. subpaleacea* Maxon
  c. The dark color of the petiole extending half or more the length of the primary costa on the abaxial side of the lamina; Mexico. (Figs. 5, 8)
                               B. Ehrenbergiana (Kl.) Underw.

Ecology and Geography (Figs. 1, 2)

Bommeria grows in xeric or semixeric regions (Fig. 1) on rocky hillsides, at the edge of boulders and on cliffs. *Bommeria hispida* and *B. pedata* usually grow in drier sites than *B. subpaleacea* and

Fig. 35.1. *Bommeria subpaleacea,* La Bufa, Chihuahua, Mexico. (Photo C. H. Haufler.)

B. Ehrenbergiana, the latter are more often in mesic canyons or in rocky places in moist forests.

Bommeria grows at 900–2400 m, from Texas to California in the southwestern United States through Mexico and sparingly in Central America to Alajuela in Costa Rica (Fig. 2).

Spores

Bommeria spores are predominantly reticulate-cristate (Figs. 9–13) or cristate (Fig. 12) with a network of strands forming the outer perispore, that is essentially similar on both proximal and distal surfaces. The strands are more or less fused in a reticulum or compact cristate structure (Fig. 10). The abraded surface in Fig. 11 shows portions of the reticulate outer stratum, the inner papillate layer and a smooth innermost layer that may be exospore. There is considerable variation in fusion of the strands and prominence of the cristate surface. Details of the stratification are reviewed by Haufler and Gastony (1978b) in their study of spore wall morphology in *Bommeria* and related genera.

Cytology

A cytological survey of the species of *Bommeria* by Gastony and Haufler (1976) shows that this group is based on $n = 30$. Three species are diploid with $n = 30$ and *B. pedata* is apogamous with both n and $2n = 90$.

Fig. 35.2. Distribution of *Bommeria*.

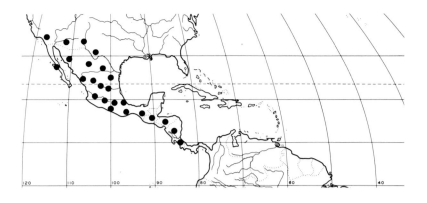

Figs. 35.3–35.6. Bommeria species, leaves and a plant, × 0.5.   **3.** *B. subpaleacea.*   **4.** *B. pedata.*   **5.** *B. Ehrenbergiana.*   **6.** *B. hispida.* (After Haufler, 1979.)

Observations

Haufler and Gastony (1978a) report the production of antherido-gen A by gametophytes of all four species of *Bommeria*. Neither antheridogen B from *Anemia phyllitidis* nor antheridogen C from *Ceratopteris* stimulated development of antheridia in the *Bommeria* gametophytes.

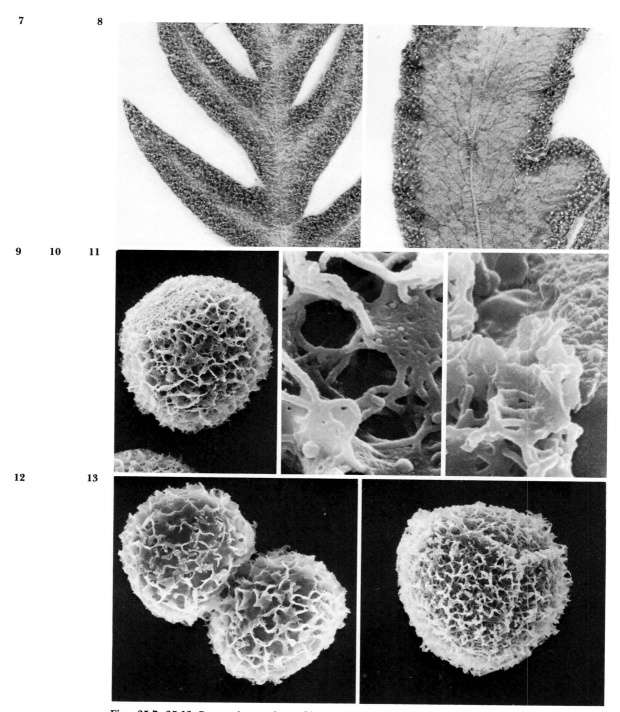

Figs. 35.7–35.13. *Bommeria*, portions of leaves, **Figs. 7, 8,** and spores, **Figs. 9–13. 7. *B. pedata*** with broad bands of sori, trichomes are especially evident on the costae, × 4. **8. *B. Ehrenbergiana*,** with a narrow marginal band of sori and partially anastomosing veins, ×4. **9–11. *B. hispida*,** reticulate-cristate spores, Mexico, *Reeder et al. 3476*. **9.** Lateral view, the proximal face left, × 1000. **10.** Detail of fused strands at base of crests, × 10,000. **11.** Detail of wall strata, the outer of fused strands, left, the inner irregularly papillate, upper right, innermost smooth surface of exospore, lower right. × 10,000. **12. *B. subpaleacea*,** cristate, lateral view, Mexico, *Knoblock 6044*, × 1000. **13. *B. Ehrenbergiana*,** reticulate-cristate proximal face, Mexico, *Fisher 37111*, × 1000.

Studies on the flavonoid constituents of *Bommeria* show the presence of quercetin and kaempferol glycosides (Haufler, 1977). *Bommeria hispida* is unique in having luteolin, while *O*-glycosides are present in both *B. hispida* and *B. subpaleacea* and absent in the other two species. The presence of quercetin and kaempferol glycosides in this group and also in *Hemionitis* was regarded by Haufler as indicating a close relationship between these genera. However, the systematic significance of the two compounds is dependent on additional analyses of other genera of cheilanthoid ferns, especially the species of *Cheilanthes* showing morphological similarities to *Bommeria*.

Literature

Gastony, G. J., and C. H. Haufler. 1976. Chromosome numbers and apomixis in the fern genus *Bommeria*.

Haufler, C. H. 1977. A Biosystematic Study of the Fern Genus *Bommeria*. Ph.D. Thesis, Indiana University, Bloomington, Indiana.

Haufler, C. H. 1979. A biosystematic revision of the fern genus *Bommeria*. Jour. Arn. Arb. 60: 445–476.

Haufler, C. H., and G. J. Gastony. 1978a. Antheridogen and the breeding system in the fern genus *Bommeria*. Canad. Jour. Bot. 56: 1594–1601.

Haufler, C. H., and G. J. Gastony. 1978b. Systematic implications of spore morphology in *Bommeria* and related fern genera. Syst. Bot. 3: 241–256.

Maxon, W. R. 1913. Notes upon *Bommeria* and related genera. Contrib. U.S. Nat. Herb. 17: 168–175.

36. *Adiantopsis*
Figs. 36.1–36.12

Adiantopsis Fée, Gen. Fil. (Mém. Fam. Foug. 5): 145. 1852. Type: *Adiantopsis paupercula* (Kze.) Fée (*Adiantum pauperculum* Kze.).

Description

Terrestrial or rupestral; stem small to moderately stout, erect or decumbent and short-creeping, rarely slender and long-creeping, bearing scales and many fibrous roots; leaves monomorphic, ca. 10–75 cm long, borne in a cluster or rarely somewhat distant, lamina pinnate, pedate or radiate, 1- to 4-pinnate, the axes strongly ridged adaxially, glabrous or finely and sparingly short-pubescent, veins free; sori single at the ends of the veins, not paraphysate, each (sometimes two) covered by a roundish to lunate indusium, moderately to well differentiated from the recurved margin; spores tetrahedral-globose, trilete, the laesurae $\frac{3}{4}$ the radius, often obscure, the surface prominently echinate and often more or less reticulate at the base of the projecting structures. Chromosome number: $2n = 60$.

Adiantopsis has the lamina axes ridged on the adaxial side (Fig. 6), asymmetrical ultimate segments (Figs. 5, 8), and echinate spores (Figs. 9–12). In addition, the ultimate segments are usually articulate with age, and the indusia are well differentiated from the margin and are discrete, usually covering a single sorus (Fig. 3).

Fig. 36.1. *Adiantopsis paupercula,* near Mason River savannah, Jamaica. (Photo W. H. Hodge.)

Systematics *Adiantopsis* is a tropical American genus of about seven species. It is closely related to *Cheilanthes,* especially the group of *C. microphylla.* Species with cristate spores such as *Adiantopsis dichotoma* and *A. regularis* are placed in *Cheilanthes,* and Old World species sometimes referred to *Adiantopsis* are also placed there.

Prantl (1883) published a synopsis of the genus, which contributed more to recognition of the group than to the knowledge of its species.

Tropical American Species

Most species of *Adiantopsis* are distinctive especially in lamina architecture. However, problems with classification of several species have proved difficult to resolve from herbarium specimens, and these require field study. *Adiantopsis ternata* Prantl and *Cheilanthes trifurcata* Baker have leaves with three main lamina branches and probably represent juvenile or depauperate plants of *A. radiata,* which has five or more main lamina branches. *Adiantopsis asplenioides* Maxon is probably a juvenile form of *A. paupercula,* and *A. rupincola* Maxon does not seem to be distinct from *A. Reesii.*

Adiantopsis minutula and *A. monticola* are apparently distinctive species of the genus, and they are included in the following key which is based on characters of mature plants.

Fig. 36.2. Distribution of *Adiantiopsis*.

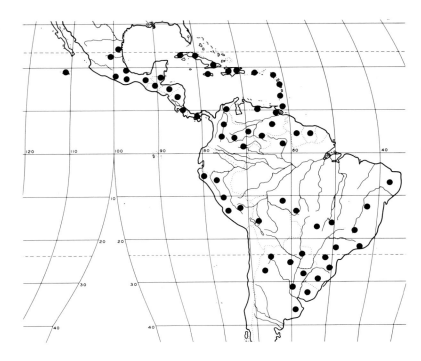

Fig. 36.3. Fertile segment of *Adiantopsis paupercula,* × 0.2.

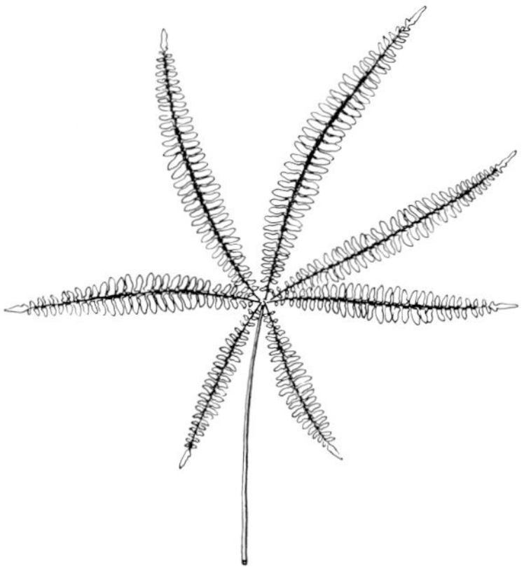

Key to Species of *Adiantopsis*

a. Lamina radiate; tropical America. (Figs. 4, 5)   *A. radiata* (L.) Fée
a. Lamina pinnate or subpedate.   b.
  b. Lamina 1-pinnate; Goias, Brazil   *A. monticola* (Hook.) Moore
  b. Lamina 2-pinnate or more complex.   c.
    c. Lamina pedate, each basal pinna strongly inequilateral, with a much-prolonged basal basiscopic segment; Greater Antilles. (Fig. 7)
                                                  A. pedata (Hook.) Moore
    c. Lamina pinnate, the basal pinnae more or less equilateral.   d.
      d. Apex of pinnae with a stalked segment.   e.
        e. Pinnae with two or three ultimate segments; Mato Grosso, Brazil.   *A. minutula* Sehnem
        e. Pinnae with several to many ultimate segments; Greater Antilles. (Figs. 1, 3, 8)   *A. paupercula* (Kze.) Fée
      d. Apex of pinnae with several confluent segments.   f.
        f. Ultimate segments mostly or partly adnate to sessile and truncate; South America.   *A. chlorophylla* (Sw.) Fée
        f. Ultimate segments mostly sessile to stalked, cuneate; Greater Antilles.   *A. Reesii* (Jenm.) C. Chr.

Ecology (Fig. 1)

Adiantopsis is a genus of forests, thickets and rocky places. *Adiantopsis paupercula* and *A. pedata* are primarily on, or perhaps confined to, limestone cliffs and calcareous rocky slopes, and *A. monticola* to acidic rocks. *Adiantopsis radiata* is a forest species, usually in lowland rain forests or also in montane forests, growing in humus or sometimes among rocks. *Adiantopsis chlorophylla* and *A. Reesii* grow most often in gallery forest, in open woods or in thickets. *Adiantopsis* grows mostly from near sea level to ca. 1000 m, sometimes up to 1600 m.

Fig. 36.4. Lamina of *Adiantopsis radiata,* × 0.25.

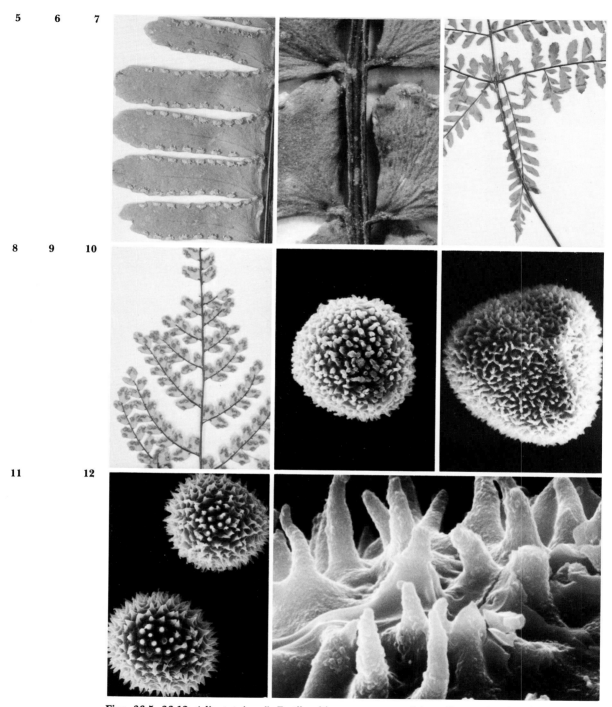

Figs. 36.5–36.12. *Adiantopsis.* **5.** Fertile ultimate segments of ***A. radiata,*** ×3. **6.** Adaxially ridged lamina branch of ***A. radiata,*** ×10. **7.** Basal pinnae of ***A. pedata*** each with an enlarged basal, basiscopic segment, ×10. **8.** Portion of apical part of lamina of ***A. paupercula,*** ×1. **9–12.** Echinate spores, ×1000; surface detail, ×10,000. **9.** ***A. paupercula,*** distal face, Cuba, *Maxon 4239.* **10.** ***A. chlorophylla,*** proximal face, Brazil, *Mexia 4827a.* **11.** ***A. Reesii,*** distal faces, Dominican Republic, *R. & E. Howard 8757.* **12.** ***A. paupercula,*** detail of echinate structure with particulate surface, Cuba, *Maxon 4239.*

Geography (Fig. 2)

Adiantopsis occurs from Mexico through Central America, in the West Indies from Cuba to St. Vincent and Trinidad; in South America it is generally distributed south to Buenos Aires in Argentina, except in the Amazon basin and the arid northeast of Brazil; it also occurs on the Revillagigedo Islands. *Adiantopsis monticola* is endemic in Goias, Brazil; and *A. paupercula*, *A. pedata* and *A. Reesii* are on two or more of the Greater Antilles. *Adiantopsis chlorophylla* occurs in Argentina and southern Brazil, and less frequently north to Colombia. *Adiantopsis radiata* encompasses the entire range of the genus, except in the south where it is absent from Uruguay and all but northeastern Argentina.

Spores

Adiantopsis spores are echinate and usually reticulate at the base of the projecing elements (Figs. 9–11). The detail (Fig. 12) shows the pelleted formation of the surface as well as portions of the reticulum below.

Cytology

The record of *Adiantopsis radiata* from Trinidad as $2n = 60$ by Walker (1973) is consistent with that of other members of the cheilanthoid alliance.

Literature

Prantl, K. 1883. *Adiantopsis alata* Prantl. Gartenfl. 32: 99–103.
Walker, T. G. 1973. Evidence from cytology in the classification of ferns, *in*: A. C. Jermy et al., The phylogeny and classification of ferns, pp. 91–110. Bot. Jour. Linn. Soc. 67, Suppl. 1.

37.   *Notholaena*
Figs. 37.1–37.28

Notholaena R. Br., Prod. Fl. Nov. Holl. 145. 1810. Type: *Notholaena trichomanoides* (L.) Desv. (*Pteris trichomanoides* L.).
Notholaena section *Argyrochosma* J. Sm., Jour. Bot. (Hook.) 4: 50. 1841. Type: *Notholaena nivea* (Poir.) Desv. (*Pteris nivea* Poir.), chosen by C. Chr. Ind. Fil. xl. 1906.
Chrysochosma (J. Sm.) Kümm., Mag. Bot. Lapok. 13: 35. 1914. *Notholaena* section *Chrysochosma* J. Sm., Hist. Fil. 279. 1875. Type: *Notholaena sulphurea* (Cav.) J. Sm. (*Pteris sulphurea* Cav.).

Description   Terrestrial or rupestral; stem small, decumbent to suberect, short- to rarely somewhat long-creeping and slender, bearing scales and usually many fibrous roots; leaves monomorphic, usually 5–25 cm, to 45 cm long, borne in a cluster, or rarely somewhat distant, lamina pinnate, 1- to 4-pinnate to rarely pedate and 2-pinnate-pinnatifid or bipinnatifid at the base, usually white, rarely yellow, farinose beneath and often also with scales and (or) trichomes, sometimes glabrous, veins free; sori usually on the somewhat modified vein-tips, sometimes of one or two

Fig. 37.1. *Notholaena Standleyi,* Camelback Mt., near Phoenix, Arizona. (Photo Timothy Reeves.)

sporangia at the vein-tips or slightly back of them, or on the apical $\frac{1}{3}$ to $\frac{1}{2}$ of the unmodified veins, or rarely nearly all along the veins, not paraphysate, exindusiate, the margin flat to strongly recurved but otherwise not or rarely slightly modified; spores somewhat globose, trilete, the laesurae $\frac{1}{2}$ to $\frac{3}{4}$ the radius, or often obscure, sometimes lacking or less than three ridges, prominently cristate and brown to tan, or black usually with a dense, granulate surface. Chromosome number: $n = 29$, ca. 30; $2n = 60$, ca. 116; ? apogamous 90.

Notholaena is characterized by the absence of an indusium, by a flat (Fig. 3) to strongly recurved (Fig. 6) but not otherwise modified margin, and the presence, with few exceptions, of a farinose indument (Fig. 4). This indument is sometimes densely covered by scales (Figs. 10, 11). In section *Argyrochosma* the stem scales are rufous and rather thin (Fig. 7) while those of section *Notholaena* are partially (Fig. 12) to wholly sclerotic and often toothed. The sori may be on modified vein-tips (Figs. 8, 9), or along much of the unmodified vein (Figs. 4, 5). Variation in the lamina architecture is shown in Figs. 13–21.

Systematics   *Notholaena* is an American genus of 39 species. In the present work two sections are recognized, especially on the basis of differences in spores. Section *Argyrochosma* has cristate spores similar to those in several related general while section *Notholaena* has spores with a unique, black surface deposit.

Notholaena section *Argyrochosma* is most closely related to the genus *Pellaea*. The apparent transition between the two sections

Fig. 37.2. Distribution of *Notholaena*.

suggests that section *Notholaena* has been derived from section *Argyrochosma* but relationships of the former to the exindusiate species of *Cheilanthes* must also be considered. Further cytological data may clarify the relations of section *Notholaena*.

The species of *Notholaena* have been revised by Tryon (1956). The present treatment places species 1–21 of that work in *Cheilanthes,* and species 22–58 in the genus *Notholaena*. Also, all Old World species of *Notholaena* are placed in *Cheilanthes*.

Notholaena trichomanoides is adopted here as the type of *Notholaena,* following the first choice by John Smith in *Historia Filicum.* The other four species included in the genus by Robert Brown we refer to *Cheilanthes.* The frequent use by earlier authors of the name "*Cincinalis* Desv." for part or all of *Notholaena* was based on an erroneous application of the name. Desvaux treated species of *Notholaena* under *Cincinalis* Gleditsch; however, that name is a synonym of *Pteridium.*

Synopsis of *Notholaena*

Section *Argyrochosma*

Stem scales lax, more or less rufous; segments mostly stalked or sometimes sessile; sori usually on the apical ¼ to ⅔ of the unmodified veins, or on nearly the whole length of the veins (*N. Palmeri*), on long clavate vein-tips (*N. bryopoda*), or with 1–2 sporangia borne slightly below the vein-tip (*N. dealbata* (Pursh) Kze., *N. Fendleri* Kze.); spores brown or tan, coarsely cristate or (in *N. incana*) the ridges often covered by a dense layer forming a slightly rugose surface.

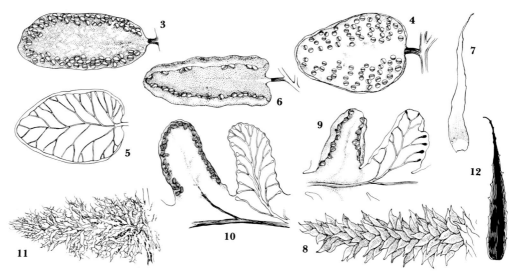

Figs. 37.3–35.12. *Notholaena* (lower surface of segments).   **3.** Fertile segment of *N. nivea* var. *oblongata*), the sporangia near the flat margins, protruding through the farinose indument, ×4.5.   **4.** Fertile segment of *N. incana*, the sporangia protruding through the farinose indument, ×4.5.   **5.** Fertile segment of *N. incana*, with sporangia and indument removed showing the unmodified fertile veins, ×4.5.   **6.** Fertile segment of *N. parvifolia*, without indument, some sporangia protruding from the strongly recurved margins, ×4.5.   **7.** Thin stem scale of *N. incana*, ×7.   **8.** Segment of *N. aurantiaca*, with imbricate scales covering the farinose indument, ×2.5.   **9.** Fertile segments of *N. Rosei*, with farinose indument and sporangia removed, right, and part of the margin bent back to show the modified fertile vein-tips, × 4.5.   **10.** Fertile segments of *N. affinis*, with farinose indument and sporangia removed, right, showing the modified fertile vein-tips, × 4.5.   **11.** Segment of *N. Galeottii*, with imbricate scales covering the farinose indument, ×4.5.   **12.** Sclerotic stem scale of *N. galapagensis*, ×9.

A section of 17 species (numbers 43–58, Tryon, 1956) and also *Notholaena Stuebeliana* (Tryon, 1961). Species that lack farinose indument are: *Notholaena formosa, N. Jonesii* Maxon, *N. Lumholtzii* Maxon & Weath., *N. parvifolia* Tryon and *N. nivea* var. *tenera* (Hook.) Griseb.

Section *Notholaena*

Stem scales rigid, dark sclerotic, or mostly with a sclerotic central band; segments mostly joined or adnate, sometimes sessile; sori borne on the somewhat modified vein-tips; spores finely cristate or not, covered by a black, more or less dense surface deposit.

A section of 22 species [numbers 22–42 of Tryon (1956) and also *Notholaena angusta* (Tryon, 1961)].

Tropical American Species

Characteristic species of *Notholaena* section *Argyrochosma* are: *Notholaena bryopoda* Maxon (Fig. 14), *N. formosa* (Liebm.) Tryon, *N. incana* Presl, *N. limitanea* Maxon, *N. nivea* (Poir.) Desv. (with four varieties), *N. Palmeri* Baker, *N. pallens* Tryon (Fig. 15), and *N. Stuebeliana* (Hieron.) Tryon. *Notholaena delicatula* Maxon & Weath. rarely has been collected and may prove to be a variety, or only a variation, of *N. incana*.

Among the species of section *Notholaena* are: *Notholaena affinis* (Mett.) Moore (Fig. 19), *N. angusta* Tryon, *N. Aschenborniana* Kl., *N. candida* (Mart. & Gal.) Hook. (with two varieties) (Fig. 21), *N. Ekmanii•* Maxon (Fig. 18), *N. Galeottii* Fée, *N. Greggii* (Kuhn)

Maxon, *N. neglecta* Maxon (Fig. 17), *N. rigida* Davenp., *N. Rosei* Maxon, *N. Schaffneri* (Fourn.) Davenp. (with two varieties), *N. sulphurea* (Cav.) J. Sm. (Fig. 20), and *N. trichomanoides* (L.) Desv. (Fig. 16).

Ecology (Fig. 1)

Notholaena usually grows in regions of sparse vegetation including grasslands, xeric scrub, or dry parklands. Species are nearly always confined to rocky slopes in ravines, at the base of boulders, on cliffs or other rocky sites; they rarely grow in clay soil. In forested areas, species may grow on cliffs or other rocky and locally xeric places. The usually compact stem is often branched and forms rather large, multicipital clumps. Most species are not restricted to a particular substrate, but some such as *Notholaena rigida*, *N. neglecta* and *N. parvifolia* are definitely calciphiles, and others such as *N. Aschenborniana* and *N. formosa* are probably restricted to basic rocks. *Notholaena Greggii* grows most often on gypsum and *N. bryopoda* is apparently confined to it. *Notholaena* grows from near sea level to ca. 3000 m or in the Andes up to 4200 m (*N. nivea*); most frequently it grows from 1000 to 2500 m.

Geography (Fig. 2)

Notholaena is an American genus ranging from Missouri and Wyoming in the central United States south through Mexico and Central America; in all of the Greater Antilles; and in South America from Colombia south to Mendoza and Buenos Aires in Argentina and Tarapacá in Chile; it is isolated in Santa Catarina and Minas Gerais in southeastern Brazil; also on the Galápagos Islands, Juan Fernandez Islands, and Revillagigedo Islands.

Section *Argyrochosma* occurs nearly throughout the range of the genus, except the Galápagos Islands, while section *Notholaena* is absent from South America except for the Andean *Notholaena sulphurea*.

The species of *Notholaena* are strongly concentrated in Mexico where 29 of the 39 species grow, 14 of them endemic. There are five species, including three endemics, in the Greater Antilles; four species, none endemic, in Central America; and three, including two endemics, in South America. The island endemics are: *Notholaena trichomanoides* (all of the Greater Antilles), *N. cubensis* Tryon and *N. Ekmanii* (Cuba), *N. galapagensis* Weath. & Svenson (Galápagos Islands), and *N. chilensis* (Juan Fernandez Islands). The general distribution is similar to that of *Pellaea*, but in *Notholaena* species endemism is well developed and more geographically restricted.

Spores

The spores are especially useful in distinguishing the sections. Those of species in section *Argyrochosma* are brown or tan with a strongly cristate surface with strands connecting the crests as in

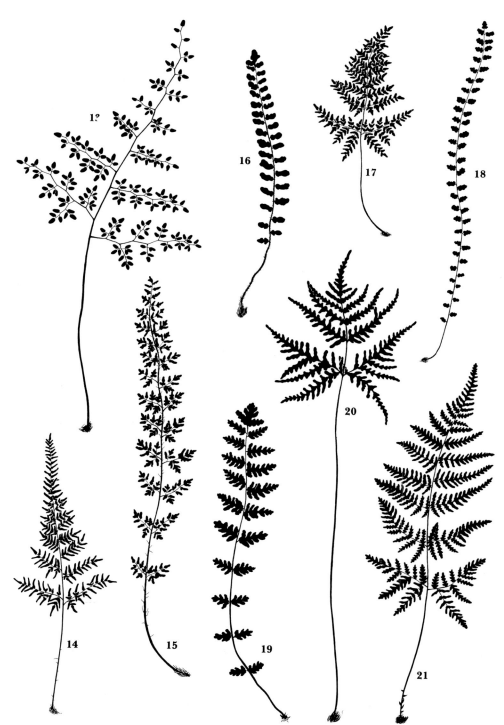

Figs. 37.13–37.21. Leaves of *Notholaena*, ×0.5. **13.** *N. parvifolia.* **14.** *N. bryopoda.* **15.** *N. pallens.* **16.** *N. trichomanoides.* **17.** *N. neglecta.* **18.** *N. Ekmanii.* **19.** *N. affinis.* **20.** *N. sulphurea.* **21.** *N. candida* (var. *candida*).

Figs. 37.22–37.28. Spores of *Notholaena,* × 1000, surface detail, × 10,000. **22.** *N. parvifolia,* cristate proximal face, lower left, and portions of distal face, Mexico, *Rollins & Tryon 58143*. **23.** *N. nivea,* from 32-spored sporangia, Peru, *Aguilar 24*. **24.** *N. trichomanoides,* proximal face, laesurae at top. cristate and slightly granulate, Jamaica, *Maxon 8753*. **25.** *N. candida,* proximal face densely granulate, Mexico, *Hinton 7395*. **26–28.** *N. Greggii.* **26.** Proximal face granulate, Mexico, *Stewart 2906*. **27.** Distal face of spore on inner, sporangium wall with granulate deposit, Mexico, *Johnston 8503*. **28.** Detail of granulate deposit, Mexico, *Stewart 2906*.

N. parvifolia (Fig. 22). They are similar to spores in some species of *Pellaea* and *Cheilanthes. Notholaena nivea,* which is an apogamous species, has large spores which are often irregular in shape and lack or have less than three laesurae (Fig. 23). Species in section *Notholaena* have black spores with granulate surface deposits usually more or less covering a cristate formation. The deposit may be slight as in *N. trichomanoides* (Fig. 24), dense as in *N. candida* (Fig. 25), or it may be formed over a relatively compact surface as in *N. Greggii* (Fig. 26). The black granulate material is also on the inner sporangium wall (Fig. 27), apparently derived from the tapetum. The surface detail of *N. Greggii* (Fig. 28) shows the outer deposit as a matrix of granulate particles and spheres.

Most species of *Notholaena* have the normal complement of 64 spores per sporangium but many have spore numbers reduced. Species may have 32 spored sporangia, or in *Notholaena trichomanoides* and *N. sulphurea,* plants may have either 32 or 64 spores. Spore number may be reduced to 16 or 32 in *N. Standleyi* Maxon and in *N. aliena* Maxon the sporangia have only 16 spores. In other allied genera reduction in spore number is associated with apogamous reproduction. Woronin (1907) reported apogamy in *Notholaena nivea, N. tenera* and *N. flavens.* All three, now treated as varieties of *N. nivea,* have 32 spored sporangia. The reduced spore number in other species of *Notholaena* may also indicate apogamy, but observations of the gametophyte are needed to confirm this condition.

Cytology

There are reports of chromosome numbers for few species of *Notholaena.* In section *Notholaena, N. rigida* from Mexico is reported as n = ca. 30, a Mexican plant of *N. Galeottii* is reported as $2n$ = 90 and presumably apogamous, and *N. Standleyi* as $2n$ = 60 from Arizona. In section *Argyrochosma* a report of n = 27 for *N. dealbata* from Kansas City, Missouri (Wagner, 1963) is inconsistent with all other records for the genus and may represent an aberrant plant. *Notholaena parvifolia* reported as n = 29 from New Mexico and $2n$ = 116 from Texas (Knobloch et al., 1973) indicates the development of polyploidy similar to that in other cheilanthoid ferns. Additional records are essential to determine the role of polyploidy and apogamy in the evolution of this genus. These will also be useful in determining whether species in section *Notholaena* are consistently based on 30, while those in section *Argyrochosma* are based on 29.

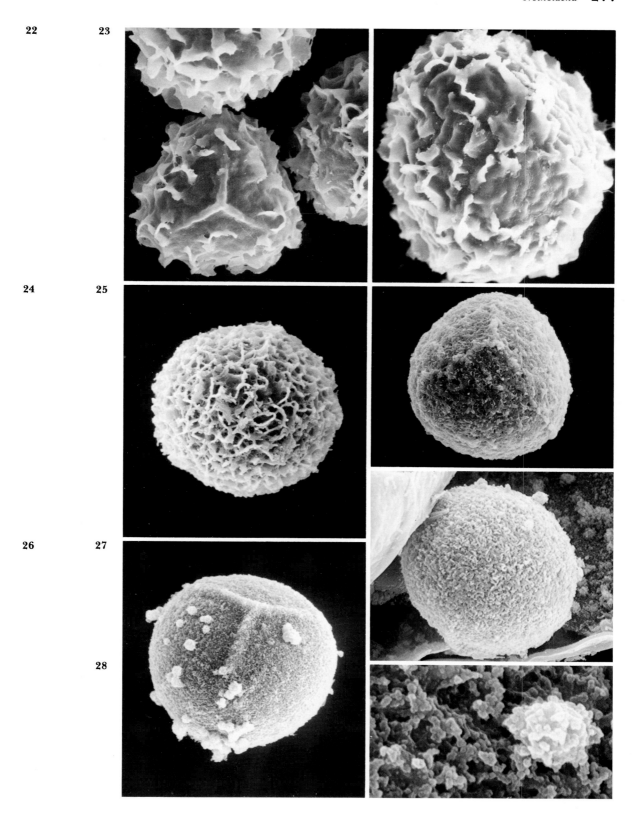

22 23

24 25

26 27

28

Observations

Studies on the chemistry of farinose indument in cheilanthoid ferns are summarized by Wollenweber (1978). There appears to be no evidence of specific compounds in these exudates that characterize genera. There are some species-specific flavonoid patterns which suggest that additional work on this group of compounds may provide further evidence for species relationships.

The farinose-secreting glands on *Notholaena* leaves are also formed on the gametophytes. They occur on the gametophytes of *Notholaena Standleyi* (Tryon, 1947) and also on gametophytes of several other species (Giauque, 1949). It is exceptional in the ferns to have a character of this type expressed in both generations.

Literature

Giauque, M. F. A. 1949. Wax glands and prothallia. Amer. Fern Jour. 39: 33–35.

Knobloch, I. W., W. Tai, and T. A. Ninan. 1973. The cytology of some species of the genus *Notholaena*. Amer. Jour. Bot. 60: 92–95.

Tryon, A. F. 1947. Glandular prothallia of *Notholaena Standleyi*. Amer. Fern Jour. 37:88.

Tryon, R. 1956. A revision of the American species of *Notholaena*. Contrib. Gray Herb. 179: 1–106.

Tryon, R. 1961. Taxonomic fern notes 1. Rhodora 63: 70–88.

Wagner, W. H. 1963. Reference under the tribe Cheilantheae.

Wollenweber, E. 1978. The distribution and chemical constituents of the farinose exudates in gymnogrammoid ferns. Amer. Fern Jour. 68: 13–28.

Woronin, H. 1907. Apogamie und Aposporie bei einigen Farnen. Flora 98: 101–162.

38. *Hemionitis*

Figs. 38.1–38.20

Hemionitis L., Sp. Pl. 1077. 1753; Gen. Pl. ed. 5, 485. 1754. Type: *Hemionitis palmata* L.

Gymnopteris Bernh., Jour. Bot. (Schrad.) 1799 (1): 297. 1799. Type: *Acrostichum rufum* L. (not *Pteris ruffa* L.) (*Gymnopteris rufa* (L.) Underw.) = *Hemionitis rufa* (L.) Sw.

Gymnogramma Desv., Berl. Mag. 5: 304. 1811, *nom. superfl.* for *Gymnopteris* Bernh. and with the same type.

Neurogramma Link, Fil. Sp. 138. 1841, *nom. superfl,* for *Gymnopteris* Bernh. and with the same type.

Gymnogramma section *Neurogramma* Kl., Linnaea 20: 410. 1847. Type: *Gymnogramma rufa* (L.) Desv. = *Hemionitis rufa* (L.) Sw. *Gymnopteris* subgenus *Neurogramma* (Kl.) C. Chr., Ind. Fil. xxxix. 1906.

Description   Terrestrial or sometimes rupestral; stem erect to decumbent or short-creeping, small, bearing scales intergrading to trichomes and usually many fibrous roots that often branch at right angles; leaves monomorphic to dimorphic (the sterile in a basal rosette, the fertile erect), ca. 5–60 cm long, borne in a cluster, the lamina rarely entire to usually shallowly to deeply 3- to 7-lobed and

Fig. 38.1. Colony of plants of *Hemionitis palmata* near Spring Garden, Tre-lawny, Jamaica. (Photo W. H. Hodge.)

subpalmate or subpedate, or imparipinnate and 1- to 2-pinnate, rarely nearly 3-pinnate, pubescent and usually with a few scales, veins free, or partially or wholly anastomosing without included free veinlets; sporangia borne on unmodified veins, either all along anastomosing veins, or in long sori on the free veins, sometimes extending below a fork, paraphyses not present, ex-indusiate; spores tetrahedral-globose or globose, trilete, the lae-surae mostly $\frac{3}{4}$ the radius or somewhat shorter, cristate, echinate, or tuberculate. Chromosome number: $n = 30$.

The stem indument of scales intergrading to trichomes is an unusual feature in ferns, most apparent in *Hemionitis* at the apex and at the base of the petiole. Sporangia are arranged along free veins (Fig. 3), or anastomosing veins (Figs. 4, 5). In three of the species the lamina architecture is pinnate (Figs. 10–12) while in the four others it is entire or variously lobed (Figs. 6–9).

Systematics   *Hemionitis* is a tropical American genus of seven species including diverse elements that seem best treated together although their evolutionary relations are not clear. Three species with en-tire to deeply lobed leaves—*H. palmata, H. pinnatifida* and *H. Levyi*—are similar while *H. elegans* differs in several characters noted in the key. *Hemionitis rufa* and *H. tomentosa* appear close and quite distinct from *H. subcordata*. Mickel (1973) recognized both *Hemionitis* and *Gymnopteris* in America and referred the Old World species to other undesignated genera. He later (1974) suggested that all of these be included in a single genus, and

Fig. 38.2. Distribution of *Hemionitis*.

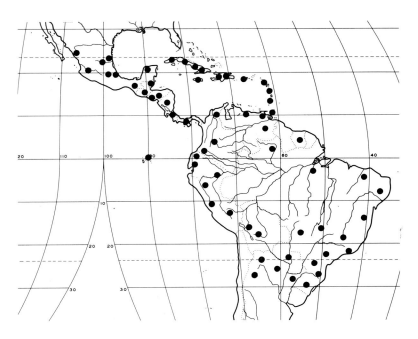

placed them in *Hemionitis* (Giannasi and Mickel, 1979). However, species of the Old World as *Hemionitis bipinnata* (Christ) Mickel and *H. vestita* (Hook.) J. Sm. are distinct from *Hemionitis* in the densely pubescent leaves, and are more closely related to *Cheilanthes*. Species with dense scales on the leaves as *Gymnopteris Marantae* (L.) Ching (*Cheilanthes Marantae* (L.) Domin), *G. Delavayi* (Baker) Underw. and *G. Muelleri* (Hook.) Underw. (*Paraceterach* Copel.) are clearly allied to *Cheilanthes*. The paleotropical species *H. arifolia* (Burm.) Moore is only superficially similar to American species and it may be related to *Syngramma* or a paleotropical element of *Doryopteris*.

Hemionitis has usually been treated with the taenitidoid genera on the basis of trichomes on the stem and sporangia along the veins. However, in other characters, especially the spores, it is related to the cheilanthoids.

Tropical American Species

The species of *Hemionitis* are usually well defined, although *H. tomentosa* is rather variable and its separation from *H. rufa* requires further study. A few intermediate plants have been reported as hybrids: *Hemionitis palmata* × *rufa* and *H. palmata* × *pinnatifida*. *Gymnopteris Gardneri* probably belongs in *Hemionitis*, but is inadequately known by a single collection from Goias, Brazil, *Hemionitis Otonis* Maxon is a synonym of *H. Levyi*.

Hemionitis subcordata was placed in *Coniogramme* by Maxon (1913), first as *C. subcordata* (Davenp.) Maxon and later as *C. americana* Maxon. Although it has some similarities with East Asian species of *Coniogramme* it is better placed in *Hemionitis* (Lellinger, 1969). The prominently cristate spores as shown in Fig. 15 differ from the relatively smooth ones of *Coniogramme*.

3   4   5

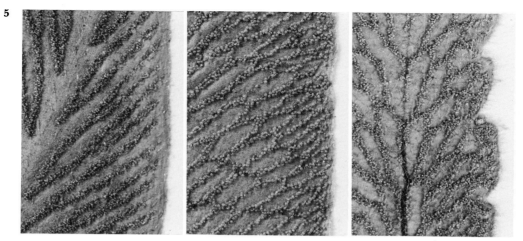

Figs. 38.3–38.5. Portions of fertile leaves of ***Hemionitis*** with sporangia along veins, × 4.0. **3. *H. rufa.* 4. *H. elegans.* 5. *H. palmata.***

Key to Species of *Hemionitis*

a. Lamina entire, or shallowly to deeply lobed, veins wholly anastomosing. b.
  b. Stem erect, leaves dimorphic, the sterile in a basal rosette, the fertile erect, petioles not fracturing. c.
    c. Lamina entire to 5-lobed. d.
      d. Lamina entire or ca. 3-lobed, with 0–3 dark primary veins beneath; Central America.   *H. Levyi* Fourn.
      d. Lamina subpalmately 5-lobed, with five dark primary veins beneath; Mexico, Central America, West Indies, South America. (Figs. 5, 7, 8)   *H. palmata* L.
    c. Lamina subpedately 7-lobed, with 7 dark primary veins beneath; Central America. (Fig. 9)   *H. pinnatifida* Baker
  b. Stem very short-creeping, leaves nearly monomorphic, the sterile not in a rosette, petioles fracturing; Mexico. (Figs. 4, 6)   *H. elegans* Davenp.
a. Lamina 1-pinnate or more complex. e.
  e. Veins partially anastomosing, petiole and rachis straw-colored, leaves glabrate; Mexico. (Fig. 10)   *H. subcordata* (Davenp.) Mickel
  e. Veins free, petiole and rachis brown to atropurpureous, leaves pubescent. f.
    f. Pinnae entire or rarely with 1 lobe, the basal short-stalked, cuneate to rounded, rarely subcordate, at the base; Mexico, Central America, Greater Antilles, Surinam to Colombia, south to Peru. (Figs. 3, 12)
                                                          H. rufa (L.) Sw.
    f. Pinnae lobed or partly 1- to rarely 2-pinnate, or entire and the basal long-stalked, cordate at the base; Peru, Bolivia, Argentina, Paraguay, Brazil. (Fig. 11)   *H. tomentosa* (Lam.) Raddi

Ecology (Fig. 1)

Hemionitis grows in open, or sometimes dense, forests, on shrubby hillsides, and in open rocky areas. It is often on stream banks, on road banks, or on old rock walls, very rarely on rotting logs. *Hemionitis palmata* may be weedy, sometimes invading coffee or banana plantations. Colonies of *Hemionitis palmata* are often formed by vegetative reproduction. Buds in the major sinuses of the lamina (Fig. 8) develop when the leaf ages and lies on the soil. *Hemionitis* usually grows between 100 and 1000 m, sometimes lower to nearly sea level, and in the Andes higher to 2800 m.

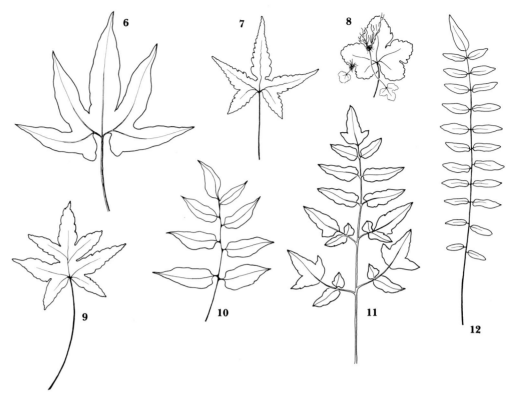

Figs. 38.6–38.12. Lamina architecture of *Hemionitis* species, all ×0.25.   **6.** *H. elegans.*   **7.** *H. palmata,* fertile.   **8.** *H. palmata,* sterile with plantlets developed from sinus buds.   **9.** *H. pinnatifida.*   **10.** *H. subcordata.*   **11.** *H. tomentosa.*   **12.** *H. rufa.*

Geography (Fig. 2)

Hemionitis occurs in Tamaulipas, Sinaloa in Mexico, Central America, the West Indes, and in South America, south to Tucumán and Santa Fé in Argentina and Rio Grande do Sul in Brazil; also on the Galápagos Islands. It is extremely rare in the Amazon basin. All of the species, except *H. tomentosa,* grow in Mexico or Central America.

Spores

Hemionitis spores are relatively small with diverse contours. The surface is cristate in *H. tomentosa* (Fig. 13), echinate in *H. palmata* (Fig. 14), and prominently cristate in *H. subcordata* (Fig. 15) with more or less fused strands. Spores of *H. rufa* have a coarse, low-tuberculate surface consisting of irregularly granulate material (Figs. 16, 17). Those of *H. pinnatifida* have smaller, compact tubercules or papillae (Fig. 18) underlaid by an inflated stratum that appears to conform to the tuberculate outer layer (Fig. 19). There are faint circular rings on the lower stratum of *H. palmata* spores (Fig. 20) that may possibly relate to the papillate elements as in *H. pinnatifida.* Spores of *Hemionitis* and *Gymnopteris* were arranged in two groups by Mickel (1974) based on whether the surface consisted of high to low tubercules or of ridges. However, spore differences in the species shown here suggest a greater diversity which cannot be readily placed in two groups.

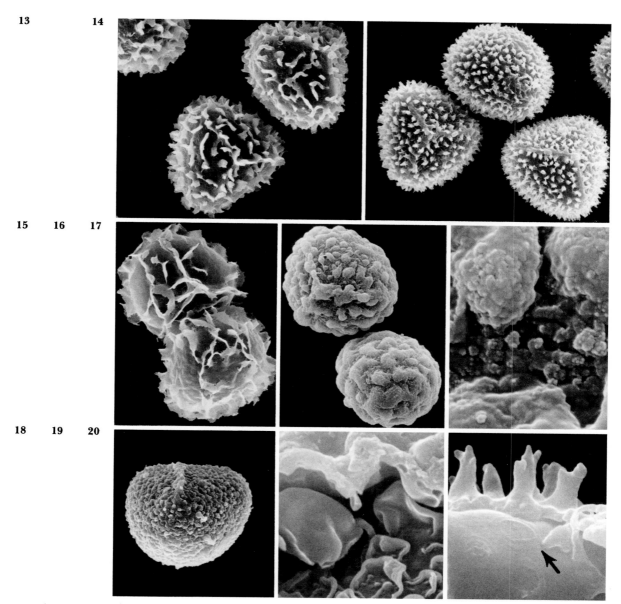

Figs. 38.13–38.20. *Hemionitis* spores, × 1000.   **13.** *H. tomentosa,* cristate, proximal faces, Peru, *Vargas 1833.*   **14.** *H. palmata,* echinate distal face above, Jamaica, *Gastony 28.*   **15.** *H. subcordata,* cristate, distal face below, Mexico, *Hinton 8063.*   **16, 17.** *H. rufa,* Jamaica, *Proctor 24564.*   **16.** Coarsely tuberculate, distal face below.   **17.** Detail of irregularly granulate surface, × 10,000.   **18, 19.** *H. pinnatifida,* Mexico, *Matuda 2128.*   **18.** Lateral view, compact tuberculate-papillate surface.   **19.** Abraded wall, the outer surface, above, conforms to the deflated, tuberculate-papillate formation below, × 10,000.   **20.** *H. palmata,* wall strata with faint rings, at arrow, that connect with outer structure, Jamaica, *Gastony 28,* × 8000.

Cytology

Hemionitis palmata and *H. rufa* (as *Gymnopteris rufa*) were both reported as *n* = 30 from St. Andrews Parish, Jamaica, by Walker (1966). The same number is recorded for both species, cultivated at Kew, by Manton and Sledge (1954) and Manton 1958). Other reports of the same number are made for *H. palmata* without documentation from Costa Rica (Gómez Pignataro, 1971) and presumably Mexico (Wagner, 1963).

Literature

Giannasi, D. E., and J. T. Mickel. 1979. Systematic implications of flavonoid pigments in the fern genus *Hemionitis* (Adiantaceae). Brittonia 31: 405–412.

Gómez Pignataro, L. D. 1971. Ricerche citologiche sulle Pteridofita della Costa Rica 1. Atti Ist. Bot. Univ. Pavia 7: 29–31.

Lellinger, D. B. 1969. The taxonomic position of *Coniogramme americana*. Amer. Fern Jour. 59: 61–65.

Manton, I. 1958. Chromosomes and fern phylogeny with special reference to "Pteridaceae." Jour. Linn. Soc. (Bot.) 56: 73–92.

Manton, I., and W. A. Sledge. 1954. Reference under the tribe Cheilantheae.

Mickel, J. T. 1973. Redefinition of the fern genus *Hemionitis*. Amer. Jour. Bot. 60 (4) Suppl.: 32.

Mickel, J. T. 1974. A redefinition of the genus *Hemionitis*. Amer. Fern. Jour. 64: 3–12.

Maxon, W. R. 1913. Notes upon *Bommeria* and related genera. Contrib. U.S. Nat. Herb. 17: 168–175.

Wagner, W. H., Jr. 1963. Reference under the tribe Cheilantheae.

Walker, T. G. 1966. Reference under the tribe Cheilantheae.

39. *Pellaea*
Figs. 39.1–39.24

Pellaea Link, Sp. Fil. 59, Sept., 1841, *nom. conserv.* Type: *Pellaea atropurpurea* (L.) Link (*Pteris atropurpurea* L.).

Platyloma J. Sm., Jour. Bot. (Hook.) 4: 160, Aug., 1841. Type: *Platyloma Brownii* J. Sm. = *Pellaea paradoxa* (R. Br.) Hook. (*Adiantum paradoxum* R. Br.). *Pellaea* section *Platyloma* (J. Sm.) Hook. & Baker, Syn. Fil. 151. 1867. *Pellaea* subgenus *Platyloma* (J. Sm.) C. Chr., Ind. Fil. xl. 1906.

Pellaea section *Holcochlaena* Hook. & Baker, Syn. Fil. 153. 1867. Type: *Pellaea articulata* (Spreng.) Baker (*Pteris articulata* Spreng.) = *Pellaea angulosa* (Willd.) Baker.

Synochlamys Fée, Mém. Fam. Foug. 7 (Mém. Soc. Hist. Nat. Strasbourg 5): 35. 1857. Type: *Synochlamys ambigua* Fée = *Pellaea ambigua* (Fée) Baker. [Section *Ormopteris*].

Ormopteris J. Sm., Hist. Fil. 281. 1875. Type: *Ormopteris gleichenioides* (Hook.) J. Sm. (*Cassebeera gleichenioides* Hook.) = *Pellaea gleichenioides* (Hook.) Christ. *Pellaea* subgenus *Ormopteris* (J. Sm.) C. Chr., Ind. Fil. xl. 1906. *Pellaea* section *Ormopteris* (J. Sm.) R. & A. Tryon, Rhodora 83: 135. 1981.

Pellaeopsis J. Sm., Hist. Fil. 289. 1875. Type: *Pellaeopsis articulata* (Spreng.) J. Sm. (*Pteris articulata* Spreng.) = *Pellaea angulosa* (Willd.) Baker. [Section *Holcochlaena*].

Pteridella Kuhn, von Decken's Reisen Ost-Afrika III (3) Bot.: 13. 1879. Type: *Pteridella Doniana* (Hook.) Kuhn = *Pellaea Doniana* Hook. *Pellaea* section *Pteridella* (Kuhn) Prantl, Engl. Bot. Jahrb. 3: 417. 1882. *Pellaea* subgenus *Pteridella* (Kuhn) C. Chr., Ind. Fil. xl. 1906. [Section *Holcochlaena*].

Description   Terrestrial or rupestral; stem small, decumbent and short, or rather slender and long-creeping, bearing scales and many fibrous roots; leaves monomorphic or nearly so, ca. 0.05–1.25 m long, borne in a cluster or somewhat distant, lamina 1- to 4-pinnate, imparipinnate, glabrous, puberulent or rarely pubescent and sometimes also with scales, rarely farinaceous among the sporangia, veins free or rarely anastomosing without included

Fig. 39.1. *Pellaea ternifolia* in crevice of volcanic rock, Pedregal las Vigas, near Jalapa, Veracruz, Mexico. (Photo W. H. Hodge.)

free veinlets; sori at the tips of unmodified veins or also extending along the apical portion of the veins, or on laterally expanded and modified vein-tips, or on a marginal commissure, not paraphysate, indusium slightly to strongly modified in texture and color from the usually recurved margin; spores tetrahedral-globose, trilete, the laesurae $\frac{3}{4}$ to nearly the radius in length, sometimes incomplete or absent, shallowly ridged to prominently cristate, or echinate or low tuberculate. Chromosome number: $n = 29, 30, 58$; $2n = 58, 60$; apogamous 87, ca. 90, 116.

Pellaea is characterized by a lamina architecture that is pinnate and imparipinnate (Figs. 3–6), segments that are mostly stalked or sessile and entire, with a continuous marginal indusium covering several to many sori (Figs. 7, 8). In section *Ormopteris* the segments are sometimes crenate or shallowly lobed and the indusium is confined to a portion of the margin (Fig. 13). Also, the fertile vein-tips are sometimes laterally expanded as in *Pellaea crenata* (Fig. 14) or form a nearly continuous commissure as in *P.*

Fig. 39.2. Distribution of *Pellaea* in America, south of lat. 35° N: section *Pellaea*, dots; section *Ormopteris*, stars.

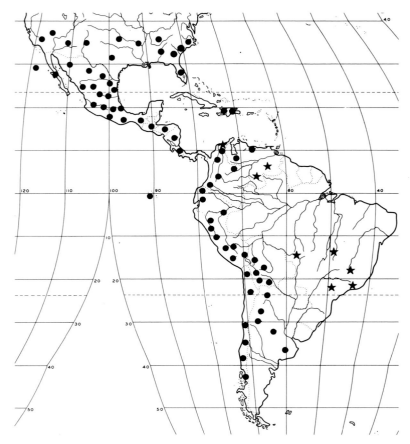

pinnata (Fig. 15). A few African species of section *Holcochlaena* also have a marginal commissure. The rachis and pinna-stalks may be terete (Figs. 9, 11) or grooved, with the groove continuous with that of the adjoining axis (Fig. 16). The stem and usually the petiole base are covered with scales, which often characterize groups of species within a section. These vary from broad, delicate, concolorous and tan scales (Fig. 10) to narrow, ruddy-brown and concolorous forms or are bicolorous with a dark, sclerotic central band (Figs. 12, 17).

Systematics

Pellaea is a genus of about 35 species, nearly pantropical and also temperate or rarely boreal. It is most closely related to *Cryptogramma, Doryopteris* and perhaps *Notholaena*.

Pellaea is treated here as a genus of four sections. It does not include several species, as the American *Pellaea allosuroides* (Mett.) Hieron, and *P. Skinneri* Hook., or African *P. hastata* (L. f.) Link, *P. pteroides* (L.) Prantl, *P. quadripinnata* (Forsk.) Prantl, and *P. viridis* (Forsk.) Prantl, or the Chinese species *P. nitidula* (Hook.) Baker and *P. Smithii* C. Chr., which we refer to *Cheilanthes* on the basis of their characters. There remain difficulties in separating *Pellaea* from some species of *Notholaena* as well as from some in *Doryopteris,* which are considered under those genera. While all groups recognized as sections have been previously treated as genera they are not considered sufficiently distinct to merit generic rank.

The species of *Pellaea* section *Pellaea* have been revised by Tryon (1957) except for *P. Bridgesii* Hook. which is now included in the section.

Synopsis of Pellaea

Section *Pellaea*

Stem short with clustered petioles, or long-creeping with the petioles more or less distant; rachis and pinna-stalks not grooved on the adaxial side, or grooved and the groove continuous with that of the adjoining axis; segments usually not articulate; indusium gradually and slightly to moderately differentiated from the margin; spores more or less ridged or cristate.

A section of 16 species, all American except *P. rufa* A. Tryon of South Africa.

Section *Ormopteris*

Stem short with clustered petioles, or rather long-creeping with the petioles more or less distant; rachis and pinna-stalks strongly grooved on the adaxial side and the groove continuous with that of the adjoining axis; segments not articulate; indusium abruptly and strongly differentiated from the margin; spores cristate.

A small group of 5–7 species of South America, all but one in Brazil.

Section *Holcochlaena*

Stem short with clustered petioles; rachis and pinna-stalks not or rarely slightly grooved on the adaxial side; segments articulate; indusium abruptly and strongly differentiated from the margin; spores with low ridges, cristate, or tuberculate.

A section of about 10 species of Africa-Madagascar with three species extending to India-Ceylon. The relations of these need further study for the diversity in spores and venation suggest that the section may not be a cohesive group. For example, free-veined species are *Pellaea Boivinii* Hook., *P. calomelanos* (Sw.) Link, *P. Doniana* Hook., *P. longipilosa* Bonap., *P. pectiniformis* Baker, and *P. tomentosa* Bonap. Species with areolate veins are *Pellaea angulosa* (Willd.) Baker, *P. dura* (Willd.) Hook., *P. prolifera* Schelpe, and *P. Schweinfurthii* (Hieron.) Diels. These areolate-veined species might merit recognition in a separate section, and those with free veins could then be placed in a section *Pteridella*.

Section *Platyloma*

Stem slender and long-creeping with the petiole distant; rachis and pinna-stalks not grooved on the adaxial side; segments articulate; indusium gradually and moderately to somewhat abruptly differentiated from the margin; spores echinate.

A small section of three species: *Pellaea falcata* (R. Br.) Fée, *P. paradoxa* (R. Br.) Hook., and *P. rotundifolia* (Forst.) Hook., ranging from New Zealand to Tasmania, Australia, New Caledonia, and New Guinea to Ceylon and India.

Tropical American Species

Species of section *Pellaea* are not sharply distinguished as tropical or temperate-boreal. Those that definitely occur in the tropics are: *Pellaea atropurpurea* (L.) Link, *P. ovata* (Desv.) Weath., *P. Pringlei* Dav., *P. sagittata* (Cav.) Link, and *P. ternifolia* (Cav.) Link.

Species of section *Ormopteris* are inadequately known for there are few collections. *Pellaea crenata* Tryon, *P. gleichenoides* (Hook.) Christ and *P. pinnata* (Kaulf.) Prantl are distinctive species. *Pellaea ambigua* (Fée) Baker, *P. brasiliensis* Baker and *P. Riedelii* Baker form a related complex, and *Pellaea flavescens* Fée of Brazil may also belong to this group.

Ecology (Fig. 1)

Pellaea usually grows in open, rocky places or in rocky soil, but also on shrubby hillsides, on cliffs and in dry woods.

In tropical America the species typically grow in relatively xeric conditions, in lava beds, in crevices and on ledges of cliffs, at the edge of large rocks, less often on dry earth banks, among shrubs or in grassy or wooded areas. Species may grow on old adobe walls, or in the joints of pre-Columbian stone work such as the terrace walls of Macchu-Picchu in Peru (*P. ternifolia*) or the pyramid of Zaculeu, Huehuetango, Guatemala (*P. atropurpurea*). Species of section *Ormopteris* are confined to sandstone and derived soils and some species of section *Pellaea* such as *P. Pringlei* and *P. atropurpurea* are calciphiles. *Pellaea ovata* has a long, fractiflex lamina that is adapted to scrambling among shrubs. On Cerro do Cipó, Minas Gerais, Brazil, *Pellea pinnata* grows in fine quartzite gravel at the base of rocks. Charred petiole bases indicate that the partially buried stem with a dense covering of scales has the capacity to survive light ground fires. Species grow from ca. 200 to 3200 m, or up to 4000 m (*P. ternifolia*) in the Andes; they occur most frequently between 1000 and 2500 m.

Geography (Fig. 2)

Pellaea occurs in North and South America, in Spain, Africa, India, and southeast to Australia and New Zealand.

In America the genus grows from Canada south through Mexico and Central America; in Hispaniola; and in South America from Venezuela to Colombia south to Valdivia in Chile and Buenos Aires in Argentina; also the Galápagos Islands. The American *P. ternifolia* is also in the Hawaiian Islands where it is probably adventive.

Species in section *Pellaea* are strongly centered in the southwestern United States (Texas to southern California) and Mexico, where 13 of the 16 species occur; there are four species in Central America, two in Hispaniola, and four in South America. There are four wide-ranging species; *Pellaea atropurpurea* occurs from Guatemala north to western and eastern Canada, *P. sagittata* ranges from the southwestern United States to Bolivia, while *P. ternifolia* and *P. ovata* with a similar range also extend to

3 **4**

5 **6**

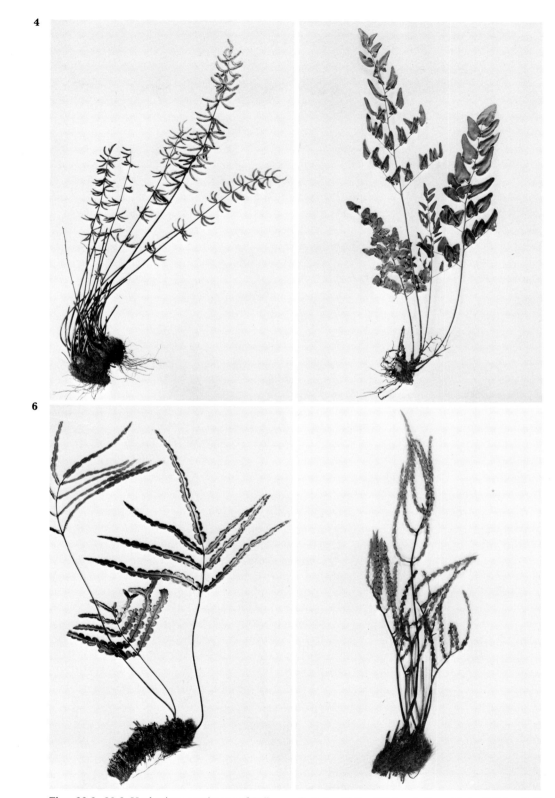

Figs. 39.3–39.6. Herbarium specimens of *Pellaea*. **3.** *P. ternifolia*, ×0.4. **4.** *P. sagittata*, ×0.4. **5.** *P. pinnata*, ×0.4. **6.** *P. gleichenioides*, ×0.5.

Figs. 39.7–39.17. *Pellaea.* **7–10.** *P. sagittata.* **7.** Pinna showing marginal indusia with protruding sporangia, × 1. **8.** Segment apex with a portion of the indusium bent to show sporangia stalks at apex of free veins, × 3.5. **9.** Part of terete rachis and pinna stalk, × 4. **10.** Concolorous stem scale, × 10. **11.** *P. atropurpurea,* part of terete rachis and pinna stalks with trichomes, × 4. **12.** *P. ternifolia,* bicolorous stem scale, × 10. **13.** *P. pinnata,* lobed pinna with mature sporangia protruding from indusia except in the central portion showing only indusia along the marginal lobes, × 1. **14.** *P. crenata,* portion of a fertile pinna with expanded terminal receptacles, × 6. **15–17.** *P. pinnata.* **15.** Apex of fertile pinna with nearly continuous marginal commissures, × 6. **16.** Part of grooved rachis and pinna stalks, adaxial surface, × 4. **17.** Bicolorous stem scale, × 10.

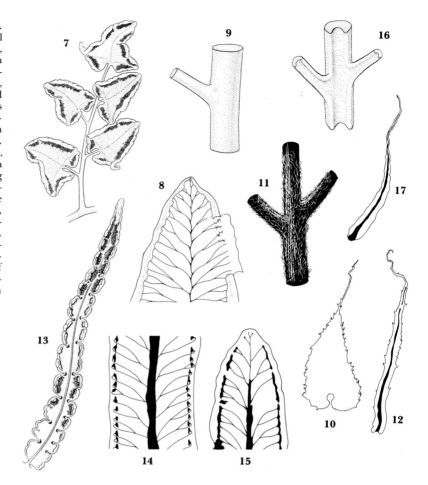

Hispaniola south to Argentina with *P. ternifolia* also in Chile. Except for *P. ternifolia,* these species are apogamous or have apogamous as well as sexual plants (Tryon and Britton, 1958; Tryon, 1968, 1972). Asexual reproduction by spores is evidently a primary reason for the broad ranges. Apogamous gametophytes do not require free water for fertilization and new plants can become established during a short moist period due to rapid germination of the spores and initiation of the sporophyte (Tryon, 1960; Whittier, 1968). There is a single extra-American species in the section, *P. rufa* of the Cape Province in South Africa. This and two others form a closely related group with an uncommon, disjunct distribution—*P. myrtillifolia* Kuhn in Chile and *P. andromedifolia* (Kaulf.) Fée in California and adjacent Mexico.

Pellaea section *Ormopteris* centers in southeastern Brazil, where all but one of the species grow. Among these, *Pellaea pinnata* is the only widely distributed one, occurring at disjunct localities in Brazil and also in the states of Amazonas and Bolivar, Venezuela. The single collection of *Pellaea ambigua* from the Sierra Nevada de Santa Marta (Colombia) may be an endemic or may represent a disjunct record of *P. brasiliensis.*

18 19 20

21 22

23 24

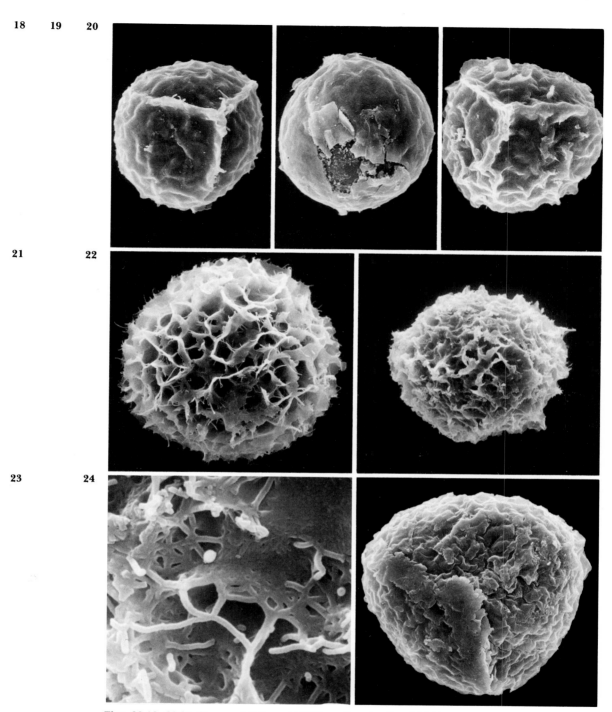

Figs. 39.18–39.24. Spores of ***Pellaea***, × 1000, surface detail, × 10,000. **18, 19. *Pellaea ternifolia,*** Dominican Republic, *R. & E. Howard 9137.* **18.** Shallowly rugose proximal face. **19.** Distal face abraded below showing inner echinate perispore. **20. *P. sagittata,*** rugose proximal face, Mexico, *Pringle 448.* **21. *P. myrtillifolia,*** distal face with prominent cristae and delicate strands connecting base of crests, Chile, *Cuming 184.* **22,23. *P. pinnata.* 22.** Cristate distal face, Brazil, *Foster 711.* **23.** Detail of cristae and more or less fused strands, Brazil, *Tryon & Tryon 6804.* **24. *P. crenata,*** proximal face, compact, rugose, Brazil, *Foster 624.*

Spores

Spores of species in section *Pellaea* are of two main types. The species with dark petioles related to *P. ternifolia* have a relatively smooth, shallowly rugose surface (Fig. 18), with an inner, echinate stratum as shown in Fig. 19. Comparisons of the strata can be made from the abraded spore illustrated in *Pellaea ternifolia*. (Tryon and Tryon, 1973). Species with tan petioles as *P. sagittata* have spores with more prominently ridged to somewhat cristate surface (Fig. 20). Species with light-colored petioles allied to *P. myrtillifolia* have strongly cristate spores as in Fig. 21. Species of section *Ormopteris* have cristate or densely rugose spores with a reticulate substructure as in *P. pinnata* and *P. crenata* (Figs. 22–24). These resemble the rugose or cristate spores in *Doryopteris*. The paleotropical *Pellaea* section *Platyloma* has echinate spores quite distinct from those of other sections in the genus. The spores of species in the largely African section *Holcochlaena* may be nearly smooth, rugose, strongly cristate, or tuberculate. The diverse types of spores in the section suggest it may be heterogeneous.

Cytology

Chromosome records of diploids $n = 29$ and tetraploids $n = 58$ are reported for *Pellaea,* mostly from American species in section *Pellaea* (Tryon, 1957, 1968; Tryon and Britton, 1958). These reports also include several apogamous triploids with 87 and tetraploids with 116 chromosomes. In the American tropics *P. ovata* and *P. sagittata* have both normal, sexual and apogamously reproducing plants. The apogamous condition is most readily recognized by sporangia bearing 32 spores rather than the normal complement of 64 per sporangia. The geography of the derived, apogamous plants can be plotted from herbarium specimens and the range of these can be compared to that of the related sexual ones. The taxa with both sexual and apogamous plants are also of interest for studies on the genetics and cytology of apogamous systems. In the paleotropical section *Platyloma* there are reports of *P. rotundifolia* with $n = 58$ and also a triploid with 87, and of *P. falcata* with $n = 58$. These suggest the presence of a polyploid complex for this group similar to those in section *Pellaea.* Cytological records for *Pellaea* seem to be consistently based on 29 with the exception of *P. Boivinii* Hook. in section *Holcochlaena* which is reported as an apogamous triploid with ca. 90 from Ceylon (Manton and Sledge, 1954). In addition there is an unverified report of $n = 30$ for this species.

Observations

The remarkable adaptations of the gametophytes and young sporophytes of *Pellaea* to the seasonally xeric conditions under which most species grow was demonstrated by the experiments of Pickett and Manuel (1926). Desiccation tests on gametophytes and sporelings of *Pellaea atropurpurea* and *P. glabella* Kuhn resulted in three-quarters of the plants recovering after three months in a greenhouse without water. Plants were also placed in closed desiccators over anhydrous calcium chloride for nine months, and others for $18\frac{1}{2}$ months. When watering was resumed some of the plants became green and continued growth in each of these tests. After five years in the calcium chloride desiccator Picket (1931) reports that five percent of the plants showed whole or partial recovery when growing conditions were restored.

Literature

Manton, I., and W. A. Sledge, 1954. Reference under the tribe Cheilantheae.

Pickett, F. L. 1931. Notes on xerophytic ferns. Amer. Fern Jour. 21: 49–57.

Pickett, F. L. and M. E. Manuel. 1926. An ecological study of certain ferns: *Pellaea atropurpurea* (L.) Link and *Pellaea glabella* Mett. Amer. Fern Jour. 53: 1–5.

Tryon, A. F. 1957. A revision of the genus *Pellaea* section *Pellaea*. Ann. Mo. Bot. Gard. 44: 125–193.

Tryon, A. F. 1960. Observations on the juvenile leaves of *Pellaea andromedifolia*. Contrib. Gray Herb. 187: 161–168.

Tryon, A. F. 1968. Comparisons of sexual and apogamous races in the fern genus *Pellaea*. Rhodora 70: 1–24.

Tryon, A. F. 1972. Spores, chromosomes and relations of the fern *Pellaea atropurpurea*. Rhodora 74: 220–241.

Tryon, A. F., and D. M. Britton. 1958. Cytotaxonomic studies on the fern genus *Pellaea*. Evolution 12: 137–145.

Tryon, R. M., and A. F. Tryon. 1973. Reference under the tribe of Cheilantheae.

Whittier, D. P. 1968. Rate of gametophyte maturation in sexual and apogamous forms of *Pellaea glabella*. Amer. Fern Jour. 58: 12–19.

40. *Doryopteris*
Figs. 40.1–40.38

Doryopteris J. Sm., Jour. Bot. (Hook.) 4: 162. 1841. Type: *Doryopteris palmata* (Willd.) J. Sm. (*Pteris palmata* Willd.) = *Doryopteris pedata* var. *palmata* (Willd.) Hicken.

Cassebeera Kaulf., Enum. Fil. 216. 1824. Type: *Cassebeera triphylla* (Lam.) Kaulf. (*Adiantum triphyllum* Lam.) = *Doryopteris triphylla* (Lam.) Christ. *Pellaea* section *Cassebeera* (Kaulf.) Prantl, Engl. Bot. Jahrb. 3: 418, 1882. [Section *Lytoneuron*].

Doryopteris section *Lytoneuron* Kl., Linnaea 20: 343. 1847. Type: *Doryopteris lomariacea* Kl.

Heteropteris Fée, Crypt. Vasc. Brésil 1: 123. 1869, not HBK., 1822 (Malpighiaceae). Type: *Heteropteris Doryopteris* Fée = *Doryopteris* hybrid. [Section *Doryopteris*].

Pellaea section *Doryopteridastrum* Prantl, Engl. Bot. Jahrb. 3: 419. 1882. Type: *Pellaea quinquelobata* Fée = *Doryopteris quinquelobata* (Fée) Diels. [Section *Lytoneuron*].

Bakeropteris O. Ktze., Rev. Gen. 2: 807. 1891, *nom. superfl.* for *Cassebeera* Kaulf. and with the same type. (*Cassebeeria* Dennst. 1818 is invalid, not an earlier homonym).

Tryonella Pic.-Ser., Webbia 29: 14. 1974, *nom. nov.* for *Heteropteris* Fée and with the same type.

Fig. 40.1. *Doryopteris crenulans* on humus covered rock in wet, forested valley, Maromba, Mt. Itatiaia, Rio de Janeiro, Brazil. (Photo Alice F. Tryon.)

Description

Terrestrial or rupestral; stem suberect to decumbent and small to moderately stout, or (in *D. ludens* and *D. papuana*) long-creeping and slender, bearing scales and many fibrous roots; leaves monomorphic to dimorphic (the fertile more erect than the sterile, with narrower segments and sometimes a more complex lamina), ca. 5–60 cm long, borne in a cluster, or rarely somewhat distant, lamina architecture diverse, entire and cordate, sagittate, hastate, 3-lobed, or usually pedate and pinnatifid to bipinnatifid, rarely tripinnatifid, or sometimes palmate, glabrous or (in *D. Rosenstockii*) with the petiole pubescent, veins free or anastomosing without included free veinlets; sori marginal, usually on a continuous vascular commissure connecting the vein-tips, or (in *D. paradoxa*) partially or mostly on modified vein-tips, or sporangia borne (in *D. Humbertii*) in a submarginal band on the outer anastomosing veins and between them, not paraphysate, the indusium abruptly and strongly differentiated from the recurved margin, covering the continuous or separate sori; spores tetrahedral-globose, trilete, the laesurae $\frac{1}{2}$ to $\frac{3}{4}$ the radius, the surface scarcely rugose to prominently cristate. Chromosome number: $n = 30, 60, 116; 2n = 60, 120, 232$.

Doryopteris is readily characterized by the pedate (Figs. 13, 22, 25, 28) or palmate (Figs. 15, 16, 27) lamina architecture and the sorus borne on a marginal commissure and covered by a well-differentiated indusium arising abruptly from the margin (Figs. 4, 5). The petiole is terete or nearly so, except in some of the paleotropical species.

Systematics

Doryopteris is a tropical genus of 25 species, which are treated in two sections, representing distinct evolutionary lines. As noted below, some species of section *Lytoneuron* have divergent characters, but these are otherwise clearly related to other species of the section. The paleotropical species of section *Doryopteris* differ

Fig. 40.2. Distribution of *Doryopteris* section *Lytoneuron.*

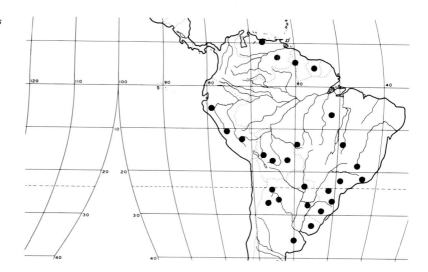

Fig. 40.3. Distribution of *Doryopteris* section *Doryopteris* in America.

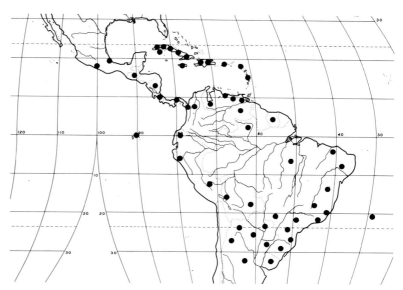

Fig. 40.4. Portion of fertile segment of *Doryopteris Rosenstockii,* showing venation, fertile commisure and indusium, × 2.0.

from the neotropical ones in characters of the stem, scales, and petiole; these may represent a convergent element derived from a different source than the American members.

Doryopteris is evidently most closely related to *Pellaea* through section *Lytoneuron. Doryopteris triphylla,* for example, resembles *Pellaea pinnata* in several characters such as the single vascular bundle in the petiole, the 3-lobed lamina of small leaves, and the deeply crenate segments with dark, semisclerotic areas in the sinuses. However, the two species are sufficiently different in petiole characters, stem scales, and lamina architecture of large leaves so they can be readily placed in their respective genera.

The type of *Pellaea* section *Doryopteridastrum* Prantl is chosen here from among the several original species. The genus *Tryonella* is based on precociously fertile juvenile plants of *Doryopteris sagittifolia* (Brade, 1965) and also probable hybrids of that species with one or more others. The name *Doryopteris* should be conserved over the seldom used *Cassebeera.*

Doryopteris has been revised by Tryon in 1942 with further

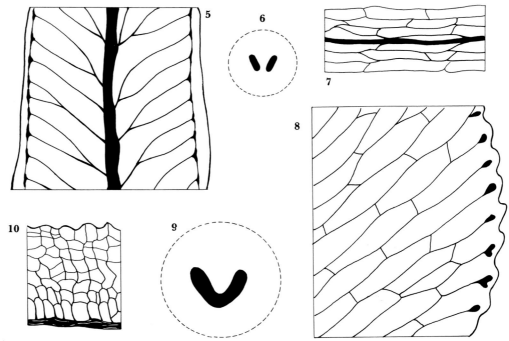

Figs. 40.5–40.10. *Doryopteris* (**Figs. 5–7,** section *Lytoneuron,* **Figs. 8–10,** section *Doryopteris*). **5.** Venation and vascular commissure of fertile segment of **D. subsimplex,** ×3.5. **6.** Section of petiole of **D. ornithopus,** with two vascular bundles, ×8.0. **7.** Cellular detail of stem scale of **D. ornithopus,** ×25.0. **8.** Venation of sterile segment of **D. nobilis,** ×2.5. **9.** Section of petiole of **D. collina,** with one vascular bundle, ×8.0. **10.** Cellular detail of stem scale of **D. rediviva,** ×25.0.

notes published in 1962. The Brazilian species were treated by Brade (1965); those of southern Brazil by Sehnem (1972); and the Argentine species by Cartaginese (1977). The genus has been altered somewhat from the treatment of Tryon by the removal of *Doryopteris concolor* and its allies to *Cheilanthes.*

Synopsis of *Doryopteris*

Section *Lytoneuron*

Stem scales long-linear; two vascular bundles at the base of the petiole (Fig. 6); veins free, except as they are connected by a commissure in fertile segments (Fig. 5).

A tropical American section of 13 species. Exceptions to the characters of the section are *Doryopteris triphylla* with a single vascular bundle, *D. ornithopus* with wholly anastomosing veins, and *D. triphylla* and *D. itatiaiensis,* with stem scales that are long-lanceolate, attenuate, or ovate-lanceolate. Most of the species have the hyaline portion of the stem scales with elongate cells about three or more times longer than broad (Fig. 7).

Section *Doryopteris*

Stem scales long-lanceolate, attenuate, to ovate-lanceolate; one vascular bundle at the base of the petiole (Fig. 9); veins partially to usually wholly anastomosing (Fig. 8).

A tropical section of 12 species, seven of them in America and five from Madagascar to New Guinea. The American species have the hyaline portion of the stem scales more or less isodia-

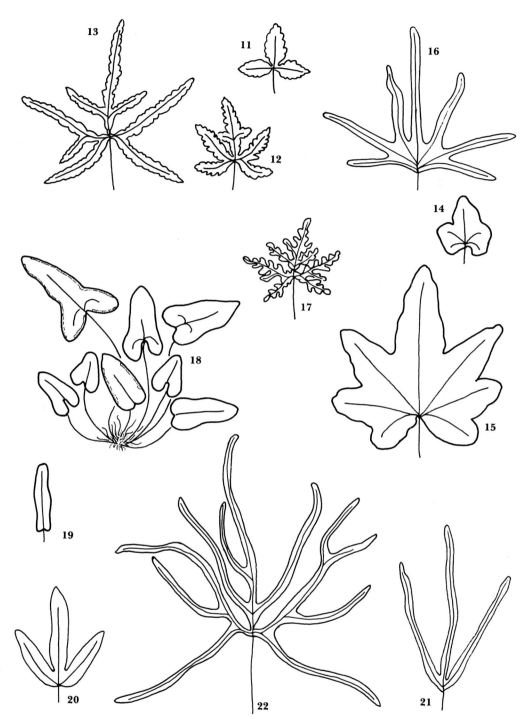

Figs. 40.11–40.22. Lamina architecture in *Doryopteris* section *Lytoneuron*, ×0.5. **11–13.** *D. triphylla.* **11, 12.** Sterile lamina. **13.** Fertile lamina. **14–16.** *D. ornithopus.* **14, 15.** Sterile lamina. **16.** Fertile lamina. **17.** *D. itatiaiensis,* sterile lamina. **18.** *D. rufa,* plant. **19–22.** *D. subsimplex.* **19, 20.** Sterile lamina. **21, 22.** Fertile lamina.

metic, or not much longer than broad (Fig. 10), the petiole terete or nearly so, and the stem short-creeping and compact. The paleotropical species have the hyaline portions of the stem scales three or more times longer than broad; and some species have a sulcate petiole and others a long-creeping, slender stem. The paleotropical species are: *Doryopteris Allenae* Tryon, *D. cordifolia* (Baker) Diels, *D. Humbertii* Tard., *D. ludens* (Hook.) J. Sm., and *D. papuana* Copel.

Tropical American Species

Doryopteris species are unusually variable in leaf architecture. Plants often have a heteroblastic leaf series (Figs. 11–16, 19–28) with gradual transitions in the form of the lamina from small, relatively simple, juvenile leaves to complexly lobed forms of large, fertile leaves on fully mature plants. Field observations have shown that juvenile leaves may be precociously fertile, a condition not always recognized in herbarium specimens. Considerably more species have been recognized by Brade (1965) and Sehnem (1972) than by Tryon (1942). The taxonomy of the group appears to be complicated by hybridization (Sehnem 1961, 1972) and polyploidy. Cytological studies are especially needed to resolve problems in species classification.

The species of section *Lytoneuron* (Tryon, 1942) are: *Doryopteris acutiloba* (Prantl) Diels, *D. conformis* Kramer & Tryon, *D. crenulans* (Fée) Christ, *D. itatiaiensis* (Fée) Christ (Fig. 17), *D. lomariacea* Kl., *D. ornithopus* (Hook. & Baker) J. Sm. (Figs. 14–16), *D. paradoxa* (Fée) Christ, *D. quinquelobata* (Fée) Diels, *D. Rosenstockii* Brade, *D. rufa* Brade (Fig. 18), *D. subsimplex* (Fée) Diels (Figs. 19–22), *D. tijucensis* Brade & Rosenst., and *D. triphylla* (Lam.) Christ (Figs. 11–13).

The American species of section *Doryopteris* (Tryon, 1942) are: *Doryopteris collina* (Raddi) J. Sm., *D. Lorentzii* (Hieron.) Diels (including *D. Juergensii* Rosenst.), *D. nobilis* (Moore) C. Chr., *D. pedata* (L.) Fée, with var. *multipartita* (Fée) Tryon (Figs. 23–25), var. *palmata* (Willd.) Hicken, and var. *pedata*, *D. rediviva* Fée (Figs. 26–28), *D. sagittifolia* (Raddi) J. Sm. (Figs. 29–31), and *D. varians* (Raddi) J. Sm.

Ecology (Fig. 1)

Doryopteris is primarily a genus of moist, or at least seasonally moist, rocky places, only a few species grow in forest or woodland humus. The paleotropical species, *Doryopteris Allenae, D. ludens* and *D. papuana,* are restricted to calcareous soils.

In the American tropics, *Doryopteris* grows most often in open rocky habitats such as cliffs, at the base of boulders, in rocky soil on steep slopes, on old rock walls along trails, and in thin soil over wet rocks. Species are most common on acidic rocks such as sandstone, quartzite and granite. Some species also grow in wet savannahs, in sphagnum swamps, on clay banks, in thickets or on shrubby hillsides. Plants of *Doryopteris itatiaiensis, D. paradoxa* and *D. crenulans* often form large clumps at the base of boulders. *Doryopteris* grows from near sea level to 2500 m.

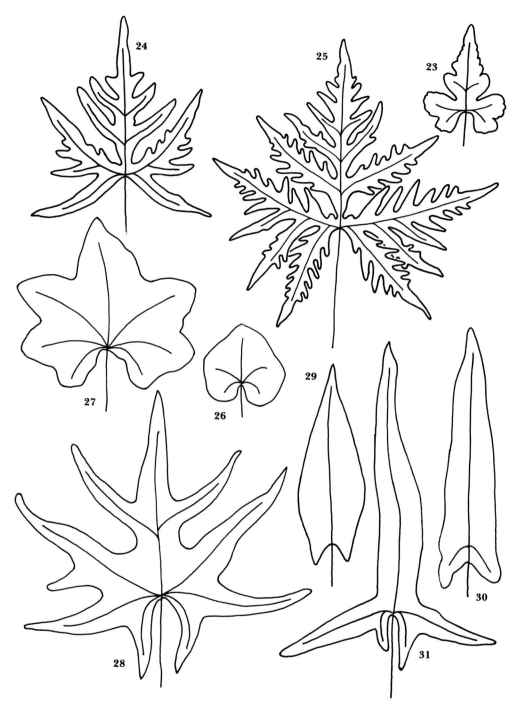

Figs. 40.23–40.31. Lamina architecture in ***Doryopteris*** section ***Doryopteris***, ×0.5. **23–25. *D. pedata* var. *multipartita*. 23.** Sterile lamina. **24, 25.** Fertile lamina. **26–28. *D. rediviva*. 26.** Sterile lamina. **27, 28.** Fertile lamina. **29–31. *D. sagittifolia*. 29.** Sterile lamina. **30, 31.** Fertile lamina.

Geography (Figs. 2, 3)

The genus *Doryopteris*, distributed through the American tropics is also found in Madagascar, and from India and China to the Philippine Islands, Java and New Guinea.

In America it occurs from southern Mexico sparingly through Central America; in the West Indies from Cuba to Martinique; and in South America from Surinam west to Colombia, south to Buenos Aires in Argentina; also on Ilha Trinidade and the Galápagos Islands.

Section *Lytoneuron* (Fig. 2) is confined to South America. Its center of diversity is southeastern Brazil; all of the species except *D. conformis* of Surinam are present there, and nine are endemic. Section *Doryopteris* (Fig. 3) is more wide ranging, occurring also in the Galápagos Islands, Central America and Mexico (*D. pedata* var. *palmata*), the West Indies (*D. pedata* var. *pedata*) and Ilha Trinidade (*D. collina*). The section also centers in southeastern Brazil, where all of the American species occur and one, *D. rediviva*, is endemic. A geographic and evolutionary history of the genus has been proposed (Tryon, 1944) involving continued speciation in southeastern Brazil and migration of some species from there to the Andes and northward.

Spores

Species in section *Lytoneuron* have relatively smooth spores with few low rugae as in *D. lomariacea* (Fig. 32) and *D. itatiaiensis* (Fig. 33). Those in section *Doryopteris* have spores mostly prominently cristate as in *D. pedata* (Figs. 35, 36), although in *D. collina* the surface is quite smooth (Fig. 34). The smooth spores of species in section *Lytoneuron* and cristate surface of those in section *Doryopteris* appear to relate to the amount of material deposited on the inner, reticulate strata of the wall. The spores of species in section *Lytoneuron* tend to be large, about 60 μm in diameter, while species in section *Doryopteris* have spores about half as large. Size differences are also evident in the varieties of *D. pedata;* spores of *D. pedata* var. *pedata* from Jamaica (Fig. 35) are smaller than those of var. *palmata* from Peru and the Galápagos Islands (Fig. 36). Spore size in these taxa appear to correlate with different ploidy levels of the varieties noted under the section on cytology.

Several strata can be recognized in spore wall sections of *Doryopteris*. In *D. sagittifolia* (Fig. 38) the surface is formed of perforated cristae and connecting strands characteristic of cristate spores. The smoother surface as in spores of *D. collina* (Fig. 34) overlays a reticulate stratum that is formed on a smoother base. Spores of *D. pedata* have similar strata as in the profile (Fig. 37).

Cytology

There is little cytological data for the genus, but *D. pedata* var. *pedata*, from Jamaica is reported as diploid with $n = 30$ (Walker, 1966), and var. *palmata* from the Galápagos Islands is tetraploid with $n = 60$ (Jarrett et al., 1968). These differences in ploidy

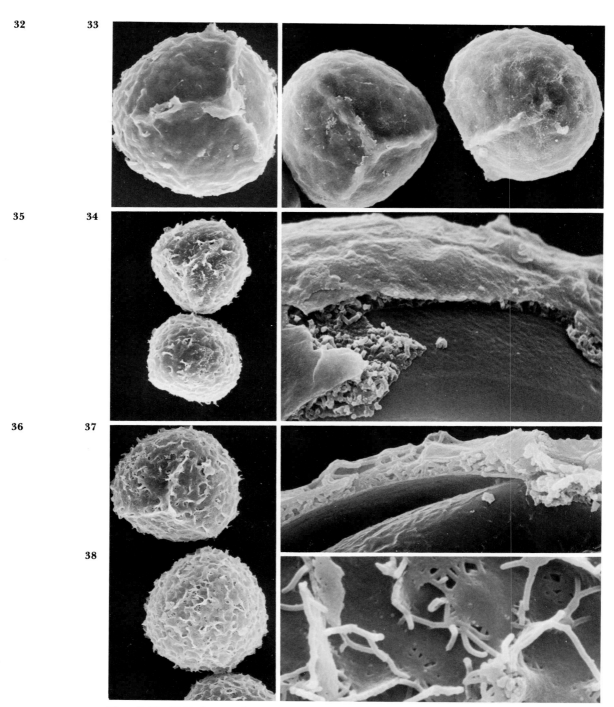

Figs. 40.32–40.38. Spores of *Doryopteris*. **32, 33.** Section *Lytoneuron*, × 1000. **32.** *D. lomariacea*, shallowly rugose proximal face, Brazil, *Mexia 5705*. **33.** *D. itatiaiensis*, shallowly rugose, Brazil, *Tryon & Tryon 6714*. **34–38.** Section *Doryopteris*. **34.** *D. collina*, abraded wall with rugose surface above reticulate layer and a smoother layer below, Brazil, *L. B. Smith 1587*, × 2500. **35–37.** *D. pedata*. **35.** var. *pedata*, cristate proximal face above, distal below, × 1000. **36, 37.** var *palmata*, Galápagos Islands, *Koford K-3*. **36.** Cristate proximal face above, distal face below, × 1000. **37.** Wall strata with cristate surface above the reticulate stratum and smoother basal layer, × 5000. **38.** *D. sagittifolia*, detail of cristate surface with partially fused strands, Venezuela, *Steyermark 89523*, × 10,000.

levels are also expressed in marked size differences of the spores as shown in Figs. 35 and 36. *Doryopteris ludens* from India has been reported as $n = 116$.

Literature

Brade, A. C. 1965. Contribução para conhecimento das espécias Brasileiras do gênero *Doryopteris*. Arq. Jard. Bot. Rio Janeiro 18: 39–72.

Cartaginese, M. A. 1977. Revisión del género "*Doryopteris*" en Argentina. Rev. Mus. Argent. Cienc. Nat. "Bernardino Rivadavia" 5: 105–122.

Jarrett, F. M., I. Manton, and S. K. Roy. 1968. Cytological and taxonomic notes on a small collection of living ferns from Galapagos. Kew Bull. 22: 475–480.

Sehnem, A. 1961. Algumas filicineas novas do Rio Grande do Sul. Pesquisas Bot. 13: 19–27.

Sehnem, A. 1972. Pteridáceas, *in*: P. R. Reitz, ed., Flora Illustrada Catariense, Pt. 1: 128–180.

Tryon, R. 1942. A revision of the genus *Doryopteris*. Contrib. Gray Herb. 143.

Tryon, R. 1944. Dynamic phytogeography of *Doryopteris*. Amer. Jour. Bot. 31: 470—473.

Tryon, R. 1962. The fern genus *Doryopteris* in Santa Catarina and Rio Grande do Sul, Brazil. Sellowia 14: 51–59.

Walker, T. G. 1966. Reference under the tribe Cheilantheae.

41. *Trachypteris*
Figs. 41.1–41.11

Trachypteris Christ, Denkschr. Schweiz. Naturfors. Gesells. 36 (Monogr. Elaphoglossum): 150. 1899. Type: *Trachypteris aureonitens* (Hook.) Christ (*Acrostichum aureonitens* Hook.) = *Trachypteris pinnata* (Hook. f.) C. Chr.

Saffordia Maxon, Smiths. Misc. Coll. 61 (4): 1. 1913. Type: *Saffordia induta* Maxon = *Trachypteris induta* (Maxon) R. & A. Tryon.

Description   Terrestrial or rupestral; stem decumbent or erect, small, bearing scales and many fibrous roots; leaves monomorphic, the lamina pinnatifid and pedate, or dimorphic (the sterile in a basal rosette, sessile to short-petiolate, narrowly obovate, or with a prolonged, proliferous apex, and the fertile long-petiolate, erect, pinnatisect to usually 1-pinnate and sometimes subpedate), ca. 10–25 cm long, borne in a close cluster, with densely imbricate scales beneath, glabrous or glabrate, or rarely thinly paleate with whitish, narrow scales above, veins anastomosing without included free veinlets; sporangia borne in a narrow to broad marginal band, or from the costa to the margin, on and between the veins, paraphyses not present, exindusiate, the margin flat to somewhat recurved; spores globose, trilete, the laesurae $\frac{3}{4}$ the radius, strongly cristate. Chromosome number: $n = 29$ or 30.

The abaxial surface of the lamina is densely covered with scales (Fig. 6) which often protrude beyond the margin onto the glabrous adaxial surface (Fig. 5). The anastomosing veins form large areolae along the costa and smaller ones at the margin. *Trachypteris induta* (Fig. 10) has a rather narrow band of sporan-

Fig. 41.1. *Trachypteris pinnata* on volcanic rock, Galápagos Islands (a rosette of sterile leaves in the center, fertile leaves at upper left). (Photo W. A. Weber.)

Fig. 41.2. Distribution of *Trachypteris* in America (dots, *T. pinnata;* stars, *T. induta*).

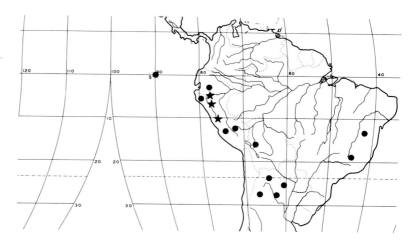

gia while *T. pinnata* (Fig. 11) has a broader one extending nearly to the costa. The sporangia mature at different times in both species and are borne between the veins as well as on them (Figs. 10, 11). This condition is rare among the cheilanthoid ferns but also occurs in *Doryopteris Humbertii.*

Systematics   *Trachypteris* is a genus of three species, two in America and one in Madagascar. *Saffordia induta,* formerly recognized as distinct from *Trachypteris* on the basis of lamina architecture and other aspects, has so many common features that there seems to be no basis for its recognition as a genus. The Madagascar endemic, *Trachypteris Drakeana* C. Chr. is similar to *T. pinnata,* but differs especially in the prolonged, proliferous apex of the sterile leaves. It has not been determined if this species has sporangia

Figs. 41.3–41.9. *Trachypteris.* **3.** Plant of *T. pinnata* with dimorphic leaves, the fertile erect, 1-pinnate, herbarium specimen, × 0.33. **4.** Plant of *T. induta* with deeply pinnatifid, pedate, monomorphic leaves, herbarium specimen, × 0.33. **5.** Portion of adaxial surface of lamina of *T. induta,* the margin with scales from the abaxial surface, × 10. **6.** Portion of abaxial surface of lamina of *T. induta,* with dense, imbricate scales, × 10. **7–9.** Spores of *Trachypteris,* × 1000, surface detail × 10,000. **7, 8.** *T. pinnata.* **7.** Cristate distal face, Galápagos Islands, *Wiggins 18369.* **8.** Detail of granulate surface, Galápagos Islands, *Snodgrass & Heller 319.* **9.** *T. induta,* proximal face, the inner echinate layer under the broken surface, Peru, *Sagástegui 0205.*

between the veins. The relationships of *Trachypteris* are not clear; however, chromosome number and spore characters place it among the cheilanthoid ferns where it undoubtedly represents a specialized genus.

Tropical American Species

The two American species, *Trachypteris induta* (Maxon) R. & A. Tryon (Fig. 4) and *T. pinnata* (Hook. f.) C. Chr. (Fig. 3) are distinctive in their lamina architecture. The very rare *Trachypteris Gilleana* (Baker) Svenson of Brazil differs from *T. pinnata* especially in the fertile lamina with three rather than five or more pinnae or segments. Until more materials are available for study, it is best considered as representing a variation of *T. pinnata.*

The pedate form of the mature lamina of *Trachypteris induta* is reached through a series of less complex leaves. The first leaves on a young plant have an orbicular lamina, with later ones oblong, 3-lobed, and finally pedate. The stem scales of this species have an exceptional pink color, especially evident at the apex in living plants. Striations on the cell walls of the scales are mentioned and figured by Ballard (1962).

Ecology (Fig. 1)

Trachypteris grows on rocks, at the base of rocks or in leaf-mold in rocky soil, or in forest humus, in open sites or more often in thickets and forest. *Trachypteris induta* grows in Peru in regions with a pronounced dry season, at which time it curls its leaves so that only the scaly undersurface is exposed; *T. pinnata* usually grows in more mesic areas. *Trachypteris pinnata* grows from 50 to 1000 m in the Galápagos Islands and at 500–2000 m in the Andes; *T. induta* occurs at 750–2900 m.

Geography (Fig. 2)

Trachypteris is local and rare in the Andes from Ecuador to northwestern Argentina and is disjunct in Minas Gerais and Bahia in eastern Brazil. In the Galápagos Islands, *Trachypteris pinnata* occurs on all five large islands and some of the smaller ones. *Trachypteris induta,* long known from only the original collection near Lima, Peru, was first recollected some 60 years later

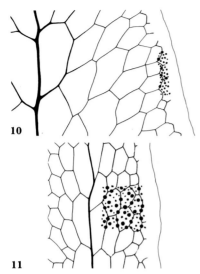

Figs. 41.10, 41.11. Portions of fertile segments of **Trachypteris,** ×6.0, veins and sporangial area (stippled); sporangia indicated by dots in part of band; size difference corresponds to degree of maturity. **10.** *T. induta.* **11.** *T. pinnata.*

at the type locality and has since also been found in La Libertad, Cajamarca, and Amazonas, Peru. The relation of *T. pinnata* with *T. Drakeana* of Madagascar is not clear. The geographic relation is similar to the disjunct distribution of *Doryopteris* section *Doryopteris,* in which species occur in Madagascar, although the section centers in Brazil.

Spores

Trachypteris spores are prominently cristate with sparse strands connecting the crests (Figs. 7, 9) and the surface detail is irregularly granulate (Fig. 8) in both species. The echinate stratum is exposed below the cristate surface in Fig. 9 of *T. induta.* The larger size of this specimen from Peru contrasts with the smaller one of *T. pinnata* from the Galápagos Islands (Fig. 7) and may reflect different polyploid levels in the two species. The similarity of spores in these American species supports their treatment in a single genus.

Cytology

The meiotic records of 29 or 30 for *Trachypteris induta* from Peru, by Manton (1958) (as *Saffordia*), is not exact but conforms with the two lowest numbers known for species of the cheilanthoid alliance.

Literature

Ballard, F. 1962. *Saffordia induta* Maxon. Hook. Icon. Pl. *t. 3599.*
Manton, I. 1958. Chromosomes and fern phylogeny with special reference to the "Pteridaceae." Jour. Linn. Soc. (Bot.) 56: 73–92.

42. *Cryptogramma*
Figs. 42.1–42.8

Cryptogramma R. Br., Addenda, Bot. Append. Franklin, Narrative Journey Polar Sea, 767 (reprint, 39). 1823. Type: *Cryptogramma acrostichoides* R. Br. = *Cryptogramma crispa* (L.) R. Br.
Phorolobus Desv., Mém. Soc. Linn. Paris 6: 291. 1827. Type: *Phorolobus crispus* (L.) Desv. (*Osmunda crispa* L.) = *Cryptogramma crispa* (L.) R. Br.
Allosorus sect. *Homopteris* Rupr., Dist. Crypt. Vasc. Ross. (Beitr. Pflanz. Russ. Reiches 3) 48. 1845. Type: *Allosorus Stelleri* (Gmel.) Rupr. (*Pteris Stelleri* Gmel.) = *Cryptogramma Stelleri* (Gmel.) Prantl. *Cryptogramma* sect. *Homopteris* (Rupr.) C. Chr., Ind. Fil. xlii. 1906.

Description Rupestral; stem decumbent, small, compact and multicipetal, or rather slender and short-creeping, bearing scales and many roots; leaves moderately to strongly dimorphic (the fertile longer than the sterile, more erect, and with entire segments), ca. 5–30 cm long, clustered to somewhat distant, lamina 2-pinnate to 3-pinnate-pinnatisect, glabrous, or sometimes the fertile with yellow farinaceous indument beneath, veins free; sori on the veins, sometimes below a fork, or only at their apex, not paraphysate, an indusium definitely differentiated from the re-

curved margin, or not; spores tetrahedral-globose, trilete with three prominent angles, the laesurae equal to the radius or nearly so, the surface strongly verrucate. Chromosome number: $n = 30, 60$; $2n = 60, 120$.

Comments on the Genus

Cryptogramma is an extratropical genus (Fig. 1) of two very distinctive species. *Cryptogramma crispa* (L.) R. Br. is widely disjunct and consists of four poorly defined varieties (Fernald, 1935). These include var. *acrostichoides* (R. Br.) C. B. Clark in the United States, Canada and northeastern Asia; var. *crispa* in Eurasia; var. *Brunoniana* (Hook. & Grev.) C. B. Clark in the mountains of southeastern Asia; and var. *chilensis* (Christ) Looser (*C. fumariifolia* (Baker) Christ) (Figs. 2–5) in southern Chile and Argentina. In this species the stem is indurated and compact with firm scales and long, fibrous roots. The fertile segments have recurved scarcely modified margins, and sometimes are yellow-farinaceous beneath.

Cryptogramma Stelleri (Gmel.) Prantl. occurs in the eastern United States, Alaska, Canada and Asia. The stem is creeping, succulent-brittle, greenish-yellow, shrivels in its second year, and bears delicate scales and few, delicate, short roots. The recurved margins of the fertile segments have the border definitely modified as an indusium. The distinctive verrucate spores show the two species to be related in spite of the differences in the stem

Fig. 42.2. Fertile segment of *Cryptogramma crispa* var. *chilensis* with reflexed margin, sori on the free veins, the lower half shows sporangia, left, and points of attachment, right, ×6.0.

Figs. 42.3–42.8. *Cryptogramma.* **3–5.** *C. crispa* var. *chilensis* **3.** Herbarium specimen, × 0.5. **4.** Sterile leaf, × 1. **5.** Fertile leaf, × 1. **6–8.** *Cryptogramma*, verrucate spores, × 1000. **6.** *C. Stelleri,* the proximal face, right, distal face, left, Vermont, *A. A. Eaton,* in 1904. **7.** *C. crispa* var. *Brunoniana,* lateral surface left, distal face right, Sikkim, *J. D. Hooker.* **8.** *C. crispa* var. *chilensis,* lateral view, Argentina, *Neumeyer 681.*

and leaves. *Cryptogramma* appears to belong with the cheilanthoid ferns on the basis of the chromosome number and general morphology but its immediate relatives are not obvious.

Spores

The verrucate elements are especially prominent at the ends of the laesurae (Fig. 6–8). There is remarkable similarity of the spores of the two species from widely disjunct populations sampled from Vermont, in the United States, Alberta in Canada, Rio Negro in Argentina and Sikkim in the Himalayas.

Cytology

The chromosome number of *Cryptogramma Stelleri,* known in plants of North America and Japan is consistently $n = 30$. Records of *C. crispa* var. *acrostichoides* from North America are uniformly $n = 30$ while report of var. *crispa* from Wales is $n = 60$ (Manton, 1950) and from Iceland $2n = 120$ (Löve, 1970). Thus diploids in both species are in North America, while in *C. crispa* derived tetraploid elements are known from the Old World (Löve et al., 1971). The chromosome number of the disjunct *C. crispa* var. *chilensis* will be of special interest in relation to these records.

Literature

Fernald, M. L. 1935. Critical plants of the upper Great Lakes region of Ontario and Michigan. *Cryptogramma crispa* and *C. acrostichoides.* Rhodora 37: 238–247.

Löve, A. 1970. Islenzk ferdaflora. 428 pp. Almenna Bokafelagid. Reykjavik.

Löve, A., D. Löve, and B. M. Kapoor. 1971. Cytotaxonomy of a century of Rocky Mountain orophytes. Arctic and Alpine Research 3: 139–165.

Manton, I. 1950. Problems of cytology and evolution in the Pteridophyta. 316 pp. Cambridge University Press, Cambridge.

43. *Llavea*
Figs. 43.1–43.6

Llavea Lag., Gen. Sp. Pl. 33. 1816. Type: *Llavea cordifolia* Lag.

Ceratodactylis Hook., Gen. Fil. t. 36. 1840. Type: *Ceratodactylis osmundioides* Hook. = *Llavea cordifolia* Lag.

Botryogramma Fée, Gen. Fil. (Mém. Fam. Foug. 5): 166. 1852. Type: *Botryogramma Karwinskii* (Kze.) Fée (*Allosorus Karwinskii* Kze.) = *Llavea cordifolia* Lag.

Description Terrestrial or rupestral; stem decumbent to nearly erect, moderately stout, bearing blackish scales and many fibrous roots; leaves partially dimorphic (the fertile with fertile pinnae apically, these with elongate, entire, narrow segments, the sterile segments broad, serrate), ca. 25–100 cm long, borne in a cluster, the petiole with large, more or less whitish, sometimes light

Fig. 43.1. Distribution of *Llavea*.

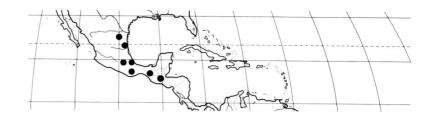

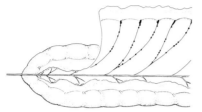

Fig. 43.2. Basal portion of fertile segment of *Llavea cordifolia*, a section bent to show the veins and points of attachment of the sporangia, the veins slightly thickened at their ends below the marginal indusium, ×6.

greenish-yellow scales at the base, lamina 2- to 4-, usually 3-pinnate, imparipinnate, glabrous, the sterile usually glaucous beneath, or the fertile sometimes short-glandular, or glandular-white farinaceous beneath, veins free; sori on the veins, not paraphysate, indusium moderately to well differentiated from the recurved margin; spores tetrahedral-globose, trilete, the laesurae $\frac{3}{4}$ the radius, the surface with coarse, more or less fused tubercules. Chromosome number: $n = 29$.

The soral condition with the sporangia along the free veins is shown in Fig. 2. The partially dimorphic fertile leaf (Fig. 3) is unusual as are the black scales on the stem and contrasting whitish ones at the base of the petiole (Fig. 4).

Systematics

Llavea is a highly distinctive, monotypic genus of Mexico and Guatemala. The single species, *Llavea cordifolia* Lag., is relatively uniform in its characters. The genus is most often placed near *Cryptogramma* and *Onychium* in spite of considerable difference from those genera. Copeland (1947) regarded it as a derivative of *Pellaea*, but there is little support for this relationship. There are resemblances to *Lygodium* in the partially dimorphic fertile leaves, the form of the sterile segments, the spores, and the chromosome number also suggests a possible distant relation to that genus.

Ecology and Geography (Fig. 1)

Llavea grows in mesic canyons, or other rocky places in pine and oak woods, or in tropical forests. Sometimes it occurs on roadsides, on rock walls or in damp soil. It is principally, perhaps always, a calciphile. The genus grows in the Sierra Madre Oriental of Mexico from Nuevo Leon southward to the mountains of Guatemala, at 800–3200 m, most frequently between 1000 and 1500 m.

Spores

Spores of *Llaeva* have rather simple, tuberculate formation with somewhat denser tubercules on the proximal face (Fig. 5), and the surface is formed of an irregular granulate material (Fig. 6). The tuberculate formation generally resembles that of spores of *Lygodium heterdoxum* or *L. venustum* than the more complex reticulate or cristate forms characteristic of the cheilanthoid ferns.

3 **5**

4 **6**

Figs. 43.3–43.6. *Llavea cordifolia.* **3.** Dimorphic leaf from herbarium specimen, the petiole bent upward, × 0.33. **4.** Blackish scales of the stem, below, and whitish scales of the petiole base, above, × 5. **5, 6.** Spores, Guatemala, *Standley 65636.* **5.** Proximal face above, distal face below, × 1000. **6.** Detail of tuberculate proximal face, × 10,000.

Cytology

Documented reports for *Llavea cordifolia* of $n = 29$ are from the states of Hidalgo (Mickel et al., 1966) and Nuevo Leon (Knobloch, 1967). This number allies *Llavea* with the cheilanthoid ferns.

Literature

Copeland, E. B. 1947. Genera Filicum. 247 pp. Chronica Botanica, Waltham, Mass.

Knobloch, I. W. 1967. Chromosome numbers in *Cheilanthes, Notholaena, Llavea, Polypodium*, Amer. Jour. Bot. 54: 461–464.

Mickle, J. T., W. H. Wagner, and K. L. Chen. 1966. Chromosome observations on the ferns of Mexico. Caryologia 19: 95–102.

14c.   Tribe Ceratopterideae

Ceratopterideae J. Sm., Hist. Fil. 170. 1875. Type: *Ceratopteris* Brongn.

Parkeriaceae Hook., Exot. Fl. 2: *t. 147*. 1825. Type: *Parkeria* Hook. = *Ceratopteris* Brongn.

Ceratopteridaceae Underw., Native Ferns, ed. 6, 78. 1900.

A single small genus, *Ceratopteris.*

44.   *Ceratopteris*
Figs. 44.1–44.14

Ceratopteris Brongn., Bull. Sci. Soc. Philom. Paris, III, 8: 186. 1821. Type: *Ceratopteris thalictroides* (L.) Brongn. (*Acrostichum thalictroides* L.).

Teleozoma R. Br., Addenda, Bot. Append. Franklin, Narrative Journey Polar Sea, 767 (reprint, 39). 1823, *nom. superfl.* for *Ceratopteris* Brongn. and with the same type.

Ellebocarpus Kaulf., Enum. Fil. 147. 1824, *nom superfl.* for *Ceratopteris* Brongn. and with the same type.

Parkeria Hook., Exotic Fl. 2: *t. 147*. 1825. Type: *Parkeria pteridoides* Hook. = *Ceratopteris pteridoides* (Hook.) Hieron.

Furcaria Desv., Mém. Soc. Linn. Paris 6: 292. 1827, *nom. superfl.* for *Ceratopteris* Brongn. and with the same type.

Description   Terrestrial, palustral, or usually aquatic; stem erect, small, hardly indurated, bearing a few scales and usually few roots, these more abundant on the petiole base; leaves dimorphic (the sterile lobed to 3-pinnate, the fertile more erect, longer, with narrower segments and 1- to 5-pinnate), ca. 5–100 cm long, borne in a rosette or cluster, glabrous, veins anastomosing without included free veinlets; sporangia solitary, borne distantly on the veins, paraphyses not present, indusium well differentiated from the recurved margin; spores globose, trilete, the laesurae $\frac{1}{2}$ the radius, or shorter, with coarse, parallel ridges extending around both faces, aligned in three units corresponding to the laesurae. Chromosome number: $n = 39, 77, 78; 2n = 80, 154.$

Fig. 44.1. *Ceratopteris pteridoides* growing with *Nymphaea* near Turrialba, Costa Rica. (Photo W. H. Hodge.)

Ceratopteris has strongly dimorphic leaves (Figs. 2, 5, 6), the fertile more complex than the sterile and bearing slender segments (Figs. 2, 5) on divergent branches (Fig. 2). Leaves have anastomosing veins, and adventitious buds are often present in marginal sinuses (Fig. 7). The fertile segments have an epidermis of lobed, isodiametric cells, except the marginal portion forming the indusium, which is composed of very different, long cells oriented with the margin (Fig. 4).

Systematics   *Ceratopteris* is a nearly pantropical genus, comprised of three species, which is morphologically isolated and consequently difficult to relate to other ferns. The Schizaeaceae, Osmundaceae and Plagiogyriaceae have all been suggested as ancestral to *Ceratopteris,* and also the pityrogrammoid and pteroid ferns. There are difficulties in distinguishing between characters modified in relation to the aquatic habit and those that reflect phyletic relations, but the general relationship appears to be with the Pteridaceae. There are similarities in the spores to the strongly ridged forms in *Anemia* and *Mohria* in the Schizaeaceae and the chromosome numbers are near 38 or 76 reported in those genera.

Ceratopteris is sometimes placed in its own family, Parkeriaceae, but it is regarded here as a specialized type rather than an isolated relict, and placed as a tribe, associating it with other genera.

Ceratopteris has been revised by Benedict (1909) and by Lloyd (1974).

Fig. 44.2. *Ceratopteris pteridoides* growing on mud near Turrialba, Costa Rica. (Photo W. H. Hodge.)

Fig. 44.3. Distribution of *Ceratopteris* in America.

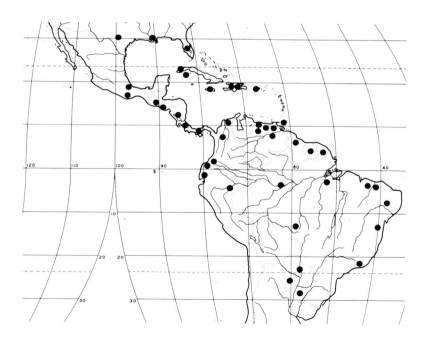

Tropical American Species

The species taxonomy of *Ceratopteris* is exceptionally difficult, for plants appear to be phenotypically plastic, a frequent condition among aquatics, with relatively few stable characters. The occurrence of morphological intermediates, including putative hybrids, complicates the delimitation of species. Also, original ranges have probably been obscured by the introduction of cultivated plants into natural habitats.

In the American tropics, two basic elements, *Ceratopteris pteridoides* and *C. thalictroides,* are recognized as species. *Ceratopteris Richardii* Brongn. (*C. deltoidea* Bened.) is included in *C. thalictroides:* the principle difference between them is that *C. thalictroides* has 32 spores in a sporangium, while *C. Richardii* has 16. Such differences in spore number occur in other species, i.e., *Notholaena Standleyi* has plants with either 32 or 16 spores per sporangium. The paleotropical *Ceratopteris cornuta* (Beauv.) Le Prieur is recognized by Lloyd and appears to be distinct from the American species.

The following key, adapted from Lloyd (1974), emphasizes differences between the species and does not account for unusual or intermediate plants.

Key to American Species of Ceratopteris

a. Plants floating, sterile lamina mostly lobed, often deeply so, or pinnate and with a broad rachis, the basal main veins or pinnae mostly opposite, annulus of few (to ca. 10) indurated cells or none, spores 32 in a sporangium; American tropics.   *C. pteridoides* (Hook.) Hieron.

a. Plants rooting above or below water, sterile lamina 1- to 3-pinnate, with a narrow rachis, the basal pinnae mostly alternate, annulus with several (ca. 15) to usually many (to 70) indurated cells, spores 32 or 16 in a sporangium; America and paleotropical.   *C. thalictroides* (L.) Brongn.

Ecology (Figs. 1, 2)

Ceratopteris is a genus of wet habitats, usually growing among other aquatic vegetation, but rarely occurring in wet grassy places. In the paleotropics it may be weedy in taro patches and in rice paddies.

In America the genus grows in ditches, lagoons, along rivers, and in lakes, ponds or marshes in fresh or sometimes in brackish water. *Ceratopteris pteridoides,* with inflated petioles, is nearly always floating, while *C. thalictroides* is rooted in wet soil or on the bottom in shallow water. Vegetative buds often occur on the leaves, especially in *C. pteridoides,* and develop vigorously to form new plants as the leaf ages and settles on the water. Donselaar (1969) reports that after the artificial Brokopondo Lake was formed in Surinam, *Ceratopteris pteridoides* greatly increased in abundance, colonizing a maximum area of 17,000 hectares. *Ceratopteris* is usually considered to be an annual, but may be more accurately described as short-lived. Many tropical aquatic habitats lack the seasonality necessary to provide a basis for an annual lifecycle. *Ceratopteris* grows mostly from sea level to ca. 300 m.

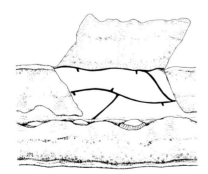

Fig. 44.4. Portion of fertile segment of ***Ceratopteris thalictroides.*** The recurved margin and indusium, below, covering sporangia, a portion raised, above, showing veins and insertion of sporangia, × 15.

Geography (Fig. 3)

Ceratopteris is widely distributed in tropical and subtropical America, in tropical Africa to southeastern Asia, Malesia, and Japan to northern Australia and the Micronesian islands; it also occurs in the Hawaiian Islands. In the paleotropics it is considered to be introduced in Fiji and the Hawaiian Islands, and its common association with taro culture suggests that it may be adventive in other parts of its range. It is also frequently grown in tropical botanical gardens and in aquaria which are likely sources for introduction.

In America the genus occurs from Texas to Florida in the southern United States, southern Mexico through Central America, the Greater Antilles, and in South America, south to Missiones in Argentina. It is adventive in Texas and perhaps elsewhere. It is probably more common and widely distributed than collections indicate, especially in the Amazon basin.

Spores

Ceratopteris spores are notable for their large size, ranging between 70 and 150 μm in diameter, the largest among homosporous ferns. Sporangia usually have 32 spores but there are only 16 in some specimens of *C. thalictroides*. Coarse parallel ridges extend around both faces in three units corresponding to the laesurae (Figs. 9, 10). The ridges are formed of a thick stratum that appears to be exospore (Fig. 8). The perispore above this usually consists of dense, projecting rods that are more or less overlaid by granulate material (Fig. 11). Spores of *C. pteridoides* differ from other species in their smaller size (Figs. 12, 13) and compact granulate surface (Fig. 14). The prominent ridged spores of *Ceratopteris* resemble those of *Mohria* and *Anemia* in the Schizaeaceae, which have coarse ridges formed by the exospore. Similarities of *Ceratopteris* spores and those of the Schizaeaceae are reviewed by Nayar (1968).

Cytology

Chromosome reports for *Ceratopteris* of $n = 39$, 78, and 77 are established with certainty. Diploids with $n = 39$ have been reported in three species by Hickok (1977), *C. pteridoides* originating from El Salvador, *C. thalictroides* (as *C. Richardii*) from Cuba, and *C. cornuta* from Ghana. Reports for *C. thalictroides* of $n = 78$ and 77 based on plants from the paleotropics have been made by Ninan (1956) and Hickok (1979). These numbers show both polyploid and aneuploid changes have occurred in *Ceratopteris*. The records of indefinite numbers as ca. 120–130 in works done prior to the use of squash methods, reported by Pal and Pal (1969), are not reliable and the record of $n = 40$ is not documented with certainty by the figure (Pal and Pal, 1963). The work on *Ceratopteris* hybrids by Hickok and Klekowski (1973, 1974) and Hickok (1977) include cytological observations of bridges, multivalents and micronuclei. From the chromosomal

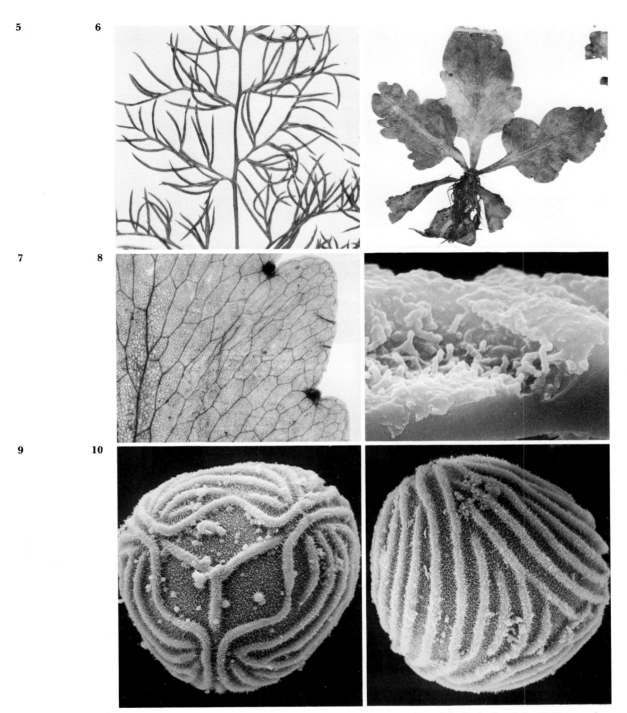

Figs. 44.5–44.10. *Ceratopteris.* **5–7.** *C. pteridoides.* **5.** Portion of fertile leaf, × 0.5. **6.** Young plant with a rosette of sterile leaves, × 0.5. **7.** A portion of a sterile leaf, with anastomosing veins and adventitious buds at the marginal sinuses, × 3.0. **8–10.** Spores of *C. thalictroides*, Isle of Pines, Cuba, *Killip 44595.* **8.** Section of wall, the perispore above, surface of more or less fused rods conforms to the dense inner exospore, below, × 5000. **9.** Proximal face with three sets of more or less parallel ridges aligned with the laesurae, and spherical deposit, × 500. **10.** Distal face tilted, the ridges continuous with those of the proximal face, × 500.

11

12

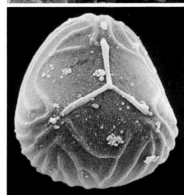

13

14

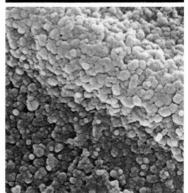

Figs. 44.11–44.14. *Ceratopteris* spores, × 500, surface detail, × 10,000. **11.** *C. thalictroides* granulate deposit densely covering rods, India, *Wight 249*. **12–14.** *C. pteridoides,* El Salvador, *Fassett 28557*. **12.** Proximal face. **13.** Distal face. **14.** Detail of granulate deposit.

associations in the diploid hybrids, they conclude that cytological differences between species involve individual chromosomes rather than the whole genome. Studies of *Ceratopteris,* especially those of Hickok on *C. thalictroides,* consider the genus to be in an active evolutionary state, based on infraspecific variation both in chromosome pairing relations and in fertility of hybrids.

Observations

Ceratopteris is useful as an experimental subject because its life cycle of about three months is one of the shortest among homosporous ferns, and also because the small plants are easily maintained in culture.

Schedlbauer (1974) identified a new antheridogen in *Ceratopteris thalictroides,* named antheridogen C, that caused potentially hermaphrodite gametophytes to become male. Tests of this substance with *Pteridium* gametophytes (antheridogen A) and with *Anemia phyllitidis* gametophytes (antheridogen B) were negative, and reciprocal tests indicated that antheridogen A and B had no effect on the development of *Ceratopteris* gametophytes.

Two inbred tetraploid plants of *C. thalictroides* were used by Hickok (1978) in studies of chromosome pairing. These provided direct evidence for a low level of homoeologous pairing, since a gametophytic mutant segregated from a sporophyte derived by selfing. In this and earlier studies (Hickok and Klekowski, 1973), emphasis is placed on the role of homoeologous pairing in homosporous, polyploid ferns as a mechanism for the release of variation following self-fertilization.

Copeland (1942) reports the development of *Ceratopteris thalictroides* as a crop plant in the Philippine Islands. Although its culture was initially successful, and there was a ready market for the leaves in Manila, a fungus disease decimated the plantings and the program was abandoned.

Literature

Benedict, R. C. 1909. The genus *Ceratopteris:* a preliminary revision. Bull. Torrey Bot. Cl. 36: 463–476.

Copeland, E. B. 1942. Edible ferns. Amer. Fern Jour. 32: 121–126.

Donselaar, J. van. 1969. On the distribution and ecology of *Ceratopteris* in Surinam. Amer. Fern Jour. 59: 3–8.

Hickok, L. G. 1977. Cytological relationship between three diploid species of the fern *Ceratopteris.* Canad. Jour. Bot. 55: 1660–1667.

Hickok, L. G. 1978. Homoeologous chromosome pairing: Frequence differences in inbred and intraspecific hybrid polyploid ferns. Science 202: 982–984.

Hickok, L. G. 1979. A cytological study of intraspecific variation in *Ceratopteris thalictroides* (L.) Brongn. Canad. Jour. Bot. 57: 1694–1700.

Hickok, L. G., and E. J. Klekowski. 1973. Abnormal reductional and nonreductional meiosis in *Ceratopteris:* Alternatives to homozygosity and hybrid sterility in homosporous ferns. Amer. Jour. Bot. 60: 1010–1022.

Hickok, L. G., and E. J. Klekowski. 1974. Inchoate speciation in *Ceratopteris:* an analysis of the synthesized hybrid *C. Richardii* × *C. pteridoides.* Evolution 28: 439–446.

Lloyd, R. M. 1974. Systematics of the genus *Ceratopteris* Brongn. (Parkeriaceae) II. Taxonomy. Brittonia 26: 139–160.

Nayar, B. K. 1968. A comparative study of the spore morphology of *Ceratopteris, Anemia* and *Mohria,* and its bearing on the relationships of the Parkeriaceae. Jour. Ind. Bot. Soc. 47: 246–256.

Ninan, C. A. 1956. Studies on the cytology and phylogeny of the Pteridophytes IV. Systematic position of *Ceratopteris thalictroides* (L.) Brongn. Jour. Ind. Bot. Soc. 35: 252–256.

Pal, N., and S. Pal. 1963. Studies on the morphology and affinity of the Parkeriaceae 2. Sporogenesis, development of gametophyte, and cytology of *Ceratopteris thalictroides.* Bot. Gaz. 124: 405–412.

Pal, N., and S. Pal. 1969. Studies on the morphology and affinity of the Parkeriaceae 3. A discussion of the systematic position of the family and specific delimitation in the genus *Ceratopteris.* Bull. Bot. Soc. Bengal 23: 17–25.

Schedlbauer, M. D. 1974. Biological specificity of the antheridogen from *Ceratopteris thalictroides* (L.) Brongn. Planta (Berl.) 116: 39–43.

14d.   Tribe Adianteae

Adianteae Presl, Tent. Pterid. 139. 1836, as Adiantaceae. Type: *Adiantum* L.

Adiantaceae (Presl) Ching, Sunyatsenia 5: 229, 1940.

This tribe consists of a single large genus, *Adiantum.*

45.   *Adiantum*
Figs. 45.1–45.35

Adiantum L., Sp. Pl. 1094. 1753; Gen. Pl., ed. 5, 485. 1754. Type: *Adiantum Capillus-Veneris* L.

Hewardia J. Sm., Jour., Bot. (Hook.) 3: 432, 1841. Type: *Hewardia adiantoides* J. Sm. = *Adiantum adiantoides* (J. Sm.) C. Chr.

Description   Terrestrial or rupestral; stem small and suberect, to stout and short-creeping or slender and long-creeping, bearing scales and usually numerous, long, fibrous roots; leaves monomorphic or nearly so, ca. 5 cm to 2 m long, borne in a cluster, or close to widely spaced, lamina simple, entire or sagittate, or 1- to 4-pinnate, rarely to 6-pinnate, or helicoid, segments usually glabrous beneath, sometimes glaucous, slightly scaly, pubescent or (in *A. Poiretii* var. *sulphureum*) farinose, veins free, rarely casually or regularly anastomosing without included free veinlets; sori borne on orbicular to very long, strongly recurved, modified marginal lobes, the sporangia confined to the veins or sometimes also on the tissue, not paraphysate, the indusium (recurved lobe) may have a membranous border; spores tetrahedral-globose, or globose, the angles sometimes prominent, trilete, the laesurae $\frac{3}{4}$ the radius, with a thin, usually fragmented surface formation above a granulate or somewhat papillate stratum. Chromosome

Fig. 45.1. *Adiantum concinnum,* Maricao, Puerto Rico. (Photo D. S. Conant.)

number: n = 29, 30, 57, 58, 60, 90, 114, 116, 150, ca. 180; $2n$ = 58, 60, 114, 116, 120, 228; apogamous 30, 60, 90, 171, ca. 175.

Diversity in the shape of ultimate segments, the length of stalk, and the form of indusia is shown in Figs. 1–13. The leaf architecture of a selection of neotropical species is illustrated in Figs. 16–28. Veins are usually free, as in Figs. 5–8, but are partially to fully anastomosing in a few species (Fig. 13). Sporangia are on the indusia, usually confined to the veins (Fig. 15) but sometimes also on the tissue (Fig. 14). In unrelated genera, similarities of leaf form through parallel development is well known in ferns (Tryon, 1964a). One of the most striking examples of such convergent evolution is found in the shape and division of the leaves between species of *Adiantum* and *Lindsaea.*

Systematics

Adiantum is a genus of about 150 species, widely distributed except in regions with extremely cold or dry climates. An evolutionary classification of *Adiantum* is much needed, but requires a thorough study of the whole genus. The groups recognized here often bring together related species, although some are doubtfully placed and relations of the groups are uncertain. Parallel and convergent evolution, which have probably occurred in leaf architecture as well as other characters, have undoubtedly obscured relationships. Species with a decompound lamina with many flabellate to cuneate segments are considered least derived, while those with branched pinnae, with dimidiate segments, or with a simpler lamina are considered more specialized.

Adiantum is an isolated and undoubtedly old genus. Its general phyletic position is among the genera of Pteridaceae on the basis of its soral characters and chromosome numbers.

There is no modern taxonomic study of the genus, although

Fig. 45.2. *Adiantum tetraphyllum,* Río Palenque, Ecuador. (Photo W. H. Hodge.)

the treatments of *Adiantum* in Costa Rica by Scamman (1960) and in Peru by Tryon (1964b) include about 50 of the American species.

The Groups of *Adiantum* and Tropical American Species

1. *Adiantum Capillus-Veneris* Group

Lamina 2- to 4-pinnate, gradually reduced to the apex, axes glabrous, ultimate segments flabellate to flabellate-cuneate, especially those at the apex and (or) base of the lamina and pinnae, short- to moderately long-stalked, with few to several indusia, veins free.

A group of about 25 species in America and nearly as many in the paleotropics including *Adiantum aethiopicum* L., *A. Davidii*

Fig. 45.3. *Adiantum Seemannii,* Turrialba, Costa Rica. (Photo W. H. Hodge.)

Franch., *A. erythrochlamys* Diels, and *A. monochlamys* D. C. Eaton. The group is undoubtedly a natural one, although a few species are divergent in some characters. *Adiantum deltoideum* and *A. sericeum* have hastate-cuneate ultimate segments, but appear to belong here on the basis of the flabellate segments of juvenile leaves. *Adiantum princeps* and *A. tenerum* have larger segments than other species and may be allied to the *Adiantum pectinatum* group. *Adiantum digittatum,* sometimes scandent with leaves to 1.5 m long and with large, deeply cleft segments, is quite different from other species.

Among the tropical American species are: *Adiantum andicola* Liebm., *A. Braunii* Kuhn, *A. Capillus-Veneris* L., *A. concinnum* Willd., *A. digittatum* Hook. (Fig. 17), *A. deltoideum* Sw., *A. Feei* Fée, *A. fragile* Sw., *A. Orbignyanum* Kuhn, *A. Poiretii* Wikstr. and its var. *sulphureum* (Kaulf.) Tryon, *A. princeps* Moore, (*A. trapezoides* Fée), *A. Raddianum* Presl (*A. cuneatum* Langsd. & Fisch.) (Fig. 16), *A. sericeum* D. C. Eaton, *A. subvolubile* Kuhn, *A. tenerum* Sw., and *A. tricholepis* Fée.

2. *Adiantum patens* Group

Lamina 1- to 3-pinnate, or helicoid, in 2- to 3-pinnate leaves with an apical 1-pinnate segment similar to the adjacent pinnae, axes glabrous or short-pubescent, ultimate segments flabellate to flabellate-cuneate, especially those at the apex and (or) base of the lamina and pinnae, mostly sessile to short-stalked, with few to several indusia, veins free.

About ten species in the American tropics, and several in the paleotropics including *Adiantum diaphanum* Bl., *A. Oatesii* Baker, and *A. Fournieri* Copel.

The group includes several species that are clearly related al-

Fig. 45.4. Distribution of *Adiantum* in America, south of lat. 35° N.

though some, especially those with a 1-pinnate lamina, are not placed here with certainty. The probable development of 1-pinnate species as *A. Ruizianum* and *A. Shepherdii* from those with a 2-pinnate lamina is suggested by species with both 1-pinnate and 2-pinnate leaves, as *A. Galeottianum*. The leaf structure in *A. lobatum* Presl and *A. sinuosum* suggests partial modification toward the helicoid form in *A. patens* and *A. pedatum*.

Among the American species are: *Adiantum Galeottianum* Hook., *A. patens* Willd. (Fig. 20), *A. pedatum* L., *A. Ruizianum* Kl. (Fig. 18), *A. sessilifolium* Hook. (Fig. 19), *A. Shepherdii* Hook., and *A. sinuosum* Gardn.

3. *Adiantum philippense* Group

Lamina 1-pinnate, axes glabrous or pubescent, ultimate segments flabellate to flabellate-cuneate, especially at the base of the lamina, short- to long-stalked, with few to several indusia, veins free.

One or more species of tropical America belong here, and 10 or more of the paleotropics including *Adiantum philippense* L. and its segregates, *A. caudatum* L. and its segregates, and *A. Balfourii* Baker. Many species have leaves specialized for vegetative reproduction with an elongate rachis that roots at the tip.

The American species in the group need to be reassessed for there are probably more names than species. *Adiantum philippense* is probably adventive in America.

Names for American species are: *Adiantum deflectens* Mart.

Figs. 45.5–45.15. *Adiantum.* **5.** Pinna, *A. scalare,* ×0.5. **6.** Pinna, *A. Seemannii,* ×0.5. **7.** Ultimate segment, *A. Poiretii* var. *Poiretii,* ×1.5. **8.** Pinnules, *A. villosum,* ×1.0. **9.** Pinnules, *A. macrocladum,* ×1.0. **10.** Pinna, *A. petiolatum,* ×1.0. **11.** Ultimate segments, *A. digitatum,* ×0.75. **12.** Pinnules, *A. sessilifolium,* ×1.0. **13.** Pinna, veins areolate, *A. olivaceum,* ×0.5. **14.** *A. Raddianum,* abaxial side of indusium, detached from the segment at the sinus, with veins and sporangia, ×10. **15.** *A. pulverulentum,* abaxial side of indusium, attached along the base, with veins, sporangia and membranous portion, above, ×10.

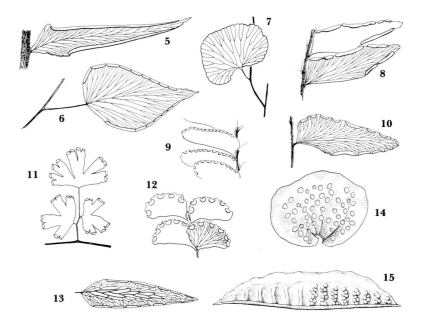

(Fig. 21), *A. delicatulum* Mart., *A. dolabriforme* Hook., *A. filiforme* Hook., *A. flagellum* Fée, *A. Fiebrigii* Hieron, *A. rhizophyton* Schrad. and *A. subaristatum* Fée.

4. *Adiantum reniforme* Group
Lamina simple, entire or nearly so, flabellate-cuneate to reniform, with few to many sori, veins free.

A paleotropical group of three species: *Adiantum flabellum* C. Chr., *A. Parishii* Hook., and *A. reniforme* L. (including *A. asarifolium* Willd.).

These species have the simplest leaf form in the genus and are regarded as highly derived forms. The chromosome number of $n = 150$ in *Adiantum reniforme* represents a decaploid, supporting its status as a derived element. Relations among the species of this group are not clear for simple leaves may have been derived from complex ones in more than one evolutionary lineage.

5. *Adiantum pectinatum* Group
Lamina 3- to 6-pinnate, at least in large leaves, with an apical 1-pinnate segment similar to the adjacent pinnae, axes scurfy-scaly or glabrous, ultimate segments acute to acuminate, especially those at the apex of the lamina and pinnae, mostly sessile to short-stalked, with few to several indusia, veins free or (in *A. Leprieurii*) anastomosing.

A group of about 12 species in the American tropics and perhaps 15 in the paleotropics including *Adiantum affine* Willd., *A. formosum* R. Br., *A. fulvum* Raoul, and *A. hispidulum* Sw.

Adiantum Mathewsianum and *A. trapeziforme,* with rather large, sometimes definitely stalked segments, are perhaps transitional between this and the group of *Adiantum platyphyllum.* Others, such as *Adiantum melanoleucum* and *A. Leprieurii,* with segments sometimes flabellate at the base of the pinnae, seem close to the *Adiantum patens* group.

Among the tropical American species are: *Adiantum brasiliense* Raddi, *A. Leprieurii* Hook., *A. macrocladum* Kl., *A. Mathewsianum* Hook., *A. melanoleucum* Willd. (Fig. 22), *A. pectinatum* Ettingsh., *A. pyramidale* (L.) Willd. (*A. cristatum* L., *A. striatum* Sw.), *A. trapeziforme* L. and *A. Wilesianum* Hook.

6. *Adiantum tetraphyllum* Group

Lamina 1- or 2-pinnate, in 2-pinnate leaves with an apical 1-pinnate segment similar to the adjacent pinnae, axes scurfy-scaly or short-pubescent, at least on the adaxial side, ultimate segments acute to acuminate, rarely obtuse, especially those at the apex of the lamina and pinnae, sessile to subsessile, or sometimes short-stalked, with few to many indusia, or a single one on the acroscopic side, veins free.

A tropical group of about 20 American species and *Adiantum Vogelii* Keys. of west Africa. The group is undoubtedly a natural one. The species taxonomy is difficult, perhaps because of hybridization. Eight probable hybrids from Surinam were listed by Kramer (1978), involving nine species of this group. In Trinidad, T. G. Walker (personal communication) has recognized the 2-pinnate form of *Adiantum lucidulum* as a hybrid between 1-pinnate *A. lucidulum* and the 2-pinnate *A. villosum;* and also the 2-pinnate form of *Adiantum petiolatum* as a hybrid between 1-pinnate *A. petiolatum* and the 2-pinnate *A. latifolium.*

Some of the American species are: *Adiantum cajennense* Kl., *A. fructuosum* Spreng, (if different from *A. tetraphyllum*), *A. fuliginosum* Fée (*A. hirtum* Splitg.), *A. humile* Kze., *A. latifolium* Lam. (Fig. 24), *A. obliquum* Willd., *A. petiolatum* Desv., *A. pulverulentum* L. (Fig. 25), *A. serratodentatum* Willd. (Fig. 23), *A. tetraphyllum* Willd., *A. tomentosum* Kl., and *A. villosum* L.

7. *Adiantum platyphyllum* Group

Lamina 1- to 4- pinnate, in 2- to 4-pinnate leaves with the apex gradually reduced or with an apical 1-pinnate segment similar to the adjacent pinnae, axes glabrous, ultimate segments acute to acuminate, rarely obtuse, especially those at the apex of the lamina and pinnae, mostly long-stalked, with several to many indusia, veins free.

A tropical American group of about six species. The species form a natural and distinctive alliance except for the 1-pinnate *Adiantum grossum,* which is placed here on the basis of its somewhat tapering apical segment and long-stalked basal pinnae. The dimidiate-flabellate pinnae, however, possibly relate it to *A. Ruizianum* of the *Adiantum patens* group.

The species are: *Adiantum anceps* Max. & Mort. (Fig. 26), *A. grossum* Mett., *A. pentadactylon* Langsd. & Fisch., *A. peruvianum* Kl., *A. platyphyllum* Sw., *A. Seemannii* Hook., and *A. subcordatum* Sw.

8. *Adiantum phyllitidis* Group

Lamina simple, entire, or 1- or 2-pinnate, with a large, entire or rarely lobed apical segment, axes scurfy-scaly or glabrous, ultimate segments acute to acuminate, sessile to usually short- to moderately long-stalked, with a single very long indusium on each side, or sometimes with a few long ones, veins free, irregularly anastomosing or fully anastomosing.

A group of about ten species in tropical America and perhaps *Adiantum phanerophlebium* of Madagascar; it is characterized by large segments and less complex leaves and may not represent a wholly natural alliance. Relations of some species, especially *A. macrophyllum* and *A. scalare,* are not clear. *Adiantum phanerophlebium* (Baker) C. Chr. is possibly a paleotropical element in the group. Its larger leaves are 1-pinnate and resemble those of *A. lucidulum* except that the apex may be proliferous.

Species with fully anastomosing veins, as *Adiantum adiantoides, A. olivaceum,* and *A. cordatum,* represent the genus *Hewardia.* However, since species such as *A. dolosum, A. lucidulum, A. phyllitidis,* and *A. Wilsonii* may have few to many irregularly anastomosing veins and appear intermediate to those with free veins, the segregate genus is not recognized. Anastomosing veins also occur in *A. Leprieurii* in the *Adiantum patens* group and suggest they have been independently derived more than once.

The tropical American species are: *Adiantum adiantoides* (J. Sm.) C. Chr., *A. cordatum* Maxon, *A. dolosum* Kze., *A. lucidulum* (Cav.) Sw., *A. macrophyllum* Sw. (Fig. 27), *A. olivaceum* Baker, *A. phyllitidis* J. Sm., *A. Poeppigianum* (Kuhn) Hieron., *A. scalare* Tryon (Fig. 28), and *A. Wilsonii* Hook.

Ecology (Figs. 1–3)

Adiantum grows in moist, shaded habitats. Species are frequently terrestrial growing in clay or humus soils in ravines, on stream banks and on montane slopes in forest. Some species are rupestral and grow on moist, shaded cliffs and ledges or in rocky places in forests or among shrubs.

In tropical America, species of groups 1–4 are often rupestral, growing on damp ledges of cliffs, in seepage crevices, near waterfalls and on rock walls or masonry. *Adiantum Braunii* grows on interior ceilings of Mayan buildings in Yucatan; this and *A. Capillus-Veneris* are usually calciphiles. Others are terrestrial in rather open, moist habitats, in forests, on stream banks, in wet ravines and shrubby areas. The coastal lomas of South America are small areas of strongly seasonal precipitation. *Adiantum Poiretii, A. subvolubile,* and *A. digittatum* grow on the lomas of Peru and *A. excisum* Kze. and *A. chilense* Kaulf. occur on those of Chile.

Species of groups 5–8 are terrestrial, mostly growing in wet montane forests. Some species, especially of the *Adiantum tetraphyllum* group may grow in seasonally inundated forests at low elevations, and also often in clearings, in old banana plantations, on roadside banks, and in thickets. The largest species in the genus, *Adiantum pectinatum,* has a shrubby habit, with leaves to 2.25 m long and stout petioles to 1 cm in diameter. Most species grow between 500 and 2500 m, some near sea level, while species as *Adiantum Raddianum, A. Poiretii, A. digittatum* and *A. Orbignyanum* may occur up to 4000 m.

Figs. 45.16–45.23. *Adiantum* leaves. **16.** *A. Raddianum,* ×0.25. **17.** *A. digittatum,* apical portion of pinna, × 0.25. **18.** *A. Ruizianum,* × 0.25. **19.** *A. sessilifolium,* × 0.25. **20.** *A. patens* × 0.25. **21.** *A. deflectens,* × 0.25. **22.** *A. melanoleucum,* base of basal pinnae, × 0.5. **23.** *A. serratodenatum,* × 0.25.

Geography (Fig. 4)

Adiantum is a very widely distributed genus, largely pantropical but also extending south to southern South America and New Zealand, and north to Newfoundland, Alaska, and northeastern Asia.

In America, *Adiantum* occurs from coastal Alaska and Newfoundland, south to Santa Cruz in Argentina and Magellanes in Chile; also on Bermuda, the Revillagigedo Islands, Cocos Island, Galápagos Islands, the Juan Fernandez Islands and the Falkland Islands.

Many species of *Adiantum* are widely distributed in the American tropics and few have limited ranges. The Andes of Colombia to Peru is probably the richest region with at least 42 species. Nearly as many occur from southern Mexico to Panama. In southeastern Brazil there are about 25 species but only 16 occur in Amazonian Brazil. There are about 25 species in the Greater Antilles and 14 in the Lesser Antilles. Among the few endemics are *Adiantum imbricatum* Tryon and *A. Ruizianum* of Peru, *A. tripteris* Kramer of Surinam, and *A. Shepherdii* of Mexico.

Four species, *Adiantum Capillus-Veneris*, *A. Jordanii* K. Müll., *A. pedatum*, and *A. tricholepis*, also occur north of the tropics, while three, *Adiantum chilense*, *A. excisum* and *A. Pearcei* Phil., are confined to south temperate regions such as Chile, southern Argentina, the Juan Fernandez Islands and the Falkland Islands.

Species of *Adiantum* have been widely cultivated and some are adventive in various parts of the tropics. *Adiantum Capillus-Veneris* is probably adventive on the sea cliffs at Lima, Peru, and elsewhere in South America. *Adiantum philippense* may be adventive in the neotropics. Sledge (1973) reports that five American species of *Adiantum* are adventive in Ceylon. Among these, *Adiantum latifolium* and *A. Raddianum* may become weeds in plantations.

Spores

Adiantum spores are exceptional in the uniformity of surface architecture of the species. The outer formation consists of a relatively thin layer that is more or less fragmented as in *A. concinnum* and *A. trapeziforme* (Figs. 29, 31), or eroded as in *A. anceps* (Fig. 32). This outer material covers a denser, granular layer as in the abraded surface of *A. concinnum* (Fig. 30) or the wall profile of *A. Shepherdii* (Fig. 35). The surface is rugose in a few species as *A. sessilifolium* (Fig. 33). There are marked size differences that may reflect distinct ploidy levels. The small spores of *A. trapeziforme* (Fig. 31) from Guerrero, Mexico, and *A. philippense* (Fig. 34) from Chiapas, Mexico, are about half the diameter of those of *A. concinnum* (Fig. 29) which is reported as a tetraploid in Nayarit, Mexico (Mickel et al. 1966), and Jamaica (Walker, 1966). Distinctive types of surfaces characteristic of species groups have not been detected in *Adiantum* spores.

Figs. 45.24–45.28. *Adiantum* leaves, ×0.25. **24.** *A. latifolium.* **25.** *A. pulverulentum.* **26.** *A. anceps.* **27.** *A. macrophyllum.* **28.** *A. scalare.*

Cytology

The number $n = 30$ is the most frequent in *Adiantum,* and polyploids formed on it reach the dodecaploid level. The occurrence of $n = 29$ or its multiples in several species indicates a second series has possibly been derived from loss or fusion of individual chromosomes. A sterile triploid hybrid with 90 chromosomes involving *A. pulverulentum* is noted by Walker (1966) from Jamaica, and other records of this number probably also relate to hybrids. One of the widely distributed species, *A. Poiretii,* is reported as $n = 57$ in Veracruz, Mexico, (Richards and Tryon, 1973) while specimens from the Cameroons in west Africa and the island of Tristan da Cunha are $n = 114$ (Manton, 1959; Manton and Vida, 1968). The Tristan material was interpreted as octoploid possibly derived from rearrangement at the tetraploid level from 58 to 57 prior to chromosome doubling. The *Adiantum caudatum* complex in Asia and Africa has been analyzed by synthesis of hybrids and examination of chromosome pairing (Sinha and Manton, 1970). Apogamy is reported in this species and others, as *A. hispidulum* (Manton and Sledge, 1954).

Observations

An analysis of the flavonoid glycosides and hydroxycinnamic acid esters of 58 species of *Adiantum* by Cooper-Driver and Swain (1977) included a wide range of the species diversity of the genus. It demonstrated that, in many cases, morphologically related species had similar chemical patterns and that these could be used to characterize certain groups of species. American species are more highly evolved chemically, with complex acylated flavonoid glucosides, sulfated cinnamic esters and no proanthocyanin, while Old World species in general have a more primitive chemistry. The study of species over a wide geographic area, as *Adiantum Capillus-Veneris,* showed uniform chemical patterns.

Adiantum is one of the most popular ferns in cultivation because of the elegant leaves with segments of diverse form and varied shades of green contrasting with the shining black or dark brown petioles. In the nineteenth century many growers were devoted to the introduction of new species to horticulture and the development of new varieties. A recent treatment of cultivated *Adiantum* (Hoshizaki, 1970) includes 28 cultivated species and 35 cultivars of *A. Raddianum,* which is the major species in the nursery trade.

Literature

Cooper-Driver, G., and T. Swain. 1977. Phenolic phytotaxonomy and phytogeography of *Adiantum.* Bot. Jour. Linn. Soc. 74: 1–21.

Hoshizaki, B. J. 1970. The genus *Adiantum* in cultivation (Polypodiaceae). Baileya 17: 97–191.

Kramer, K. U. 1978. The Pteridophytes of Suriname. Nat. Stud. Suriname en Nederland. Antill. 93. 198 pp.

Manton, I. 1959. Cytological information, *in*: Alston, A. H. G., The ferns and fern allies of West Tropical Africa, pp. 75–81. Flora of West Tropical Africa, Suppl. 89 pp. Millbank, London.

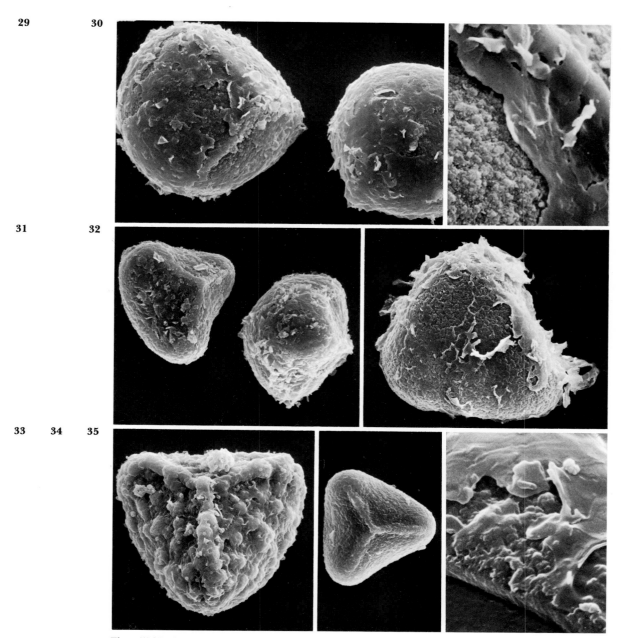

Figs. 45.29–45.35. ***Adiantum*** spores × 1000. **29, 30.** ***A. concinnum.*** **29.** Proximal face, portion of distal face, right, Mexico, *Pringle 15715.* **30.** Detail of abraded wall, the surface, right, inner granulate deposit, left, Costa Rica, *Scamman 7063,* ×5000. **31.** ***A. trapeziforme,*** fragmented surface, proximal face, left, lateral view, right, Mexico, *Palmer 519.* **32.** ***A. anceps,*** distal face with eroded outer deposit, Peru, *Tryon & Tryon 5288* (NY). **33.** ***A. sessilifolium,*** rugose proximal face, Peru, *Bryan 428.* **34.** ***A. philippense,*** rugulose proximal face, *Breedlove 27004* (NY). **35.** ***A. Shepherdii,*** detail of abraded wall, the thin perispore surface, above, papillate-granulate lower perispore, below, exospore at lower left, Mexico, *Mexia 8804,* ×10,000.

Manton, I., and W. A. Sledge. 1954. Observations on the cytology and taxonomy of the pteridophyte flora of Ceylon. Phil. Trans. Roy. Soc. London. 238: 127–185.

Manton, I., and G. Vida.1968. Cytology of the fern flora of Tristan da Cunha. Proc. Roy. Soc. (B) 170: 361–379.

Mickel, J., W. H. Wagner, and K. Chen. 1966. Chromosome observations on the ferns of Mexico. Caryologia 19: 95–102.

Richards, J., and A. Tryon. 1973. Cytology, in: R. Tryon et al., Fern biology in Mexico. A class field program. pp. 30, 31. BioScience 23: 28–33.

Scamman, E. 1960. The maidenhair ferns (*Adiantum*) of Costa Rica. Contrib. Gray Herb. 187: 3–22.

Sinha, B. M., and I. Manton. 1970. Cytotaxonomic studies in the *Adiantum caudatum* complex of Africa and Asia. III. Bot. Jour. Linn. Soc. 63: 247–264.

Sledge, W. A. 1973. Native and naturalized species of *Adiantum* in Ceylon. Ceylon Jour. Sci. (Biol. Sci.) 10: 144–154.

Tryon, R. 1964a. Evolution in the leaf of living ferns. Mem. Torrey Bot. Cl. 21: 73–85.

Tryon, R. 1964b. The ferns of Peru. Contrib. Gray Herb. 194. 253 pp.

Walker, T. G. 1966. A cytotaxonomic survey of the pteridophytes of Jamaica. Trans. Roy. Soc. Edinburgh 66: 169–237.

14e.   Tribe Pterideae

Acrostichaceae Frank, Syn. Pflanzenkr. (Leunis) ed. 2, 3: 1458. 1877. Type: *Acrostichum* L.

A tribe of seven genera; two of them *Afropteris* Alston and *Ochropteris* J. Sm. are found only in the paleotropics and are doubtfully distinct from *Pteris*. The genus *Idiopteris* T. G. Walker of India and Ceylon with a chromosome report of $n = 27$ may belong to this tribe.

46.   *Pteris*
Figs. 46.1–46.30

Pteris L., Sp. Pl. 1073. 1753; Gen. Pl. ed. 5, 484. 1754. Type: *Pteris longifolia* L.

Campteria Presl, Tent. Pterid. 146. 1836. Type: *Campteria Rottleriana* Presl, *nom. superfl.* for *Pteris nemoralis* Willd. = *Campteria nemoralis* (Willd.) J. Sm. = *Pteris biaurita* L. *Pteris* section *Campteria* (Presl) Hook., Sp. Fil. 2: 202. 1858.

Litobrochia Presl, Tent. Pterid. 148. 1836. Type: *Litobrochia denticulata* (Sw.) Presl = *Pteris denticulata* Sw. *Pteris* section *Litobrochia* (Presl) Hook., Sp. Fil. 2: 207. 1858.

Pteris subgenus *Pteridopsis* Link, Fil. Sp. 49. 1841. Type:·*Pteris longifolia* L.

Pycnodoria Presl, Epim. Bot. 100. 1849. Type: *Pycnodoria opaca* (J. Sm.) Presl = *Pteris opaca* J. Sm.

Heterophlebium Fée, Gen. Fil. (Mém. Fam. Foug. 5): 139. 1852. Type: *Heterophlebium grandifolium* (L.) Fée = *Pteris grandifolia* L. *Pteris* section *Heterophlebium* (Fée) Hook., Sp. Fil. 2: 201. 1858.

Pteris section *Parapteris* Keys., Polypod. Cyath. Herb. Bung. 4. 1873. Type: *Pteris quadriaurita* Retz. is an adequate choice.

Fig. 46.1. *Pteris longifolia,* near Retreat, St. Mary, Jamaica. (Photo W. H. Hodge.)

Description Terrestrial; stem erect or short- to long-creeping, small to stout, bearing scales and few to many, usually long, fibrous roots; leaves usually monomorphic, or dimorphic (the fertile taller and more erect and with usually narrower segments), ca. 10 cm to 6 m long, borne in a cluster or at intervals, lamina 1- to 5-pinnate, of diverse architecture, gradually reduced apically to imparipinnate, the basal pinnae often elaborated basally, segments usually glabrous beneath to pubescent or sparingly scaly, veins free or anastomosing without included free veinlets; sori on a marginal commissure connecting the vein ends, with filamentous paraphyses (sometimes few), indusium strongly differentiated from the recurved margin; spores tetrahedral or globose, trilete, the laesurae $\frac{1}{2}$ to $\frac{3}{4}$ the radius, usually with an equatorial

Fig. 46.2. *Pteris deflexa,* Pedregal Esquilón, Veracruz, Mexico. (Photo W. H. Hodge.)

flange, and low tubercules more or less fused into ridges, or prominently tuberculate, or reticulate. Chromosome number: $n = 29, 58, 116$; $2n = 58, 116$, ca. 232; apogamous 58, 87, ca. 90, 116, ca. 120.

The veins may be free (Fig. 13) or anastomosing to form areolae only along the costa (Fig. 14), or they may be fully anastomosing. The costal areolae may be prominent (Fig. 15), the marginal ones much smaller (Fig. 16), or the aerolae may be of fairly uniform size (Fig. 17). Many of the species have the costa ridged on the upper side, with a portion of the ridge extended at the base of main veins to form an awn (Fig. 19). The lamina is rarely densely pubescent as in *Pteris Lechleri* (Fig. 18). The characteristic marginal sorus of *Pteris* is shown in Fig. 4.

Systematics

Pteris is a genus of about 200 species, with some 55 in America. These can be grouped into alliances based mainly on the leaf architecture and venation, but the natural subgeneric or sectional groups are not certain. The classical division of the genus, on the basis of venation, into *Pteris,* with free veins, *Campteria,* with partially areolate venation, and *Litobrochia,* with completely areolate venation, is certainly artificial.

Shieh (1966) proposed a new infrageneric classification of two subgenera, four sections and seven subsections for the 34 species of Japan, Ryukyu Islands and Taiwan. This is based on new characters such as those of the vein-tips, the scales, and false-veins, which merit investigation on a much broader geographic basis.

The relations of *Pteris* are obscure, although in general they

Fig. 46.3. Distribution of *Pteris* in America.

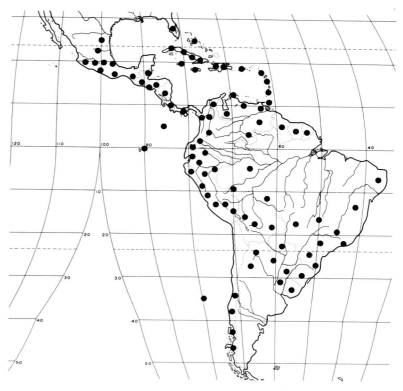

are associated with the cheilanthoid ferns in soral characters and with the taenitoid ferns in spore characters.

The nomenclature for *Pteris,* presented above, includes only names applicable to America. There is no modern revision of the species of *Pteris.*

Tropical American Species

The following synopsis includes a selection of the American species. Species groups with the least specialized leaf architecture (Figs. 5, 6) are placed first. The groups that follow, with a more specialized architecture (Figs. 7–12), are considered as advanced. This arrangement seems to bring together related species, but further study is essential to confirm the naturalness of the groups. Evidence that hybridization and polyploidy have been operative in the genus suggests that relations may be reticulate. The species of *Pteris* often have a heteroblastic leaf development, and for this reason, only the characters of large leaves are employed.

1. *Pteris chilensis* Group

Basal pinnae 1-pinnate or more complex beyond the basal pinnules, the basal pinnae and the lamina gradually reduced to the apex, veins free.

Pteris Berteroana Ag., *P. chilensis* Desv., *P. leptophylla* Sw. (Fig. 5), and *P. tremula* R. Br. (adventive).

2. *Pteris deflexa* Group

Basal pinnae 1-pinnate beyond the basal pinnules, the basal pinnules and (or) the lamina moderate to strongly reduced at the apex, veins free or anastomosing.

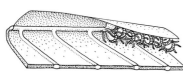

Fig. 46.4. Sorus of *Pteris stridens,* with marginal indusium, vascular commissure, sporangia and paraphyses, diagrammatic, × 10.

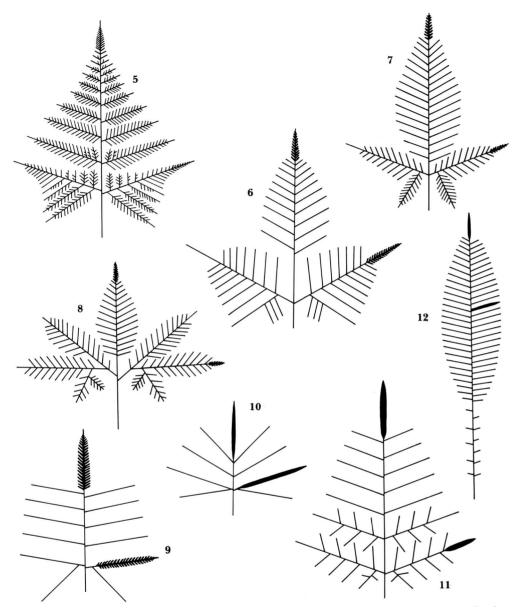

Figs. 46.5–46.12. Diagrams of leaf architecture in *Pteris* species, including all axes except costules, lamina apex and one pinna are indicated as pinnatifid or entire in solid black. **5.** *P. leptophylla*, × 0.1. **6.** *P. deflexa*, × 0.05. **7.** *P. muricella*, × 0.05. **8.** *P. podophylla*, × 0.025. **9.** *P. quadriaurita*, × 0.05. **10.** *P. cretica*, × 0.05. **11.** *P. Haenkeana*, × 0.05. **12.** *P. longifolia*, × 0.125.

Pteris altissima Poir., *P. coriacea* Desv., *P. deflexa* Link (Fig. 6), *P. gigantea* Willd., *P. livida* Mett., *P. muricata* Hook., *P. muricella* Fée (Fig. 7), *P. podophylla* Sw. (Fig. 8), *P. propinqua* Ag., *P. semiadnata* Phil., *P. stridens* Ag., and *P. tripartita* Sw. (adventive).

3. *Pteris quadriaurita* Group

Basal pinnae pinnatisect beyond the basal pinnules (or with a second staked pinnule), lamina with a pinnatifid, conform apical segment, veins free or anastomosing.

Pteris biaurita L., *P. decurrens* Presl, *P. horizontalis* (Fée) Rosenst., *P. Lechleri* Mett., *P. pungens* L., *P. quadriaurita* Retz. (Fig. 9), and *P. sericea* (Fée) Christ.

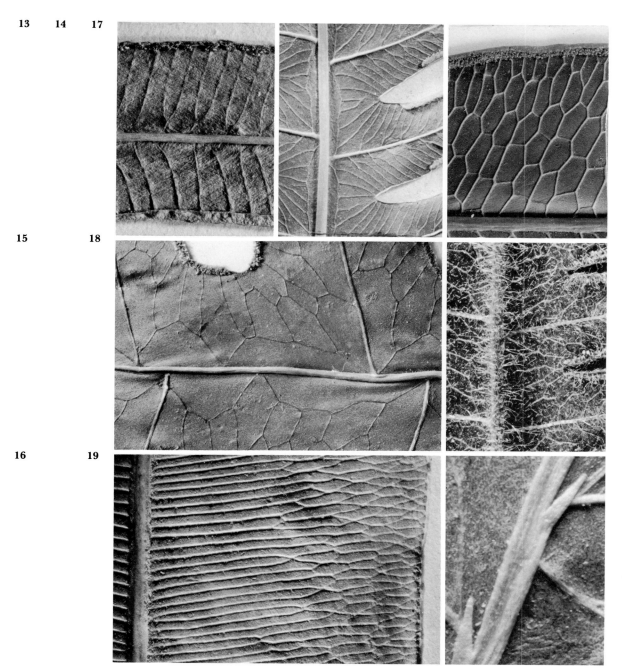

Figs. 46.13–46.19. *Pteris*. 13. *Pteris semiadnata,* free veins, ×5. **14. *P. biaurita,*** costal areolae, ×5. **15. *P. horizontalis,*** anastomosing veins, ×5. **16. *P. grandifolia,*** anastomosing veins, ×5. **17. *P. splendens,*** indusium, sporangia and venation, ×5. **18. *P. Lechleri,*** segments pubescent beneath, ×5. **19. *P. deflexa,*** awns on upper side of costa, ×20.

Figs. 46.20–46.30. *Pteris* spores, × 1000, surface detail, × 5000. **20.** *P. biaurita,* proximal face with tubercules coalescent into ridges, the equatorial flange peripheral, Peru, *Tryon & Tryon 5246.* **21.** *P. altissima,* proximal face above, rugose distal face, below, Peru, *Ridoutt s. n.* **22.** *P. podophylla,* proximal face slightly tuberculate to rugose, part of rugose distal face, below, Colombia, *Killip & Smith 20351.* **23, 24.** *P. ensiformis,* Jamaica, *Walker 4515.* **23.** Lateral view, the equatorial flange projecting more than tubercules. **24.** Proximal face, coarsely tuberculate, equatorial flange peripheral. **25.** *P. quadriaurita,* detail of tuberculate, proximal face, laesura across top, Peru, *Aguilar 508.* **26.** *P. grandifolia,* proximal face, right, equatorial flange peripheral, distal face tilted, left, the flange along top, Panama, *Allen 2673.* **27.** *P. mutilata* spore tetrad, Haiti, *Leonard 8306.* **28–30.** *P. longifolia,* Mexico, *Durantes 197.* **28.** Proximal face with large areolae in equatorial area. **29.** Distal face coarsely reticulate. **30.** Detail of papillate surface of large areola and ridges.

4. *Pteris Haenkeana* Group
Pinnae or ultimate segments large, entire, veins anastomosing.
Pteris grandifolia L., *P. Haenkeana* Presl (Fig. 11), *P. mexicana* (Fée) Fourn., *P. Schwackeana* Christ, and *P. splendens* Kaulf.

5. *Pteris longifolia* Group
Lamina more or less pectinately 1-pinnate, with long, narrow, entire pinnae, imparipinnate, veins free. The spores are quite distinct from other species of *Pteris* and indicate the relations of this group need investigation.
Pteris longifolia L. (Fig. 12), *P. Purdoniana* Maxon (perhaps a variety of *P. longifolia*), and *P. vittata* L. (adventive).

6. *Pteris cretica* Group
Small plants, usually with dimorphic leaves, the lamina with an elongate, entire, terminal segment, veins free or anastomosing.
Pteris ciliaris D. C. Eaton, *P. cretica* L. (Fig. 10) (probably adventive), *P. denticulata* Sw. and probable hybrids, *P. ensiformis* Burm. (adventive in Jamaica), *P. multifida* Poir. (adventive in southern United States), and *P. mutilata* L.

Ecology (Figs. 1, 2)

Pteris is a genus of diverse ecology, although most species occur in forest, frequently secondary forests, in openings or along rocky stream banks.

In tropical America, species grow in wet forests, at the edge of clearings, in thickets, sometimes in cloud forests or in gallery forests, or occasionally on cliffs. Disturbed or artificial habitat such as road banks, rocks or old brick walls are also invaded. *Pteris cretica* and *P. longifolia* are calciphiles usually found in limestone sinks and other locally moist places in pine woods or deciduous forests, and also on masonry. Although the genus is generally one of lower altitudes, from sea level to 2000 m, *Pteris coriacea,* a subpáramo species of rocky and shrubby sites ranges between 2800 and 3500 m in Costa Rica and the Andes.

Geography (Fig. 3)

Pteris is a pantropic genus with extensions into temperate regions in Chile, the Mediterranean region, South Africa, Korea, Japan, Tasmania and New Zealand.

20 21 22

23 24 25

26 27

28 29 30

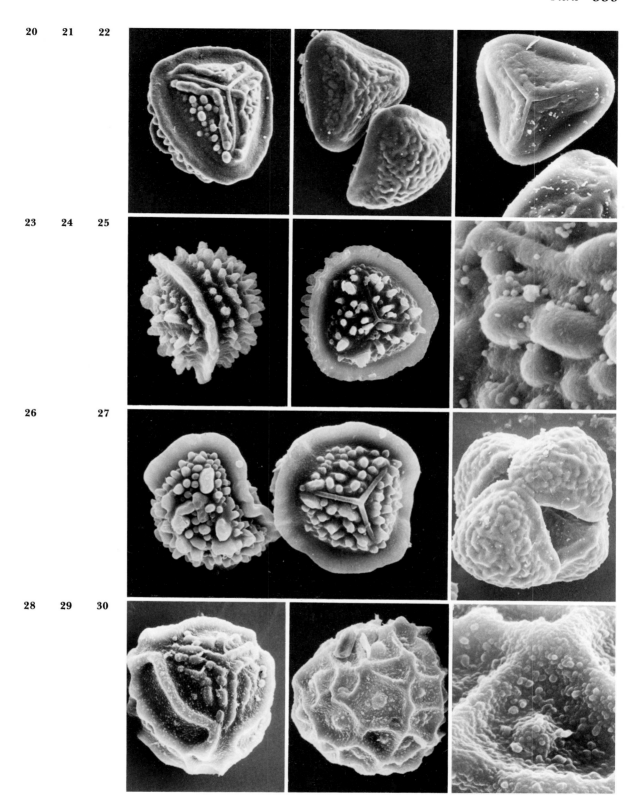

In America, native species of *Pteris* occur from Nuevo Leon in Mexico and Florida in the United States, south to La Rioja and Entre Ríos in Argentina and Aisén in Chile; also on Cocos Island, the Galápagos Islands and Juan Fernandez Islands. The American center of diversity of *Pteris* is in the Andes from Venezuela to Bolivia, where nearly half of the neotropical species occur. The genus is rare in the Amazon Basin, being represented there by only four or five species.

Spores

The American species of *Pteris* generally have uniform spores characterized by the strong equatorial flange, coarse tubercules more or less fused into ridges and spherical deposit (Fig. 22–27). The faces are unequal with the flange closer to the apex as in *Pteris altissima* (Figs. 21), *P. ensiformis* (Fig. 23), and *P. grandifolia* (Fig. 26). The larger distal face often is rugose as in *P. podophylla,* and *P. mutilata* (Figs. 22, 27) and the tubercules of the proximal face are coalescent in ridges (Fig. 21). Spores of *P. mutilata* tend to be retained in tetrads joined at the equator (Fig. 27). The reticulate surface (Figs. 28–30) in *P. longifolia* differs from the formation in other American species and resembles that of the paleotropical *Pteris vittata.*

The strong equatorial flange and coarsely ridged and tuberculate spores of *Pteris* resemble those of genera allied to *Pityrogramma.* This relationship is also supported by a basic chromosome number of 29, common to these genera.

Cytology

Cytologically *Pteris* is one of the best known fern genera due to the studies of Walker (1958, 1960, 1962) on species over a broad geographic range. Polyploidy, apogamy and hybridization are evidently widespread systems in *Pteris* and have been important in speciation. The chromosome numbers range from diploid $n = 29$ to octoploid $2n = 232$. Aside from a few uncertain reports of 90, the numbers in *Pteris* are uniformly based on 29 and there are no definite records of 30. About 55% of the cytologically known species are polyploid and apogamy is known in about a third of these. American tropical species are at diploid, tetraploid and octoploid levels and there are apogamous diploids, triploids and tetraploids. The irregular lamina architecture often associated with hybrids, as in the *Pteris quadriaurita* complex in Ceylon (Walker, 1958), is also evident in specimens of some American species as *Pteris denticulata* and *P. petiolulata* Tryon. Comparison of the geographic ranges of the different cytotypes show that the sexual diploids have a more restricted distribution than the higher polyploids or apogamous forms. The study of karyotypes of *Pteris,* including a tetraploid and two diploid species of Japan (Kawakami, 1971), indicates the basic number is ten. The diploid chromosomes are arranged in nine sets of six and one set of four, while in the tetraploid they are in nine sets of twelve and one set of eight chromosomes.

Observations

The elaborate branching pattern of the leaves of most species of *Pteris* (Figs. 5–7, 9, 10) is unusual among ferns. Field observations of *Pteris* leaves indicate that in species such as *P. deflexa* (Fig. 6), *P. muricella* (Fig. 7) and *P. podophylla* (Fig. 8), the central portion of the lamina is in one plane, while the basal pinnae and their enlarged segments are oriented in different planes. These characters possibly relate to the most effective use of light in the shaded situations where these species grow. The leaves of juvenile plants of some species are also unusual because, as in *Pteris altissima,* the lamina is strongly dissected into slender, linear segments, quite unlike those of the adult leaves.

Walker (1958) reported fertile hybrid swarms arising in Ceylon from crosses of *Pteris quadriaurita* and *P. multiaurita* Ag. The only other example of a fertile hybrid swarm in ferns is in populations of the hybrid *Alsophila* × *Nephelea,* in Puerto Rico.

Literature

Kawakami, S. 1971. Karyological studies of Pteridaceae l. Karyotypes of three species in *Pteris*. Bot. Mag. (Tokyo) 84: 180–186.

Shieh, W-C., 1966. A synopsis of the fern genus *Pteris* in Japan, Ryukyu, and Taiwan. Bot. Mag. (Tokyo) 79: 283–292.

Walker, T. G. 1958. Hybridization in some species of *Pteris* L. Evolution 12: 82–92.

Walker, T. G. 1960. The *Pteris quadriaurita* complex in Ceylon. Kew Bull. 14. 321–332.

Walker, T. G. 1962. Cytology and evolution in the fern genus *Pteris* L. Evolution 16: 17–43

47. *Anopteris*
Figs. 47.1–47.11

Anopteris (Prantl) Diels, Nat. Pflanz. 1 (4): 288. 1899. *Cryptogramma* sect. *Anopteris* Prantl, Engl. Bot. Jahrb. 3: 414. 1882. Type: *Cryptogramma heterophylla* (L.) Prantl (*Pteris heterophylla* L., 1759) = *Anopteris hexagona* (L.) C. Chr.

Description Terrestrial or rupestral; stem ascending to usually erect, small, bearing scales and long, slender, fibrous roots; leaves usually dimorphic (the fertile longer and the form of the segments distinct from the sterile), ca. 15–80 cm long, borne in a cluster, lamina 2- to usually 3- to 5-pinnate, gradually reduced at the apex to imparipinnate, glabrous, veins free; sori short to long, the sporangia borne on a marginal vascular commissure connecting the vein ends, with numerous filamentous paraphyses, indusium strongly differentiated from the recurved margin; spores tetrahedral-globose, somewhat compressed with prominent angles, the laesurae $\frac{3}{4}$ the radius, the distal face coarsely tuberculate, or rugose, the proximal face slightly rugose to nearly smooth. Chromosome number: $n = 58$; $2n = 116$.

The stem scales are thin to slightly sclerotic, rather broad and with a dark brown center and lighter margins.

Fig. 47.1. *Anopteris hexagona*, Fern Gully, near Ocho Rios, Jamaica. (Photo W. H. Hodge.)

Fig. 47.2. Distribution of *Anopteris*.

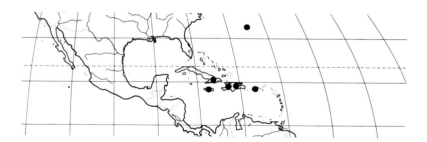

Anopteris is an American genus of one species, *A. hexagona* (L.) C. Chr. in the Greater Antilles and Bermuda. It is closely related to *Pteris,* differing primarily in the lack of an equatorial flange on the spores. Further study of the relationship of these genera is needed.

Tropical American Species

The single species *Anopteris hexagona* (*Adiantum hexagonum* L., 1753) is highly variable in lamina complexity and segment form. The variant with a highly dissected lamina was recognized as *Onychium strictum* Kze. (*O. multifidum* Fée) and questionably included in *Onychium* by Copeland (1947) as the only American

Figs. 47.3–47.5. *Anopteris hexagona,* leaf architecture, × 0.5. **3.** Ssp. *hexagona,* sterile leaf. **4.** Ssp. *hexagona,* fertile leaf. **5.** Ssp. *multifida,* basal pinna.

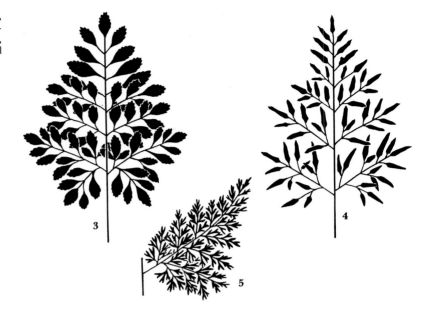

Figs. 47.6, 46.7. *Anopteris hexagona,* fertile segments, × 4. **6.** Ssp. *hexagona.* **7.** Ssp. *multifida.*

representative of a small Old World genus. Morton (1957) corrected this interpretation, pointing out that *Onychium* was quite different in its strongly ornamented spores and absence of paraphyses. He also recognized the unusual variation that connected the different leaf forms. The following key separates the two extremes of the species. Plants of intermediate form, that occur in Hispaniola, may be placed in subspecies *intermedia* Morton.

8 9

10 11

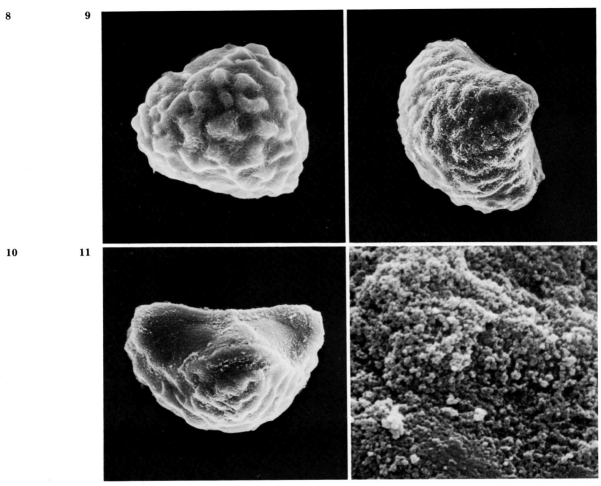

Figs. 47.8–47.11. Spores of ***Anopteris hexagona***, Cuba, *Maxon 4110,* × 1000, surface detail, × 10,000. **8.** Distal face tuberculate-rugose. **9.** Distal face tilted with one of the three prominent lobes, center right. **10.** Lateral view, proximal face above. **11.** Detail of granulate surface, distal face.

Key to Subspecies of *Anopteris hexagona*

a. Lamina 2- to 3-pinnate, ultimate segments rather large, with 8 or more veins, toothed, indusia not approaching the costa; Bermuda, Jamaica, Hispaniola. (Figs. 3, 4, 6)   ssp. *hexagona*

a. Lamina 4- to 5-pinnate, ultimate segments with 1–2 veins, entire, indusia closely approaching the central vein or overlapping; Cuba, Hispaniola, Puerto Rico. (Figs. 5, 7)   ssp. *multifida* (Fée) Morton

Ecology and Geography (Figs. 1, 2)

Anopteris hexagona occurs in Bermuda, Cuba, Jamaica, Hispaniola and Puerto Rico (Fig. 2). It grows on limestone cliffs, in sinkholes, and among limestone rocks in thin to dense shade in usually wet forests. Most commonly it grows in rocky places in dense woods (Fig. 1) at or near sea level to ca. 1000 m. It is apparently restricted to limestone. The stem may branch with age to produce a dense clump of plants.

Spores

The coarsely tuberculate to rugose spores have more prominent contours on the distal than the proximal face (Figs. 8–10). The surface, of more or less compact granulate deposit, may be relatively smooth especially on the proximal face. These spores have coarse tubercules and ridges similar to those of *Pteris* and *Onychium* but lack the equatorial ridges characteristic of those genera.

Cytology

Two populations of *Anopteris hexagona* from Jamaica are reported as $n = 58$ by Walker (1966) and regarded as sexual tetraploids. These records support the alliance of the genus with *Pteris,* in which the chromosome number is also based on 29.

Literature

Morton, C. V. 1957. The fern genus *Anopteris*. Bull. Jard. Bot. L'Etat, Bruxelles 27: 579–584.

Walker, T. G. 1966. A cytotaxonomic survey of the pteridophytes of Jamaica. Trans. Roy. Soc. Edinburgh 66: 169–237.

48. *Neurocallis*
Figs. 48.1–48.7

Neurocallis Fée, Acrost. (Mém. Fam. Foug. 2): 19, 89, 1845. Type: *Neurocallis praestantissima* Fée.

Description

Terrestrial; stem decumbent-ascending, rather stout, bearing scales and many long, somewhat soft roots; leaves dimorphic (the fertile pinnae narrower than the sterile), to 2 m long, borne in a cluster, lamina 1-pinnate, imparipinnate, the pinnae entire, glabrous or minutely pubescent beneath, veins coarsely anastomosing without included free veinlets; sporangia borne on and also between the veins, from the margin to the costa, or in a broad marginal band, mixed with numerous, mostly capitate paraphyses, partly covered by a narrow marginal indusium; spores tetrahedral-globose to somewhat flat on the distal face trilete, the laesurae extending $\frac{3}{4}$ or more of the radius, rugulose, with low ridges more prominent between the laesurae and on the distal face. Chromosome number: $n = 58$.

The stem scales vary from small, blackish, sclerotic and narrowly thin-margined to long, very narrow and dark brown or reddish-brown. The long-petioled, dimorphic leaves are shown in Figs. 3 and 4 and the areolate venation in Fig. 5.

Systematics

Neurocallis is a genus of one species, *N. praestantissima* Fée of the American tropics. It is related to *Pteris,* and resembles species such as *P. splendens,* except for the difference in soral condition and its thin-walled spores without an equatorial flange. The

Fig. 48.1. *Neurocallis praestantissima,* Cerro de Zurqui, Costa Rica. (Photo W. H. Hodge.)

Fig. 48.2. Distribution of *Neurocallis*.

plants of Costa Rica have been distinguished as var. *subcaudata* Gómez (Rev. Trop. Biol. 20: 183. 1972). They have the pinnae long-petiolulate and usually with long, often subcaudate apices.

Ecology and Geography (Figs. 1, 2)

Neurocallis grows in wet, usually dense rain forest from c. 800–1200 m in Guadeloupe, Dominica, Martinique, St. Vincent, and in Aragua and Sucre in Venezuela, and Cartago in Costa Rica. It is evidently very rare except in Guadeloupe and Dominica, where it has been frequently collected.

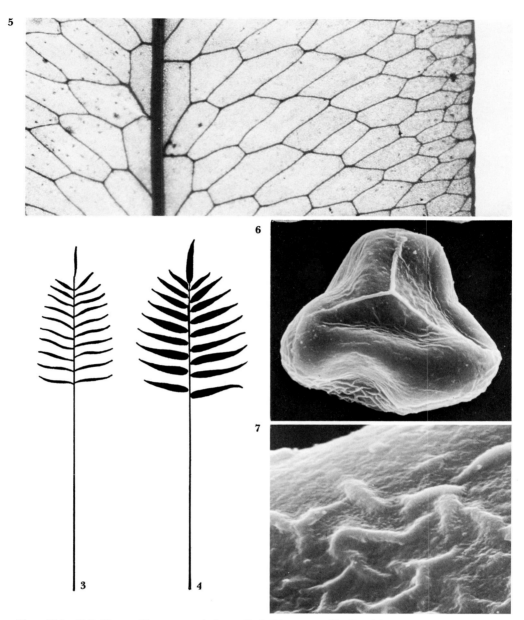

Figs. 48.3–48.7. *Neurocallis praestantissima.* **3, 4.** Diagrams of leaf architecture, ×0.05. **3.** Fertile leaf. **4.** Sterile leaf. **5.** Portion of a pinna showing anastomosing veins, ×5. **6, 7.** Spores, Venezuela, *Steyermark 94966.* **6.** Proximal face, part of rugulose distal face, below, ×1000. **7.** Detail of rugulose equatorial area, distal face, ×5000.

Spores

The spores are somewhat flattened, and mature, fully developed specimens have concave areas between the laesurae (Fig. 6). These areas seem to contract in dry spores and suggest that the walls are thin. Although the genus is regarded as related to *Pteris,* the spores differ in their shallow, rugulose to generally even surface (Fig. 7) and in the absence of an equatorial flange.

Cytology

Neurocallis praestantissima is reported as $n = 58$ from Costa Rica by Wagner (1980). This number is also known in several other genera of the Pteridaceae.

Literature

Wagner, F. S. 1980. New basic chromosome numbers for genera of neotropical ferns. Amer. Jour. Bot. 67: 733–738.

49. *Acrostichum*
Figs. 49.1–49.13

Acrostichum L. Sp. Pl.: 1067. 1753; Gen. Pl. ed. 5, 484. 1754. Type: *Acrostichum aureum* L.

Chrysodium Fée, Acrost. (Mém. Fam. Foug. 2): 33, 97. 1845. *nom. superfl.* for *Acrostichum* L. and with the same type.

Description

Terrestrial-palustral; stem decumbent-ascending to erect, stout, bearing scales and large spongy roots; leaves slightly dimorphic (the fertile pinnae usually somewhat narrower and shorter than the sterile), up to 2–4 m long, borne in a cluster, lamina 1-pinnate, imparipinnate, the pinnae entire, glabrous or minutely pubescent beneath, veins closely anastomosing without included free veinlets; sporangia borne over the surface on and between the veins, interspersed with short, capitate paraphyses, exindusiate, the margin coriaceous; spores tetrahedral-globose, trilete, the laesurae $\frac{1}{2}$ to $\frac{3}{4}$ the radius, the surface coarsely diffuse papillate or tuberculate with rods or sparse strands denser on the distal face. Chromosome number: $n = 30$; $2n = 60$.

The large stem of *Acrostichum* is usually decumbent, with an ascending apex (Figs. 3, 4). The leaf architecture (Fig. 5) and the venation pattern (Figs. 7, 8) are characteristic of the genus. There are abundant paraphyses among the sporangia (Fig. 6). The stem scales (Fig. 9) are usually large, up to ca. 4 cm long, brown to dark brown, centrally thickened, and with thin margins. *Acrostichum aureum* has abortive petiolules, varied in form but usually rather spinelike, on the upper part of the petiole. Occasionally there are also much reduced basal pinnae.

Systematics

Acrostichum is a pantropical genus of three or perhaps more species. *Acrostichum aureum* is pantropical, *A. speciosum* Willd. is paleotropical and *A. danaeifolium* is neotropical. *Acrostichum aureum*

Fig. 49.1. *Acrostichum aureum,* Playa de la Mancha, coastal Veracruz, Mexico. (Photo W. H. Hodge.)

Fig. 49.2. Distribution of *Acrostichum* in America.

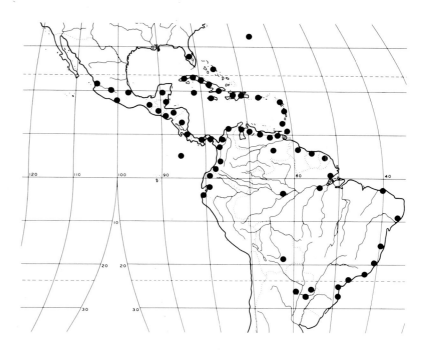

is variable in the Old World and requires further study to assess the taxonomic status of the geographic variations.

The genus is distinctive and morphologically isolated. It is probably rather distantly related to *Pteris*.

Tropical American Species

The two American species are distinctive, although more adequate specimens and also field studies are needed to confirm their characters. Both are sexual diploids with the same chromosome number; the basis of their morphological divergence merits further study. A hybrid has been reported from the Dominican Republic by López (1978).

Acrostichum danaeifolium has been known as *A. excelsum* Maxon (*A. lomarioides* Jenm., not Bory).

Adams and Tomlinson (1979) present a detailed account of *Acrostichum* in Florida and a general review of the genus in America.

Key to American Species of *Acrostichum*

a. Pinnae spaced, coriaceous, fertile pinnae confined to the apical portion of the lamina, abortive pinnae represented by spinelike structures on the apical portion of the petiole, paraphyses with the apex somewhat globose, irregularly lobed.   *A. aureum* L.
a. Pinnae crowded, chartaceous, fertile pinnae extending to the base of the lamina or nearly so, petiole without abortive pinnae, paraphyses with the apex laterally extended, slightly lobed.   *A. danaeifolium* Langsd. & Fisch.

Ecology (Fig. 1)

Acrostichum grows in brackish or salt water and is a characteristic member of mangrove communities; rarely it is found on sea cliffs.

In America it is also common in open salt marshes, on alluvial banks of estuaries, on the shores of embayments and along ditches. Occasionally, *Acrostichum* grows in lightly shaded places such as thickets and it has been reported in fresh water somewhat above the sea. Large clumps of plants 1 m or more in diameter may develop from the branching stem.

Geography (Fig. 2)

Acrostichum grows in the tropics and subtropics in America, Africa, and east to tropical southeastern Asia and Queensland, Australia, north to the Ryukyu Islands, and through the Pacific to Tahiti and the Austral Islands.

In America, the genus grows in Bermuda, in Florida in the United States, in Mexico to Panama, the West Indies, and along the Gulf and Atlantic coasts of South America to Santa Catarina in southeastern Brazil, with inland extensions along the Amazon River, the Río Paraná, and the Río Uruguay in Argentina and in Paraguay and Bolivia. It has a less extensive distribution on the Pacific coast, south to Tumbes in northernmost Peru; also on Cocos Island.

In Florida and in the Lesser Antilles, where the distribution of

Figs. 49.3, 49.4. Stem of *Acrostichum danaeifolium,* × 0.33. **3.** Trimmed stem surface, showing petiole bases (P), scales (S), and spongy roots (R). **4.** Longisection through stem, showing portions of the vascular system (V), and below, several partially decayed roots.

the two species is well known, *A. danaeifolium* is more common and widespread than *A. aureum.*

Spores

Spores of tropical American *Acrostichum aureum* are slightly rougher than those of *A. danaeifolium* and have strands or rods mostly associated with coarsely papillate structures (Figs. 10–

Figs. 49.5–49.8. Acrostichum. 5. Diagram of sterile leaf of **A. aureum,** ×0.05.   **6.** Portion of a fertile pinna of **A. aureum,** with sporangia among the darker paraphyses, ×10.   **7.** Venation, portion of sterile pinna of **A. aureum,** ×2.25.   **8.** Venation, portion of sterile pinna of **A. danaeifolium,** ×2.25.

13). Spores of collections of *Acrostichum aureum* from west Africa, Australia and the Philippines vary in density and prominence of the papillate formation but are generally similar. The profile of the spore wall of *A. aureum* from Madagascar has a thin, papillate surface and a thicker, columellate strata below (Tardieu-Blot, 1963). The spores of *Acrostichum* differ from those of other genera and represent a discrete type among the pteroid ferns.

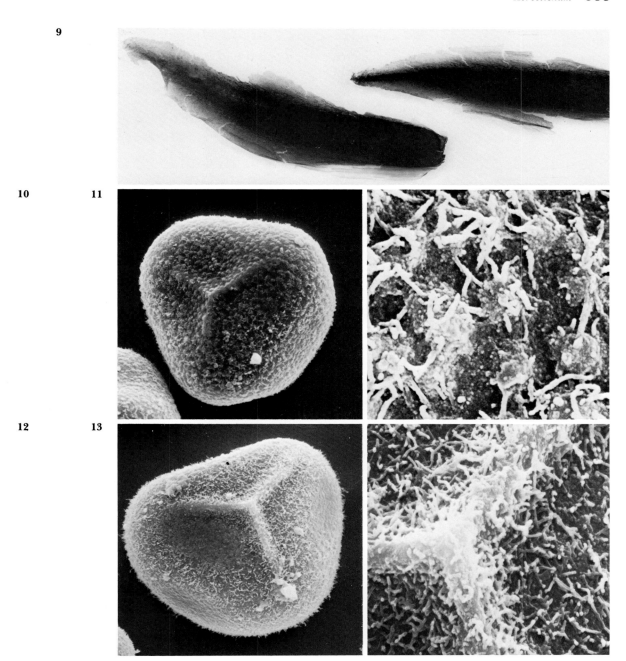

Figs. 49.9–49.13. *Acrostichum*. 9. Stem scales of *A. aureum*, ×2.25. **10–13.** Spores. **10, 11.** *A. aureum*, French Guiana, *Broadway 297*. **10.** Proximal face, ×1000. **11.** Detail of strands on diffuse papillae, proximal face, ×10,000. **12, 13.** *A. danaeifolium*, Mexico, *Alava & Cook 1703*. **12.** Proximal face, ×1000. **13.** Detail of strands at apex, ×5000.

Cytology

Both American species of *Acrostichum* are reported as $n = 30$ by Walker (1966) based on plants from Jamaica. The same number is also reported for *A. aureum* from Ceylon (Manton and Sledge, 1954). These cytological samples from widely distant regions suggest the species may be cytologically uniform. Polyploidy, which is so frequent in the genera of tropical ferns, apparently has not developed in these semiaquatic species.

Observations

Acrostichum danaeifolium and *A. aureum* have a sex-organ ontogeny, when grown in culture, characteristic of gametophytes adapted to outcrossing (Lloyd and Gregg, 1975; Lloyd, 1980). However, they also have a low genetic load, which implies frequent selfing. If gametophytes are regularly self-fertilized in nature, this may account for the separation of the two species in spite of their overlapping distribution.

Crotty (1955) studied the heteroblastic leaf development of *Acrostichum danaeifolium* as related to the leaf primordia and the size of the stem. He concluded that the sequence from young plants producing elongate simple leaves to mature ones with 1-pinnate leaves with many pinnae, was based on a combination of factors such as size of the stem apex, auxin inhibition, nutritional levels and the time available for leaf maturation.

Literature

Adams, D. C., and P. B. Tomlinson. 1979. *Acrostichum* in Florida and tropical America. Amer. Fern Jour. 69: 42–46.

Crotty, W. J. 1955. Trends in the pattern of primordial development with age in the fern *Acrostichum danaeaefolium*. Amer. Jour. Bot. 42: 627–636.

Lloyd, R. M. 1980. Reproductive biology and gametophyte morphology of New World populations of *Acrostichum aureum*. Amer. Fern Jour. 70: 99–110.

Lloyd, R. M., and T. L. Gregg. 1975. Reproductive biology and gametophyte morphology of *Acrostichum danaeifolium* from Mexico. Amer. Fern Jour. 65: 105–120.

López, I. G. 1978. Revisión del género *Acrostichum* en la República Dominicana. Moscosoa 1: 64–70.

Manton, I., and W. A. Sledge. 1954. Observations on the cytology and taxonomy of the pteridophyte flora of Ceylon. Phil. Trans. Roy. Soc. London 238: 127–185.

Tardieu-Blot, M. 1963. Sur les spores de Pterideae Malgaches. Pollen et Spores 5: 337–353.

Walker, T. G. 1966. A cytotaxonomic survey of the pteridophytes of Jamaica. Trans. Roy. Soc. Edinburgh 66: 169–237.

Family 15.   Vittariaceae

Vittariaceae (Presl) Ching, Sunyatsenia 5: 232. 1940.
Vittarieae Presl, Tent. Pterid. 164. 1836, as Vittariaceae. Type: *Vittaria* Sm.
Antrophyaceae Link, Fil. Sp. 140. 1841, as suborder. Type: *Antrophyum* Kaulf.
Antrophyaceae (Link) Ching, Acta Phytotax. Sinica 16: 11. 1978.

Description   Stem nearly erect to short-creeping, small, or long-creeping and rather slender, with a protostele, siphonostele or dictyostele, not indurated, bearing scales; leaves ca. 3 cm to 1 m long, simple, entire, or (in *Hecistopteris*) furcate, or (in *Rheopteris*) pinnate, circinate in the bud, petiole without stipules; sporangia borne in abaxial, often sunken, sori, or (in *Anetium*) the sporangia mostly in scattered groups on and between the veins and superficial, exindusiate; sporangia rather short-stalked, the stalk 1- to 2-rowed below its apex, the vertical annulus interrupted by the stalk; homosporous, spores lacking chlorophyll. Gametophyte epigeal, with chlorophyll, elongate and irregularly branched, sometimes with gemmae, archegonia borne scattered on the lower surface toward the margins, antheridia 3-celled, borne on the lower surface, mostly apart from the archegonia.

Comments on the Family

The Vittariaceae are a closely knit family of six genera, four of them American, of which two are pantropical, and *Monogramma* Schkuhr (*Vaginularia* Fée) of the paleotropics and *Rheopteris* Alston of New Guinea.

The Vittariaceae are predominently epiphytes with clathrate scales on the stem and, except in *Hecistopteris,* with entire, pendent leaves. The plants are anatomically unusual in lacking sclerenchyma and in having elongate sclerids (spicular idioblasts) in the epidermis of the leaf. The sporangium (Wilson, 1959) is of delicate and precise construction, with four well differentiated stomium cells. The basic morphological and anatomical study of the genera is by Williams (1927).

The family is a distinctive one with no known fossil record and of uncertain close relation to other ferns. The chromosome number of $n = 60$ or multiples relates the family, at least in a general way, to the Pteridaceae often with $n = 30$ or multiples. The same relationship is suggested by the spicular cells which also occur in the leaf of *Adiantum.* Nearly all recent authors agree with this relationship, especially to genera of the taenitoid ferns. Copeland (1947), however, was in doubt and unwilling to specify a relation for the family.

Key to Genera of Vittariaceae in America

a. Veins free, lamina more or less dichotomously lobed or furcate.
$\qquad$ 50. *Hecistopteris*
a. Veins anastomosing, lamina entire. b.
$\qquad$ b. Sporangia on the veins, contiguous in rather short to very long, sometimes branched, more or less sunken sori. c.
$\qquad\qquad$ c. Several to many sori on each side of the costa. 51. *Antrophyum*
$\qquad\qquad$ c. One very long sorus or one line of sori near each margin.
$\qquad\qquad\qquad$ 52. *Vittaria*
$\qquad$ b. Sporangia single or in scattered groups on the veins and between them, not in definite sori, superficial. 53. *Anetium*

Literature

Copeland, E. B. 1947. Genera Filicum. 247 pp. Chronica Botanica, Waltham, Mass.

Williams, S. 1927. A critical examination of the Vittarieae with a view to their systematic comparison. Trans. Roy. Soc. Edinburgh 55: 127–217.

Wilson, K. A. 1959. Sporangia of the fern genera allied with *Polypodium* and *Vittaria.* Contrib. Gray Herb. 185: 97–127.

50. *Hecistopteris*
Figs. 50.1–50.4

Hecistopteris J. Sm., Lond. Jour. Bot. 1: 193. 1842. Type: *Hecistopteris pumila* (Spreng.) J. Sm. (*Gymnogramma pumila* Spreng.).

Description

Epiphytic, rarely rupestral; stem short-creeping, very small, bearing scales and a few long roots; leaves monomorphic, or the fertile usually larger, ca. 1–3 cm long, borne in a cluster, lamina furcate and broader apically, or rarely nearly entire, glabrous, veins free; sori superficial, elongate along the distal veins but not extending to their apex, sometimes joined below a fork of a vein, with simple or branched paraphyses with an enlarged apical cell, exindusiate; spores tetrahedral-globose, trilete, the laesurae ¾ the radius, the surface layer somewhat granulate with irregular, scattered spheres on a somewhat granulate to papillate surface. Chromosome number: not reported.

The small, usually flabellate-cuneate leaves (Fig. 1) are sometimes nearly entire. The roots, which sometimes have been misinterpreted as stems, are often long and may have proliferous buds that produce new plants.

Systematics

Hecistopteris is a distinctive genus of the American tropics with a single species, *H. pumila* (Spreng.) J. Sm. Characters such as free venation, superficial sori and trilete spores indicate that it is one of the least specialized in the tribe.

Ecology and Geography (Fig. 2)

Hecistopteris grows especially in rain forests and montane forests, mostly below 1000 m. It is commonly epiphytic, frequently growing among small clumps of moss on tree trunks. It also grows on rotten tree logs and sometimes on wet rocks.

The genus has a wide distribution (Fig. 2), based on relatively

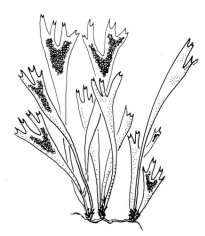

Fig. 50.1. Plants of *Hecistopteris pumila,* connected by proliferous roots, × 2.

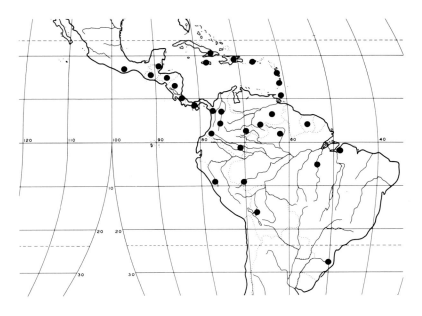

Fig. 50.2. Distribution of *Hecistopteris.*

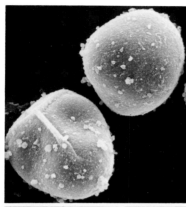

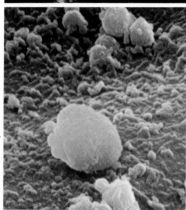

few records. The diminutive plants may be rare, or perhaps are often overlooked.

Hecistopteris ranges from Oaxaca in Mexico, through Central America, in the Greater Antilles, in Guadeloupe and Martinique in the Lesser Antilles, in Trinidad, and in South America south to Bolivia and Santa Catarina in southeastern Brazil.

Spores

Hecistopteris spores are trilete with sparse, somewhat spherical deposit on both faces (Fig. 3) overlaying a granulate to papillate formation (Fig. 4). The surface is generally similar to that of *Vittaria graminifolia* and differs from the stranded formation on spores of *Anetium* and *Antrophyum*.

Figs. 50.3, 50.4 *Hecistopteris* spores, Brazil, *Conant 916.* **3.** Distal face above the proximal below, × 1000. **4.** Detail of irregular spheres on granulate-papillate surface, × 10,000.

51. *Antrophyum*
Figs. 51.1–51.13

Antrophyum Kaulf., Enum. Fil. 197. 1824. Type: *Antrophyum plantagineum* (Cav.) Kaulf. (*Hemionitis plantaginea* Cav.) [Chosen by J. Smith, Hist. Fil. 1875].

Polytaenium Desv., Mém. Soc. Linn. Paris 6: 218. 1827. Type: *Polytaenium lanceolatum* (Sw.) Desv. (*Vittaria lanceolata* Sw.) = *Antrophyum lineatum* (Sw.) Kaulf. *Antrophyum* subgenus *Polytaenium* (Desv.) Bened., Bull. Torrey Bot. Cl. 34: 447. 1907.

Scoliosorus Moore, Ind. Fil. xxix. 1857. Type: *Scoliosorus ensiformis* (Hook.) Moore = *Antrophyum ensiforme* Hook. *Antrophyum* subgenus *Scoliosorus* (Moore) Bened., *ibidem* 34: 447. 1907. [Subgenus *Antrophyum*]

Antrophyum subgenus *Antrophyopsis* Bened., *ibidem* 34: 447. 1907. Type: *Antrophyum Boryanum* (Willd.) Spreng.

Antrophyum subgenus *Bathia* C. Chr., Notes Ptérid. (Bonap.) 16: 110. 1925. Type: *Antrophyum bivittatum* C. Chr.

Description Epiphytic or sometimes rupestral; stem short-creeping, small, bearing scales and many roots; leaves monomorphic, ca. 3–50 cm long, borne in a cluster or somewhat spaced, lamina entire, linear, oblanceolate to suborbicular, glabrous, veins anastomosing without included free veinlets; or (in *A. bivittatum* C. Chr. and *A. trivittatum* C. Chr.) free; sori superficial or in grooves, two or more very long sori between the costa and margin, or usually

Fig. 51.1. *Antrophyum ensiforme,* Pedregal Las Vigas, above Jalapa, Veracruz, Mexico. (Photo W. H. Hodge.)

many on the anastomosing veins, with simple or usually branched paraphyses with or without an enlarged apical cell, exindusiate, spores tetrahedral-globose, trilete, the laesura $\frac{1}{2}$ to $\frac{3}{4}$ the radius, or somewhat ellipsoidal, monolete, the laesura nearly equal the spore length, usually with irregular strands or long echinate rods on a somewhat papillate surface. Chromosome number: $n = 60, 120, 180; 2n = $ ca. 180.

In some species of *Antrophyum* the lamina has no costa, while in others it is well developed to the apex (Fig. 3). Several to many sori may be along portions of the veins (Fig. 1) or there may be a few very long sori (Fig. 5) in deep grooves (Fig. 6), or the sporangia may follow most of the veins (Fig. 4). Small plants of *Antrophyum lineatum* may have single long sorus on each side of the costa.

Fig. 51.2. Distribution of *Antrophyum* in America.

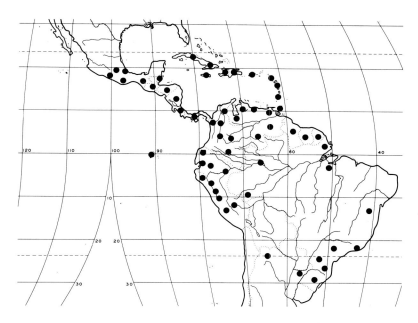

Systematics

Antrophyum is a pantropical genus of 40 more species, with perhaps ten in America. It is closely related to *Vittaria*.

The subgenus *Polytaenium* is recognized primarily on the basis of the lack of paraphyses and its occurrence only in America. Although sometimes recognized as a genus, it is not a very distinctive taxon. The paleotropical subgenera *Bathia* and *Antrophyopsis* need to be assessed on the basis of a modern study of the whole genus.

The classification of the genus and the American species has been studied by Benedict (1907, 1911).

Synopsis of American *Antrophyum*

Subgenus *Antrophyum*

Paraphyses present, costa absent or partially developed, rarely (in *A. ensiforme*) extending to the apex of the lamina, spores trilete or monolete.

One species, *Antrophyum ensiforme* Hook. of Mexico and Central America, and thirty or more species in the paleotropics.

Subgenus *Polytaenium*

Paraphyses absent, costa extending to the apex of the lamina, or rarely (in *A. anetioides*) only partially developed, spores trilete.

An American subgenus of ten or fewer species, the commonly recognized ones are: *Antrophyum anetioides* Christ, *A. brasilianum* (Desv.) C. Chr., *A. cajenense* (Desv.) Spreng. (*A. discoideum* Kze.), *A. Dussianum* Bened., *A. guayanense* Hieron., *A. Jenmanii* Bened., *A. lanceolatum* (L.) Kaulf. (*Polytaenium Feei* (Fée) Maxon), *A. lineatum* (Sw.) Kaulf. (*Vittaria intramarginalis* Baker), and *A. Urbanii* Brause.

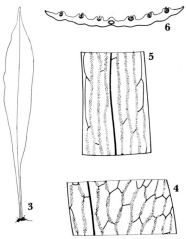

Fig. 51.3–51.6. *Antrophyum*. 3. Leaf of *A. cajenense,* ×0.25. **4.** Venation, the sori stippled along veins, *A. cajenense,* ×1. **5.** Venation, the sori stippled along veins, *A. lineatum,* ×1.5. **6.** Section of fertile leaf of *A. lineatum* the sori in grooves, ×3.

Tropical American Species

Three of the species of subgenus *Polytaenium* are distinctive. *Antrophyum lineatum* has several long, sunken sori parallel to the costa; *A. lanceolatum* has the areolae oriented parallel to the costa; and *A. anetioides* has a partially developed costa. The remaining have a complete costa, sori irregularly divergent from the costa, or sporangia along many of the veins and most areolae divergent from the costa. The exact number of these species is uncertain because variation of the plants has not been adequately surveyed. Characters such as lamina thickness, marginal venation and depth of the soral grooves require assessment in relation to environmental conditions.

Prior to the work of Tryon (1964) the names of three species were confused due to inadequate study of their types: *A. cajenense* was commonly called *A. brasilianum, A. brasilianum* was named *A. discoideum* and *A. guayanense* was named *A. cajenense.*

Ecology (Fig. 1)

Antrophyum usually grows in moist or wet forests, often in deep shade. It is commonly epiphytic, but also grows on mossy logs and rotten wood, and sometimes on boulders or moist cliffs.

In tropical America it grows in rain forests or cloud forests, sometimes in disturbed sites, or (in Mexico) in caverns in lava beds. It is rare in less mesic habitats such as savannah forests. *Antrophym* usually grows at 100–1500 m, sometimes it occurs near sea level, or up to 2500 m.

Geography (Fig. 2)

Antrophyum is pantropic, in America, Africa, and eastward to Malesia and through the Pacific to Tahiti, north to Japan and south to Australia.

In America it occurs from Hidalgo in Mexico, through Central America, in the Greater and Lesser Antilles, and in South America south to Salta and Missiones in northern Argentina and Rio Grande do Sul in southeastern Brazil; also in the Galápagos Islands.

Antrophyum anetioides is endemic to Costa Rica and *A. Urbanii* to Hispaniola. *Antrophyum ensiforme* occurs from Mexico to Panama, and *A. Dussianum* is in the West Indies and Trinidad. Other species are rather widely distributed, and there is no significant center of diversity.

Spores

Most *Antrophyum* species have trilete spores (Figs. 7, 8) with a uniform papillate surface upon which spheres, strands, and echinate elements are formed. A few species have monolete spores, as *A. ensiforme* (Figs. 9, 10), and a papillate surface with sparse or no spherical deposit (Fig. 11). Marked size differences between and within species (Figs. 7–10) probably reflect different ploidy levels. Details of the echinate rods and surface

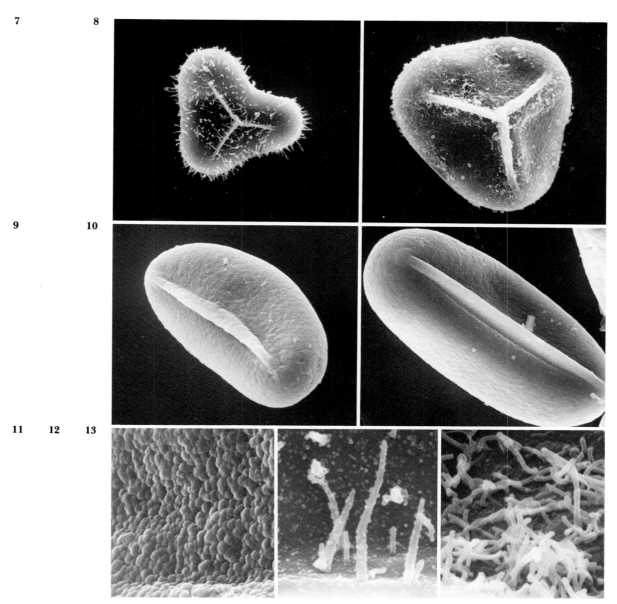

Figs. 51.7–51.13. *Antrophyum* spores, × 1000, surface detail, × 10,000. **7.** *A. guayanense,* trilete, echinate, Brazil, *Prance et al. 1339.* **8.** *A. lanceolatum,* trilete, with sparse strands, Brazil, *Egler & Irwin 46693.* **9–11.** *A. ensiforme,* monolete. **9.** Veracruz, Mexico, *Dressler & Jones 147.* **10.** Chiapas, Mexico, *Ghiesbreght 218.* **11.** Detail of papillate surface, laesura at base, *Dressler & Jones 147.* **12.** *A guayanense,* surface detail with echinate rods, portion of laesura at base, *Prance et al. 1339.* **13.** *A. lanceolatum,* detail of surface, *Egler & Irwin 46693.*

strands show the particulate formation (Figs. 12, 13). These spores have been reported as lacking perispore however the stranded and echinate structure appears to be perispore.

Cytology

Chromosome numbers from Jamaica, reported by Walker (1966), include *Antrophyum lineatum* with $n = 60$ (as *Polytaenium lineatum*) considered a tetraploid and *A. Dussianum* with $n = 120$

(as *P. discoideum*) an octoploid. The record of *A. lanceolatum* with n = ca. 60, from Puerto Rico, listed in Fabbri (1965) is likewise tetraploid. The few chromosome reports for paleotropical species are also n = 60 or 180, and an unidentified specimen from Ceylon with $2n$ = ca. 180 was regarded as a hexaploid by Manton and Sledge (1954). These ploidy levels are based on 30, known in the related genus, *Monogramma* (reported as *M. trichoidea* (Fée) Hook.) by Manton (1954).

Literature

Benedict, R. C. 1907. The genus *Antrophyum* 1. Synopsis of subgenera, and the American species. Bull. Torrey Bot. Cl. 34: 445–458.

Benedict, R. C. 1911. The genera of the fern tribe Vittarieae. Their external morphology, venation, and relationships. Bull. Torrey Bot. Cl. 38: 153–190.

Fabbri, F. 1965. Secondo supplemento alle Tavole Cromosomiche delle Pteridophyta do Alberto Chiarugi. Carylogia 18: 675–731.

Manton, I. 1954. Cytological notes on one hundred species of Malayan ferns, *in*: R. E. Holttum, A revised Flora of Malaya, Vol. 2. Ferns of Malaya, pp. 623–628. Government Printing Office. Singapore.

Manton, I., and W. A. Sledge 1954. Observations on the cytology and taxonomy of the Pteridophyte flora of Ceylon. Phil. Trans. Roy. Soc. London, Ser. B 238: 127–185.

Tryon, R. M. 1964. Taxonomic fern notes. 4. Some American Vittarioid ferns. Rhodora 66: 110–117.

Walker, T. G. 1966. Reference under the family.

52.   *Vittaria*
Figs. 52.1–52.17

Vittaria Sm., Mém. Acad. Turin 5: 413. 1793. Type: *Vittaria lineata* (L.) Sm. (*Pteris lineata* L.).

Haplopteris Presl, Tent. Pterid. 141. 1836. Type: *Haplopteris scolopendrina* (Bory) Presl (*Pteris scolopendrina* Bory) = *Vittaria scolopendrina* (Bory) Thwait. *Vittaria* subgenus *Haplopteris* (Presl) C. Chr., Ind. Fil. xlvi. 1906. *Vittaria* section *Haplopteris* (Presl) Ching, Sinensia 1: 176. 1931.

Taeniopsis J. Sm., Jour. Bot. (Hook.) 4: 67. 1841, *nom. superfl.* for *Haplopteris* Presl and with the same type.

Taeniopteris Hook., Gen. Fil. text, *t. 76B*. 1842. Type: *Taeniopteris Forbesii* Hook. = *Vittaria scolopendrina* (Bory) Thwait.

Oetosis Greene, Pittonia 4: 105. 1900, (not O. Ktze. 1891 = *Drymoglossum*), *nom. superfl.* for *Vittaria* Sm. and with the same type.

Ananthacorus Underw. & Maxon, Contrib. U.S. Nat. Herb. 10: 487. 1908. Type: *Ananthacorus angustifolius* (Sw.) Underw. & Maxon (*Pteris angustifolia* Sw.) = *Vittaria costata* Kze.

Vittaria subgenus *Radiovittaria* Bened., Bull. Torrey Bot. Cl. 38: 166. 1911. Type: *Vittaria remota* Fée.

Vittaria section *Pseudotaenitis* Ching, Sinensia 1: 176. 1931. Type: five species were included in the section.

Description   Epiphytic or rarely rupestral, very rarely terrestrial: stem suberect to short-creeping, small, bearing scales and a few spongy roots; leaves monomorphic ca. 5–100 cm long, borne in a cluster or somewhat spaced, lamina entire, filiform to linear or rarely broader, glabrous, veins anastomosing without included

Fig. 52.1. *Vittaria lineata,* Marshals Pen, Mandeville, Jamaica. (Photo W. H. Hodge.)

free veinlets; sori more or less sunken, one very long sorus or a line of sori at or near each margin, with simple or branched paraphyses with an enlarged apical cell, exindusiate, spores tetrahedral-globose, trilete, the laesurae $\frac{3}{4}$ the radius, or ellipsoidal, monolete, the laesura $\frac{3}{4}$ to nearly the length of the spore, with spherical particles, or strands on a papillate surface. Chromosome number: $n = 60, 120$; $2n = 120$, ca. $180, 240$.

Vittaria is distinguished from *Antrophyum* by the single long sorus, or line of sori somewhat back of the margin or apparently marginal. The lamina has a single row of areolae on each side of the costa (Figs. 3, 7) or two or more rows in *Vittaria costata* (Fig. 10). The clathrate stem scales that are typical of the family are

Fig. 52.2. Distribution of *Vittaria* in America.

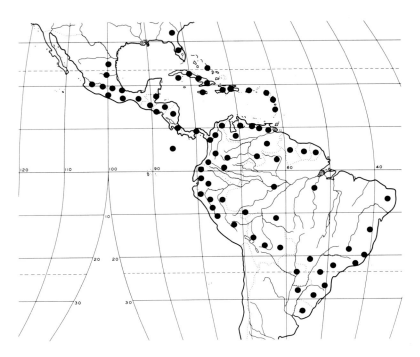

shown in Figs. 5 and 6. The characteristic simple paraphysis structure is shown in Fig. 4, and the sunken sori in Figs. 8 and 9.

Systematics

Vittaria is a pantropic genus of perhaps 50 or more species, with about nine in America. The infrageneric classification is uncertain due to the lack of a modern treatment of the genus. For example, subgenus *Radiovittaria* and the sections recognized by Ching (1931) are not adopted because of doubt as to their distinctiveness and relation to other groups in the genus.

The species *Vittaria costata* (*Ananthacorus angustifolius* (Sw.) Underw. & Maxon) is often maintained as a genus on the basis of more than one row of areolae on each side of the costa. However, this character does not provide sufficient reason to place it in a separate genus or subgenus.

The generic name *Pteropsis* Desv. has, in the past, been cited as a synonym of *Vittaria*. According to the International Code of Botanical Nomenclature (Stafleu et al. 1978) this is now a synonym of *Drymoglossum*.

A classification of *Vittaria* and a treatment of the American species were presented by Benedict (1911, 1914).

Tropical American Species

The taxonomy of the American species is adequately known although some problems remain. It is not clear that *Vittaria dimorpha* K. Müll. is separable from *V. graminifolia,* and *Vittaria minima* (Baker) Bened. appears to represent small, fertile plants of *V. remota.*

Key to American Species of *Vittaria*

a. Two to several rows of areolae on each side of the costa. *V. costata* Kze.
a. A single row of areolae on each side of the costa. b.
  b. Petiole lighter in color than the green lamina, or concolorous with the brownish lamina when darkened with age or drying; stem dorsiventral. c.
    c. Paraphyses slender, tan to light reddish-brown, the apical cell not or only slightly enlarged; spores reniform, monolete. *V. lineata* (L.) Sm.
    c. Paraphyses stout, reddish-brown to dark reddish-brown, the apical cell enlarged; spores tetrahedral-globose, trilete. *V. graminifolia* Kaulf. (*V. filifolia* Fée)
  b. Petiole usually reddish-brown to atropurpureous, darker than the lamina; stem radial. d.
    d. Petiole terete or oval, or flattened only at the very base, not alate, hard. e.
      e. Stem scales 2 cells wide (Fig. 6) or sometimes 3 cells wide at the base, sori in deep grooves near the margin. *V. stipitata* Kze.
      e. Stem scales mostly 5 or more cells wide (Fig. 5); sori in shallow grooves back from the margin. *V. Moritziana* Mett.
    d. Petiole flattened and two-angled throughout, wholly or mostly narrowly alate, firm or rather soft, usually irregularly wrinkled in drying. f.
      f. Stem scales mostly dark brown to atropurpureous, with sclerotic lateral cell walls, the margin with short, stout teeth. g.
        g. Lamina narrowly elliptic to linear-elliptic, often falcate, a coastal ridge on the upper surface from the base to the center or beyond. h.
          h. Long axis of the areolae oblique to the costa (a lateral vein reaches the margin above the point where the next one on the same side arises from the costa). *V. remota* Fée
          h. Long axis of the areolae nearly parallel to the costa (a lateral vein reaches the margin below the point where the next one on the same side arises from the costa). *V. Gardneriana* Fée (*V. Karsteniana* Mett.)
        g. Lamina narrowly linear, or narrower, rather straight, lacking a costal ridge on the upper surface at the base (one may be somewhat developed beyond the base). *V. Ruiziana* Fée
      f. Stem scales light brown, with slightly sclerotic lateral cell walls, the margin with long, slender cilia. *V. latifolia* Bened.

Ecology (Fig. 1)

Vittaria is primarily a genus of rain forests, cloud and mossy forests, less often of partially shaded or more open habitats. It is usually epiphytic, and occurs less often on rotten wood and or on boulders or cliffs, and rarely on earth banks.

In America, the genus grows in lowland rain forests, montane, cloud, and elfin forests usually in especially shaded and moist habitats as along streams and in deep ravines. Less often it will grow in lightly shaded sites. It is usually found below 1000 m, with some species such as *Vittaria graminifolia* growing as high as 3300 m.

Geography (Fig. 2)

Vittaria is pantropic, from America to Africa and eastward through the Pacific to the Hawaiian Islands, Pitcairn and Easter Island, and north to the southeastern United States, Japan, and south to Australia.

In America it occurs in Georgia and Florida in the southeastern United States, in the Bahamas, and from Neuvo Leon in Mexico through Central America, in the West Indies, and in

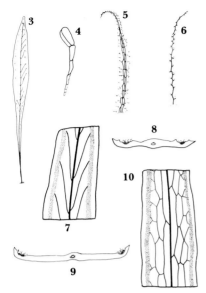

Fig. 52.3–52.10. *Vittaria.* **3.** Venation in leaf of *V. latifolia,* ×0.25. **4.** Detached paraphysis of *V. graminifolia,* enlarged. **5.** Stem scale of *V. Moritziana,* ×6. **6.** Stem scale of *V. stipitata,* ×6. **7.** Venation, the sori stippled along veins, *V. remota,* ×1. **8.** Section of fertile leaf of *V. stipitata,* with sori near margins, ×4. **9.** Section of fertile leaf of *V. costata,* with sori near margins, ×3. **10.** Venation, the sori stippled along veins, *V. costata,* ×1.

South America south to Bolivia, Paraguay, northeastern Argentina, and Uruguay; also on Cocos Island.

Vittaria latifolia occurs in Peru and Bolivia, while other species are more widely distributed. The center of diversity is in the Andes from Venezuela to Bolivia where nine species occur. Five species occur in Costa Rica, five in the Greater Antilles, and fewer elsewhere.

Spores

Most species of *Vittaria* have monolete spores but a few are trilete (Figs. 12, 13). The spore studies of Columbia Pteridophyta (Murillo and Bless, 1974, 1978) include six species, five monolete and one trilete. In the monolete forms the laesura is often prominent as in *V. remota* (Fig. 11). The surface in both types of spores is finely papillate with large spherical deposits as in *V. graminifolia* (Fig. 14), or smaller particles as in *V. Ruiziana* and *V. Gardneriana* (Figs. 15, 17), or strands as in *V. latifolia* (Fig. 16). The surface especially of monolete spores is uniformly papillate. *Vittaria* spores are described as lacking perispore (Erdtman and Sorsa, 1971). However, the abraded surface of *V. Gardneriana* (Fig. 17) has a particulate perispore that overlays a relatively smooth exospore. The small spores of *V. graminifolia* from Costa Rica (Fig. 12) probably reflect a lower ploidy level than the larger ones from Venezuela (Fig. 13).

Cytology

Reports of Jamaican specimens by Walker (1966) include *Vittaria costata* (as *Ananthacorus angustifolius*) with $n = 120$, regarded as octoploid, and an unidentified specimen, resembling *V. lineata,* with $2n = $ ca. 180, forming 60 bivalents and 60 univalents, determined as a hybrid. The chromosome numbers in two species of *Vittaria* in tropical America were used by Gastony (1977) to identify vegetatively reproducing gametophytes. These were identified as *Vittaria lineata* on the basis of the corresponding numbers $n = 120$ in gametophytes from Kentucky and $2n = 240$ in sporophytes of *V. lineata,* from Florida, rather than $2n = 120$ in sporophytes of *V. graminifolia* from Veracruz, Mexico. However, marked differences in spore size as in *V. graminifolia* (Figs. 12, 13) suggests that different ploidy levels exist within species thus the chromosome number may not be conclusive for identification.

Observations

Colonies of gametophytes in the southeastern United States, forming large mats on noncalcareous rocks from Alabama to Ohio, have been identified as *Vittaria.* Reproduction is from gemmae, although both antheridia and archegonia are formed. Small sporophytes have been found in association with the gametophytes and numerous, apogamous embryos have been produced in laboratory cultures. Identification of these gametophytes as *Vittaria lineata* is supported on the basis of

11   12

13   14

15 16 17

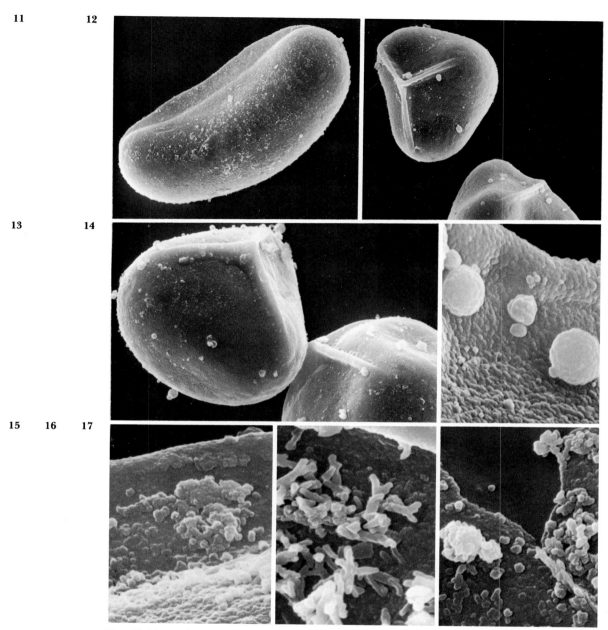

Figs. 52.11–52.17. *Vittaria* spores, × 1000; surface detail, × 10,000.   **11.** *V. remota* ellipsoidal, mono-lete, the laesura at top, Surinam, *Maguire 2418*.   **12–14.** *V. graminifolia,* tetrahedral-globose, tri-lete.   **12.** Costa Rica, *Scamman & Holdridge 8110*.   **13.** Venezuela, *Pittier 9971*.   **14.** Detail of papillate surface and spherical deposit, Costa Rica, *Scamman & Holdridge 8110*.   **15.** *V. Ruiziana,* detail of fused spheres, Peru, *Vargas 11157*.   **16.** *V. latifolia,* detail of surface strands, Bolivia, *Williams 1337*.   **17.** *V. Gardneriana,* abraded surface, the papillate perispore, below, smoother exospore, above, Ecuador, *André K 81*.

chromosome numbers (Gastony, 1977) as noted under cytology. Farrar (1978) questions the identity of these gametophytes on the basis of morphological differences between them and game-tophytes of *V. lineata* from Florida. He suggests that they may represent a species from which they have been long separated or perhaps one that is now extinct.

Literature

Benedict, R. C. 1911. The genera of the fern tribe Vittarieae. Bull. Torrey Bot. Cl. 38: 153–190.

Benedict, R. C. 1914. A revision of the genus *Vittaria* J. E. Smith. Bull. Torrey Bot. Cl. 41: 391–410.

Ching, R. C. 1931. The studies of Chinese ferns, VI. Genus *Vittaria* of China and Sikkim-Himalaya. Sinensia 1: 175–192.

Erdtman, G. and P. Sorsa. 1971. Pollen and Spore Morphology/Plant Taxonomy. Pteridophyta. 302 pp. Almqvist & Wiksell, Stockholm.

Farrar, D. R. 1978. Problems in the identity and origin of the Appalachian *Vittaria* gametophyte, a sporophyteless fern of the eastern United States. Amer. Jour. Bot. 65: 1–12.

Gastony, G. J. 1977. Chromosomes of the independently reproducing Appalachian gametophyte: A new source of taxonomic evidence. Syst. Bot. 2: 43–48.

Murillo, M. T., and J. M. Bless. 1974. Spores of recent Colombian Pteridophyta. 1. Trilete spores. Rev. Palaeobot. Palynol. 18: 223–269.

Murillo, M. T., and J. M. Bless. 1978. Spores of recent Colombian Pteridophyta. 2. Monolete spores. Rev. Palaeobot. Palynol. 25: 319–365.

Stafleu, F. A. et al. (eds.). 1978. XI: Pteridophyta, pp. 294–297, *in*: International Code of Botanical Nomenclature. 457 pp. Bohn, Scheltema & Holkema, Utrecht.

Walker, T. G. 1966. Reference under the family.

53. *Anetium*
Figs. 53.1–53.6

Anetium (Kze.) Splitg., Tijdsch. Nat. Gesch. 7: 395. 1840. *Acrostichum* section *Anetium* Kze., Flora, Beibl. 1839, I: 47. Type: *Acrostichum citrifolium* L. = *Anetium citrifolium* (L.) Splitg.

Pteridanetium Copel., Gen. Fil. 224. 1947, *nom. superfl.* for *Anetium* Splitg. and with the same type.

Description

Epiphytic or rarely rupestral; stem long-creeping, rather slender, bearing scales and many spongy roots; leaves monomorphic, ca. 10–100 cm long, spaced, lamina entire, usually oblanceolate, glabrous, veins anastomosing without included free veinlets; sporangia superficial, mostly in scattered groups on and between the veins, no paraphyses, exindusiate; spores tetrahedral-globose, trilete, the laesurae $\frac{3}{4}$ the radius, with strands more or less covering the papillate surface often fused in compact sheaths. Chromosome number: $n = 60$.

The leaves are succulent (fleshy-coriaceous) when fresh and the sporangia are readily detached from the leaf.

Systematics

Anetium is a tropical American genus of one species, *A. citrifolium* Splitg. (Fig. 2). It is distinguished by the subacrostichoid sporangia arrangement (Fig. 4). This character along with the special-

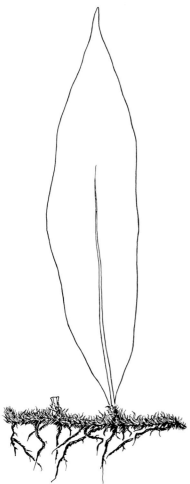

Fig. 53.1. *Anetium citrifolium,* on palm trunk, Río Bobonaza, Ecuador. (Photo B. Øllgaard.)

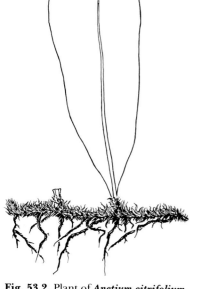

Fig. 53.2. Plant of *Anetium citrifolium,* ×0.5.

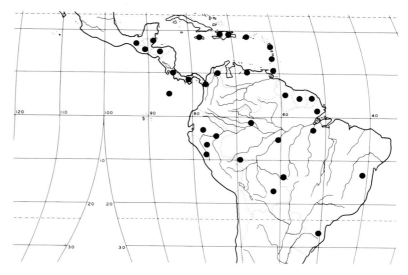

Fig. 53.3. Distribution of *Anetium.*

ized spore architecture evidently indicate a derived condition in the tribe.

Copeland proposed a new name for *Anetium,* but since the earlier *Anetia* Endl. is not a homonym as he supposed, the name is superfluous.

Ecology and Geography (Figs. 1, 3)

Anetium grows in wet forests, from sea level to ca. 1200 m. It is usually epiphytic, especially on palm trunks (Fig. 1), but rarely it grows on rocks.

Fig. 53.4. Venation, groups of sporangia on and between the veins, stippled, *A. citrifolium,* × 0.75.

The genus occurs (Fig. 3) from Chiapas in southern Mexico through Central America, in the West Indies, and is widely scattered in South America south to Bolivia, and Santa Catarina in southeastern Brazil; also on Cocos Island.

Spores

Anetium spores (Fig. 5) have a relatively uniform papillate surface as in other vittarioid genera but additional deposits are formed on this layer as in Fig. 6. The surface is partly to entirely covered with strands that become fused into a dense, sheathlike formation as in the spores of *Doryopteris* or *Pellaea.*

Cytology

Anetium citrifolium is reported (as *Pteridanetium*) from Trinidad with $n = 60$, and regarded as a tetraploid by Walker (1966).

Literature

Walker, T. G. 1966. Reference under the family.

Figs. 53.5, 53.6. *Anetium citrifolium* spores, Peru, *Schunke 3203.* **5.** Spore with strand remnants, × 1000. **6.** Detail of the strands and part of the outer sheathlike formation lower left, papillate surface of laesura, above, × 10,000.

5   6

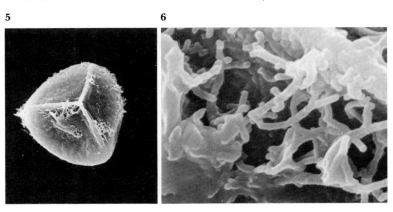

Family 16.   Dennstaedtiaceae

Dennstaedtiaceae Pic.-Ser., Webbia 24: 704. 1970. Type: *Dennstaedtia* T. Moore.

Family synonyms are placed under the tribes, where they correspond more closely to the taxonomy.

Description   Stem erect, rarely arborescent, to short- or very long-creeping (to ca. 60 m or more), small to stout or slender, protostelic with internal phloem, sphonostelic or rarely dictyostelic, usually indurated or (in *Lonchitis*) succulent, bearing trichomes or scales, or both; leaves ca. 20 cm to 7 m long, usually pinnate or (in *Lindsaea*) rarely simple and cordate to sagittate, circinate in the bud, petiole without stipules; sporangia borne in marginal, submarginal or rarely abaxial sori, at a vein end, receptacle elevated or not, or on a vascular commissure connecting vein ends, indusiate, the indusium cup- or purse-shaped, or formed by the modified, recurved margin, or abaxial and laterally more or less extended, or rarely (in *Paesia* and sometimes in *Pteridium*) both a marginal and abaxial indusium present, or (in *Hypolepis*) exindusiate; sporangia short- to usually long-stalked, the stalk 1- to 3-rowed below its apex, the annulus vertical to slightly oblique, at least the indurated portion interrupted by the stalk, homosporous, spores lacking chlorophyll. Gametophyte epigeal, with chlorophyll, obcordate to reniform, slightly thickened centrally, glabrous, archegonia borne on the lower surface, usually in the central region, antheridia 3-celled, borne on the lower surface, mostly apart from the archegonia.

Comments on the Family

The Dennstaedtiaceae are a pantropical family of 17 genera, with a few of them extending to boreal and (or) south temperate regions. The group is characterized by long-creeping stems bearing large, decompound leaves with marginal or submarginal, indusiate sori. Many species have epipetiolar buds which develop into branches. The chromosome numbers are diverse and usually high, ranging from $n = 26$ to $n = 220$. Since there is little information on the gametophytes, they cannot be adequately described.

Fossil records are meager and represent rather late periods as *Dennstaedtia americana* Knowlton from the Paleocene and Miocene and *Dennstatedtiopsis* Arnold & Daugherty from the Eocene.

The family undoubtedly includes a number of evolutionary lines from an ancient complex, as shown by the several distinctive genera without close affinities to others. There are similarities to several primitive families but it is difficult to assess their phyletic value.

Each of the genera of the Dennstaedtiaceae have distinctive types of spores except for the Lindsaeeae, and *Hypolepis* and *Blotiella*. The chromosome numbers are also diverse including a large range indicative of both polyploid and aneuploid changes. The discordant reports known for *Dennstaedtia, Hypolepis* and *Lindsaea* suggest cytological complexities that possibly involve more than a single base number in each of these genera.

The tribe **Dennstaedtieae** is characterized by a typical although sometimes polycyclic siphonostele, and the indusium is either purse- or cup-shaped as in *Dennstaedtia*, or formed by the more or less modified, reflexed margin, or an abaxial as well as a marginal indusium is present as in *Paesia* and *Pteridium*. The spores have various but usually prominent architecture. Spores, chromosome numbers and other characters indicate that the relationships of the genera are complex and that they represent isolated elements.

The tribe **Lindsaeeae** is characterized by a special type of stele, a protestele with internal phloem. Found in nearly all members of the tribe it is often referred to as the Lindsaea-type. The abaxial indusium is laterally short to very long and the opposed leaf margin is little if at all modified. The surface architecture of the spores is generally less elaborate than in the spores of the Dennstaedtieae. Within the tribe, *Odontosoria* and *Lindsaea* are closely related and are not clearly separable because of intermediate species. *Ormoloma* is clearly allied to *Lindsaea*. The other two paleotropical genera *Tapeinidium* and *Xyropteris* are not closely related to each other nor to other genera of the tribe.

The tribe **Monachosoreae** (Ching) R. & A. Tryon, Rhodora 84:126. 1982 includes *Monachosorum* Kze., an unusual small genus of uncertain affinity. It has a dictyostele, unique, small trichomes on the stem, two vascular bundles in the petiole, an exindusiate sorus, irregularly tuberculate spores and a chromosome number of $n = 56$ or 84. Its morphology and relations have been reviewed by Nair and Sen (1974). It has been allied to the dennstaedtioid ferns, the davallioids, thelypteroids and taenitoids, and Ching (1978) has placed it in a separate family, Monachosoraceae Ching, Acta Phytotax. Sinica 14 (4): 17. 1978. Type: *Monachosorum* Kze.

Key to Genera of Dennstaedtiaceae in America

a. Indusium cup- or purse-shaped, the abaxial and adaxial portions joined, recurved, spores tuberculate-verrucate, reticulate, or coarsely ridged.
          55. *Dennstaedtia,* p. 377
a. Indusium formed only by the modified, recurved margin, or a separate abaxial indusium may also be present.   b. (see third a).
    b. Sori 1-nerved   59. *Hypolepis,* p. 398
    b. Sori few to many nerved.   c.
        c. Sterile segments and sterile portions of fertile segments with a modified, recurved margin similar to the indusium but not as broad.
            57. *Pteridium,* p. 387
        c. Only the fertile portion of segments with a modified, recurved margin (indusium).   d.
            d. Abaxial indusium present.   58. *Paesia,* p. 395
            d. Abaxial indusium absent.   e.
                e. Lamina glabrous, or with a very few scales, usually glaucous beneath, stem long-creeping, slender, indurated, spores monolete.   62. *Histiopteris,* p. 410
                e. Lamina pubescent, not glaucous beneath.   f.
                    f. Stem decumbent to erect, indurated, veins fully anastomosing, spores monolete.   60. *Blotiella,* p. 403
                    f. Stem short-creeping, succulent, veins free or casually anastomosing, spores trilete.   61. *Lonchitis,* p. 406
a. Indusium abaxial, the opposed margin not or hardly modified, flat or nearly so.   g.
    g. Stem erect, spores dichotomously ridged.   56. *Saccoloma,* p. 383
    g. Stem creeping to ascending.   h.
        h. Stem with trichomes, sori at or near sinuses of the fertile segments, spores regularly and prominately echinate.   54. *Microlepia,* p. 373
        h. Stem with scales and sometimes also with trichomes, or if only with trichomes then the sori terminal on the fertile segments, spores smooth or with granulate or spherical deposition, or with strands forming an irregular, low echinate surface.   i.
            i. Lamina 1-pinnate and the sori 1-nerved.   65. *Ormoloma,* p. 429
            i. Lamina entire, or 2-pinnate or more complex, or 1-pinnate and the sori many nerved.   j.
                j. Sori laterally short, with the indusium attached at the base and at least partially on the sides.   63. *Odontosoria,* p. 414
                j. Sori laterally short with the indusium attached only at the base, or sori laterally elongate.   64. *Lindsaea,* p. 421

Literature

Ching, R. C. 1978. The chinese fern families and genera: Systematic arrangement and historical origin (cont.). Acta Phytotax. Sinica 16 (4): 16–37.

Lugardon, B. 1974. La structure fine de l'exospore et de la périspore des Filicinées isosporées, 2. Filicales, Commentaries. Pollen et Spores 16: 161–226.

Nair, G. B. and U. Sen. 1974. Morphology and anatomy of *Monachosorum subdigittatum* (Bl.) Kze. with a discussion of its affinities. Ann. Bot. 38: 749–756.

Smith, A. R., and J. T. Mickel. 1977. Chromosome counts for Mexican ferns. Brittonia 29: 391–398.

Walker, T. G. 1966. A cytotaxonomic survey of the pteridophytes of Jamaica. Trans. Roy. Soc. Edinburgh 66: 169–237.

Walker, T. G. 1973. Evidence from cytology in the classification of ferns, *in:* A. C. Jermy et al., The phylogeny and classification of the ferns, pp. 91–110. Bot. Jour. Linn. Soc. 67, Suppl. 1.

16a. Tribe Dennstaedtieae

Hypolepidiaceae Pic.-Ser., Webbia 24: 705. 1970. Type: *Hypolepis* Bernh.

Pteridiaceae Ching, Acta Phytotax. Sinica 13: 96. 1975. Type: *Pteridium* Scop.

Nine genera of America and the paleotropics; also *Leptolepia* Diels of New Zealand, Queensland and New Guinea, *Coptidipteris* Nakai & Momose of eastern Asia and Japan, and *Oenotrichia* Copel. of New Caledonia. *Coptidipteris* has coarsely tuberculate spores and a chromosome number of $n = 31$. The only species, *C. Wilfordii* (Moore) Nakai & Momose, has been placed in both *Dennstaedtia* and in *Microlepia* but may not be closely related to either. *Oenotrichia maxima* (Fourn.) Copel, the type species, of *Oenotrichia* and *O. MacGillivrayi* (Fourn.) Brownlie, both of New Caledonia, have trilete, coarsely ridged spores that are clearly dennstaedtioid. Spores of *Oenotrichia tripinnata* (F. v. Muell.) Copel. of Queensland and an apparently undescribed species of New Caledonia are bilateral and echinate. These two species also have scales rather than trichomes on the stem; their relation is uncertain, they do not belong in *Oenotrichia* and the spores suggest an alliance with *Thelypteris* or *Ctenitis*.

54. *Microlepia*
Figs. 54.1–54.9

Microlepia Presl, Tent. Pterid. 124. 1836. Type: *Microlepia polypodioides* (Sw.) Presl (*Dicksonia polypodioides* Sw.) = *Microlepia speluncae* (L.) Moore.

Dennstaedtia Bernh., Jour. Bot. (Schrad.) 1800 (2): 124. 1802. Type: *Dennstaedtia flaccida* (Forst.) Bernh. (*Trichomanes flaccida* Forst.) = *Microlepia flaccida* (Forst.) Fée.

Scypholepia J. Sm., Hist. Fil. 261. 1875. Type: *Scypholepia Hookeriana* (Hook.) J. Sm. (*Davallia Hookeriana* Hook.) = *Microlepia Hookeriana* (Hook.) Presl.

Description

Terrestrial, very rarely rupestral; stem long-creeping to some-times short-creeping, slender to rather stout, bearing trichomes or rarely bristles and few to many fibrous roots; leaves mono-morphic, ca. 20 cm to 3 m, rarely to 7 m long, usually distant, or sometimes closely spaced, lamina 1- to 4-pinnate, usually pubes-cent, veins free; sori abaxial to marginal, 1-nerved, receptacle not or somewhat elevated, sometimes paraphysate, indusium half cup-shaped, attached at its base and sides or rarely only along a broad base, directed outward, or rarely cup-shaped and directed downward; spores tetrahedral-globose, prominently lobed, trilete, the laesurae short, usually about $\frac{1}{2}$ the radius, the surface of slender rods or strands forming a fine, echinate struc-ture. Chromosome numbers: $n = 43$, 44, 84–87, 86, 129; $2n = 84$, 86, 160, ca. 170, 172.

The sori of *Microlepia* are at or near the margin (Fig. 2) and are typically covered by an arching indusium. The leaves vary from decompound (Fig. 3) to 1-pinnate in the Old World *Micro-lepia Hookeriana* (Hook.) Presl.

Systematics

Microlepia is a pantropical and somewhat extratropical genus of about 45 species with only one, *M. speluncae* (L.) Moore, in America. It is usually distinct from *Dennstaedtia* in characters of the sorus and indusia but in the paleotropics it grades into that genus except in spore architecture. *Microlepia flaccida* (Forst.) Fée has the sorus and indusium cup-shaped and directed down-ward as in *Dennstaedtia* and species such as *M. concinna* R. & A. Tryon and *M. Hooveri* Christ sometimes have a somewhat re-curved, shallow, cup-shaped indusium. These species have large, decompound leaves and, except for their delicately echi-nate spores, resemble species of *Dennstaedtia* more than some *Microlepia* species as *M. strigosa* (Thunb.) Presl and *M. marginata* (Panzer) C. Chr.

The earliest name for the genus, *Dennstaedtia* Bernh., is not adopted in order to avoid confusing changes of the species names.

Tropical American Species

The pantropical *Microlepia speluncae* (L.) Moore is variable through its range, and *Microlepia jamaicensis* (Hook.) Fée is one of the many local variations. It differs from typical *M. speluncae* in that the indusium is only partly attached at the sides rather than fully fused, a condition that also occurs in plants of the Old World.

Ecology

Microlepia is usually a genus of moist to wet forests, where it grows on mountain slopes, in ravines or along streams. It also grows in dry forests, on lightly shaded hillsides, less often in thickets, along roadsides or in open places.

In America it grows in shady forested ravines and on moun-tain sides in forests at ca. 400–850 m.

Fig. 54.1. Distribution of *Microlepia* in America.

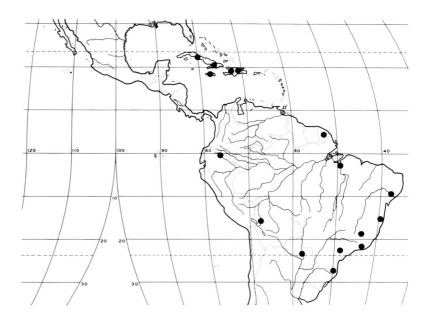

Geography (Fig. 1)

Microlepia grows in the American tropics, in tropical Africa and Madagascar, India, Ceylon, China, and eastward through Malesia across the Pacific to the Hawaiian Islands and Easter Island, north to Korea and Japan, and south to Queensland, Australia. It is adventive in the Hawaiian Islands and perhaps elsewhere.

In America the distribution of *Microlepia* is widely scattered. It grows in the Greater Antilles except Puerto Rico, and in South America in eastern and southeastern Brazil, also Paraguay, Bolivia, Ecuador, and French Guiana. It lacks a coherent range in America which suggests that it may be adventive.

Spores

Microlepia spores are relatively small and finely echinate. The characteristic surface is best developed on the lowest spore (Fig. 5) and on the proximal surface of the spore at higher magnification (Fig. 6). Young spores are usually strongly 3-lobed and each lobe appears to fit into depressed areas of adjacent spores of the tetrad. These associations are evident in spores somewhat displaced from the tetrad (Fig. 4). *Microlepia* spores are described as lacking perispore by Erdtman and Sorsa (1971) but SEM micrographs of abraded surfaces show three types of perispore structure. The outer echinate formation overlays a granulate stratum (Fig. 7) and below this is a meshwork of strands (Fig. 8). *Microlepia* spores are consistently echinate as in *M. speluncae* and in species of the Old World as *M. marginata* from Japan (Fig. 9). These clearly differ from the coarsely ridged or reticulate spores of *Dennstaedtia*.

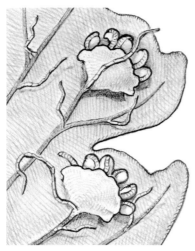

Fig. 54.2. Sori of *Microlepia speluncae*, × 15.

3   4

5   6

7   8   9

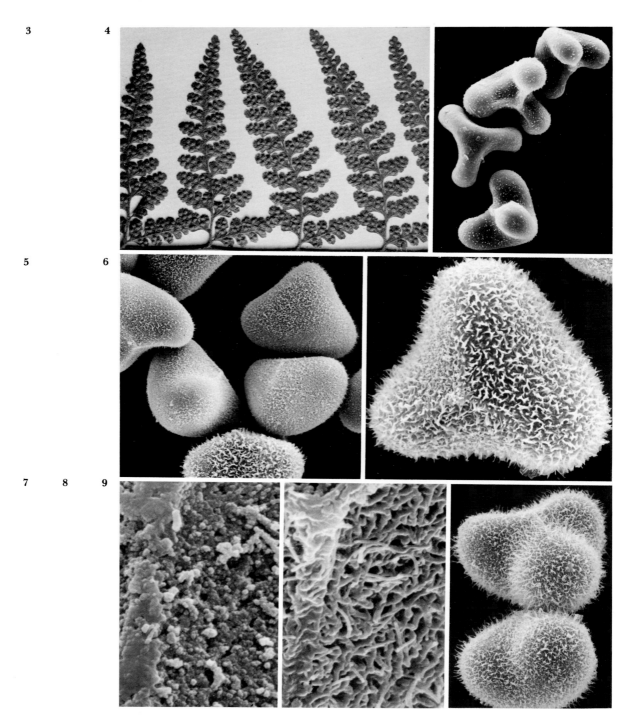

Fig. 54.3–54.9. *Microlepia.*   **3.** Portion of a pinna of *M. speluncae,* ×1.5.   **4–9.** Spores.   **4.** *M. denn-staedtioides* Copel, four spores of a young tetrad, strongly lobed with concave areas adjacent to laesurae, Philippines. *Edaño,* in 1932, × 700.   **5–8.** *M. speluncae.*   **5, 6.** Cuba, *Smith & Hodgdon 3241.*   **5.** The complete echinate surface on spore at base, × 1000.   **6.** Echinate, proximal face, × 2000.   **7, 8.** Lower perispore deposition, Paraguay, *Rojas 1051,* × 10,000.   **7.** Granulate perispore deposit below echinate surface, part of laesura at left.   **8.** Fused strands of lowest perispore deposit, part of laesura at left.   **9.** *M. marginata,* proximal face, above, lateral view, below, Taiwan, *Chuang 4766,* × 1000.

Cytology

There are no reports of chromosome numbers for *Microlepia* in America. Records of *M. speluncae* from India and Ceylon include several ploidy levels ranging from $n = 43, 84–87, 86$, and 129, while other species of the Old World are mostly 43 or 86.

Literature

Erdtman, G., and P. Sorsa. 1971. Pollen and Spore Morphology/Plant Taxonomy. Pteridophyta. Almqvist & Wiksell, Stockholm.

55. *Dennstaedtia*
Figs. 55.1–55.13

Dennstaedtia Moore, Ind. Fil. xcvii. 1859. Type: *Dennstaedtia cicutaria* (Sw.) Moore (*Dicksonia cicutaria* Sw.).

Sitobolium Desv., Mém. Linn. Soc. Paris 6: 262. 1827, altered to *Sitolobium* by J. Sm., Jour. Bot. (Hook.) 3: 418. 1841. Type: *Sitobolium punctilobulum* (Michx.) Desv., as *punctilobum* (*Nephrodium punctilobulum* Michx.) = *Dennstaedtia punctilobula* (Michx.) Moore.

Patania Presl, Tent. Pterid. 137. 1836. Type: *Patania obtusifolia* (Willd.) Presl (*Dicksonia obtusifolia* Willd.) = *Dennstaedtia obtusifolia* (Willd.) Moore.

Adectum Link, Fil. Sp. 42. 1841. Type: *Adectum pilosiusculum* (Willd.) Link (*Dicksonia pilosiuscula* Willd.) = *Dennstaedtia punctilobula* (Michx.) Moore.

Litolobium Newm., Phytologist 5: 236. 1854, *nom. superfl.* for *Sitobolium* Desv. and with the same type.

Costaricia Christ, Bull. Soc. Bot. Genève, II, 1: 229. 1909. Type: *Costaricia Werckleana* Christ = *Dennstaedtia* sp.

Fuziifilix Nakai & Momose, Cytologia, Fujii Jub. Vol. 365. 1937. Type: *Fuziifilix pilosella* (Hook.) Nakai & Momose (*Davallia pilosella* Hook.) = *Dennstaedtia hirsuta* (Sw.) Mett.

Paradennstaedtia Tagawa, Jour. Jap. Bot. 27: 213. 1952. Type: *Paradennstaedtia glabrata* (Ces.) Tagawa (*Dicksonia glabrata* Ces.) = *Dennstaedtia glabrata* (Ces.) C. Chr.

Emodiopteris Ching, Acta Phytotax. Sinica 16: 21. 1978. Type: *Emodiopteris appendiculata* (Hook.) Ching (*Dicksonia appendiculata* Hook.) = *Dennstaedtia appendiculata* (Hook.) J. Sm.

Description   Terrestrial or rarely rupestral; stem long-creeping or rarely short-creeping, slender to moderately stout, bearing trichomes and a few to many long fibrous roots; leaves monomorphic, ca. 20 cm to 4 m, rarely to 7 m long, borne at intervals to sometimes loosely clustered, lamina 1-pinnate, imparipinnate to 4-pinnate-pinnatifid, glabrous or pubescent, veins free; sori marginal, 1-nerved, the receptacle more or less elevated, sometimes paraphysate, indusium cup- or purse-shaped, directed more or less

Fig. 55.1. *Dennstaedtia circutaria,* central cordillera, Puerto Rico. (Photo D. S. Conant.)

downward, sometimes slightly bilabiate; spores tetrahedral-globose with prominent lobes, usually somewhat compressed, trilete, the laesurae $\frac{3}{4}$ the radius, often obscured by adjacent surface features, the surface prominently verrucate-tuberculate, reticulate or ridged. Chromosome number: $n = 30, 33–34, 34, 46, 47, 60, 64, 65, 94$; $2n = 60$, ca. $94, 120, 128$.

Dennstaedtia typically has the indusium bent downward (Figs. 3–6) and it is sometimes bilabiate as in Fig. 3. The stem is usually slender but may be up to 2.5 cm in diameter and rarely is nearly devoid of indument. The pinnules of *D. bipinnata* in Fig. 7 show details of leaf architecture typical of many species.

Systematics   *Dennstaedtia* is a genus of tropical and extratropical regions with about 45 species, 12 of them in America.

Similarities in many characters indicate a close relation to *Microlepia* although there are differences in the indusium, and the verrucate-tuberculate, reticulate or coarsely ridged spores are quite different from the echinate ones of *Microlepia*. In tropical America the genus is relatively homogeneous in characters of the spores and leaves except for *D. Wercklei* with a 1-pinnate lamina. The morphological diversity is greater in the paleotropics, where species as *Dennstaedtia scandens* (Bl.) Moore have spines on

Fig. 55.2. Distribution of tropical species of *Dennstaedtia* in America.

the petiole and rachis, and others as *D. glabrata* (Ces.) C. Chr. have articulate pinnae.

The most divergent elements in the genus are extratropical species as *Dennstaedtia appendiculata* (Hook.) J. Sm., *D. puncti-lobula* (Michx.) Moore and *D. hirsuta* (Sw.) Mett. These are sometimes segregated as genera but seem better placed in *Dennstaedtia;* further study may indicate that an infrageneric classification is needed.

Costaricia was based on sterile specimens that have trichomes on the stem and elongate leaves similar to juvenile ones of *Dennstaedtia.* They undoubtedly represent juvenile plants of *Dennstaedtia dissecta* or *D. obtusifolia.* Fertile specimens collected later and referred to *Costaricia* (Tryon, 1961) represent a variant of *D. obtusifolia.*

Scyphofilix Thouars, Nov. Gen. Madagas. (Melanges Bot. 2): 1. 1806, of Madagascar was published without an included species. Kunze (1837, p. 38) suggested it might be *Dicksonia madagas-cariensis* Kze. (= *Dennstaedtia madagascariensis* (Kze.) Tard.). It could as well be a *Davallia* (*D. chaerophylloides* (Poir.) Steud.) as Thouars indicated and it is best placed there until an authentic specimen is found, since *Davallia* is the older name.

Since the type of *Dennstaedtia* Bernh. is a *Microlepia,* the name *Dennstaedtia* is taken from a later publication and with a different type in order to avoid confusing name changes.

The American species of *Dennstaedtia* have been revised by Tryon (1960).

Fig. 55.3. Sori of *Dennstaedtia globulifera*, × 10.

Tropical American Species

Tropical American species of *Dennstaedtia* are relatively well known, but additional field work is needed to determine the relationship between *D. dissecta* and *D. obtusifolia*. They evidently differ only in quantitive characters of the indusium and sorus and may be variations of a single species. Further study is also needed to adequately separate *D. arborescens* from *D. Kalbreyeri* which was long known from limited material but has recently been collected several times in Ecuador.

The large leaves in most species are developed through heteroblastic series from elongate leaves of juvenile plants to broadly ovate or deltoid leaves on adult plants. Study of these developmental series may provide new species characters and better evidence for their relations.

The following key has been adapted from the more detailed one in Tryon (1960).

Key to Species of *Dennstaedtia* in Tropical America

a. Upper surface of axis of the penultimate segments bordered on each side by a pronounced herbaceous wing perpendicular to the plane of the segment, the wing on the basiscopic side decurrent onto the axis of the next order as a wing or a pronounced ridge. b.
  b. Basal segments of the pinnules of the central pinnae usually subopposite to nearly opposite, rather or quite equal in size, the inferior not or slightly ascending.   *D. globulifera* (Poir.) Hieron.
  b. Basal segments of the pinnules of the central pinnae definitely alternate, quite unequal in size, the inferior slightly to strongly ascending.
          D. bipinnata (Cav.) Maxon.
a. Upper surface of axis of the penultimate segments lacking perpendicular herbaceous wings, or if these present, the one on the basiscopic side not decurrent onto the axis of the next order. c.
  c. Sterile vein-tips on the upper surface slender, ending well back of the glabrous margin. d.
    d. Pinnae alternate, the lower ones stalked with the basal pinnules not or hardly reduced. e.
      e. Many or most of the sori borne in a sinus, pinnules more or less pubescent beneath, lamina deltoid.   *D. cicutaria* (Sw.) Moore.
      e. All or most of the sori terminal on lobes, lamina ovate or deltoid-lanceolate.   *D. glauca* (Cav.) Looser.
    d. Pinnae opposite or rarely subopposite, the lower ones sessile with a pair of usually much reduced basal pinnules.
          D. distenta (Kze.) Moore.
  c. Sterile vein-tips on the upper surface enlarged, clavate to punctate or rarely (in *D. obtusifolia*) slender. f.
    f. Lamina 1-pinnate, the pinnae entire.
          D. Wercklei (Christ) Tryon.
    f. Lamina 1-pinnate-pinnatifid or more complex. g.
      g. Sterile veins nearly reaching the persistently pubescent margin.
          D. Sprucei Moore.
      g. Sterile veins ending well back of the glabrous margin. h.
        h. Pinnules entire to deeply pinnatifid; apical segments of the pinnae confluent back of the prolonged apex. i.
          i. Segments persistently and prominently pubescent on both surfaces.   *D. Kalbreyeri* Maxon.
          i. Segments glabrous to glabrate on both surfaces.
          D. arborescens (Willd.) Maxon.
        h. Pinnules 1-pinnate to 1-pinnate-pinnatifid, apical segments of the pinnae distinct, or nearly so, up to the prolonged apex. j.
          j. Mature sori mostly 1.0–1.5 mm broad and about half as thick, especially the acroscopic one on a segment.
          D. dissecta (Sw.) Moore.
          j. Mature sori mostly 0.5–1.0 mm broad and about as thick.
          D. obtusifolia (Willd.) Moore.

Fig. 55.4. Sori of *Dennstaedtia Wercklei,* × 10.

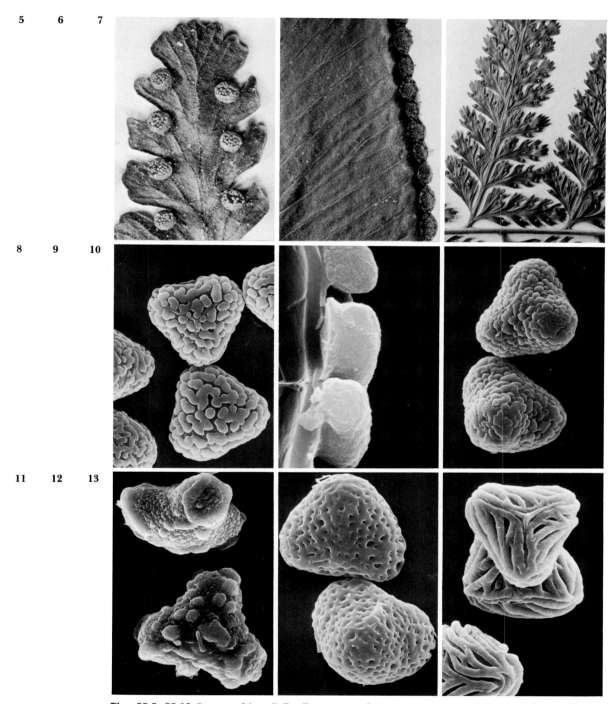

Figs. 55.5–55.13. *Dennstaedtia.* **5.** Fertile segment of *D. obtusifolia* ×5.0. **6.** Portion of a fertile pinna of *D. Wercklei*, ×5.0. **7.** Pinnules of *D. bipinnata*, ×1.5. **8–13.** Spores, ×1000. **8–9.** *D. cicutaria,* Panama, *Burch 218.* **8.** Verrucate elements more or less fused into ridges, proximal face above. **9.** Wall profile with a thin outer perispore over the verrucate exospore, ×10,000. **10.** *D. obtusifolia,* verrucate, proximal faces tilted, Colombia, *Pennell et al. 8665.* **11.** *D. bipinnata* somewhat compressed, irregularly papillate and tuberculate, Colombia, *Killip 5003.* **12.** *D. Smithii,* coarsely reticulate, Moluccas, *Buwalda 5734.* **13.** *D. samoensis,* coarsely ridged proximal face, above, part of distal face lower left, New Hebrides, *Braithwaite 2479.*

Ecology (Fig. 1)

Dennstaedtia is predominantly a genus of wet mountain forests, forest openings and secondary vegetation. The northern species such as *D. hirsuta* grow in woods, pastures, rocky slopes, on shaded banks and roadsides, and along streams. *Dennstaedtia punctilobula* grows in a variety of usually moist habitats, such as woods and forest borders, rocky hillsides, meadows; in the southwestern portion of its range it is restricted to sandstone cliffs.

In tropical America, *Dennstaedtia* usually grows in rain forests or cloud forests on mountain slopes, in ravines or on stream banks. It may also grow in forest borders or clearings and rarely in rocky places, on road banks, or open sunny habitats. It is sometimes a weed or persists in coffee plantations.

Species are usually colonial from the long-creeping, frequently branching stems. The leaves may be semiscandent in tall shrubs or among the branches of small trees.

Dennstaedtia grows from near sea level to 3200 m, usually from 500 to 2500 m.

Geography (Fig. 2)

Dennstaedtia occurs in America, in Madagascar and the Mascarenes, and eastward through Malesia and in the Pacific to Samoa and Tonga, south to eastern Australia and north to Japan and the adjacent mainland.

In America the genus occurs in the eastern United States and adjacent Canada (*D. punctilobula*) and from southern Florida (*D. bipinnata*) and Texas (*D. globulifera*) through Mexico and Central America, the West Indies, and in South America from French Guiana west to Colombia, south to Bolivia, central Chile (*D. glauca*) and Argentina to southeastern Brazil and Uruguay; also on Cocos Island and in the Galápagos Islands.

Most of the species have typical pan-Andean ranges, from the Greater Antilles, southern Mexico and Central America, Venezuela, Colombia to Bolivia and sometimes in southeastern Brazil. A few, as *D. obtusifolia,* are somewhat more widely distributed and also occur in the Lesser Antilles. Several have rather restricted ranges as *D. distenta* from Mexico to Panama and in the Greater Antilles, *D. glauca* from Chile, northwest Argentina to southern Peru, *D. Sprucei* in Ecuador and Peru, *D. Kalbreyeri,* Ecuador and Colombia, and *D. Wercklei* in Costa Rica, Colombia and Peru.

Spores

Three main types of spores in *Dennstaedtia* are generally correlated with geography. The American species have tuberculate to verrucate spores as *D. obtusifolia* (Fig. 10) or the verrucate elements may be more or less fused into rugae as in *D. cicutaria* (Fig. 8). The spores are often somewhat compressed rather than spheriodal as in *D. obtusifolia* (Fig. 10) and in *D. bipinnata* (Fig. 11) which has an unusually irregular surface. Spore wall sections of *D. bipinnata* by Lugardon (1974) show a thin perispore formed above a compact exosore layer which is traversed by exceptionally large canals. The profile of the wall of *D. cicutaria* (Fig. 9) shows a thin outer perispore layer conform to the veruccate exospore. The paleotropical species have spores with two main types of surface—reticulate as in *D. Smithii* (Hook.) Moore (Fig. 12), or ridged as in *D. samoensis* (Brack.) Moore (Fig. 13). Temperate species as *D. punctilobula* have shallowly tuberculate spores with a slightly raised equatorial ridge.

Cytology

There appear to be two distinct cytological groups in *Dennstaedtia*. The American tropical species with $n = 46$ or 47, and 94 based on records from Jamaica and Trinidad (Walker, 1966, 1973) and Chiapas, Mexico (Smith and Mickel, 1977), represent one group. The record from the paleotropics of the *D. glabrata* complex with $n = 46$, reported from New Guinea by Walker (1973) may also belong here. The temperate species *D. punctilobula* with $n = 34$ represents a second group with lower chromosome numbers between 30 and 34 that appear to center in Asia. A scheme involving polyploid and aneuploid changes to accommodate the many chromosome numbers based on 15, 16 and 17 was proposed by Walker (1973). However, it was noted that further sampling of species from all parts of the range is required to clarify the cytological complexities of the genus.

Literature

Kunze, G. 1837. Analecta pteridographica. pp. 1–50. Leopold Voss, Lipsiae.

Lugardon, B. 1974. Reference under the family.

Smith, A. R., and J. Mickel. 1977. Reference under the family.

Tryon, R. 1960. A review of the genus *Dennstaedtia* in America. Contrib. Gray Herb. 187: 23–52.

Tryon, R. 1961. Taxonomic fern notes, I. Rhodora 63: 70–88.

Walker, T. G. 1966. Reference under the family.

Walker, T. G. 1973. Reference under the family.

56. *Saccoloma*
56.1–56.12

Saccoloma Kaulf., Berl. Jahrb. Pharm. 1820: 51. Type *Saccoloma elegans* Kaulf.

Neuropteris Desv., Mém. Linn. Soc. Paris 6: 292. 1827. Type: *Neuropteris elegans* Desv. = *Saccoloma elegans* Kaulf.

Orthiopteris Copel., Bishop Mus. Bull. 59: 14. 1929. Type: *Orthiopteris ferulacea* (Moore) Copel. (*Davallia ferulacea* Moore, Ind. Fil. 294. 1861, *nom. nov.* for *D. trichomanoides* Hook. 2nd Cent. *t. 64.* 1861, not Bl., 1828) = *Saccoloma ferulaceum* (Moore) R. & A. Tryon.

Ithycaulon Copel., Univ. Cal. Publ. Bot. 16: 79, 1929. Type: *Ithycaulon moluccanum* (Bl.) Copel. (*Davallia moluccana* Bl.) = *Saccoloma moluccanum* (Bl.) Kuhn.

Fig. 56.1. *Saccoloma inaequale,* Toro Negro, west of Barranquitas, Puerto Rico. (Photo D. S. Conant.)

Description

Terrestrial; stem erect to decumbent, moderately stout, bearing scales and many fibrous roots; leaves monomorphic, ca. 50 cm to 2.5 m long, borne in a cluster, lamina 1-pinnate, imparipinnate, to 5-pinnate, glabrous or nearly so, veins free; sori marginal or submarginal, 1-nerved, receptacle not to slightly elevated, not paraphysate, indusium half-conical, attached at its cuneate base and sides, or half cup-shaped, or purse-shaped and slightly bilabiate and more or less directed downward; spores tetrahedral-globose, trilete, the laesurae $\frac{2}{3}$ to $\frac{3}{4}$ the radius, the surface with generally parallel, dichotomous ridges usually partly obscured by spherical deposit. Chromosome number: $n = 188$; $2n = 63 \pm 2$ II, ca. 376.

Characteristic sori of *Saccoloma* are shown in Fig. 3, and those of the 1-pinnate *S. elegans* in Fig. 4. In the latter, simplification of the lamina to pinnae with entire margins brings the sori in close alignment. Stem scales with a basal peltate attachment have been described in *S. elegans* (Nair, 1979).

Systematics

Saccoloma is a tropical genus of about 10 species with three present in America. It is distinguished by its usually erect or sometimes decumbent stem bearing scales and the unusual delicately ridged spores. *Saccoloma elegans* is sometimes distinguished as a monotypic genus. However, except for the 1-pinnate lamina architecture it is similar to other species. A 1-pinnate lamina has also developed in other genera as in *Microlepia Hookeriana* (Hook.) Presl and *Dennstaedtia Wercklei.*

Ithycaulon has been considered a synonym of *Tapeinidium,* but the type specimen of *Davallia moluccana* Bl. is a *Saccoloma* (Kramer, 1967).

Fig. 56.2. Distribution of *Saccoloma* in America.

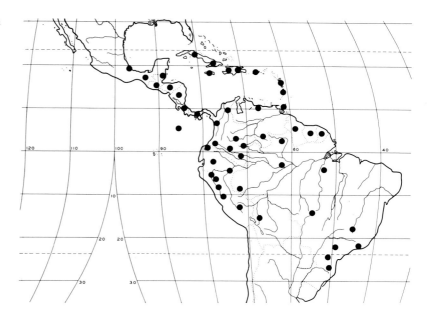

Tropical American Species

Saccoloma inaequale and *S. domingense* differ in chromosome number and evidently are valid species. They are also distinguished by differences in lamina architecture, sorus and indusium, but there are some specimens that appear intermediate. Some of these may represent hybrids, but others are from areas where only one species is known.

The following key has been adapted from the treatment of the genus by Tryon (1962).

Key to American Species of *Saccoloma*

a. Lamina 1-pinnate (Figs. 4, 5)   *S. elegans* Kaulf.
a. Lamina 2-pinnate or more complex.   b.
    b. Ultimate segments mostly large, entire to shallowly toothed, apices of lamina and of the larger pinnae pinnatifid, sori mostly close on adjacent veintips, indusium broadly cuneate (Fig. 6)   *S. domingense* (Spreng.) Prantl
    b. Ultimate segments mostly small, strongly toothed or lobed, apices of lamina and of the larger pinnae more or less pinnatisect, sori distant, indusium narrowly cuneate (Figs. 3, 7)   *S. inaequale* (Kze.) Mett.

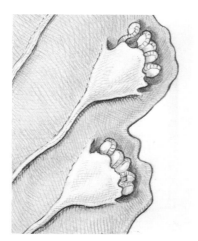

Fig. 56.3. Sori of *Saccoloma inaequale,* × 15.

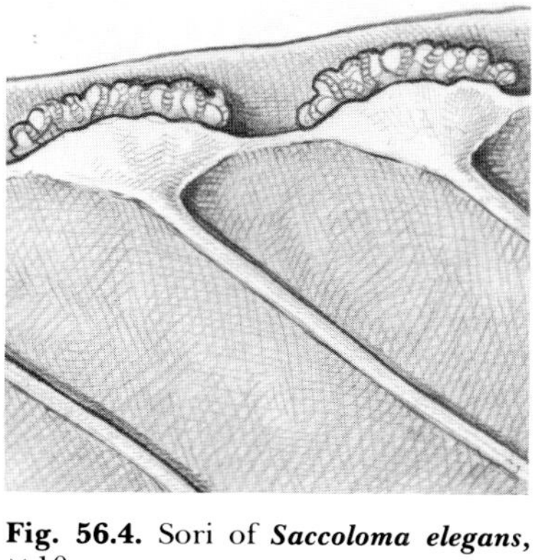

Ecology (Fig. 1)

Saccoloma is a forest genus, often growing in dense shade. In America it grows in rain forest and cloud forest on steep mountain slopes, in ravines, less often in secondary forest, or rarely in swamps or on rocks; the genus grows from ca. 50 to 2000 m altitude.

Geography (Fig. 2)

Saccoloma occurs in the American tropics, in Madagascar and in Malesia eastward to Samoa. In America it ranges from Veracruz in Mexico, through Central America, the West Indies, and in northern South America to Bolivia and Santa Catarina in Brazil; also on Cocos Island.

Fig. 56.4. Sori of *Saccoloma elegans,* × 10.

5   6   7

8   9

10   11   12

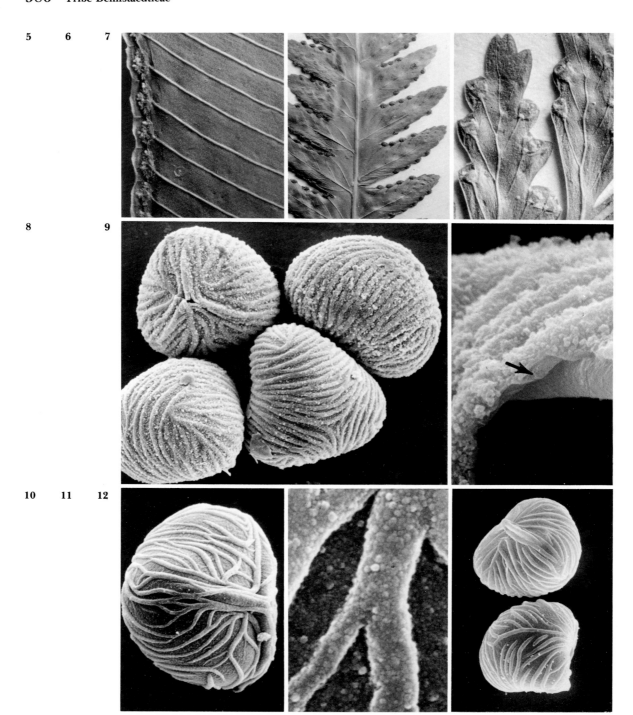

Figs. 56.5–56.12. Saccoloma. **5.** Portion of a pinna of **S. elegans,** × 5. **6.** Portion of a pinnule of **S. domingense,** × 1.5. **7.** Pinnules of **S. inaequale,** × 5. **8–12.** Spores. **8, 9. S. elegans,** Colombia, *Schultes & Idrobo 795.* **8.** Ridged spores with granulate deposit, proximal face upper left, distal face, lower right, lateral view, lower left and upper right, × 1500. **9.** Wall strata, the granulate perispore above, ridges formed by lower perispore (arrow) above the exospore, lower right, × 10,000. **10, 11. S. elegans,** British Honduras, *Schipp 90.* **10.** Lateral view, ridges with sparse granulate deposit, × 2000. **11.** Surface detail of ridges and granulate deposition, × 20,000. **12. S. moluccanum,** the proximal face, above, distal below, Solomon Islands, *Brass 2914,* × 1000.

Spores

The dichotomous ridges enveloping the spores of *Saccoloma* form an unusual type of surface that is consistent in American tropical species such as *S. elegans* (Fig. 8) and those of the paleotropics as *S. moluccanum* (Bl.) Kuhn (Fig. 12). The ridges are formed by the lower perispore, exposed as in Fig. 10, and are underlaid by a thick exospore (Fig. 9). An upper, dense granulate perispore deposit forms the surface (Figs. 9, 11). The wall is often peculiarly cracked or chipped suggesting it may be brittle. The thin, scarcely raised ridges differ from those of other ridged spores among the dennstaedtioid genera. *Saccoloma* spores resemble those of the paleotropical genus *Cystodium* of the Dicksoniaceae and suggest a possible link with that family.

Cytology

Reports of $n = 188$ and $2n = 376$ for *Saccoloma domingense,* (as *Orthiopteris*) from Jamaica were regarded by Walker (1966) as probably an octoploid level based on 47. Another high but discordant record of $2n = $ ca. 63 ± 2 II or ca. 63 for *S. inaequale* is reported from Chiapas, Mexico, by Smith and Mickel (1977). On the basis of this record, they question the octoploid level proposed for the Jamaican plants and the close relationship of *Saccoloma* to the dennstaedtioid ferns. The high chromosome numbers known in *Saccoloma* and the complex ridged spore architecture indicate that the genus represents a specialized element among the dennstaedtioids. It is perhaps best treated in this alliance until there is clear evidence for placing it elsewhere.

Literature

Kramer, K. U. 1967. A revision of *Tapeinidium.* Blumea 15: 545–556.
Nair, G. B. 1979. Peltate scales in *Saccoloma.* Fern Gaz. 12: 53–55.
Smith, A. R., and J. T. Mickel. 1977. Reference under the family.
Tryon, R. 1962. Taxonomic fern notes, III. Contrib. Gray Herb. 191: 91–107.
Walker, T. G. 1966. Reference under the family.

57. *Pteridium*
Figs. 57.1–57.12

Pteridium Scop., Fl. Carn. 169. 1760. Type: *Pteridium aquilinum* (L.) Kuhn (*Pteris aquilina* L.).

Cincinalis Gled., Syst. Pl. 290. 1764. Type: "*Cincinalis aquilina* Gled." (Scop. Fl. Carn. 2: 290. 1772) = *Pteridium aquilinum* (L.) Kuhn.

Eupteris Newm., Phytologist 2: 278. 1845. Type: *Eupteris aquilina* (L.) Newm. (*Pteris aquilina* L.) = *Pteridium aquilinum* (L.) Kuhn.

Ornithopteris (Ag.) J. Sm., Hist. Fil. 297. 1875, not Bernh. 1806 (= *Anemia*). *Pteris* section *Ornithopteris* Ag., Rec. Gen. Pterid. 45. 1839. Type: *Pteris aquilina* L. = *Pteridium aquilinum* (L.) Kuhn.

Fig. 57.1. *Pteridium aquilinum* var. *Feei,* north of Jalapa, Veracruz, Mexico. (Photo W. H. Hodge.)

Description

Terrestrial; stem long-creeping, rather slender, bearing trichomes and a few fibrous roots; leaves monomorphic, ca. 40 cm to 3 m, rarely to 7 m long, borne at intervals, lamina 2-pinnate-pinnatifid to 4-pinnate, pubescent to rarely glabrous, veins free; sori marginal, the sporangia on a vascular commissure between the apex and sinus of a segment, not paraphysate, the adaxial indusium formed by the recurved, modified margin, the abaxial indusium membranous, or hardly to not developed, the sterile margins modified similarly to the adaxial indusium; spores tetrahedral-globose, the distal face well rounded, the laesurae $\frac{2}{3}$ the radius, irregularly granulate with particles fused into masses, or more or less fused into short rods. Chromosome number: $n = 52$; $2n = 52, 208$.

Pteridium has several anatomical and morphological features that are unusual in ferns. There are true vessels in the stem, which rarely occur in pteridophytes. There are two types of stems: one is a long shoot that branches but does not bear leaves, the other is a short shoot that usually bears a single leaf during a

Fig. 57.2. Distribution of ***Pteridium aquilinum*** in America, south of lat. 35° N.

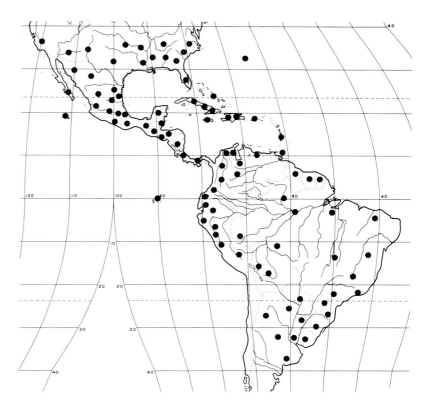

growing season. Glandular areas, often called nectaries, occur at the base of the pinnae and exude a substance attractive to ants. *Pteridium* is unique in having the sterile segments with a modified margin (Fig. 6) similar to that of the adaxial indusium of the fertile margins. This may be narrower or the same width as the adaxial indusium. The abaxial indusium is variously developed, is sometimes a continuous membrane, or is only partially developed as in Fig. 4, or it may be absent.

Systematics

Pteridium is a genus of one species, *P. aquilinum* (L.) Kuhn, of worldwide distribution except in regions that are extremely dry or cold. It is morphologically isolated, but its closest affinities are with the dennstaedtioid ferns. *Pteridium aquilinum* is a complex species consisting of 12 geographic varieties. Subspecies *aquilinum* has eight varieties and ssp. *caudatum* has four. While these have distinctive characters, they are difficult to define precisely because they are both phenotypically and genetically variable.

Seven varieties occur in the Old World: subspecies *aquilinum* var. *aquilinum* in Europe and Africa, var. *africanum* Bonap. in central Africa, var. *Wightianum* (Ag.) Tryon in southeastern Asia to Ceylon and New Guinea, var. *decompositum* (Gaud.) Tryon in the Hawaiian Islands, and var. *latiusculum* of America which extends from northern Europe across Asia to China and Kamchatka; in subspecies *caudatum* there are var. *esculentum* (Forst.) Kuhn, distributed from Australia to Tahiti, and var. *yarrabense* Domin from northern India to Australia.

Pteridium has been revised by Tryon (1941).

Fig. 57.3. General distribution of the tropical varieties of *Pteridium aquilinum* in America: solid area, var. *Feei;* dashed line, var. *arachnoideum;* hatched area, var. *caudatum.*

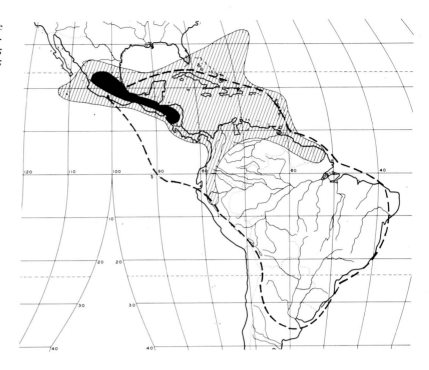

Key to Varieties of *Pteridium aquilinum* in America

a. Ultimate segments not decurrent at the base, or decurrent and equally surcurrent; subspecies *aquilinum.* b.
    b. Adaxial indusium and modified sterile margin ciliate and (or) pubescent on the adaxial surface. c.
        c. Modified sterile margin ca. 0.25 mm broad, the adaxial indusium broader on the same segment.   var. *pubescens* Underw.
        c. Modified sterile margin ca. 0.4 mm or more broad, the adaxial indusium at least as broad on the same segment. (Fig. 6)
                              var. *Feei* (Fée) Maxon
    b. Adaxial indusium and modified sterile margin glabrous. d.
        d. Ultimate segments moderately pubescent at the base of the indusium, the longest about four times as long as broad.
                          var. *latiusculum* (Desv.) Heller.
        d. Ultimate segments glabrous or subglabrous at the base of the indusium, the longest about nine times as long as broad.
                    var. *pseudocaudatum* (Clute) Heller.
a. Ultimate segments, at least some of them, more strongly decurrent than surcurrent at the base; subspecies *caudatum* (L.) Bonap. e.
    e. Free lobes present on the axes between the ultimate or penultimate segments. (Fig. 5)   var. *arachnoideum* (Kaulf.) Brade.
    e. Free lobes absent on the axes between the ultimate and penultimate segments. (Fig. 7)   var. *caudatum* (L) Sadeb.

Ecology (Fig. 1)

Pteridium grows in a variety of habitats as woods, forests and scrublands, in thickets and along open borders of woods, and especially in disturbed places as pastures, abandoned fields and burned or cut-over land. It usually grows in poor, more or less acid soils but also in fertile and sometimes calcareous soils.

In tropical America, var. *Feei* grows most commonly in oak and pine woods in mountaneous areas, often in calcareous soils, in rocky places, old pastures or clearings. The varieties *caudatum* and *arachnoideum* grow in open forests especially along the bor-

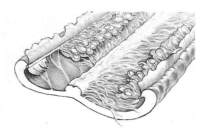

Fig. 57.4. Sorus and indusia of *Pteridium aquilinum* var. *caudatum* × 8. The adaxial indusium is bent to show the sorus, and the sporangia removed, at left, to show the partly developed abaxial indusium.

ders, in thickets, grasslands, and rarely in wetlands. Variety *caudatum* rarely grows around limestone sinks. *Pteridium* grows from sea level to ca. 3000 m.

The species can be a pernicious weed in tropical America, as in most parts of the world, for it invades lands cleared for agriculture or pasture and often develops into pure stands. Allelopathy appears to be involved in the promotion and maintenance of this dense growth as shown in the studies of Gliessman and Muller (1978).

The leaves of *Pteridium* are usually one to two meters tall but those growing in thickets or under small trees may be semiscandent on the branches and reach a length of 4 or rarely up to 7 m. The stem is wide creeping, it frequently branches, usually well below the soil surface, and it can be very long lived. Stems examined by Watt (1940) were 60 m long including the branches, but this is probably not the maximum for a fully developed living system. A study of Oinonen (1967a, 1967b) on bracken clones in Finland estimated the age of the oldest to be about 1500 years. Established plants may survive considerable change and a major environmental alteration is probably necessary to completely exterminate a colony. The capacity for survival of the plants at one locality over a long period reflects the broad ecological tolerance of the species.

There are few reports of *Pteridium* gametophytes and sporelings in nature but these stages appear to occur especially in areas that have been burned (Oininen, 1967a; Gliessman, 1978).

Geography (Figs 2, 3)

Pteridium grows in North and South America, Europe and Africa, and eastward across Asia, through Malesia and in the Pacific to Tahiti and the Hawaiian Islands; it extends northward to Scandinavia and southward to Tasmania and New Zealand.

In America it occurs from Newfoundland to southern Alaska, south to Buenos Aires in Argentina, and Uruguay (Fig. 2); also on the Revillagigedo Islands, and the Galápagos Islands.

The general distribution of the tropical American varieties is shown in Fig. 3. In addition to these, var. *pubescens* is in the western United States, north to southern Alaska, and south in the mountains of Mexico to Hidalgo; var. *latiusculum* of eastern North America and the Rocky Mountains occurs at Monterrey, Nuevo Leon, Mexico; and var. *pseudocaudatum* ranges from Massachusetts to southern Indiana and Illinois, south to Florida and Texas.

Five of the six American varieties occur in eastern Mexico, but this is probably not a center of evolution since all of the varieties are evidently quite old.

Spores

The granulate surface deposit is relatively uniform in spores of several varieties of *Pteridium* from widely distant areas (Figs. 8–12). The often partly eroded surface, as on the distal face in Fig.

8, exposes a compact stranded formation. The section in Fig. 10 shows that the exceptionally thin wall consisting of a somewhat thicker exospore below the granulate perispore surface. In this section of the wall the surface appears to be quite uniformly granulate. This is also shown at greater magnification in the TEM study of the wall by Lugardon (1974). The detail, in Fig. 12, shows the meshwork of rods or strands is formed below the granulate surface deposit.

Cytology

A meiotic chromosome number of 52 is consistently reported for plants of North America, Europe, Asia and New Zealand. However there is a record of $2n = 52$ for a plant from Spain by Löve and Kjellquist (1972) that was proposed as possibly an ancient element surviving in calcareous areas of southern Europe. The range of this cytotype and its age in relation to other elements in the species need to be substantiated. A plant from the Galápagos Islands with $2n = 208$ (Jarrett et al., 1968) was considered tetraploid but a second collection had the usual number 52. The base number of 26 established for *Pteridium* is an uncommon one in ferns, but this or its multiples are also known in *Paesia* and *Hypolepis*. A close relationship of these three genera has been proposed on the basis of the chromosome number, but marked differences in the spores do not confirm a close alliance.

Observations

There are more studies covering a diversity of topics on *Pteridium* than on any other fern. This is due in part to the wide distribution of the species and its abundance as well as unusual anatomical and morphological features. It is also the most economically important fern and there is considerable literature pertaining to its useful and detrimental qualities. The latter seem to predominate at present but formerly the fern was employed for many purposes as a source of potash for making glass and soap, for thatch, swine food, bedding material for animals and man, and as packing material for fruits. It is still used in rural regions of Venezuela for packing and especially for wrapping heads of curing cheese.

A major undesirable aspect of bracken is its poisonous effect when consumed as fodder by animals or when eaten by man. The young leaves contain a nerve poison and are also carcinogenic (Evans, I. A., 1976; Evans, W. C., 1976). In regions where large quantities of bracken croziers are consumed, as in Japan (Hodge, 1973), they may be implicated in the high incidence of stomach cancer. Bracken is regarded as a pernicious weed in some areas where it takes over land intended for cultivation or grazing. It often forms dense colonies over large areas excluding all or most other plants. Eradication of bracken has met with only limited success because of the high cost of effective control measures on the poorer land that it usually invades.

Two major publications on bracken, including about 1000 references covering most of the extensive work, are by Braid (1959)

5   **6**   **7**

8   **9**

10   **11**   **12**

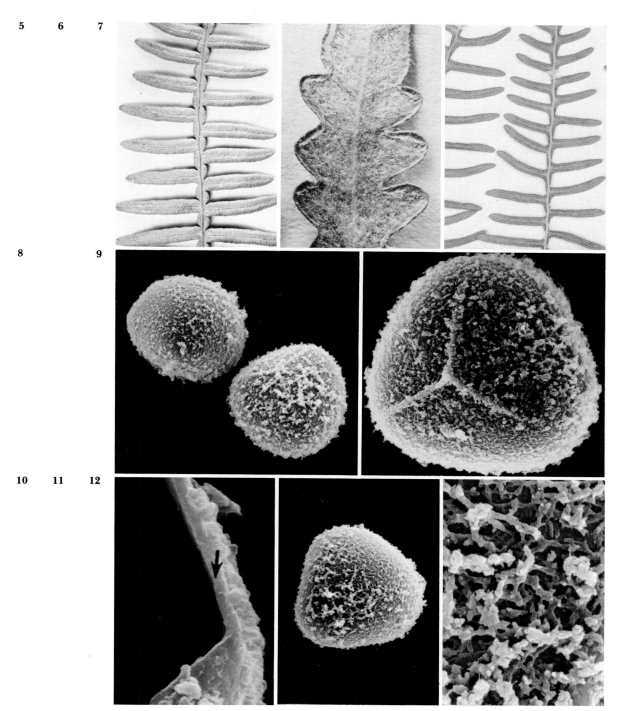

Fig. 57.5–57.12. *Pteridium aquilinum.*   **5.** Portion of a pinnule of var. ***arachnoideum,*** × 1.5.   **6.** Sterile pinnule of var. ***Feei*** with modified margin, × 5.   **7.** Portion of pinnules of var. ***caudatum,*** × 1.5.   **8–12.** Spores.   **8.** var. ***arachnoideum,*** the distal face above with surface partially eroded showing lower reticulate stratum, Colombia, *Little 8781,* × 1000.   **9.** var. ***aquilinum,*** granulate proximal face, Uganda, *Loverage 473,* × 2000.   **10.** var. ***arachnoideum,*** wall section, the exospore (at arrow) below granulate perispore deposit, Brazil, *Irwin & Soderstrom 5301,* × 10,000.   **11.** var. ***caudatum,*** distal face, Dominican Republic, *Fuertes 1053,* × 1000.   **12.** var. ***esculentum,*** detail of more or less granulate deposit on proximal face, New Zealand, *Hunnewell 13360,* × 6000.

and Perring and Gardner (1976). Some special areas of research on the species include genetics, by Wilkie (1956), Klekowski (1973) and Chapman et al. (1979), and flavonoid chemistry by Cooper-Driver (1976). Studies of ecdysones by Kaplanis et al. (1967) have found two major insect molting compounds in bracken that may explain the general resistence of the plants to insect predation. Some basic studies on the development and growth of *Pteridium* are by Webster and Steves (1958), O'Brien (1963) and Watt (1940).

Literature

Braid, K. W. 1959. Bracken: A review of the literature. Commonwealth Bureau Pastures Field Crops. Mimeogr. Publ. No. 3/1959, pp. 1–69.

Cooper-Driver, G. 1976. Chemotaxonomy and phytochemical ecology of bracken. Bot. Jour. Linn. Soc. 73: 35–46.

Chapman, R. H., E. J. Klekowski, and R. K. Selander. 1979. Homoeologous heterozygosity and recombination in the fern *Pteridium aquilinum*. Science 204: 1207–1209.

Evans, I. A. 1976. Relationship between bracken and cancer. Bot. Jour. Linn. Soc. 73: 105–112.

Evans, W. C. 1976. Bracken thiaminase-mediated neurotoxic syndromes. Bot. Jour. Linn. Soc. 73: 113–131.

Gliessman, S. R. 1978. The establishment of bracken following fire in tropical habitats. Amer. Fern. Jour. 68: 41–44.

Gliessman, S. R., and C. H. Muller. 1978. The allelopathic mechanisms of dominance in bracken (*Pteridium aquilinum*) in southern California. Jour. Chem. Ecol. 4: 337–362.

Hodge, W. H. 1973. Fern foods of Japan and the problems of toxicity. Amer. Fern Jour. 63: 77–80.

Jarrett, F. M., I. Manton, and S. K. Roy. 1968. Cytological and taxonomic notes on a small collection of living ferns from Galápagos. Kew Bull. 22: 475–480.

Kaplanis, J. N., M. J. Thompson, W. E. Robbins, and B. M. Bryce. 1967. Insect hormones: Alpha ecdysone and 20-hydroxyecdysone in bracken fern. Science 157: 1436–1438.

Klekowski, E. J. 1973. Genetic endemism of Galápagos *Pteridium*. Bot. Jour. Linn. Soc. 66: 181–188.

Löve, A., and E. Kjellquist. 1972. Cytotaxonomy of Spanish plants. I. Lagascalia 2: 23–25.

Lugardon, B. 1974. Reference under the family.

O'Brien, T. P. 1963. The morphology and growth of *Pteridium aquilinum* var. *esculentum* (Forst.) Kuhn. Ann. Bot. n.s. 27: 253–267.

Oinonen, E. 1967a. Sporal regeneration of bracken (*Pteridium aquilinum* (L.) Kuhn) in Finland in the light of the dimensions and age of its clones. Acta Forest. Fennica 83 (1): 1–96.

Oinonen, E. 1967b. The correlation between the size of Finnish bracken (*Pteridium aquilinum* (L.) Kuhn) clones and certain periods of its site history. Acta Forest. Fennica 83 (2): 1–51.

Perring, F. H., and B. G. Gardiner (eds.). 1976. The biology of bracken. Bot. Jour. Linn. Soc. 73: 1–302.

Tryon, R. M. 1941. A revision of the genus *Pteridium*. Rhodora 43: 1–31, 37–67. (Contrib. Gray Herb. 134.)

Watt, A. S. 1940. Contributions to the ecology of bracken (*Pteridium aquilinum*), I. The rhizome. New Phytol. 39: 401–422.

Webster, B. D., and T. A. Stevens. 1958. Morphogenesis in *Pteridium aquilinum* (L.) Kuhn-general morphology and growth habit. Phytomorph. 8: 30–41.

Wilkie, D. 1956. Incompatibility in bracken. Heredity 10: 247–256.

58. *Paesia*

Figs. 58.1–58.10

Paesia St.-Hil., Voy. Distr. Diamans 1: 381. 1833. Type: *Paesia viscosa* St.-Hil. = *Paesia glandulosa* (Sw.) Kuhn.

Description
Terrestrial, rarely rupestral; stem long-creeping, rather slender, bearing trichomes and few to many fibrous roots; leaves monomorphic, ca. 20 cm to 2.5 m long, borne at intervals, lamina to 4-pinnate-pinnatifid, glabrate, glandular-pubescent or pubescent, veins free; sori marginal, the sporangia on a short to long vascular commissure, not paraphysate, the adaxial indusium formed by the modified, recurved margin, the abaxial indusium firmly membranous; spores somewhat ellipsoidal, monolete, the laesura ca. $\frac{3}{4}$ the spore length, usually in a furrow formed by adjacent rugae, the surface coarsely rugose. Chromosome number: $n = 26, 104$.

Paesia is characterized by a well-developed abaxial indusium as well as a marginal indusium (Fig. 3). The abaxial one is invariably present and of a firm, membraneous texture, contrasting with the structure in *Pteridium* that is often poorly developed or absent. The rachis, especially in large leaves, is usually somewhat fractiflex and appears to be adapted for a scandent or scrambling habit. The segment architecture is shown in Figs. 4 and 5.

Systematics
Paesia is a predominantly tropical genus of about 12 species, two of them in tropical America. It has been related to *Pteridium* on the basis of characters such as the vascular commissure, abaxial indusium and the chromosome number based on $n = 26$. The

Fig. 58.1. *Paesia anfractuosa,* south of Cartago, Costa Rica. (Photo W. H. Hodge.)

Fig. 58.2. Distribution of *Paesia* in America.

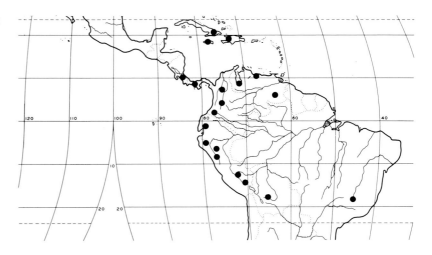

rugose, monolete spores suggest that the alliance may not be close.

Hypolepsis bivalvis vAvR. with an abaxial indusium, is correctly placed as *Paesia bivalvis* (vAvR.) R. & A. Tryon.

Tropical American Species

The marked variability of *Paesia glandulosa* has led to its recognition under several names which are based on local variants not sufficiently constant for taxonomic recognition. The name *Paesia viscosa* has been commonly used for this species, but examination of the type material of *Cheilanthes glandulosa* Sw. at Stockholm showed this to be an earlier name, contrary to Tryon (1964).

Key to American Species of *Paesia*

a. Basal segment of the pinnae and secondary segments on the acroscopic side.
P. glandulosa (Sw.) Kuhn
a. Basal segment of the pinnae and secondary segments on the basiscopic side.
P. anfractuosa (Christ) C. Chr.

Ecology (Fig. 1)

Paesia is a genus of montane forests and occurs in natural forest openings. In America, *Paesia* most often grows in cloud forests or wet, montane forests, usually in shrubby places, at the edge of woods, and in rocky woods; occasionally it invades road banks in forest zones. It occurs at altitudes between 1400 and 3600 m. *Paesia* may form large colonies and probably has a long, creeping and branching stem; however, data are not available on the stem system. The leaves are semiscandent, on and among branches when plants grow among shrubs and small trees.

Geography (Fig. 2)

Paesia is present in the American tropics and in the Old World tropics from Malesia to Tahiti, and south to New Zealand.

In America the genus occurs in Costa Rica and Panama, Cuba, Jamaica, Hispaniola, Venezuela to Colombia, south to Bolivia,

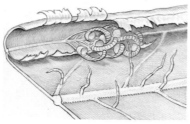

Fig. 58.3. Sorus of *Paesia glandulosa*, showing the adaxial and abaxial indusia, × 20.

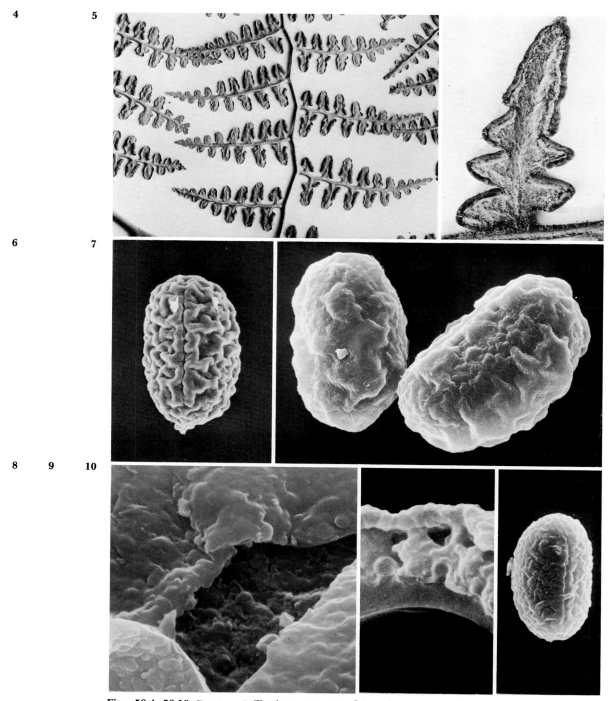

Figs. 58.4–58.10. *Paesia.* **4.** Tertiary segments of *P. anfractuosa,* × 1.5. **5.** Abaxial surface of segment of *P. glandulosa,* × 5. **6–10.** Spores, × 1000. **6–9.** *P. glandulosa.* **6.** Proximal face with rugae obscuring the central laesura, Colombia, *Daniel 660.* **7–9.** Peru, *Pennell 14024.* **7.** Coarsely rugose, laesura across top of the spore, right. **8.** Part of abraded perispore and the lower papillate exospore, × 10,000. **9.** Wall profile with thick, ridged perispore above a denser, more or less papillate exospore, × 10,000. **10.** *P. scaberula,* rugose proximal face the rugae obscuring the central laesura, New Zealand, *Craig 157.*

and disjunct in Minas Gerais in Brazil; this is also the range of *Paesia glandulosa*. The other American species, *P. anfractuosa* is endemic to Costa Rica.

Spores

The spores of *Paesia* are consistently rugose but vary in density and prominence of the rugae (Figs. 6–10). There are marked size differences between specimens of *Paesia glandulosa* from Colombia (Fig. 6) and those from Peru (Fig. 7). The species is reported as octoploid in Jamaica thus the larger spored material from Peru may represent a higher ploidy level. The abraded surface (Fig. 8) and section of the wall (Fig. 9) show that the thick outer rugose perispore overlays a more or less papillate exospore. Spores of *P. scaberula* (A. Rich.) Kuhn from New Zealand (Fig. 10) are rugose but smaller than those of the American tropics. The alliance of *Paesia, Pteridium* and *Hypolepis* is proposed on the basis of similar chromosome numbers. However, marked differences in shape and surface of the spores suggest the relationships may not be close.

Cytology

The record of $n = 26$ in *Paesia anfractuosa* from Costa Rica (Gómez Pignataro, 1971) is consistent with that of *P. scaberula* from New Zealand (Brownlie, 1957). The report of $n = 104$ for *P. glandulosa,* (as *P. viscosa*) from Jamaica was regarded as an octoploid by Walker (1966). Differences in spore size of specimens from distant portions of the range may reflect the changes in ploidy levels within the species.

Literature

Brownlie, G. 1957. Cytotaxonomic studies on New Zealand Pteridaceae. New Phytol. 57: 207–209.

Gómez Pignataro, L. 1971. Ricerche citologiche sulle Pteridofite della Costa Rica 1. Atti. Ist. Bot. Univ. Pavia 7: 29–31.

Tryon, R. 1964. The ferns of Peru. Polypodiaceae (Dennstaedtieae to Oleandreae). Contrib. Gray Herb. 194: 1–253.

Walker, T. G. 1966. Reference under the family.

59. *Hypolepis*
Figs. 59.1–59.12

Hypolepis Bernh., Neues Jour. Bot. (Schrad.) 1 (2) 34. 1806. Type: *Hypolepis tenuifolia* (Forst.) Presl (*Lonchitis tenuifolia* Forst.).

Description   Terrestrial; stem long-creeping, rather slender, bearing trichomes and long, fibrous roots; leaves monomorphic, ca. 10 cm to 3 m, rarely to 7 m long, borne at intervals, lamina 2-pinnate-pinnatifid to 4-pinnate-pinnatifid, usually pubescent, or nearly glabrous, veins free; sori submarginal, 1-nerved, usually borne in a sinus, not paraphysate, indusium slightly to strongly dif-

Fig. 59.1. *Hypolepis repens,* north of Jalapa, Veracruz, Mexico. (Photo W. H. Hodge.)

ferentiated from a recurved marginal lobe, or lacking and the lobe flattened and unmodified; spores somewhat ellipsoidal, monolete, the laesura $\frac{2}{3}$ the length, the surface echinate, often with more or less reticulate strands connecting the echinate elements. Chromosome number: n = 29, 39, 51–53, 52, ca. 92, 98, ca. 100, 104; $2n$ = ca. 92, 104, ca. 150, 208.

Hypolepis has a marginal or submarginal sorus (Fig. 4) usually covered by a modified, recurved, marginal indusium (Figs. 3). The lamina is usually 3-pinnate or more complex (Figs. 1, 5), but the smallest species in the genus, *H. obtusata,* may be only 2-pinnate-pinnatifid at the base and 2-pinnate above (Fig. 6). The margin is unmodified in some species, as *H. rugulosa* and *H. pulcherrima.*

Systematics   *Hypolepis* is a tropical and extratropical genus of perhaps 40 species with about 15 in America. Relationships of *Hypolepis* with other dennstaedtioid ferns are not close. There are similarities in chromosome numbers to *Paesia* and *Pteridium,* and to *Blotiella* in the spores, but other characters do not support a close alliance with these genera. Species lacking an indusium, as *H. pulcherrima* suggest a possible distant relationship to the taenitioid ferns of the Pteridaceae.

Fig. 59.2 Distribution of *Hypolepis* in America.

Tropical American Species

Over 20 species have been recognized in America but there are probably fewer; a modern revision of these and of the paleotropical species is needed. Some of the American species are: *Hypolepis bogotensis* Karst., *H. crassa* Maxon, *H. hostilis* (Kze.) Presl, *H. melanochlaena* A. R. Smith, *H. nigrescens* Hook., *H. obtusata* (Presl) Hieron., *H. parallelogramma* (Kze.) Presl, *H. pulcherrima* Maxon, *H. repens* (L.) Presl, *H. rugulosa* (Labill.) J. Sm., *H. Stuebelii* Hieron., and *H. viscosa* Karst.

Ecology (Fig. 1)

Hypolepis usually grows in forests or along their borders, in clearings and thickets, sometimes in moist to wet, open habitats.

In America it grows in rain or montane forests, páramillo thickets, brushy places, forest borders or clearings, in fields, open pastures, and less often in open, rocky places or on cliffs. A few species in the Andes may extend into the páramo.

Hypolepis usually grows from sea level to 2500 m; with *H. crassa* at altitudes to 3900 m and *H. obtusata* to 4200 m.

Hypolepis may form dense colonies in shrubby vegetation and among small trees. The leaves may be long and subscandent and the stem may be very long and frequently branched. The excavated stem system of a plant in *Hypolepis repens* studied by Gruber (1981) is about 22 m in length, although incomplete (Fig. 12). The extensive development and numerous apices show the potential for growth of these plants.

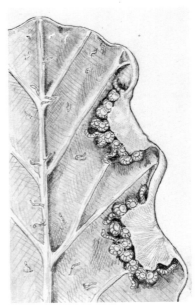

Fig. 59.3. Sorus of *Hypolepis parallelogramma*, ×15.

Figs. 59.4–59.11. ***Hypolepis.*** **4.** Fertile segments of ***H. parallelogramma,*** ×5.   **5.** Pinnules of ***H. repens,*** × 1. **6. Segments of** *H. obtusata* (central portion of lamina), × 1. **7–11.** Spores, × 1000, surface detail, ×10,000.   **7–8.** ***H. crassa,*** reticulate, Ecuador, *Øllgaard & Balslev 8155.*   **7.** Echinate-reticulate surface, the laesura at top of upper spore.   **8.** Detail of particulate strands connecting echinate elements.   **9.** ***H. hostilis,*** with dense echinate-reticlate elements, Costa Rica, *Berger & Liesner 7038.*   **10.** *H. nigrescens,* detail of echinate surface, Jamaica, *Maxon & Killip 1139.*   **11.** *H. bogotensis,* echinate-reticulate elements and papillate surface, part of laesura at base, Colombia, *Pennell & Hazen 10068.*

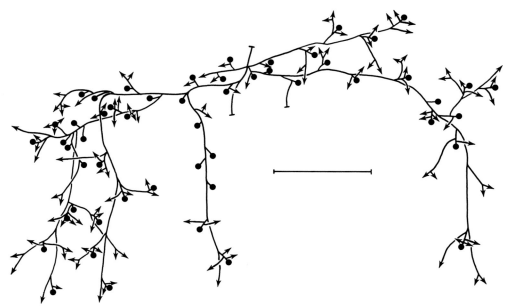

Fig. 59.12. Stem system of *Hypolepis repens,* mapped in situ in a cloud forest north of Jalapa, Mexico, (Gruber, 1981). Solid dots indicate the position of leaves, arrows the stem apices, and short bars the decayed stem ends; the scale is shown by a 1 m bar.

Geography (Fig. 2)

Hypolepis grows in tropical America and temperate South America, in Africa eastward to southeastern Asia and Malesia, east through the Pacific to Pitcairn Island and the Hawaiian Islands, south to Tasmania, New Zealand and Auckland Island.

In America *Hypolepis* grows from Puebla and Veracruz in Mexico to Panama, southern Florida, the West Indies, and in South America from Surinam, to Colombia, south to southern Chile, northeastern Argentina and southeastern Brazil; also on Cocos Island, the Galápagos Islands and the Juan Fernandez Islands.

The greatest diversity of species is in the Andes from Colombia to Peru. There are several species in Central America and in the Greater Antilles, and fewer elsewhere. *Hypolepis rugulosa* occurs in Chile, the Juan Fernandez Islands, and also in Australia and New Zealand.

Spores

Hypolepis spores are basically echinate-reticulate, with slender strands forming a reticulum incorporating the echinate formation as in *H. crassa* (Figs. 7, 8) and *H. bogotensis* (Fig. 11). The echinate elements may be dense as in *H. hostilis* (Fig. 9). Spores of *H. nigrescens* lack the reticulate strands (Fig. 10) and the species is also distinguished by its divergent chromosome number. *Hypolepis* spores have one of the most complex types of surface structure among the dennstaedtioid ferns, suggesting that the genus may represent a derived element in the group.

Cytology

The large range of chromosome numbers reported for *Hypolepis* indicates complex cytological changes, although several species have meiotic numbers of 52 or 104. Reports of putative hybrids involving *H. viscosa* from Oaxaca, Mexico, and *H. repens* from Costa Rica are based upon study of chromosome pairing in triploid plants (Smith and Mickel, 1977). The record of $n = 29$ in *H. nigrescens* from Jamaica (Walker, 1966), and from Oaxaca, Mexico (Smith and Mickel, 1977), is a divergent number for the genus. However, Walker (1973) suggests that the basic numbers 26 and 29 may represent a short aneuploid series.

Literature

Gruber, T. M. 1981. The branching pattern of *Hypolepis repens*. Amer. Fern Jour. 71: 41–47.

Smith, A. R., and J. Mickel. 1977. Reference under the family.

Walker, T. G. 1966. Reference under the family.

Walker, T. G. 1973. Reference under the family.

60. *Blotiella*

Figs. 60.1–60.11

Blotiella Tryon, Contrib. Gray Herb. 191: 96. 1962. Type: *Blotiella glabra* (Bory) Tryon (*Lonchitis glabra* Bory).

Description

Terrestrial; stem erect, to 4.5 m tall, to decumbent, stout, bearing trichomes and many long, fibrous roots; leaves monomorphic, ca. 50 cm to 6 m long, borne in a crown or cluster, lamina 1-pinnate-lobed to 2-pinnate-pinnatifid, more or less pubescent, veins partially to usually wholly anastomosing, without included free veinlets; sori marginal, continuous around base of narrow sinuses, sporangia borne on a vascular commissure, paraphysate, indusium well differentiated from the recurved margin; spores somewhat ellipsoidal, monolete, the laesura $\frac{1}{2}$ to $\frac{2}{3}$ the spore length, the surface coarsely echinate, sometimes with connecting reticulate strands, or granulate. Chromosome number: $n = 38, 76$.

The lamina indument is somewhat variable but is usually dense as in Fig. 4. It may be sparse and, especially with age, the lamina may be glabrate. Reduced (stipular) pinnules may occur at the base of the lower pinnae as in *Histiopteris*. The marginal sori are continuous around the sinuses of the ultimate segments as in Fig. 3 and the veins are anastomosing as in Figs. 4 and 5.

Systematics

Blotiella is a genus of about 15 species, one in America and the others in Africa and Madagascar. The American *Blotiella Lindeniana* (Hook.) Tryon is variable especially in the color, density and distribution of the lamina indument. However, there seems to be no basis for the recognition of more than one taxon. The African and Madagascar species require additional study be-

Fig. 60.1. *Blotiella Lindeniana.* Tapanti, Cartago, Costa Rica, pinnae of a scandent leaf. (Photo Alice F. Tryon.)

Fig. 60.2. Distribution of **Blotiella** in America.

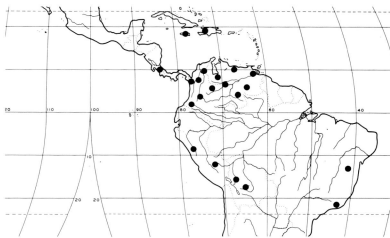

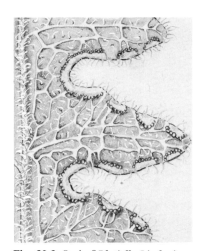

Fig. 60.3. Sori of **Blotiella Lindeniana,** ×2.

cause of variability of the lamina architecture and precocious fertility.

Although they are very distinctive, the genera *Blotiella, Lonchitis,* and *Pteris* have been confused (Copeland, 1947) primarily because of similarities of the sorus and indusium. *Blotiella* is a new name for some species previously called *Lonchitis* (Kümmerle, 1915). The type of *Lonchitis* has erroneously (Proctor, 1977) been considered to be *Lonchitis aurita* L. It is actually *Lonchitis hirsuta* L. which belongs in the genus often called *Anisosorus* (Lellinger, 1977), while *Lonchitis aurita* is a synonym of *Pteris arborea* L.

Ecology and Geography (Figs. 1, 2)

Blotiella grows in wet forest, in ravines and along streams. In America *B. Lindeniana* grows in cloud forests, wet, montane forests (Fig. 1) and sometimes in more open, wet sites. The leaf is

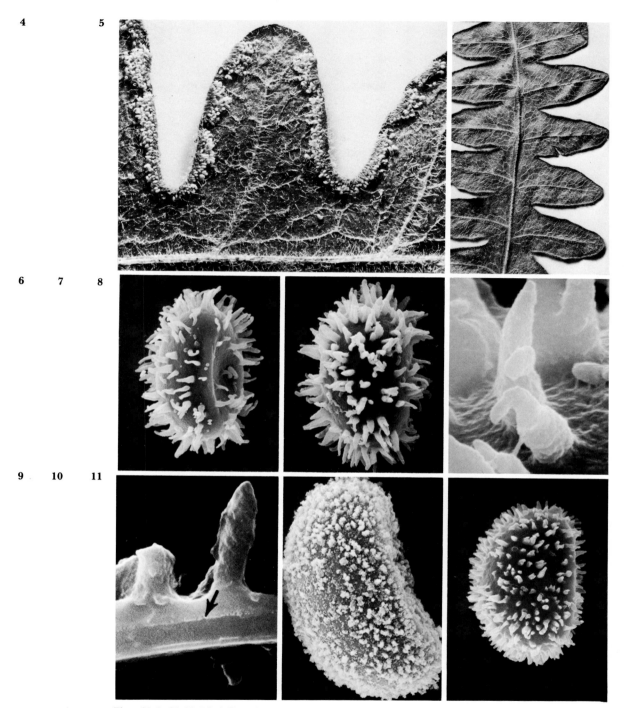

Figs. 60.4–60.11. *Blotiella.* **4.** *B. Lindeniana* sori with lobed, marginal indusia, abaxial leaf surface pubescent, × 5.0. **5.** *B. Lindeniana* areolate venation of sterile segment, × 1.5. **6–11.** Spores, × 1000. **6 –8.** *B. Lindeniana,* Brazil, *Brade 18285.* **6.** Echinate proximal face with laesura. **7.** Distal face. **8.** Surface detail with strand connecting echinate elements, × 10,000. **9.** *B. Lindeniana,* wall profile, the coarse echinate elements formed by upper perispore, lower perispore at arrow, above the exospore. Costa Rica, *Burger & Stolze 5698,* × 10,000. **10.** *B. glabra,* granulate spore, laesura, right, Tanganyika, *Richards 6320.* **11.** *B. glabra,* echinate spore, laesura, right, Swaziland, *Schelpe 6320* (US).

sometimes scandent on shrubs and small trees. *Blotiella* ranges in altitude from 1500 to 2300 m.

The genus is in tropical America, tropical and subtropical Africa and in Madagascar and the Mascarenes. In America it occurs uncommonly in Costa Rica, Jamaica and Hispaniola, and in South America from Venezuela and Colombia south to Bolivia and eastern Brazil.

Spores

Spores of the American *Blotiella Lindeniana* are coarsely echinate (Figs. 6, 7) often with slender strands connecting the echinate elements (Fig. 8). The wall profile of this species (Fig. 9) shows the echinate structure is formed by an outer perispore, and below this a slightly rougher, thin granulate stratum overlays a thick exospore. Spores of the African species vary from granulate to echinate as in *B. glabra* (Bory) Tryon (Figs. 10, 11). The echinate material appears to be organized from the less structured granulate form. Variance in size of these spores indicates possible differences in ploidy levels. The echinate-reticulate spores of *B. Lindeniana* are similar to those of *Hypolepis* and are considered to represent a specialized, derived form among the dennstaedtioid ferns.

Cytology

Chromosome numbers are not reported for the American *Blotiella Lindeniana* but two diploids and a tetraploid based on $n = 38$ are known for species in Africa. This number is quite different from that of $n = 100$ reported for *Lonchitis* from Jamaica and supports recognition of the two as distinct genera.

Literature

Copeland, E. B. 1947. Genera Filicum. 247 pp. Chronica Botanica, Waltham, Mass.

Kümmerle, J. B. 1915. Monographie generis Lonchitidis prodromus. Botan. Közlem. 14: 166–188.

Lellinger, D. B. 1977. The identity of *Lonchitis aurita* and the generic names *Anisosorus* and *Lonchitis*. Taxon 26: 578–580.

Proctor, G. R. 1977. Pteridophyta, *in:* R. A. Howard, Flora of the Lesser Antilles. 2. 414 pp. Arnold Arboretum, Harvard University, Cambridge, Mass.

61. *Lonchitis*
Figs. 61.1–61.9

Lonchitis L., Sp. Pl. 1078. 1753; Gen. Pl. ed. 5, 485. 1754. Type: *Lonchitis hirsuta* L. (Chosen by Brongn., Dict. Class. Hist. Nat. (Bory) 9: 490. 1826).

Anisosorus Maxon, Sci. Surv. Porto Rico & Virg. Isls. 6: 429. 1926, *nom. superfl.* for *Lonchitis* L. and with the same type.

Fig. 61.1. *Lonchitis hirsuta,* a small plant, Spring Garden, Jamaica. (Photo W. H. Hodge.)

Description

Terrestrial or sometimes rupestral; stem more or less short-creeping, moderately stout, bearing trichomes and many long, fibrous roots; leaves monomorphic, ca. 40 cm to 2.5 m long, rather closely spaced, lamina 2-pinnate-pinnatifid to 3-pinnate-pinnatifid, more or less pubescent, veins free or partly anastomosing without included free veinlets; sori marginal, sporangia on a vascular commissure between the apex and sinus of segment or sometimes at the base of the sinus, sometimes paraphysate, the indusium well differentiated from the recurved margin; spores spheroidal, trilete with broad, raised laesurae $\frac{3}{4}$ the radius with diffuse, spherical material irregularly deposited on a slightly granulate surface. Chromosome number $n = 50, 100$.

The stem of *Lonchitis* is crisp and fleshy, and bears succulent leaves in two ranks. It is often reported as bearing scales but these are true trichomes that may be flattened and laterally coherent appearing as narrow scales. The sorus and modified marginal indusium are shown in Figs. 3–5, and the scattered lamina pubescence on the abaxial surface in Fig. 5.

Systematics

The two species of *Lonchitis* are geographically distinct with *L. hirsuta* L. in America and *L. occidentalis* Baker in tropical Africa and Madagascar. *Lonchitis* has often been allied to *Pteris,* and Copeland (1947) placed it in that genus. Several characters including the spores and chromosome numbers, clearly relate it to

Fig. 61.2. Distribution of *Lonchitis* in America.

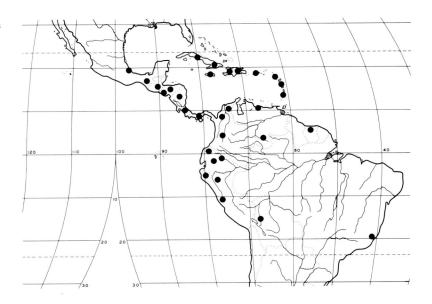

the dennstaedtioid rather than the pterioid ferns. The name *Lonchitis* has usually been applied to the previous genus, *Blotiella,* while the genus *Lonchitis* has usually been called *Anisosorus.*

The typification of *Lonchitis* was conclusively established by Lellinger (1977).

Ecology (Fig. 1)

Lonchitis grows in wet forests, especially in ravines, on stream banks and near waterfalls, it grows less often on wet rocks or ledges of wet cliffs. In America it occurs between about 100 to 1800 m.

Geography (Fig. 2)

Both of the species are wide ranging, one in tropical America and the other in tropical Africa and Madagascar.

In America *Lonchitis* occurs from Veracruz in Mexico, south to Panama; in the Greater Antilles and many of the Lesser Antilles; in South America from Surinam to Colombia, south to Bolivia and also disjunct in Rio de Janeiro, Brazil.

Spores

Lonchitis spores are spheroidal with a relatively simple granulate surface as in Figs. 6–8. Slender strands form a basal meshwork (Fig. 9) on which a granulate deposit is coalesced into a thick layer forming the main part of the wall, and coarse, irregular spheres are formed on the surface as in Fig. 7. Size differences are evident in spores from distant geographic areas. Those from Veracruz, Mexico, (Fig. 6) are smaller than spores from Ecuador (Fig. 8), possibly reflecting different ploidy levels within the species. These trilete spores are unlike those of other genera of the dennstaedtioids, especially the complex, monolete, echinate spores of *Blotiella,* a genus formerly associated with *Lonchitis.*

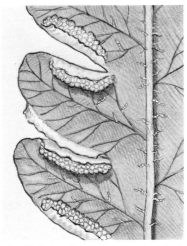

Fig. 61.3. Fertile segments of *Lonchitis hirsuta* with free veins, sori and marginal indusia, × 4.

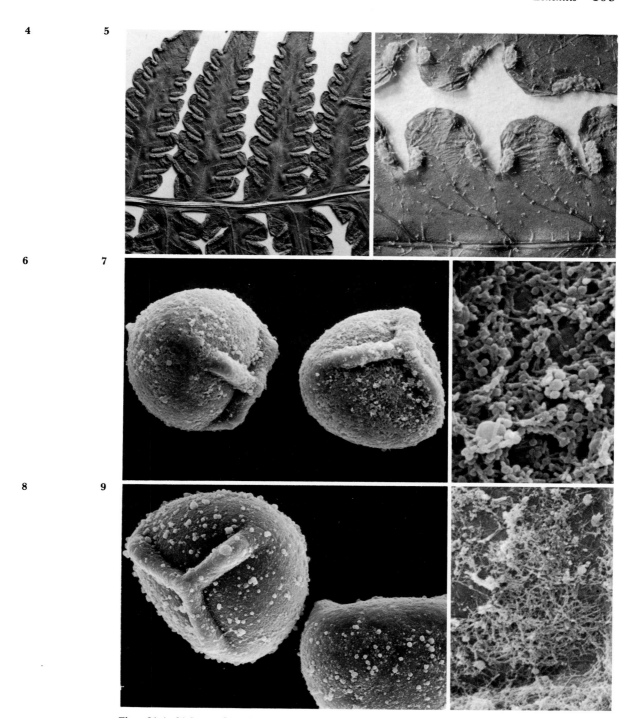

Figs. 61.4–61.9. *Lonchitis hirsuta.* **4.** Portion of fertile pinna with marginal sori, × 1.5. **5.** Sori and trichomes on abaxial surface of fertile segments, × 5.0. **6–9.** *Lonchitis hirsuta,* spores, × 1000. **6–7.** Mexico, *Purpus 2928.* **6.** Granulate spores with prominent laesurae. **7.** Detail of spherical surface deposit, × 10,000. **8.** Proximal face left, and part of distal face right, Ecuador, *Holm-Nielson 547.* **9.** Detail of slender strands below the spherical surface deposit, Guatemala, *Skutch 1861,* × 5000.

The prominent, raised laesurae and spherical surface deposit most closely resemble that of *Odontosoria* spores. These similarities provide a connection between the tribes Dennstaedtieae and Lindseeae that otherwise have discrete kinds of spores.

Cytology

The report of $n = 100$ for *Lonchitis* from Jamaica was considered a tetraploid by Walker (1966). The record of $n =$ ca. 50 for *L. occidentalis* (as *Anisosorus occidentalis*) from Ghana by Manton (1958) is consistent with this. The distinction between *Lonchitis* and *Blotiella* is reinforced by the meiotic chromosome reports of 38 and 76 in African species of *Blotiella*.

Literature

Copeland, E. B. 1947. General Filicum. 247 pp. Chronica Botanica. Waltham, Mass.

Lellinger, D. B. 1977. The identity of *Lonchitis aurita* and the generic names *Anisosorus* and *Lonchitis*. Taxon 26: 578–580.

Manton, I. 1958. Chromosomes and fern phylogeny with special reference to "Pteridaceae." Jour. Linn. Soc. Bot. 56: 73–92.

Walker, T. G. 1966. Reference under the family.

62. *Histiopteris*
Figs. 62.1–62.8

Histiopteris (Ag.) J. Sm., Hist. Fil. 294. 1875. *Pteris* section *Histiopteris* Ag., Rec. Gen. Pterid. 76. 1839. Type: *Pteris vespertilionis* Labill. (as *Histiopteris vespertilionis* (Labill.) J. Sm.) = *Histiopteris incisa* (Thunb.) J. Sm.

Lepidocaulon Copel., Univ. Cal. Publ. Bot. 8: 218. 1942. Type: *Lepidocaulon caudatum* Copel. = *Histiopteris caudata* (Copel.) Holtt.

Description Terrestrial; stem long-creeping, slender to rather stout, bearing scales and sometimes also trichomes, or only trichomes, and long, fibrous roots; leaves monomorphic, ca. 50 cm to 3 m, rarely to 6 m or more long, borne at intervals, lamina 2- to 4-pinnate, glabrate, veins free to usually anastomosing without included free veinlets; sori marginal, the sporangia on a vascular commissure, sometimes around a sinus, paraphysate, the indusium well differentiated from the recurved margin; spores ellipsoidal, monolete, the laesura $\frac{1}{2}$ to $\frac{2}{3}$ the spore length, the surface tuberculate to rugose, usually less prominent near the laesura. Chromosome number: $n = 48, 96$.

The position and structure of the sorus in *Histiopteris* (Figs. 3, 5) resembles that of *Pteris* and the sometimes fully anastomosing veins (Fig. 6) are likewise similar to *Pteris*. The leaves are unusual in the development of stipule-like reduced pinnules (Fig. 4) especially at the base of large pinnae. These may also be formed at the base of pinnules. The lamina is usually glaucous beneath and glabrous or sometimes with a few large trichomes.

Fig. 62.1. *Histiopteris incisa,* near El Empalme, Costa Rica, (Photo W. H. Hodge.)

Fig. 62.2. Distribution of *Histiopteris incisa* in America.

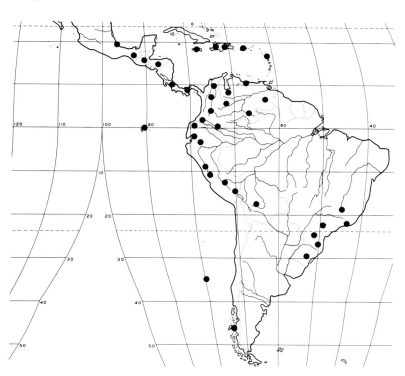

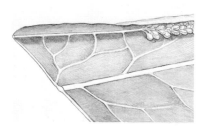

Fig. 62.3. Sorus of *Histiopteris incisa* with continuous commissure and marginal indusium, × 12.

Fig. 62.4. Laminar stipules at base of pinna, × 0.5.

5

6

Figs. 62.5, 62.6. *Histiopteris incisa.* **5.** Fertile segments, × 5. **6.** Sterile pinnule with anastomosing veins, × 0.5.

Systematics

Histiopteris is a pantropical and temperate genus of one or perhaps a few species. *Histiopteris incisa* (Thunb.) J. Sm. occurs throughout the range of the genus. Other species sometimes recognized in the paleotropics are: *Histiopteris caudata* (Copel.) Holtt., *H. estipulata* vAvR., *H. sinuata* (Brack.) J. Sm., and *H. stipulacea* (Hook.) Copel. Revisionary work on the group is necessary to confirm the status of these segregate species.

Ecology (Fig. 1)

Histiopteris is a genus of wet forests and especially occurs along the borders, in clearings and natural openings in forested areas.

In tropical America it grows in montane forests, cloud forests, and higher páramillo thickets, in clearings of dense forests, open woodlands, and grassy places, it is less often on rocky slopes and rarely on cliffs. *Histiopteris* usually grows from 1500 to 3000 m, less often to 3600 m.

Geography (Fig. 2)

Histiopteris grows in tropical and south temperate America, in Africa and east to China and Australia, and in the Pacific to Tahiti and Rapa, north to Japan and south to Tasmania; it is also in the Auckland Islands.

In America, *Histiopteris* is distributed from Veracruz in Mexico, through Central America, in the Greater Antilles, except Cuba, and some of the Lesser Antilles (Guadaloupe, Dominica), and in South America from Venezuela and Colombia south to Bolivia; it is disjunct in southeastern Brazil, and in Aisén in Chile; also on the Galápagos and on the Juan Fernandez Islands.

7

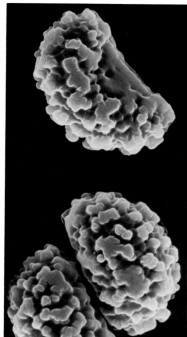

8

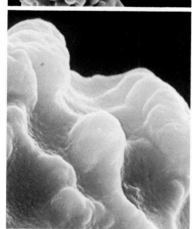

Figs. 62.7, 62.8. *Histiopteris incisa* spores, Costa Rica, *Allen 623*. **7.** Tuberculate the laesura, right, upper spore, ×1000. **8.** Detail of tuberculate surface, ×5000.

Spores

The coarsely tuberculate, monolete spores of *Histiopteris* (Fig. 7) have a coalescent, papillate surface deposit (Fig. 8). The contours appear to be formed by a lower layer of the perispore, although Erdtman and Sorsa (1971) indicate that the perine is lacking. The tuberculate surface resembles that of the trilete spores in the American tropical species of *Dennstaedtia*. The monolete form of *Histiopteris* spores and the relatively high chromosome numbers in the American plants suggest the species represents a derived element among the dennstaedtioid ferns.

Cytology

Plants of *Histiopteris* are reported with $n = 96$ from Jamaica and there are tentative records probably representing this number from Tristan da Cunha, and for species of Ceylon and New Zealand. A record of $n = 48$ for plants of the Bonin Islands is reported and a base number of 48 or 24 suggested by Mitui (1973).

Literature

Erdtman, G., and P. Sorsa. 1971. Pollen and Spore Morphology/Plant Taxonomy. 4. Pteridophyta. 302 pp. Almqvist & Wicksell, Stockholm.
Mitui, K. 1973. A cytological survey on the Pteridophytes of the Bonin Islands. Jour. Jap. Bot. 48: 247–253.

16b. Tribe Lindsaeeae

Lindsaeeae Hook., Sp. Fil. 1: 202. 1846. Type: *Lindsaea* Sm.
Lindsaeaceae Pic.-Ser., Webbia 24: 707. 1970. Type: *Lindsaea* Sm.

Three American genera, two of them also paleotropical, and *Xyropteris* Kramer and *Tapeinidium* (Presl) C. Chr. confined to the paleotropics.

This treatment of the genera of the tribe Lindsaeeae is based on the work of Karl U. Kramer, and his review of our manuscript. In reference to indument there are two differences between the terminology used here and that in the work of Kramer. The uniseriate stem indument that is referred to as scales in the work of Kramer is designated here as trichomes. The minute, usually 2- to 3-celled trichomes, often born on the receptacle, are considered as paraphyses by Kramer, while the sorus is considered as nonparaphysate here since similar 2- to 3-celled trichomes are also borne on other parts of the lamina.

63. *Odontosoria*
Figs. 63.1–63.18

Odontosoria Fée, Mém. Fam. Foug. 5 (Gen. Fil.): 325. 1852. Type: *Odontosoria uncinella* (Kze.) Fée (*Davallia uncinella* Kze.).

Stenoloma Fée, Mém. Fam. Foug. 5 (Gen. Fil.): 330. 1852. Type: *Stenoloma dumosa* (Sw.) Fée (*Davallia dumosa* Sw.) = *Odontosoria aculeata* (L.) J. Sm. (Internat. Code Bot. Nomencl. ed. 8, 220. 1956). [Incorrectly changed in ed. 9, 231. 1961 to *Stenoloma clavata* (L.) Fée = *Odontosoria clavata* (L.) J. Sm; see Kramer 1971, p. 179.].

Odontosoria (Presl) J. Sm., Hist. Fil. 263. 1875, not Fée, 1852. *Davallia* section *Odontosoria* Presl, Tent. Pterid. 129. 1836. Type: *Davallia tenuifolia* (Lam.) Sw. (*Odontosoria tenuifolia* (Lam.) J. Sm., *Adiantum tenuifolium* Lam.) = *Odontosoria chinensis* (L.) J. Sm.

Sphenomeris Maxon, Jour. Wash. Acad. Sci. 3: 144. 1913. Type: *Sphenomeris clavata* (L.) Maxon (*Adiantum clavatum* L.) = *Odontosoria clavata* (L.) J. Sm.

Description Terrestrial or rupestral; stem short- to long-creeping, slender to rather stout, bearing scales that sometimes intergrade to trichomes, or (in *O. clavata*) often only trichomes, and few to many, long, fibrous roots; leaves monomorphic, ca. 20 cm to 6 m long, borne in a loose cluster or at intervals, lamina 2- to 5-pinnate, glabrous, veins free; sori marginal, 1-nerved or sometimes 2- to 8-nerved and then borne on a vascular commissure connecting the vein-ends, not paraphysate, covered by an abaxial indusium attached at the base and at least partly on the sides, the opposed margin not or slightly modified; spores usually spheroidal, trilete, the laesurae $\frac{2}{3}$ to $\frac{3}{4}$ the radius (usually broad and raised), or sometimes ellipsoidal, monolete with laesura $\frac{3}{4}$ the length (broad and raised), the surface granulate, with spherical deposit, or nearly smooth. Chromosome number: $n = 38, 39, 47, 48$, ca. 88, 94, ca. 96, 100, 145–147; $2n = 76$; apogamous 47, 48, 94.

The spines that are borne along the axes of most species (Fig. 6) are an unusual development in ferns. The terminal sori at the ends of single veins (Fig. 5) are characteristic of the genus, although in a few species two or more veins extend to the sorus (Fig. 9).

Systematics *Odontosoria* is a pantropical and partly extratropical genus of about 20 species, with some 12 of them in America. The genus is related to *Lindsaea,* especially section *Schizoloma.*

Although *Sphenomeris* has been widely accepted as a genus since its recognition by Maxon, the difficulties of separating it from *Odontosoria* were pointed out by Kramer (1972) and since then he has expressed the opinion that a single genus should be recognized. The species with large leaves with periodic growth

Fig. 63.1. *Odontosoria gymnogrammoides,* Tapanti, Cartago, Costa Rica. (Photo W. H. Hodge.)

and dormancy of the apex and with spines represent a specialized element in the genus, but one that is not clearly distinguished from the species with smaller and normally expanded leaves.

The American species of *Odontosoria* were revised by Maxon (1913) and the ones previously placed in *Sphenomeris* by Kramer (1957).

Fig. 63.2. *Odontosoria clavata,* Retreat, St. Mary, Jamaica. (Photo W. H. Hodge.)

Tropical American Species

The revision of Maxon recognized 10 species. However, specimens collected since his study indicate that most of the characters used to distinguish the species, such as segment size and shape, and type and orientation of the spines are sometimes variable in a species and do not always correlate. Several of the species need a modern assessment employing new characters. *Odontosoria Wrightiana* Maxon (Fig. 7), for example, is surely only a narrow-segmented form of *O. aculeata,* and *O. fumarioides* of Jamaica needs to be more clearly distinguished from *O. gymnogrammoides* of Costa Rica.

There are three species in Colombia: *O. colombiana, O. Killipii* (*Spehnomeris Killipii* (Maxon) Kramer), and *O. spathulata* (*Sphenomeris spathulata* (Maxon) Kramer). These are all very rare and poorly known and their status should be reassessed when adequate materials are available.

The following key has been adapted and simplified from Maxon (1913) and Kramer (1957).

Fig. 63.3. Distribution of *Odontosoria* in America.

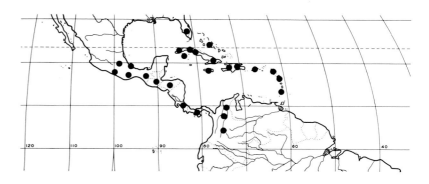

Geographic-Morphological Key to American Species of *Odontosoria*

a. Sori 2- to 4-nerved (Fig. 9) at least on the larger segments, stem mostly or wholly with trichomes. *O. clavata* (L.) J. Sm.
a. Sori 1-nerved, stem with scales, sometimes intergrading to trichomes. b.
  b. Mexico and Central America. c.
    c. Leaves lacking spines on the axes. d.
      d. Ultimate lobes long and narrow (Fig. 4), hardly broader than the alate axes to which they are attached; Mexico to Honduras. *O. Schlectendahlii* (Presl) C. Chr.
      d. Ultimate lobes mostly rather short, many somewhat to rather deeply bifid, mostly broader than the alate axes to which they are attached; Guatemala. *O. guatemalensis* Christ.
    c. Leaves with some of the axes spiny; Costa Rica and Panama. *O. gymnogrammoides* Christ.
  b. Greater Antilles. e.
    e. Axes with long often acicular, mostly spreading spines (Fig. 6). *O. aculeata* (L.) J. Sm.
    e. Axes with stout, conical, mostly retrorse spines, these sometimes few. f.
      f. Pinnae with a few, elongate, 1-pinnate pinnules and an elongate conform terminal segment. *O. uncinella* (Kze.) Fée.
      f. Pinnae with many 2- to 3-pinnate pinnules, the apex more or less gradually reduced. g.
        g. Ultimate segments deeply flabellate or bifid (Fig. 8), nearly all fertile lobes with one sorus. *O. fumarioides* (Sw.) J. Sm.
        g. Ultimate segments mostly with shallow lobes, many fertile lobes with 2–3 sori. *O. Jenmanii* Maxon.
  b. Lesser Antilles. *O. flexuosa* (Spreng.) Maxon.
  b. Colombia. h.
    h. Rachis spiny. *O. colombiana* Maxon.
    h. Rachis lacking spines. i.
      i. Ultimate segments spathulate. *O. spathulata* (Maxon) R. & A. Tryon.
      i. Ultimate segments linear or slightly cuneate. *O. Killipii* (Maxon) R. & A. Tryon.

Ecology (Figs. 1, 2)

Odontosoria most commonly grows in regions of rain forests, montane forests, and cloud forests, but also in oak and pine woods. It especially grows in secondary vegetation, on brushy slopes, in thickets and clearings and in disturbed sites such as roadsides, landslips and sometimes in plantations. Rarely it grows in swampy places or on mossy rocks near streams or waterfalls. *Odontosoria clavata* (Figs. 2, 9–11) grows in limestone sinks and on limestone outcrops, especially in pinelands, also along streams and on river banks in shaded situations. It sometimes also grows on serpentine rocks. *Odontosoria* occurs from near sea level to ca. 2500 m, most commonly from 500 to 1500 m.

Many species of *Odontosoria* in America are thicket-forming ferns with long, scandent leaves scrambling on other vegetation. The pinnae are usually opposite and often flexuous, and the axes spiny; these characters appear to be adaptations for a scandent habit. In addition, the apical leaf-bud is dormant during the time that the last pair of pinnae expand, following which growth is resumed to produce another pair of pinnae. This periodic growth pattern of the leaf is probably also an adaptation that allows the pinnae to gain support from surrounding plants prior to further development of the leaf. The leaf of *Odontosoria* is often reported to be indeterminate, but a determinate apex is apparently eventually produced.

Geography (Fig. 3)

Odontosoria is distributed in tropical and subtropical America, in Africa, and eastward to southeastern Asia, Malesia, and through the Pacific to Rapa, Tahiti, the Marquesas and the Hawaiian Islands, and northward to Korea and Japan. It is notably absent from Australia and New Zealand.

In America it occurs from Puebla in Mexico, south to Panama, southern Florida and the Bahamas, in all of the Greater Antilles and some of the Lesser Antilles, and in Colombia.

Spores

Odontosoria spores are generally larger than those of *Lindsaea* and the trilete spores of some species are the largest in the family. The surface may be relatively smooth as in the large spores of *O. guatemalensis* at half the usually magnification (Fig. 12) and also in the monolete spores of the Old World *O. chinensis* (L.) J. Sm. (Fig. 18). The granulate surface may be sparsely and irregularly deposited as in *O. aculeata* (Fig. 13) or denser as in *O. Killipii* (Figs. 15, 16) but without surface strands as in spores of *Lindsaea* or *Ormoloma*. The profile of a portion of the wall of *O. aculeata* (Fig. 14) shows two perispore strata consisting of a thick outer layer appressed to a thinner, somewhat denser lower one adjacent to the exospore. The spores of *O. clavata* (Fig. 17) differ from other species in having a somewhat rougher surface with coarse, spherical deposit.

Cytology

Records of $n = 38$ and $2n = 76$ for *Odontosoria clavata* (as *Sphenomeris*) from Jamaica (Walker, 1966) clearly document these numbers. A report of $n = 39$ from Florida (Wagner, 1963) may also represent 38. *Odontosoria fumarioides* and *O. Jenmanii* are reported with $n = $ ca. 96 from Jamaica (Walker, 1966). There are records for several paleotropical species of numbers ranging between $n = 47$ and $n = 147$ that have been considered to be based on 47, 48 or 50. The discrete numbers of neotropical species as compared to those from the paleotropics suggests that more than one lineage may be included in *Odontosoria*.

4   5   6

7   8

9   10   11

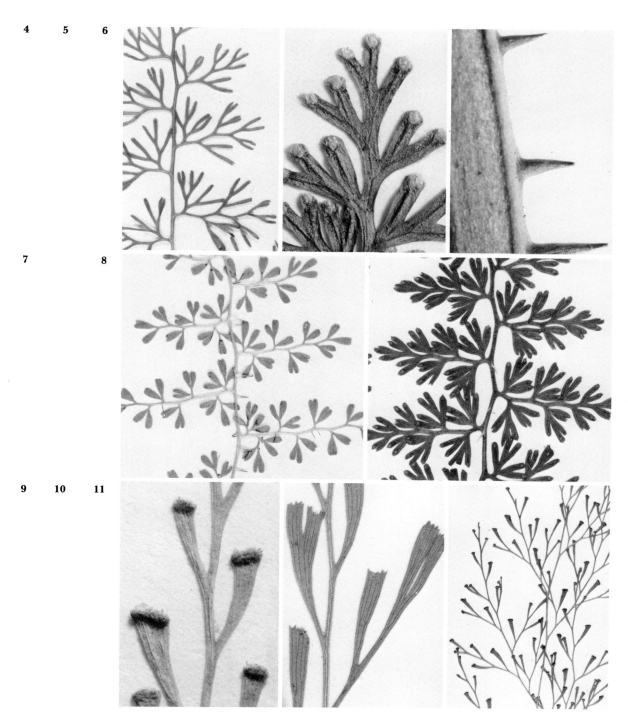

Figs. 63.4–63.11. ***Odontosoria.***   **4.** Segments of *O. Schlechtendahlii,* ×2.   **5.** Ultimate segments of ***O. fumarioides*** with marginal sori, ×5.   **6.** Spines on rachis of ***O. aculeata,*** ×8.5.   **7.** Narrow segments of ***O. aculeata*** (the form described as *O. Wrightiana*), ×2.   **8.** Segments of ***O. fumarioides,*** ×2.   **9.** Fertile segments of ***O. clavata,*** ×4.   **10.** Sterile segments of ***O. clavata,*** ×3.   **11.** Portion of a leaf of ***O. clavata,*** ×0.75.

12   13   14

15   16

17   18

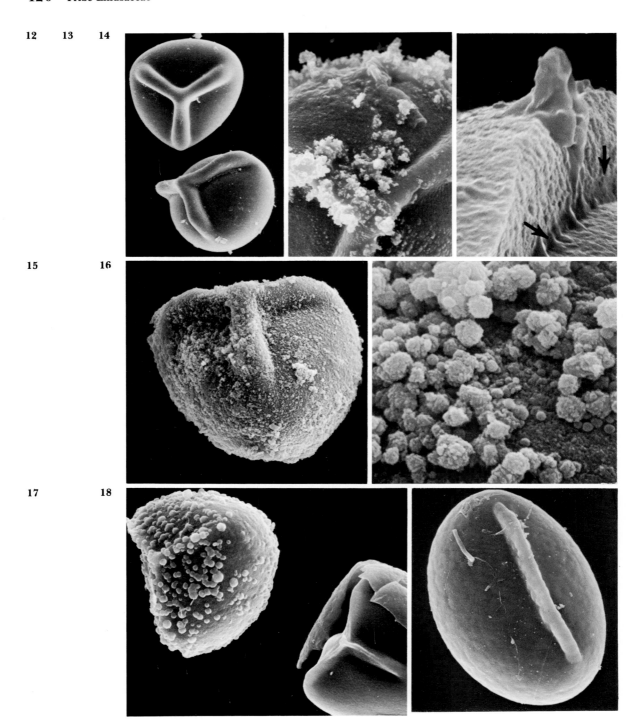

Figs. 63.12–63.18. *Odontosoria* spores.   **12. *O. guatemalensis,*** nearly smooth with raised laesurae, Guatemala, *Muenscher 12130* (F), × 500.   **13, 14. *O. aculeata.*   13.** Detail of granulate proximal face, part of laesura, below, Dominican Republic, *Valeur 291,* × 1600.   **14.** Wall profile with thick outer and thinner inner perispore strata (arrow), the exospore below, right, Cuba, *Maxon 4131,* × 10,000.   **15, 16. *O. Killipii,*** Colombia *Killip 7947* (isotype).   **15.** Proximal face, tilted, granulate, the laesurae at top, × 1000.   **16.** Detail of granulate surface, × 10,000.   **17. *O. clavata,*** spherical surface deposit on spore, at left, part of abraded spore with smooth exospore exposed, right, Nassau, Bahama Islands, *A. Wight 137,* × 1000.   **18. *O. chinensis,*** monolete spore, shallowly rugose, Hong Kong, *C. Wright in 1853–56,* × 1000.

Literature

Kramer, K. U. 1957. See reference under *Lindsaea*.

Kramer, K. U. 1972. The Lindsaeoid ferns of the Old World, IX. Africa and its islands. Bull. Jard. Bot. Nat. Belg. 42:305–345.

Maxon. W. R. 1913. Studies of tropical American ferns, 4. Contrib. U.S. Nat. Herb. 17: 133–179.

Wagner, W. H. 1963. A biosystematic study of United States ferns. Preliminary abstract. Amer. Fern Jour. 53: 1–16.

Walker, T. G. 1966. Reference under the family.

64. *Lindsaea*
Figs. 64.1–64.24

Lindsaea Sm. Mém. Acad. Turin 5: 413. 1793. Type: *Lindsaea trapeziformis* Dryand. = *Lindsaea lancea* (L.) Bedd.

Schizoloma Gaud., Ann. Sci. Nat. 3: 507. 1824. Type: *Schizoloma Billardieri* Gaud. = *Lindsaea ensifolia* Sw. *Lindsaea* subgenus *Schizoloma* (Gaud.) Hook. Sp. Fil. 1: 219. 1846. *Lindsaea* section *Schizoloma* (Gaud.) Kramer, Acta Bot. Neerland. 15: 571. 1967. [Subgenus *Lindaea*].

Isoloma J. Sm., Jour. Bot. (Hook.) 3: 414. 1841. Type: *Isoloma divergens* (Hook. & Grev.) J. Sm. = *Lindsaea divergens* Hook. & Grev. [Subgenus *Lindsaea*].

Odontoloma J. Sm., *ibidem* 3: 415. 1841, not HBK., 1820 (Compositae). Type: *Odontoloma Boryana* (Presl) J. Sm. (*Davallia Boryana* Presl) = *Lindsaea repens* (Bory) Thwaites.

Synaphlebium Hook., Gen. Fil. *t. 101.* 1842. Type: *Synaphlebium recurvatum* Hook. = *Lindsaea cultrata* (Willd.) Sw. [Subgenus *Lindsaea*].

Davallia subgenus *Odontoloma* Hook., Sp. Fil. 1: 174. 1846. Type: the same as that of *Odontoloma* J. Sm. *Lindsaea* subgenus *Odontoloma* (Hook.) Kramer, Blumea 15: 561. 1968.

Lindsaenium Fée, Mém. Mus. Hist. Nat. Strassburg 4: 201. 1850. Type: (Fée, Mém. Fam. Foug. 5 (Gen. Fil.): 333. 1852.) *Lindsaenium rigidum* (Hook.) Fée (as *Lindsaynium*) = *Lindsaea rigida* Hook. [Subgenus *Odontoloma*].

Lindsaea section *Paralindsaea* Keys., Polypod. Cyath. Herb. Bung. 3, 21, 70. 1873. Type: *Lindsaea linearis* Sw. [Subgenus *Lindsaea*].

Sambirania Tard., Mém. Instit. Sci. Madagas. ser. B, 7: 34. 1956. Type: *Sambirania plicata* (Baker) Tard. = *Lindsaea plicata* Baker [Subgenus *Lindsaea*].

Humblotiella Tard., *ibidem* 7: 38. 1956. Type: *Humblotiella odontolabia* (Baker) Tard. (*Davallia odontolabia* Baker) = *Lindsaea odontolabia* (Baker) Kramer. [Subgenus *Odontoloma*].

Lindsaea section *Pseudosphenomeris* Kramer, Acta Bot. Neerland. 6: 165. 1957. Type: *Lindsaea bifida* (Kaulf.) Mett. (*Davallia bifida* Kaulf.) = section *Schizoloma*.

Lindsaea section *Crematomeris* Kramer, *ibidem* 171. 1957. Type: *Lindsaea pendula* Kl. [Subgenus *Lindsaea*].

Lindsaea section *Temnolindsaea* Kramer, *ibidem* 176. 1957. Type: *Lindsaea Klotzschiana* Ettingsh. [Subgenus *Lindsaea*].

Lindsaea section *Haplolindsaea* Kramer, *ibidem* 260. 1957. Type: *Lindsaea sagittata* (Aubl.) Dryand. (*Adiantum sagittatum* Aubl.) [Subgenus *Lindsaea*].

Lindsaea section *Tropidolindsaea* Kramer, *ibidem* 267. 1957. Type: *Lindsaea Seemannii* J. Sm. [Subgenus *Lindsaea*].

Description   Terrestrial, rupestral or epiphytic; stem decumbent and very short-creeping, moderately stout to slender, or long-creeping and slender, bearing scales that usually intergrade to stiff trichomes, and usually many, long, fibrous roots; leaves monomorphic or (in *L. cubensis*) somewhat dimorphic (with the fertile larger and more erect), ca. 15–100 cm long, borne in a cluster or at intervals, lamina simple to ca. 4-pinnate, glabrous, veins free or sometimes anastomosing without included free veinlets; sori marginal, rarely 1-nerved, usually 2- to many-nerved and borne on a vascular commissure connecting the vein-ends, not paraphysate, covered by an abaxial indusium attached at the base and sometimes the sides, the opposed margin unmodified; spores usually trilete, spheroidal or 3-lobed, the laesurae $\frac{2}{3}$ to $\frac{3}{4}$ the radius, often broad and raised or sometimes ellipsoidal, monolete, and the laesura nearly as long as the spore, broad and raised, the surface often smooth or granulate or with curled strands. Chromosome number: n = 34, ca. 40, 42, 44, 44 or 45, 47, ca. 50, ca 84, ca. 87, 88, ca. 100, ca. 150, ca. 153, 155, ca. 220; $2n$ = ca. 100, ca. 176; ? apogamous 47, 80, 94, ca. 130.

The sorus of *Lindsaea* is sometimes short laterally, as on narrow segments of species such as *L. bifida* (Fig. 3). More typically it extends along the upper and outer sides of a segment as in *L. guianensis* (Fig. 4). In species with long, entire pinnae or a simple lamina, the sorus extends along nearly all of the margin (Fig. 6). When the sorus is mature, the indusium becomes more or less bent backward (Fig. 5). The lamina architecture is diverse in *Linsaea,* and a selection of the types among the American species is shown in Figs. 11–24. The minute trichomes, mentioned by Kramer (1957) as probably occurring on the leaves of all *Lindsaea* species, are too small to be recognized as pubescence.

Systematics   *Lindsaea* is a pantropical and extratropical genus of about 150 species, with about 45 in America.

The genus is a complex one, with two subgenera and 23 sections (Kramer, 1971). This classification provides a basis for further detailed studies, especially of species relations and of the polyploid complexes that undoubtedly exist. Section *Schizoloma* is closest to *Odontosoria* and is probably the most primitive element in the genus.

Lindsaea has a series of lamina forms that often closely resemble some of those in *Adiantum*.

The American species of *Linsaea* have been revised by Kramer (1957).

Synopsis of Subgenera of *Lindsaea*

Subgenus *Lindsaea*

Terrestrial or only casually epiphytic, stem short- to only moderately long-creeping, the stele radially symmetrical or nearly so.

A subgenus of 17 sections, seven of them America. The subgenus *Schizoloma* in Kramer (1957) was later altered to a section of subgenus *Lindsaea*.

Fig. 64.1. *Lindsaea portoricensis,* Mason River savannah, Jamaica. (Photo W. H. Hodge.)

Subgenus *Odontoloma*

Epiphytic, stem very long-creeping, the stele strongly dorsiventral. A subgenus of six sections, all paleotropical.

Tropical American Species

The following enumeration provides some of the characteristics of the American sections of subgenus *Lindsaea* and includes all or a selection of the species.

Section *Schizoloma*

Lamina usually gradually reduced apically and often decompound. It has about 40 species, with four of them American: *L. bifida* (Kaulf.) Kuhn (Fig. 11), *L. macrophylla* Kaulf., with a 1-pinnate lamina and some anastomosing veins, *L. sphenomeridopsis* Kramer, and *L. virescens* Sw. (Fig. 12).

Fig. 64.2. Distribution of **Lindsaea** in America.

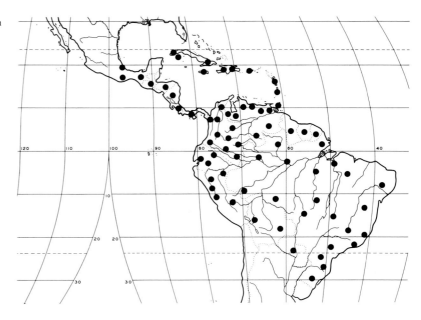

Section *Crematomeris*
Twisted axes that result in pendulous segments. It has two species: *L. meifolia* (HBK.) Kuhn and *L. pendula* Kl. (Fig. 16), both American.

Section *Temnolindsaea*
2-pinnate lamina and incised pinnules. It has eight species, five of them American, including *L. Klotzschiana* Ettingsh., *L. tetraptera* Kramer, and *L. Parkeri* (Hook.) Kuhn. The three Old World species may not belong with the American ones.

Section *Lindsaea*
1- to 2-pinnate lamina and a conform terminal pinna-segment. It has 28 species, 26 of them American, including: *L. arcuata* Kze., *L. botrychioides* St.-Hil. (Fig. 14), *L. dubia* Spreng. (Fig. 13), *L. guianensis* (Aubl.) Dryand., *L. hemiglossa* Kramer (Fig. 21), *L. lancea* (L.) Bedd. (Figs. 17, 18), *L. portoricensis* Desv., *L. quadrangularis* Raddi, *L. Schomburghkii* Kl., *L. stricta* (Sw.) Dryand. (Figs. 19, 20), and *L. Ulei* Hieron. (Fig. 22). The two Old World species may not belong with those in America.

Section *Haplolindsaea*
Reniform, cordate or sagittate lamina. It has three species: *L. cyclophylla* Kramer, *L. reniformis* Dryand. (Fig. 23), and *L. sagittata* (Aubl.) Dryand. (Fig. 24), all American.

Section *Paralindsaea*
Small, somewhat dimorphic leaves. It has four species, with one of them, *L. cubensis* Underw. & Maxon (Fig. 15), American. This species is only doubtfully related to the Old World ones.

3   **4**   **5**

6   **7**   **8**

9   **10**

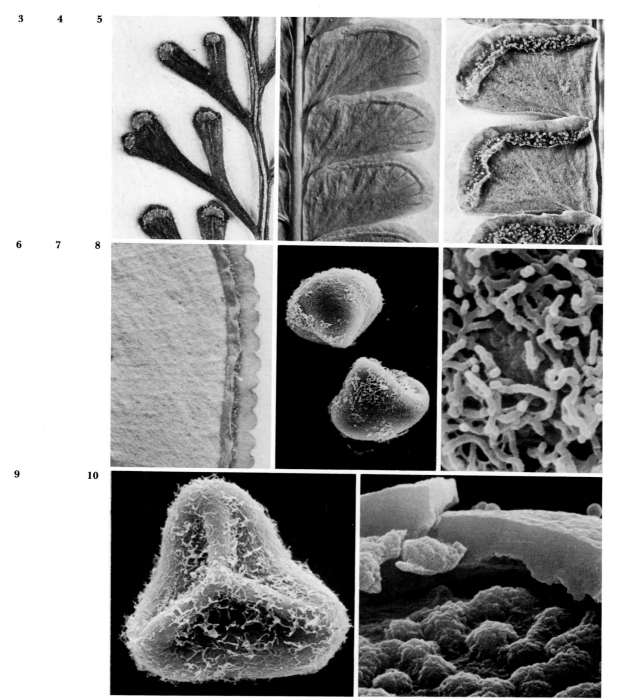

Figs. 64.3–64.10. *Lindsaea.* **3–6.** Segments and sori. **3.** Fertile ultimate segments of *L. bifida*, ×8. **4.** Fertile pinnules of *L. guianensis* with immature indusium, ×5. **5.** Fertile pinnules of *L. stricta* with mature sori and indusia, ×8. **6.** Portion of fertile lamina of *L. cyclophylla* with continuous sorus and indusium, ×5. **7–10.** Spores. **7.** *L. guianensis,* lateral view of two young spores with fused strands, Brazil, *Dusén* in 1908, ×1000. **8.** *L. lancea* var. *lancea,* surface detail of strands, Dominican Republic, *Valeur 548,* ×10,000. **9.** *L. portoricensis* proximal face of young spore with sparse strands, Jamaica, *Gastony 77,* ×2000. **10.** *L. philippensis,* part of wall section with thick outer perispore, above, and tuberculate inner perispore, below, Philippines, *Elmer 12416,* ×10,000.

Section *Tropidolindsaea*

1-pinnate lamina reduced both apically and basally. It has three species: *L. pratensis* Maxon, *L. protensa* C. Chr., and *L. Seemannii* J. Sm., all American. The Philippine *L. adiantoides* Hook. is probably closely related.

Ecology (Fig. 1)

Lindsaea is primarily a genus of moist or wet forests, but species also grow in grasslands, marshes, and open rocky places.

In America the genus grows in low rain forests, in montane and in elfin forests, sometimes in pine lands, in savannahs, in thickets and low scrub, and in shaded or exposed rocky sites. Sometimes it grows in clearings or on exposed clay road banks. The genus grows on a great variety of soil and rock types. Several species are sometimes low epiphytes on mossy tree trunks or branches but these (subgenus *Lindsaea*) are not adapted, as subgenus *Odontoloma* is, for a truly eiphytic habit. *Lindsaea* grows from near sea level up to about ca. 2300 m, most commonly below 1500 m.

Geography (Fig. 2)

Lindsaea occurs through the American tropics, in Africa and eastward to Malesia and southeastern Asia, north to Japan and south to Tasmania and New Zealand, it extends eastward in the Pacific to the Marquesas and the Hawaiian Islands.

In American *Lindsaea* grows from Veracruz in Mexico, through Central America, in the West Indies, and generally through South America south to Bolivia, Paraguay and Rio Grande do Sul in Brazil.

A few species have ranges that extend nearly throughout that of the genus in America: *L. guianensis, L. portoricensis, L. stricta* and *L. lancea*. Most have moderate-sized ranges, but seven are local endemics: *L. pratensis* of Costa Rica; *L. stenomeris* Kramer of Cerro Neblina, Amazonas, Venezuela; *L. Spruceana* Kuhn of Mt. Guayrapurima, Tarapoto, Peru; *L. Herminieri* Fée of Guadeloupe, *L. taeniata* Kramer of Antioquia, Colombia; *L. cubensis* of Cuba and the Isle of Pines; and *L. protensa* of Hispaniola.

The center of diversity and endemism of *Lindsaea* is in the area of the Roraima sandstone formation in northern South America where 27 of the 45 species occur and 14 of them are endemic. Southeastern Brazil is a secondary center, with 13 species, three of them endemic, and also Panama and Andean Colombia with 12 species and one endemic. *Lindsaea* is one of three fern genera with a dominant geographic center in Guayana; the others are *Hymenophyllopsis* and *Pterozonium* which are endemic or nearly endemic to the region.

Spores

Lindsaea spores that have strands or dense granulate surface deposits are usually strongly lobed (Fig. 7) and have concave areas between the laesurae. Those with a nearly smooth surface are

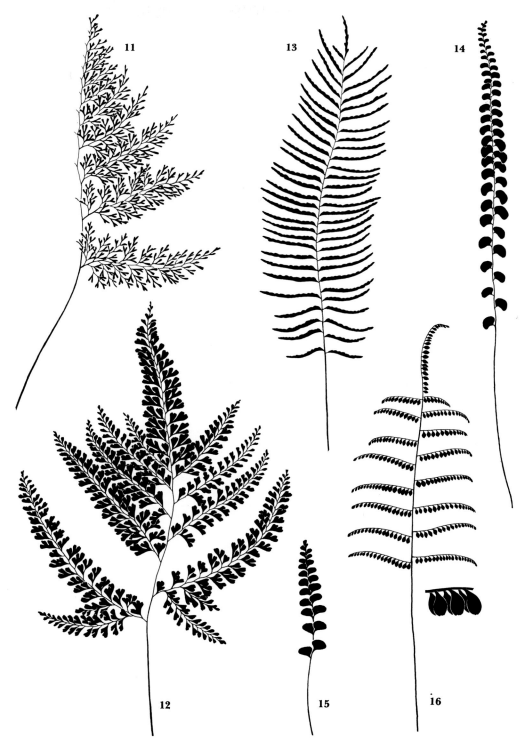

Figs. 64.11–64.16. Lamina architecture of *Lindsaea.* **11.** *L. bifida,* ×0.5. **12.** *L. virescens,* ×0.5. **13.** *L. dubia,* ×0.5. **14.** *L. botrychioides,* ×2.5. **15.** *L. cubensis,* fertile leaf, ×0.5. **16.** *L. pendula,* ×0.5, detail enlarged.

Figs. 64.17–64.24. Lamina architecture of *Lindsaea*. **17.** *L. lancea,* 2-pinnate form, var. *lancea,* × 0.25. **18.** *L. lancea* var. *falcata* (Dryand.) Ros., × 0.25. **19.** *L. stricta,* 2-pinnate form, × 0.25. **20.** *L. stricta,* 1-pinnate form, × 0.25. **21.** *L. hemiglossa,* × 0.25. **22.** *L. Ulei,* × 0.25. **23.** *L. reniformis,* × 0.25. **24.** *L. sagittata,* × 0.25.

usually spheroidal or ellipsoidal. The surface strands (Figs. 8, 9) appear to be incorporated in the outer part of the wall as the spores develop. Although the surface structure of *Lindsaea* spores is relatively smooth the wall stratification is complex. A section of the wall of *Lindsaea philippensis* Kramer (Fig. 10) shows the thick, outer perispore layer is closely appressed to a lower thin, tuberculate perispore above the exospore. The lobed form of the spores and surface strands are similar to those of *Ormoloma*. The paleotropical genera *Tapeinidium* and *Xyropteris* have monolete spores with smooth, thick outer walls, prominent laesurae and generally resemble the ellipsoidal spores of *Lindsaea*.

Cytology

Species of *Lindsaea* from the American tropics are reported as $n =$ ca. 88 from Jamaica (Walker, 1966); $2n =$ ca. 88 II, from Chiapas, Mexico (Smith and Mickel, 1977), and $n = 42$, $n =$ ca. 84 from Brazil (Tryon et al., 1975). The diverse numbers of *Lindsaea*, between $n = 34$ and 220 are assessed by Kramer (1971) and several approximate records are considered as equal to or derived from 44 to 47 or possibly 94. At least three discrete series of numbers involving 34, 44 and 47 are also evident in the arrangement made by Walker (1973) based on reports of species in six sections of the genus. The Old World species with monolete spores including *Lindsaea viridis* Colenso and *L. odorata* Roxb. have high numbers ranging from $n = 88$ to 220 that suggest they represent derived elements.

Literature

Kramer, K. U. 1957. A revision of the genus *Lindsaea* in the New World with notes on allied genera. Acta Bot. Neerland. 6: 97–290.

Kramer, K. U. 1971. Lindsaea-group. Fl. Malesiana 1:177–254. (This is the author's major paper, among six, on paleotropical *Lindsaea*.)

Smith, A. R., and J. T. Mickel. 1977. Reference under the family.

Tryon, A. F., H. P. Bautista, and I. Araujo. 1975. Chromosome studies of Brazilian ferns. Acta Amazonica 5: 35–43.

Walker, T. G. 1966. Reference under the family.

Walker, T. G. 1973. Reference under the family.

65. *Ormoloma*
Figs. 65.1–65.6

Ormoloma Maxon, Proc. Biol. Soc. Wash. 46: 143. 1933. Type: *Ormoloma Imrayanum* (Kze.) Maxon (*Saccoloma Imrayanum* Kze.).

Description

Terrestrial or casually epiphytic; stem rather long-creeping, slender, bearing scales and many, long, fibrous roots; leaves monomorphic, ca. 20–60 cm long, borne at intervals, lamina 1-pinnate, glabrous, veins free; sori marginal, 1-nerved, not paraphysate, covered by an abaxial indusium attached only at the base, the opposed margin unmodified; spores trilete, somewhat 3-lobed, the laesurae $\frac{2}{3}$ the radius, with slender, more or less coalescent strands. Chromosome number: $n = 42$.

Fig. 65.1. *Ormoloma Imrayamum,* Cloud forest, Monteverde, Costa Rica, (Photo L. D. Gómez.)

Ormoloma is a distinctive genus among the lindsaeoid ferns in having 1-nerved sori (Fig. 4) borne on nearly entire pinnae. These are not connected by a vascular commissure except occasionally where a vein branches very near the margin (Fig. 3).

Systematics

Ormoloma is a monotypic genus based on *O. Imrayanum* (Kze.) Maxon. While it is clearly related to *Lindsaea* its affinity to a particular group within the genus is uncertain.

Maxon (1933) described a second species. *Ormoloma Standleyi* of Costa Rica, on the basis of the color of the petiole, the shape of the stem scales, and size of the indusium, but later collections show that all of these characters are variable.

Ecology (Fig. 1)

Ormoloma grows in moist or wet forests, in deep shade or in somewhat open forests. It is terrestrial and also grows on old logs and tree stumps and rarely is a low epiphyte on mossy tree trunks. *Ormoloma* grows from ca. 750 to 1800 m.

Fig. 65.2. Distribution of ***Ormoloma.***

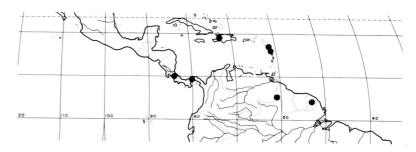

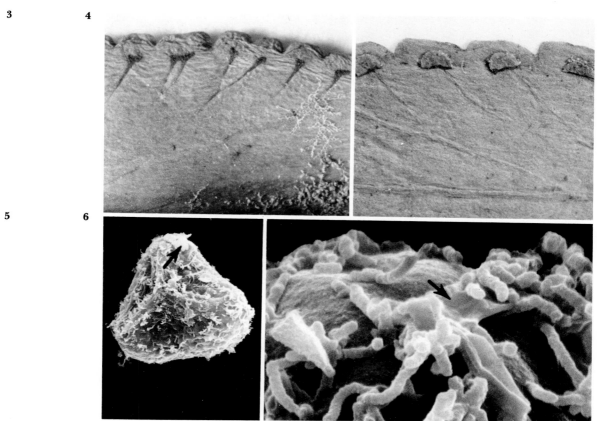

Figs. 65.3–65.6. *Ormoloma Imrayanum.* **3.** Portion of a fertile pinna, adaxial side, with discrete fertile vein-ends, except where a vein has branched close to the margin, × 8. **4.** Portion of a fertile pinna, abaxial side, × 8. **5–6.** Spores. Dominica, *Lellinger 485.* **5.** Proximal face slightly tilted, with strands more or less coalescent in plates (at arrow), × 1000. **6.** Detail of strands, coalescent at arrow, × 10,000.

Geography (Fig. 2)

Ormoloma occurs in Haiti, Guadeloupe, and Dominica, in French Guiana and on Mount Roraima, and in Costa Rica and Panama.

Spores

The strands on the spores of *Ormoloma Imrayanum* appear to co-alesce and form compact plates, at the arrow near the top of the whole spore in Fig. 5. This and other compact fragments suggest that the strands become incorporated into a denser surface. At higher magnification (Fig. 6) the sparse strands are more or less fused and form a denser plate, at the arrow. The general form of the spores and especially the stranded elements composing the wall most closely resemble spores of *Lindaea* species such as *L. lancea* (Fig. 64.8).

Cytology

The record of $n = 42$ for *Ormoloma* from Costa Rica (Wagner, 1980) is the same number as in some species of *Lindsaea* and is consistent with the close relationship of these genera as noted in the spores as well as other characters reviewed by Kramer (1957).

Literature

Maxon, W. R. 1933. A second species of *Ormoloma*. Proc. Biol. Soc. Wash. 46: 157–158.

Kramer, K. U. 1957. Reference under *Lindsaea*.

Wagner, F. S. 1980. New basic chromosome numbers for genera of neotropical ferns. Amer. Jour. Bot. 67: 733–738.

Family 17.   Thelypteridaceae

Thelypteridaceae Pic.-Ser., Webbia 24: 711. 1970. Type: *Thelypteris* Schmidel.

Description   Stem erect, decumbent or to very long-creeping, stout to slender, with a dictyostele, usually indurated, bearing scales; leaves ca. 10 cm to 5 m long, entire or usually pinnate, circinate in the bud, petiole without stipules; sporangia borne in roundish to elongate abaxial sori or rarely also on the leaf-tissue, indusiate, the indusium usually reniform, or exindusiate, mostly rather short-stalked, the stalk 2- to 3-rowed below its apex, the vertical annulus interrupted by the stalk; homosporous, spores lacking cholorphyll. Gametophyte epigeal, with chlorophyll, more or less obcordate, slightly thickened centrally, often with unicellular, rarely multicellular trichomes and (or) glands, archegonia borne on the lower surface, mostly in the central region, antheridia 3-celled, borne on the lower surface, mostly among the archegonia.

Comments on the Family   The Thelypteridaceae are essentially a worldwide family. The generic classification has been diverse, some authors recognizing a single genus *Thelypteris,* as we do, while others such as Iwatsuki (1964) recognize three genera, and some recognize many, for example, Pichi-Sermolli (1977) has 32 genera, and Holttum (1971) has 23 genera in the Old World. These ferns were formerly placed in the large genus *Dryopteris* and much of the literature concerning them uses that name.

The family is characterized by two vascular bundles at the base of the petiole, by leaves that bear unicellular acicular, or variously branched, trichomes, by the pinna-rachis and rachis not being connected by a groove, and a series of chromosome numbers based on $n = 27$ to 36. Loyal (1977) reports four bundles in the petiole of one species. The spores are generally monolete and ellipsoidal but are trilete, and spheroidal, in the few species of the segregate genus *Trigonospora* Holtt.

Aspidistes Thomasii Harris of the Jurassic may be an early thelypteroid fern (Lovis, 1977). The family is an isolated one and probably old. It has been related to the Gleicheniaceae (Holttum, 1947), to the Cyatheaceae (Holttum, 1971), and to the Dennstaedtiaceae (Pichi-Sermolli, 1977). These divergent views suggest that evidence for its phyletic relation is equivocal and that further study of the problem is required.

Literature

Holttum, R. E. 1947. A revised classification of Leptosporangiate ferns. Jour. Linn. Soc. Bot. 53: 123–158.

Holttum, R. E. 1971. Studies in the family Thelypteridaceae, III. A new system of genera in the Old World. Blumea 19: 17–52.

Iwatsuki, K. 1964. Taxonomy of the Thelypteroid ferns, with special reference to the species of Japan and adjacent regions. Mem. Coll. Sci. Kyoto, Ser. B, 31: 11–40.

Lovis, J. D. 1977. Evolutionary patterns and processes in ferns. Advances Bot. Research 4: 229–415.

Loyal, D. S. 1977. Two unusual features in Thelypteroid ferns and their evolutionary significance. Amer. Fern Jour. 67: 53–57.

Pichi-Sermolli, R. E. G. 1977. Tentamen Pteridophytorum genera in taxonomicum ordinem redigendi. Webbia 31: 313–512.

66. *Thelypteris*
Figs. 66.1–66.56

Thelypteris Schmidel, Icon. Pl. (ed. Keller) 3rd page, *t. XI* text. Oct. 1763, *nom. conserv.* Type: *Acrostichum Thelypteris* L. = *Thelypteris palustris* Schott.

Thelypteris Adans., Fam. Plantes 2: 20. July–Aug. 1763. Type: *Acrostichum Thelypteris* L. (chosen by Schmidel, *op. cit.*) = *Thelypteris palustris* Schott.

Meniscium Schreber, Gen. Pl. (Linn.), ed. 8 (Schreber), 2: 757. 1791. Type: *Polypodium reticulatum* L. = *Meniscium reticulatum* (L.) Sw. = *Thelypteris reticulata* (L.) Proctor. *Phegopteris* subgenus *Meniscium* (Schreber) Christ, Farnkr. Erde 269. 1897. *Thelypteris* subgenus *Meniscium* (Schreber) Reed, Phytologia 17: 254. 1968.

Lastrea Bory, Dict. Class. Hist. Nat. 6: 588. 1824, *nom. superfl.* for *Thelypteris* Adans. and with the same type.

Stegnogramma Blume, Enum. Pl. Jav. 172. 1828. Type: *Stegnogramma aspidioides* Blume = *Thelypteris Stegnogramma* (Blume) Reed (*Gymnogramma Stegnogramma* Blume, *nom. nov.* for *Stegnogramma aspidioides* Blume, not *Gymnogramma aspidioides* (Willd.) Kaulf. (=*Thelypteris aspidioides* (Willd.) Tryon). *Gymnogramma* subgenus *Stegnogramma* (Blume) Clarke, Trans. Linn. Soc. II, 1: 569. 1880. *Thelypteris* subgenus *Stegnogramma* (Blume) Reed, Phytologia 17: 254. 1968.

Cyclosorus Link, Hort. Reg. Bot. Berol. 2: 128. 1833. Type: *Cyclosorus gongylodes* (Schkr.) Link (*Aspidium gongylodes* Schkr., as *goggilodus*) = *Thelypteris gongylodes* (Schkr.) Small = *Thelypteris interrupta* (Willd.) Iwats. *Dryopteris* subgenus *Cyclosorus* (Link) C. Chr., Ind. Fil. xxi. 1906. *Thelypteris* subgenus *Cyclosorus* (Link) Morton, Amer. Fern Jour. 53: 153. 1963.

Polypodium section *Phegopteris* Presl, Tent. Pterid. 179. 1836. Type: *Polypodium Phegopteris* L. = *Phegopteris connectilis* (Michx.) Watt. = *Thelypteris Phegopteris* (L.) Slosson. *Phegopteris* (Presl) Fée, Gen. Fil. (Mém. Fam. Foug. 5): 242. 1852. *Polypodium* subgenus *Phegopteris* (Presl) Clarke, Trans. Linn. Soc. II, 1: 542. 1880. *Thelypteris* subgenus *Phegopteris* (Presl) Ching, Bull. Fan Mem. Inst. Biol. Bot. 6: 250. 1936.

Goniopteris Presl, Tent. Pterid. 181. 1836. Type: *Goniopteris crenata* (Sw.) Presl (*Polypodium crenatum* Sw.) = *Thelypteris Poiteana* (Bory) Proctor. *Polypodium* subgenus *Goniopteris* (Presl) Clarke, Trans. Linn. Soc. II, 1: 547. 1880. *Thelypteris* subgenus *Goniopteris* (Presl) Duek, Adansonia II, 11: 720. 1971.

Leptogramma J. Sm., Jour. Bot. (Hook.) 4: 51. 1841. Type: *Leptogramma totta* (Schlect.) J. Sm. (*Gymnogramma totta* Schlect., nom. nov. for *Polypodium tottum* Willd., not Thunb.) = *Thelypteris Pozoi* (Lag.) Morton. *Gymnogramma* subgenus *Leptogramma* (J. Sm.) Clarke, Trans. Linn. Soc. II, 1: 567. 1880. *Thelypteris* subgenus *Leptogramma* (J. Sm.) Reed, Phytologia 17: 254. 1968. [Subgenus *Stegnogramma*].

Amauropelta Kze., Farnkr. 1: 109. 1843. Type: *Amauropelta Breutelii* Kze. = *Thelypteris limbata* (Sw.) Proctor. *Thelypteris* subgenus *Amauropelta* (Kze.) A. R. Sm., Amer. Fern Jour. 63: 121. 1973.

Hemestheum Newm., Phytologist 4: append. xxi. 1851. (not Lev. 1915 = *Polystichum*), nom. superfl. for *Thelypteris* Adans. and with the same type.

Oochlamys Fée, Gen. Fil. (Mém. Fam. Foug. 5): 297. 1852. Type: *Oochlamys Rivoirei* Fée = *Thelypteris opposita* (Vahl) Ching [Subgenus *Amauropelta*].

Glaphyropteris Fée, Crypt. Vasc. Brésil 2: 40. 1873. [Not validly published by Presl as often cited]. Type: *Glaphyropteris decussata* (L.) Fée (*Polypodium decussatum* L.) = *Thelypteris decussata* (L.) Proctor. *Dryopteris* subgenus *Glaphyropteris* (Fée) C. Chr., Biol. Arb. til. Eug. Warming 80. 1911. *Thelypteris* subgenus *Glaphyropteris* (Fée) Alston, Jour. Wash. Acad. Sci. 48: 234. 1958. *Thelypteris* section *Glaphyopteris* (Fée) Morton, Amer. Fern Jour. 51: 31. 1961. [Subgenus *Steiropteris*].

Nephrodium subgenus *Lastrea* Hook., Sp. Fil. 4: 5, 84. 1862. Type: *Nephrodium Oreopteris* (Ehrh.) Desv. (*Polypodium Oreopteris* Ehrh.) = *Thelypteris limbosperma* (All.) Fuchs. *Thelypteris* subgenus *Lastrea* (Hook.) Alston, Jour. Wash. Acad. Sci. 48: 232. 1958.

Dryopteris subgenus *Steiropteris* C. Chr., Biol. Arb. til. Eug. Warming 81. 1911. Type: *Dryopteris deltoidea* (Sw.) C. Chr. (*Polypodium deltoideum* Sw.) = *Thelypteris deltoidea* (Sw.) Proctor. *Thelypteris* subgenus *Steiropteris* (C. Chr.) Iwats., Mem. Coll. Sci. Kyoto, Ser. B, 31: 31. 1964. *Steiropteris* (C. Chr.) Pic.-Ser., Webbia 28: 449. 1973.

Christella Lév., Fl. Kouy-tchéou 472. 1915. Type: *Christella parasitica* (L.) Lév. (*Polypodium parasiticum* L.) = *Thelypteris parasitica* (L.) Tard. [Subgenus *Cyclosorus*].

Thelypteris section *Parathelypteris* H. Ito, Nov. Fl. Jap. (Nakai & Honda), 4: 127. 1939. Type: *Thelypteris glanduligera* (Kze.) Ching (*Aspidium glanduligerum* Kze.). *Parathelypteris* (H. Ito) Ching, Acta Phytotax. Sinica 8: 300. 1963. *Thelypteris* subgenus *Parathelypteris* (H. Ito) R. & A. Tryon, Rhodora 84: 128. 1982.

Thelypteris section *Macrothelypteris* H. Ito, Nov. Fl. Jap. (Nakai & Honda) 4: 141. 1939. Type: *Thelypteris oligophlebia* (Baker) Ching (*Nephrodium oligophlebium* Baker). *Macrothelypteris* (H. Ito) Ching, Acta Phytotax. Sinica 8: 308. 1963. *Thelypteris* subgenus *Macrothelypteris* (H. Ito) A. R. Sm., Phytologia 34: 233. 1976.

Fig. 66.1. *Thelypteris serra,* Retreat, St. Mary, Jamaica. (Photo W. H. Hodge.)

Thelypteris subgenus *Cyclosoriopsis* Iwats., Mem. Coll. Sci. Kyoto, Ser. B, 31: 28. 1964. Type: *Thelypteris dentata* (Forsk.) E. P. St. John (*Polypodium dentatum* Forsk.) [Subgenus *Cyclosorus*].

Oreopteris Holub, Fol. Geobot. Phytotax. Praha 4: 46. 1969. Type: *Oreopteris limbosperma* (All.) Holub (*Polypodium limbospermum* All.) = *Thelypteris limbosperma* (All.) Fuchs [Subgenus *Lastrea*].

Amphineuron Holtt., Blumea 19: 45. 1971. Type: *Amphineuron opulentum* (Kaulf.) Holtt. (*Aspidium opulentum* Kaulf.) = *Thelypteris opulenta* (Kaulf.) Fosberg. [Subgenus *Cyclosorus*].

Description   Terrestrial or rupestral; stem erect, decumbent or short- to very long-creeping, stout to slender, bearing scales and usually many fibrous or sometimes rather few thick roots; leaves monomorphic to rarely dimorphic (the fertile erect and with smaller segments than the sterile), ca. 10 cm to 5 m long, borne in a crown, or clustered to widely spaced, lamina simple to 3-pinnate-pinnatifid, usually 1-pinnate or 1-pinnate-pinnatifid, pubescent,

Fig. 66.2. *Thelypteris sagittata*, Mt. Albion, St. Ann, Jamaica. (Photo W. H. Hodge.)

glandular, sparingly scaly or glabrate beneath, veins free to fully anastomosing, areolae with or without included free veinlets; sori borne on the veins on the lower surface of the lamina, nearly round to elongate, sometimes arcuate, very rarely (in subgenus *Meniscium*) the sporangia also on the leaf tissue, not paraphysate or rarely so (in subgenera *Steiropteris* and *Meniscium*), with a well-developed and reniform to small and spathulate indusium or exindusiate; spores ellipsoidal, monolete, the laesura $\frac{2}{3}$ or more the spore length, sometimes obscured by the perispore, the surface often with an appressed or raised reticulum and (or) with more or less connected, winglike, often perforate ridges, or with coarse ridges or irregularly verrucate to papillate, or echinate, (in the Old World *Trigonospora* Holtt., spores spheroidal and trilete). Chromosome number: $n = 27$, 29, 30, 31, 32, 34, 35, 36, 58, 60, 62, 64, 70, 72, 93, 116, ca. 136; $2n = 68$, 70, 72, 124, 128, 144, 246; apogamous 90, ? apogamous 108.

Some of the diversity of lamina structure of *Thelypteris* is shown in Figs. 19–35; and some venation types in Figs. 5, 7, 8. Details of the indument afford important taxonomic characters and some of the principal types are shown in Figs. 11–18. Stalked or sessile stellate trichomes (Fig. 11), or furcate ones (Fig. 12) occur especially in subgenus *Goniopteris* which also sometimes has anchor-shaped trichomes (Fig. 17). Trichomes

Fig. 66.3. *Thelypteris sancta,* Mt. Albion, St. Ann, Jamaica. (Photo W. H. Hodge.)

that are fasciculate (Fig. 13) or hooked (Fig. 16) may occur in subgenus *Amauropelta.* Sessile glands (Fig. 18) or stipitate ones occur in several subgenera. Sporangia may have setae on the capsule (Fig. 14) or stalk, or they may bear stellate trichomes (Fig. 15) or glands.

Systematics

Thelypteris is a very large and diverse genus of about 800 species and its distribution is nearly worldwide. The group is nevertheless relatively homogeneous and few of the segregate genera that are sometimes adopted are especially distinctive. The species of *Thelypteris* can be placed in about 30–35 groups but these are not readily combined into a few genera or subgenera. All of the major groups in America are treated here as subgenera. Those with a chromosome number of $n = 36$ or multiples are placed first and those with lower numbers follow. The principal systems of classification are those of Christensen (1911, 1913), Ching (1963), Morton (1963), Iwatsuki (1964) and Holttum (1971).

The treatment of this genus has been prepared in close collaboration with Alan R. Smith.

Only American groups are included in the nomenclature and taxonomy. The names of sections of *Thelypteris* adopted by Morton (1963) and A. R. Smith (1974) for American species may be found in those papers. *Athyrium decurtatum* (Link) Presl is a species of *Thelypteris,* perhaps belonging to subgenus *Amauropelta.*

The generic name *Thelypteris* Adanson has been dated prior to *Thelypteris* of Schmidel. Adanson did not include species in his genus; the first to be included as a modern binomial was *Acrostichum Thelypteris* L. by Schmidel (as Linn. Spec. Pl. 1071, no. 21). This serves to typify Adanson's name. The treatment of the genus *Lastrea* Bory included the same species (as *Polypodium Thelypteris*) and is therefore superfluous. The correct author for the name *Thelypteris* evidently is Adanson, but since the name has been conserved with the authorship of Schmidel, we follow that designation.

The basic revisions of Christensen (1907, 1909, 1913, 1920) covered all tropical American species. Recent revisions include those of Maxon and Morton (1938), on subgenus *Meniscium* and A. R. Smith's on the subgenera *Cyclosorus* (1971), *Amauropelta* (1974) and *Steiropteris* (1980), and *Amauropelta* and *Goniopteris* in Mexico (1973).

Key to Tropical American Subgenera of *Thelypteris*

a. Lamina 2-pinnate-pinnatifid or more complex.   9. *Macrothelypteris*
a. Lamina 1-pinnate-pinnatifid or less complex, or rarely 2-pinnate. b.
   b. Five or more pairs of united veins between the primary veins, these cross-veins often producing a usually free, excurrent veinlet, i.e. venation meniscioid (Fig. 5); pinnae, when present, entire or nearly so. c.
      c. Stellate or furcate trichomes (Figs. 11, 12) present, especially in the adaxial groove of the rachis, or if absent, then the sori in two rows between the primary veins (Fig. 7); indusium present or absent.
          4. *Goniopteris*
      c. Stellate or furcate trichomes absent; sori oblong, often arcuate, in a single row between the primary veins (Fig. 6); indusium absent.
          5. *Meniscium*
   b. Veins free, or one to rarely four pairs of veins between the primary veins, uniting (Fig. 7) to form a common vein extending to the sinus; pinnae, when present, shallowly or more deeply lobed, or rarely 1-pinnate. d.
      d. Stellate and (or) furcate trichomes (Figs. 11, 12) present, especially on the scales at the stem apex, on the rachis, especially in the adaxial groove and often on other parts of the lamina, or rarely absent.
          4. *Goniopteris*
      d. Stellate or furcate trichomes absent (fasciculate trichomes, Fig. 13, rarely present in subgenus *Amauropelta*). e.
         e. Sporangia setose (as in Fig. 14) and the sori elongate.
          2. *Stegnogramma*
         e. Sporangia not setose, the sori nearly round, oblong, or elongate, or if setose, then the sori nearly round. f.
            f. Lamina with all veins extending to the margin above the sinus (Fig. 8), or the segments without sinuses. g.
               g. Several pairs of lower pinnae strongly reduced, or if not, then the veins simple and the stem compact or short-creeping.   11. *Amauropelta*
               g. Lower pinnae not reduced. h.
                  h. Many veins of the sterile lamina forked; stem slender, long-creeping.   6. *Thelypteris*
                  h. Veins simple; stem stout, suberect.   3. *Steiropteris*
            f. Lamina with at least some of the basal veins extending to, connivent to, or united below the sinus. i.
               i. Base of the pinna usually with an aerophore, this sometimes also present at the base of the segment (as in Fig. 9); usually with a more or less pubescent keel on the lower surface below the sinus (Fig. 10); lamina eglandular. 3. *Steiropteris*
               i. Base of the pinna without an aerophore; without a keel or with a poorly developed one below the sinus; lower surface of the lamina often with sessile (as in Fig. 18) or stipitate glands.   1. *Cyclosorus*

Fig. 66.4. Distribution of *Thelypteris* in America, south of lat. 35° N.

American Subgenera and Species of *Thelypteris*

The description of subgenus *Cyclosorus* includes only the native American species; the other subgeneric descriptions are intended to include all of the species. The three boreal American subgenera are included here but not in the above key.

1. Subgenus *Cyclosorus*

Lamina 1-pinnate-pinnatifid, the apex gradually reduced, the base not or rarely reduced, without proliferous buds and aerophores, with acicular trichomes, with sessile or usually stipitate glands, or sometimes eglandular, one or more of the basal veins connivent to the sinus, or the basal veins joined below the sinus, indusiate, without paraphyses, sporangia without trichomes; spores with few prominent winglike ridges and (or) shorter, disconnected ones, or somewhat echinate. Chromosome number: $n = 36, 72$.

A pantropical subgenus of about 70 species, 16 native in the American tropics, with five of them extending into the southern United States.

Subgenus *Cyclosorus* is adopted here in a rather broad sense, since comparative study of the American and paleotropical species is needed for an understanding of its natural groups of species.

Thelypteris dentata (Forsk.) E. P. St. John and *T. opulenta* (Kaulf.) Fosberg (*T. extensa* (Blume) Morton) are adventive in the American tropics.

5   6

7   8   9

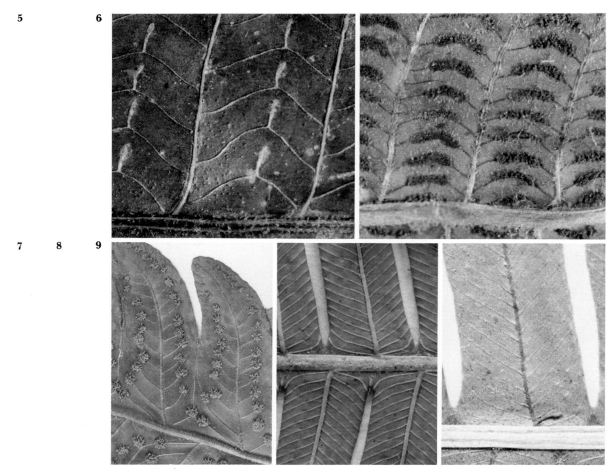

Figs. 66.5–66.9. *Thelypteris.*   **5.** Meniscioid venation, sterile pinna of *T. macrophylla,* ×6.   **6.** Arcuate sori of *T. serrata,* in a single row between the primary veins, ×6.   **7.** Sori in two rows between the primary veins, the basal veins united to form a common vein extending to the sinus, *T. pennata,* ×2.   **8.** Basal veins extending to the margin above the sinus, *T. decussata,* ×5.   **9.** Aerophores at the base of pinna-segments of *T. decussata,* ×8.

Some of the native American species are *Thelypteris grandis* A. R. Sm., *T. interrupta* (Willd.) Iwats. (*T. totta* (Thunb.) Schelpe), *T. Kunthii* (Desv.) Morton (*T. normalis* (C. Chr.) Moxley) (Fig. 19), *T. patens* (Sw.) Small, *T. hispidula* (Decne.) Reed (*T. quadrangularis* (Fée) Schelpe) and *T. serra* (Sw.) R. P. St. John (Fig. 1).

2. Subgenus *Stegnogramma*

Lamina 1-pinnate-pinnatifid, the apex gradually reduced, the base not or only slightly reduced, without proliferous buds and aerophores, with acicular trichomes, eglandular, the basal veins connivent to the sinus or anastomosing, exindusiate, without paraphyses, sporangia setose; spores echinate. Chromosome number: $n = 36, 72$.

A subgenus of about 12 species, with *Thelypteris pilosa* (Mart. & Gal.) Crawf. (Fig. 20) in the southeastern United States (Alabama) and Mexico to Honduras. Other species occur in the Azores, South Africa and eastward to Malesia and north to Japan, with most of them Asian or Malesian. Elongate sori are characteristic of this subgenus.

10   11   12

13   14   15

16   17   18

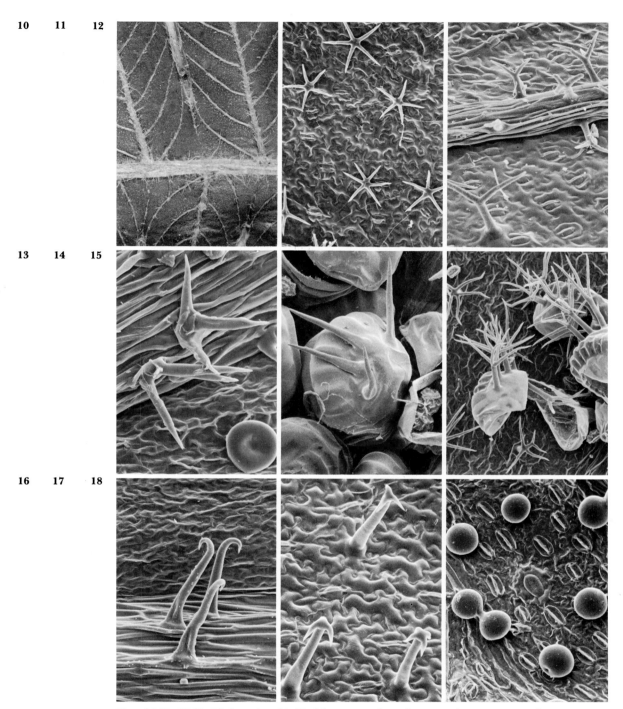

Figs. 66.10–66.18. *Thelypteris.*   10. Keel below the sinus, between pinna-segments of *T. Leprieurii,* ×8. **11.** Stellate trichomes of *T. nephrodioides,* ×100. **12.** Furcate trichomes of *T. lugubris,* ×150. **13.** Fasciculate trichomes of *T. Thomsonii,* ×250. **14.** Setose sporangia of *T. Poiteana,* × 200. **15.** Stellate trichomes on sporangia of *T. asterothrix,* × 100. **16.** Hooked trichomes of *T. heteroclita,* × 150. **17.** Anchor-shaped trichomes of *T. Biolleyi,* ×250. **18.** Sessile glands of *T. opposita,* ×125.

Two American species placed in this group by Christensen (1913), *Thelypteris polypodioides* (Raddi) Reed and *Dryopteris dasyphylla* C. Chr., do not belong here, but are of uncertain affinity.

3. Subgenus *Steiropteris*

Lamina 1-pinnate-pinnatifid, or (in *T. hottensis* and *T. Wrightii*) pinnatifid with a few pinnae at the base, the apex gradually reduced or sometimes with a nearly conform apical segment, the base usually not reduced, or (in *T. deltoidea*) strongly reduced, without or (in *T. Seemannii*) with proliferous buds, usually with aerophores, with acicular trichomes, eglandular, or (in *T. decussata*) with sessile glands, or (sometimes in *T. Leprieurii*) with stipitate glands on the indusium, the basal veins (in section *Glaphyropteris*) extending to the margin above the sinus or usually (in section *Steiropteris*) connivent to the sinus and, on the lower surface, with an often pubescent and sometimes veinlike keel extending from the sinus toward the costa, indusiate or (in *T. decussata* and *T. glandulosa*) exindusiate, without paraphyses or (in *T. valdepilosa*) paraphysate with stalked, globose glands on the receptacle, sporangia without trichomes, or (in *T. setulosa*) setose, spores with few to many prominent, connected, rugae or (in section *Glaphyropteris*) the ridges thicker, denser, and more or less connected. Chromosome number: $n = 36, 72; 2n = 72$.

A tropical American subgenus of about 20 species, treated in two sections. Section *Glaphyropteris* (Fée) Morton, treated by Christensen (1913) as a subgenus, included diverse elements, some properly placed in his subgenus *Lastrea* (treated here as *Amauropelta* except for a few species), and the remainder to be placed here. These are *Thelypteris comosa* (Morton) Morton, *T. decussata* (L.) Proctor (Fig. 23) (with four varieties), *T. Hatschbachii* A. R. Sm., *T. Mexiae* (Copel.) Ching, and *T. polyphlebia* (C. Chr.) Morton.

Some of the species of the larger section *Steiropteris* are *Thelypteris clypeolutata* (Desv.) Proctor (Fig. 21), *T. deltoidea* (Sw.) Proctor (Fig. 24), *T. glandulosa* (Desv.) Proctor, *T. hottensis* (C. Chr.) A. R. Sm., *T. Leprieurii* (Hook.) Tryon, *T. Seemannii* (J. Sm.) A. R. Sm., *T. setulosa* A. R. Sm., *T. valdepilosa* (Baker) Reed, and *T. Wrightii* (D. C. Eaton) Reed (Fig. 22).

4. Subgenus *Goniopteris*

Lamina pinnatifid, 1-pinnate-pinnatifid, or simple (in *T. Cumingiana*), the apex gradually reduced or sometimes with a conform apical segment, the base reduced or not, often with proliferous buds, without aerophores, with acicular, stellate, and (or) anchor-shaped trichomes, eglandular or rarely with stipitate glands, the basal veins connivent to or united below the sinus, indusiate and the sporangia without trichomes or exindusiate with setose sporangia; spores with prominent, short, connected, winglike ridges with erose edges and usually short-echinate processes. Chromosome number: $n = 36, 72;$? apogamous 108.

Goniopteris is a tropical American subgenus of about 60 species. Stellate trichomes on the segments and especially on the scales at the stem apex and in the adaxial groove of the rachis

Figs. 66.19–66.29. *Thelypteris* leaves. **19.** Subgenus *Cyclosorus.* *T. Kunthii,* × 0.5. **20.** Subgenus *Stegnogramma.* *T. pilosa,* × 0.5. **21–24.** Subgenus *Steiropteris.* **21.** *T. clypeolutata,* × 0.5. **22.** *T. Wrightii,* × 0.5. **23.** *T. decussata,* × 0.5. **24.** *T. deltoidea,* × 0.5. **25–29.** Subgenus *Goniopteris.* **25.** *T. cordata,* × 1.0. **26, 27.** *T. scolopendrioides.* **26.** Sterile leaf, × 0.5. **27.** Fertile leaf, × 0.5. **28.** *T. reptans,* × 0.5. **29.** *T. Poiteana,* × 0.5.

are a distinctive feature of the group. They are lacking, however, in a few species such as *T. macrotis, T. Ghiesbreghtii, T. semihastata,* and sometimes in *T. Poiteana.* Another characteristic of the subgenus is the presence of relatively few thick roots.

Some of the species of subgenus *Goniopteris* are *Thelypteris asterothrix* (Fée) Proctor, *T. Biolleyi* (Christ) Proctor, *T. blanda* (Fée) Reed, *T. cordata* (Fée) Proctor (Fig. 25), *T. Cumingiana* (Kze.) Reed, *T. Francoana* (Fourn.) Reed, *T. Ghiesbreghtii* (Hook.) Morton, *T. lugubris* (Mett.) R. & A. Tryon, *T. macrotis* (Hook.) Tryon, *T. nephrodioides* (Kl.) Proctor, *T. obliterata* (Sw.) Proctor, *T. pennata* (Poir.) Morton (*T. megalodus* (Schkuhr) Proctor), *T. Poiteana* (Bory) Proctor (Fig. 29), *T. reptans* (J. F. Gmel.) Morton (Fig. 28), *T. sagittata* (Sw.) Proctor, *T. scolopendrioides* (L.) Proctor (Figs. 26, 27), *T. semihastata* (Kze.) Ching, *T. tetragona* (Sw.) Small and *T. tristis* (Kze.) Tryon.

5. Subgenus *Meniscium*

Lamina 1-pinnate, with nearly entire pinnae, the apex with a conform apical segment, the base not reduced or (in *T. gigantea* and *T. minuscula*) simple, without buds or (in *T. reticulata* and sometimes in *T. membranacea*) with proliferous buds, without aerophores, glabrate or with acicular trichomes, eglandular, or (sometimes in *T. longifolia*) with stipitate glands, with meniscioid venation, exindusiate, without paraphyses or sometimes paraphysate, sporangia without trichomes, or setose; spores with an appressed reticulum or with prominent, perforate, winglike ridges, or echinate. Chromosome number: $n = 36, 72$.

Meniscium is a tropical American subgenus of about 20 species. The paleotropical *Thelypteris prolifera* (Retz.) Reed is excluded, more study is needed to determine its affinities. The paraphysate species *Thelypteris Andreana* and *T. arcana* have elongate glands, and *T. gigantea* has stipitate glands among the sporangia. The species of subgenus *Meniscium* have relatively few, thick roots.

Some of the species are *Thelypteris Andreana* (Sod.) Morton, *T. angustifolia* (Willd.) Proctor (Fig. 31), *T. arborescens* (Willd.) Morton (*Dryopteris permollis* Maxon & Morton), *T. arcana* (Maxon & Morton) Morton, *T. falcata* (Liebm.) Tryon, *T. gigantea* (Mett.) Tryon (Fig. 30), *T. longifolia* (Desv.) Tryon, *T. macrophylla* (Kze.) Morton, *T. membranacea* (Mett.) Tryon, *T. minuscula* (Maxon) Morton, *T. reticulata* (L.) Proctor, *T. Salzmannii* (Fée) Morton, and *T. serrata* (Cav.) Alston.

6. Subgenus *Thelypteris*

Lamina 1-pinnate-pinnatifid, the apex gradually reduced, the base not reduced, without proliferous buds and aerophores, with acicular trichomes, usually with sessile or stipitate glands, the basal veins extending to the margin above the sinus, indusiate, without paraphyses, sporangia without trichomes; spores (in *T. palustris*) with a more or less raised reticulum, or sometimes also coarsely verrucate to papillate or somewhat echinate, or (in *T. confluens*) coarsely echinate. Chromosome number: $n = 35; 2n = 70$.

Figs. 66.30–66.35. *Thelypteris* leaves. **30, 31.** Subgenus *Meniscium.* **30.** *T. gigantea*, × 0.25. **31.** *T. angustifolia*, ×0.5. **32–35.** Subgenus *Amauropelta*. **32.** *T. consimilis*, ×0.5. **33.** *T. sancta*, ×1.0. **34.** *T. pusilla*, ×1.0. **35.** *T. pteroidea*, ×0.5.

The two species included in subgenus *Thelypteris* have forked veins, especially in the sterile lamina. *Thelypteris palustris* Schott occurs in eastern North America and is disjunct across Eurasia to Japan. The record of *Dryopteris tremula* Christ from near Morelia, Michoacán, Mexico, is questionable. The only collections are those by Arsène and these resemble very closely some specimens of *Thelypteris palustris* from Louisiana, where he also collected. *Thelypteris confluens* (Thunb.) Morton occurs in Buenos Aires, Argentina (*T. Cabrerae* (Weath.) Abbiatti), Africa, and from India to New Zealand and New Guinea.

7. Subgenus *Parathelypteris*

Lamina 1-pinnate-pinnatifid, the apex gradually reduced, the base reduced or not, without proliferous buds and aerophores, with acicular trichomes, usually with sessile glands, the basal veins extending to the margin above the sinus, indusiate, without paraphyses, sporangia without trichomes: spores with a raised reticulate surface and few prominent ridges. Chromosome number: $n = 27, 31, 62, 64$.

Parathelypteris is a subgenus of about 12 species. Ten are distributed in continental southeastern Asia, Ceylon, Malesia and north to Japan. The remaining species are *T. nevadensis* (Baker) Morton of the western United States and adjacent Canada, and *T. noveboracensis* (L.) Nieuwl. in the eastern United States and adjacent Canada.

8. Subgenus *Lastrea*

Lamina 1-pinnate-pinnatifid, the apex gradually reduced, the base reduced, without proliferous buds and aerophores, with acicular trichomes, with sessile glands, the basal veins extending to the margin above the sinus, indusiate, without paraphyses, sporangia without trichomes; spores (in *T. limbosperma*) with few, prominent, short, or more or less connected, winglike ridges, or (in *T. quelpartensis*) with an appressed basal reticulum and prominent echinate elements. Chromosome number: $n = 34$, ca. 136.

Lastrea is a small, disjunct subgenus of three species. The veins, especially those of the sterile lamina, are often forked. *Thelypteris limbosperma* (All.) Fuchs occurs in Europe, the near East, northwestern North America and Newfoundland; *T. quelpartensis* (Christ) Ching in eastern Asia; and *T. Elwesii* (Baker) Ching in the Himalayas.

9. Subgenus *Macrothelypteris*

Lamina 2-pinnate to 3-pinnate-pinnatifid, the apex gradually reduced, the base not reduced, without proliferous buds and aerophores, with acicular trichomes, eglandular or with stipitate glands, the basal veins not extending to the margin or sinus, usually indusiate, without paraphyses, sporangia often with short-stipitate glands; spores reticulate with coarse, more or less connected, perforate ridges. Chromosome number: $n = 31, 62, 93$.

Macrothelypteris is a paleotropical subgenus of about nine species, distributed from the Mascarene Islands east to the Pacific. A single species, *Thelypteris Torresiana* (Gaud.) Alston is widely adventive in tropical America, where it has also been known as *T. setigera* (Blume) Ching or *T. uliginosa* (Kze.) Ching. It is readily distinguished from all other species in America by its 2-pinnate-pinnatifid to 3-pinnate-pinnatifid lamina.

Leonard (1972) reported the spread of *Thelypteris Torresiana* in the southeastern United States, from early records in central Florida to recent ones in South Carolina, northern Mississippi and eastern Texas.

10. Subgenus *Phegopteris*

Lamina bipinnatifid, the apex gradually reduced, the base not reduced or (in *T. decursivepinnata*) reduced, without proliferous buds and aerophores, with acicular trichomes and stipitate glands, with a basal vein extending to the sinus or the basal veins extending to the margin above the sinus, exindusiate, without paraphyses, sporangia often with short, stipitate glands, spores with an appressed reticulum with few, slightly raised, perforate ridges, or coarsely tuberculate to somewhat papillate. Chromosome number: $n = 30, 60$; apogamous 90.

Subgenus *Phegopteris* is a group of three species: *Thelypteris Phegopteris* (L.) Slosson of circumboreal, disjunct distribution, *T. hexagonoptera* (Michx.) Weath. of eastern North America, and *T. decursivepinnata* (van Hall) Ching of eastern Asia.

11. Subgenus *Amauropelta*

Lamina 1-pinnate or usually 1-pinnate-pinnatifid, or (in *T. pteroidea*) to 2-pinnate, the apex gradually reduced, the base usually reduced, without proliferous buds or (in *T. Linkiana*) with buds, aerophores present or not, with acicular, hooked, or fasciculate trichomes, eglandular or with sessile or stipitate glands, the basal veins extending to the margin above the sinus, indusiate or exindusiate, without paraphyses, sporangia without trichomes or (in *T. concinna*) setose; spores with a raised reticulum, the surface forming a fairly uniform network, or sometimes with prominent coarse ridges with basal areas more or less coalescent. Chromosome number: $n = 29, 58, 116$.

Amauropelta is a large, primarily tropical American subgenus of about 200 species. Five species are in Africa, Madagascar and the Mascarene Islands, and three in the Pacific, including the Hawaiian Islands. The subgenus has been treated in nine sections by Smith (1974). These are characterized by a combination of features such as the orientation of the stem, the type and distribution of trichomes and glands, the presence or absence of aerophores, the position of the sori, and the presence or absence of an indusium.

A few of the species are *Thelypteris aspidioides* (Willd.) Tryon, *T. brachypus* (Sod.) R. & A. Tryon, *T. cheilanthoides* (Kze.) Proctor, *T. concinna* (Willd.) Ching, *T. consimilis* (Baker) Proctor (Fig. 32), *T. deflexa* (Presl) Tryon, *T. Funckii* (Mett.) Alston, *T. heteroclita* (Desv.) Ching, *T. Linkiana* (Presl) Tryon, *T. Mettenii* (Copel.) Abbiatti, *T. opposita* (Vahl) Ching, *T. Pavoniana* (Kl.) Tryon, *T. ptarmica* (Mett.) Reed, *T. pteroidea* (Kl.) Tryon (Fig. 35), *T. pusilla* (Mett.) Ching (Fig. 34), *T. resinifera* (Desv.) Proctor, *T. sancta* (L.) Ching (Fig. 33), and *T. Thomsonii* (Jenm.) Proctor.

Ecology (Figs. 1–3)

Primarily a genus of wet to moist forested regions, *Thelypteris*, because of the numerous species, occupies a diversity of other habitats.

In tropical American most species grow in low elevation rain

Figs. **66.36–66.47.** *Thelypteris* spores, × 1000, surface detail, × 5000 or × 10000. **36–40.** Subgenus *Amauropelta.* **36.** *T. cheilanthoides,* the laesura central, Colombia, *Barkley & Juajibioy 73110.* **37.** *T. Linkiana,* detail of raised reticulum, Costa Rica, *Skutch 2330,* × 10,000. **38.** *T. Mettenii,* abraded surface, the underlaying papillate surface, above, reticulate outer stratum, below, Brazil, *Rambo 31221,* × 10,000. **39, 40.** *T. deflexa,* Ecuador, *Holm-Nielsen 4405.* **39.** Ridged-perforate surface, thae laesura central. **40.** Surface detail of reticulum, fused, × 5000. **41–43.** Subgenus *Steiropteris.* **41.** *T. decussata,* winglike ridges with erose edges. Guadeloupe, *Duss 4108.* **42.** *T. glandulosa,* winglike ridges, Ecuador, *Holm-Nielsen 4615.* **43.** *T. Leprieurii,* detail of rugulose surface, Peru, *Tryon & Tryon 5279,* × 5000. **44–47.** Subgenus *Goniopteris.* **44.** *T. tristis,* winglike ridges, Ecuador, *Pinkley 27.* **45, 46.** *T. Poiteana,* Jamaica, *Gastomy 34.* **45.** Dense winglike ridges. **46.** Detail of surface with echinate elements. × 10,000. **47.** *T. tristis,* surface detail of papillate and echinate elements, Ecuador, *Pinkley 27,* × 5000.

forests, montane forests, or in cloud forests. They are also found in secondary forests especially along the borders, in gallery forests, thickets, swamps, wet or moist savannahs, cliffs and rocky places.

Some species especially of subgenera *Cyclosorus* and *Meniscium* are pioneers in open, disturbed sites such as landslides and roadside banks. Subgenus *Goniopteris* frequently occurs in neglected coffee, cacao and banana plantations. *Steiropteris* is mainly confined to shade in forests. Several species are calciphiles, especially in subgenus *Goniopteris* in the Greater Antilles and in subgenus *Cyclosorus.* Some may be obligate calciphiles but additional data is needed to confirm their ecology.

Thelypteris interrupta, with slender creeping stems to 2 m or more long, is often in boggy places or occasionally grows on floating mats of aquatic plants. Species as *Thelypteris angustifolia, T. aspidioides, T. ptarmica, T. sancta,* and *T. Francoana* may be rheophytes on rocks in streams or on rocky banks where the plants are subject to periodic inundation. These species all have rather streamlined leaves compared to their relatives. *Thelypteris reptans* reproduces vigorously by its gemmiferous leaves forming colonies. A few species, as *T. pteroidea,* have scandent leaves to ca. 3 m long. Others, as *T. brachypus,* are subarborescent with an erect stem to 40 or 50 cm tall.

Thelypteris grows from sea level to ca. 4100 m. *Thelypteris Pavoniana* and *T. Funckii* are at the highest sites in the Andes. Species most commonly occur at 500–2500 m and are infrequent above that altitude.

Geography (Fig. 4)

The distribution of *Thelypteris* is nearly worldwide, but with relatively few species in boreal and south-temperate regions.

In America, *Thelypteris* occurs from Newfoundland and the Aleutian Islands south to the United States to Maule in Chile and San Luis, Mendoza, and Buenos Aires in Argentina; also on Bermuda, Cocos Island and the Galápagos Islands.

In tropical America, the greatest diversity of species and the highest endemism is in the Andes from western Venezuela to

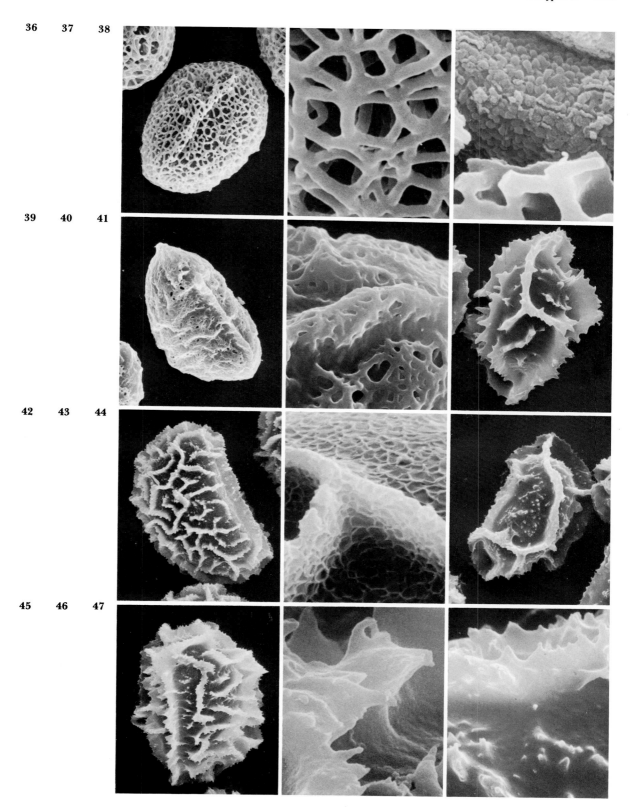

Bolivia. Secondary centers of diversity and endemism are in southern Mexico to Costa Rica, in the Greater Antilles, and in southeastern Brazil. Distributions of many species of subgenus *Amauropelta* dominate the range of the genus. Notable geographic features of other subgenera include *Cyclosorus* with no endemics in the Andes, *Steiropteris* with poor representation in Mexico and Central America, *Goniopteris* with centers in the eastern slopes of the Andes and the Greater Antilles, and *Meniscium* with greatest diversity in the low, east slopes of the Andes and eastward through Amazonia and the Guayana region.

The primarily extratropical subgenera *Parathelypteris, Lastrea* and *Phegopteris* do not have close relationships with species of tropical America, but appear to be closest to elements in southeastern Asia and in Malesia. Subgenus *Thelypteris* is amphitropical in America.

Spores

Contours of ridged and winged spores of *Thelypteris* are derived from outer perispore folds that are compressed and more or less protruding as in spores of several species of *Dryopteris*. They represent a more derived type of wall structure than that in which contours are based on the exospore. Spores provide new data useful in characterizing subgenera of *Thelypteris* and suggest possible relationships within the genus. There are three main types of surface architecture in spores of the American species. The ridged and reticulate forms were recognized in *Thelypteris* by Wood (1973). In addition to these, an echinate type is characteristic of subgenus *Stegnogramma*. Spores of *Amauropelta* are consistently reticulate but vary in the prominence and degree of fusion of the network. In *T. cheilanthoides* and *T. Linkiana* there is a uniform reticulum (Figs. 36, 37). The abraded surface of *T. Mettenii* spores (Fig. 38) shows an irregular, papillate stratum underlying the reticulum. The surface is undulate with a more or less fused reticulum in *T. deflexa* (Figs. 39, 40). Spores of subgenus *Steiropteris* consistently have prominent winglike ridges as in *T. glandulosa* (Fig. 42). The surface especially between the ridges is rugulose in *T. Leprieurii* (Fig. 43) or fairly smooth in *T. decussata* (Fig. 41). Spores of subgenus *Goniopteris* have prominent winglike ridges and sparse projections as in *T. tristis* (Figs. 44, 47) and *T. Poiteana* (Figs. 45, 46), but lack low wrinkles. The single American species of subgenus *Stegnogramma, T. pilosa,* has distinctive echinate spores (Fig. 48) similar to those of Old World species of the subgenus. Spores of subgenus *Cyclosorus* are not as consistent as those of other groups. In *T. dentata* there are sparse, prominent, winglike ridges (Fig. 49), while in *T. interrupta* (Fig. 50) the ridges are less prominent and disconnected and also there are papillae. In subgenus *Meniscium* the spores usually have an appressed reticulum and coarse, low or raised, perforate ridges as in *T. Salzmannii* (Figs. 51, 52). The ridges may be more or less obscured by dense, papillate to echinate elements as in *T. falcata* (Figs. 53, 56). *Meniscium* spores resemble those of subgenera *Amauropelta, Parathelypteris, Phegopteris* and

48   49   50

51   52   53

54   55   56

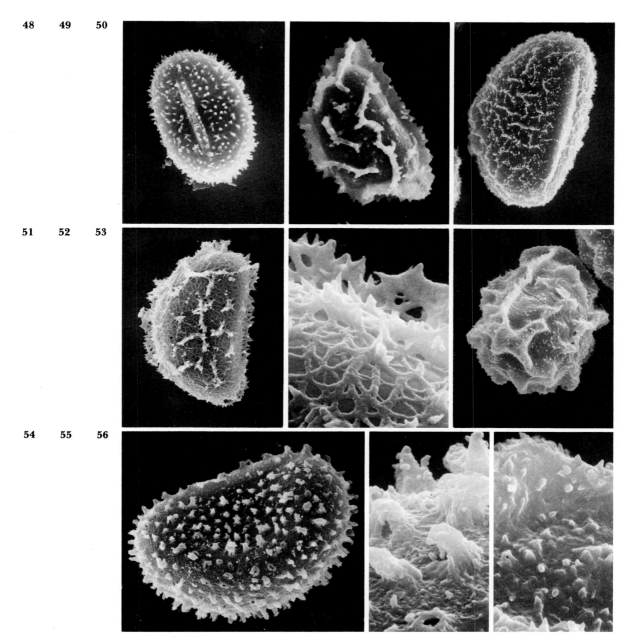

Figs. 66.48–66.56. *Thelypteris* spores, × 1000, surface detail, × 5000.   **48.** Subgenus *Stegnogramma*. *T. pilosa,* echinate, the laesura central, Mexico, *Reineck* in 1899.   **49, 50.** Subgenus *Cyclosorus*.   **49.** *T. dentata,* winglike ridges with erose edges, St. Lucia, *Proctor 17533*.   **50.** *T. interrupta* with low ridges and papillae, British Guiana, *De la Cruz 3811*.   **51–53, 56.** Subgenus *Meniscium*.   **51, 52.** *T. Salzmannii,* Ecuador, *Holm-Nielsen 1050*.   **51.** Appressed reticulum and perforate ridges.   **52.** Surface detail.   **53.** *T. falcata* coarsely ridged, papillate, Nicaragua, *C. L. Smith 2030*.   **54, 55.** Subgenus *Thelypteris*. *T. confluens,* Argentina, *Schulz 831*.   **54.** Coarsely echinate surface.   **55.** Surface detail.   **56.** *T. falcata,* detail of papillate and echinate elements, *C. L. Smith 2030*.

Lastrea, which also have a basically reticulate surface often with raised, perforate ridges.

Spores are helpful in assessing relationships of species in subgenus *Thelypteris* (Tryon, 1971). There are two main types of surface, one primarily reticulate and the other echinate. *Thelypteris palustris* spores examined in eight collections from widely disjunct areas in the eastern United States and also from Bermuda, Japan and Europe have a basic, appressed reticulate surface that may have coarse ridges, tubercules, or papillate to echinate elements. The spores of *T. confluens* from Argentina (Figs. 54, 55) have a prominent, coarse echinate architecture similar to those of collections from Africa, India and New Zealand.

Cytology

Chromosome numbers in *Thelypteris* range from $n = 27$ to 136 including aneuploid series between 36 to 34 and 32 to 29, as well as several polyploid levels. Octoploids occur on bases of 29 and 34 and hexaploids on 35. Most reports are either 72 or the diploid 36, which is the most common number reported and is regarded by Smith (1971) as a probable ancestral base number. The subgenera *Cyclosorus, Goniopteris, Steiropteris* and *Meniscium* are uniformly 36 or multiples on this base. Subgenus *Amauropelta,* including several reports from the American tropics, is uniformly based on 29. The lowest number known for the genus, 27, in subgenus *Parathelypteris* along with 31 and 64 represent an unusual diversity of numbers among the American subgenera. Subgenus *Thelypteris* is clearly 35 based on records for *T. palustris* from North America, Europe, and Japan. Other reports of 35 or its multiples, in two Old World subgenera, are in species not closely allied to subgenus *Thelypteris.*

The chromosome number correlates with the type of spore architecture in several subgenera and provides useful characters for assessing relations of the groups. The subgenera *Cyclosorus, Goniopteris* and *Steiropteris,* based on 36, have spores with prominent winglike ridges and appear to represent less specialized elements in the genus. In contrast to these, subgenera *Amauropelta, Phegopteris* and *Parathelypteris* have lower chromosome numbers which are evidently derived by reduction, and they also have a more complex, reticulate spore architecture.

Observations

Davidonis and Ruddat (1973) demonstrated that the roots of *Thelypteris Kunthii* (as *T. normalis*) produced allelopathic substances, called Thelypterin A and B. A later study (Davidonis, 1976) reported that leaves and gametophytes of this species also contained Thelypterin A; and that the roots of *Thelypteris dentata* contained Thelypterin A and B. Considering the more or less weedy nature of these two species, the compounds may be advantageous in reducing intraspecific competition in the immediate vicinity of parental plants.

Literature

Christensen, C. 1907. Revision of the American species of *Dryopteris* of the group of *D. opposita.* Danske Vidensk. Selsk. Skrift. VII, 4: 249–336.

Christensen, C. 1909. The American ferns of the group of *Dryopteris opposita* contained in the U.S. National Museum. Smiths. Misc. Coll. 52: 365–396.

Christensen, C. 1911. On a natural classification of the species of *Dryopteris.* Biol. Arb. til. Eug. Warming. 73–85.

Christensen, C. 1913. A monograph of the genus *Dryopteris,* Part I, the tropical American pinnatifid-bipinnatifid species. Danske Vidensk. Selsk. Skrift. VII, 10: 55–282.

Christensen, C. 1920. A monograph of the genus *Dryopteris,* Part II, the tropical American bipinnate-decompound species. Danske Vidensk. Selsk. Skrift. VIII, 6: 1–32.

Davidonis, G. H. 1976. The occurrence of Thelypterin in ferns. Amer. Fern Jour. 66: 107–108.

Davidonis, G. H., and M. Ruddat. 1973. Allelopathic compounds, Thelypterin A and B in the fern *Thelypteris normalis.* Planta (Berl.) 111: 23–32.

Holttum, R. E. 1971. Reference under the family.

Iwatsuki, K. 1964. Reference under the family.

Leonard, S. W. 1972. The distribution of *Thelypteris Torresiana* in the southeastern United States. Amer. Fern Jour. 62: 97–99.

Maxon, W. R., and C. V. Morton. 1938. The American species of *Dryopteris* subgenus *Meniscium.* Bull. Torrey Bot. Cl. 65: 347–376.

Morton, C. V. 1963. The classification of *Thelypteris.* Amer. Fern Jour. 53: 149–154.

Smith, A. R. 1971. Systematics of the neotropical species of *Thelypteris* section *Cyclosorus.* Univ. Cal. Publ. Bot. 59: 1–136.

Smith, A. R. 1973. The Mexican species of *Thelypteris* subgenera *Amauropelta* and *Goniopteris.* Amer. Fern Jour. 63: 116–127.

Smith, A. R. 1974. A revised classification of *Thelypteris* subgenus *Amauropelta.* Amer. Fern Jour. 64: 83–95.

Smith, A. R. 1980. Taxonomy of *Thelypteris* subgenus *Steiropteris,* including *Glaphyropteris.* Univ. Cal. Publ. Bot. 76: 1–38.

Tryon, A. 1971. Structure and variation in spores of *Thelypteris palustris.* Rhodora 73: 444–460.

Tryon, A. F., R. Tryon, and F. Badré. 1980. Classification, spores, and nomenclature of the Marsh Fern. Rhodora 82: 461–474.

Wood, C. C. 1973. Spore variation in the Thelypteridaceae, *in*: A. C. Jermy et al., The phylogeny and classification of the ferns, pp. 191–202. Bot. Jour. Linn. Soc. 67, Suppl. 1.

Family 18.   Dryopteridaceae

Dryopteridaceae Herter, Rev. Sudam. Bot. 9: 15. 1949; description from Baker, Mart. Fl. Brasil. 1(2): 458, sub tribe Aspidieae. Type: *Dryopteris* Adans.

Family synonyms of American Dryopteridaceae are placed under the tribes, where they correspond more closely with the taxonomy.

Description

Stem erect, very rarely arborescent, decumbent, or to long-creeping or long-scandent, stout to very slender, with a dictyostele, usually indurated, bearing scales; leaves ca. 1 cm to 3 m long, entire or lobed to usually pinnate, very rarely pedate or flabellate, circinate in the bud, petiole without stipules, rarely laterally expanded; sporangia borne in round to very long-linear abaxial sori, the receptable on a vein or at a vein-tip, elevated or not, very rarely stalked, indusiate, the indusium squamiform, patelliform, reniform, peltate, elongate to long-linear, or very rarely globose, or exindusiate and sometimes the sporangia borne over the abaxial surface, sporangia short- to long-stalked, the stalk 2- to 3-rowed below its apex, the annulus vertical or nearly vertical and interrupted by the stalk; homosporous, spores lacking chlorophyll or rarely with chlorophyll (green). Gametophyte epigeal, with chlorophyll, more or less obcordate, to ligulate, rarely branched, slightly to definitely thickened centrally, glabrous or often with unicellular glands or trichomes, rarely with multicellular trichomes, archegonia borne on the lower surface, mostly in the central region, antheridia 3-celled, borne on the lower surface, among or mostly apart from the archegonia.

Comments on the Family

The Dryopteridaceae are a family of worldwide distribution with over 50 genera, 30 of them American. It is a large and diverse group, especially characterized by monolete, spheroidal spores and chromosome numbers based on $n = 40$ or 41.

The number of genera recognized differs widely among authors, especially with respect to the representatives in eastern Asia which is the center of diversity of the family. Study, especially on a worldwide basis, is essential before many of the genera and their relationships are adequately known.

The publication of the family name Dryopteridaceae by Herter was very brief but it is a valid and legitimate new name for a new family. The protologue refers to the latin description of the Aspidieae by Baker in Flora Brasiliensis, which is sufficient to conform with Arts. 32 and 36 of the International Code of Botanical Nomenclature. The earlier name Aspidiaceae is illegitimate, being based on *Aspidium,* an illegitimate generic name.

The fossil record of the Dryopteridaceae is meagre. One of the early records is *Onoclea hebridica* (Forbes) Bell from the late Cretaceous of British Columbia.

The groups we recognize as tribes of Dryopteridaceae are

generally considered to be related, except for the Oleandreae which are sometimes placed with the Davalliaceae. They are treated as families by Pichi-Sermolli (1977) and as subfamilies by Crabbe et al. (1975). Some are more distinctive than others, but together they form a large evolutionary group. Relationships of the family are uncertain but the Dryopteridaceae appear to be more closely related to the Aspleniaceae than to other families.

The tribe **Peranemeae** Presl, Tent. Pterid. 64. 1836, as Peranemaceae, has three or more vascular bundles in the petiole, the lamina is usually large and decompound, and bears stout, multicellular trichomes, especially on the adaxial side, and the receptacle is elevated. It is an Old World tribe of four genera: *Acrophorus* Presl, *Nothoperanema* (Tagawa) Ching, *Peranema* Don (*Diacalpe* Blume) and *Stenolepia* vAvR.

Peranemaceae (Presl) Ching, Sunyatsenia 5: 246. 1940, is a latter homonym of Peranemaceae Klebs, 1892, an illegitimate name for a family of Euglenophyta, based on *Peranema* Dujardin, 1841, not Don, 1825.

The tribe **Dryopterideae** is terrestrial, rupestral or rarely the stem is scandent-epiphytic, the leaves are usually monomorphic, although strongly dimorphic in a few genera, they have three or more vascular bundles in the petiole and the pinnae or pinnules are rarely articulate. It is a large tribe centering on *Ctenitis, Tectaria, Dryopteris,* and *Polystichum,* with many divergent lines of evolution.

The tribe **Physematieae** has two vascular bundles in the petiole, these uniting close to the base or above it (more than two vascular bundles are reported in the petiole of *Diplazium pseudo-Doederleinii* Hayata of Formosa) and the leaves are monomorphic or nearly so. It is a large group, comprising the traditional genera *Athyrium* and *Diplazium* and their numerous proposed segregates.

The tribe **Onocleeae** has two vascular bundles in the petiole, these joining in the rachis, the leaves are strongly dimorphic, the sori are enclosed by modified fertile segments, the receptacle is elevated, the indusium is very fragile, when present, and the spores contain chlorophyll and are green. It is a small tribe of three distinctive genera.

The tribe **Oleandreae** has a long, often epiphytic stem, the leaves are monomorphic and articulate, and the sporangia are borne in usually indusiate sori. It is a small but distinctive tribe of two or three genera.

The tribe **Bolbitideae** is often epiphytic and the stem is sometimes scandent, the stem and stele are dorsiventral, the leaves are usually dimorphic, the fertile portions reduced, the leaves and pinnae are sometimes more or less articulate, and the sporangia are borne over the abaxial surface (acrostichoid). It is a rather large group, with many species especially in *Elaphoglossum.*

Key to Genera of Dryopteridaceae in America

a. Sporangia acrostichoid, generally distributed over the surface of the fertile segment.   b.
  b. Fertile lamina 1-pinnate-pinnatifid or more complex.
                                                 80. *Polybotrya,* p. 536
  b. Fertile lamina 1-pinnate or less complex, or 1-pinnate-pinnatifid only at the basal portion of the lamina.   c.
    c. Sterile lamina with anastomosing veins, rarely simple, entire and then acute to acuminate and with prominent primary veins.   d.
      d. Stem scandent-epiphytic, pinnae of the sterile lamina without lateral veins, paraphyses present.   95. *Lomagramma,* p. 613
      d. Stem terrestrial or rupestral, or scandent-epiphytic and the pinnae of the sterile lamina with prominent lateral veins, paraphyses absent.   93. *Bolbitis,* p. 600
    c. Sterile lamina with free veins, or if rarely anastomosing then simple, entire, obtuse and without definite primary veins.   e.
      e. Fertile lamina 1-pinnate.   94. *Lomariopsis,* p. 607
      e. Fertile lamina entire, slightly lobed or deeply pedately lobed.   96. *Elaphoglossum,* p. 617
a. Sporangia in sori, although not always on a receptacle.   f.
  f. Fertile and sterile leaves strongly dimorphic and the veins of the sterile lamina free.   g.
    g. Sori enclosed by the indurated, brown pinnae.   89. *Matteuccia,* p. 585
    g. Sori exposed on alate axes of the fertile lamina.   68. *Atalopteris,* p. 468
    g. Sori covered by a large, persistent indusium.   79. *Maxonia,* p. 533
  f. Fertile and sterile leaves monomorphic to somewhat dimorphic, or strongly dimorphic and the veins of the sterile lamina anastomosing.   h.
    h. Indusium inferior, attached completely around the receptacle.   i.
      i. Lamina simple, entire or with basal lobes, veins anastomosing.   70. *Hypoderris,* p. 481
      i. Lamina 1-pinnate or more complex, veins free.   87. *Woodsia,* p. 577
    h. Indusium superior, attached at one point or on one side, or on both sides of a vein, or absent.   j.
      j. Petiole jointed and disarticulate at the joint.   k.
        k. Lamina entire.   92. *Oleandra,* p. 593
        k. Lamina 1-pinnate   91. *Arthropteris,* p. 590
      j. Petiole continuous.   l.
        l. Lamina 1-pinnate, the pinnae with a large basal auricle on the basiscopic side.   71. *Cyclopeltis,* p. 484
        l. Lamina simple to decompound, if 1-pinnate then the pinnae rounded to tapering at the base on the basiscopic side.   m.
          m. Three or more vascular bundles in the petiole.   n.
            n. Costa of the penultimate segments raised on the adaxial side.   o.
              o. Indusium peltate, rachis with a central ridge on the adaxial side.   72. *Rumohra,* p. 487
              o. Indusium reniform to subpeltate or absent, only lateral ridges on the adaxial side of the rachis.   p.
                p. Groove of the pinna-stalk continuous with the groove on the adaxial side of the rachis, basiscopic margin of the ultimate or penultimate segments continuous with an adaxial ridge of the axis below.   73. *Lastreopsis,* p. 491
                p. Groove of the pinna-stalk, when present, not continuous with the groove on the adaxial side of the rachis, basiscopic margin of the ultimate or penultimate segments not decurrent, or decurrent as a lateral wing on the axis below.   67. *Ctenitis,* p. 459
            n. Costa of the penultimate segments flat or grooved on the adaxial side or with slightly raised edges.   q.
              q. Lamina 1-pinnate, the pinnae entire, with the margins sharply toothed or with short to long spinules, especially toward the apex of the pinna, sori multiseriate.   75. *Cyrtomium,* p. 510
              q. Lamina simple to decompound, if the lamina 1-pinnate with entire pinnae and the sori multiseriate, then the margins of the pinnae smooth.   r.
                r. Stem dorsiventral, very long-creeping, usually scandent-epiphytic and densely scaly.   79. *Maxonia,* p. 533
                r. Stem radially symmetrical, erect, decumbent, or rather short-creeping (to a few dm), terrestrial.   s.

s. Indusium peltate, persistent or fugacious.   t.

 t. Veins free or if anastomosing the sterile, and usually the fertile, pinnae with one triangular areola between the primary veins.   u.

  u. Sori in two or more rows on each side of the costa, or if in one row then the stem creeping.

    77. *Stigmatopteris,* p. 519

  u. Sori in one row on each side of the costa, rarely a few present beyond the row, stem erect or decumbent. 78. *Polystichum,* p. 524

 t. Veins anastomosing, one or more long areolae parallel to the costa between the primary veins.

    69. *Tectaria,* p. 470

s. Indusium subpeltate, reniform, elongate-cordate, elongate or linear, or absent.   v.

 v. Leaf-tissue pellucid-punctate, with transparent glandular areas evident when observed by transmitted light.

    77. *Stigmatopteris,* p. 519

 v. Leaf-tissue not pellucid-punctate.   w.

  w. Veins anastomosing, or if free the lamina mostly to wholly pinnatifid or pinnatisect.

    69. *Tectaria,* p. 470

  w. Veins free, the lamina 1-pinnate or more complex.   x.

   x. Sorus indusiate, or if exindusiate the lamina pubescent.   y.

    y. Sori and indusia elongate.

     76. *Didymochlaena,* p. 515

    y. Sori and indusia more or less round, or the sori exindusiate. 74. *Dryopteris,* p. 496

   x. Sorus exindusiate, the lamina glabrate or scaly, not pubescent. 78. *Polystichum,* p. 524

m. Two vascular bundles in the petiole, at least basally, these uniting above.   z.

 z. Leaves strongly dimorphic, the sori enclosed by a modified portion of the fertile segment.   aa.

  aa. Sterile lamina 1-pinnate, the pinnae subcordate at the base, fertile leaf herbaceous, withering. 88. *Onocleopsis,* p. 582

  aa. Sterile lamina deeply pinnatifid or 1-pinnate basally and the pinnae attenuate at the base, fertile leaf indurated, persistent. 90. *Onoclea,* p. 587

 z. Leaves more or less monomorphic, sori covered, or not, by an indusium, not by a modified portion of the fertile segment.   bb.

 bb. Indusium peltate, lamina glandular on both surfaces.

    84. *Adenoderris,* p. 565

 bb. Indusium attached basally or laterally or absent.   cc.

  cc. Indusium attached basally, arching over the sorus.

    85. *Cystopteris,* p. 568

  cc. Indusium attached laterally or absent.   dd.

   dd. Lamina triangular, indusium absent, stem very slender and long-creeping. 86. *Gymnocarpium,* p. 574

   dd. Lamina elongate, or if triangular then the sori indusiate.   ee.

    ee. Lamina 1-pinnate, the pinnae entire, veins anastomosing toward the margin, pinnae glabrous beneath, sori not branched. 83. *Hemidictyum,* p. 561

    ee. Lamina simple and entire to decompound, veins usually free, if the lamina 1-pinnate and the pinnae entire and the veins anastomosing, then the pinnae minutely scaly beneath and the sori branched with the veins.   ff.

     ff. Sori more or less elongate to linear, usually indusiate, some or most of them on both sides of a vein and the indusia separate distally, spores usually with prominent winglike folds.

      81. *Diplazium,* p. 543

     ff. Sori roundish and exindusiate, or somewhat elongate, on one side of a vein or some hook-shaped and the indusium continuous distally, spores usually with coarse, rugose folds. 82. *Athyrium,* p. 555

Literature

Crabbe, J. A., A. C. Jermy, and J. T. Mickel. 1975. A new generic sequence for the pteridophyte herbarium. Fern Gaz. 11: 141–162.

Pichi-Sermolli, R. E. G. 1977. Tentamen Pteridophytorum genera in taxonomicum ordinem redigendi. Webbia 31: 313–512.

18a.　Tribe Dryopterideae

Aspidiaceae Frank, Syn. Pflanzenf. (Leunis), ed. 2, 3: 1469. 1877, *illegit.*, based on *Aspidium* Sw. *illegit.*

A tribe of 20 or more genera, with 14 in America. The genera confined to the Old World are: *Hemigramma* Christ, *Heterogonium* Presl (*Ctenitopsis* Ching), *Lithostegia* Ching, *Pleocnemia* Presl, *Psomiocarpa* Presl, and *Pteridrys* C. Chr. & Ching, and of less certain status, *Arcypteris* Underw. (*Dictyopteris* Presl), *Cyrtogonellum* Ching, *Cyrtomidictyum* Ching, *Dryopolystichum* Copel., *Luerssenia* Luers., *Stenosemia* Presl, and *Tectaridium* Copel.

The grooves and ridges on the adaxial surface of the rachis and pinna-rachises have been emphasized as important characters, especially among genera of the Dryopterideae, by Holttum (1960), Tindale (1965), and Pichi-Sermolli (1977). While these structures undoubtedly provide characters of systematic value, they do vary depending on the complexity of the lamina and the position of the pinnae. Additional study, especially of fresh material, is required to refine the characters of the major axes among the genera.

The ridges and grooves of the costa, or axis of the next lower order, often provide more useful characters than those of the major axes. In some genera, as *Lastreopsis,* the penultimate segments have the costa raised on the adaxial surface, and their basal segments have the margin continuous with a ridge of the axis that bears the costa (Fig. 73.3). In other genera, as *Dryopteris,* the penultimate segments have the costa grooved on the adaxial surface and the basal segments have a basiscopic ridge that is continuous with a ridge of the axis that bears the costa (Fig. 74.4).

Literature

Christensen, C. 1913. A monograph of the genus *Dryopteris,* part I. The tropical American pinnatifid-bipinnatifid species. Danske Vidensk. Selsk. Skrift. VII, 10: 55–282.

Christensen, C. 1920. A monograph of the genus *Dryopteris,* part II. The tropical American bipinnate-decompound species. Danske Vidensk. Selsk. Skrift, VIII, 6: 3–132.

Holttum, R. E. 1960. Vegetative characters distinguishing the various groups of ferns included in *Dryopteris* of Christensen's Index Filicum and other ferns of similar habit and sori. Gard. Bull. Singapore 17: 361–367.

Manton, I. 1950. Problems of Cytology and Evolution in the Pteridophyta. 316 pp. Cambridge Univ. Press, Cambridge, England.

Pichi-Sermolli, R. E. G. 1977. Fragmenta Pteridologiae, VI. Webbia 31: 237–259.

Smith, A. R., and J. T. Mickel. 1977. Chromosome counts for Mexican ferns. Brittonia 29: 391–398.

Tindale, M. D. 1965. A monograph of the genus *Lastreopsis* Ching. Contrib. N. S. Wales Nat. Herb. 3: 249–339.

Walker, T. G. 1966. A cytotaxonomic survey of the pteridophytes of Jamaica. Trans. Roy. Soc. Edinburgh 66: 169–237.

Walker, T. G. 1973. Additional cytotaxonomic notes on the pteridophytes of Jamaica. Trans. Roy. Soc. Edinburgh 69: 109–135.

67. *Ctenitis*

Figs. 67.1–67.29

Ctenitis (C. Chr.) C. Chr., Verdoorn, Man. Pterid. 543. 1938. *Dryopteris* subgenus *Ctenitis* C. Chr. Biol. Arb. til. Eug. Warming. 77. 1911. Type (Ching, Fan Mem. Instit. Biol. Bot. 8: 275. 1938): *Dryopteris ctenitis* (Link) O. Ktze. (*Aspidium ctenitis* Link) = *Ctenitis distans* (Brack.) Ching.

Description

Terrestrial or rupestral; stem erect, rarely arborescent to 3 m tall, or decumbent or (in *C. protensa*) creeping, rather small to stout, bearing scales and usually many, fibrous roots; leaves monomorphic or nearly so, ca. 20 cm to 4 m long, borne in a crown or cluster or spaced, lamina 1-pinnate-pinnatifid to 4-pinnate, with various kinds of 1- to few- to many-celled trichomes, also often scaly, sometimes with bullate scales, veins free; sori roundish, borne on the veins or at their tip, receptacle somewhat elevated, not paraphysate, trichomes similar to those on other parts of the lamina rarely in the sorus, covered by a reniform to nearly orbicular indusium with a narrow sinus, sometimes small or fugacious, or exindusiate; spores somewhat ellipsoidal, monolete, the laesura $\frac{1}{2}$ to $\frac{3}{4}$ the spore length, echinate with few, coarse projections, or spinulose, or with parallel ridges with sharply erose margins, or saccate, or with longer inflated folds, or with winglike projections and the surface reticulate-echinate to nearly smooth. Chromosome number: $n = 41, 82$.

Ctenitis is characterized by the raised costa on the adaxial surface of the segments (Fig. 3). The sori may be indusiate as in *C. protensa* (Fig. 5) or exindusiate as in *C. pulverulenta* (Fig. 4). The petiole scales are of diverse form and often characterize species. Many species have simple veins and medial sori (Fig. 6). Scales are long, narrow, pale colored and form a dense mat at the base of the petiole in *C. Sloanei* (Fig. 7), and they are broader and dark colored on the petiole of *C. atrogrisea* (Fig. 8). The leaves are generally large with complex architecture. The lamina or portions of it, illustrated for several species in Figs. 9–21, show the diversity in the form of the leaves.

Systematics

Ctenitis is a tropical and temperate genus of some 130 species, with about 75 in America. It is characterized by the costae that are raised on the upper surface of the segments, by short, multicellular trichomes, especially on the major axes of the lamina, and by the groove of the pinna-stalk not connecting to the

Fig. 67.1. *Ctenitis equestris,* north of Jalapa, Veracruz, Mexico. (Photo W. H. Hodge.)

groove of the rachis. On the basis of Asiatic species, two subgenera were recognized by Ching (1938) and two sections, each with subsections, by Ito (1939). The American elements of the genus were arranged into five species-groups by Christensen (1920). A suitable infrageneric classification remains to be worked out for the whole genus. The spores especially need to be reviewed in relation to such a classification, as considerable diversity has been discovered since Christensen considered echinate spores as a generic character.

Ctenitis appears to be closely related to *Tectaria* and also to *Lastreopsis,* while *Atalopteris* is an obvious derivative from it. A relationship of *Ctenitis* with the Cyatheaceae is suggested by the arborescent stem, the large, decompound leaves, and the bullate lamina scales that occur in some species. However, the type of indusia, the form of the spore and the chromosome number differ in the two groups.

The name *Ctenitis,* originally published as a subgenus of *Dryopteris,* was treated as a genus in three separate publications in

Fig. 67.2. Distribution of *Ctenitis* in America.

1938. The earliest, by Christensen, in Verdoorn, "Manual of Pteridology" (copy received at the Gray Herbarium, May 28, 1938), the next by Tardieu and Christensen in Not. Syst., Oct. 1938, and the last by Ching in Fan Mem. Instit. Biol. Bot. 8. Nov. 1938. Ching was the first author to specifically designate a type.

The American species of *Ctenitis* have been revised by Christensen (1913, 1920).

Tropical American Species

The numerous American species of *Ctenitis* need to be revised to accommodate new species and the abundant materials assembled since 1920. Christensen arranged the American species into five groups based on characters of the stem, lamina architecture, type of scales and trichomes, and details of venation. Characters of the spores correlate with these groups to a considerable extent, but his group of *Ctenitis hirta* is diverse and it seems better to place the 1-pinnate-pinnatifid species (nos. 299–304 of C. Chr., 1920) with those of the group of *C. submarginalis* which have similar spores. The decompound-leaved species of the *C. hirta* group (nos. 305–312) have spores similar to those of species in the *C. ampla* group and are thus transferred to the latter. The following synopsis is simplified and altered, as noted above, from Christensen (1920, pp. 31, 32).

Synopsis of American *Ctenitis*

a. Stem erect or decumbent, bearing the leaves in a crown or cluster.   b.
    b.   Trichomes on the lamina all short, multicellular.   c.
        c. Lamina 1-pinnate-pinnatifid or predominently 1-pinnate-pinnatisect, at least above the basal pinnae.

Group of *C. submarginalis*

A group of about 20 species in America. The spores are saccate (as in Fig. 27) in *Ctenitis cirrhosa* (Schum.) Ching, *C. falciculata* (Raddi) Ching, *C. strigillosa* (Davenp.) Copel., *C. submarginalis* (Langsd. & Fisch.) Ching (Fig. 27), *C. Tonduzii* (Christ) R. & A. Tryon and *C. velata* (Mett.) R. & A. Tryon (Fig. 14). They are regularly and coarsely echinate (Fig. 25) in *C. vellea* (Willd.) Proctor and are irregularly and coarsely echinate (as in Fig. 22) in *C. nigrovenia* (Christ) Copel.

Among other species are *Ctenitis distans* (Brack.) Ching, *C. Hemsleyana* (Baker) Copel., and *C. refulgens* (Mett.) Vareschi.

        c. Lamina 2-pinnate-pinnatifid or more complex above the basal pinnae.

Group of *C. ampla*

A group of about 15 species in America. The spores are irregularly and coarsely echinate (as in Fig. 22) in *Ctenitis ampla* (Willd.) Ching (*C. catocarpa* (Kze.) Morton, *C. nemophila* (Kze.) Ching), *C. cheilanthina* (C. Chr.) Morton (Fig. 21), *C. crystallina* (Kze.) Proctor (Fig. 18), *C. hirta* (Sw.) Ching (Fig. 22), *C. oophylla* (C. Chr.) Ching, and *C. Sloanei* (Spreng.) Morton (Fig. 13), (*Ctenitis ampla* of Christensen, 1920 and other authors, not of Willd.).

Among other species are *Ctenitis equestris* (Kze.) Ching, *C. melanosticta* (Kze.) Copel, and *C. meridionalis* (Poir.) Ching.

    b. Other kinds of trichomes on the lamina, usually in addition to short, multicellular ones, lamina 1-pinnate-pinnatifid or more complex.

Group of *C. subincisa*

A group of about 35 species in America. The spores are spinulose (as in Fig. 23) in *Ctenitis inaequalifolia* (Colla) Ching and *C. spectabilis* (Kaulf.) Kunkel. The spores have parallel ridges that are sharply erose on the margin (as in Fig. 23) in *Ctenitis atrogrisea* (C. Chr.) Ching, *C. biserialis* (Hook. & Baker) Lell., *C. connexa* (Kaulf.) Copel., *C. grandis* (Presl) Copel. (Fig. 20), *C. honesta* (Kze.) R. & A. Tryon, *C. pulverulenta* (Poir.) Copel. (Fig. 24), (*Dryopteris Karsteniana* (Kl.) Hieron), and *C. subincisa* (Willd.) Ching (Figs. 15, 16).

a. Stem creeping, bearing the leaves at intervals.

Group of *C. protensa*

Ctenitis protensa (Sw.) Ching (Figs. 9–11) of America and Africa-Madagascar has spores with winglike ridges and the surface reticulate-echinate (Fig. 29). A few other species of Africa and Madagascar are sometimes recognized in this group.

3 4 5

6 7 8

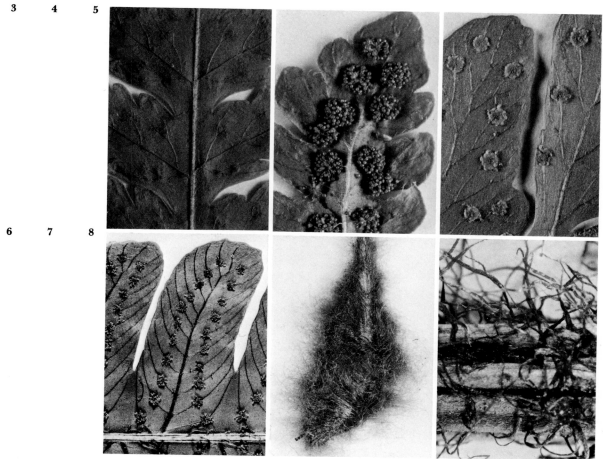

Figs. 67.3–67.8. *Ctenitis.* **3.** Raised costa on adaxial side of segment of *C. Sloanei,* ×5. **4.** Exindusiate sori of *C. pulverulenta,* ×8. **5.** Indusiate sori of *C. protensa,* ×5. **6.** Sori on simple veins of *C. refulgens,* ×4. **7.** Mat of narrow scales at base of petiole of *C. Sloanei,* ×0.5. **8.** Petiole scales of *C. atrogrisea,* ×6.

Ecology (Fig. 1)

Ctenitis is primarily a genus of wet primary forests, sometimes also growing in secondary forests, at forest borders or in thickets.

In America *Ctenitis* usually grows in forests at low elevations or in montane or cloud forests, where it occurs especially in ravines, along watercourses or on mountain slopes. It sometimes grows in gallery or swamp forests, in cutover areas or in thickets. It often grows in rocky places and rarely on cliffs. The species grow in many soil types and some, as *Ctenitis hirta,* are evidently confined to limestone. *Ctenitis* grows from near sea level to 2800 m, most often at 500–1800 m.

Arborescent species include *Ctenitis atrogrisea, C. subincisa* and *C. pulverulenta.* In these plants the crown forms part of the understory rather than part of the canopy since their stems do not exceed 3 m in height.

Figs. 67.9–67.14 *Ctenitis* leaves. **9–11.** *C. protensa,* variation in leaves, ×0.5. **9.** Half of a lamina. **10.** Base of a basal pinna. **11.** Base of a basal pinna. **12.** Central pinnae of *C. submarginalis,* × 0.5. **13.** Pinnules at basal pinna of *C. Sloanei,* × 0.5. **14.** Leaf of *C. velata,* × 0.25.

Geography (Fig. 2)

Ctenitis grows in tropical, subtropical and south temperate America, in Africa and Madagascar and eastward to southern India, Ceylon and to the Philippine Islands and New Guinea; also it is in the Himalayas, China and north to Japan. In the Pacific it is on the Bonin Islands, the Hawaiian Islands, Fiji to Tahiti, Rapa and Pitcairn Island and in New Zealand.

In America *Ctenitis* occurs in southern Florida in the United States, in central Mexico through Central America, in the Antilles and New Providence in the Bahamas, and in South

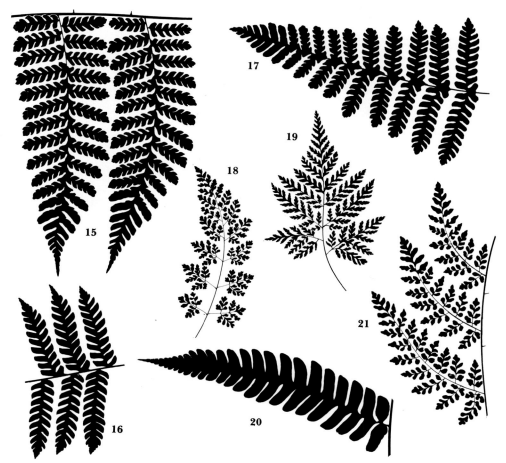

Figs. 67.15–67.21. *Ctenitis* leaves. **15, 16.** *C. subincisa,* ×0.5. **15.** Pinnules of basal pinna. **16.** Portion of an apical pinna. **17.** Apical portion of a pinna of *C. atrogrisea,* × 0.5. **18.** Leaf of *C. crystallina,* ×0.5. **19.** Leaf of *C. hirta,* ×0.5. **20.** Pinna of *C. grandis,* × 0.25. **21.** Leaf of *C. cheilanthina,* ×0.5.

America it occurs from the north to Uruguay, and in Chile south to Aisén; also on Bermuda, Cocos Island, the Galápagos Islands and the Juan Fernandez Islands.

Spores

The spores of *Ctenitis* are of two basic forms—either echinate or saccate—but with unusual diversity in the details of each of these. Although most species are echinate, especially those of tropical America, several different forms can be recognized. The coarsely echinate type in *Ctenitis vellea* (Fig. 25) has large, diffuse projections. Several species with coarsely echinate spores have more compact, irregular elements as in *C. hirta* (Fig. 22) and this resembles the echinate formation in other genera as in *Hypoderris.* This type of asymmetrical formation suggests the possibility of irregularity in sporogenesis perhaps involving hybridity. Other echinate spores in *Ctenitis* have regular, spinelike elements as in *C. inaequalifolia* (Fig. 23) of the Juan Fernandez Islands and *C. spectabilis* of southern South America. The same kind of spore occurs in *Ctenitis aquilina* (Thouars) Pic.-Ser. of

22   23   24

25   26

27   28   29

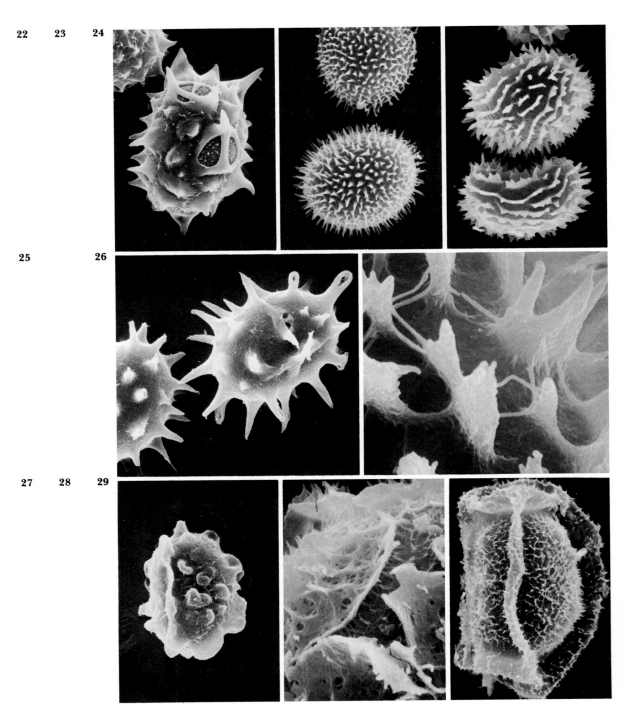

Figs. 67.22–67.29. *Ctenitis* spores, × 1000.   **22.** *C. hirta,* irregularly echinate, the lower rugose layer apparent in perforate areas, Jamaica, *Maxon 10370.*   **23.** *C. inaequalifolia,* spinulose, Juan Fernandez, *Chapin 1091.*   **24.** *C. pulverulenta,* parallel ridges with erose margins, Ecuador, *Holm-Nielsen & Jeppesen 1053.*   **25.** *C. vellea,* regularly echinate folds, Jamaica, *Webster 5280.*   **26.** *C. connexa,* detail of strands connecting erose ridges, Brazil, *Handro 2193,* × 10,000.   **27.** *C. submarginalis,* saccate, the laesura, left, Brazil, *Sehnem 2308.*   **28, 29.** *C. protensa.*   **28.** Abraded spore, the reticulate-echinate surface with lower reticulate stratum, British Guiana, *Maguire & Fanshawe 22871,* × 5000.   **29.** Winglike folds and reticulate-echinate surface, Brazil, *Conant 1005.*

Tristan da Cunha and in the Chinese *Ctenitis apiciflora* (Mett.) Ching. The most unusual spores among the American species have ridges aligned mostly parallel, with erose margins as in *Ctenitis pulverulenta* (Fig. 24). Slender strands connect the echinate projections forming the erose margins as in *Ctenitis connexa* (Fig. 26). This type of spore is apparently unique among those of the Dryopteridaceae.

Several species have spores with inflated folds or saccate structures as in *C. submarginalis* (Fig. 27) which are characteristic of other genera of the family especially *Dryopteris*. *Ctenitis* spores of New Zealand are of this type according to Harris (1955) and they are also in some of the species of China figured by Chang et al. (1976). A few *Ctenitis* species have spores with folds elaborated into prominent wings and a reticulate-echinate surface (Fig. 29) with a lower echinate perispore stratum (Fig. 28) as in *C. protensa*. Spores of African specimens of this species, described by Erdtman and Sorsa (1971), correspond to the American material illustrated here. These are similar to prominently winged spores in several genera of the Dryopteridaceae, especially *Tectaria*.

Cytology

Reports of chromosome numbers for several species of *Ctenitis* from Jamaica (Walker, 1966, 1973) and Mexico (Smith and Mickel, 1977) are largely diploid $n = 41$ with fewer tetraploids. Cytological sampling of the genus from distant regions as New Zealand, Ceylon, and Japan also indicates mostly diploids and some tetraploids. Hybrids are not reported among eight cytologically sampled, of the 11 species in Jamaica, nor are they indicated in other cytological accounts of the genus. The relatively few records for the large genus *Ctenitis* are based on 41 and consistent with those of other genera in the Dryopteridaceae.

Literature

Chang, Y. L., I. C. Hsi, C. T. Chang, K. C. Kao, N. C. Tu, H. C. Sun and C. C. Kung. 1976. Sporae Pteridophytorum Sinicorum. 455 pp. Academica Sinica, Science Press, Peking.

Ching, R. C. 1938. A revision of the Chinese and Sikkim-Himalayan *Dryopteris* with reference to some species from neighbouring regions. Fan Mem. Instit. Biol. Bot. 8: 275–334.

Christensen, C. 1913. Reference under the tribe Dryopterideae.

Christensen, C. 1920. Reference under the tribe Dryopterideae.

Erdtman, G., and P. Sorsa. 1971. Pollen and Spore Morphology/Plant Taxonomy. 302 pp. Almqvist & Wiksell, Stockholm.

Harris, W. F. 1955. A manual of the spores of New Zealand Pteridophyta. Bull. 116 N. Dept. Sci. & Ind. Res.

Ito, H. 1939. Polypodiaceae—Dryopteridoideae, I, Nakai & Honda, Nov. Fl. Jap. 4: 1–243.

Smith, A. R., and J. T. Mickel. 1977. Reference under the tribe Dryopterideae.

Walker, T. G. 1966. Reference under the tribe Dryopterideae.

Walker, T. G. 1973. Reference under the tribe Dryopterideae.

68.   *Atalopteris*
Figs. 68.1–68.8

Atalopteris Maxon, Contrib. U.S. Nat. Herb. 24: 55. 1922. Type: *Atalopteris aspidioides* (Griseb.) Maxon (*Polybotrya aspidioides* Griseb.).

Description

Terrestrial; stem decumbent to nearly erect, small, bearing scales and a few fibrous roots; leaves dimorphic (the fertile narrower than the sterile and its lamina reduced to narrowly alate axes), ca. 25–50 cm long, borne in a cluster, mostly 1-pinnate-pinnatisect, somewhat scaly and short-pubescent, veins free; sori borne, often in pairs, on the abaxial side of the tertiary axes near the veins ends, usually extending to the margin and sometimes to the adaxial side, not paraphysate, exindusiate; spores ellipsoidal to somewhat spheroidal, monolete, the laesura $\frac{3}{4}$ or less the spore length, echinate with prominent, coarse spines that are basally rugulose and perforate. Chromosome number: not reported.

The petiole and rachis are more or less densely scaly with elongate, brownish scales. Few-celled, glandular trichomes are borne among the sporangia, but these also occur elsewhere on the lamina thus they are not considered to be paraphyses. The fertile leaves are soft and undoubtedly are fugacious. The axis of the pinnae is raised on the adaxial surface. The architecture of the sterile lamina is shown in Fig. 3 and the free venation in Fig. 2. The sori, borne on alate fertile axes, are shown in Fig. 4.

Systematics

Atalopteris is an American genus of the Greater Antilles. It is related to *Ctenitis* from which it differs especially in the extreme dimorphism of the leaves. While formerly allied with the Philippine endemic *Psomiocarpa,* the study of that genus by Zamora and Chandra (1977) indicates that the two genera are probably distantly, if at all, related.

Fig. 68.1. Distribution of *Atalopteris.*

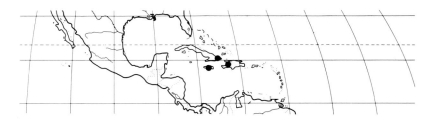

Figs. 68.2–68.4. *Atalopteris aspidioides.* **2.** Segment of sterile leaf with free venation, × 2. **3.** Pinnae of sterile leaf, × 0.5. **4.** Fertile segment, two sori one with sporangia and one with capsules removed, × 12.

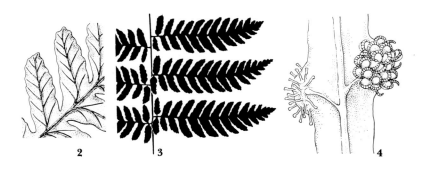

2   3   4

5   **6**

7   **8**

Figs. 68.5–68.8. Spores of *Atalopteris*, × 1000.   **5–7. A. Ekmanii,** Haiti, *Ekman 5189.*   **5.** Spores with compact, irregular spines.   **6.** Detail of spines with ridged, perforate bases, × 5000.   **7.** Surface spine cavity with exposed lower papillate deposit, × 10,000.   **8. A. Maxonii,** spinose, Jamaica, *Maxon 2228.*

Tropical American Species

Three species of *Atalopteris* have been described: *A. aspidioides* (Griseb.) Maxon of Cuba, *A. Maxonii* (Christ) Maxon of Jamaica, and *A. Ekmanii* Maxon (1924) of Haiti. However, there are so few collections of these that it is not possible to assess their status.

Ecology and Geography (Fig. 1)

Atalopteris grows in dense woods on rocky slopes at 600–900 m. The genus occurs in eastern Cuba, in Jamaica, and in several localities in Haiti.

Spores

Spores are coarsely echinate as in *Atalopteris Ekmanii* (Fig. 5) and *A. Maxonii* (Fig. 8) with the base of the spines and adjacent area ridged to reticulate and perforate (Fig. 6). A lower granulate layer is formed below the echinate structure as shown within the base of a broken spine (Fig. 7). These spores most closely resemble some of the coarsely echinate types in *Ctenitis* as those of *C. vellea*. The apex of the spines may be looped appearing as if this develops from folds as in the ridged spores. *Atalopteris* spores are clearly distinct from those of the Philippine genus, *Psomicarpa* which has very prominently winged spores, and do not support the close relationship between these genera, proposed by Christ (1911).

Literature

Christ, H. 1911. On *Psomiocarpa*, a neglected genus of ferns. Smiths. Misc. Coll. 56: 1–4.

Maxon, W. R. 1924. A third species of *Atalopteris*. Proc. Biol. Soc. Wash. 37: 63–64.

Zamora, P. M., and S. Chandra. 1977. Morphology of the sporophyte of *Psomiocarpa* (Aspidiaceae). Kalikasan 6: 217–228.

69. *Tectaria*
Figs. 69.1–69.37

Tectaria Cav., Anal. Hist. Nat. 1: 115. 1799. Type: *Polypodium trifoliatum* L. = *Tectaria trifoliata* (L.) Cav. 1802.

Aspidium Sw., Jour. Bot. (Schrad.) 1800 (2): 4, 29. 1802, *nom. superfl.* for *Tectaria* and with the same type.

Sagenia Presl, Tent. Pterid. 86. 1836. Type: *Sagenia hippocrepis* (Jacq.) Presl (*Polypodium hippocrepis* Jacq.) = *Tectaria cicutaria* (L.) Copel.

Aspidium section *Bathmium* Presl, *ibidem* 88. 1836. Type: *Aspidium singaporianum* Hook. & Grev. = *Tectaria singaporiana* (Hook. & Grev.) Copel.

Amphiblestra Presl, *ibidem* 150. 1836. Type: *Amphiblestra latifolia* (Willd.) Presl (*Pteris latifolia* Willd.) = *Tectaria Amphiblestra* R. & A. Tryon.

Faydenia Hook., Gen. Fil. *t. 53 B*. 1840. Type: *Faydenia prolifera* Hook. (*Aspidium proliferum* Hook. & Grev. 1828, not. R. Br. 1810) = *Faydenia Hookeri* (Sweet) Maxon (*Aspidium Hookeri* Sweet) = *Tectaria prolifera* (Hook.) R. & A. Tryon.

Dictyoxiphium Hook., *ibidem t. 62*. 1840. Type: *Dictyoxiphium panamense* Hook. = *Tectaria panamensis* (Hook.) R. & A. Tryon.

Bathmium Link, Fil. Sp. 114. 1841, *nom. superfl.* for *Tectaria* and with the same type.

Microbrochis Presl, Epim. Bot. 51. 1851. Type: *Microbrochis apiifolia* (Schkuhr) Presl (*Aspidium apiifolium* Schkuhr) = *Tectaria apiifolia* (Schkuhr) Copel.

Polydictyum Presl, *ibidem* 52. 1851. Type: *Polydictyum menyanthidis* (Presl) Presl (*Aspidium menyanthidis* Presl) = *Tectaria menyanthidis* (Presl) Copel.

Dryomenis Fée, Mém. Fam. Foug. 5 (Gen. Fil.): 225. 1852. Type: *Dryomenis phymatodes* Fée = *Tectaria siifolia* (Willd.) Copel.

Podopeltis Fée, *ibidem* 5 (Gen. Fil.): 286. 1852. Type: *Podopeltis singaporiana* (Hook. & Grev.) Fée (*Aspidium singaporianum* Hook. & Grev.) = *Tectaria singaporiana* (Hook. & Grev.) Copel.

Camptodium Fée, *ibidem* 5 (Gen. Fil): 289. 1852. Type: *Camptodium pedatum* (Desv.) Fée (*Aspidium pedatum* Desv. = *Tectaria pedata* (Desv.) R. & A. Tryon.

Phlebiogonium Fée, *ibidem* 5 (Gen. Fil): 314. 1852. Type: *Phlebiogonium impressum* Fée = ? *Tectaria variolosa* (Hook.) C. Chr.

Cardiochlaena Fée, *ibidem* 5 (Gen. Fil.): 314. 1852, *nom. superfl.* for *Polydictyum* and with the same type.

Fig. 69.1. *Tectaria rheosora,* Río Reventazón, Turrialba, Costa Rica. (Photo W. H. Hodge.)

Cionidium Moore, Ind. Fil. xcviii. April, 1857 (Gard. Comp. 143. 1852, *nom. nud.*). Type: *Cionidium Moorii* (Hook.) Moore (*Deparia Moorii* Hook.) = *Tectaria Moorii* (Hook.) C. Chr.

Trichiocarpa (Hook.) J. Sm., Cult. Ferns 68. June, 1857, *nom. superfl.* for *Cionidium* and with the same type. *Deparia* section *Trichiocarpa* Hook., Jour. Bot. Kew Misc. 4: 55. 1852. Type: *Deparia Moorii* Hook. = *Tectaria Moorii* (Hook.) C. Chr.

Grammatosorus Regel, Gartenfl. 15: 335. 1866. Type: *Grammatosorus Blumeanus* Regel = *Tectaria Blumeana* (Regel) Morton.

Quercifilix Copel., Phil. Jour. Sci. 37: 408. 1928. Type: *Quercifilix zeilanica* (Houtt.) Copel. (*Ophioglossum zeilanicum* Houtt.) = *Tectaria zeilanica* (Houtt.) Sledge.

Pleuroderris Maxon, Jour. Wash. Acad. Sci. 24: 550. 1934. Type: *Pleuroderris Michleriana* (D. C. Eaton) Maxon (*Lindsaea Michleriana* D. C. Eaton) = *Tectaria incisa* × *panamensis*.

Pseudotectaria Tard., Not. Syst. 15: 87. 1955. Type: *Pseudotectaria Decaryana* (C. Chr.) Tard. = *Tectaria Decaryana* C. Chr.

Fig. 69.2. *Tectaria heracleifolia,* Puente Nacional, Veracruz, Mexico. (Photo W. H. Hodge.)

Description

Terrestrial or rupestral; stem erect or decumbent and often stout, including the more or less persistent petiole bases, to moderately long-creeping and slender, bearing scales, especially at the apex and rarely some trichomes (in *T. draconoptera*) and many long, fibrous roots; leaves usually monomorphic, to dimorphic (the fertile often erect, usually longer and less expanded than the sterile), ca. 10 cm to 2 m long, borne in a cluster to widely spaced, lamina simple, entire to deeply lobed, ranging to 1-pinnate to 3-pinnate-pinnatifid or bipinnatifid to tripinnatifid, glabrate, scaly or pubescent, veins usually anastomosing, usually with included, free veinlets, the veins sometimes free or mostly so; sori roundish to elongate to linear, usually on the veins, or rarely on a marginal commissure that is continuous or interrupted, rarely on a marginal projection, or rarely with the sporangia partly acrostichoid (in *T. zeilanica*), not paraphysate, covered by a usually reniform or peltate, or less often an elongate to linear indusium, or exindusiate; spores somewhat ellipsoidal, monolete, the laesura $\frac{1}{2}$ to $\frac{3}{4}$ the spore length, with winglike folds, cristate, or echinate, the areas between the folds

Fig. 69.3. *Tectaria panamensis,* Río San Carlos, Boca Arenal, Costa Rica. (Photo W. H. Hodge.)

usually reticulate-echinate or spinulose. Chromosome number: $n = 40, 80, 120$, ca. 160; $2n = 80, 160$.

Some of the variation in leaf architecture is shown in Figs. 1–3 and 12–22. The venation is usually fully anastomosing with included free veinlets (Fig. 5), but it varies to mostly free as in *T. pedata* (Fig. 6). The indusia are usually reniform or peltate (Fig. 9); in *T. panamensis* they are linear and attached to the costal side of the marginal sorus (Fig. 10). The costae are flat or only slightly raised, sometimes with slightly raised edges, or definitely raised on the adaxial surface.

Systematics

Tectaria is a large pantropical and subtropical genus of perhaps 150 species, with about 35 in America.

Tectaria varies in characters that are usually stable and often

Fig. 69.4. Distribution of *Tectaria* in America.

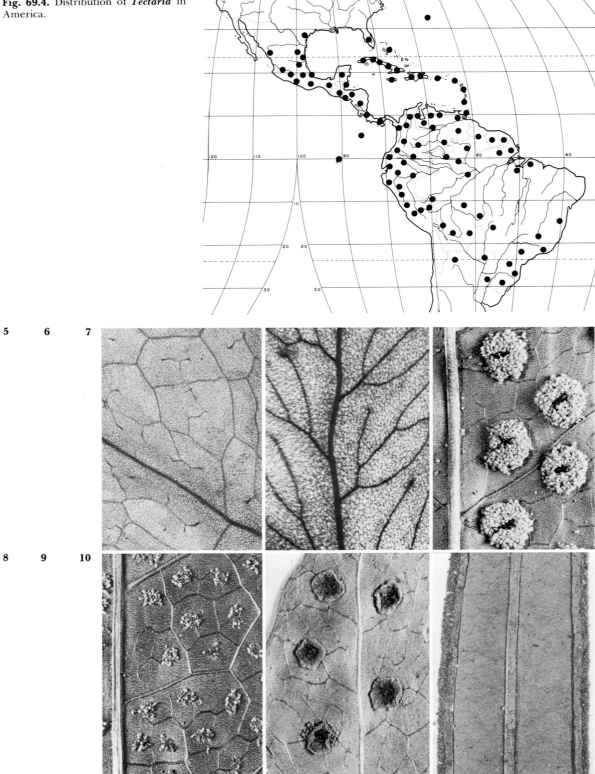

Figs. 69.5–69.10. *Tectaria*. 5. Anastomosing veins with included free veinlets, *T. heracleifolia*, ×8. **6.** Predominently free veins, *T. pedata*, ×8. **7.** Mature sori of *T. incisa*, ×4.5. **8.** Exindusiate sori of *T. draconoptera*, ×5. **9.** Young sori with peltate indusia, *T. heracleifolia*, ×4.5. **10.** Portion of fertile leaf of *T. panamensis*, ×4.

Fig. 69.11. Heteroblastic leaf series, ***Tectaria heracleifolia,*** the smallest leaf from the youngest, the largest from the oldest plant, × 0.25.

afford distinguishing generic features as the venation, soral position, the type of stem and the indument. Several small genera are sometimes recognized on the basis of these and other characters as leaf architecture and degree of leaf dimorphism. A number of these genera are clearly minor evolutionary developments within *Tectaria,* and in America only *Hypoderris* seems sufficiently divergent for recognition. A modern monographic treatment of *Tectaria* is much needed to evaluate the variation in characters, to relate the smaller satellite groups and provide an infrageneric classification. *Tectaria* is clearly allied to genera of the Dryopterideae, especially to *Ctenitis.*

The complex leaf form in *Tectaria* is undoubtedly less derived and evolutionary lines appear to have developed simpler forms. The heteroblastic series of leaves that occur in several species, as *T. heracleifolia* (Fig. 11) suggest that simpler leaf architecture may have evolved through the process of neotony.

Unusual leaf architecture in collections of *Tectaria* suggests that hybridization may be of some frequency although only a few hybrids have been proposed. Walker (1966) reports a hybrid involving *T. coriandrifolia* and perhaps *T. apiifolia* in Jamaica and Gómez (1977) reports *T. draconoptera* (as *T. nicotianifolia*) × *nicaraguensis* from Nicaragua. *Tectaria Amesiana* A. A. Eaton is doubtless a hybrid of *T. coriandrifolia* and *T. lobata* (*T. minima*) as Eaton suggested. Another long suspected hybrid, *Pleuroderris Michleriana* has recently been recognized as a cross between *T. incisa* (Figs. 12, 13) and *T.* (*Dictyoxiphium*) *panamensis* (Figs 3, 19) (Wagner et al. 1978). The lamina architecture of this hybrid is highly variable as is the sorus and indusium. The spores are irregular in size and surface structure but some may be fairly well developed. It occurs throughout the range of *Tectaria panamensis* and although it is locally frequent there is no evidence that it is fertile. Analysis of the characters of hybrids will be of special interest in reference to the role of hybridization in the evolution of the remarkable morphological divergence of species of *Tectaria.*

The Mexican species of *Tectaria* were revised by Morton (1966).

Tropical American Species

About twenty American species of *Tectaria* are readily characterized by several common features.

Stem decumbent and very short-creeping to erect, bearing nonclathrate scales, leaves monomorphic or nearly so, with a petiole, lamina 1-pinnate to 3-pinnate-pinnatifid, at least basally in larger leaves, veins anastomosing, with included free veinlets or not, sori few to many, borne on the veins, with a reniform or peltate indusium.

Representative species are: *Tectaria andina* (Baker) C. Chr., *T. apiifolia* (Schkuhr) Copel., *T. cicutaria* (L.) Copel., *T. coriandrifolia* (Sw.) Underw. (Fig. 18), *T. heracleifolia* (Willd.) Underw. (Fig. 2), *T. incisa* Cav. (*T. martinicensis* (Spreng.) Copel.) (Figs. 12, 13), *T. mexicana* (Fée) Morton (Mexican and Central American plants previously called *T. dilacerata* (Kze.) Maxon, a synonym of

12

13

14

Figs. 69.12–69.14. *Tectaria* leaves, ×0.125. **12.** Sterile leaf of *T. incisa.* **13.** Fertile leaf of *T. incisa.* **14.** *T. draconoptera.*

T. cicutaria) (Fig. 17), *T. rheosora* (Baker) C. Chr. (Fig. 1), *T. subebenea* (Christ) C. Chr., and *T. trifoliata* (L.) Cav.

In addition, a number of American species are related to these on the basis of several of their characters. However, they are divergent in some structures such as the stem, leaf, sorus, indusium or scales. These are listed below in alphabetical order since relationship among them and with the species of the first group are uncertain.

Tectaria Amphiblestra R. & A. Tryon (*Amphiblestra latifolia* (Willd.) Presl. is known only from four early collections in Colombia and Venezuela. The lamina is similar to that of *T. heracleifolia* and to broad forms of *T. incisa,* the sori are marginal and borne on a vascular commissure as in *T. panamensis* but are irregularly interrupted and exindusiate.

Tectaria Brauniana (Karst.) C. Chr. of Colombia to Bolivia. Stem moderately long-creeping, bearing clathrate scales, lamina bipinnatifid or tripinnatifid, veins free or rarely anastomosing.

Tectaria draconoptera (D. C. Eaton) Copel. (*T. euryloba* (Christ) Maxon, *T. myriosora* (Christ) C. Chr., *T. nicotianifolia* (Baker) C. Chr.) (Fig. 14) is distributed from Nicaragua to Colombia and Ecuador, east to Venezuela and Trinidad. Stem long- to moderately long-creeping, leaves borne at intervals, lamina deeply pinnatifid or pinnatisect into a few large pinna-segments, sori small, ca. 1 mm in diameter (Fig. 8), very numerous and exindusiate. The more typical mature sori of *Tectaria* are ca. 2–3 mm in diameter (Fig. 7).

Tectaria lobata (Presl) Morton (Fig. 21) (*T. minima* Underw.) is a diminutive species with the fertile leaves often only 5–10 cm long. The stem has clathrate scales and the lamina is lobed to pinnatifid, rarely pedate. The spores have folds and a perforate surface similar to those of *T. incisa.*

Tectaria nicaraguensis (Fourn.) C. Chr. of Nicaragua is rather similar to *T. plantaginea* but differs in a truncate rather than attenuate base of the entire lamina and also in having smaller spores with the surface folds dissected into a cristate form. These differences suggest that the entire leaves of these two species have developed independently.

Tectaria panamensis (Hook.) R. & A. Tryon (Figs. 3, 19) of southern Mexico to Panama and northern Colombia. Sterile leaf nearly sessile, lamina entire, narrowly lanceolate, the fertile leaf nearly erect, narrower and with a continuous, marginal sorus borne on a vascular commissure. An indusium arising from the costal, rather than the marginal side, covers the sorus.

Tectaria pedata (Desv.) R. & A. Tryon (*Camptodium pedatum* (Desv.) Fée) (Fig. 16) and *Tectaria pinnata* (C. Chr.) R. & A. Tryon (*Camptodium pinnatum* C. Chr.) (Fig. 20). *Tectaria pedata* occurs in Cuba and Jamaica, and *T. pinnata* in Cuba (*Wright 997,* GH, in part) and Hispaniola. The veins are free or rarely a few are anastomosing without included free veinlets, the lamina is pedate, deeply pinnatifid or 1-pinnate-lobed basally.

Tectaria plantaginea (Jacq.) Maxon (Fig. 15) is distributed from British Honduras to Panama, Colombia south to Peru, east to Surinam and north to northern Amazonas, Brazil; also Puerto

Figs. 69.15–69.22. *Tectaria* leaves, ×0.25. **15.** *T. plantaginea*, an old prostrate leaf with proliferous apex. **16.** *T. pedata.* **17.** Basal portion of basal pinna, *T. mexicana.* **18.** *T. coriandrifolia.* **19.** *T. panamensis*, sterile leaf, left; fertile leaf, right. **20.** *T. pinnata.* **21.** *T. lobata.* **22.** *T. prolifera*, plant, two sterile leaves with attenuate, proliferous apex, fertile leaf with sori white.

Rico, the Lesser Antilles and Trinidad. Stem rather short-creeping, bearing clathrate scales, leaves clustered to slightly spaced, lamina entire or broadly crenate with an acute to broadly obtuse apex bearing a proliferous bud, with a peltate indusium or exindusiate.

Tectaria prolifera (Hook.) R. & A. Tryon (Fig. 22) grows in all of the Greater Antilles. Sterile leaf nearly sessile, lamina entire and often with a prolonged tip with a proliferous apical bud, fertile erect and narrower, often with a single row of areolae on each side of the costa.

None of the four earlier names for this species are available in *Tectaria. Asplenium proliferum* Sw. 1788, not Lam. 1786 and *Aspidium proliferum* Hook. & Grev. 1828, not R. Br. 1810 are both later homonyms. *Aspidium Hookeri* Sweet 1830 provides the earliest epithet under *Faydenia* as *Faydenia Hookeri* (Sweet) Maxon, however it is not available because of *Tectaria Hookeri* Brownlie, 1977. *Polystichum Grevillanum* Presl, 1838 is a superfluous name for *Aspidium Hookeri* Sweet.

Ecology (Figs. 1–3)

Tectaria is predominently a genus of wet forests, where it grows especially on stream banks, in rocky places and on forest slopes. A few species grow in open forests, along road or trail banks in the forest and some are confined to limestone habitats.

In America *Tectaria* usually grows in rain forests on mountain slopes, on stream banks or in wet ravines, occasionally on wet rocks by waterfalls. Sometimes it occurs in more exposed habitats as the borders of forests or in disturbed sites such as muddy stream banks, in coffee plantations and along roadsides. *Tectaria lobata* and *T. pedata* are evidently confined to limestone, and other species frequently grow on calcareous rocks or old masonry. *Tectaria* is most common from sea level to 1000 m, and rarely occurs up to 2500 m.

Tectaria prolifera and *T. plantaginea* have buds at the leaf apex that produce new plants as the leaf ages and lies on the soil. *Tectaria incisa* and rarely *T. coriandrifolia* have proliferous buds in the axils of the pinnae.

Geography (Fig. 4)

Tectaria is widely distributed in tropical and subtropical America, and in tropical Africa eastward to China, Malesia and Queensland in Australia, and in the Pacific east to the Hawaiian Islands, Tahiti, and the Austral Islands.

In America it grows in Texas and Florida in the southern United States and is widely distributed in Mexico, the Antilles and in South America south to Salta and Missiones in Argentina, and Rio Grande do Sul in Brazil; also in Bermuda, Cocos Island and the Galápagos Islands. The center of species diversity is in the Andes, where about 18 species occur from Colombia to Bolivia. *Tectaria incisa* is the only widely distributed species occurring nearly throughout the range of the genus in America.

Spores

Spores of *Tectaria* have winglike folds as in *T. plantaginea* (Fig. 23) or the folds may be more prominent with spinulose surfaces as in *T. heracleifolia* (Fig. 24) or with smaller crests as in *T. trifoliata* (Fig. 25). The surface between the folds is usually reticulate-echinate as in *T. Lizarzaburui* (Sod.) C. Chr. (Fig. 26) often with echinate processes protruding from a lower stratum as shown in the abraded spore of *T. subebenea* (Fig. 27). The surface may be slightly perforate as in *T. incisa* (Fig. 28) or strongly so in *T. lobata* (Fig. 29). The sporoderm appears to be formed of three strata in *T. angulata* (Willd.) Copel. (Fig. 34) or two layers can be distinguished above a fairly smooth exospore in *T. draconoptera* (Fig. 30) and *T. trifoliata* (Fig. 31). The folds are hollow with thin walls as shown in section near the top of Fig. 30. The echinate structure in *T. Brauniana* is irregularly fused forming a more or less cristate surface (Fig. 33) is quite distinct from the regular echinate-reticulate formation in paleotropical species as *T. decurrens* (Presl) Copel. (Fig. 32). The spore architecture supports the inclusion of *T. prolifera* (Fig. 35), *T. pedata* (Fig. 36) and *T. panamensis* (Fig. 37) in *Tectaria*. The spores also suggest a close relationship between *Tectaria* and segregate genera of the paleotropics as *Tectaridium* and *Heterogonium*.

Cytology

Chromosome numbers for American species of *Tectaria* are mostly diploid with $n = 40$ or tetraploid with $n = 80$. Four of six Jamaican species are diploid (Walker, 1966, 1973). Both diploid and tetraploid plants of *T. incisa* occur there, the tetraploid at two disjunct localities and there also are reports of diploids from Trinidad and Peru. Plants of *T. heracleifolia* from Jamaica

23 24 25

26   27

28 29 30

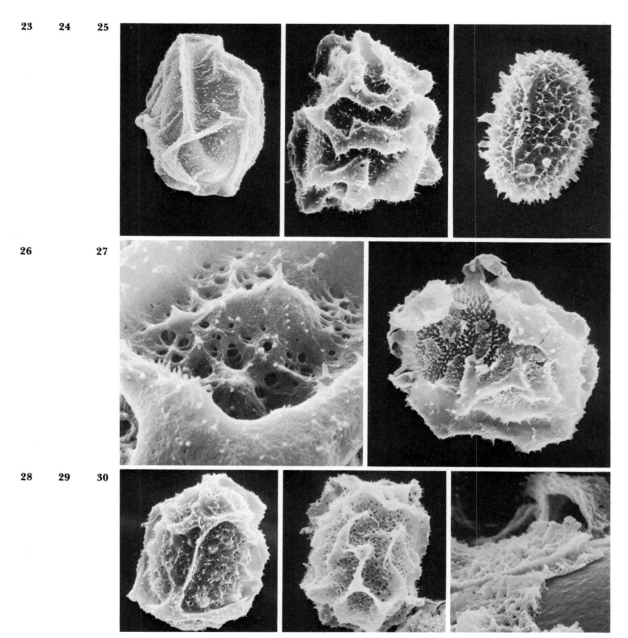

Figs. 69.23–69.30. *Tectaria* spores, × 1000, surface detail, × 5000. **23.** *T. plantaginea* with low folds, laesura at center, Puerto Rico, *Scamman 6538.* **24.** *T. heracleifolia* spinulose folds, Colombia, *Haught 4234.* **25.** *T. trifoliata,* cristate, laesura near center, Guadaloupe, *Proctor 20382.* **26.** *T. Lizarzaburui,* detail of reticulate-echinate area between folds, Ecuador, *Harling et al. 6977.* **27.** *T. subebenea,* abraded surface at left, lower echinate structure exposed, Costa Rica, *Burger & Stolze 5874.* **28.** *T. incisa,* surface perforate, laesura center, Peru, *Cerrate 2830.* **29.** *T. lobata,* winglike folds and perforate surface, Bahamas, *O'Neill 7802.* **30.** *T. draconoptera,* abraded perispore surface, section of a fold at top, smooth exospore, lower right, Costa Rica, *Burger & Stolze 5767.*

and Florida are tetraploid, while diploids are known from Mexico and Costa Rica. *Tectaria prolifera* is reported as triploid with 40 pairs and 40 univalents, by Walker (1966) (as *Faydenia*). Reports of chromosome numbers of Old World tectarias are also mostly diploid and tetraploid. *Tectaria Vieillardii* (Fourn.) C. Chr. is reported as a hexaploid with $n = 120$ from New Caldonia (Fabbri, 1965) and suggests possible derivation through hybrid-

31 **32** **33**

34 **35**

36 **37**

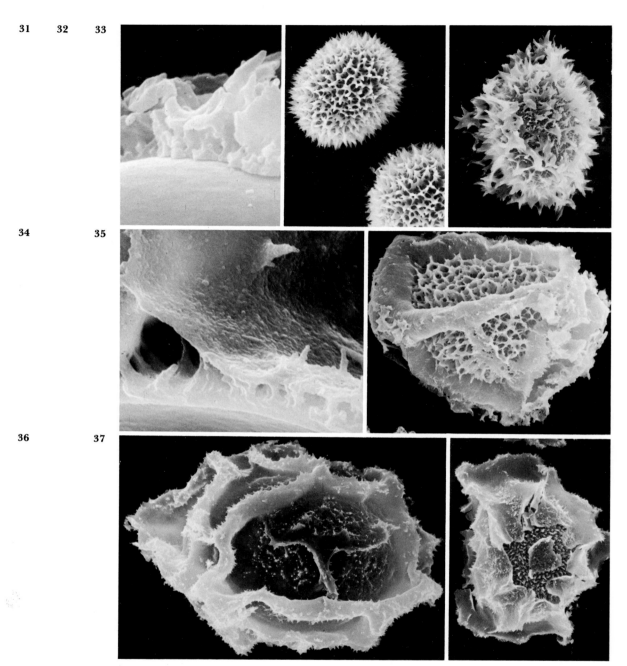

Figs. 69.31–69.37. *Tectaria* spores, × 1000, wall sections, × 10,000. **31.** *T. trifoliata,* wall section with reticulate perispore above smooth exospore, Guadeloupe, *Proctor 20382.* **32.** *T. decurrens,* echinate-reticulate spores, New Guinea, *Brass 13855A.* **33.** *T. Brauniana,* cristate-echinate spores, Colombia, *Haught 1939.* **34.** *T. angulata,* wall section with three perispore strata, a spine attached to surface of fold, the smooth exospore below, left, New Guinea, *Brass 8836.* **35.** *T. prolifera* with broad folds, surface abraded at right, Cuba, *Wright 844.* **36.** *T. pedata,* with prominent winglike folds, Hispaniola *Ekman 5311.* **37.** *T. panamensis,* a portion of the surface abraded exposing the lower echinate formation, Nicaragua, *Bunting & Licht 597.*

ization. There are also diploid and tetraploid complexes in Old World species as *T. devexa* (Kze.) Copel. that extend over a wide geographic range. The highest chromosome number known in the genus is $n = 160$, an octoploid, in *T. angelicifolia* (Schum.) Copel. from Ghana.

Literature

Fabbri, F. 1965. Secundo supplemento alle tavole chromosomiche delle pteridophyta de Alberto Chiarugi. Caryologia 18: 675–731.

Gómez, L. D. 1977. Contribuciones a la pteriodologia Centroamericana 2. Novitates. Brenesia 10/11: 115–119.

Morton, C. V. 1966. The Mexican species of *Tectaria*. Amer. Fern Jour. 56: 120–137.

Wagner, W. H. Jr., F. S. Wagner, and L. D. Gómez P. 1978. The singular origin of a Central American fern, *Pleuroderris Michleriana*. Biotropica: 254–265.

Walker, T. G. 1966. Reference under the tribe Dryopterideae.

Walker, T. G. 1973. Reference under the tribe Dryopterideae.

70.   *Hypoderris*
Figs. 70.1–70.9

Hypoderris R. Br., Wall. Icon. Pl. Asiat. Rar. 1: 16 (sub *Matonia*) 1830, and Hook., Gen. Fil. *t. l.* 1838. Type: *Hypoderris Brownii* Hook.

Description   Terrestrial or rupestral; stem rather long-creeping, rather small, bearing scales and many fibrous roots; leaves monomorphic, ca. 25–75 cm long, borne somewhat distantly, lamina simple, entire, lanceolate or with two large, divergent lobes at the base, somewhat scaly, veins anastomosing with included free veinlets; sori very numerous, borne on the minor veins on a

Fig. 70.1. *Hypoderris Brownii* on an earth bank, North Range, Trinidad. (Photo L. D. Gómez.)

Fig. 70.2. Distribution of *Hypoderris*.

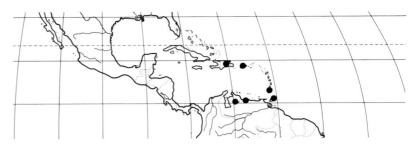

more or less oval receptacle, not paraphysate, with a rather irregular, membranous, ciliate indusium appressed to the leaf surface beneath the sporangia; spores somewhat ellipsoidal, to somewhat spheroidal, monolete, the laesura $\frac{3}{4}$ the spore length, obscured by the cristate-echinate surface with prominent, compact, often irregular projections. Chromosome number: $n = 40$; $2n = 80$.

The stem scales are brown, rather thin and subclathrate. The small, appressed indusium (Fig. 5) evidently encloses the sorus only when the sporangia are very young. Small leaves have a simple lamina (Fig. 3) and are usually sterile or only partially fertile while large leaves are lobed at the base (Fig. 4) and are usually fertile. The complex anastomosing vein systems are shown in Figs. 6 and 7 and the association of these and the sori in Fig. 7. The main axis of the lamina is raised on the adaxial surface.

Systematics

Hypoderris is a monotypic genus of tropical America based on *H. Brownii* Hook. The South American *H. Stuebelii* Hieron. is *Bolbitis oligarchia* while four species of Central America described under *Hypoderris* are hybrids between *Tectaria incisa* and *T. panamensis*.

An affinity with *Tectaria,* especially species such as *T. euryloba,* is indicated by the rather long-creeping stem, larger leaves with the lamina deeply lobed, numerous small sori and cristate spores. However, the unique basal indusium of *Hypoderris* (Fig. 5) sets the genus apart from the minor evolutionary lines within *Tectaria.* The indusium structure is so different from other tectarioid ferns that it was possibly independently derived from an exindusiate sorus.

Ecology and Geography (Figs. 1, 2)

Hypoderris grows in wet, shaded habitats, in ravines, in woods, on cliffs or in rocky places, from about 200 to 1100 m.

It grows in Hispaniola, Puerto Rico, Grenada, Trinidad, where it is most common, and in the states of Falcon and Miranda in Venezuela.

Spores

The spores are unusually varied in size and shape but are generally ellipsoidal and prominently cristate-echinate (Fig. 8). They often adhere in tetrads and are then more spheroidal (Fig.

3

4

Figs. 70.3, 70.4. *Hypoderris* leaves. **3.** Sterile leaf, ×0.25. **4.** Fertile leaf, ×0.25.

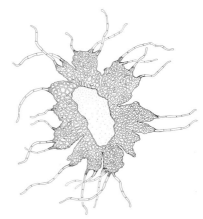

Fig. 70.5. Indusium of *Hypoderris*, surrounding a central receptacle and closely appressed to the leaf surface at maturity, ×25.

9). The surface architecture consists of more or less fused, irregular processes with basal perforations. The asymmetry of this formation and varability in size and shape of these spores suggests possible irregularities in sporogenesis. The surface is generally similar but notably more irregular that the cristate formation in *Tectaria* spores especially those of *T. Brauniana* or *T. trifoliata*.

Cytology

The reports of $n = 40$ and $2n = 80$ for *Hypoderris Brownii* from Trinidad by Walker (1973) support the association of the genus with *Tectaria*.

Literature

Walker, T. G. 1973. Evidence from cytology in the classification of ferns, *in*: A. C. Jermy et al., The phylogeny and classification of the ferns, pp. 91–110. Bot. Jour. Linn. Soc. 67, Suppl. 1.

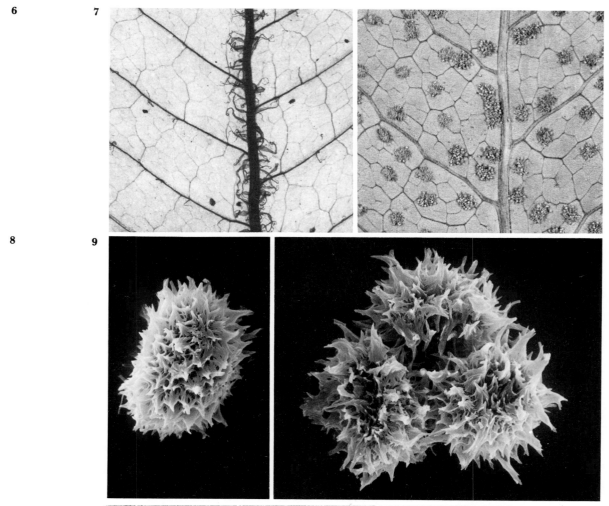

Figs. 70.6–70.9. *Hypoderris Brownii.* **6.** Venation of a young leaf with a few immature sori (small, dark areas), ×3. **7.** Venation of fertile leaf with sori, ×3. **8, 9.** Spores, Grenada, *Proctor 17247*, ×1000. **8.** Cristate-echinate, ellipsoidal. **9.** Three somewhat spheroidal spores of a tetrad.

71. *Cyclopeltis*
Figs. 71.1–71.8

Cyclopeltis J. Sm., Curtis Bot. Mag. 72 (III,2), Comp., first of two pages 36. 1846. Type: *Cyclopeltis semicordata* (Sw.) J. Sm. (*Aspidium semicordatum* (Sw.) Sw., *Polypodium semicordatum* Sw.)

Hemicardion Fée, Mém. Fam. Foug. 5 (Gen. Fil.): 282. 1852, *nom. superfl.* for *Cyclopeltis* and with the same type.

Description

Terrestrial or rarely rupestral; stem decumbent, moderately stout, including the more or less persistent petiole bases, rather short-creeping, bearing scales and numerous, long, fibrous roots; leaves monomorphic, ca. 20 cm to 1.5 m long, borne in a loose cluster, lamina 1-pinnate, usually imparipinnate, with the pinnae entire or nearly so, somewhat scaly and short-pubescent, veins free, the acroscopic one often extending about halfway to the margin; sori roundish, abaxial, borne on the veins in 1 to 3 regular or rather irregular ranks on each side of the costa, not paraphysate, covered by a peltate, fugacious to persistent indusium; spores more or less spheroidal, monolete, the laesura $\frac{1}{2}$ to $\frac{2}{3}$ the spore length with prominent winglike folds, the surface more or less spinulose. Chromosome number: $n = 41$.

Fig. 71.1. *Cyclopeltis semicordata,* Rose Craig, Portland, Jamaica. (Photo W. H. Hodge.)

Fig. 71.2. Distribution of *Cyclopeltis* in America.

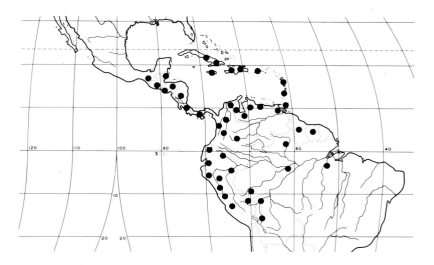

The pinnae of *Cyclopeltis* are articulate and deciduous with age, their base is strongly inequilateral in *C. semicordata* (Fig. 4) and more or less equally truncate in *C. crenata*. The stem scales are long and brown. The indument on the rachis is of short, 1- to few-celled trichomes, these mostly with an enlarged apical cell, and often also of long, filiform scales. A lateral groove of the rachis is interrupted by the base of the pinna (Fig. 3), which has the costa raised on the adaxial surface. The characteristic sori and indusia are shown in Fig. 6.

Systematics

Cyclopeltis is a tropical genus of four to six species, with only *C. semicordata* (Sw.) J. Sm. in America. The genus is readily identified in America by the well-developed basal auricle of the pinnae that overlies the rachis (Figs. 4, 5). In the paleotropics are *C. crenata* (Fée) C. Chr., *C. Cumingiana* (Fée) Ching (*C. Presliana* Berkl.), *C. mirabilis* Copel., and perhaps others although most of the species have not been adequately defined. The genus is distinctive and probably related to *Ctenitis*.

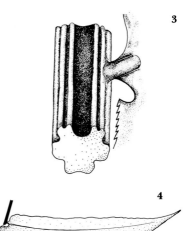

3

4

Figs. 71.3, 71.4. *Cyclopeltis semicordata.* 3. Rachis and pinna base, enlarged. **4.** Pinna with basal auricle, ×0.5.

Ecology (Fig. 1)

Cyclopeltis is primarily a genus of wet primary forests, where it grows on mountain slopes, in ravines, on stream banks, and less often on limestone rocks.

In America *Cyclopeltis semicordata* grows in dense rain forests or in wet thickets, usually in well-drained sites such as rocky ravines, stream banks, or hillsides. It sometimes grows on limestone, especially in the Greater Antilles and rarely it grows on old stone walls. *Cyclopeltis semicordata* occurs from sea level to 400 m or less often, in the Andes, to 750 m.

Geography (Fig. 2)

Cyclopeltis occurs in the American tropics and from Burma and the Malay Peninsula to the Philippine Islands, Borneo, New Guinea and the Caroline Islands (Truk).

In America it occurs from Chiapas in southern Mexico,

5 6

7 8

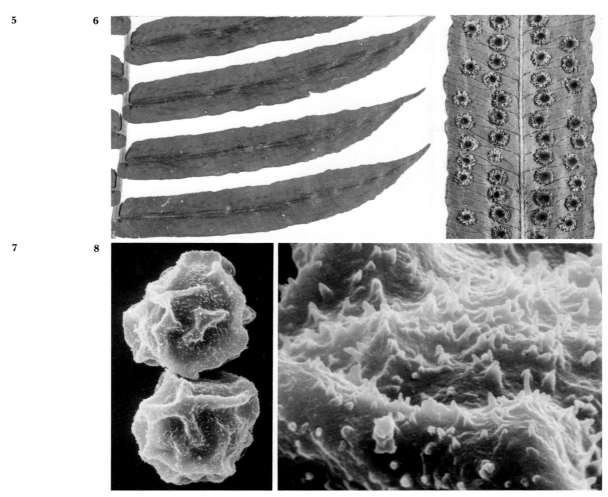

Fig. 71.5–71.8. *Cyclopeltis semicordata.* **5.** Portion of a lamina, × 1. **6.** Portion of a fertile pinna, × 3. **7, 8.** Spores, Guatemala, *Ortiz 2683*. **7.** Two spores with winglike folds, × 1000. **8.** Detail of irregularly echinate surface, × 10,000.

through Central America, in all of the Greater Antilles and some of the Lesser Antilles, and in South America south to Bolivia, and the Amazon basin of Brazil.

Spores

The inflated folds and spinulose surface (Fig. 7) are similar to spores of several dryopteroid genera especially *Dryopteris*. The short, echinate elements are fairly diffuse with sharp apices (Fig. 8) but vary in density and prominence. This formation represents one of the most unspecialized among the spores of the dryopteroids and is less complex than the echinate types in other genera in the tribe as *Ctenitis, Hypoderris* and *Atalopteris*.

Cytology

The report of $n = 41$ for *Cyclopeltis semicordata* from Jamaica, by Walker (1966), along with the characteristics of the spores, noted above, helps to establish the position of the genus among the dryopteroids.

Literature

Walker, T. G. 1966. Reference under the tribe Dryopterideae.

72. *Rumohra*
Figs. 72.1–72.9

Rumohra Raddi, Opusc. Sci. Bolog. 3: 290. 1819. Type: *Rumohra aspidioides* Raddi = *Rumohra adiantiformis* (Forst.) Ching.

Description

Terrestrial, rupestral or rarely epiphytic; stem long-creeping, rather stout, bearing a dense covering of scales and many fibrous roots; leaves monomorphic, ca. 0.1–1 m long, borne distantly to rather closely spaced, lamina 2-pinnate-pinnatifid to 4-pinnate, glabrous or somewhat scaly, veins free; sori roundish, borne on the veins or at the vein-tips, not paraphysate, covered by a peltate indusium that is usually fugacious from the persistent stalk; spores somewhat ellipsoidal, monolete, the laesura $\frac{1}{2}$ the spore length, the surface with rounded, saccate projections or long folds and small, low ridges. Chromosome number: $n = 41$.

The scaly stem of *Rumohra* (Fig. 4) bears mostly cordate but some peltate scales. The stem is dorsiventral as is the stele. The deltoid to ovate lamina varies in size, division and architectural details (Fig. 3). The rachis has two grooves on the adaxial side and a continuous central ridge, and the pinna-rachis is similar with lateral grooves continuous with those of the rachis (Fig. 2), and the axis of the penultimate segments is raised on the adaxial surface. The usually large indusium is peltate (Fig. 5).

Systematics

Rumohra is a distinctive genus of two to six species with a primarily circum-austral distribution. It is now restricted to the widespread *Rhumora adiantiformis* and a few related species that were earlier placed in *Polystichum*. The relation of the genus appears to be with the dryopteroids in some characters and with the davallioids in others. Pérez (1928) and Holttum (1947) place special emphasis on the dorsiventral stem and stele as indicating an affinity with the Davalliaceae. Kato (1974) considered that the sorus, indusium and petiole anatomy demonstrated an affinity with the group we recognize as Dryopteridaceae. The surface formation of the spores especially in *Rumorha Berteriana* resemble the folded formation in many genera of the Dryopteridaceae although the coarse projections in *R. adiantiformis* are unusual.

American Species

There are two species of *Rumohra* in America, the widely distributed *R. adiantiformis* (Forst.) Ching and *R. Berteriana* (Colla) Duek & Rodrig. of the Juan Fernandez Islands. The latter was distinguished by Christensen (1910) and may be maintained

Fig. 72.1. Distribution of **Rumohra** in America.

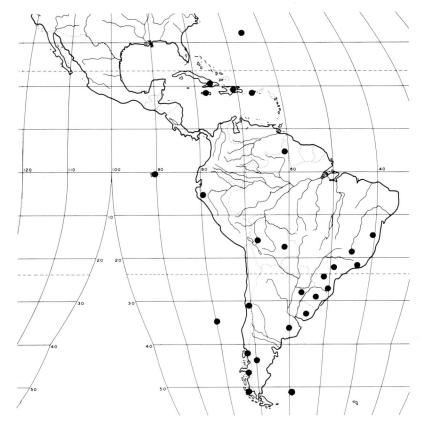

Fig. 72.2. Rachis and base of pinna stalk of **Rumohra adiantiformis,** enlarged.

especially on the basis of the narrower, dark colored indurated stem scales, saccate spores, and a reddish exudate which usually forms small dots on both surfaces of the lamina. The stem scales of *R. adiantiformis* are broad, light brown and thin and the exudate is absent.

Ecology

Rumohra grows in a variety of habitats, in open sandy soil, in shrubby areas, on rocks and in forest.

In America *Rumohra adiantiformis* grows on beach strand, on dunes, on rocks in streams, in shrubby places, in forests, and rarely on cliffs. It is sometimes an epiphyte, growing on trees (Cuba) and on *Acrostichum* (Bermuda). It occurs from sea level up to 2400 m in the Andes of Peru.

Fig. 72.3. Portions of pinnae from different plants of **Rumohra adiantiformis,** × 0.5.

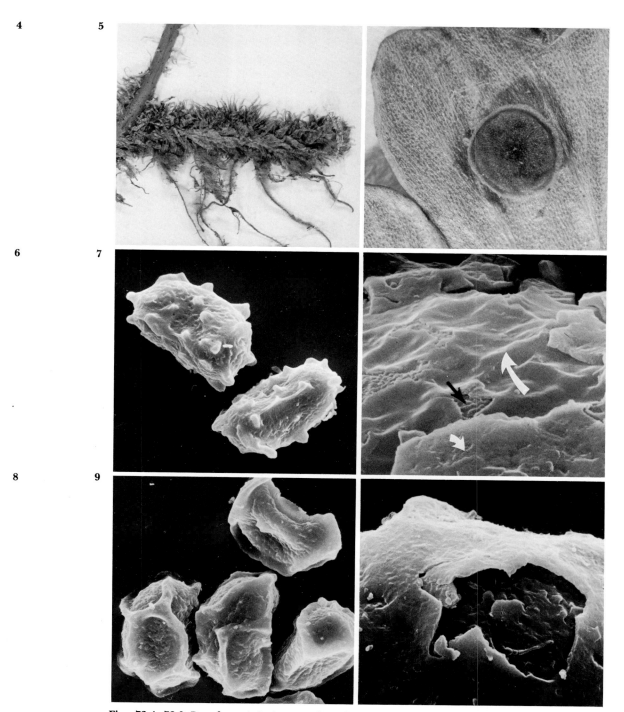

Figs. 72.4–72.9. ***Rumohra.*** **4.** Scaly stem of ***R. adiantiformis,*** × 1.25. **5.** Peltate indusium of ***R. adiantiformis,*** × 15. **6–9.** Spores of ***Rumohra.*** **6, 7.** ***R. adiantiformis.*** **6.** Tuberculate-saccate spores, the lower
with central laesura, Peru, *Wurdack 1470,* × 1000. **7.** Abraded surface lowest papillate stratum at black
arrow, small white arrow on surface, and large white arrow on the central ridged stratum, Jamaica,
Maxon 9515, × 10,000. **8, 9.** ***R. Berteriana,*** Juan Fernandez, *Bertero 1529.* **8.** Spores with long, coarse
folds, × 1000. **9.** Abraded surface with part of lower papillate stratum exposed, × 5000.

There is considerable range in leaf size in *Rumohra adianti-formis* with the fertile ones from ca. 10 cm to 1 m in length. The differences in size appear to be related to environmental conditions. Plants from exposed sites in southern South America may have leaves with strongly ascending and closely imbricate pinnae.

Geography (Fig. 1)

Rumohra occurs in America, on Tristan da Cunha and Gough Islands, in southern Africa, Madagascar, the Mascarenes and Seychelles, in New Guinea and southeastern Australia, Tasmania and New Zealand. In the Old World the range of the genus is largely based on *R. adiantiformis,* except for a few endemics in Madagascar.

In America *Rumohra adiantiformis* occurs in the Greater Antilles, in northern South America on Mt. Roraima, in northern Peru to Bolivia, southeastern Brazil, and southward to Argentina and Chile; also on Bermuda, the Galápagos and the Falkland Islands. *Rumohra Berteriana* is endemic to the Juan Fernandez Islands.

Spores

Rumohra spores are relatively small, about 30–38 μm long. Specimens of *R. adiantiformis* examined from Jamaica, Peru and New Zealand have relatively uniform spores with rounded saccate projections (Fig. 6). An abraded spore shows three strata (Fig. 7), the lowest papillate formation (black arrow), an outer layer (short, white arrow), and a slightly ridged central stratum (long, white arrow). The spores of *R. Berteriana* (Fig. 8) have long folds and a papillate lower stratum is also shown in an abraded spore (Fig. 9).

Cytology

Rumohra adiantiformis is reported as $n = 41$ from Jamaica by Walker (1966) (as *Polystichum adiantiforme*). The same number is known for the species in New Zealand and Tristan da Cunha.

Literature

Christensen, C., 1910. On some species of ferns collected by Dr. Carl Skottsberg in temperate South America. Ark. för Bot. 10 (2): 1–32.

Holttum, R. E. 1947. A revised classification of Leptosporangiate ferns Jour. Linn. Soc. Bot. 53: 123–158.

Kato, M. 1974. A note on the systematic position of *Rumohra adianti-formis*. Acta Phytotax. Geobot. 26: 52–57.

Pérez A., E. 1928. Die natüraliche Gruppe der Davalliaceen (Sm.) Kaulf. Bot. Abhandl. Goebel 14: 1–96.

Walker, T. G. 1966. Reference under the tribe Dryopterideae.

73. *Lastreopsis*

Figs. 73.1–73.13

Lastreopsis Ching, Bull. Fan Mem. Instit. Biol. Bot. 8: 157. 1938. Type: *Lastreopsis recedens* (T. Moore) Ching (*Lastrea recedens* T. Moore) = *Lastreopsis tenera* (R. Br.) Tindale.

Parapolystichum (Keys.) Ching, Sunyatsenia 5: 239. 1940. *Polystichum* section *Parapolystichum* Keys., Polypod. Cyath. Herb. Bung. 11, 45. 1873. Type: *Polystichum effusum* (Sw.) Keys. (*Polypodium effusum* Sw.) = *Parapolystichum effusum* (Sw.) Ching = *Lastreopsis effusa* (Sw.) Tindale.

Description

Terrestrial, less often rupestral or rarely (in *L. davallioides*) epiphytic; stem long- to short-creeping to erect, usually moderately stout including the sometimes persistent petiole bases, bearing scales and many long, fibrous roots; leaves monomorphic, ca. 50 cm to 2 m rarely to 3 m, long, borne in a crown or a cluster to widely spaced, lamina usually 2-pinnate-pinnatifid to 4-pinnate, less often 2-pinnate or to 5-pinnate-pinnatifid, pubescent, glandular-pubescent and often scaly, veins free; sori roundish, borne on a vein to marginally at a vein-tip, on a slightly to moderately raised receptacle, not paraphysate, covered by a reniform or rarely subpeltate indusium or exindusiate; spores more or less ellipsoidal, monolete, the laesura $\frac{3}{4}$ the spore length, rugulose-saccate or with long folds and an echinulate surface. Chromosome number: $n = 41, 82; 2n = 82$.

The costae are raised on the adaxial side (Fig. 4) and the basal, ultimate or penultimate segments have the basiscopic margin continuous with an adaxial ridge of the axis below (Fig. 3). The lamina is catadromic (Fig. 12) in most species, at least above the base, but in a few—*Lastreopsis amplissima, L. Killipii* (Fig. 13) and *L. davallioides* (Brack.) Tindale—it is wholly anadromic.

Systematics

Lastreopsis is a pantropical and south temperate genus of about 35 species with five in America. It appears to be related to *Ctenitis* by the multicellular trichomes on the adaxial side of the axis and the raised axis of the penultimate segments. However, the spores with long folds and those with saccate surface have counterparts among those of some *Dryopteris* species and suggest possible alliances with that genus.

Lastreopsis has been revised by Tindale (1965).

Tropical American Species

Two of the American species of *Lastreopsis* are rather widely distributed and have pronounced geographic variation. Variants have been taxonomically recognized by Tindale and these include four more or less sympatric subspecies of *L. effusa* and three partially sympatric subspecies of *L. exculta. Lastreopsis exculta* ssp. *guatemalensis* (Bak.) Tindale has recently been treated as a species, *L. chontalensis* (Fourn.) Lell., based on its narrow lamina and closely placed central pinnae.

Fig. 73.1. *Lastreopsis effusa,* near Albion, St. Ann, Jamaica. (Photo W. H. Hodge.)

Fig. 73.2. Distribution of *Lastreopsis* in America.

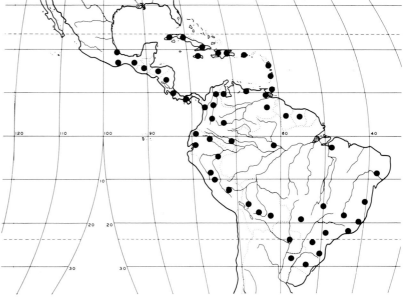

3   4

5   6

Figs. 73.3–73.6. *Lastreopsis*. 3. Basal penultimate segment of *Lastreopsis effusa* with basiscopic margin continuous with an adaxial ridge (arrow) of the axis below, ×8. **4.** Raised costae on adaxial surface of segments of *L. effusa*, ×5. **5.** Indusiate sori of *L. exculta*, ×12. **6.** Exindusiate sori of *L. effusa*, ×12.

Key to American Species of *Lastreopsis*

a. The basal ultimate segment of a penultimate segment with a midvein arising from the same axis as the costa of the penultimate segment.
            L. acuta (Hook.) Tindale

a. The basal ultimate segment of a penultimate segment with a midvein arising from the costa of the penultimate segment.   b.

  b. Lamina anadromic throughout (Fig. 13).   c.

    c. Petiole and rachis with few persistent scales; Bolivia to southeastern Brazil.   *L. amplissima* (Presl) Tindale

    c. Petiole and rachis persistently scaly; Costa Rica to Colombia.
            L. Killipii (Maxon) Tindale

  b. Lamina catadromic (Fig. 12) at least above the basal pinnae.   d.

    d. Sori indusiate. (Fig. 5)   *L. exculta* (Mett.) Tindale

    d. Sori exindusiate. (Fig. 6)   *L. effusa* (Sw.) Tindale

Ecology (Fig. 1)

Lastreopsis is a genus of tropical to temperate rain forests. A single species, *L. davallioides* of the Pacific, is sometimes epiphytic.

In America *Lastreopsis* usually occurs in wet montane forests, or in lowland forests, especially along streams and in ravines. It usually grows on the forest floor but may also be on clay stream

7 8 9

10 11

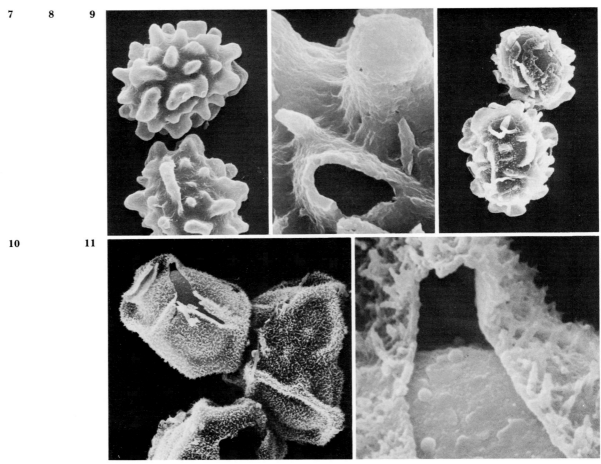

Figs. 73.7–73.11. Spores of *Lastreopsis,* × 1000, **7, 8.** *L. effusa,* Cuba, *Jack 7223.* **7.** Rugulose-saccate inflated. **8.** Detail slightly rugulose or papillate surface, × 5000. **9.** *L. exculta* saccate elements compressed, Costa Rica, *Scamman 5934.* **10, 11.** *L. Killipii,* Panama, *Killip 5360.* **10.** Long folds and echinulate surface. **11.** Detail of abraded spore in Fig. 10, a raised, echinate fold, above, rugulose lower stratum, below, × 10,000.

banks, or on rocky banks. Sometimes it grows in secondary vegetation, in thickets, or along roadsides. *Lastreopsis* occurs from nearly sea level to 1500 m, rarely higher to 2300 m.

Lastreopsis effusa and *L. exculta* commonly have a scaly proliferous bud on the rachis near the lamina apex, by which the plants can reproduce vegetatively.

Geography (Fig. 2)

Lastreopsis is distributed in tropical America, tropical Africa and Madagascar and the Seychelles east to Taiwan, the Philippine Islands and Timor; also in eastern Australia and Tasmania and in the Pacific from New Caledonia, and New Zealand east to Tahiti.

In America the genus ranges from Veracruz in Mexico, south to Panama, in all of the Greater Antilles, in Martinique, St. Vincent and Grenada in the Lesser Antilles, and in South America south to Bolivia and to Missiones in northeastern Argentina, and Rio Grande do Sul in Brazil.

Figs. 73.12, 73.13. Lamina architecture of *Lastreopsis.* **12.** *Lastreopsis effusa,* catadromic: the basal pinnule (arrow) is on the basiscopic side, that is, the side of the pinna-rachis toward the base of the lamina, ×0.5. **13.** *Lastreopsis Killipii,* anadromic: the basal pinnule (arrow) is on the acroscopic side, the side of the pinna-rachis toward the apex of the lamina, ×0.25.

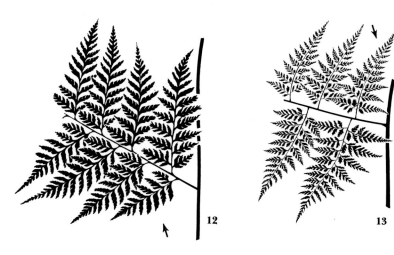

Lastreopsis effusa is distributed throughout the range of the genus, *L. exculta* occurs from southern Mexico to Panama, Ecuador and Venezuela. *Lastreopsis amplissima* occurs from Mt. Roraima and Ecuador to Bolivia and southeastern Brazil, *L. Killipii* from Costa Rica to Peru, and *L. acuta* is confined to southeastern Brazil.

Spores

Lastreopsis spores are mostly rugulose-saccate as in *L. effusa* (Fig. 7) but may have long folds and a densely echinate surface as in *L. Killipii* (Fig. 10). The saccate formation is formed of inflated, hollow folds (Fig. 8) and may be compressed and appear somewhat cristate as in *L. exculta* (Fig. 9). The echinate spores have a relatively thin outer layer that is underlayed by a rugulose stratum (Fig. 11). Spores are described as ruguloso-saccate, cristate and rarely echinate in the monograph of the genus by Tindale (1965). Both saccate and echinate spores occur in the American species. The saccate type of spore is characteristic of several related genera as *Dryopteris,* and *Ctenitis* but densely echinate spores are uncommon.

Cytology

The report of *n* = 41 for *Lastreopsis* in the American tropics, based on *L. effusa* from Jamaica (Walker, 1966) is also known for paleotropical species. The cytological work on the genus, reviewed in Tindale's treatment (1965), includes diploids from a wide geographic sample of species in Australia, New Caledonia, Ghana and Ceylon, while in New Zealand tetraploids as well as diploids have been detected.

Literature

Tindale, M. D., 1965. A monograph of the genus *Lastreopsis* Ching. Contrib. N. S. Wales Nat. Herb. 3: 249–339.
Walker, T. G. 1966. Reference under the tribe Dryopterideae.

74. *Dryopteris*

Figs. 74.1–74.25

Dryopteris Adans., Fam. Plantes 2: 20, 551. 1763, *nom. conserv.* Type: *Dryopteris Filix-mas* (L.) Schott (*Polypodium Filix-mas* L.).

Nephrodium Michx., Fl. Bor. Amer. 2: 266. 1803. Type: *Nephrodium marginale* (L.) Michx. (*Polypodium marginale* L.) = *Dryopteris marginalis* (L.) A. Gray. [Subgenus *Dryopteris*].

Arachniodes Bl., Enum. Pl. Jav. 241. 1828. Type: *Arachniodes aspidioides* Bl. = ? *Dryopteris aristata* (Forst.) O. Ktze. [Subgenus *Polystichopsis*].

Lophodium Newm., Phytol. 4: 371. 1851. Type: *Lophodium Foenisecii* (R. T. Lowe) Newm. (*Nephrodium Foenisecii* R. T. Lowe) = *Dryopteris aemula* (Ait.) O. Ktze. [Subgenus *Dryopteris*].

Dichasium Fée, Mém. Fam. Foug. 5 (Gen. Fil): 302. 1852. Type: *Dichasium parallelogrammum* (Kze.) Fée (*Aspidium parallelogramma* Kze.) = *Dryopteris paleacea* (Sw.) Hand.-Mazz. [Subgenus *Dryopteris*].

Pycnopteris Moore, Gard. Chron. 1855: 468. Type: *Pycnopteris Sieboldii* Moore = *Dryopteris Sieboldii* (Moore) O. Ktze. [Subgenus *Dryopteris*].

Diclisodon Moore, Ind. Fil. xcv. 1857. Type: *Diclisodon deparioides* Moore = *Dryopteris deparioides* (Moore) O. Ktze. [Subgenus *Dryopteris*].

Acrorumohra (H. Ito) H. Ito, Fl. Nov. Jap. (Nakai & Honda) 4: 101. 1939. *Rumohra* section *Acrorumohra* H. Ito, Jour. Jap. Bot. 11: 583. 1935. Type: *Rumohra diffracta* (Baker) Ching (*Nephrodium diffractum* Baker) = *Dryopteris diffracta* (Baker) C. Chr. [Subgenus *Dryopteris*].

Leptorumohra (H. Ito) H. Ito, *ibidem* 4: 118. 1939. *Rumohra* section *Leptorumohra* H. Ito, *ibidem* 11: 579. 1935. Type: *Rumohra Miqueliana* (Franch. & Sav.) Ching (*Aspidium Miqueliana* Franch. & Sav.) = *Dryopteris Miqueliana* (Franch. & Sav.) C. Chr. [Subgenus *Polystichopsis*].

Polystichopsis (C. Chr.) Morton, Amer. Fern Jour. 50: 147–148. 1960. *Dryopteris* subgenus *Polystichopsis* C. Chr., Danske Vidensk. Selsk. Skrift. Nat. Math. VIII, 6 (1): 101. 1920. Type: *Dryopteris pubescens* (L.) O. Ktze. (*Polystichopsis pubescens* (L.) Morton).

Byrsopteris Morton, Amer. Fern Jour. 50. 149. 1960. Type: *Byrsopteris aristata* (Forst.) Morton (*Polypodium aristatum* Forst.) = *Dryopteris aristata* (Forst.) O. Ktze. [Subgenus *Polystichopsis*].

Phanerophlebiopsis Ching, Acta Phytotax. Sinica 10: 115. 1965. Type: *Phanerophlebiopsis Tsiangiana* Ching = *Dryopteris Tsiangiana* (Ching) R. & A. Tryon. [Subgenus *Dryopteris*].

Description   Terrestrial, rupestral or very rarely (as in *D. subarborea*) epiphytic; stem erect to ca. 30 cm tall, or usually decumbent small or stout including the persistent leaf bases, or (in *D. angustifrons*) long-creeping, bearing scales and many long, fibrous roots; leave monomorphic or slightly dimorphic, ca. 10 cm to 2 m long, borne usually in a crown or cluster, lamina 1-pinnate to usually 2-pinnate-pinnatifid or 3-pinnate, to 6-pinnate, glabrous, glandular, pubescent or scaly, veins free; sori roundish, borne on the veins or at a vein-tip, with a flat to slightly raised receptacle, not paraphysate, covered by a usually reniform or rarely subpeltate indusium or exindusiate; spores more or less ellipsoidal, monolete, the laesura $\frac{1}{2}$ to $\frac{3}{4}$ the spore length, rugose-saccate or with folds or wings, the surface slightly rugulose, echinate or spinulose. Chromosome number: $n = 41, 82, 123$; $2n = 82, 164$; apogamous 82, 123, 164.

The costa of the penultimate segments is grooved on the adaxial surface (Fig. 5) and the basal, ultimate or penultimate,

Fig. 74.1. *Dryopteris paleacea*, Villa Mills, Cerro de la Muerte, Costa Rica. (Photo W. H. Hodge.)

segments have the basiscopic ridge continuous with an adaxial ridge of the axis below (Fig. 4). Most of the species have a reniform indusium, which is rarely glandular (Fig. 7). In subgenus *Dryopteris* the lamina is typically catadromic, at least above the base (Fig. 10), and in subgenus *Polystichopsis* it is usually anadromic (Fig. 12). These types of branching are also illustrated in *Lastreopsis,* the catadromic in Fig. 73.12 and the anadromic in Fig. 73.13.

Systematics   *Dryopteris* is a nearly worldwide genus of about 150 species, with some 25 in America and 11 of these in the American tropics. It is most diverse in eastern Asia, where further study is needed to clarify species as well as generic relationships. Serizawa (1976) published a classification of the dryopteroid ferns of Japan and adjacent areas and reviewed previous classifications of Asian species. An assessment of these classifications with reference to the entire genus is needed. The several segregates of *Dryopteris* frequently recognized are *Acrorumohra, Leptorumohra, Polystichopsis, Phanerophlebiopsis,* and *Arachniodes.* The last is most often maintained as a genus distinct from *Dryopteris* as by Holttum (1954) the first to recognize it as a natural group, and Tindale (1961, 1965). The analysis of characters by Tindale include features such as orientation of the stem, the stem scales, lamina shape and texture that are not clearly distinctive for each genus.

Fig. 74.2. *Dryopteris denticulata,* Cordillera de Talamanca, Costa Rica. (Photo W. H. Hodge.)

Some, such as the segment teeth may vary within a species-group, for example in *Dryopteris denticulata* (*sens. lat.*) the teeth vary from cuspidate as in *Arachniodes* to rounded as in *Dryopteris.* Sledge (1973) and Serizawa (1976) also note other features of the genera. However, all these authors recognize difficulties in separating the groups, as Sledge noted in reference to *Arachniodes:* "Even after the removal of discordant species, its boundaries remain somewhat ill-defined though its maintenance seems to be justified if only on the grounds of convenience". Other characters such as lamina architecture, soral position and the indusium do not provide a basis for a satisfactory classification.

Recognition of genera in this complex is difficult, for it appears to involve a series of intergrading groups with reticulate relationships of characters, as well as intermediate species. An inclusive genus is adopted here, with the recognition of the two major groups as subgenera. Discrete evolutionary elements within this complex may be detected as a thorough review is made of all species and additional data on anatomy, chemistry and spores is incorporated.

Dryopteris differs from *Polystichum* in its reniform or rarely sub-peltate, rather than peltate, indusium and by the structure of the spores in which a lower perispore layer is not prominently echinate as in *Polystichum*. Subgenus *Dryopteris* usually has the segments not as sharply toothed as in *Polystichum* and subgenus *Polystichopsis,* which often does have sharply toothed segments, often has a creeping stem rather than a decumbent or erect stem as in *Polystichum*.

The name *Polystichopsis* has never been satisfactorily published except as a subgenus of *Dryopteris* by Christensen. John Smith used the name earlier but without rank or proper subordination under *Lastrea* section *Thelypteris.* Tindale (1961) considered Christensen to be the author of subgenus *Polystichopsis,* and (1965) Morton to be the author at generic rank. We agree, although Morton's publication is clearly inadvertent.

Christensen (1913, 1920) has revised the tropical American species of subgenus *Dryopteris* and of subgenus *Polystichopsis* (*D. Trianae* and *D. amplissima* are to be excluded).

Synopsis of *Dryopteris*

The description of the subgenera includes the central characters and some exceptional ones are also noted.

Subgenus *Dryopteris*

Stem erect to decumbent, with internal glands, lamina ovate to ovate-lanceolate or narrower, catadromic, at least above the basal pinnae, the basal pinnae not or not much larger than the next pair, ultimate segments nearly entire to toothed, the teeth blunt to acuminate.

A subgenus of about 100 species, with ca. 20 in America and eight of these in the American tropics.

Some of the species with exceptional characters are: *Dryopteris angustifrons* (Moore) O. Ktze. with a long-creeping stem, *D. fragrans* (L.) Schott without internal glands in the stem, *D. deparioides* (Moore) O. Ktze., with the lamina strongly anadromic, *D. formosana* (Christ) C. Chr. and *D. chinensis* (Bak.) Koidz. with the lamina pentagonal and the basal pinnae much larger than the next pair, and *D. expansa* (Presl) Fras.-Jenk. & Jermy and *D. campyloptera* Clarkson with the ultimate segments toothed, the teeth cuspidate.

Subgenus *Polystichopsis*

Stem short- to long-creeping, without internal glands, lamina pentagonal, anadromic, the basal pinnae larger than the next pair, ultimate segments toothed, the teeth cuspidate.

A subgenus of about 50 species, three of them in tropical America.

Some of the species with exceptional characters are: *Dryopteris denticulata* with erect stems, *D. Hasseltii* (Bl.) C. Chr. with internal glands in the stem, *D. pubescens* with sometimes a predominently catadromic lamina, *D. Standishii* (Moore) C. Chr. with the basal

Fig. 74.3. Distribution of *Dryopteris* in America, south of lat. 35° N.

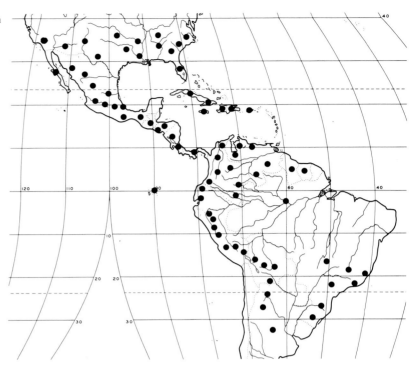

pinnae not much larger than the next pair, and *D. macrostegia* with bluntly toothed ultimate segments. The spores are generally more prominently cristate than those of subgenus *Dryopteris*.

The *Dryopteris pubescens* complex is sometimes recognized as a small genus *Polystichopsis* (*sens. strict.*) with affinities to *Lastreopsis*. However, it appears that the grooved costae on the adaxial surface of the lamina ally it with *Dryopteris* and *D. macrostegia* is transitional between elements of *D. pubescens* and more typical species of subgenus *Polystichopsis*.

Tropical American Species

Most tropical American species of *Dryopteris* are sufficiently distinctive but three species complexes are especially in need of study. Christensen considered each of these to include several species and his treatments have been generally followed, although more adequate collections accumulated since his work raise questions as to their distinctiveness.

Dryopteris patula sens. lat. includes *D. cinnamomea* (Cav.) C. Chr., *D. glandulifera* (Liebm.) C. Chr., *D. indecora* (Liebm.) C. Chr., and *D. mexicana* (Presl) C. Chr.

Dryopteris denticulata sens. lat. includes *D. formosa* (Fée) Maxon, *D. leucostegioides* C. Chr., and *D. rigidissima* (Hook.) C. Chr.

Dryopteris pubescens sens. lat. includes *D. argillicola* Proctor, *D. chaerophylloides* (Poir.) C. Chr., *D. lurida* Slosson, and *D. muscosa* (Vahl) Proctor. This complex consists of several more or less distinctive leaf forms with few other, variable characters supporting their status as species. The variation is similar in some ways to that in the leaf forms of *Anopteris hexagona,* which are partly correlated with the geography.

Key to Tropical American *Dryopteris*

a. Lamina anadromic throughout and eglandular, or if catadromic above the base then the abaxial surface of the major axes with abundant long trichomes. (Subgenus *Polystichopsis*).   b.

  b. Lamina long-pubescent on the adaxial surface, fertile leaf ca. 20–30 cm long; West Indies. (Fig. 9)   *D. pubescens* (L.) O. Ktze. *sens. lat.*

  b. Lamina glabrous on the adaxial surface or rarely somewhat pubescent in the grooves of the major axes.   c.

    c. Stem erect to decumbent, the leaves clustered, most ultimate segments stalked to sessile; tropical America except the Lesser Antilles. (Figs. 2, 12)   *D. denticulata* (Sw.) O. Ktze. *sens. lat.*

    c. Stem creeping, the leaves borne at intervals, many ultimate segments partially to wholly adnate; Jamaica, Panama, northern South America. (Fig. 5)   *D. macrostegia* (Hook.) O. Ktze.
                    (*D. ochropteroides* (Baker) C. Chr.)

a. Lamina catadromic, at least above the base, and glabrous, glandular or scaly, or if anadromic then minutely glandular. (Subgenus *Dryopteris*).   d.

  d. Indusium vaulted over the sporangia and concealing them, of the same form in old sori.   e.

    e. Lamina eglandular; Mexico, western United States to southern British Columbia. (Fig. 11)   *Dryopteris arguta* (Kaulf.) Watt.
                    (*D. Maxonii* C. Chr.)

    e. Lamina with abundantly cylindrical glands; Mexico to Costa Rica. (Fig. 7)   *Dryopteris Karwinskyana* (Mett.) O. Ktze.

  d. Indusium flat or nearly so, the sporangia extending beyond except in very young sori, variously distorted in age.   f.

    f. Lamina eglandular, 1-pinnate-pinnatifid to 2-pinnate.   g.

      g. Lamina herbaceous, with usually few tan scales, ultimate segments toothed below the rounded to subacute apex; North America, south to Guatemala.   *Dryopteris Filix-mas* (L.) Schott

      g. Lamina coriaceous, with many brown to dark reddish-brown scales, especially on the rachis, ultimate segments entire or nearly so below the truncate apex; Mexico, Central America, Greater Antilles, South America to Bolivia and southeastern Brazil. (Figs. 1, 6, 8)   *Dryopteris paleacea* (Sw.) Hand.-Mazz.
                    (*D. parallelogramma* (Kze.) Alston)

    f. Lamina more or less minutely glandular, or if eglandular, then 3-pinnate or more complex at least basally.   h.

      h. Pinnae stalked, or if sessile then the lamina 2-pinnate or more complex.   i.

        i. Lamina long-triangular to ovate to lanceolate or narrowly so, basal pinnae long-triangular to lanceolate, usually equilateral or nearly so, ultimate segments ascending; southwestern United States, Mexico, Central America, Greater Antilles, South America south to Argentina and southeastern Brazil. (Fig. 10)   *Dryopteris patula* (Sw.) Underw., *sens. lat.*

        i. Lamina and basal pinnae deltoid, strongly inequilateral, ultimate segments patent.   j.

          j. Pinnae ascending, indusium ca. 1 mm in diameter; southern Mexico.   *Dryopteris futura* A. R. Sm.

          j. Pinnae patent, indusium ca. 0.5 mm in diameter; Guatemala.   *Dryopteris nubigena* Maxon

      h. Pinnae, or most of them, sessile, the lamina mostly 1-pinnate-pinnatifid, reduced at the base, ca. 10–30 cm long; Peru.   *Dryopteris Saffordii* C. Chr.

Ecology (Figs. 1, 2)

Dryopteris is primarily a genus of wet forests, where it grows in ravines, or shaded slopes, often among rocks which may be igneous, sandstone, limestone or shale. It also occurs in thickets, in swamps, in grassy areas and on cliffs. In the paleotropics, *D. arborescens* (Bak.) O. Ktze. and *D. subarborea* (Bak.) C. Chr. are exceptional in sometimes being epiphytic.

In tropical America *Dryopteris* grows most often in wet montane forests, also in pine and oak woods (Mexico to Honduras), in cloud forests and in lowland rain forests. In xeric regions it occurs in mesic canyons. Species may grow in moist niches on lava beds, on earth banks, on shrubby hillsides or in coffee plantations. At the highest elevations, in the Andes, *Dryopteris denticulata* and *D. Saffordii* are among rocks in grassy páramos. Species occur from near sea level to nearly 4000 m, mostly at 1000–2500 m. In South America, *Dryopteris denticulata* grows from 50 m in the Amazon basin up to 3200 m in the Andes. *Dryopteris Saffordii* also has a broad altitudinal range, from 500 m on Loma Lachay, north of Lima, Peru to 3900 m in the Andes of Peru.

At high altitudes, *Dryopteris denticulata* may develop a congested mound of multicipital stems up to 50 cm broad and 15 cm tall.

Geography (Fig. 3)

Dryopteris is widely distributed, on all continents and in the Pacific islands east to the Hawaiian Islands and Easter Island. It is absent from southern South America, central and western Australia and New Zealand.

Dryopteris occurs throughout most of the United States and Canada, with a center of 12 species in eastern North America. *Dryopteris fragrans* extends the range northward to nearly lat. 80° N. Some of these temperate and boreal species also occur in Europe and Eastern Asia and others are most closely related to species of those regions.

In tropical America the genus occurs in Mexico, Central America, the West Indies, and in South America south to Bolivia, Córdoba in central Argentina, and southeastern Brazil; also in the Galápagos Islands. The two subgenera have nearly the same range in tropical America but subgenus *Dryopteris* is absent from the Lesser Antilles and also from the Amazon Basin, where subgenus *Polystichopsis* is rare.

Spores

Spores of subgenus *Dryopteris* usually are coarsely rugose often with a finely rugulose surface as in the tropical American species *D. patula* (Fig. 13), *D. paleacea* (Fig. 16) and *D. Saffordii* (Fig. 15). The surface is generally similar to many Asiatic species such as *D. Championii* (Benth.) Ching (Fig. 17). The perispore is formed of two strata, a slightly rugose one between the smooth exospore and an inflated outer layer as in *D. patula* (Fig. 14). Spores of the *D. Carthusiana* complex of temperate regions in Europe, America and eastern Asia are distinguished from the others by a more or less dense spinulose surface. These may also have the characteristic saccate formation as in *D. cristata* (L.) A. Gray (Fig. 18) but usually have longer folds as in *D. intermedia* (Willd.) A. Gray (Fig. 19) and *D. Carthusiana* (Vill.) Fuchs (Fig. 20). These spinulose types have been figured in several studies of this complex in Europe and England by Jermy (1968), Crabbe et al. (1970) and in America by Britton (1972a, 1972b); and several related species lacking spinules have been reviewed by Britton and Jermy (1974). A SEM survey of 45 species of Japan includes

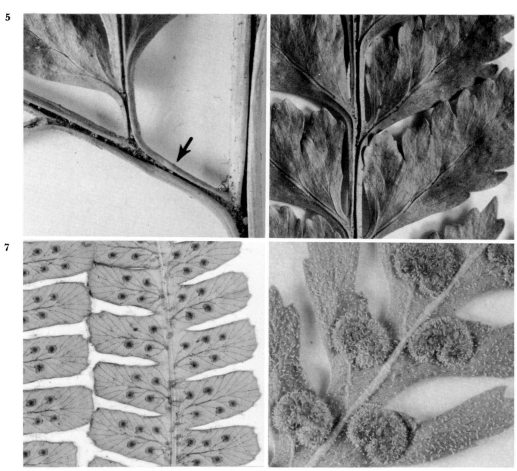

Figs. 74.4–74.7. *Dryopteris.* **4.** Basal penultimate segment of *D. macrostegia* with the basiscopic ridge continuous with an adaxial ridge (arrow) of the axis below, rachis at right, × 8. **5.** Grooved costa on adaxial surface of segment of *D. macrostegia,* × 5. **6.** Portion of fertile pinna of *D. paleacea,* × 3. **7.** Glandular surface of segment and reniform indusia of *D. Karwinskyana,* × 12.

predominantly saccate types classified into several groups on the basis of the surface (Mitui, 1972).

Spores of subgenus *Polystichopsis* are generally rugose-saccate and similar to those of subgenus *Dryopteris* as in *D. macrostegia* (Fig. 21). Those of *D. pubescens* are prominently saccate-tuberculate with strands and connecting folds (Figs. 24, 25). Abraded spores have two perispore strata, a thin, rugose layer below an inflated one as in *D. macrostegia* (Fig. 22) characteristic of subgenus *Dryopteris.* Spores of the tropical American *D. denticulata sens. lat.* are distinctive with slender, prominent winglike folds (Fig. 23) and generally resemble spores of *Thelypteris* subgenus *Steiropteris.*

Cytology

In subgenus *Dryopteris,* diploids with $n = 41$ are more than twice as frequent as tetraploids. Hybrids and apogamous forms are frequent in the genus although the hexaploid, the highest level known, has been established only rarely. The few cytological reports for American tropical species are mostly diploids as in

Dryopteris patula and *D. arguta* (*D. Maxonii*). *Dryopteris paleacea* (*D. parallelogramma*) is reported as diploid and triploid. A hexaploid, from Mexico, recognized as "*Dryopteris* sp., aff. *filix-mas* (L.) Schott, $2n = 123$ II," by Smith and Mickel (1977), probably also involves *D. paleacea,* and denotes complexities in the tropical American species. Subgenus *Polystichopsis* has been reported as $n = 41$ and $n = 82$ from both America and the Old World. These records are consistent with those in subgenus *Dryopteris* and related genera. *Dryopteris denticulata* (as *Arachniodes*) is reported as tetraploid with $n = 82$ from Jamaica by Walker (1966), while plants from Chiapas, Mexico, are diploid with $n = 41$ (Smith and Mickel, 1977). Diploids are somewhat more frequent than polyploids among Old World species of subgenus *Polystichopsis.* The mitotic number 82, for the subgenus, is clearly documented by a photograph of a root tip squash of *Arachniodes chinensis* (Rosenst.) Ching (Roy and Holttum, 1965).

Relations of temperate species of the *Dryopteris Carthusiana* complex have been most thoroughly studied in Britain and North America. In eastern North America this group includes four diploids, four tetraploids and the hexaploid, *D. Clintoniana* (D. C. Eaton) Dowell. The origin of the hexaploid, central to a genomic interpretation of the complex, has not been clearly established. Genome relations of the species, were proposed by Wagner (1971) based on morphological characters and chromosome pairing, and included an unknown diploid taxon. An alternative explanation, proposed by Hickok and Klekowski (1975) regards the origin of the polyploids wholly from extant species, some with a genome partially differentiated from the parental species. Cytological relations in this complex studied by Gibby (1977) are based on chromosome pairing and have utilized synthesized material from controlled crosses. The species of Europe and Macaronesia are the main focus of this work but it does not include the critical American hexaploid, *Dryopteris Clintoniana.* Five basic genomes, including three that occur in North America, *Dryopteris intermedia, D. expansa* (*D. assimilis* S. Walker) and *D. ludoviciana* (Kze.) Small are recognized, as well as three American tetraploids, *D. cristata, D. campyloptera* and *D. Carthusiana* (*D. spinulosa* (O. F. Muell.) Watt.

A cytogeographic analysis of *Dryopteris* of Japan by Hirabayshi (1974) includes numerous meiotic figures documenting the cytological work. The cytotypes are assessed in relation to six phytogeographic patterns based on mean isotherm lines for Japan. Apogamous cytotypes are the most common and include one tetraploid, four diploids and 26 triploids. Twenty-two diploids are reported among normal reproducing species and also five tetraploids along with 15 hybrids for which the putative parents are proposed.

The cytological modifications involved in the apogamous life cycle of ferns were first delineated by the work of Döpp (1932) in *Dryopteris remota* (A. Br.) Druce. It was developed in further detail in other ferns, and especially in *D. affinis* (Lowe) Fras.-Jenk. (*D. pseudomas* (Wollast.) Holub, *D. Borreri* Newm.) by Manton (1950).

Figs. 74.8–74.12. Leaves of ***Dryopteris*** species. **8. *D. paleacea,*** × 0.25. **9. *D. pubescens,*** × 0.5. **10. *D. patula,*** catadromic above base: the basal pinnule is on the side of the pinna-rachis directed toward the base of the lamina, × 0.5. **11. *D. arguta,*** × 0.25. **12. *D. denticulata,*** anadromic: the basal pinnule is on the side of the pinna-rachis directed toward the apex of the lamina, × 0.5.

13   14

15 16 17

18 19 20

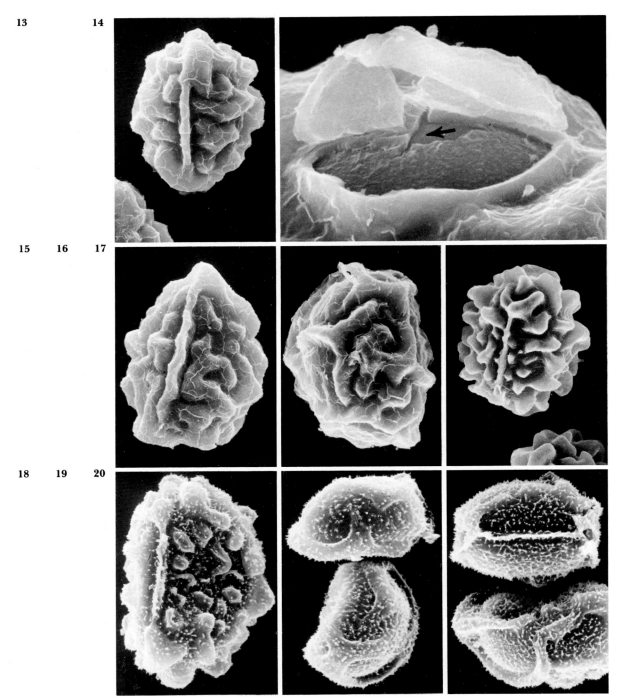

Figs. 74.13–74.20. Spores of *Dryopteris* subgenus *Dryopteris*, × 1000. **13. *D. patula,*** *sens. lat.* (*D. mexicana*), rugate-saccate surface, laesura central, Mexico, *Pringle 11773.* **14. *D. patula,*** *sens. lat.* (*D. cinnamomea*), abraded wall, (arrow) at exospore, the inner perispore layer above and outer perispore raised, Mexico, *Correll & Gentry 23002,* × 5000. **15. *D. Saffordii,*** rugose-saccate, Peru, *Vargas 1112.* **16. *D. paleacea,*** rugose-saccate, Mexico, *Anderson & Laskowski 3997.* **17. *D. Championii,*** saccate, laesura central, China, *Fan & Li 85.* **18. *D. cristata,*** saccate-spinulose, laesura left, New Hampshire, *Kennedy in 1890.* **19. *D. intermedia,*** spinulose, Massachusetts, *Faxon in 1874.* **20. *D. Carthusiana,*** spinulose, Massachusetts, *Weatherby et al. Gray Herb. Exsicc. 803.*

21   23   24

22   25

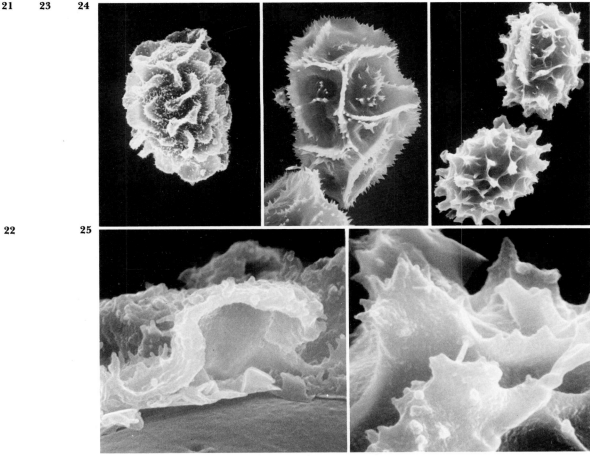

Figs. 74.21–74.25. Spores of *Dryopteris* subgenus *Polystichopsis*, ×1000. **21, 22. *D. macrostegia*,** Colombia, *Schultes & Cabrera 16711*. **21.** Elongated saccate folds. **22.** Section of perispore through fold and lower perispore layer, the exospore, at base, ×10,000. **23. *D. denticulata*** winged-echinate, Colombia, *Hodge 6573*. **24, 25. *D. pubescens*,** Jamaica, *Maxon 9103*. **24.** Saccate-tuberculate, laesura, left, upper spore. **25.** Detail of dentate saccate elements and connecting strands, ×10,000.

Observations

Early recognition of "Filix-mas" of the apothecaries as an effective vermifuge initiated a long history of work on the compounds produced in *Dryopteris*. The active principles in the Male Fern are acylphloroglucinols (phloroglucinols) and a large number of these have been identified in *Dryopteris* species. New data on systematics and relationships have resulted from a series of collaborative studies by biochemists and botanists.

The presence of phloroglucinols is always associated with glands. In subgenus *Dryopteris* these are most often within the stem but there are rarely only external glands on the stem and leaf. In subgenus *Polystichopsis* the glands are usually external and only rarely present internally in the stem.

Taxa are sometimes characterized by a unique set of acylphloroglucinol compounds which aid in their taxonomic evaluation. Similarities of the sets of compounds establish relations among species, and in a few examples, the parentage of alloploids has been clarified by their particular combination of acylphloroglucinols (Britton and Widen, 1974; Widen, 1969; Widen and Britton, 1971a–d; Widén et al., 1975; Widén et al., 1978; Widén et al. 1976a; Widén et al., 1976b). The tropical American *Dryopteris patula* has been related to the *Dryopteris Carthusiana* (*D. dilatata* (Hoffm.) A. Gray) complex through the presence of aspidin, and *D. paleacea* (*D. parallelogramma*) related to the *D. Filix-mas* complex, especially to *D. affinis* (*D. Borreri*) by the presence of filixic acid (Tryon et al., 1973).

Study of the flavonoids of Japanese species of *Dryopteris* (Hiraoka, 1978) demonstrated that there were groups of species with distinctive sets of compounds and also a few with unique sets.

Literature

Britton, D. M. 1972a. Spore ornamentation in the *Dryopteris spinulosa* complex. Canad. Jour. Bot. 50: 1617–1621.

Britton, D. M. 1972b. The spores of *Dryopteris clintoniana* and its relatives. Canad. Jour. Bot. 50: 2027–2029.

Britton, D. M., and A. C. Jermy. 1974. The spores of *Dryopteris filix-mas* and related taxa in North America. Canad. Jour. Bot. 52: 1923–1926.

Britton, D. M., and C-J. Widén. 1974. Chemotaxonomic studies on *Dryopteris* from Quebec and eastern North America. Canad. Jour. Bot. 52: 627–638.

Christensen, C. 1913. Reference under the tribe Dryopterideae.

Christensen, C. 1920. Reference under the tribe Dryopterideae.

Crabbe, J. A., A. C. Jermy, and S. Walker. 1970. The distribution of *Dryopteris assimilis* S. Walker in Britain. Watsonia 8: 3–15.

Döpp, W. 1932. Die apogamie bei *Aspidium remotum* A. Br. Planta 17: 86–152.

Gibby, M. 1977. A Cytogenetic and Taxonomic Study of the *Dryopteris Carthusiana* Complex. Ph.D. Thesis, Department of Genetics, University of Liverpool, England.

Hickok, L. G., and E. J. Klekowski. 1975. Chromosome behavior in hybrid ferns: A reinterpretation of Appalachian *Dryopteris*. Amer. Jour. Bot. 62: 560–569.

Hirabaysahi, H. 1974. Cytogeographic studies on *Dryopteris* of Japan. 176 pp. Hara Shobo, Tokyo, Japan.

Hiraoka, A. 1978. Flavonoid patterns in Athyriaceae and Dryopteridaceae. Biochem. Syst. Ecol. 6: 171–175.

Holttum, R. E. 1954. Flora of Malaya 2: Ferns. 643 pp. Singapore, Government Printing Office.

Jermy, A. C. 1968. Two new hybrids involving *Dryopteris aemula*. Brit. Fern Gaz. 10: 9–12.

Manton, I. 1950. Reference under the tribe Dryopterideae.

Mitui, K. 1972. Spore ornamentation of Japanese species of *Dryopteris* Bull. Nippon Dental Coll. Gen. Ed. 1: 99–116.

Roy, S. K., and R. E. Holttum. 1965. Cytological observations on ferns from southern China. Amer. Fern Jour. 55: 154–158.

Serizawa, S. 1976. A revision of the dryopteroid ferns in Japan and adjacent regions. Sci. Report Tokyo Kyoiku Daigaku, Sec. B, 16: 109–148.

Sledge, W. A. 1973. The dryopteroid ferns of Ceylon. Bull. Brit. Mus. (Nat. Hist.), Bot. 5: 1–43.

Smith, A. R., and J. T. Mickel. 1977. Reference under the tribe Dryopterideae.

Tindale, M. D. 1961. Pteridophyta of south eastern Australia. Contrib. N. S. Wales Nat. Herb., Flora Series 208–211.

Tindale, M. D. 1965. A monograph of the genus *Lastreopsis* Ching. Contrib. N. S. Wales Nat. Herb. 3: 249–339.

Tryon, R., C-J. Widén, A. Huhtikangas, and M. Lounasmaa. 1973. Phloroglucinol derivatives in *Dryopteris parallelogramma* and *D. patula*. Phytochemistry 12: 683–687.

Wagner, W. H. 1971. Evolution of *Dryopteris* in relation to the Appalachians, *in*: The distributional history of the biota of the southern Appalachians. Part 2: Flora. P. Holt, ed, Research Div. Monograph 2, Va. Polytech. and State Univ., Blacksburg, Virginia.

Walker, T. G. 1966. Reference under the tribe Dryopterideae.

Widén, C-J. 1969. Chemotaxonomic investigations on Finnish *Dryopteris* species and related North American taxa. Ann. Acad. Sci. Fennicae Ser. A, IV. Biol. 143: 3–19.

Widén, C-J., and D. M. Britton. 1971a. A chromatographic and cytological study of *Dryopteris dilatata* in North America and eastern Asia. Canad. Jour. Bot. 49: 247–258.

Widén, C-J., and D. M. Britton. 1971b. Chemotaxonomic investigations on *Dryopteris fragrans*. Canad. Jour. Bot. 49: 989–992.

Widén, C-J., and D. M. Britton. 1971c. Chemotaxonomic investigations on the *Dryopteris cristata* complex in North America. Canad. Jour. Bot. 49: 1141–1154.

Widén, C-J., and D. M. Britton. 1971d. A chromatographic and cytological study of *Dryopteris filix-max* and related taxa in North America. Canad. Jour. Bot. 49: 1589–1600.

Widén, C-J., D. M. Britton, W. H. Wagner, Jr., and F. S. Wagner. 1975. Chemotaxonomic studies on hybrids of *Dryopteris* in eastern North America. Canad. Jour. Bot. 53: 1554–1567.

Widén, C-J., A. Huure, J. Sarvela, and K. Iwatsuki. 1978. Chemotaxonomic studies on *Arachniodes* (Dryopteridaceae), II. Phloroglucinol derivatives and taxonomic evaluation. Bot. Mag. (Tokyo) 91: 247–254.

Widén, C-J., M. Lounasmaa, A. C. Jermy, J. von Euw, and T. Reichstein. 1976a. Die Phloroglucide von zwei Farnhybriden aus England und Schottland, von authentischem *"Aspidium remotum"* A. Braun und von *Dryopteris aemula* (Aiton) O. Kuntze aus Ireland. Helvet. Chimica Acta 59: 1725–1744.

Widén, C-J., J. Sarvela, and K. Iwatsuki. 1976b. Chemotaxonomic studies on *Arachniodes* (Dryopteridaceae), I. Phloroglucinol derivatives of Japanese species. Bot. Mag. (Tokyo) 89: 277–290.

75. *Cyrtomium*

Figs. 75.1–75.13

Cyrtomium Presl, Tent. Pterid. 86. 1836. Type: *Cyrtomium falcatum* (L. f.) Presl (*Polypodium falcatum* L. f.).

Phanerophlebia Presl, *ibidem* 84. 1836. Type: *Phanerophlebia nobilis* (Schlect. & Cham.) Presl (*Aspidium nobile* Schlect. & Cham.) = *Cyrtomium nobile* (Schlect. & Cham.) Moore. A synonym of *Cyrtomium* by Moore, Ind. Fil. lxxxii. 1857.

Amblia Presl. *ibidem* 184. 1836. Type: *Amblia juglandifolia* (Willd.) Presl (*Polypodium juglandifolium* Willd.) = *Cyrtomium juglandifolium* (Willd.) Moore. A synonym of *Cyrtomium* by Moore, Ind. Fil. lxxxii. 1857.

Description Terrestrial or rupestral; stem short, nearly erect to usually decumbent, sometimes rather stout including the persistent petiole bases, bearing scales and numerous fibrous roots; leaves monomorphic, ca. 15 cm to 1.5 m long, borne in a crown or cluster, lamina 1-pinnate, usually imparipinnate, or (in *C. hemionitis* Christ) entire or somewhat lobed, deeply cordate and acuminate, more or less fibrillose-scaly, sometimes minutely so, to glabrate, veins free or regularly anastomosing with included free veins, or rarely casually anastomosing and without included veins, sori roundish, borne on the veins or at their tip in two to several rows on each side of the costa, receptacle slightly to somewhat elevated, not paraphysate, covered by a persistent to fugacious, peltate indusium or (in *C. dubium*) exindusiate; spores ellipsoidal to spheroidal, monolete, the laesura $\frac{3}{4}$ the spore length, with irregular rugose-saccate projections or folds, these sometimes winglike, the surface smooth or finely rugulose. Chromosome number: $n = 41, 82$; apogamous 123.

The round sorus and persistent indusium of *Cyrtomium umbonatum* is shown in Fig. 3, the characteristic multiseriate sori of the genus in Fig. 4, and the anastomosing veins of *C. remotisporum* in Fig. 5. Most species have an imparipinnate lamina (Fig. 6). The margins of the pinnae usually bear prominent spinules (Fig. 5) and are often cartilaginous.

Systematics *Cyrtomium* is a temperate and tropical genus of perhaps 25 species, with 15 or more primarily in eastern Asia, and nine or fewer in America. All Old World species have pinnae with anastomosing veins with free included veinlets, while most American species have free venation. Underwood (1899) recognized the American species as a distinct genus, *Phanerophlebia,* on the basis of differences in venation. However, the American *Cyrtomium juglandifolium* has the same type of venation as species of the Old World and the American *C. dubium* and *C. remotisporum* vary in their venation from anastomosing without free included veinlets to nearly free-veined. Therefore, there is neither a clear geographic nor morphological basis for the recognition of the mostly free-veined American species as a separate group.

Cyrtomium is related, in a number of characters, to Old World species of the large genera *Polystichum* and *Dryopteris.* It is distinguished from *Polystichum* by the sori that are in two or more rows on each side of the costa rather than a single row, by the usually imparipinnate lamina rather than gradually reduced lamina

Fig. 75.1. *Cyrtomium remotisporum,* *Liquidambar* cloud forest north of Jalapa, Veracruz, Mexico. (Photo W. H. Hodge.)

Fig. 75.2. Distribution of *Cyrtomium* in America.

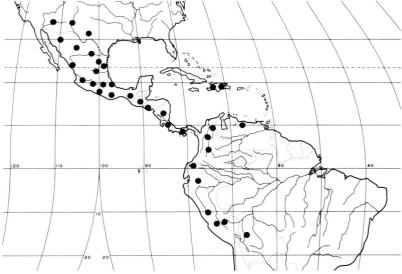

apex, and by the spores with a compact, nonperforate perispore rather than a perforate or reticulate perispore. *Cyrtomium* is distinguished from *Dryopteris* by its peltate, rather than reniform to subpeltate indusium.

The American species of *Cyrtomium* were revised by Underwood (1899) and by Maxon (1912). Morton (1957) provided correct names for the American species and a key to the species cultivated in America.

Tropical American Species

Maxon's revision of American *Cyrtomium* was based on relatively few specimens that provided a limited scope of variation. The species require a new study because of the variability of many characters employed by Maxon in the definition of species. The following synopsis includes all of the names of American species currently recognised and will aid in the identification of the more clearly defined one.

a. Lamina abruptly to gradually reduced at the apex: *C. dubium* (Karst.) R. & A. Tryon. The veins are regularly to casually anastomosing and the sori are exindusiate. The character of the lamina apex is unique among American species, but occurs in some of the Old World such as *C. Fortunei* J. Sm.; the exindusiate condition is unique in the genus. The chromosome number of $n = 41$ places the species in the dryopteroid alliance and the structure of the perispore and multiseriate sori support placing it in *Cyrtomium* rather than *Polystichum*.

a. Lamina imparipinnate, the apex with a conform terminal segment.   b.
   b. Veins regularly to casually anastomosing: *C. juglandifolium* (Willd.) Moore and *C. remotisporum* (Fourn.) Morton may represent the same species.
   b. Veins free.   c.
      c. Indusium indurated, persistent: *C. umbonatum* (Underw.) Morton. The indusia are often raised at the center (umbonate), but less so in the northern part of the range.
      c. Indusium rather thin, fugacious.   d.
         d. Pinnae, especially the basal ones, acroscopically auriculate: *C. auriculatum* (Underw.) Morton. The auricles vary in their degree of development and shape.
         d. Pinnae not auriculate: *C. macrosorum* (Baker) Morton (*C. guatemalense* (Underw.) Morton), *C. nobile* (Schlect. & Cham.) Moore (*C. haitiense* (C. Chr.) Morton; *Phanerophlebia Lindenii* Fourn.), and *C. pumilum* (Mart. & Gal.) Morton are all very similar and characters such as the number of pinnae, the shape of their base, the nature of the margins and the size of the sori are all variable. A critical species revison of this group is especially needed.

Ecology (Fig. 1)

Cyrtomium usually grows in forests, especially in ravines, on rocky banks and by streams. It also grows in thickets, on road banks and on moist cliffs.

In the Americas *Cyrtomium* grows in locally moist situations in canyons, ravines and on shaded cliffs in the northern part of its range, while in southern Mexico and southward it usually grows in oak or pine woods, in montane forests and cloud forests. It most frequently grows in rocky places and is often associated with limestone. It usually grows at elevations of 1000–3000 m.

Cyrtomium juglandifolium sometimes has a proliferous bud in the axil of the pinnae just below the apical segment, from which new plants may be produced.

3   4   5

Figs. 75.3–75.5. *Cyrtomium.* **3.** Sorus and persistent peltate indusium of *C. umbonatum*, × 10. **4.** Multiseriate sori of *C. macrosorum*, × 2.5. **5.** Anastomosing veins and marginal spinules of *C. remotisporum*, × 3.5.

Fig. 75.6. Lamina of *Cyrtomium macrosorum*, × 0.25.

Geography (Fig. 2)

Cyrtomium occurs in America, Africa, Madagascar and southern India, north and eastward to the Himalayas, China, northern Viet Nam, Formosa, Japan and Korea; also in the Hawaiian Islands.

In America, *Cyrtomium* occurs in Texas, New Mexico and Arizona in the southwestern United States, and south through Mexico and Central America to Panama, in Hispaniola, and in South America in northern Venezuela and Colombia, south to Bolivia. Old World species of *Cyrtomium*, especially *C. falcatum* (L. f.) Presl and *C. Fortunei* are frequently cultivated in America and are sometimes adventive. Specimens of *Cyrtomium nobile* from Porto Alegre, Rio Grande do Sul, Brazil, represent an unusual case of a American species that is cultivated and may be adventive.

All of the Old World species occur in China and Japan, and all American species are found in Mexico, except *Cyrtomium dubium* from Costa Rica to Bolivia.

Spores

Cyrtomium spores are characterized by prominent, inflated folds as in *C. auriculatum* and *C. dubium* (Figs. 7, 9, 10). The folds are slender, winglike especially in less fully developed spores as *C. macrosorum* (Fig. 11), or shorter, saccate, in *C. juglandifolium* (Fig. 12) and the Asian *C. caryotideum* (Hook & Grev.) Presl (Figs. 12, 13). The larger size of the latter may reflect the triploid level of that species. Abraded spores of *C. nobile* (Fig. 8) show the wall is formed of a finely rugulose surface above a tuberculate layer with a smooth exospore below. The compact form of the perispore of *Cyrtomium* spores is similar to that of *Dryopteris* and differs from the perforate and reticulate formation characteristic of *Polystichum*.

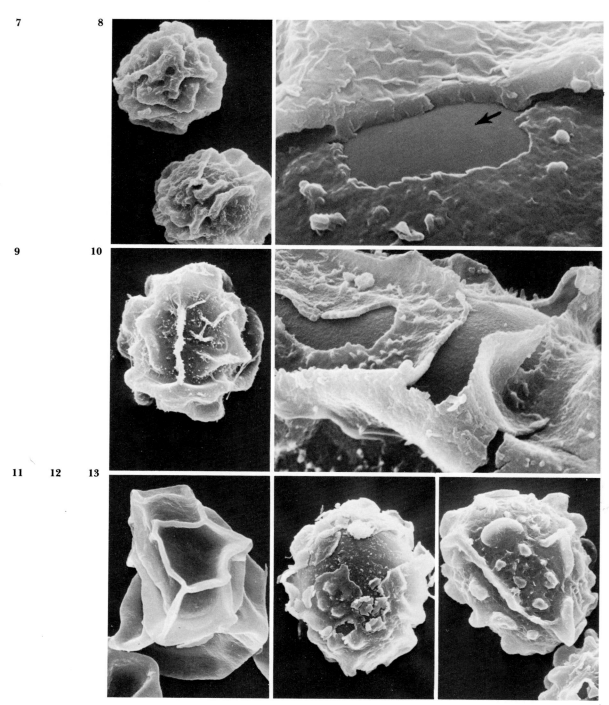

Figs. 75.7–75.13. *Cyrtomium* spores, × 1000. **7.** *C. auriculatum,* rugose-saccate folds, Mexico, *Pringle 831.* **8.** *C. nobile,* abraded surface with outer rugulose perispore above, the lower, irregularly papillate below, and smooth exospore (at arrow), Haiti, *Ekman 7793,* × 10,000. **9, 10.** *C. dubium,* Panama, *Davidson 340.* **9.** Surface folds partly inflated. **10.** Abraded spore with papillate perispore surface and portions of smooth exospore, below, section of fold, right, × 5000. **11.** *C. macrosorum,* slender winglike folds of incompletely developed spore, Guatemala, *Honeywell 15499.* **12.** *C. juglandifolium,* abraded surface with exposed inner perispore, Venezuela, *Fendler 233.* **13.** *C. caryotideum,* saccate, China, *Fan & Class 21.*

Cytology

Cyrtomium umbonatum is reported as $n = 41$ (as *Phanerophlebia*) from San Luis Potosi, Mexico (Mickel et al., 1966), and *C. dubium* with $n = 41$ (as *Polystichum*) from Costa Rica (Wagner, 1980). These records of American species are consistent with those of Asia and also common to several other allied genera. Diploid and tetraploid populations of *Cyrtomium falcatum* are known from Japan and also triploid apogamous plants. Five species of Japan are apogamous triploids suggesting that this is a common mode of reproduction in the genus although it has not been detected among the tropical American species. The old classical work on apogamy by de Bary included observations on *Cyrtomium* and detailed cytological examination of the modifications in apogamous plants of *Cyrtomium* were investigated by Manton (1950).

Literature

Manton, I. 1950. Reference under the tribe Dryopterideae.

Maxon, W. R. 1912. Notes on the North American species of *Phanerophlebia*. Bull. Torrey Bot. Cl. 39: 23–28.

Mickel, J., W. H. Wagner, and K. Chen. 1966. Chromosome observations on the ferns of Mexico. Caryologia 19: 95–102.

Morton, C. V. 1957. Observations on cultivated ferns, II. The proper generic name of the Holly Fern. Amer. Fern Jour. 47: 52–55.

Underwood, L. M. 1899. American Ferns, II. The genus *Phanerophlebia*. Bull. Torrey Bot. Cl. 26: 205–216.

Wagner, F. S. 1980. New basic chromosome numbers for genera of neotropical ferns. Amer. Jour. Bot. 67: 733–738.

76. *Didymochlaena*
Figs. 76.1–76.7

Didymochlaena Desv., Berl. Mag. 5: 303. 1811. Type: *Didymochlaena sinuosa* Desv. = *Didymochlaena trunculata* (Sw.) J. Sm.

Tegularia Reinw., Sylloge Plant. Nov. 2: 3. 1824. Type: *Tegularia adianthifolia* Reinw., *nom. superfl.* for *Aspidium trunculatum* Sw. = *Didymochlaena trunculata* (Sw.) J. Sm.

Monochlaena Gaud., Bot. Freycinet Voy. 340. 1828, *nov. superfl.* for *Didymochlaena* and with the same type.

Description

Terrestrial; stem erect or sometimes decumbent, rather stout including the more or less persistent petiole bases, bearing scales and many rather soft roots; leaves monomorphic, ca. 75 cm to 2 m, rarely to nearly 3 m long, borne in a cluster, lamina 2-pinnate, imparipinnate, scaly and pubescent to nearly glabrous, veins free; sori elongate, borne on the veins-tips well back of the margin, not paraphysate, covered by an elongate indusium attached along the center, continuous around the receptacle at the

Fig. 76.1. *Didymochlaena trunculata,* south of Misantla, Veracruz, Mexico, (Photo Alice F. Tryon.)

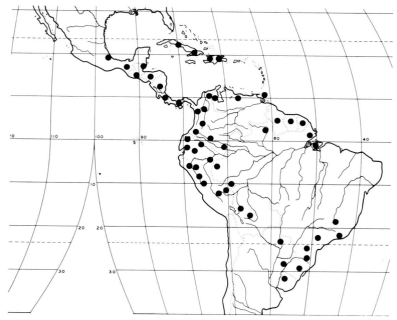

Fig. 76.2. Distribution of *Didymochlaena* in America.

apex but not at the base; spores ellipsoidal to somewhat spheroidal, monolete, the laesura about half or less the spore length, with prominent winglike folds usually more or less inflated, sometimes shorter, and rugose-saccate, the surface spinulose. Chromosome number: $n = 41$.

The stem of *Didymochlaena* bears long, narrow, brown scales and smaller ones that grade into trichomes, the petiole base is scaly and also covered by long, slender, tortuous, more or less matted trichomes and similar indument may be borne on the axes of the lamina. The rachis is grooved on the adaxial side and the ridges are continuous across the base of the grooved pinnastalk (Fig. 4). The pinnae and the dimidiate pinnules (Fig. 3) with elongate sori (Fig. 5) are mostly subarticulate and deciduous with age. At the base of the pinnae and pinnules there usually is a tuft of scales or cluster of spinescent processes. The latter appear to represent a reduced and thickened form of scale.

Systematics

Didymochlaena is best regarded as a monotypic, tropical genus based on *D. trunculata* (Sw.) J. Sm. Various segregates have been proposed but these do not seem to be clearly discrete. Review of the genus is needed to evaluate the variation in leaf architecture (Fig. 3) in plants over the broad geographic range. Those in Madagascar, recognized as *D. microphylla* (Bonap.) C. Chr., seem to represent a small-pinnuled form.

Didymochlaena is a highly distinctive genus allied to the dryopteroid ferns by the elongate indusium attached at the center, and the absence of ctenitoid trichomes.

Ecology (Fig. 1)

Didymochlaena usually grows in wet forests, often in gullies and on stream banks.

In America it grows in rain forests, cloud forests, mossy woods, wet thickets, in ravines and on rocky banks. It rarely occurs in cutover areas. It usually occurs at altitudes ranging between 100 and 1000 m, but occasionally up to 2300 m.

Geography (Fig. 2)

Didymochlaena occurs in tropical America, in Africa and Madagascar, and from Assam to Java, New Guinea and the Philippine Islands and eastward to Fiji.

In America it grows from Veracruz in Mexico, south to Panama, in the Greater Antilles (except Jamaica), and in South America south to Bolivia, southeastern Brazil and Uruguay. It is absent from most of the Amazon basin and northeastern Brazil.

Spores

Specimens from the neo- and paleotropics have rather uniform spores with prominent, inflated folds or saccate formation (Fig. 6). The irregularly echinate elements are especially dense on the

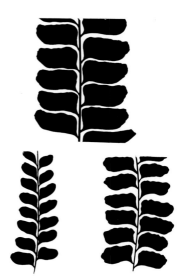

Fig. 76.3. *Didymochlaena trunculata,* portions of pinnae showing variation in the shape and size of the pinnules, $\times 0.5$.

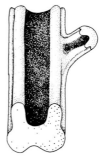

Fig. 76.4. *Didymochlaena trunculata,* channeled rachis and pinna-stalk, enlarged.

inflated folds. The abraded surface shows a fairly thin outer perispore and fragments of a rugose formation overlay the smooth exospore (Fig. 7). The spores generally resemble those of several dryopteroid genera especially *Dryopteris.*

Cytology

The chromosome number $n = 41$ reported in *Didymochlaena truncatula* from Malaya (Manton, 1954), and also from Costa Rica (Gómez-Pignataro, 1971) suggests the diploid extends over a large part of the range. The number supports the disposition of the genus in the tribe Dryopterideae.

Literature

Gómez-Pignataro, L. D. 1971. Ricerche citologiche sulle Pteridofite della Costa Rica 1. Atti Ist. Bot. Univ. Pavia 7: 29–31.

Manton, I. 1954. Cytological notes on one hundred species of Malayan ferns *in*: R. E. Holttum Flora of Malaya 2. Ferns of Malaya. 643 pp. Government Printing Office, Singapore.

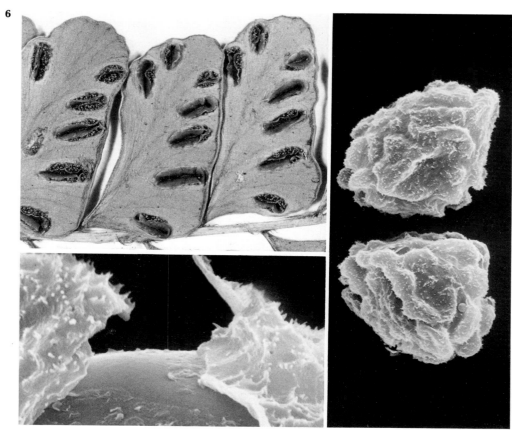

Figs. 76.5–76.7. *Didymochlaena trunculata.*   **5.** Fertile pinnules, ×4.   **6–7.** Spores.   **6.** Long inflated rugose-saccate folds and echinate surface, Nicaragua, *Molina 20581,* ×1000.   **7.** Detail of abraded surface, a portion of lower perispore at base, smooth exospore, above, Uganda, *Longfield 86,* ×5000.

77. *Stigmatopteris*
Figs. 77.1–77.17

Stigmatopteris C. Chr., Bot. Tidsskr. 29: 292. 1909. Type: *Polypodium flavopunctatum* Kaulf. = *Stigmatopteris rotundata* (Willd.) C. Chr. (*Aspidium rotundatum* Willd.). *Dryopteris* subgenus *Stigmatopteris* (C. Chr.) C. Chr., Danske Visdensk. Selsk. Skrift. VII. 10: 73. 1913.

Cyclodium Presl, Tent. Pterid. 85. 1836. Type: *Cyclodium meniscioides* (Willd.) Presl (*Aspidium meniscioides* Willd.) = *Stigmatopteris meniscioides* (Willd.) Kramer.

Peltochlaena Fée, Mém. Fam. Foug. 5 (Gen. Fil): 289. 1852, invalid, (*nom. provis.*) Based on "*Peltochlaena nephrodiiformis* Fée" = *Stigmatopteris guianensis* (Kl.) C. Chr.

Description   Terrestrial or rarely rupestral or epiphytic; stem erect, decumbent or rarely moderately long-creeping, rather small to somewhat stout, bearing scales and scattered to numerous, long, fibrous roots; leaves monomorphic to rarely dimorphic (the fertile longer than the sterile and with shorter and narrower pinnae), ca. 40 cm to 2 m long, clustered to somewhat spaced, 1-pinnate with entire pinnae or 1-pinnate basally and pinnatifid above, to usually 1-pinnate-pinnatifid, or rarely to 2-pinnate-pinnatifid, rarely imparipinnate, glabrate to somewhat scaly, or also with few-celled trichomes on the adaxial surface of the axes, usually with pellucid-punctate leaf-tissue, veins free or casually to regularly anastomosing with or without included free veins; sori roundish to somewhat elongate, borne on the veins or at their tip, on a somewhat elevated receptacle, not paraphysate, or (in *S. Michaelis*) paraphysate, covered by a peltate indusium or exindusiate; spores ellipsoidal to somewhat spheroidal, monolete, the laesura $\frac{3}{4}$ or less the spore length, with inflated folds and relatively smooth surface or winglike with a reticulate-echinate surface. Chromosome number: $n = 41$.

Most species of *Stigmatopteris* have exindusiate sori as in *S. prionites* (Fig. 3), while a few have indusiate sori (Fig. 4). The lamina characteristically has pellucid glands in the leaf-tissue (Fig. 5). Anastomosing veins are uncommon but occur in a few species such as *S. meniscioides* (Fig. 6).

Systematics   *Stigmatopteris* is a tropical American genus of perhaps 20 species, generally characterized by a costa that is grooved on the adaxial side, absence of trichomes on the lamina, and large, pellucid glands in the leaf-tissue. These internal glands are quite evident in many species but are small, obscure, and perhaps absent in others.

The deciduous capsule and persistent stalk of the sporangium were indicated by Christensen (1909) as generic characters, but these are not consistent features, for the capsule may be retained and the stalk deciduous in *Stigmatopteris*. These characters of the sporangium also occur in other genera such as *Ctenitis* and *Polystichum*. *Stigmatopteris Michaelis* (Baker) C. Chr. has sori with long, multicellular paraphyses, a character that is evidently unique in the tribe Dryopterideae.

The relationships of *Stigmatopteris* are uncertain. The occurrence of short, few-celled trichomes on the adaxial surface of the

Fig. 77.1. *Stigmatopteris contracta,* Tapanti, Costa Rica. (Photo W. H. Hodge.)

costa in a few species such as *S. alloeoptera, S. opaca* and *S. meniscioides* suggests a relationship with *Ctenitis* or *Tectaria*. *Stigmatopteris meniscioides* has the venation pattern, peltate indusia, multiseriate sori and imparipinnate lamina similar to *Cyrtomium* species. However, it seems to be a specialized species of *Stigmatopteris* and similarities with *Cyrtomium* are convergent.

Stigmatopteris has been consistently used for this segregate genus of *Dryopteris sens. lat.* It is adopted rather than the earlier *Cyclodium* in order to avoid changing the species names, especially since *Cyclodium,* with a peltate indusium, may merit generic rank.

Stigmatopteris was revised by Christensen (1909, with additions in 1913 and 1920) and the species with anastomosing veins by Morton (1939).

Tropical American Species

Stigmatopteris is of uncommon occurrence, although it is widely distributed in the American tropics. Variation in many species has not been adequately assessed because of limited materials and there are probably fewer species than currently recognized.

Most species of *Stigmatopteris* are exindusiate. Many of these have free veins and the lamina is 1-pinnate to 1-pinnate-pinnatifid, as in *Stigmatopteris contracta* (Christ) C. Chr., *S. longicaudata* (Liebm.) C. Chr., *S. nothochlaena* (Maxon) C. Chr. (Fig. 13), *S.*

Fig. 77.2. Distribution of ***Stigmatopteris.***

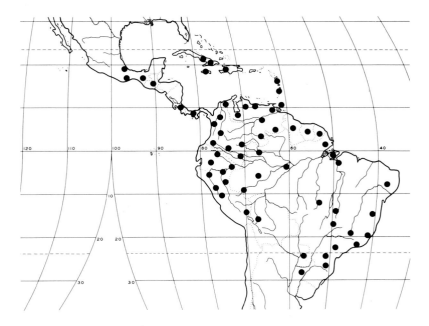

prionites (Kze.) C. Chr., and *S. rotundata* (Willd.) C. Chr. (Fig. 17).
Stigmatopteris cyclocolpa (Christ) C. Chr. is similar but has a 2-pin-
nate-pinnatifid lamina. A few species have anastomosing veins
and the lamina mostly pinnatifid to pinnatisect with large segments as in *Stigmatopteris alloeoptera* (Kze.) C. Chr. and *S. opaca*
(Baker) C. Chr.

The other species have a peltate indusium and among these
Stigmatopteris meniscioides (Willd.) Kramer and *S. paludosa* (Morton) R. & A. Tryon have anastomosing veins and an imparipinnate lamina. The former species is sometimes dimorphic (Figs.
15, 16). Species such as *Stigmatopteris guianensis* (Kl.) C. Chr. (Fig.
14) and *S. sancti-gabrielii* (Hook.) C. Chr. (Fig. 12) have the lamina apex gradually to abruptly reduced.

Ecology (Fig. 1)

Stigmatopteris is usually a genus of primary rain forests, where it
occurs on river banks, along streams, in ravines and on mountain slopes. It often grows on bare clay soil, sometimes on rocks
and it may also grow on road banks and rarely in banana plantations. Less often it grows at forest borders or in open forests.
Stigmatopteris meniscioides occurs as a low epiphyte in varzea forests near Manaus, Brazil. The genus grows from near sea level to
ca. 1800 m, most commonly at 500–1200 m.

Stigmatopteris hemiptera (Maxon) C. Chr. and *S. gemmipara* C.
Chr. have scaly proliferous buds in the axils of the pinnae
toward the apex of the lamina.

Geography (Fig. 2)

Stigmatopteris is widely distributed in tropical America, from
Veracruz in Mexico, south through Central America, in the Antilles, and in South America south to Bolivia, Paraguay, Corrientes in Argentina, and Paraná in Brazil.

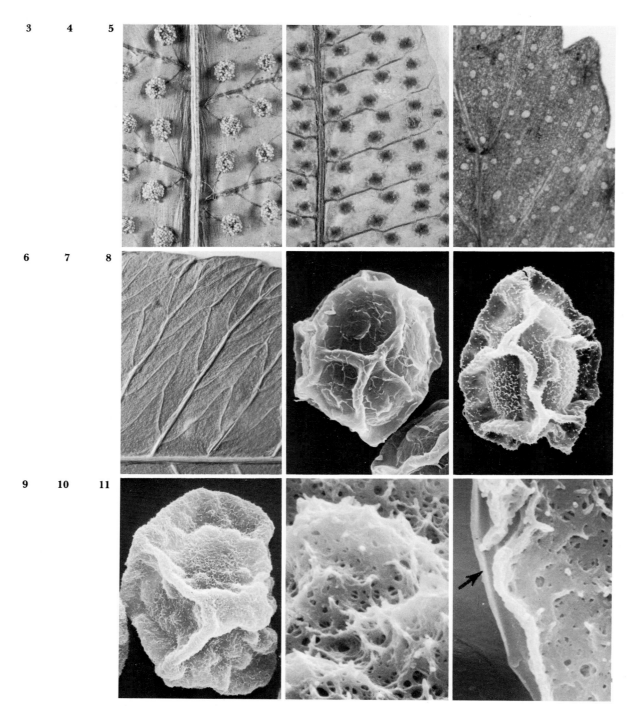

Figs. 77.3–77.11. *Stigmatopteris.* **3–6.** Lamina and sorus characters. **3.** Exindusiate sori of *S. prionites*, ×5. **4.** Immature sori with indusia appressed to the leaf-tissue, *S. menisciodes,* ×3. **5.** Pellucidpunctate leaf-tissue, *S. nothochlaena,* ×20. **6.** Anastomosing veins, *S. meniscioides,* ×3. **7–11.** Spores, ×1000, detail of wall structure, ×10,000. **7.** *S. nothochlaena,* inflated slightly rugulose folds, Jamaica, *Maxon 8898.* **8.** *S. guianensis* winglike papillate folds, Brazil, *Blanchet 2208.* **9–11.** *S. meniscioides,* Colombia, *Schultes & Cabrera 17354.* **9.** Inflated saccate folds. **10.** Detail of reticulate-echinate surface. **11.** Detail of lower perispore (at arrow) between exospore at left and reticulate surface, right.

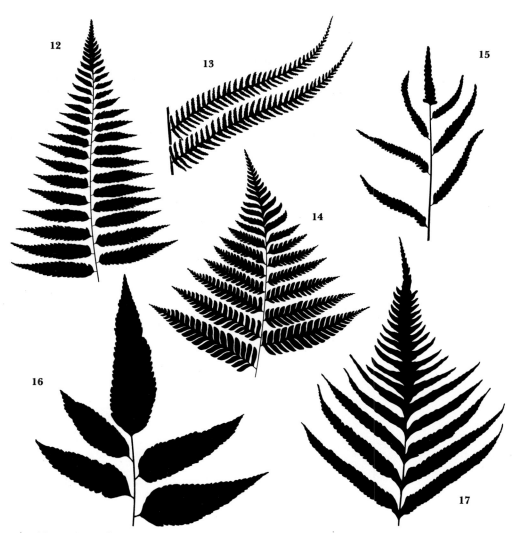

Figs. 77.12–77.17. Lamina architecture of ***Stigmatopteris***, × 0.25. **12.** Upper part of lamina ***S. sancti-gabrielii***. **13.** Pinnae of ***S. nothochlaena***. **14.** Upper part of lamina ***S. guianensis***. **15.** Apex of fertile lamina, ***S. meniscioides***. **16.** Apex of sterile lamina, from same plant as 15. **17.** Upper part of lamina ***S. rotundata***.

Spores

Stigmatopteris spores usually have inflated folds and a slightly rugulose surface as *S. nothochlaena* (Fig. 7). The folds may be elaborated in prominent wings, as *S. guianense* (Fig. 8), or somewhat saccate in *S. meniscoides* (Fig. 9). The perispore is formed of at least two main strata as in the abraded spore (Fig. 11) consisting of a smooth lower perispore between the exospore and the reticulate-echinate surface (Fig. 10). Marked size differences indicate possible diverse ploidy levels. The spores are generally similar to those of other genera as *Ctenitis* and *Tectaria* in the dryopteroids.

Cytology

Stigmatopteris nothochlaena is reported as diploid with $n = 41$ from Jamaica (as *Dryopteris* subgenus *Stigmatopteris*) by Walker (1966). The record is helpful in the general alignment of the genus among others with this number in the tribe Dryopterideae.

Literature

Christensen, C. 1909. On *Stigmatopteris,* a new genus of ferns with a review of its species. Bot. Tidsskr. 29: 291–304.

Christensen, C. 1913. Reference under the tribe Dryopterideae.

Christensen, C. 1920. Reference under the tribe Dryopterideae.

Morton, C. V. 1939. On the genus *Cyclodium.* Bull. Torrey Bot. Cl. 66: 47–52.

Walker, T. G. 1966. Reference under the tribe Dryopterideae.

78. *Polystichum*
Figs. 78.1–78.29

Polystichum Roth, Tent. Fl. Germ. 3: 31, 69. 1799. Type: *Polystichum Lonchitis* (L.) Roth (*Polypodium Lonchitis* L.)

Hypopeltis Michx., Fl. Bor. Amer. 2: 266. 1803. Type: *Hypopeltis lobulata* Bory, Exped. Sci. Morée 3 (2): 286. 1832. = *Polystichum aculeatum* (L.) Roth.

Plecosorus Fée, Mém. Fam. Foug. 5 (Gen. Fil.): 150. 1852. Type: *Plecosorus mexicanus* Fée, *nom. superfl.* for *Cheilanthes speciosissima* Kze. = *Plecosorus speciosissimus* (Kze.) Moore = *Polystichum speciosissimum* (Kze.) R. & A. Tryon.

Sorolepidium Christ, Bot. Gaz. 51: 350. 1911. Type: *Sorolepidium glaciale* (Christ) Christ (*Polystichum glaciale* Christ) = *Polystichum Duthei* (Hope) C. Chr.

Hemesteum Lev., Fl. Kouy-tchéou 450, 496. 1915, (not Newm. 1851 = *Thelypteris*), several species of *Polystichum* are listed.

Aetopteron House, Amer. Fern Jour. 10: 88. 1920, invalid (Barnhart, *ibidem* 10: 111. 1920).

Papuapteris C. Chr., Brittonia 2: 300. 1937. Type: *Papuapteris linearis* C. Chr. = *Polystichum lineare* (C. Chr.) Copel.

Acropelta Nakai, Bull. Nat. Sci. Mus. Tokyo 33: 5. 1953. Type: *Acropelta omeiensis* (C. Chr.) Nakai (*Polystichum omeiense* C. Chr.) = *Polystichum carvifolium* Diels, based on *Aspidium carvifolium* Baker, not Kze.

Description   Terrestrial or rupestral; stem erect or decumbent, small to sometimes stout, including the persistent petiole bases, bearing scales and many, long, fibrous roots; leaves monomorphic or rarely partially dimorphic with the apical fertile portion somewhat contracted, or (in *P. dimorphophyllum* Hayata) dimorphic (the fertile leaves taller than the sterile and with smaller segments), 5 cm to usually 0.5–1 m long, to rarely 3 m long, borne in a crown or cluster, lamina rarely entire to pinnatifid, usually 1-pinnate or 2-pinnate, to 3-pinnate-pinnatifid, more or less persistently and often densely scaly, veins free; sori round, borne on the veins on a nearly flat receptacle, not paraphysate, covered by a peltate, persistent or caducous indusium or exindusiate; spores ellipsoidal to somewhat spheroidal, monolete, the

Fig. 78.1. *Polystichum muricatum,* Veracruz, Mexico. (Photo W. H. Hodge.)

laesura $\frac{2}{3}$ to $\frac{3}{4}$ the spore length, with irregular, saccate or winglike folds, and the surface more or less spinulose, often perforate. Chromosome number: $n = 41, 82, 164; 2n = 82$, ca. 160, ca. 164; apogamous 82, 123.

The petioles of *Polystichum* are usually persistently scaly, especially at the base and rarely they are densely scaly to the apex of the petiole (Fig. 7). Most species have the pinnae or pinnules auricled on the acroscopic side (Fig. 9), and the ultimate segments sharply toothed. The costa of a segment is grooved on the adaxial side. The characteristic peltate indusium is shown in Fig. 4, and the less common exindusiate sorus in Fig. 5. The diversity of lamina architecture in American *Polystichum* is shown in Figs. 10–23.

Fig. 78.2. *Polystichum orbiculatum,* Cotopaxi, Ecuador. (Photo W. H. Hodge.)

Systematics

Polystichum is a relatively homogeneous, nearly worldwide genus of perhaps as many as 160 species, with about 55 in tropical America. Four segregate genera, sometimes recognized, are based on single rather extreme species. *Polystichum lineare* (C. Chr.) Copel. (*Papuapteris*) is an alpine species of New Guinea with narrow, elongate leaves and very reduced pinnae and pinnules. Forms of *Polystichum papuanum* C. Chr. and *P. cheilanthoides* Copel. are transitional between *P. lineare* and other New Guinea species of lower altitudes with broad, expanded leaves. Similar variation, although not as extreme, occurs in the Andean *Polystichum pycnolepis* (Fig. 8). *Polystichum Duthei* (Hope) C. Chr. (*Sorolepidium*) has a narrow lamina that is densely and persistently scaly. *Polystichum speciosissium* (*Plecosorus*) is exindusiate and with recurved segment margins that are thin, and light brown to

Fig. 78.3. Distribution of *Polystichum* in America, south of lat. 35° N.

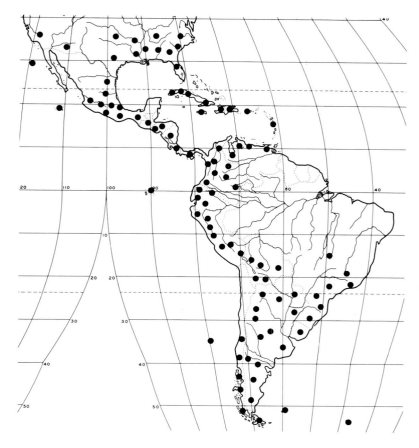

white (Fig. 6). *Polystichum carvifolium* Diels (*Acropelta*) has the pinnae finely dissected.

There is no satisfactory infrageneric classification for the genus. Tagawa (1940) proposed eight sections in *Polystichum* for the area including Japan, Korea and Formosa, and Daigobo (1972) had 16 sections for the genus within a similar area. These classifications need to be assessed for their application to the whole genus.

The alliance of *Polystichum* and *Dryopteris* is generally regarded as a close one. However, some specialized characters of *Polystichum,* as the spores with a strongly perforate lower perispore stratum in contrast to the nonperforate lower layer in *Dryopteris,* suggest that their relationship may more distant.

The West Indian species of *Polystichum* have been revised by Maxon (1909, 1912) and those of Cuba by Morton (1967).

Tropical American Species

Polystichum is one of the taxonomically difficult genera of the American tropics. The species are poorly known, the accurate application of early names is often uncertain, and species are exceptionally variable. Studies of the genus, as those by Maxon on the West Indian species, need revision for they are outdated by the accumulation of additional collections.

Cytotaxonomic studies of *Polystichum* in Japan, Europe and

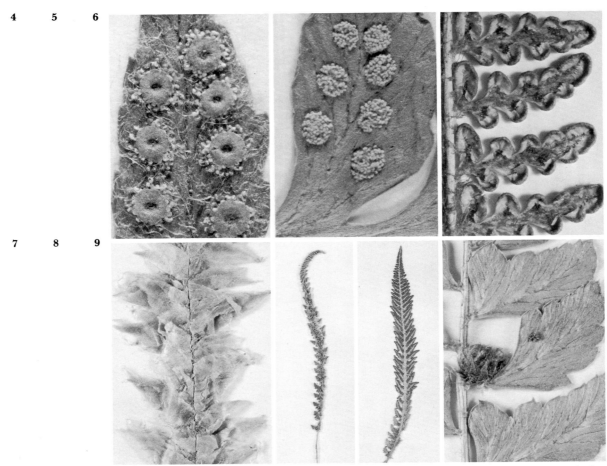

4 5 6

7 8 9

Figs. 78.4–78.9. *Polystichum.* **4.** Peltate indusium of *P. mucronatum,* ×8. **5.** Exindusiate sori of *P. platyphyllum,* ×8. **6.** Fertile segments of *P. speciosissimum* with recurved modified segment margins, ×5. **7.** Apical portion of petiole of *P. dictyophyllum* with large, dense scales, ×2. **8.** High altitude leaf forms of *P. pycnolepis:* left from 3750 m, ×0.3, right from 3400 m, ×0.15. **9.** Portion of lamina apex of *P. platyphyllum,* a scaly bud in the axil of a pinna, ×3.3.

the western United States and Canada have shown that some of the taxonomic problems arise from extensive hybridization that has given rise to alloploids and sterile hybrids. It is probable that the tropical American species have also hybridized freely but cytological studies are needed to interpret the variation of cytotypes and their hybrids. Variation in leaves is illustrated by specimens (Figs. 20–22) sampled from a colony of *Polystichum platphyllum* growing near the railroad at Macchu Picchu, near Cuzco, Peru (*Tryon & Tryon 5404*). The plants have proliferous buds on the leaves thus it is possible that the colony may have formed from one or few plants. Variation is also documented in a collection of *Polystichum polyphyllum* from Páramo El Gavilán, Mérida, Venezuela (*Tryon & Tryon 5861*). The sample includes plants from an exposed site with leaves ca. 25 cm long, 2–3 cm broad with imbricate pinnae and also plants from a nearby sheltered site with leaves 1 m long, 5–6 cm broad and distant pinnae.

Among the many tropical American species of *Polystichum* are the following: *Polystichum boboense* Hieron. (Fig. 12), *P. Christianae* Maxon (Fig. 13), *P. decoratum* Maxon, *P. dictyophyllum*

Figs. 78.10–78.23. Lamina architecture of *Polystichum*, all × 1 except as noted. **10.** Pinna of *P. distans.* **11.** Pinna of *P. yungense,* × 0.5. **12.** Pinna of *P. boboense.* **13.** Pinna of *P. Christianae.* **14.** Leaf of *P. nudicaule.* **15.** Leaf of *P. Harrisii,* rooting at the tip, × 0.5. **16.** Pinna of *P. muricatum.* **17.** Leaf of *P. Plaschnickianum,* rooting at the tip. **18.** Apical portion of lamina of *P. echinatum.* **19.** Apical portion of lamina of *P. tridens.* **20–22.** Apex of lamina from plants in one colony of *P. platyphyllum.* **23.** Leaf of *P. rhizophyllum,* rooting at the tip.

(Kuhn) C. Chr., *P. distans* Fourn. (*P. pallidum* Fourn., not (Bory) Todaro) (Fig. 10), *P. echinatum* (Gmel.) C. Chr. (Fig. 18), *P. Harrisii* Maxon (Fig. 15), *P. machaerophyllum* Slosson, *P. mucronatum* (Sw.) Presl, *P. muricatum* (L.) Fée (Figs. 1, 16), *P. nudicaule* Rosenst. (Fig. 14), *P. orbiculatum* (Desv.) Gay (Fig. 2), *P. Plaschnickianum* (Kze.) Moore (Fig. 17), *P. platyphyllum* (Willd.) Presl (Figs. 20–22), *P. polyphyllum* (Presl) Presl, *P. pycnolepis* (Kl.) Moore, *P. rhizophorum* (Jenm.) Maxon, *P. rhizophyllum* (Sw.) Presl (Fig. 23), *P. speciosissimum* (Kze.) R. & A. Tryon, *P. triangulum* (L.) Fée, *P. tridens* (Hook.) Fée (Fig. 19), and *P. yungense* Rosenst. (Fig. 11).

Ecology (Figs. 1, 2)

Polystichum is primarily a temperate genus in northern and southern hemispheres and in mountainous regions of the tropics. It grows especially in wet forests and also, at higher altitudes or latitudes, in more open shrubby or grassy places, particularly among rocks.

In the American tropics, *Polystichum* usually grows in wet montane forests or cloud forests, in ravines, along stream banks, in rocky woods, on cliffs or talus slopes. Sometimes species grow in less mesic habitats such as oak or pine woods, in rough pastures or along road banks. In the Greater Antilles species such as *P. echinatum, P. Christianae, P. Harrisii* and *P. triangulum* are strongly associated with calcareous rocks, and *P. tridens* sometimes grows on serpentine. In the Andes species such as *P. pycnolepis, P. orbiculatum* and *P. polyphyllum* grow in páramos and in grassy or rocky alpine situations. The genus occurs from sea level to 4600 m, most frequently at 1000–4000 m.

Vegetative reproduction by buds is frequent in *Polystichum*, particularly in species of the Greater Antilles. Buds are formed at the tip of a long, naked rachis in *P. decoratum, P. Harrisii* and *P. rhizophorum*, at the apex of a long-attenuate lamina in *P. rhizophyllum*, (Fig. 23), and on the rounded apex in *P. Plaschnickianum* (Fig. 17). Scaly buds may develop along the apical part of the rachis in a few species such as the widely distributed *P. platyphyllum* (Fig. 9).

Geography (Fig. 3)

Polystichum is widely distributed throughout most of the world, on all continents, and in the Pacific east from New Zealand to Tahiti, Rapa and Easter Island and north to the Hawaiian Islands; it also occurs on the southern islands of South Georgia, Gough, Marion, Amsterdam and Auckland.

In America *Polystichum* ranges from Alaska and Canada southward through the United States to Central America, the Greater Antilles, Guadeloupe in the Lesser Antilles, and in South America from Venezuela to Colombia, southward to Tierra del Fuego; also on Isla Guadelupe, the Revillagigedo Islands, the Galápagos Islands, the Juan Fernandez Islands, the Falkland Islands and South Georgia. The genus is absent from the Amazon basin, the Guianas and from the xeric northeast of Brazil.

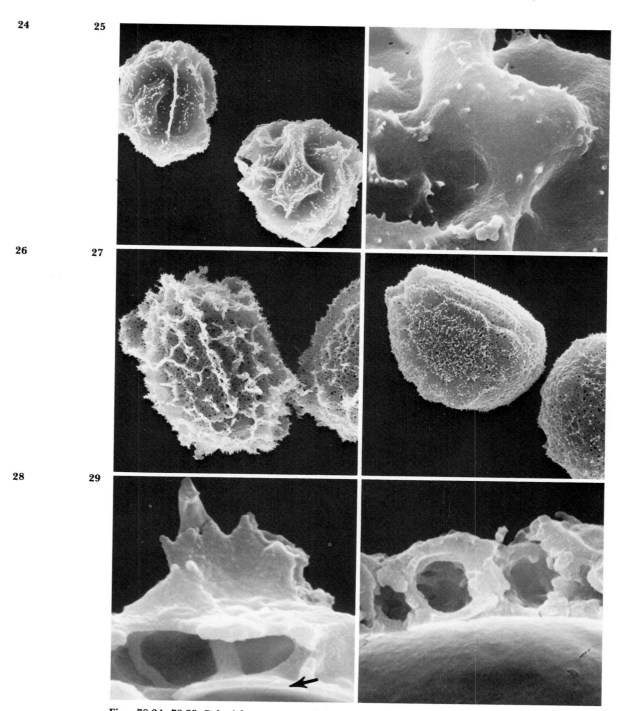

Figs. 78.24–78.29. *Polystichum* spores, × 1000. **24.** *P. mucronatum,* inflated, saccate folds, laesura at center upper spore, Jamaica, *Proctor 4559.* **25.** *P. rhizophyllum,* detail of folds with sparse spinules, Jamaica, *Moore* in 1896, × 5000. **26.** *P. muricatum,* perforate winglike ridges, Mexico, *Copeland 37.* **27.** *P. distans,* low perforate, folds, Mexico, *Hinton 13454.* **28.** *P. machaerophyllum,* profile of outer dentate perispore, above, lower perispore (at arrow) below, and central columnar structure, Haiti, *Leonard 4779,* × 10,000. **29.** *P. echinatum,* wall profile, outer perispore, above, lower perispore overlying smooth exospore, below, and central columnar structure, Haiti, *Leonard 37772,* × 10,000.

It is evident that there are two centers of diversity of American *Polystichum* although the species and their distributions are not adequately known.

Twenty or more species are currently recognized in the Greater Antilles and about the same number in the Andes of Colombia, Ecuador and Peru. Nearly all of the Greater Antillean species are endemic, and few have close relations outside of the region.

Spores

Spores of tropical American species of *Polystichum* have prominent folds in *P. mucronatum* (Fig. 24) and sparse spinules in *P. rhizophyllum* (Fig. 25), or may be reticulate in *P. distans* (Fig. 27) or with perforate ridges in *P. muricatum* (Fig. 26). Spores with folds are generally smaller than reticulate forms.

The wall consists of a lower perispore layer in *P. machaerophyllum* (Fig. 28, arrow) with columns between this and an outer perispore stratum in *P. echinatum* (Fig. 29). This wall structure in *Polystichum* appears more complex than the compact perispore formation in spores of *Dryopteris or Cyrtomium*. SEM studies of *Polystichum* spores of Japan and Taiwan (Mitui, 1973, 1974) show predominately perforate forms that are considered to be quite distinct from those of *Dryopteris*. Spores of western North American polystichums, examined by Wagner (1979), are also mainly reticulate or perforate. Variation of surface structure noted in that work appears to relate to differences in the density of deposition over the perforate formation.

Cytology

Chromosome numbers, known for about a third of the species of *Polystichum,* range from diploid with $n = 41$ to octoploid $n = 164$. Tetraploids are about as frequent as diploids except in the American tropics where the relatively few records are mostly diploids. There are few hexaploids, and a single species, *Polystichum falcinellum* (Sw.) Presl, an endemic of Madeira, is octoploid. Accounts of sterile hybrids, where species grow together, are made by Daigobo (1974), Manton (1950), Nakaike (1973), Wagner (1979), and Wagner (1973). Experimental studies of triploid hybrids in Europe by Vida and Reichstein (1975) demonstrate the occurrence of fertile spores in *Polystichum* × *illyricum* (Borbás) Hahne and of fertile hexaploid plants from the triploid gametophytes. This work shows that higher polyploid levels can be attained by fusion of gametes as well as by chromosome doubling in the sporophyte.

Literature

Daigobo, S. 1972. Taxonomical studies on the fern genus *Polystichum* in Japan, Ryukyu and Taiwan. Sci. Report Tokyo Kyoiku Daigaku, Sect. B, 15: 57–80.

Daigobo, S. 1974. Chromosome numbers of the fern genus *Polystichum.* Jour. Jap. Bot. 49: 371–378.

Manton, I. 1950. Reference under the tribe Dryopterideae.

Maxon, W. R. 1909. A revision of West Indian species of *Polystichum*. Contrib. U.S. Nat. Herb. 13: 25–39.

Maxon, W. R. 1912. Further notes on the West Indian species of *Polystichum*. Contrib. U.S. Nat. Herb. 16: 49–51.

Mitui, K. 1973. Characteristics of the perispore of the genus *Polystichum* by the scanning electron microscope. Bull. Nippon Dental Coll., Gen. Ed. 2: 103–123.

Mitui, K. 1974. On the perispore formation of some *Polystichum* species (Fillicales). Bull. Nippon Dental Coll. Gen. Ed. 3: 103–118.

Morton, C. V. 1967. Studies of fern Types, I., Key to the Cuban species of *Polystichum*. Contrib. U.S. Nat. Herb. 38: 68.

Nakaike, T. 1973. Studies in the fern genus *Polystichum*, I. Observations on the section *Metapolystichum* at Gobô-sawa, Pref. Chiba. Bull. Nat. Sci. Mus. Tokyo 16: 437–457.

Vida, G., and T. Reichstein. 1975. Taxonomic problems in the fern genus *Polystichum* caused by hybridization, *in*: S. M. Walters, ed., European floristic and taxonomic studies, pp. 125–136. Conf. Rep. Bot. Soc. Brit. Isl. 15.

Wagner, D. 1979. Systematics of *Polystichum* in western North America north of Mexico. Pteridologia 1: 1–64.

Wagner, W. H. 1973. Reticulation of Holly Ferns (*Polystichum*) in western United States and adjacent Canada. Amer. Fern Jour. 63: 99–115.

79. *Maxonia*

Figs. 79.1–79.8

Maxonia C. Chr., Smiths. Misc. Coll. 66 (9): 3. 1916. Type: *Maxonia apiifolia* (Sw.) C. Chr. (*Dicksonia apiifolia* Sw.).

Dryopteris subgenus *Peisomopodium* Maxon, Contrib. U. S. Nat. Herb. 13: 39. 1909. Type: *Dryopteris apiifolia* (Sw.) O. Ktze. (*Dicksonia apiifolia* Sw.) = *Maxonia apiifolia* (Sw.) C. Chr.

Description Terrestrial and epiphytic; stem long-creeping to very long-creeping, rather slender or moderately stout, bearing scales and rather few, long, fibrous roots; leaves moderately to strongly dimorphic (the fertile with somewhat to strongly reduced segments), ca. 50 cm to 1 m or more long, borne at usually distant intervals, lamina 3- to 4-pinnate, short-pubescent and with a few narrow scales, veins free; sori roundish, borne on the vein-tips, well back of the margin, not paraphysate, covered by a large, cordate to peltate indusium; spores more or less spheroidal, monolete, the laesura $\frac{1}{4}$ to $\frac{1}{2}$ the spore length with prominent folds, the surface spinulose or echinate to rugulose. Chromosome number: $n = 41$.

Maxonia is characterized by a dorsiventral stem that is thick and densely scaly in scandent forms (Fig. 6) and slender with sparse scales in terrestrial plants (Fig. 5). The leaves are dimorphic, the sterile expanded (Fig. 4) and the fertile with a reduced lamina although the indusia are large and conspicuous (Fig. 3). The axis of the penultimate segments is grooved on the adaxial surface.

Fig. 79.1. *Maxonia apiifolia,* Albion, St. Ann, Jamaica. (Photo W. H. Hodge.)

Fig. 79.2. Distribution of *Maxonia.*

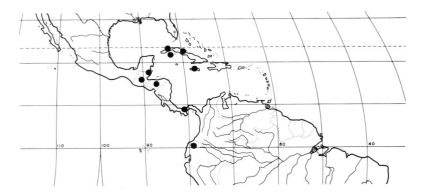

Systematics

Maxonia is a monotypic, tropical American genus based on *M. apiifolia* (Sw.) C. Chr. In the shape and dimorphic condition of the leaves it seems most closely related to *Polybotrya.* The plants of the Greater Antilles referred to *M. apiifolia* var. *apiifolia* are dubiously distinct from those of the continent that have been recognized as *M. apiifolia* var. *dualis* (Donn. Sm.) C. Chr. The Antillean specimens have the stem scales finely toothed, especially at the apex, while those of the continent have scales with smooth margins but this distinction is not wholly constant.

Ecology (Fig. 1)

Maxonia grows in shrubby savannahs, on open hillsides, at the edge of forests, in wet ravines and in dense forests. It occurs from ca. 75 to 1000 m. The plants are initially terrestrial with slender rhizomes with rather small, scattered scales, and bear only small sterile leaves (Walker, 1972). Older plants with stems climbing on trees, especially on palms or tree ferns, have stems

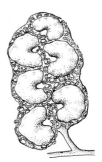

Fig. 79.3. Fertile segment of *Maxonia apiifolia* with mature sori and indusia, ×5.

4   5   6

7   8

Figs. 79.4–79.8. *Maxonia apiifolia.* **4.** Portion of sterile lamina, × 1. **5.** Portion of slender, sparsely scaly, terrestrial stem, × 1. **6.** Portion of stout, densely scaly, climbing stem, × 1. **7, 8.** Spores, Jamaica, *Maxon 10437.* **7.** Inflated folds and rugulose surface, × 1000. **8.** Detail of rugulose, echinate surface, × 10,000.

three or more times stouter than terrestrial ones and have a dense cover of long scales as well as both sterile and fertile leaves.

Geography (Fig. 2)

Maxonia occurs in Cuba and Jamaica, and in British Honduras (Belize) to Panama; also in Ecuador. It has been collected in south Florida (Morton, 1961), where it is evidently not an established species. It may have been a hurricane waif, or perhaps originated from a cultivated plant.

Spores

The major contours are formed by prominent, sparse, long folds (Fig. 7). The surface of these and adjacent areas are rather uniformly covered with smaller rugulose structures. Detail of the rugulose formation consists of more or less reticulate, low strands and echinate elements (Fig. 8). These spores are generally similar to those of *Polybotrya* but with somewhat longer strands.

Cytology

Walker (1966) reports the chromosome number of *Maxonia* as $n = 41$, based on plants from Mt. Diabolo, Jamaica, with comments on the paucity of countable cells although numerous sporangia were formed. A photograph of the chromosomes at diakinesis, with strongly contracted bivalents, documents the report. The number is characteristic of most dryopteroid genera and established the general alignment of the genus.

Literature

Morton, C. V. 1961. Another genus of ferns new to the United States. Amer. Fern Jour. 51: 81–83.
Walker, T. G. 1966. Reference under the tribe Dryopterideae.
Walker, T. G. 1972. The anatomy of *Maxonia apiifolia:* A climbing fern. Brit. Fern Gaz. 10: 241–250.

80.   *Polybotrya*
Figs. 80.1–80.30

Polybotrya Willd., Sp. Pl. 5: 99. 1810. Type: *Polybotrya osmundacea* Willd.
Olfersia Raddi, Opusc. Sci. Bolog. 3: 283. 1819. Type: *Olfersia corcovadensis* Raddi = *Polybotrya cervina* (L.) Kaulf.
Soromanes Fée Mém. Fam. Foug. 2 (Hist. Acrost.): 16, 82. 1845. Type: *Soromanes serratifolium* Fée = *Polybotrya serratifolia* (Fée) Kl.
Dorcapteris Presl, Epim. Bot. 166. 1851. Type: *Dorcapteris cervina* (L.) Presl (*Osmunda cervina* L.) = *Polybotrya cervina* (L.) Kaulf.

Description   Terrestrial or sometimes rupestral, usually becoming scandent-epiphytic; stem short- to very long-creeping, or scandent to 8 m or more, moderately slender to stout, bearing a usually dense covering of scales and few to many long, fibrous roots; leaves strongly dimorphic (the fertile of usually greater complexity than the sterile and with much reduced segments) ca. 40 cm to 1.5 m long, borne at usually distant intervals, the sterile 1-pinnate to 3-pinnate-pinnatifid, the fertile rarely 1-pinnate-pinnatifid to usually 2-pinnate to 4-pinnate, glabrate, pubescent or somewhat scaly, veins free, or (in *P. cervina*) connected by a marginal strand, or sometimes anastomosing without included free veins; sporangia borne over the abaxial surface of the segments often on the margins and sometimes also on the adaxial surface, no paraphyses or indusium; spores more or less ellipsoidal, monolete, the laesura about $\frac{1}{2}$ the spore length with few, prominent somewhat inflated folds and slightly to densely echinate surface. Chromosome number: $n = 41$; $2n = 82$.

The terrestrial stem of *Polybotrya caudata* is rather slender (Fig. 8) in comparison to the stouter scandent one with a dense cover of scales (Fig. 9). Sporangia are generally distributed over the

Fig. 80.1. *Polybotrya cervina,* Río Reventazón, Turrialba, Costa Rica. (Photo W. H. Hodge.)

abaxial surface (Fig. 4) and on the margins but not on the adaxial surface in *P. osmundacea* (Fig. 3). In others as *P. cervina* sporangia also cover the adaxial surface (Fig. 5). Free veins are characteristic of the genus; they are connected by a marginal strand in *P. cervina* (Fig. 6), and are anastomosing in a few species as *P. suberecta* (Fig. 7).

Systematics *Polybotrya* is a tropical American genus of perhaps 40 species. It is characterized by acrostichoid sporangia and strongly dimorphic leaves, with the fertile usually more complex than the ster-

Fig. 80.2. Distribution of *Polybotrya*.

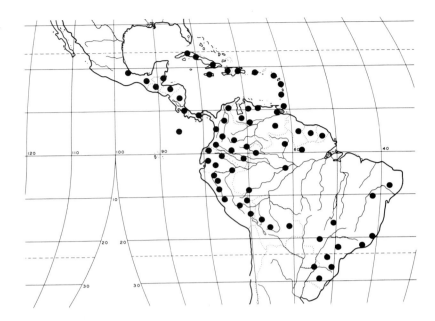

ile and ephemeral. The stem is somewhat dorsiventral, at least in the scandent phase. The costa is grooved on the upper surface, or rarely, as in *Polybotrya Lechleriana*, it is slightly grooved or rounded. This species also has rather long, multicellular trichomes borne among the sporangia as well as on other parts of the fertile lamina. The sporangia are described as being borne on the margins, and in some species on the adaxial surface, as well as on the abaxial surface, but this is not true in an anatomical sense. Schumann (1915) showed that the sporangiferous tissue of *Polybotrya osmundacea* extends beyond the true margins of the segment and that in *P. cervina* the tissue occupies the upper surface, the margins being obsolete.

Polybotrya is close to *Maxonia* in the architecture of the leaves, the surface structure of the spores, and in stem dimorphism, as in *P. caudata*.

Tropical American Species

There is no systematic treatment of *Polybotrya* and a modern revision is much needed to assess the numerous species and establish their relationships. Species with a 1-pinnate lamina have probably been derived from those with a more complex lamina. Several of them have anastomosing veins, a condition that often is correlated with a reduction in lamina architecture. Leaves become increasingly complex from juvenile to adult plants as in *Polybotrya cervina* (Figs. 10–14) and probably other species. Study of developmental change in form of these leaves is needed to understand the ontogeny of the plants as well as species relationships.

Among the species with a 1-pinnate lamina, some have free veins, as *Polybotrya crassirhizoma* Lell., *P. fractiserialis* (Baker) J. Sm., and *P. plumbicaulis* (Baker) J. Sm.; *Polybotrya cervina* (Figs. 1, 14, 15) has long free veins connected at their apex by a marginal strand. Others have anastomosing veins as *Polybotrya Aucuparia*

3   4   5

6   7   8

9

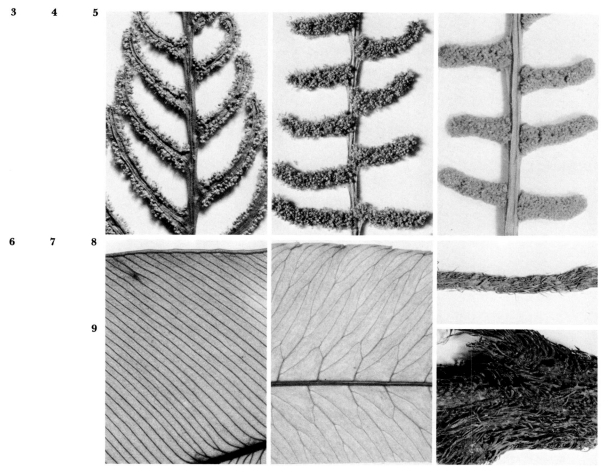

Figs. 80.3–80.9. *Polybotrya.*   **3.** Fertile segments of *P. osmundacea*, adaxial side, × 3.   **4.** Fertile segments of *P. osmundacea*, abaxial side, × 3.   **5.** Fertile segments of *P. cervina* with immature sporangia, adaxial side, × 3.   **6.** Portion of sterile pinna of *P. cervina*, the free veins connected by a marginal strand, × 2.   **7.** Portion of a sterile pinna of *P. suberecta* with anastomosing veins, × 2.   **8.** Terrestrial stem of *P. caudata*, × 1.   **9.** Scandent-epiphytic stem of *P. caudata*, × 1.

Christ, *P. Kalbreyeri* C. Chr., *P. serratifolia* (Fée) Kl. (Figs. 18, 19), and *P. suberecta* (Baker) C. Chr.

Species with the sterile lamina 2-pinnate-pinnatifid or more complex have free veins as in *Polybotrya canaliculata* Kl. (Figs. 16, 17), *P. caudata* Kze. (Figs. 22, 23), *P. Lechleriana* Mett., *P. osmundacea* Willd. (Figs. 20, 21), and *P. scandens* (Raddi) Christ.

Ecology (Fig. 1)

Polybotrya is a genus of wet primary forests and sometimes cloud forests. Less often it grows in secondary forests or thickets and is rarely on wet cliffs. It grows from near sea level to about 2500 m but mostly not exceeding 1500 m.

The plants are initially terrestrial with the stem short- to long-creeping on soil, fallen tree trunks or among rocks. The stem becomes scandent on tree trunks, attached by roots, but retains its terrestrial connection. In *Polybotrya caudata* the terrestrial stem bears smaller, less complex leaves than the scandent one.

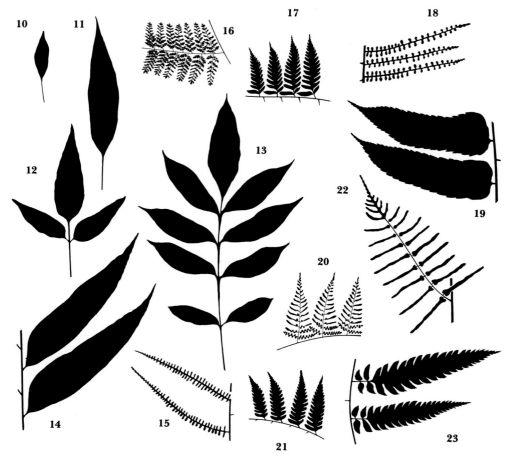

Figs. 80.10–80.23. Lamina architecture of *Polybotrya*, ×0.25. **10–13.** Leaves from juvenile plants of *P. cervina*, Fig. 10 is from the smallest plant, Fig. 13 from the largest. **14, 15.** Pinnae from leaves of adult plants of *P. cervina*. **14.** Sterile pinnae. **15.** Fertile pinnae. **16, 17.** Portions of pinnae of *P. canaliculata*. **16.** Fertile pinna. **17.** Sterile pinna. **18, 19.** Pinnae of *P. serratifolia*. **18.** Fertile pinnae. **19.** Sterile pinnae. **20, 21.** Portions of pinnae of *P. osmundacea*. **20.** Fertile pinna. **21.** Sterile pinna. **22, 23.** Pinnae of *P. caudata*. **22.** Fertile pinna. **23.** Sterile pinnae of leaf borne on a terrestrial stem, the leaves of scandent-epiphytic stems are similar to those of *P. osmundacea* (Fig. 21).

Geography (Fig. 2)

Polybotrya occurs from Veracruz in Mexico, south through Central America, in the Greater and Lesser Antilles, and in South America south to Bolivia, Missiones in Argentina, and Rio Grande do Sul in Brazil.

Spores

Polybotrya spores often have relatively broad, inflated folds, as in *P. fractiserialis* (Fig. 26), or more prominent and winglike folds, as in *P. serratifolia* (Fig. 24). The surface varies from a compact echinate form in *P. serratifolia* and *P. scandens* (Figs. 24, 25, 28, 29), diffuse in *P. fractiserialis* (Figs. 26, 27), or sparse in *P. cervina* (Fig. 30). The density of the echinate elements depends to a certain extent on the age or development of the spores but is generally consistent in species. The wall appears to be relatively thin with two perispore strata above the smooth exospore and the

24 **25**

26 **27**

28 **29** **30**

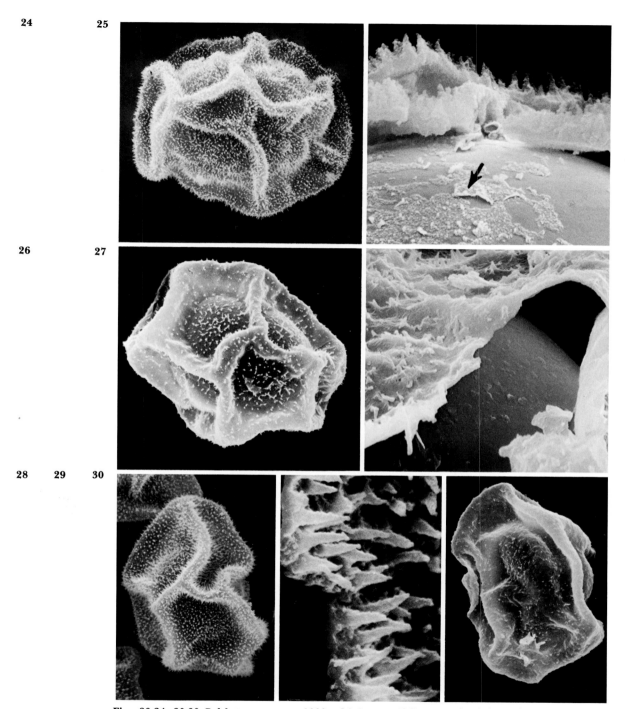

Figs. 80.24–80.30. *Polybotrya* spores, × 1000.   **24.** *P. serratifolia,* somewhat inflated folds and compact, echinate surface, Ecuador, *Madison et al. 13356.*   **25.** *P. scandens,* abraded surface, compact outer echinate perispore, above, and thin lower perispore (at arrow) on smooth exospore surface, below, Colombia, *Hodge 6592,* × 5000.   **26, 27.** *P. fractiserialis,* Peru, *Tryon & Tryon 5221.*   **26.** Inflated folds and moderately dense echinate surface.   **27.** Abraded spore with echinate surface and section of inflated fold, × 5000.   **28, 29.** *P. osmundacea,* Jamaica, *Maxon 9500.*   **28.** Folds and compact echinate surface.   **29.** Section of echinate perispore, the smooth exospore left, × 10,000.   **30.** *P. cervina,* incompletely developed prominent folds and sparse echinate surface, Guatemala, *Skutch 1807.*

outer surface more or less echinate (Fig. 25). *Polybotrya* spores are relatively large in comparison to other genera as *Dryopteris* in the tribe Dryopterideae. The longest dimension is generally between 50 and 60 μm but may be nearly to 70 μm in species such as *P. scandens.* The folded contours and echinate surface of *Polybotrya* spores resemble those of *Maxonia.* They are usually less elaborate than the prominently winged spores of the bolbitid genera.

Cytology

The first record of the chromosome number for *Polybotrya* was made in two species from Jamaica by Walker (1966). Meiotic cells were examined in *P. osmundacea* and these as well as mitotic cells of the root tips with $2n = 82$ were studied in *P. cervina.* The latter, also reported as $n = 41$ from Trinidad (Smith and Mickel, 1977), indicates a uniform chromosome level extends over a considerable area of the Caribbean range.

Literature

Schumann, E. 1915. Die Acrosticheen und ihre Stellung im System der Farne. Flora 108: 201–260 (reprint pp. 1–60).
Smith, A. R., and J. T. Mickel. 1977. Reference under the tribe Dryopterideae.
Walker, T. G. 1966. Reference under the tribe Dryopterideae.

18b.   Tribe Physematieae

Physematieae J. Sm., Hist. Fil. 235. 1875. Type: *Physematium* Kaulf. = *Woodsia.*
Woodsieae Diels, Nat. Pflanz. 1 (4): 159. 1899. Type: *Woodsia* R. Br. Woodsiaceae (Diels) Herter, Rev. Sudam. Bot. 9: 14. 1949.
Athyriaceae Alston, Taxon 5: 25. 1956. Type: *Athyrium* Roth.

A tribe of at least nine genera, with seven in America and *Hypodematium* Kze. and *Trichoneuron* Ching confined to the Old World. *Cheilanthopsis* Hieron. is a poorly known and dubious genus. Additional genera should be recognized in the *Athyrium-Diplazium* alliance which includes 500 or more species. This major group still lacks a reasonably satisfactory classification. It is not possible to assess segregate genera in the absence of a modern review of the whole complex. Thus two genera, *Athyrium* and *Diplazium,* are adopted for the American species.

Ching (1978) lists 17 genera in this complex in China, while Kato (1977) has four genera, *Athyrium, Diplazium, Cornopteris* and *Deparia* Hook. & Grev. (*sens. lat.*), with a total of 11 sections or species groups for Japan and adjacent eastern Asia. These classifications utilize new characters but unfortunately they have restricted geographic scope and include only a portion of all species.

Among the segregate genera recognized by Pichi-Sermolli (1977) and by Ching (1978), the following are excluded from the treatments of *Athyrium* and *Diplazium.* Those with a chromosome number of $n = 40$ or multiples are: *Anisocampium* Presl, *Athyriopsis* Ching, *Diplaziopsis* C. Chr., *Dryoathyrium* Ching, *Kuniwatsukia* Pic.-Ser. (*Microchlaena* Ching, not Wight & Arnott), and *Rhachidosorus* Ching. The genus *Callipteris* has $n = 41$, and *Cornopteris* has $n = 40$, and 41 or multiples. The cytology is not known in *Allantodia* R. Br. (*sens. str.*), *Cystoathyrium* Ching, *Dictyodroma* Ching, *Monomelangium* Hayata and *Triblemma* Ching.

Although the difference between *Athryium* with $n = 40$ and *Diplazium* with $n = 41$ has been repeatedly emphasized as significant, in the group as a whole these numbers do not correlate very well with other characters. Future classifications probably will recognize natural groups with both $n = 40$ and 41, as currently known in *Cornopteris*.

Literature

Chang, Y. L., I. C. Hsi, C. T. Chang, K. C. Kao, N. C. Tu, H. C. Sun, and C. C. Kung. 1976. Sporae Pteridophytorum Sinicorum. 451 pp. Academic Sinica Science Press. Peking.

Ching, R. C. 1978. The Chinese fern families and genera: systematic arrangement and historical origin. Acta Phytotax. Sinica 16 (3): 1–19.

Kato, M. 1977. Classification of *Athyrium* and allied genera of Japan. Bot. Mag. (Tokyo) 90: 23–40.

Lovis, J. D. 1977. Evolutionary patterns and processes in ferns. Advances of Botanical Research 4: 229–440.

Lugardon, B. 1971. Contribution à la connaissance de la morphogenese et de la structure des parois sporales chez las filicinées isosporées. 257 pp. Thése, Univ. Paul Sabatier, Toulouse.

Manton, I. 1950. Problems of cytology and evolution in the Pteridophyta. 316 pp. Cambridge Univ. Press, Cambridge, England.

Pichi-Sermolli, R. E. G. Tentamen Pteridophytorum genera in taxonomicum ordinem redigendi. Webbia 31: 313–512.

Sledge, W. A. 1962. The athyrioid ferns of Ceylon. Bull. Brit. Mus. (Nat. Hist.) 2 (11): 277–323.

Tatuno, S. and H. Okada. 1970. Karyological studies in the Aspidiaceae 1. Bot. Mag. (Tokyo) 83: 202–210.

Walker, T. G. 1966. A cytotaxonomic survey of the pteridophytes of Jamaica. Trans. Roy. Soc. Edinburgh 66: 169–237.

Walker, T. G. 1973. Additional cytotaxonomic notes on the pteridophytes of Jamaica. Trans. Roy. Soc. Edinburgh 69: 109–135.

81. *Diplazium*
Figs. 81.1–81.39

Diplazium Sw., Jour. Bot. (Schrad.) 1800 (2): 61. 1802. Type: *Diplazium plantagineum* Sw. (*Asplenium plantagineum* L., nom. superfl. for *Asplenium plantaginifolium* L.) = *Diplazium plantaginifolium* (L.) Urban.

Anisogonium Presl, Tent. Pterid. 115. 1836. Type: *Anisogonium fraxinifolium* (Presl) Presl = *Diplazium fraxinifolium* Presl.

Digrammaria Presl, *ibidem* 116. 1836. Type: *Digrammaria ambigua* (Sw.) Presl (*Asplenium ambiguum* Sw.) = *Diplazium esculentum* (Retz.) Sw.

Oxygonium Presl, *ibidem* 117. 1836. Type: *Oxygonium ovatum* Hook. = *Diplazium cordifolium* Blume.

Lotzea Kl. Linnaea 20: 358. 1847. Type: *Lotzea diplazioides* Kl. = *Diplazium diplazioides* (Kl.) Alston.

Brachysorus Presl, Epim. Bot. 70. 1851. Type: *Brachysorus woodward-ioides* Presl = *Diplazium woodwardioides* (Presl) Holtt.

Microstegia Presl, *ibidem* 90. 1851. Type: *Microstegia esculenta* (Retz.) Presl (*Hemionitis esculenta* Retz.) = *Diplazium esculentum* (Retz.) Sw.

Ochlogramma Presl, *ibidem* 93. 1851. Type: *Ochlogramma Cumingii* Presl = *Diplazium Cumingii* (Presl) C. Chr.

Pteriglyphis Fée, Mém. Fam. Foug. 5 (Gen. Fil.): 219. 1852, *nom. superfl.* for *Ochlogramma* Presl and with the same type. The earlier publication of the name by Fée, (X Congr. Sci. France 1: 178. 1843) is without description, and invalid.

Description

Terrestrial, sometimes rupestral, or very rarely epiphytic; stem usually erect, to 1 m tall, or decumbent to rather long-creeping, small to stout, bearing scales and usually many long, fibrous roots; leaves monomorphic or nearly so, ca. 25 cm to 3 m long, borne in a crown, a cluster, or more or less distant, lamina simple and entire to 3-pinnate-pinnatifid, glabrous or glabrate, finely or slightly pubescent beneath, or somewhat scaly, veins free or anastomosing; sori elongate, borne on the veins, on one or often on both sides of a vein on a scarcely to slightly raised receptacle, very rarely following the course of anastomosing veins, not paraphysate, covered by an indusium or rarely exindusiate; spores, more or less ellipsoidal, monolete, the laesura $\frac{1}{2}$ to $\frac{3}{4}$ the spore length, with prominent, winglike folds, the surface often smooth or with irregular echinate elements, or irregularly papillate or echinate. Chromosome number: $n = 41, 82$, ca. $121, 123, 164$; $2n = 82, 164$; apogamous $123, 164$, ca. $200, 328$.

The stem is erect in most species but the lamina structure is diverse as shown in the American species (Figs. 10–25). The elongate sori (Figs. 4–9) are often on both sides of the vein (Figs. 6, 8) and rarely follow the course of anastomosing veins (Fig. 9). The indusium may be so incompletely and irregularly developed that the sorus is considered to be exindusiate.

Systematics

Diplazium is nearly a worldwide genus of about 300 species. Segregate genera, or an infrageneric classification, are not adopted for American species since an adequate classification has not been developed for the whole group. At the present time it is not feasible to suggest relationships of the tropical American groups to each other or to the paleotropical elements.

The similarity of spores of *Diplazium* to those of species in several genera of the Dryopteridaceae as *Ctenitis, Bolbitis, Lomagramma* and *Stigmatopteris* reinforces other common features within the family including the chromosome numbers. In surface formation of the spores as well as the diversity of form, *Diplazium* spores have marked similarities to those of *Tectaria* (Figs. 69.24–69.39), suggesting possible close relationships of these genera.

A small group of tropical American species is of uncertain affinity. This includes species such as *Diplazium Bradeorum* Rosenst., *D. Lehmannii* Hieron., *D. Wilsonii* (Baker) Diels, *Athyrium ferulaceum* (Hook.) Christ, and perhaps *A. ordinatum* Christ. All of these have small ultimate segments with one to usually a few sori, and most have a 3-pinnate lamina. Although they are in-

Fig. 81.1. *Diplazium striatum,* Río Palenque, Ecuador. (Photo W. H. Hodge.)

cluded here as a group, their relationships are uncertain. The chromosome number $n = 40$ indicated for *D. Bradeorum,* as *"Athyrium"* by Gómez-Pignataro (1971) and the sori which are rarely hook-shaped suggest an affinity with *Athyrium,* but the surface folds of the spores (Fig. 34) are similar to species of *Diplazium,* such as *D. unilobum* (Fig. 33). The strongly winged-echinate spores of *D. Lehmanii* (Fig. 38) are distinct from those of *D. Bradeorum* but are similar to those of other diplaziums.

Two species of the eastern United States and adjacent Canada that have long been referred to *Diplazium* are sometimes now placed in segregate genera. *Diplazium pycnocarpon* (Spreng.) Broun has been recognized as the monotypic genus *Homalosorus* Pic.-Ser. (Pichi-Sermolli, 1977). Except for its free rather than anastomosing veins, it bears a close resemblance to *Diplaziopsis Cavaleriana* (Christ) C. Chr. of eastern Asia. Ching treated *Diplazium acrostichoides* (Sw.) Butters in the genus *Lunathyrium* Koidz. as *L. acrostichoides* (Sw.) Ching (1964) while Kato (1977, 1980) treated it in the genus *Deparia* Hook. & Grev. as *D. acrostichoides* (Sw.) Kato.

Oxygonium Presl is correctly typified by *Oxygonium ovatum* Hook. rather than by *Oxygonium alismifolium* (Presl) Presl which is a species of *Syngramma* (Holttum and Tryon, 1981).

Tropical American Species

Biosystematic studies are needed in addition to morphological, anatomical and cytological analyses to develop a modern classification of the American species of *Diplazium.* Particular attention should be given to the modes of reproduction and cytology. Complete collections are essential, especially of the large species

Fig. 81.2. Distribution of *Diplazium* in America.

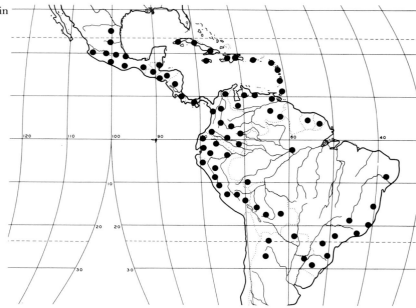

and for the interpretation of the ontogenetic series of leaf forms that appear to be frequent in these species.

Most of the 100 or more species of American *Diplazium*, excluding the small group discussed above, may be allied, by their lamina architecture, lamina indument, and venation into four groups. In addition, there are a few highly distinctive species without clear affinity to any others. The alliance of species by these characters will undoubtedly bring some related species together. However, some of those with a simple or 1-pinnate lamina have probably been derived from different sources among species with a more complex lamina and these affinities need to be determined.

1. *Diplazium umbrosum* Group

Lamina 2-pinnate-pinnatifid to 3-pinnate-pinnatifid, pinnae glabrate, slightly scaly or short-pubescent beneath, veins free. A group of about 30 species including *D. diplazioides* (Kl.) Alston, (*D. Klotzschii* (Mett.) Moore), *D. herbaceum* Fée (Fig. 10), *D. hians* Kl., *D. Lilloi* (Hicken) R. & A. Tryon, with prominently winged spores, and *D. umbrosum* Willd.

2. *Diplazium striatum* Group

Lamina 1-pinnate with entire pinnae to 1-pinnate-pinnatifid, the pinnae equilateral at the base or nearly so, more or less glabrous beneath or finely pubescent, veins free. A group of about 45 species, including *D. brasiliense* Rosenst. (Fig. 13), *D. celtidifolium* Kze. (Fig. 16), *D. flavescens* (Mett.) Christ, *D. Riedelianum* (Kuhn) C. Chr. (Fig. 25), *D. Roemerianum* (Kze.) Presl (Fig. 23), *D. striatum* (L.) Presl (Fig. 1), (*D. crenulatum* Liebm., *D. striatastrum* Lell.), and *D. ternatum* Liebm. (Fig. 22).

In this group there are about 10 species that appear to be intermediate with the next one in having small brown scales on the lamina beneath. However, the species with lobed or pinnatifid pinnae differ by the small ultimate segments as compared to the large segments of group 3.

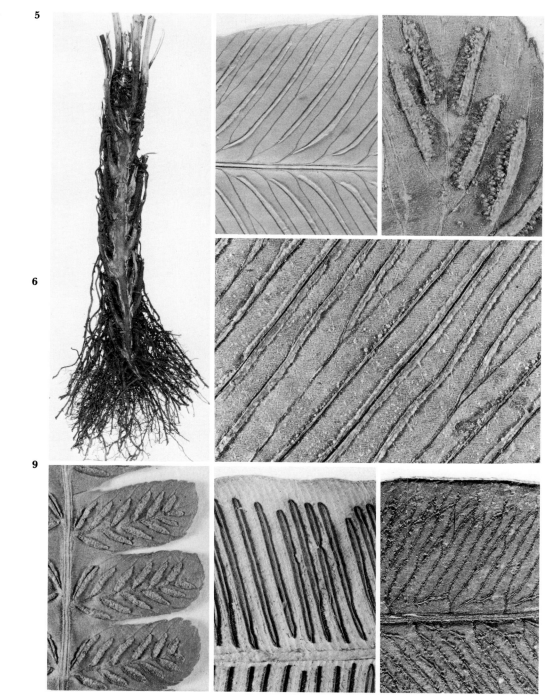

Figs. 81.3–81.9. *Diplazium*. 3. Erect stem of *D. umbrosum*, ×0.3. **4.** Indusiate sori of *D. flavescens*, ×2. **5.** Indusiate sori of *D. cristatum*, ×6. **6.** Indusiate sori of *D. Trianae*, several on both sides of a vein, ×5. **7.** Indusiate sori of *D. striatum*, the basal ones on both sides of the vein, ×3. **8.** Indusiate sori of *D. Lechleri*, on both sides of the vein, ×2. **9.** Nearly exindusiate sori of *D. pinnatifidum* following the anastomosing veins, ×2.

Some of these species are *D. costale* (Sw.) Presl, *D. Eggersii* (Sod.) C. Chr. and *D. Lindbergii* (Mett.) Christ.

3. *Diplazium pinnatifidum* Group

Lamina pinnatifid, 1-pinnate-pinnatifid or 2-pinnate, pinnae equilateral at the base, with small brown scales beneath, especially along the veins, ultimate segments large, veins free or anastomosing. A group of about 10 species, including those with free veins such as *D. macrophyllum* Desv. (Fig. 14), *D. macropterum* (Sod.) C. Chr. (Fig. 15), and *D. Trianae* (Mett.) C. Chr. (Fig. 11), and those with anastomosing veins such as *D. ceratolepis* (Christ) Christ and *D. pinnatifidum* Kze. (Fig. 20).

In the latter species the less complex leaves of young plants are shown in Figs. 19 and 21. In species with anastomosing veins the sori may branch and rejoin following the veins.

4. *Diplazium cristatum* Group

Lamina 1-pinnate, with nearly entire pinnae to partly 2-pinnate, the pinnae inequilateral at the base with the acroscopic side more developed, glabrate, veins free. A group of about 10 species, including *D. cristatum* (Desr.) Alston (Fig. 12), *D. Franconis* Liebm., *D. lonchophyllum* Kze., and *D. unilobum* (Poir.) Hieron. (Fig. 24).

5. Highly Distinctive Species of Uncertain Affinity

Diplazium Lechleri (Mett.) Moore has a large 1-pinnate lamina with entire pinnae, the veins are free except at their tip where they are connected by a slightly inframarginal strand.

Diplazium plantaginifolium (L.) Urban (Fig. 17) has a simple, entire lamina with free veins.

Diplazium praestans (Copel.) Morton (Fig. 18) has a simple, often irregularly lobed lamina with anastomosing veins and linear sori.

Diplazium aberrans Max. & Mort. has a simple entire lamina, long tapering to the base, anastomosing veins, and the sori and indusia branch and rejoin, following the veins.

Ecology (Fig. 1)

Diplazium is primarily a genus of wet montane forests, where it grows in ravines, by streams, or on rocky slopes. The plants occur in thickets, in open places in wet ground, along forest borders or in mesic woods.

In America *Diplazium* grows in primary rain forests, in oak woods, in cloud forests and also in secondary forests and in thickets. Species grow on stream banks, on rocky slopes or ravines, in wet canyons and less often on cliffs; also in clearings, on roadside banks, and rarely on old walls; *D. cristatum* is rarely epiphytic. *Diplazium* occurs from sea level to about 3000 m, most commonly at 500–2000 m.

Some species have proliferous buds on the rachis toward the apex of the lamina that may produce new plants, as in *Diplazium obscurum* Christ, *D. Werckleanum* Christ, and *D. gemmiferum* Christ.

Figs. 81.10–81.16. Portions of *Diplazium* leaves, × 0.25. **10.** Portion of a pinna of *D. herbaceum*. **11.** Pinna of *D. Trianae*. **12.** Upper portion of lamina of *D. cristatum*. **13.** Pinnae of *D. brasiliense*. **14.** Pinna of *D. macrophyllum*. **15.** Pinna of *D. macropterum*. **16.** Pinna of *D. celtidifolium*.

Figs. 81.17–81.25. Leaves of *Diplazium*, ×0.25. **17.** *D. plantaginifolium.* **18.** *D. praestans.* **19–21.** *D. pinnatifidum.* **19.** Leaf of juvenile plant. **20.** Leaf of adult plant. **21.** Leaf of young plant. **22.** *D. ternatum.* **23.** *D. Roemerianum.* **24.** *D. unilobum.* **25.** *D. Riedelianum.*

26 27 28

29 30

31 32 33

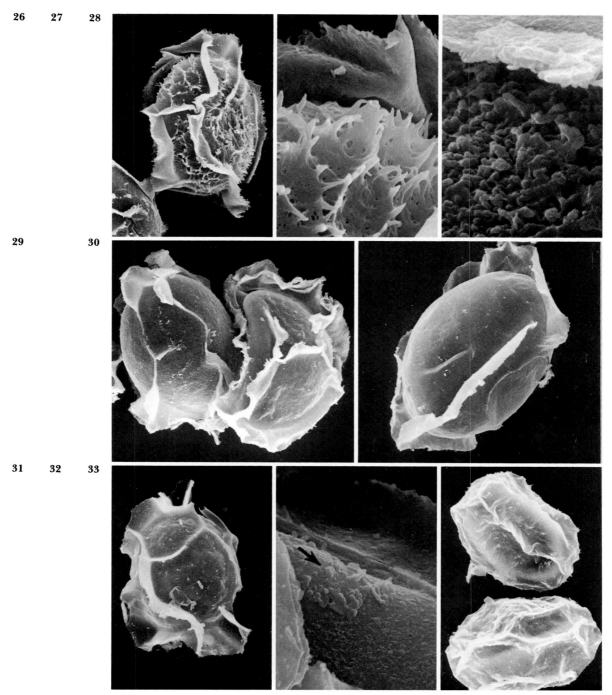

Figs. 81.26–81.33. *Diplazium* spores, × 1000. **26, 27.** *D. hians*, Venezuela, *Steyermark 109943.* **26.** Echinate surface and prominent wings. **27.** Detail of echinate-perforate surface, × 5000. **28.** *D. flavescens*, abraded wall with outer perispore surface, top, lower, coarse papillate layer, below, Peru, *Killip & Smith 23553*, × 10,000. **29.** *D. Eggersii*, surfaces of two spores with winglike folds, Peru, *Schunke 281.* **30.** *D. flavescens*, surface, prominent winglike folds, Peru, *Killip & Smith 23553.* **31, 32.** *D. pinnatifidum*, Ecuador, *Mexia 6865.* **31.** Surface, prominent winglike folds. **32.** Abraded wall, the surface left, below, a central coarse papillate layer (arrow), and lowest, diffuse papillae layer, below and right, × 5000. **33.** *D. unilobum*, low folds, Dominican Republic, *Gastony et al. 571.*

Geography (Fig. 2)

Diplazium is a pantropical genus with a few extensions to subtemperate regions. It occurs in America, Africa-Madagascar, and eastward to China, Japan, New Guinea and eastern Australia; also on the Pacific islands east to Pitcairn Island.

In America it is distributed from Nuevo Leon in Mexico through Central America, the Antilles, and in South America south to Tucumán and Missiones in Argentina and Rio Grande do Sul in Brazil; also in the Galápagos Islands. The genus is nearly absent from the Amazon basin. The center of species diversity is in the Andes from Colombia to Bolivia.

Diplazium lonchophyllum Kze. of Mexico and Central America has been reported from the southern United States on Avery Island, Louisiana, perhaps as a hurricane waif. Two Old World diplazioid ferns have been reported as adventive also in the southern United States, *Diplazium esculentum* (Retz.) Sw. in Florida and *Diplazium* (*Athyriopsis*) *japonicum* (Thumb.) Bedd. in Alabama and Florida. *Diplazium* (*Athyriopsis* or *Lunathyrium*) *Petersenii* (Kze.) Christ has been collected in the state of Guanabara (Rio de Janeiro) (*Smith*, 1932) and São Paulo (*Tryon & Tryon 6578*, GH), Brazil. Both of these collections were made in native vegetation without a nearby cultivated source.

Spores

Diplazium spores are generally characterized by prominently projecting winglike folds as in *D. Eggersii* (Fig. 29), *D. flavescens* (Fig. 30), and *D. pinnatifidum* (Fig. 31). The surface may be relatively smooth or with irregular echinate elements projecting from a compact, perforate base as in *D. hians* (Figs. 26, 27). One or two perispore strata are evident below the surface as in *D. flavescens* (Fig. 28) or in *D. pinnatifidum* (Fig. 32) which has a lower layer of smaller papillae. The surface varies from few, low folds as in *D. unilobum* (Fig. 33) to very prominent, elaborate wings as in *D. diplazioides* (Figs. 36, 37). The winged folds that are characteristic of the American species are also prominently developed in many of the Old World ones as shown in the work on the species of China, mostly treated under *Allantodia, sens. lat.* (Chang et al. 1976) and the species of Ceylon (Sledge, 1962). Some species of these regions as *D. zeylandicum* (Hook.) Moore have spores with folds forming short, truncate projections. The echinate spores as in *D. cognatum* (Hieron.) Sledge (Fig. 39) represent a distinct form that is not represented among the neotropical species.

Figs. 81.34–81.39. *Diplazium* spores, ×1000. **34. D. Bradeorum,** coarse, low folds, Costa Rica, *Scamman 7668*. **35. D. macropterum,** echinate and prominent wings, Colombia, *Ewan 16071*. **36, 37. D. diplazioides,** Venezuela, *Fendler 149*. **36.** Dense, fimbriate wings. **37.** Surface detail of wings, ×5000. **38. D. Lehmannii,** prominently echinate surface and wings, Costa Rica, *Scamman 7141*. **39. D. cognatum,** echinate, Ceylon, *Beckett 825*.

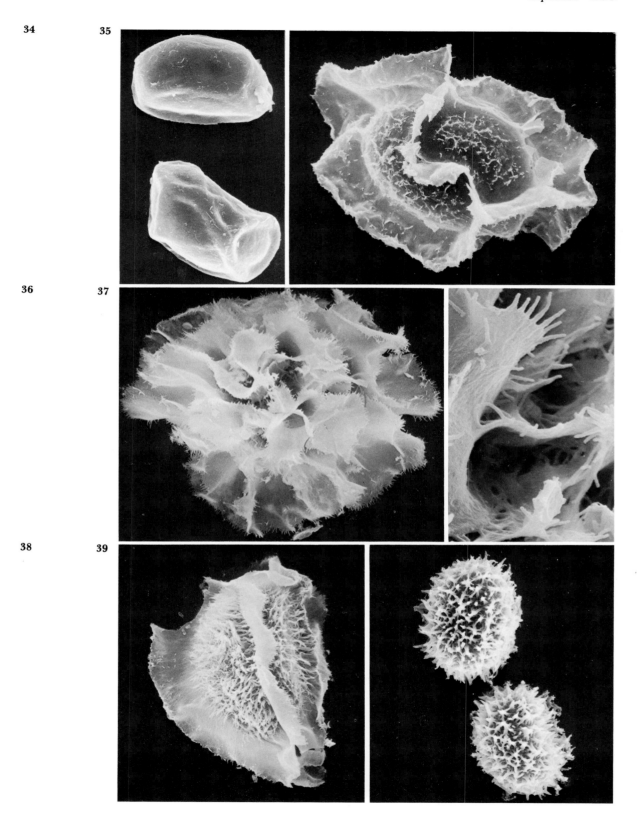

The surface structure and diversity of *Diplazium* spores resemble those of other genera of the Dryopteridaceae, especially those of *Tectaria*. There are resemblances in the general spore contours of the American *Diplazium macropterum* (Fig. 35) and those of *Tectaria pedata* (Fig. 69.37) and in the echinate forms of *Diplazium cognatum* (Fig. 39) and *Tectaria decurrens* (Fig. 69.33) of the paleotropics. The similarity of spores of several species of *Diplazium* and *Tectaria* and consistency of divergent forms is suggestive of basic affinities among these genera.

Cytology

The American species of *Diplazium* are largely tetraploid or octoploid with relatively few diploids. Three diploids are listed among reports of nine species from Mexico and Costa Rica (Smith and Mickel, 1977). A diploid, two tetraploids and eight octoploids are reported from Jamaica by Walker (1966, 1973). *Diplazium unilobum* is commonly diploid in Jamaica but a tetraploid was found among plants periodically submerged along a stream bank. Plants of *D. striatum* with $n = 164$, reported from Jamaica and Veracruz, Mexico, suggest that the species may be a wide-ranging polypoid. The highest number for the genus, $n = 328$ in *D. hians,* is an apogamous octoploid from Jamaica. The relatively high base numbers in *Diplazium* and *Athyrium,* differ by a single bivalent, posing difficulty in accurately establishing the chromosome number; thus photographic documentation is especially critical for these genera. Extratropical species of *Diplazium* are mostly $n = 41$ or 82 but a few apogamous triploids are known from Japan. Karyotypes of two species, analyzed by Tatuno and Okada (1970), have chromosomes arranged into six sets A–F: sets B–E each with eight bivalents, set F four, and set A with five. They propose that 41 in *Diplazium* is an octoploid based on 11 with loss of three bivalents from set A.

Literature

Chang, Y. L., et. al. 1976. Reference under the tribe Physematieae.

Ching, R. C. 1964. On some confused genera of the family Athyriaceae. Acta Phytotax. Sinica 9: 41–48.

Gómez-Pignataro, L. D. 1971. Ricerche citologiche sulle Pteridofite della Costa Rica, I. Atti Ist. Bot. Univ. Pavia 7: 29–31.

Holttum, R. E. and R. M. Tryon. 1981. The lectotypification of *Oxygonium* Presl (Pterid.) Taxon 30: 823–824.

Kato, M. 1977. Reference under the tribe Physematieae.

Kato, M. 1980. A new combination in *Athyrium* (Athyriaceae) of North America. Ann. Carnegie Mus. 49: 177–179.

Pichi-Sermolli, R. E. G. 1977. Fragmenta Pteridologiae, VI. Webbia 31: 237–259.

Sledge, W. A. 1962. Reference under the tribe Physematieae.

Smith, A. R., and J. T. Mickel, 1977. Chromosome counts for Mexican ferns. Brittonia 29: 391–398.

Smith, L. B. 1932. *Diplazium Petersenii* as an escape. Amer. Fern Jour. 22: 99–100.

Tatuno, S., and H. Okada. 1970. Reference under the tribe Physematieae.

Walker, T. G. 1966. Reference under the tribe Physematieae.

Walker, T. G. 1973. Reference under the tribe Physematieae.

82. *Athyrium*
Figs. 82.1–82.19

Athyrium Roth, Tent. Fl. Germ. 3: 58. 1799. Type: *Athyrium Filix-femina* (L.) Roth (*Polypodium Filix-femina* L.).

Pseudathyrium Newm., Phytol. 4: 370. 1851. Type: *Pseudathyrium alpestre* (Hoppe) Newm. (*Aspidium alpestre* Hoppe) = *Athyrium alpestre* (Hoppe) Clairv.

Pseudocystopteris Ching, Acta Phytotax. Sinica 9: 76. 1964. Type: *Pseudocystopteris spinulosa* (Maxim.) Ching (*Cystopteris spinulosa* Maxim.) = *Athyrium spinulosum* (Maxim.) Milde.

Systematics

Terrestrial or rupestral; stem nearly erect to usually decumbent, very short-creeping and usually moderately stout, to long-creeping and slender, bearing scales, which (in *A. Skinneri*) intergrade into trichomes, and many long, fibrous or (in *A. Skinneri*) often soft roots; leaves monomorphic or nearly so, ca. 20 cm to 2 m long, borne in a crown, loose cluster, or at intervals, lamina 1-pinnate-pinnatifid to partly 3-pinnate-pinnatifid, more or less glabrate or somewhat scaly or slightly pubescent, veins free; sori round or usually somewhat elongate on one side of a vein, or sometimes also developed distally on the other side, on a scarcely raised receptacle, not paraphysate, usually covered by a reniform to somewhat elongate or hook-shaped indusium corresponding to the extent of the sorus, or rarely exindusiate; spores more or less ellipsoidal, monolete, the laesura $\frac{2}{3}$ to $\frac{3}{4}$ the spore length with moderately prominent folds forming a coarse, rugose surface or sometimes with prominent winglike folds. Chromosome number: $n = 40, 80, 120$, ca. 160; $2n = 80, 160, 240$; apogamous 120.

The indusium in *A. alpestre* is so small and incompletely developed that the sorus is considered to be exindusiate. *Athyrium Filix-femina* has a compact stem with persistent petiole bases (Fig. 2), while *A. Skinneri* has an elongate stem with spaced petiole bases that soon decay (Fig. 3). The typical elongate sori and indusia of *Athyrium,* some of them hook-shaped, are shown in Figs. 4–6.

Description

Athyrium is primarily a north temperate genus of 100 or more species, with three of them in America. The genus as treated here includes the well-known *Athyrium Filix-femina* and its immediate allies. Sledge (1962) treated *Athyrium* of Ceylon in the same sense; however, Kato (1977) enlarged it to include *Anisocampium* and *Kuniwatsukia*. It is expected that the generic circumscription may change with further studies on the relationships with other athyrioid or diplazioid groups.

Some tripinnate American species that may be referable to *Athyrium* are discussed under *Diplazium.*

American Species

There are three basic species of *Athyrium* in America. The widespread *Athrium Filix-femina,* which is represented by three geographic varieties in the United States and Canada, has not been studied critically in the tropics. Several Mexican segregates such as *Athyrium Martensii* (Kze.) Moore, *A. paucifrons* C. Chr. and *A.*

Fig. 82.1. Distribution of *Athyrium* and of *A. Filix-femina* in America, south of lat. 35° N.

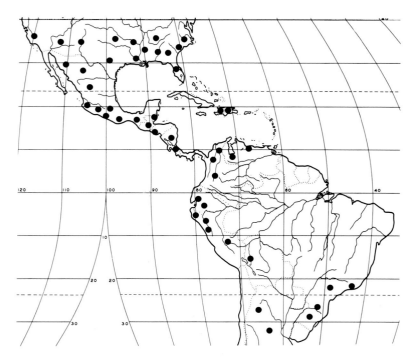

Bourgaei Fourn. undoubtedly do not represent species, while *Athyrium Dombeyi* Desv. of Central and South America is probably a geographic variety of *Athyrium Filix-femina*. *Athyrium Skinneri* is the most distinctive species although it varies in the size and division of the lamina (Figs. 7, 8) especially if *A. palmense* (Christ) Lell. is included.

The third species, *Athyrium alpestre,* is boreal to subarctic. It has recently been called *Athyrium distentifolium* Opiz (Fuchs, 1974), but *Athyrium alpestre* (Hoppe) Clairville (Man. Herbor, Suisse Valais, 1811) is the correct, earlier name.

Athyrium decurtatum (Link) Presl of Argentina, Uruguay and southeastern Brazil is a species of *Thelypteris*.

Butters (1917) published a critical study of the American *Athyrium Filix-femina* group, although some of his species are now recognized at an infraspecific rank.

Key to Species of *Athyrium* in America

a. Sori exindusiate. *A. alpestre* (Hoppe) Clairv.
a. Sori indusiate. b.
  b. Stem short-creeping to nearly erect, bearing the leaves in a crown or cluster, the old petiole bases adjacent, persistent.
    A. Filix-femina (L.) Roth (*sens. lat.*)
  b. Stem elongate, sometimes moderately long-creeping, bearing the leaves at intervals, the old petiole bases more or less distant, ephemeral.
    A. Skinneri (Baker) Diels (*sens. lat.*)

Ecology

Athyrium is a genus of forests and thickets, open, shrubby habitats, grassy and rocky places, and stream sides.

In tropical America, *Athyrium* grows in rain forests, in montane forests, in pine and oak woods, often in canyons or ravines,

2

4

3

5

6

7

Figs. 82.2–82.7. *Athyrium.* **2.** Stem longisection (apex, right) and persistent petiole bases of *A. Filix-femina,* ×0.75. **3.** Stem (apex, right) and decayed petiole bases of *A. Skinneri,* ×0.75. **4.** Fertile segments of *A. Filix-femina,* ×4. **5.** Sori and indusia of *A. Filix-femina,* ×10. **6.** Sori and indusia of *A. Skinneri,* ×10. **7.** Lamina of *A. Skinneri,* ×0.3.

on stream banks or sometimes on cliffs. At high altitudes it grows in low thickets, often among rocks, in wet meadows or along watercourses. *Athyrium* grows most often at 1000–2500 m; *A. Skinneri* is usually at lower altitudes while *A. Filix-femina* has a broad ecological distribution extending to 3500 m.

Geography (Fig. 1)

Athyrium occurs in America, Eurasia and southward to Africa-Madagascar, the Himalayas and Ceylon, Japan, the Philippine Islands, Java and New Guinea.

In America it grows from Laborador to Alaska and southward through Mexico, Central America and South America to southeastern Brazil and Córdoba in central Argentina; it is disjunct on Hispaniola.

Athyrium Filix-femina grows through the range of the genus in tropical America while *A. Skinneri* occurs in Mexico south to the Sierra Nevada de Santa Marta in northern Colombia.

Spores

The low, coarsely rugose surface as in *Athyrium Dombeyi* (Fig. 11) and *A. Filix-femina* (Figs. 13, 15) is characteristic of the genus. However, this outer surface is often detached exposing a lower, shallowly rugulose formation as in *A. Dombeyi* (Fig. 12) and *A. Filix-femina* (Fig. 16). The surface detail of this lower perispore stratum (Fig. 17) shows the particulate material is fused into irregular plates less compact than the perispore surface (Fig. 14). The wall sections of *Athyrium Filix-femina* and *A. alpestre* studied by Lugardon (1971) indicate three perispore strata. The lowest is adherent to the exospore while the upper are less strongly attached and tend to become dislodged along a line of dark staining granules. The spores of *Athyrium Skinneri* differ from those of other American species in their strongly protruding winglike folds, echinate especially along the margins (Figs. 18, 19). Several Old World species have coarsely rugose spores similar to those of *A. Dombeyi* and the outer surface is also readily detached. Spore photographs of athyriums of China in Chang et al. (1976) and drawings in the work of Sledge (1962) on athyrioid ferns of Ceylon are mostly rugose, or rugulose to smooth. A few species have well-developed wings, but ridged spores are characteristic of *Athyrium* and distinct from the predominately winged spores of the diplaziums.

Cytology

The chromosome number $n = 40$ in *Athyrium* is largely based on reports for species of India and Japan. The records are mostly diploid or tetraploid but an octoploid and also a few hexaploids with $n = 120$ are known. Reports of *Athyrium alpestre* and *A. Filix-femina* from North America and Europe (Schneller, 1979) are consistently diploid. The widest-ranging species, *A. Filix-femina*, is also diploid in Asia but reports are lacking for the tropical

Figs. 82.8–82.10. Central pinnae of *Athyrium*, ×1. **8.** *A. Skinneri.* **9, 10.** *A. Filix-femina.*

American plants. Cytologically, *Athyrium* differs from *Diplazium* in having one less bivalent. The somatic chromosomes of *Athyrium niponicum* (Mett.) Hance, figured by Tatuno and Okada (1970) with $2n = 80$, generally appear similar in size and form to those of *Diplazium,* but the karyotype has not been analyzed as in the latter genus.

11 12 13

14 15 16

17 18 19

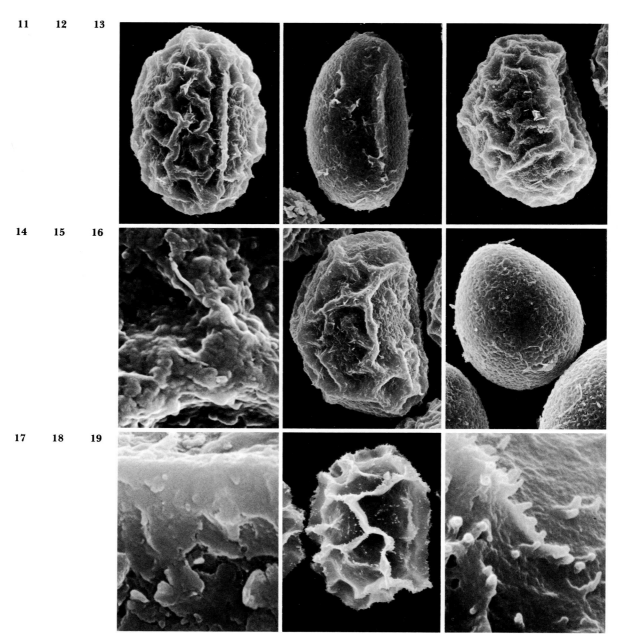

Figs. 82.11–82.19. *Athyrium* spores, × 1000, surface detail, × 10,000. **11, 12. *A. Dombeyi.* 11.** Outer rugose surface, laesura, right, Mexico, *Riba et al. 283.* **12.** Lower rugulose perispore, the outer surface detached. Peru, *Sanchez 318.* **13–17. *A. Filix-femina.* 13, 14.** Mexico, *Nicholas* in 1910. **13.** Outer rugose perispore surface. **14.** Particulate surface deposition of Fig. 13. **15.** Outer rugose perispore surface, Mexico, *Moore 5378.* **16, 17.** Lower, rugulose perispore layer, Mexico, *Ghiesbreght 229.* **16.** Eroded distal surface. **17.** Detail lower perispore with particulate deposit fused into irregular plates, the laesura at top. **18, 19. *A. Skinneri,*** Mexico, *R. & C. Wilbur 2344.* **18.** Lateral surface, winglike folds. **19.** Echinate surface and edge of fold.

Observations

Biosystematic investigations on *Athyrium Filix-femina* in Europe by Schneller (1979) include observations on growth and development of plants in nature as well as in culture. The analyses of spore longevity, germination and dispersal relate to the large annual production estimated at 75 million spores for a single plant. An antheridogen system influences development of the gametophytes and also enables the spores to germinate in the dark. No crossing barriers were found between populations either from close or far distant localities. Relatively little evidence of successful intragametophytic crossing was detected while intergametophytic crossing was the predominant system and apparently accounts for the great variability of the species.

Literature

Butters, F. K. 1917. The genus *Athyrium* and the North American ferns allied to *A. Filix-femina*. Rhodora 19: 170–207.

Chang, Y. L., et al. 1976. Reference under the tribe Physematieae.

Fuchs, H. P. 1974. The correct name for the alpine lady fern. Candollea 29: 181–205.

Kato, M. 1977. Reference under the tribe Physematieae.

Lugardon, B. 1971. Reference under the tribe Physematieae.

Schneller, J. J. 1979. Biosystematic investigations on the Lady Fern (*Athyrium Filix-femina*). Pl. Syst. Evol. 132: 255–277.

Sledge, W. A. 1962. Reference under the tribe Physematieae.

Tatuno, S. and H. Okada. 1970. Reference under the tribe Physematieae.

83. *Hemidictyum*
Figs. 83.1–83.9

Hemidictyum Presl, Tent. Pterid. 110. 1836. Type: *Hemidictyum marginatum* (L.) Presl (*Asplenium marginatum* L.).

Description

Terrestrial; stem decumbent to erect, stout, including the more or less persistent petiole bases, bearing scales and many large fibrous roots; leaves monomorphic, ca. 1–3 m long, borne in a crown, lamina 1-pinnate, with entire pinnae, imparipinnate, glabrous, veins free except toward the margin where they are fully anastomosing without included free veinlets; sori elongate along the veins, on a somewhat raised receptacle, not paraphysate, covered by an equally elongate, membranous indusium; spores more or less ellipsoidal, monolete, the laesura $\frac{3}{4}$ the spore length, with perforate folds, saccate prominences and somewhat rugulose surface. Chromosome number: $n = 31$.

The stem of *Hemidictyum* is massive and becomes erect (Fig. 3) producing a crown of large leaves. The pinnae are unusually large and thin (Fig. 4), in comparison to other ferns with entire pinnae, attaining a size of about 40 cm long and 8 cm wide. Sori are borne on long veins (Fig. 5), while the veins beyond anastomose and connect in a continuous strand somewhat back of the margin (Fig. 6).

Fig. 83.1. *Hemidictyum marginatum*, Luquillo Mountains, Puerto Rico. (Photo D. S. Conant.)

Systematics

Hemidictyum is a monotypic, apparently isolated, American genus based on *H. marginatum* (L.) Presl. It is sometimes placed in *Diplazium* (*D. marginatum* (L.) Diels, or correctly as *D. limbatum* (Willd.) Proctor) but it is not clearly related to that genus. The chromosome number *n* = 31 is unusual in the family. This inconsistency, and differences noted below in spores, raise doubt as to its position in the Dryopteridaceae. The spores and cytology as well as the two vascular bundles in the petiole suggest a relationship to the thelypteroid ferns (Lovis, 1977) but other characters are not consistent with that alliance. A second species, *Hemidictyum Purdieanoides* (Karst.) Vareschi is *Asplenium Purdieanum*.

Ecology (Fig. 1)

Hemidictyum grows in wet forests, especially along streams, in ravines, or on mountain slopes. it grows from sea level to ca. 1300 m, most often at 100–500 m.

Geography (Fig. 2)

Hemidictyum is distributed from Oaxaca in southern Mexico, through the Antilles, to South America and from French Guiana, west to Colombia and south in the Andes to Bolivia; it is disjunct in southeastern Brazil from Bahia to Santa Catarina.

Spores

The surface contours are formed by relatively few, long, low folds or short, saccate projections (Fig. 7). The folds are more or less perforate and shallowly rugulose (Fig. 8). The relatively thin

Fig. 83.2. Distribution of *Hemidic-tyum.*

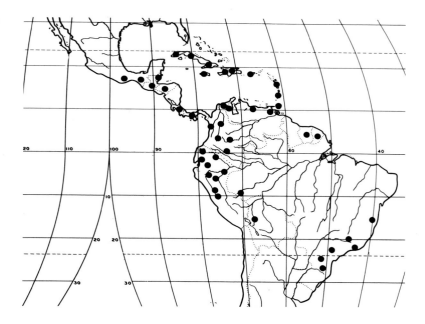

Fig. 83.3. Longisection of stem of *Hemidictyum marginatum* (P, petiole; S, stem), ×0.33.

Fig. 83.4. Pinna of *Hemidictyum mar-ginatum,* ×0.25.

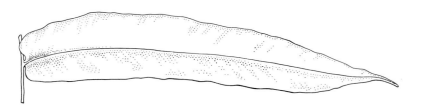

5   8

6

7   9

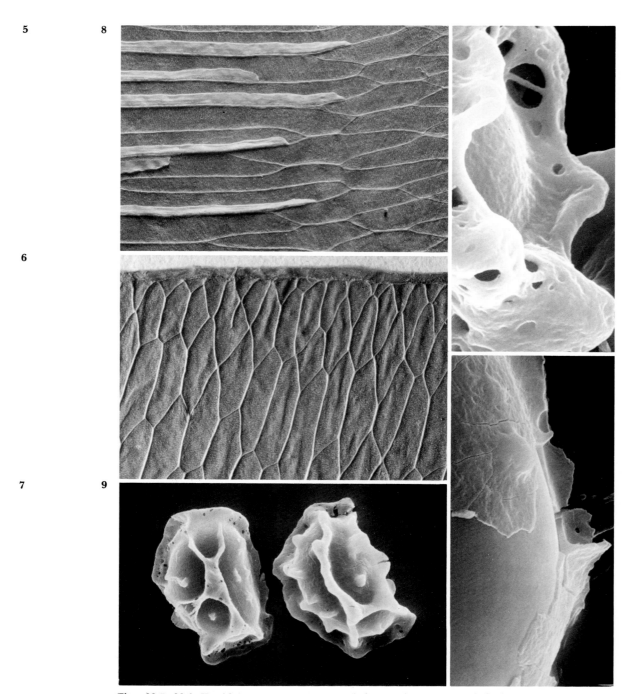

Figs. 83.5–83.9. *Hemidictyum marginatum*. **5.** Apical portions of young indusiate sori and basal portion of the region of anastomosing veins (pinna margin toward right), ×5. **6.** Marginal portion of a pinna, anastomosing veins and connecting strand, ×5. **7–9.** Spores, Peru, *Wurdack 1890*. **7.** Lateral surface of two spores with long, perforate and saccate folds, ×1000. **8.** Detail of perforate folds, ×5000. **9.** Abraded surface, thin, perispore above, smooth exospore at left, below, ×5000.

outer perispore overlays a smooth exospore (Fig. 9) without an elaborate lower perispore layer as in *Tectaria, Oleandra* or *Diplazium*. The spores of these genera usually have a more complex surface structure than those of *Hemidictyum*. The thin, perforate perispore generally resembles some spores of *Thelypteris*.

Cytology

The report of $n = 31$ for *Hemidictyum* from Trinidad (Walker, 1973), confirmed by another from Costa Rica (Wagner, 1980), is an exceptional one in the Dryopteridaceae. Alignment of this number with other dryopteroid genera that are mostly $n = 40$ or 41 would require drastic aneuploid reduction.

Literature

Lovis, J. D. 1977. Reference under the tribe Physematieae.
Wagner, F. S. 1980. New basic chromosome numbers for genera of neotropical ferns. Amer. Jour. Bot. 67: 733–738.
Walker, T. G. 1973. Reference under the tribe Physematieae.

84. *Adenoderris*
Figs. 84.1–84.8

Adenoderris J. Sm., Hist. Fil. 222. 1875. Type: *Adenoderris glandulosa* (Presl) J. Sm. (*Polystichum glandulosum* Presl, *nom. nov.* for *Aspidium glandulosum* Hook. & Grev. 1829, not Blume, 1828).

Description Terrestrial or rupestrial; stem erect, small, bearing scales and many fibrous roots; leaves monomorphic, ca. 5–20 cm long, borne in a crown, lamina pinnatisect or 1-pinnate-pinnatifid, viscid-glandular, veins free; sori round, borne on the veins or at their tips on a somewhat raised receptacle, not paraphysate, covered by a peltate to somewhat cordate indusium; spores more or less ellipsoidal, monolete, the laesura $\frac{1}{2}$ to $\frac{3}{4}$ the spore length, with inflated rugose-saccate folds or wings, the surface rugulose or echinate. Chromosome number: not reported.

The stem scales are relatively large, broad and light brown. The leaves are short-petioled or in some plants are nearly sessile. The indusiate sori are shown in Fig. 3 and the viscid glands, that are especially dense along the margin, in Fig. 4.

Systematics *Adenoderris* is a little-known genus of two species in the American tropics. Its systematic position is uncertain and it has been placed near *Polystichum* or *Dryopteris*. The two vascular bundles

Fig. 84.1. Distribution of *Adenoderris*.

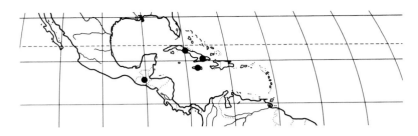

in the petiole relate it to the tribe Physematieae, although the spores indicate a closer relation to the Dryopterideae. Marked differences of spores in the two species raise a question as to their relationships. More adequate specimens and cytological information are essential to properly place the genus.

It has evidently been overlooked that *Polystichum glandulosum* Presl was a new name for *Aspidium glandulosum* Hook. & Grev., not Blume. It provides an earlier epithet than *Aspidium viscidulum* Mett. for this species under *Andenoderris*.

Maxon (1905) provided a synopsis of the genus.

Key to the Species of *Adenoderris*

a. Lamina pinnatisect, the pinna-lobes adnate (Fig. 2) veins mostly forked, ending near the margin, sori borne on the veins.   *A. glandulosa* (Presl) J. Sm.
(*A. viscidula* (Mett.) Maxon)

a. Lamina 1-pinnate-pinnatifid, the pinnae sessile to stalked, veins mostly simple, ending well back of the margin, sori borne at the vein-ends.
A. sororia Maxon

Ecology and Geography (Fig. 1)

Adenoderris grows on or among rocks, in ravines and on mountain slopes. Sometimes it grows on limestone. The genus occurs at elevations of 450–1200 m. *Adenoderris glandulosa* grows in Cuba and Jamaica and *A. sororia* in Guatemala.

Spores

Spores of the two species of *Adenoderris* differ in the type of surface structure, folds and size (Figs. 5, 7), but have a similar thin development of the perispore (Figs. 6, 8). The spores are of a relatively simple structure and generally similar to those of several dryopteroid genera as *Dryopteris, Cyrtomium* and *Didymochlaena*.

Cytology

The chromosome number has not been reported; however, the marked diversity in spore size suggest that different ploidy levels may be represented.

Literature

Maxon, W. R. 1905. *Adenoderris*, a valid genus of ferns. Bot. Gaz. 39: 366 –369.

Fig. 84.2. Leaf of *Adenoderris glandulosa*, ×0.5.

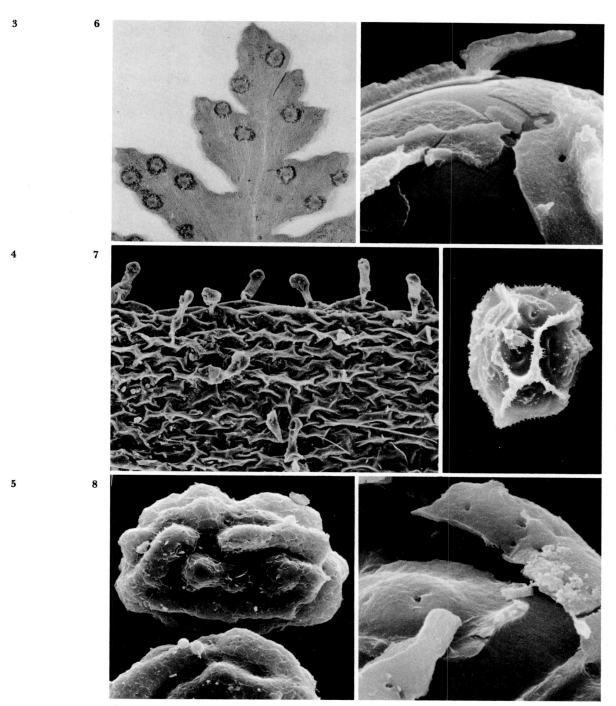

Figs. 84.3–84.8. *Adenoderris.* **3.** Lamina apex with indusiate sori, × 3. **4.** Lamina glands, × 200. **5–8.** Spores, × 1000. **5.** *A. sororia,* rugose-saccate, Guatemala, *Tuerckheim 868.* **6.** *A. glandulosa,* detail of abraded surface, the lower perispore stratum and exospore, below, Cuba, *Hodge & Howard 4584,* × 5000. **7.** *A. glandulosa,* winged, echinate folds, Cuba, *Morton 4078.* **8.** *A. sororia,* abraded perispore, over smooth exospore surface, *Tuerckheim 868,* × 3000.

85. *Cystopteris*
Figs. 85.1–85.12

Cystopteris Bernh., Neues Jour. Bot. (Schrad.) 1 (2): 26. 1806, *nom. conserv.* Type: *Cystopteris fragilis* (L.) Bernh. (*Polypodium fragile* L.)

Filix Adans., Fam. Plts. 2: 20, 558. 1763. Type: *Cystopteris fragilis* (L.) Bernh.

Cystea Sm., English Fl. 4: 297. 1828, *nom. superfl.* for *Cystopteris* Bernh. and with the same type.

Cyste Dulac, Fl. Dept. Haute-Pyrenees 33. 1867, *nom. superfl.* for *Cystopteris* Bernh. and with the same type.

Acystopteris Nakai, Bot. Mag. (Tokyo) 47: 180. 1933. Type: *Acystopteris japonica* (Luerss.) Nakai = *Cystopteris japonica* Luerss. *Cystopteris* subgenus *Acystopteris* (Nakai) Blasdell, Mem. Torrey Bot. Cl. 21 (4): 49. 1963.

Description Terrestrial or rupestral; stem short, decumbent and small to long-creeping and slender, bearing scales and rarely some trichomes, and usually many slender, fibrous roots; leaves monomorphic, ca. 10–80 cm long, borne in a cluster to rather widely spaced, lamina 2-pinnate to 3-pinnate-pinnatifid, glabrous, glandular or slightly pubescent, veins free; sori round, borne on the veins on a scarcely to somewhat raised receptacle, not paraphysate, indusiate, the indusium scalelike, broadly ovate, lanceolate to half-cup-shaped; spores more or less ellipsoidal, monolete, the laesura $\frac{3}{4}$ the spore length, echinate, reticulate, verrucate, or with inflated, bulbous processes. Chromosome number $n = 42$, 84, 126, 168; $2n = 126$, 168, 252.

The indusium of *Cystopteris* arches over the sorus (Fig. 4) although it is sometimes small and concealed beneath the mature sporangia. There are many variations in the form of the leaves in *C. fragilis* including small, abundantly fertile forms (Fig. 5) and more complex leaves (Fig. 6). Pichi-Sermolli (1968) reports glandular trichomes mixed with the scales on the stem of *C. nivalis.*

Systematics *Cystopteris* is a nearly worldwide genus of about six species with the variable *C. fragilis* in the American tropics. The indusium, basally attached and arching over the sporangia, at least while they are immature, is the most distinctive feature of the genus. In other characters it is closely related to *Athyrium* and perhaps to *Gymnocarpium.* The stout multicellular trichomes and complex lamina of subgenus *Acystopteris,* which may merit generic rank, suggest a possible relationship to *Stenolepia* in the tribe Peranemeae.

The name *Filicula* Séguier, 1754, sometimes cited as a synonym of *Cystopteris,* is invalid; although two species are described, a generic description is lacking.

Cystopteris has been revised by Blasdell (1963). His treatment recognized three segregate species in the *C. fragilis* complex—*C. protrusa* (Weath.) Blasdell, *C. diaphana* (Bory) Blasdell and *C. Douglasii* Hook.—and numerous widely distributed hybrids of the first two with *C. fragilis.* On the basis of the variability of the taxa and the integradation between them, the *C. fragilis* complex is treated here as a single species, and *C. Dickeana* Sim is also in-

Fig. 85.1. *Cystopteris fragilis,* Pedregal Las Vigas, above Jalapa, Veracruz, Mexico. (Photo W. H. Hodge.)

cluded. *Cystopteris pellucida* (Franch.) C. Chr. is included with *C. sudetica* as these are too close for specific distinction.

Synopsis of *Cystopteris*

Subgenus *Cystopteris*

Multicellular trichomes, when present on the lamina, glandular, spores echinate, verrucate or granulose.

Cystopteris fragilis (L.) Bernh. (Fig. 3), *C. montana* (Lam.) Bernh., *C. sudetica* A. Br. & Milde, and *C. bulbifera* (L.) Bernh. A circumboreal group, also in the mountains of the tropics, and widely distributed in austral regions. *Cystopteris nivalis* (Pirotta) Pic.-Ser. of Ruwenzori and *C. membranifolia* Mickel of Mexico cannot be adequately assessed on the present limited material. Since local endemism is not a feature of the genus, it seems likely that these are not distinct species.

Subgenus *Acystopteris*

Lamina with stout, nonglandular, multicellular trichomes especially on the adaxial surface, spores with prominent bulbous processes.

Cystopteris japonica Luerss. and *C. tenuisecta* (Bl.) Mett. of eastern and southeast Asia and Malaysia.

Tropical American Species

A single variable species, *Cystopteris fragilis,* occurs in the American tropics.

Fig. 85.2. Distribution of *Cystopteris* and of *C. fragilis* in America, south of lat. 35° N.

Ecology (Fig. 1)

Cystopteris is primarily a genus of mesic, rocky habitats but also is terrestrial in forests and alluvial stream valleys. It is often on old masonry and rarely grows on the base of trees.

In the American tropics it grows in montane rain forests, in pine and oak woodlands, in elfin forests, and also in thickets, alpine scrub and páramo. It is nearly always associated with rocks of all types and may be among boulders, on lava beds, cliffs, talus slopes and rocky hillsides and rocky stream banks. It grows from 1500 to 4500 m in the tropics and north and south at lower elevations, descending to sea level.

Geography (Fig. 2)

Cystopteris is widely distributed in America, in Africa and its islands, in Eurasia and Malesia. It also occurs in southeastern Australia, Tasmania, New Zealand and the Hawaiian Islands as well as the austral islands of Kerguelen and South Georgia.

In America, *Cystopteris fragilis* grows from lat. 83° N south to Tierra del Fuego and is also on the Falkland Islands, the Juan Fernandez islands and South Georgia. It is confined to mountainous regions in the tropics. The other American species, *C. montana,* is boreal and arctic and *C. bulbifera* is north temperate and boreal.

Fig. 85.3. Plant of *Cystopteris fragilis;* the central leaf is immature ×0.5.

Spores

Spores of *Cystopteris* are unusually diverse for a small genus. The exceptional variation in spores of *C. fragilis* (Figs. 7–10) is similar to that shown by Pearman (1976). The echinate type most frequent in the species (Fig. 7) is formed of coalescent strands (Fig. 8). Sections of the wall examined by Lugardon (1971) show a thin outer layer and an echinate formation of perispore overlay a dense exospore. Verrucate spores are also common to the species (Fig. 9) and have been the basis for distinguishing *C. Dickieana* (Jermy and Harper, 1971). Another type of spore within the *Cystopteris fragilis* complex, with a diffuse reticulum and small echinate structures (Fig. 10), has been used to distinguish *C. diaphana*. Variation in the spores of *Cystopteris fragilis* in North America studied by Hagenah (1961) showed that non-echinate spores are largely in the northern and western regions and there are intermediates between the echinate and verrucate forms. This variation is similar to that in spores of *Thelypteris palustris* and developmental studies of specimens over a broad geo-

4   5   6

7   8   9

10   11   12

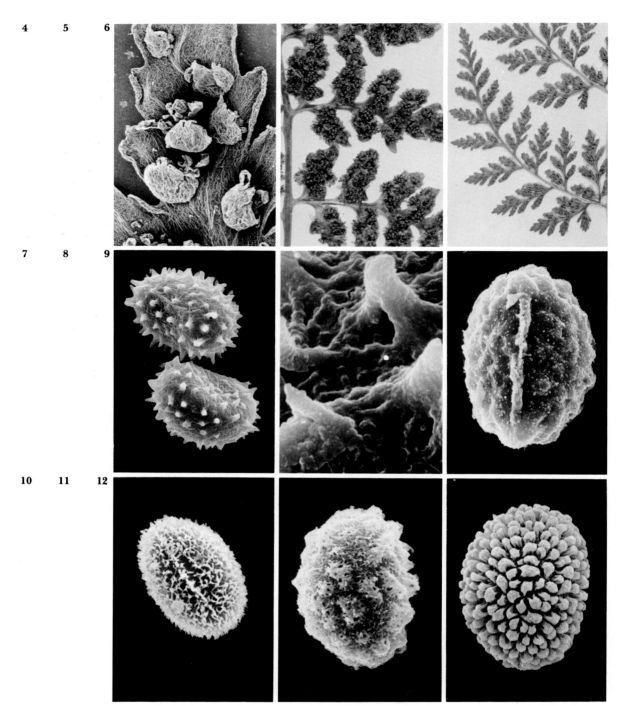

Figs. 85.4–85.12. *Cystopteris.* **4.** Sori and indusia of *C. fragilis,* ×20. **5.** Pinnae of heavily fertile leaf of *C. fragilis,* 9 cm long, growing in rock crevices at 4300 m, Peru, ×5. **6.** Central pinnae of leaf of *C. fragilis,* 65 cm long ×1. **7–12.** Spores, ×1000, surface detail, ×10,000. **7–10.** *C. fragilis.* **7.** Echinate spores, Mexico, *Conzatti & Gonzales 696½.* **8.** Detail of echinate surface, Argentina, *Cantino 207.* **9.** Verrucate, laesura at center, Alaska, *Argus 1121.* **10.** Echinate-reticulate surface, the laesura at center, Costa Rica, *Scamman 5920.* **11.** *C. montana,* verrucate-reticulate, Ontario, Canada, *Hosie et al. 50.* **12.** *C. japonica* with inflated, bulbous projections, Japan, *Shiota 6029.*

graphic range of these species are essential for interpretation of the exceptional variation. The spores of *Cystopteris montana* are more consistently verrucate with a reticulate surface similar to the echinate forms of *C. fragilis* (Fig. 11). The most distinctive spores are those of subgenus *Acystopteris*. *Cystopteris tenuisecta* and *C. japonica* (Fig. 12) have spores with fused strands forming bulbous apical structures that protrude from the surface.

Cytology

Chromosome numbers in *Cystopteris* range from diploid $n = 42$ to octoploid $n = 168$, with the tetraploid level reported most frequently. Records of tropical American species include polyploids in *C. fragilis,* a tetraploid in Jamaica (Walker, 1966) and two hexaploids, one from Peru and another from Costa Rica (as *C. diaphana,* Blasdell, 1963). Cytological analyses of *Cystopteris fragilis* in Europe (Vida, 1974) are based on study of pairing relationships. In that work a synthesized apogamous diploid in which somatic doubling of cells was detected suggested the possibility of somatic polyploidization in this complex. Crosses made between diploids and tetraploids resulted in formation of few bivalents, while those involving hexaploids or crosses between tetraploids yielded greater chromosome pairing. It is possible that increased pairs in crosses involving higher polyploids was due to homeologous pairing. However, Vida's analyses are based on the premise that lack of pairing is indicative of unrelated genomes. The explanation for the origin of four tetraploids belonging to the *Cystopteris* complex in Europe by Vida involves three unknown diploids. Cytological examination of the American elements in the complex may reveal the missing diploids and provide a broader base for interpreting relationships in the genus.

Literature

Blasdell, R. F. 1963. A monographic study of the fern genus *Cystopteris.* Mem. Torrey Bot. Cl. 21 (4): 1–102.

Hagenah, D. J. 1961. Spore studies in the genus *Cystopteris* 1. The distribution of *Cystopteris* with non-spiny spores in North America. Rhodora 63: 181–193.

Jermy, A. C., and L. Harper. 1971. Spore morphology of the *Cystopteris fragilis* complex. Brit. Fern Gaz. 10: 211–213.

Lugardon, B. 1971. Reference under the tribe Physematieae.

Pearman, R. W. 1976. A scanning electron microscope investigation of the spores of the genus *Cystopteris.* Brit. Fern Gaz. 11: 221–230.

Pichi-Sermolli, R. E. G. 1968. Fragmenta Pteridologiae I. Webbia 23: 159–207.

Vida, G. 1974. Genome analysis of the European *Cystopteris fragilis* complex. Acta Bot. Acad. Sci. Hungaricae 20: 181–193.

Walker, G. T. 1966. Reference under the tribe Physematieae.

86. *Gymnocarpium*
Figs. 86.1–86.5

Gymnocarpium Newm., Phytol. 4: 371. 1851. Type: *Gymnocarpium Dryopteris* (L.) Newm. (*Polypodium Dryopteris* L.).
Currania Copel., Phil. Jour. Sci. 4: 112. 1909. Type: *Currania gracilipes* Copel. = *Gymnocarpium gracilipes* (Copel.) Ching = *Gymnocarpium oyamense* (Baker) Ching.

Description

Terrestrial or rupestral; stem long-creeping, slender, bearing scales and usually few slender, fibrous roots; leaves monomorphic, ca. 10–50 cm long, borne at intervals, lamina deeply pinnatifid or 2-pinnate-pinnatifid to 3-pinnate-pinnatifid, glabrous to glandular, veins free; sori round to elongate, borne on an unmodified vein, near or well back of the tip, not paraphysate, exindusiate; spores more or less ellipsoidal, monolete, the laesura $\frac{3}{4}$ the spore length, irregularly reticulate, or with low, cristate or winglike folds. Chromosome number: $n = 40, 80$; $2n = 80, 160$.

A peculiarity of the genus, emphasized by Ching (1933) is the jointed, although not disarticulate, basal pinnae that evidently accommodates for the unusual orientation of the lamina which is more or less at right angles to the petiole.

Comments on the Genus

Gymnocarpium is largely circumboreal and includes three to six species, with two or three in the United States and Canada. It is characterized by its long, slender stem, exindusiate sori (Fig. 3) and chromosome number based on 40. The relation of the genus is with the tribe Physematieae, on the basis of the two vascular bundles in the petiole, and it appears to be near *Athyrium* and *Cystopteris*.

Sarvela (1978) recognizes six species in two subgenera: subgenus *Gymnocarpium* with two main species, *G. Dryopteris* (L.) Newm. (Fig. 2) and *G. Robertianum* (Hoffm.) Newm., and three others, *G. jessoense* (Koidz.) Koidz., *G. Fedtschenkoanum* Pojark and *G. remote-pinnatum* (Hayata) Ching, that are closely related and not markedly distinct. Subgenus *Currania* (Copel.) Sarvela is treated as monotypic based on *G. oyamense* (Baker) Ching. Four hybrids are reported (Sarvela, 1980; Sarvela et al., 1981) involving *Gymnocarpium Dryopteris, G. jessoense* and *G. Robertianum*.

Gymnocarpium grows in rocky habitats, talus slopes and at the base of cliffs or rocks and is terrestrial in forests, or sometimes in wet woods, or in sphagnum, in both calcareous and acidic substrates. It is broadly distributed in North America (Fig. 1) extending southward in the west to the mountains of Arizona and New Mexico and in the east in the mountains to North Carolina. It is wide ranging in Eurasia extending south and east to the Himalayas, Japan, Formosa, the Philippine Islands and New Guinea.

The irregular, verrucate, or low winglike contours of *Gymnocarpium* spores have a reticulate surface (Fig. 4). A rugose formation that underlays the reticulum is apparent in the eroded spore at the upper left. The detail of the surface structure (Fig. 5) shows the complex reticulate formation and the granulate

Fig. 86.1. Distribution of *Gymnocarpium* in America.

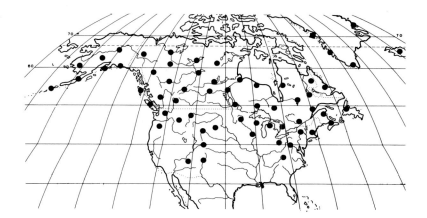

Fig. 86.2. Plant of *Gymnocarpium Dryopteris,* × 0.3.

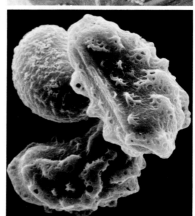

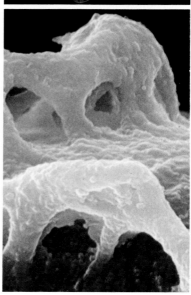

composition of the surface. The large spores of the Asiatic species, *Gymnocarpium oyamense,* with verrucate contours and reticulate surface shown in Chang et al. (1976) are similar to the species in America and support treatment of the species here rather than as a distinct genus *Currania.*

The initial cytological studies by Manton (1950) showing *Gymnocarpium* species of Europe with $n = 80$, provided new evidence for recognition of this as one of the distinct genera formerly treated as *Dryopteris (sens. lat.).* Additional cytological work reports diploids as well as tetraploids in two species. In *G. jessoense* the diploid, subspecies *jessoense* is reported from Japan and the circumboreal tetraploid has been distinguished as subspecies *parvulum* Sarvela. In *G. Dryopteris* the diploid, $n = 40$, subspecies *disjunctum* (Schur.) Sarvela is known from the American Pacific northwest while the tetraploid subspecies *Dryopteris* is wide ranging. Hybridization and apparently apogamy are also involved in these species complexes. The cytological records of *Gymnocarpium Robertianum* are consistently tetraploid. *Gymnocarpium oyamense* is also tetraploid with $n = 80$ and this supports its position in the genus rather than in *Currania* (Kurita, 1962).

Literature

Chang, Y. L., et al. 1976. Reference under the tribe Physematieae.

Ching, R. C. 1933. On the nomenclature and systematic position of *Polypodium Dryopteris* L. and related species. Contrib. Biol. Lab. Sci. Soc. China 9: 30–43.

Kurita, S. 1962. Chromosome numbers of some Japanese ferns 3. Jour. Col. Arts & Sci. Chiba Univ. 3: 463–468.

Manton, I. 1950. Reference under tribe Physematieae.

Sarvela, J. 1978. A synopsis of the fern genus *Gymnocarpium.* Ann. Bot. Fennici 15: 101–106.

Sarvela, J. 1980. *Gymnocarpium* hybrids from Canada and Alaska. Acta Bot. Fennici 17: 292–295.

Sarvela, J., D. M. Britton, and K. Pryer. 1981. Studies on the *Gymnocarpium Robertianum* complex in North America. Rhodora 83: 421–431.

Figs. 86.3–86.5. *Gymnocarpium Dryopteris.*   3. A segment with exindusiate sori and free veins, × 8.   **4–5.** Spores.   **4.** Proximal surface at upper right with central laesura, lateral view below, eroded spore with rugose lower perispore, upper left, Nova Scotia, *Bartram & Long 23028,* × 1000.   **5.** Detail of particulate perispore deposit, Maine, *Kennedy* in 1902, × 10,000.

87. *Woodsia*
Figs. 87.1–87.9

Woodsia R. Br., Prod. Fl. Nov. Holl. 158, obs. iv sub *Alsophila*. 1810, as *Woodia;* also, and correct in Trans. Linn. Soc. London 11: 171, 173. Type: *Polypodium ilvense* (L.) Vill. (*Acrostichum ilvense* L.) = *Woodsia ilvensis* (L.) R. Br.

Physematium Kaulf., Flora 12: 341. 1829. Type: *Physematium molle* Kaulf. = *Woodsia mollis* (Kaulf.) J. Sm.

Hymenocystis C. A. Meyer, Verz. Pl. Caucas. 229, 1831. Type: *Hymenocystis caucasica* C. A. Meyer = *Woodsia caucasica* (C. A. Meyer) J. Sm. = *Woodsia fragilis* (Trev.) Moore.

Trichocyclus Dulac, Fl. Dept. Haute-Pyrenees 31. 1867, *nom. superfl.* for *Woodsia* R. Br. and with the same type.

Protowoodsia Ching, Lingnan Sci. Jour. 21: 36. 1945. Type: *Protowoodsia manchuriensis* (Hook.) Ching = *Woodsia manchuriensis* Hook.

Description Terrestrial or usually rupestral; stem erect to decumbent and usually very short-creeping, small, bearing scales and many, slender fibrous roots; leaves monomorphic, ca. 5–50 cm long, borne in a cluster, 1-pinnate to 1-pinnate-pinnatisect or rarely 2-pinnate-pinnatifid, usually glandular and/or pubescent and/or scaly or sometimes glabrous, veins free; sori rather round, borne on a vein back of or near its tip, on a hardly to somewhat raised receptacle, not paraphysate, covered or not by an indusium surrounding the receptacle and of diverse form, globose, or of a few large to small segments, or composed of a few trichomes; spores ellipsoidal, to spheroidal, monolete, the laesura $\frac{1}{2}$ to $\frac{3}{4}$ the spore length with winglike folds that may be short, anastomosing and more or less echinate, or the surface with a regular reticulum and few folds. Chromosome number: $n = 33, 38, 39, 41, 76, 82$; $2n = 66, 78, 82, 152, 156, 164$.

Several species, including the well-known northern ones, *Woodsia ilvensis* (L.) R. Br., *W. alpina* (Bolton) S. F. Gray, and *W. glabella* R. Br., have a jointed petiole at which the leaf disarticulates.

Systematics *Woodsia* is primarily a circumboreal genus of about 25 species with nearly half in America and two or four in the American tropics. The most distinctive feature of *Woodsia* is the indusium that surrounds the receptacle. In its greatest development it is globose and more or less encloses the sorus, even at maturity of the sporangia.

Woodsia was previously considered to have an affinity with the Cyatheaceae, a relationship that is not supported by the different form and surface features of the spores and the cytology. Holttum (1949) proposed a relationship with *Dennstaedtia*, which was supported by Kurita (1965) on the basis of the cytology. However, that affinity is not indicated by the form or surface formation of the spores, the structure of the indusium and petiole, and other features. The spore characters, chromosome numbers and two vascular bundles in the petiole suggest alliances in the Dryopteridaceae perhaps close to *Cystopteris*.

Three subgenera were recognized by Hooker (1846) on the basis of the degree of development of the indusium, and Ching

Fig. 87.1. *Woodsia montevidensis,* east of Zumbagua, Cotopaxi, Ecuador. (Photo W. H. Hodge.)

(1932) recognized three sections and two subsections on the same basis and the articulate or continuous petiole. However, the indusium evidently is not a reliable character for species relationships and the petiole articulation does not correlate sufficiently with other characters to provide a basis for an infrageneric classification.

Woodsia has been revised by Brown (1964).

Tropical American Species

The species of *Woodsia* in tropical America are difficult to distinguish with precision due to variation of the indusium and the indument.

Woodsia Plummerae Lemmon of the southwestern United States and Mexico has an indusium with few or no multicellular trichomes on the indusial segments, while *W. mexicana* Fée, with a similar range, has several to many long trichomes on the indusium, easily visible among the mature sporangia. These partly sympatric species are very close and both are similar to *Woodsia oregana* D. C. Eaton that has a more northern distribution but is also in the southwestern United States. The three clearly form a closely related complex in the genus.

Woodsia mollis (Kaulf.) J. Sm. (Fig. 3), of Mexico and Guatemala, has a globose indusium (Fig. 4), sometimes split into sev-

Fig. 87.2. Distribution of *Woodsia* in America, south of lat. 35° N.

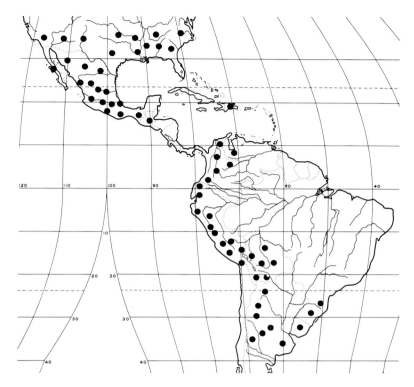

eral segments, more or less enclosing the mature sorus, while *W. montevidensis* (Spreng.) Hieron., of Guatemala, the Dominican Republic and South America, usually has smaller indusial segments, mostly concealed by the mature sorus. However, in northern South America, *W. montevidensis* sometimes has a well-developed indusium. Together these differ from the *Woodsia oregana* complex by the presence of rather long, multicellular trichomes on the lamina, although these sometimes also occur on the segment margins of *W. mexicana*. The combined range of *Woodsia mollis* and *W. montevidensis* is similar to that of several xeric ferns with a Mexican-Andean distribution, suggesting that, in the absence of distinctive characters, they may be a single, partly differentiated, species.

Ecology (Fig. 1)

Woodsia grows in forests, in woodlands, on shrubby hillsides and in open sparsely vegetated habitats nearly always associated with rocks, on cliffs, talus slopes, on ledges, on boulders, old stone walls and in rocky soil. It rarely grows in soil not associated with rocks. It grows in moist but often seasonally cold or dry habitats.

In tropical America, *Woodsia* grows in the mountains of Mexico and Guatemala, Hispaniola, and the Andes south to Bolivia, usually at elevations of 1500–3000 m, although sometimes ascending to 4300 m in Peru and Bolivia. It grows nearer sea level at 300 m or somewhat higher on the costal lomas of Peru, and in the southern part of its range to 200 m in Argentina; also in the north in the United States and Canada it descends to near sea level.

Fig. 87.3.. Leaf of *Woodsia mollis* and part of the stem, ×0.5.

Geography (Fig. 2)

Woodsia is widely distributed in North and South America, in northern and central Europe and eastward to the Caucasus and across Asia to Kamchatka, and from the Himalayas and China to Japan; also isolated in southern Africa.

In America, *Woodsia* grows from about lat. 78° N south to Mexico and Guatemala, in the Dominican Republic, and in South America from western Venezuela and Colombia south to central Argentina.

Three American species are more or less circumboreal in their distribution, *Woodsia alpina*, *W. ilvensis* and *W. glabella*. Seven endemics occur in America, 11 from the Himalayas to Siberia, two in southern Africa, and *Woodsia fragilis* (Trev.) Moore is endemic to the Caucasus.

Spores

Woodsia spores tend to have a somewhat spheroidal form as in *W. montevidensis* (Fig. 6) in comparison to most monolete spores. The surface of low, largely anastomosing, somewhat echinate folds (Figs. 5–7, 9) is relatively uniform in most species. A well-developed reticulate formation overlays a fairly smooth exospore (Fig. 8). The spores of *W. Andersonii* (Bedd.) Christ, of China, have a reticulate surface and few, low folds. The surface and wall structure of *Woodsia* spores is generally similar to those with winglike folds in other genera of the Dryopteridaceae.

Cytology

The range of chromosome numbers reported for *Woodsia* is unusually wide although the frequency of 41 or its multiples support its treatment among the Dryopteridaceae. Records of 38 and 39 in the genus are regarded by Lovis (1977) as evidence of aneuploid reduction, although the derivation of $n = 33$, in *W. manchuriensis* Hook., would involve more complex changes. This species along with *Woodsia elongata* Hook. was placed in a group of Asiatic species with chromosome numbers of $n = 41$ (Kurita, 1965).

Literature

Brown, D. F. M. 1964. A monographic study of the fern genus *Woodsia*. Beihefte Nova Hedwigia 16: 1–154.

Ching, R. C. 1932. The studies of Chinese ferns, VIII, 1. *Woodsia* R. Brown. Sinensia 3: 131–154.

Holttum, R. E. 1949. The classification of ferns. Biol. Reviews 24: 267–296.

Hooker, W. J. 1846. Species Filicum. 1: 59–64. William Pamplin, London.

Kurita, S. 1965. Chromosome numbers and systematic position of the genus *Woodsia*. Jour. Jap. Bot. 40: 358–362.

Lovis, J. D. 1977. Reference under the tribe Physematieae.

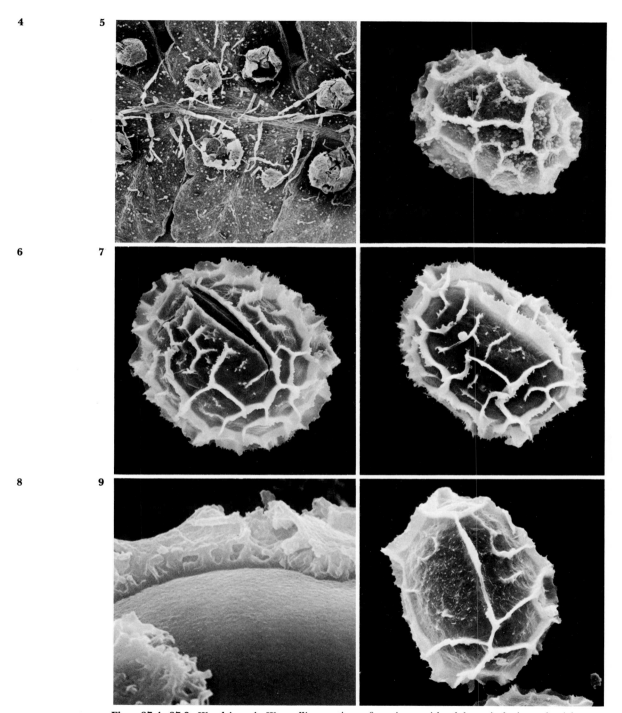

Figs. 87.4–87.9. Woodsia.   4. W. mollis, portion of a pinna with globose indusia and trichomes, ×12.   **5–9.** Spores ×1000, profile, ×5000.   **5. W. Plummerae,** lateral view anastomosing folds, Mexico, *Pringle 836.*   **6, 7. W. montevidensis,** Peru.   **6.** Proximal surface split along the laesura, *Correll & Smith P188.*   **7.** Lateral surface with echinate folds, *Woytkowski 5252.*   **8. W. elongata,** profile of reticulate perispore and smooth exospore below, India, *Strachey & Winterbottom 1.*   **9. W. polystichoides** D. C. Eaton, elongate, low folds, Japan, *Kodama 6769.*

18c.   Tribe Onocleeae

Onocleeae Tod., Syn. Pl. Vasc. Sicil. 29. 1866 and Giorn. Sci. Nat. Econ.
Palmero 1: 234. 1866, as Onocleideae. Type: *Onoclea* L.
Onocleaceae Pic.-Ser., Webbia 24: 708. 1970. Type: *Onoclea* L.

A tribe of three distinctive genera, all American and two also
in the Old World. Kato and Sahashi (1980) have revised the classification of the onocleoid ferns by including *Onocleopsis* in *Matteuccia* and *M. orientalis* in *Onoclea,* primarily on the basis of stem
and leaf characters. However, differences in complexity and in-
duration of the fertile leaves of *Onocleopsis* and *Matteuccia Struthiopteris,* and differences in chromosome numbers of *Onoclea* and
Matteuccia orientalis are not considered. The chlorophyll bearing
spores of the onocleoids have remarkably rigid walls with a fairly
well-developed perispore of two strata. Differences in prominence and fusion of the surface elements are indicated in the generic treatments. The surface structure is basically similar in the
tribe, and to some extent, differences as observed by Kato and
Sahashi may be due to the developmental condition of the material. Present data supports the classification of Lloyd (1971)
rather than the proposed changes of Kato and Sahashi.

Literature

Kato, M., and N. Sahashi. 1980. Affinities in the onocleoid ferns. Acta
Phytotax. Geobot. 31: 127–138.
Lloyd, R. M. 1971. Systematics of the onocleoid ferns. Univ. Cal. Publ.
Bot. 61: 1–81.

88.   *Onocleopsis*
Figs. 88.1–88.8

Onocleopsis Ballard, Amer. Fern Jour. 35: 1. 1945. Type: *Onocleopsis
Hintonii* Ballard.

Description   Terrestrial; stem erect, stout, to ca. 1 m tall, sometimes stoloniferous, bearing a few brownish scales and, basally, long fibrous
roots; leaves strongly dimorphic (the sterile ca. 0.5–1.75 m long,
1-pinnate with the pinnae slightly to rather deeply lobed, the
fertile shorter, 3-pinnate or nearly so, with much reduced laminar tissue), borne in a crown, slightly scaly and short-pubescent,

Fig. 88.1. Distribution of *Onocleopsis*.

the sterile pinnae with anastomosing veins without included veinlets; sori borne on free veins, wholly enclosed by the papyraceous fertile segments which irregularly rupture at maturity, the sporangia borne at different levels on an erect receptacle, enclosed when young by a very thin, hyaline, fugacious indusium; spores ellipsoidal, monolete, the laesura $\frac{1}{2}$ the spore length, with prominent folds, the surface minutely echinate-rugose, more or less fused at the base. Chromosome number $n = 40$.

The sterile lamina (Fig. 2) is gradually reduced at the base, the pattern of the anastomosing veins (Fig. 3) differs from *Onoclea,* and the margins of the pinnae are serrate. The fertile leaves are herbaceous, not indurated or long-persistent as in *Matteuccia* and *Onoclea.* The highly modified fertile pinnules (Fig. 4) enclose the sori (Fig. 5) until maturity.

Systematics

Onocleopsis is a distinctive tropical American genus of one species, *O. Hintonii* Ballard. It was first collected in Mexico by Pringle in 1896 and by Conzatti and Gonzáles in 1897, but was not recognized as a significant new kind of fern until the George B. Hinton collections were made and studied about 40 years later.

As the least dimorphic of the onocleoid ferns, *Onocleopsis* is evidently the most primitive, but it is sufficiently specialized that it does not provide evidence for a close relationship with other dryopteroid ferns.

Ecology and Geography (Fig. 1)

Onocleopsis grows in deeply shaded, small canyons along streams, either in wet soil or in running water (Ballard, 1948), at ca. 2000–2500 m. It occurs in southern Mexico in the states of Mexico, Morelos, Guerrero and Oaxaca, and in the Departments of San Marcos and Jalapa in Guatemala.

Fig. 88.2. Sterile leaf of **Onocleopsis Hintonii,** × 0.125.

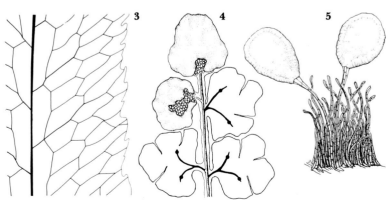

Figs. 88.3–88.5. Onocleopsis Hintonii. 3. Portion of sterile pinna, × 2.5. **4.** Apical portion of a fertile pinnule with three segments open to show the veins and insertion of the sori (black dots), × 8. **5.** Sorus, only two capsules shown, × 40.

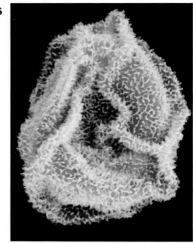

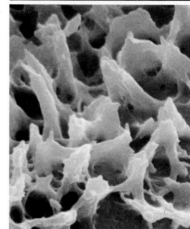

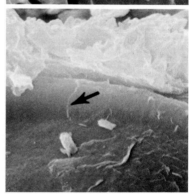

Spores

Onocleopsis spores have prominent, coarse folds (Fig. 6) and minutely echinate-rugose surface elements fused at the base (Fig. 7). Short ridges form somewhat more connected units than the echinate forms in *Onoclea* or the compact, cristate ones in *Matteuccia*. Kato and Sahashi (1980) describe this as muri-like ridges interrupted at random. The lower perispore layer is thin and slightly rugose (Fig. 8) as in the spores of the other onocleoid genera.

Cytology

A report of 40 bivalents at metaphase I (Lloyd, 1971) is based on a collection from Oaxaca, Mexico. The same number is known for several collections of *Matteuccia* and suggests there is a closer relationship to this genus than to *Onoclea* with $n = 37$.

Literature

Ballard, F. 1948. Further notes on *Onocleopsis*. Amer. Fern Jour. 38: 125–132.

Kato, M., and N. Sahashi. 1980. Reference under the tribe Onocleeae.

Lloyd, R. M. 1971. Reference under the tribe Onocleeae.

Figs. 88.6–88.8 *Onocleopsis Hintonii* spores. **6.** Coarse folds and minutely echinate-rugose surface, Oaxaca, Mexico, *Hinton 16314*, × 1000. **7.** Detail of echinate-rugose surface, Mexico, *Hinton 16314*, × 10,000. **8.** Abraded surface showing the thin lower perispore (arrow) over part of the laesura, outer perispore at top, Oaxaca, Mexico, *Conzatti & Gonzales 480*, × 5000.

89. *Matteuccia*

Figs. 89.1–89.9

Matteuccia Tod., Syn. Pl. Vasc. Sicil. 30. 1866, and Giorn. Sci. Nat. Econ. Palmero 1: 235. 1866, *nom. conserv.* Type: *Matteuccia Struthiopteris* (L.) Tod. (*Osmunda Struthiopteris* L.).

Struthiopteris Willd., Berl. Mag. 3: 160, 1809, not Scop. 1760 (=*Blechnum*) nor Bernh. 1802 (=*Osmunda*). Type: *Osmunda Struthiopteris* L. = *Struthiopterus germanica* Willd. = *Matteuccia Struthiopteris* (L.) Tod.

Pteretis Raf., Amer. Monthly Mag. & Crit. Rev. 2: 268. 1818, *nom. nov.* for *Struthiopteris* Willd. and with the same type.

Pentarhizidium Hayata, Bot. Mag. Tokyo 42: 345. 1928. Type: *Pentarhizidium japonicum* Hayata = *Matteuccia orientalis* (Hook.) Trev.

Description

Terrestrial; stem erect, rather stout, to ca. 20 cm tall, and (in *M. Struthiopteris*) stoloniferous, bearing a few brown scales and, basally, many, long, fibrous roots; leaves strongly dimorphic (the sterile ca. 0.5–2 m, or more, long, 1-pinnate-pinnatifid, the fertile somewhat to much shorter, 1-pinnate, or 1-pinnate-pinnatifid at maturity, with indurated pinnae), borne in a crown, finely pubescent, with a few scales, veins free, sori borne on the veins, wholly enclosed by the strongly modified pinna which opens and sometimes ruptures at maturity, not paraphysate, the sporangia are borne at different levels on an erect receptacle, more or less enclosed by a very thin, hyaline, somewhat persistent, indusium, or exindusiate; spores ellipsoidal, monolete, the laesura $\frac{1}{2}$ the spore length, with few large folds, the surface minutely cristate to somewhat echinate, the elements often fused forming a compact surface. Chromosome number: $n = 39, 40, 80; 2n = 78, 80$.

Comments on the Genus

Matteuccia is a distinctive temperate and boreal genus of two or three species. *Matteuccia Struthiopteris* (L.) Tod. is disjunctly circumboreal, and *M. orientalis* (Hook.) Trev. occurs in eastern Asia southward to the Himalayas. *Matteuccia intermedia* C. Chr. may be a distinct species or it may be a small form of *M. orientalis*.

The leaf bases are so closely placed on the erect stem that it is not clear whether it bears scales. Large, thin brown scales are especially abundant on the petiole base and cover the young crozier. The sterile lamina has free veins (Fig. 5) and regularly pinnatifid pinnae. The fertile leaves are indurated and brown at

Fig. 89.1. Distribution of Matteuccia in America.

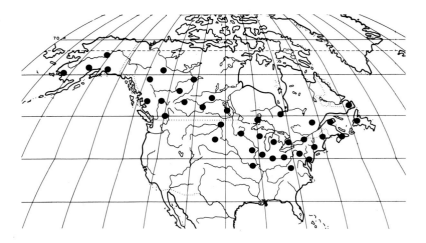

6

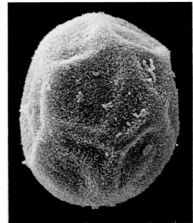

7

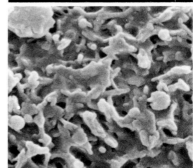

8

9

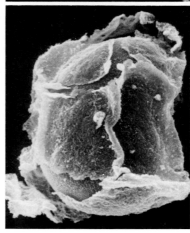

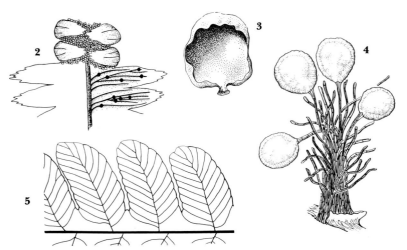

Figs. 89.2–89.5. *Matteuccia Struthiopteris.* **2.** Portion of fertile pinna opened to show the veins and insertion of the sori (black dots), ×4. **3.** Indusium from immature sorus, ×12. **4.** Sorus and base of the indusium, only four capsules shown, ×30. **5.** Portion of a sterile pinna, ×2.5.

maturity; they may persist, in *M. Struthiopteris,* for about two years. The highly modified fertile pinnae (Fig. 2) enclose the sori (Fig. 4) until maturity. The sorus is sometimes partially enclosed by a rather persistent indusium (Fig. 3).

In America *Matteuccia* occurs from Labrador and Newfoundland to Virginia and west to British Columbia and Alaska (Fig. 1). It grows in wet habitats or in moist, shady places, sometimes in periodically inundated streams valleys where the erect stems may be nearly covered by alluvial deposition.

Matteuccia spores have few, coarse folds and obscure, minute, cristate to echinate surface elements (Fig. 6) generally similar but more compact than those of *Onoclea* or *Onocleopsis* (Fig. 7). The perispore is formed of two distinct strata—a thicker outer surface overlays a thin, rugulose layer—above a fairly smooth exospore (Fig. 8). The spore of *Matteuccia orientalis* (Fig. 9), at lower magnification, may represent a tetraploid for it is nearly twice as large as the spore of *M. Struthiopteris.* Differences in wall structure shown in the work of Kato and Sahashi (1980) appear to represent different developmental stages than the more fully formed spores figured here.

Figs. 89.6–89.9. *Matteuccia* spores. **6–8.** *M. Struthiopteris.* **6.** Spore with coarse, low folds and minutely cristate to echinate surface, Massachusetts, *Tryon & Tryon* May, 1978, × 1000. **5.** Detail of compact cristate-echinate surface, *Tryon & Tryon* May, 1978, ×10,000. **8.** Abraded wall, the compact outer surface, right, the thin slightly rugulose lower perispore, left, smooth exospore at center, Massachusetts, *Fernald* in 1910, × 5000. **9.** *M. orientalis,* few prominent folds the echinate surface broken with lower rugulose perispore below, China, *Wang 20657,* ×600.

Chromosome numbers of *Matteuccia Struthiopteris* from Canada, Europe, and Japan are mostly $n = 40$ but there is record of a cultivated specimen with $n = 39$ and a collection with $2n = 78$ from Canada (Löve and Löve, 1976). Several reports of *M. orientalis* from Asia are diploid and a tetraploid $n = 80$ is based on a collection from Japan (Lloyd, 1971).

Literature

Kato, M., and N. Sahashi. 1980. Reference under the tribe Onocleeae.
Lloyd, R. M. 1971. Reference under the tribe Onocleeae.
Löve, A., and D. Löve, 1976. Reports, *in*: A. Löve (ed.), IOBP Chromosomes number reports, LIII. Taxon 25: 484–487.

90. *Onoclea*
Figs 90.1–90.7

Onoclea L., Sp. Pl. 1062. 1753; Gen. Pl. ed. 5, 484. 1754. Type: *Onoclea sensibilis* L.
Angiopteris Adans., Fam. Plts. 2: 21, 518. 1763, not Hoffm. (Marattiaceae) *nom. superfl.* for *Onoclea* and with the same type.
Calypterium Bernh., Jour. Bot. (Schrad.) 1801 (1): 22. 1802, *nom. superfl.* for *Onoclea* and with the same type.
Riedlea Mirb., Hist. Nat. Pl. 4: 65. 1803, and Hist. Nat. Veg. 5: 71. 1803, *nom. superfl.* for *Onoclea* and with the same type.
Ragiopteris Presl, Tent. Pterid. 95, *t. 3, f. 9, 10*. 1836. Type: *Ragiopteris onocleoides* Presl = *Onoclea sensibilis* L. (*Ragiopteris onocleoides*, type: Willd. Herb. 19835, sheet 3, the fertile leaf *Onoclea sensibilis*, the sterile pinna = *Dryopteris Filix-mas*).
Pteridinodes O. Ktze., Rev. Gen. Pl. 2: 819. 1891, *nom. superfl.* for *Onoclea* and with the same type.

Description Terrestrial; stem long-creeping, rather slender, bearing a few broad, thin scales and numerous, long, fibrous roots; leaves strongly dimorphic (the sterile ca. 0.2–1 m long, or more, deeply pinnatifid, bipinnatifid or 1-pinnate-pinnatifid, most complex basally, the fertile somewhat to much shorter, 2-pinnate, with small, globose, indurated pinnules), borne in a loose cluster to rather distantly, with a few scales, hardly pubescent, the sterile segments with anastomosing veins without included veinlets; sori borne on free veins, wholly enclosed by the pinnules which rupture at maturity into ca. 5 segments, not paraphysate, the sporangia borne at different levels on an erect receptacle, more or less enclosed by a very thin, hyaline somewhat persistent indusium; spores ellipsoidal, monolete, the laesura $\frac{1}{2}$ the spore length, with a few prominent folds, and minutely echinate-cristate. Chromosome number: $n = 37$; $2n = 74$.

Fig. 90.1. Distribution of *Onoclea* in America.

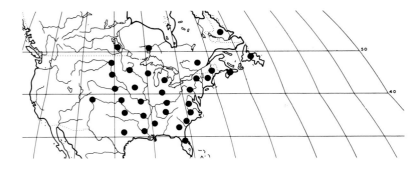

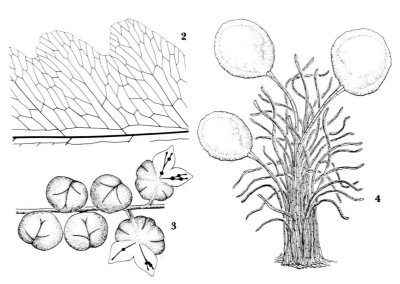

Figs. 90.2–90.4. *Onclea sensibilis.* **2.** Portion of a sterile pinna, ×3. **3.** Portion of a fertile pinna, two pinnules partly open to show veins and insertion of the sori (black dots), ×4. **4.** Sorus, only three capsules shown, ×40.

Comments on the Genus

Onoclea is a distinctive temperate and boreal genus of one species, *O. sensibilis* L. (Lloyd, 1971).

The sterile lamina has anastomosing veins (Fig. 2) of a different pattern than *Onocleopsis* and the margins of the segments are nearly smooth. The fertile leaves are indurated, brown at maturity and persist for a year or more after the spores are shed. The sori (Fig. 4) are enclosed in small, globose pinnules (Fig. 3) that rupture into segments to release the spores. In the northern part of the range the spores are shed in late winter.

Onoclea occurs in eastern Asia and in North America (Fig. 1) from Newfoundland and Laborador to Florida and west to Manitoba, Colorado and Texas. It grows in a variety of habitats but most commonly in marshes, swamps, and wet woods, less often in drier places.

The spores of *Onoclea* often have large particles scattered over the surface and few prominent folds (Fig. 5). The surface is generally similar to that of *Onocleopsis* or *Matteuccia* spores but with somewhat more discrete, minute, echinate-cristate elements (Fig. 6). The lower perispore is slightly rugulose (Fig. 7).

Recent records of chromosome numbers of collections from the northern United States and southern Canada and Japan are consistently $n = 37$ or $2n = 74$. Earlier discrepant records are

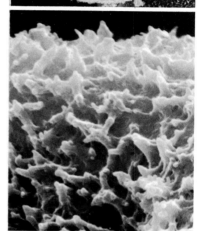

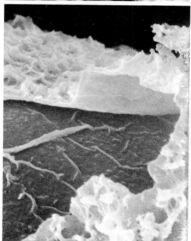

evidently inaccurate. The difference between this number and reports of *n* = 40 for *Matteuccia* indicate evolutionary divergence in chromosome numbers of these genera as well as in other characters.

The special fertile leaves of *Onoclea* supply a convenient source of spores for experimental work. The spores have been used to study stored substances, primarily lipids and proteins, and the role of these in providing energy during spore germination and early gametophyte development (DeMaggio and Stetler, 1980). *Onoclea* gametophytes have been used to study the effect of light on the growth of cells, and the causes of the switch from 1-dimensional, filamentous growth to 2-dimensional growth (Cooke and Paolillo, 1979; Miller, 1968, 1980).

Literature

Cooke, T. J., and D. J. Paolillo. 1979. The photobiology of fern gametophytes. 2. The photocontrol of filamentous growth and its implication for the photocontrol of the transition to two-dimensional growth. Amer. Jour. Bot. 66: 376–385.

DeMaggio, A. E., and D. A. Stetler. 1980. Storage products in spores of *Onoclea sensibilis* L. Amer. Jour. Bot. 67: 452–455.

Lloyd, R. M. 1971. Reference under the tribe Onocleeae.

Miller, J. H. 1968. Fern gametophytes as experimental material. Bot. Rev. 34: 361–440. (Includes numerous references.)

Miller, J. H. 1980. Orientation of the plane of cell division in fern gametophytes: The role of cell shape and stress. Amer. Jour. Bot. 67: 534–542.

Figs. 90.5–90.7. ***Onoclea sensibilis*** spores. **5.** Surface with prominent folds, few large, scattered, irregular particles and minute, echinate-cristate deposit, New Hampshire, *Harris & Bean 19562*, ×10,000. **6.** Detail of echinate-cristate structure basally fused, *Harris & Bean 19562*, ×10,000. **7.** Abraded surface with lower rugulose perispore (center darker) below a surface fold, Massachusetts, *Kennedy* in 1886, ×5000.

18d.   Tribe Oleandreae

Oleandreae J. Sm., Ferns Brit. For. 73. 1866. Type: *Oleandra* Cav. Oleandraceae (J. Sm.) Pic.-Ser., Webbia 20: 745. 1965.

A tribe of two or perhaps three genera with two in America. *Psamiosorus* C. Chr. is endemic to Madagascar and questionably distinct from *Arthropteris*.

91.   *Arthropteris*
Figs. 91.1–91.9

Arthropteris Hook. f., Fl. Nov.-Zealand. 2: 43. 1854. Type: *Arthropteris tenella* (Forst.) Hook. f. (*Polypodium tenellum* Forst.).

Description

Terrestrial, rupestral or usually scandent-epiphytic, or epiphytic; stem slender, very long, to 10 m or more, bearing scales, rarely a few trichomes, and few, slender, fibrous roots; leaves monomorphic, ca. 15–50, to 90 cm long, borne at intervals, lamina 1-pinnate, gradually reduced at the apex, or imparipinnate, with multicellular trichomes, veins free; sori roundish, borne on the abaxial side of the pinna, at the vein-tips or sometimes at the fork of a vein, on a slightly to definitely raised receptacle, not paraphysate, covered by an orbicular indusium with a deep sinus, or a reniform indusium, or exindusiate; spores spheroidal to ellipsoidal, monolete, the laesura $\frac{1}{2}$ to $\frac{3}{4}$ the spore length with irregular winglike or cristate folds and reticulate surface. Chromosome number: $n = 41$, ca. 42.

Comments on the Genus

Arthropteris is a genus of about 15 species, widely distributed although absent from continental America. A single species, *Arthropteris altescandes* (Colla) J. Sm. is endemic on the Juan Fernandez Islands (Fig. 2).

The stem is dorsiventral in *Arthropteris* and usually bears only scales but in *A. Palisotii* (Desv.) Alston some long trichomes are mixed among the scales. The leaf of *A. altescandens* is shown in Fig. 3 and the exindusiate sori in Fig. 4. The petiole is jointed close to or well beyond the stem and the leaves disarticulate at that place. The pinnae also disarticulate at basal joints and in some species with an imparapinnate lamina, the terminal segment is also jointed, suggesting the abortion of an original apex, as in some species of *Lomariopsis*.

The relationship of this genus with others is not clear. It has been placed in the davallioids near *Nephrolepis* but marked differences in the spores of the two genera do not support a close relationship. The prominent, irregular folds forming the spore surface are similar to those of *Lomariopsis* and these also are scandent plants with articulate pinnae, but differing in chromosome number. *Arthropteris* belongs in an alliance with the dryopteroids rather than the davallioides but its precise generic relationship is uncertain.

Arthropteris grows in wet or locally moist habitats, on the ground, on rotten logs, on rocks or on cliffs; it is usually scan-

Fig. 91.1. *Arthropteris altescandens,* Mas Afuera, Juan Fernandez Islands. (Photo F. G. Meyer.)

Fig. 91.2. Distribution of *Arthropteris* in America.

dent-epiphytic on tree trunks, sometimes wholly epiphytic. On the Juan Fernandez Islands it grows in dense woods, or rocks in woods, on fallen trees, and especially climbs on the Myrtaceous genus *Myrceugenia* (Fig. 1). It occurs at rather low altitudes about 100 to 400 m. *Arthropteris* occurs in tropical Africa and adjacent Arabia east to Ceylon, southern China and the Philippine Islands, south to eastern Australia and New Zealand and eastward through the Pacific islands to Tahiti and the Juan Fernandez Islands.

The irregular cristate surfaces of the four spores in Fig. 5 have connections between the cristate structure of adjacent spores, and also the slightly projecting laesura. A thin-stranded, reticulate perispore stratum below the cristate formation overlays a smooth exospore (Fig. 7). Spores of *A. altescandens* (Figs. 5–7) are generally similar to those of the Australian *A. tenella* (Forst.) Hook. f. (Figs. 8, 9) but with less elaborated surface structure. Several species of Asia have spores with intact wing-like folds rather than crests. Similarities of these spores with those of *Oleandra* and *Nephrolepis* are discussed by Liew (1977). However, the structure of the lower perispore, which is echinate in *Oleandra,* relatively even and compact in *Nephrolepis,* and reticulate in *Arthropteris,* is not considered.

The chromosome number *n* = ca. 42 is reported for *Arthropteris tenella* of New Zealand, by Brownlie (1958), and also several counts of *n* = 41 for species of Australia and Africa by Manton (1959). The number was considered by Manton as relevant to the position of *Arthropteris* near *Nephrolepis* in the system of Copeland. However, 41 is known in a number of genera in the Dryopteroidaceae and some in the Davalliaceae.

Literature

Brownlie, G. 1958. Chromosome numbers in New Zealand ferns. Trans. Roy. Soc. N. Z. 85: 213–216.

Liew, F. S. 1977. Scanning electon microscopical studies on the spores of Pteridophytes. XI. The family Oleandraceae (*Oleandra, Nephrolepis* and *Arthropteris*) Gard. Bull. Singapore. 30: 101–110.

Manton, I. 1959. Cytological information on the ferns of west tropical Africa. 75–81, *in*: A. H. G. Alston, The Ferns and Fern Allies of West Tropical Africa, pp. 1–89. Crown Agents, London.

Fig. 91.3. Leaf attached to stem of *Arthropteris altescandens,* × 0.5.

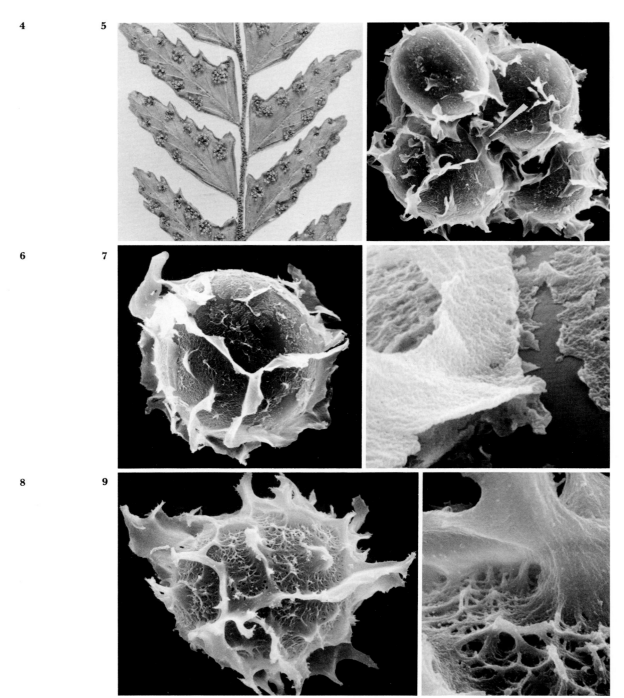

Figs. 91.4–91.9. *Arthropteris.* **4.** Pinnae and exindusiate sori of *A. altescandens,* ×2.5. **5–9.** Spores. **5–7.** *A. altescandens,* Juan Fernandez, *Solbrig, et al 3780.* **5.** Four spores, connections of cristae (arrow) on adjacent spores, the laesura across spore at top, left, ×600. **6.** Irregular cristate wings and reticulate surface, ×1000. **7.** Surface detail of finely reticulate strata overlaying smooth exospore, ×10,000. **8, 9.** *A. tenella,* Australia, *Constable P 6545.* **8.** Irregular cristate wings and reticulate surface, ×1000. **9.** Detail of reticulate structure at base of crest, ×5000.

92. *Oleandra*
Figs. 92.1–92.15

Oleandra Cav., Anal. Hist. Nat. 1: 115. 1799. Type: *Oleandra neriformis.* Cav.

Ophiopteris Reinw., Sylloge Plant. 2: 3. 1824. Type: *Ophiopteris verticillata* Reinw. = *Oleandra pistillaris* (Sw.) C. Chr.

Neuronia Don, Prod. Fl. Nepal. 6. 1825. Type: *Neuronia asplenioides* Don, *nom. superfl.* for *Aspidium Wallichii* Hook. = *Oleandra Wallichii* (Hook.) Presl.

Description

Terrestrial, rupestral, scandent-epiphytic or epiphytic; stem erect-branching or climbing or long-creeping, to 5 m long, slender, bearing scales, and few to sometimes many, often very long, roots; leaves monomorphic or (in *O. Werneri* Rosenst.) dimorphic (the fertile longer than the sterile and much narrower), ca. 10–50 cm long, entire, glabrate, scaly and/or pubescent, veins free; sori roundish, borne on the veins on a somewhat raised receptacle, not paraphysate, covered by a reniform or orbicular-peltate or orbicular indusium with a deep sinus; spores more or less spheroidal, monolete, the laesura $\frac{1}{2}$ to $\frac{3}{4}$ the spore length, with prominent winglike folds, these and adjacent areas echinate, or with low folds and rugulose. Chromosome number: $n = 40, 41,$ ca. 80.

Oleandra has indusiate sori borne on the free, usually evenly spaced veins (Fig. 3). The sori are sometimes in a regular row on each side of the costa or often are irregularly placed on the veins (Fig. 13). *Oleandra* also has well-developed phyllopodia (Fig. 4). Some of the diversity in the shape of the lamina is shown in Figs. 13–15.

Some specimens of *Oleandra distenta* Kze. have long, slender, tortuous trichomes which are numerous among the sporangia but relatively sparse on the lamina. These are not considered to be paraphyses since they are similar to the lamina trichomes, but possibly may represent an example of incipient development of paraphyses.

Systematics

Oleandra is a pantropical genus of perhaps 40 species. The articulate petioles, entire lamina and indusiate sori clearly distinguish it from all others. It is similar to *Arthropteris* in its cytology, the spores of some species, the scandent stem and articulate petioles. A relationship to *Tectaria* is suggested by the chromosome number $n = 40$ and the spore characters.

Many of the species of *Oleandra* have a set of characters that place them in one of two groups. *Oleandra articulata*, *O. hirta* and *O. tricholepis* Kze., for example, have a long-creeping, epiphytic, somewhat flexible stem with leaves at intervals, and spreading, rather thin, reddish brown scales with few or no long cilia. *Oleandra costaricensis* (Fig. 12), *O. pistillaris* (Sw.) C. Chr. and *O. oblanceolata* Copel. are examples of species with an erect to climbing rigid stem often bearing leaves in pseudowhorls and appressed, thickened, basally dark brown to blackish scales with many long, caducous, marginal cilia. In spite of these, and other differences, these groups have not been recognized as infrageneric

Fig. 92.1. *Oleandra articulata,* Turrialba, Costa Rica. (Photo W. H.
Hodge.)

taxa. In some species of the paleotropics the characters are variable, and in others such as *Oleandra Cumingii* J. Sm., *O. hirtella*
(Miq.) Kze., and *O. Wallichii* Presl, they do not correlate.

Morton (1968) changed the application of the name *Oleandra
articulata* from an American species to one of Africa and used
the name *O. nodosa* for the American species. De Joncheere
(1969) and others have shown that Morton's conclusion was
wrong and that *O. articulata* is correctly applied to the species of
the American tropics.

The American species of *Oleandra* were revised by Maxon
(1914).

Tropical American Species

The American species of *Oleandra* need reassessment on the
basis of the considerable materials obtained since Maxon's study
and the several species described since then. Quantitative characters and those of venation and indument are variable and
there may be few, widely distributed, American species, rather
than the several local ones Maxon recognized. The study of

Fig. 92.2. Distribution of *Oleandra* in America.

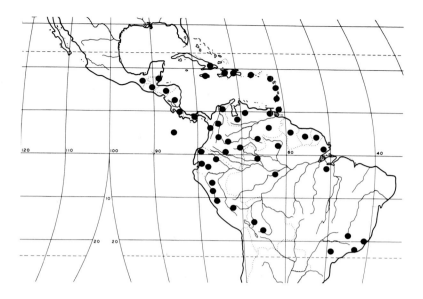

Pichi-Sermolli (1965) on *Oleandra distenta* of Africa points out the variation in many characters of that species and Harmata and Kornas (1978) demonstrate the variability in spore size.

In addition to the six species included in the key, the following are often recognized: *Oleandra Archeri* Maxon, *O. duidae* A. C. Smith, *O. guatemalensis* Maxon, and *O. Urbanii* Brause—allied to *O. Lehmannii; Oleandra decurrens* Maxon, *O. dura* Maxon, *O. panamensis* Maxon, and *O. trujillensis* Karst.—allied to *O. pilosa; Oleandra Baetae* Damaz.—allied to *O. hirta;* and *Oleandra zapatana* Lell.—allied to *O. articulata.*

Key to Major Species of American *Oleandra*

a. Stem long-creeping, with widely spreading scales, those at the stem apex rather loose with few if any long cilia (Fig. 5). b.
    b. Lamina glabrous or nearly so, costal scales may be present beneath. c.
        c. Stem with a whitish deposit.   *O. Bradei* Christ
        c. Stem brownish, without a whitish deposit.
            O. articulata (Sw.) Presl (*O. nodosa* (Willd.) Presl)
    b. Lamina pubescent, especially at the margins and base.   *O. hirta* Brack.
a. Stem ascending to erect-climbing, with closely appressed scales, those at the apex tightly appressed, with many very long, slender, whitish, tortuous, tangled, cilia (Fig. 6). d.
    d. Phyllopodia ca. 10–20 mm long, the barren ones obvious on the stem.
            O. Lehmannii Maxon
    d. Phyllopodia ca. 1–5 mm long, the barren ones hardly evident on the stem. e.
        e. Lamina glabrous, costal scales may be present beneath.
            O. costaricensis Maxon
        e. Lamina more or less pubescent, especially on the costa beneath.
            O. pilosa Hook.

Ecology (Fig. 1)

Oleandra grows in wet forests in humus, on rotten tree stumps and trunks, among boulders and on cliffs, frequently in acidic substrates. It is often scandent-epiphytic or epiphytic, sometimes in the crowns of trees. *Oleandra* also grows in forest clearings, in seasonally dry forests and in elfin forests.

In America, *Oleandra* grows in similar habitats, from sea level to ca. 2000 m, mostly from 500 to 1500 m.

Oleandra has long, branched stems with diverse habits. They may be terrestrial and creeping, or these may produce erect, branched stems to 1 or 2 m tall, forming a thicket. This shrubby habit is unique among ferns except for *Oleandropsis ferrea* (Brause) Copel. of New Guinea. Terrestrial plants may also produce scandent stems climbing on tree trunks to a height of at least 5 m. Some species are epiphytic and in these the stems are long-creeping among branches or rarely long-pendent, or erect.

Geography (Fig. 2)

Oleandra is distributed in tropical America, Africa and Madagascar, east to Ceylon and from the Himalayas and southern China south to Malaysia and eastward to the Philippine Islands, New Guinea and Queensland in Australia, and through the Pacific islands in New Hebrides, Fiji, Samoa and Tahiti.

In America, *Oleandra* occurs from Chiapas in southern Mexico through Central America, in the Antilles, in northern South America and in the Andes south to Bolivia, and is disjunct in southeastern Brazil; also on Cocos Island.

Spores

The winglike folds as in spores of *Oleandra Archeri* (Fig. 7) and relatively uniformly echinate, lower formation, are characteristic of the genus. The outer surface is lacking in the spores of *O. pilosa* (Fig. 9) but usually forms an intact surface of low folds as in *O. costaricensis* (Fig. 8). Connections between the echinate elements and the surface layer (Fig. 11), and the particulate formation (Fig. 10) show the complex elaboration of the spore wall. The spores of several paleotropical species, drawn by Braggio (1966) illustrate stratification similar to that of American species. Spores of *O. distenta* of Africa, examined with SEM (Harmata and Kornas, 1978), are also similar but show marked difference in size. The similarities of these spores and those of *Ctenitis, Asplenium* and especially *Tectaria* support the treatment of *Oleandra* among the dryopteroid rather than the davalloid genera.

Cytology

Oleandra articulata is reported as $n = 41$ from Jamaica and Trinidad (Walker, 1966) and there are reports of this same number for several species of India and Africa. There are less certain records of $n = $ ca. 80 for a cultivated plant of *O. distenta* and $n = $ ca. 40 for *O. musifolia* (Bl.) Presl from Ceylon (Manton and Sledge, 1954), but 40 is confirmed in a report of this species from India (Abraham et al., 1962). Aneuploid sequences as $n = 40$ and 41 in *Oleandra* are less frequent in genera of the Dryopteridaceae than in Thelypteridaceae or Dennstaedtiaceae.

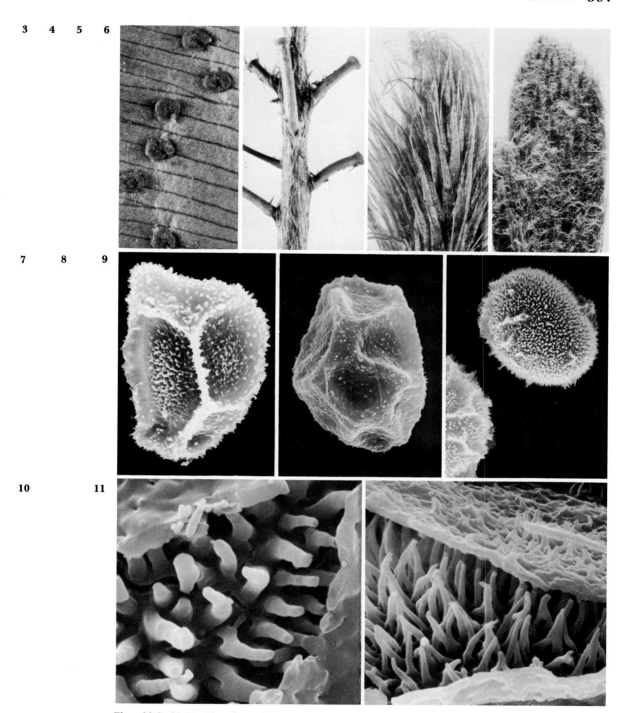

Figs. 92.3–92.11. *Oleandra*. 3. *O. articulata*, sori and veins, ×8. **4. *O. Lehmannii*,** phyllopodia on stem, ×2.5. **5. *O. articulata*,** stem apex, ×7. **6. *O. costaricensis*,** stem apex, ×7. **7–11. *Oleandra*** spores, ×1000. **7. *O. Archeri*** prominent winglike folds and echinate surface, Colombia, *Killip 11728*. **8. *O. costaricensis*** compact, low folds, Costa Rica, *Skutch 2961*. **9, 10. *O. pilosa*,** Peru, *Schunke 5194*. **9.** Inner echinate perispore, lacking the outer compact layer. **10.** Detail of inner echinate formation, the outer surface (right, upper left), ×10,000. **11. *O. articulata*,** portions of the rugulose-papillate outer surface (above and below) and lower echinate formation, Jamaica, *Maxon & Killip 544*, ×5000.

Figs. 92.12–92.15. *Oleandra.* **12.** Diagram of apical portion, 1 m long, of stem of *O. costaricensis;* roots are thin lines, leaf bases toward apex are solid triangles. **13.** Leaf of *O. articulata,* and irregularly placed sori, ×0.5. **14.** Leaf of *O. pilosa,* ×0.5. **15.** Leaf of *O. Lehmannii,* ×0.5.

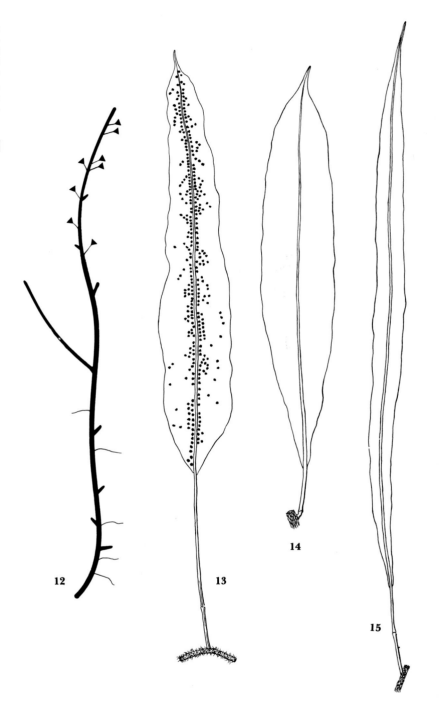

Observations

The roots of *Oleandra* may be long, extending at least 35 cm before branching in the substrate. Ogura (1972) considered the long, unbranched portion a "rhizophore" on the basis of anatomical differences in comparison to the branch roots. Wetter (1951), however, did not find significant differences. Further investigations should be focused on the influence of the humidity and soil substrate on the anatomical structure of this organ.

Literature

Abraham, A., C. Ninan, and P. Mathew. 1962. Studies on the cytology and phylogeny of the pteridophytes. Jour. Ind. Bot. Soc. 41: 339–421.

Braggio, G. 1966. Morfologio delle spore e sistematica delle Davalliales. Webbia 21: 725–767.

Harmata, K., and J. Kornas. 1978. Spore morphology of two varieties of *Oleandra distenta* (Davalliaceae, Filicopsida) from southern tropical Africa. Zesz. Nauk. Univ. Jagiellonskiego 193. Prac. Bot. 6: 7–14.

Joncheere, G. J. de 1969. The typification of *Oleandra articulata* (Filic.). Taxon 18: 538–541.

Manton, I., and W. A. Sledge. 1954. Observations on the cytology and taxonomy of the pteridophyte flora of Ceylon. Phil. Trans. Roy. Soc. Lond. Ser. B 238: 127–185.

Maxon, W. R. 1914. Studies of tropical American ferns, 5. The American species of *Oleandra*. Contrib. U.S. Nat. Herb. 17: 392–398.

Morton, C. V. 1968. The correct name of a common tropical American *Oleandra*. Amer. Fern Jour. 58: 105–107.

Ogura, Y. 1972. Comparative anatomy of vegetative organs of the pteridophytes, ed. 2. 502 pp. Borntraeger. Berlin.

Pichi-Sermolli, R. E. G. 1965. Adumbratio florae Aethiopicae. 11. Oleandraceae. Webbia 20: 745–769.

Walker, T. G. 1966. A cytotaxonomic survey of the pteridophytes of Jamaica. Trans. Roy. Soc. Edinburgh 66: 169–237.

Wetter, C. 1951. Über die Luftwurzeln von *Oleandra*. Planta (Berl.) 39: 471–475.

18e. Tribe Bolbitideae

Bolbitideae Pic.-Ser., Webbia 23: 381. 1969. Type: *Bolbitis* Schott.
Lomariopsidaceae Alston, Taxon 5: 25. 1956. Type: *Lomariopsis* Fée.
Elaphoglossaceae Pic.-Ser., Webbia 23: 209. 1968. Type: *Elaphoglossum* J. Sm.
Bolbitidaceae (Pic.-Ser.) Ching, Acta Phytotax. Sinica 16(3): 15. 1978.

A tribe of five genera, four in America. *Teratophyllum* Kuhn (*Arthrobotrya* J. Sm.) is primarily Malesian. *Thysanosoria* Gepp of New Guinea is considered to be an atavistic form of *Lomariopsis* by Hennipman (1977).

Literature

Hennipman, E. 1977. Reference under *Bolbitis*.

Holttum, R. E. 1978. *Lomariopsis* group, Flora Malesiana, Ser. II. Pteridophyta, 1(4): 255–330.

Walker, T. G. 1966. A cytotaxonomic survey of the pteridophytes of Jamaica. Trans. Roy. Soc. Edinburgh 66: 169–237.

93. *Bolbitis*
Figs. 93.1–93.17

Bolbitis Schott, Gen. Fil. *t. 13* (fasc. 3). 1835. Type: *Bolbitis serratifolia* (Kaulf.) Schott (*Acrostichum serratifolium* Kaulf.).

Egenolfia Schott, Gen. Fil. *t. 16* (fasc. 4). 1835 or 1836. Type: *Egenolfia Hamiltoniana* Schott = *Bolbitis appendiculata* (Willd.) Iwats.

Campium Presl, Tent. Pterid. 238. 1836. Type: *Campium virens* (Hook. & Grev.) Presl (*Acrostichum virens* Hook. & Grev.) = *Bolbitis virens* (Hook. & Grev.) Schott.

Poecilopteris Presl, Tent. Pterid. 241. 1836. Type: *Poecilopteris punctulata* (Sw.) Presl (*Acrostichum punctulatum* Sw.) = *Bolbitis auriculata* (Lam.) Alston.

Jenkinsia Hook., Gen. Fil. *t. 75B.* 1841. Type: *Jenkinsia undulata* Hook. = *Bolbitis virens* (Hook. & Grev.) Schott.

Cyrtogonium J. Sm., Jour. Bot. (Hook.) 4: 154. 1841, *nom. superfl.* for *Bolbitis* and with the same type.

Heteroneuron Fée, Mém. Fam. Foug. 2 (Acrost.): 20, 91. 1845. Type: *Heteroneuron argutum* Fée = *Bolbitis heteroclita* (Presl) C. Chr.

Lacaussadea Gaud., Voy. Bonite, *t. 118–120.* 1846. Type: *Lacaussadea montana* Gaud. = *Bolbitis appendiculata* (Willd.) Iwats.

Edanyoa Copel., Phil. Jour. Sci. 81: 22. 1852. Type: *Edanyoa difformis* Copel. = *Bolbitis heteroclita* (Presl) C. Chr.

Description   Terrestrial or usually rupestral, sometimes becoming scandent-epiphytic; stem short- to long-creeping, or scandent to ca. 15 m, slender to moderately stout, bearing scales and few to very many, long, sometimes slender, fibrous roots; leaves dimorphic (the fertile more erect than the sterile and with narrower segments), ca. 25 cm to 2 m long, borne in a cluster to widely spaced, lamina simple and entire to pinnatifid, 1-pinnate, 1-pinnate-pinnatifid, or (in *B. appendiculata*) the fertile 2- to 3-pinnate, sometimes imparipinnate, glabrate to somewhat scaly, veins usually anastomosing, with or without free included veins, or rarely free; sporangia usually borne over the abaxial surface of the segment, rarely only on and near the veins, no paraphyses, no indusium; spores more or less ellipsoidal, monolete, the laesura $\frac{3}{4}$ or less the spore length, usually with prominent, winglike to somewhat echinate folds, or reticulate. Chromosome number: $n = 41, 82; 2n = 82$.

The sporangia in *Bolbitis* are usually borne over the abaxial surface of the segments (Fig. 3) or rarely the border of the segment is sterile; or they are largely confined to the veins. The usually complex venation of the fertile segments (Fig. 4) and of the sterile segments may have all of the free veins directed toward the margin, the excurrent condition (Fig. 12), or some of the free veins may be recurrent, that is, directed toward the costa (Fig. 11). The diversity of the lamina and dimorphism of the fertile and sterile leaves are illustrated in Figs. 13–17.

Systematics   *Bolbitis* is a pantropical genus of 44 species including 14 in America. It has greater morphological diversity than other genera in the tribe Bolbitideae, especially in the paleotropical species. In America it is characterized by nonarticulate pinnae, by prominent lateral veins, and the absence of paraphyses, as well

Fig. 93.1. *Bolbitis portoricensis,* Turrialba, Costa Rica. (Photo W. H. Hodge).

as by a dorsiventral stem and stele. Hennipman (1977) arranged most of the species into 10 series with an additional 10 species of uncertain position. The general relation of *Bolbitis* is with *Lomariopsis* and *Lomagramma* and the usually terrestrial or rupestral habit suggests that it is less derived than either of those two genera.

This treatment of *Bolbitis* is based on the monograph of Hennipman (1977) and his review of our manuscript.

Tropical American Species

The lamina architecture in *Bolbitis* species is usually variable, depending on the age and size of the plant and on the growing conditions. However, a typical range of variation in the lamina architecture provides characters for recognition of species. Other important characters of the species are those of the stem, scales, bulbils and venation pattern.

Hennipman proposes twelve hybrids in *Bolbitis* and several others that are less certain. A single one in America, *Bolbitis nicotianifolia* × *portoricensis* is based on a collection from Guatemala, but other examples of *Bolbitis* hybrids will surely be recognized as additional field studies are made in tropical America.

Key to American Species of *Bolbitis*

a. High climbing plants, sterile leaves with the tertiary veins prominent throughout. b.
   b. Stem scales with entire margins.   *B. Lindigii* (Mett.) C. Chr.
   b. Stem scales with denticulate margins.   *B. Bernoullii* (Christ) C. Chr.
a. Terrestrial plants, or sometimes at the base of a tree, sterile leaves with the tertiary veins prominent only near the base of the costa or near the secondary veins. c.
   c. Sterile leaves entire to pinnatifid, or pinnate with the terminal segment usually conform to the pinnae, at least in shape, rarely narrowly triangular and much shorter than the remaining lamina. d.
      d. Sterile leaves pinnatifid. e.
         e. Sterile leaves with a subterminal bulbil, Mexico.
                                B. hastata (Fourn.) Hennip. (Fig. 15)
         e. Sterile leaves without a bulbil, South America.
                             B. semipinnatifida (Fée) Alston.
      d. Sterile leaves entire or pinnate. f.
         f. Sterile leaves entire and with a subterminal bulbil, or pinnate. g.
           g. Venation without recurrent, included free veins. h.
              h. Lamina with 4–8 pinnae, rachis narrowly winged.
                              B. serrata (Kuhn) C. Chr.
              h. Lamina with 9–30 or more pinnae, rachis not winged.
          B. serratifolia (Kaulf.) Schott (*B. crenata* (Presl) C. Chr.) (Fig. 13)
           g. Venation with recurrent and excurrent included free veins. i.
              i. Bulbils absent. j
                j Scales of the stem soft, rusty brown, bullate.
                         B. pergamentacea (Maxon) C. Chr.
                j. Scales of the stem firm, dark brown or blackish, flat, appressed.   *B. nicotianifolia* (Sw.) Alston
              i. Bulbils present. k.
                k. Bulbils subterminal.
                       B. pandurifolia (Hook.) C. Chr. (Fig. 16)
                k. Bulbils axillary. l.
                  l. Base of the terminal segment of the sterile leaves definitely asymmetrical, Trinidad and Venezuela.
                                B. hemiotis (Maxon) C. Chr.
                  l. Base of the terminal segment of the sterile leaves symmetrical or nearly so, Costa Rica to Bolivia.
                          B. oligarchica (Baker) Hennip. (Fig. 14)
         f. Sterile leaves entire, without a subterminal bulbil. m.
           m. Scales of the stem rusty brown, bullate.   *B. pergamentacea*
           m. Scales of the stem blackish, flat. n.
              n. Base of the sterile lamina truncate to subcordate.
                                    B. hemiotis
              n. Base of the sterile lamina more or less acute.
                                    B. nicotianifolia
   c. Sterile leaves pinnate with the terminal segment triangular and often longer than the remaining lamina. o.
      o. Venation without included free veins, or nearly so. p.
         p. Leaves clustered, margins of the sterile pinnae with teeth in the sinuses.   *B. umbrosa* (Liebm.) C. Chr. (Fig. 17)
         p. Leaves spaced, margins of the sterile pinnae without teeth in the sinuses when present.   *B. aliena* (Sw.) Alston
      o. Venation with many recurrent and excurrent included free veins. q.
         q. Scales of the stem up to 10 mm long, basally attached, bulbils present.
         B. portoricenis (Spreng.) Hennip. (*B. cladorrhizans* (Spreng.) C. Chr.)
         q. Scales of the stem up to 4 mm long, pseudopeltate, bulbils absent.
                                     B. semipinnatifida

Ecology (Fig. 1)

Bolbitis grows in wet to seasonally dry forests in shaded habitats, and mostly along watercourses where it is often on rocks. *Bolbitis Heudelotii* (Fée) Alston of Africa is sometimes submerged in streams.

Fig. 93.2. Distribution of *Bolbitis* in America.

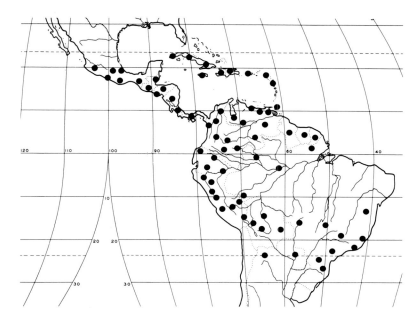

In America *Bolbitis* usually grows in lowland rain forests or montane forests, especially along streams, in ravines, and sometimes in streambeds. It frequently grows on sandstone, limestone, or other rocks; also in soil or on the base of trees. *Bolbitis Bernoullii* and *B. Lindigii* are the only high-climbing species of the genus. In these and some other scandent species, observations are needed to ascertain whether scandent stems retain a terrestrial connection. *Bolbitis* grows from sea level to 1800 m, most often below 1000 m.

The sterile leaf of *Bolbitis portoricensis* may have a long-attenuate apex which roots at the tip, and several species, such as *B. nicotianifolia* (Fig. 5) have scaly bulbils on the lamina that can produce new plants.

Geography (Fig. 2)

Bolbitis occurs in the American tropics, tropical Africa, eastward to southern India and Ceylon, north and eastward to China and southern Japan, and south to Queensland in Australia and across the Pacific to Tahiti and Rapa.

In America it grows in Veracruz to Jalisco in Mexico, south through Central America, in all of the Greater Antilles, and many of the Lesser Antilles, and in South America south to Salta in Argentina, Paraguay and Santa Catarina in southeastern Brazil.

Twelve of the 14 American species occur in the Cordillera from Bolivia north to southern Mexico; 10 are in the Andes and 10 from Panama to Mexico. Many Andean genera have strong secondary centers in the Greater Antilles and southeastern Brazil. However, these are not evident in *Bolbitis*, which has only four species in the Greater Antilles and a single one in Brazil. There is little local endemism in America. *Bolbitis umbrosa* is known only from Veracruz, Mexico, *B. hastata* from southern

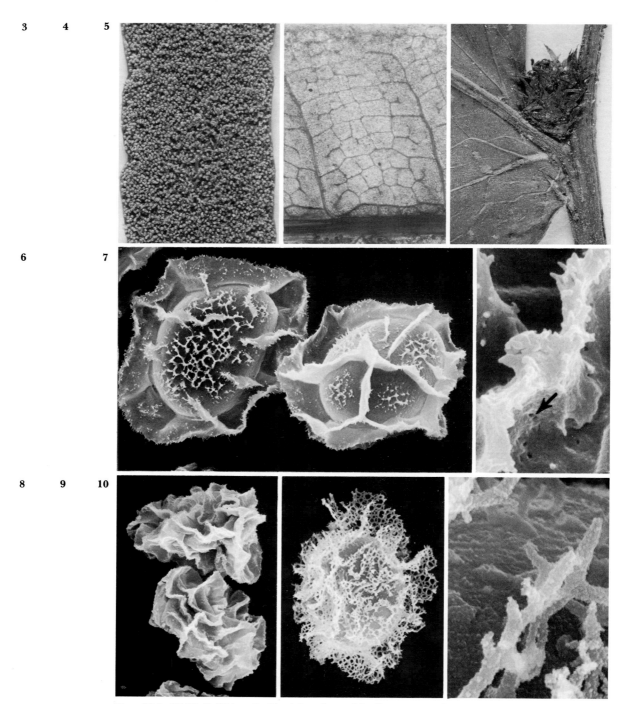

Figs. 93.3–93.10. *Bolbitis.* **3.** Abaxial surface of fertile segment of *B. portoricensis*, ×5. **4.** Abaxial surface of fertile segment of *B. pergamentacea,* with sporangia removed to show venation, ×8. **5.** Scaly bulbil of *B. hemiotis,* ×4. **6–10.** Spores, ×1000, surface detail, ×10,000. **6.** *B. Lindigii,* prominent winglike folds and adjacent echinate areas, Colombia, *Hodge 7072.* **7, 8.** *B. pergamentacea,* Costa Rica, *Scamman 6162.* **7.** Surface detail of echinate, reticulate folds, and perforations (at arrow). **8.** Spores with winged folds and undeveloped central cell. **9, 10.** *B. appendiculata,* Thailand, *Tagawa 1079.* **9.** Reticulate surface folds. **10.** Detail of reticulate structure.

Figs. 93.11–93.12. Venation of sterile segments of *Bolbitis* **11. *B. nicotianifolia***, ×2. **12. *B. Bernoullii***, ×2. (After Hennipman, 1977.)

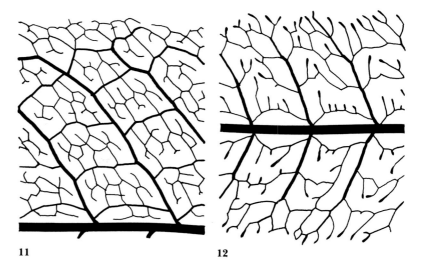

11   12

Mexico, and *B. pandurifolia* from Ecuador and Peru. Other species have moderate to rather wide distributions.

Spores

The compact surface is formed over a reticulate lower formation as in *B. pergamentacea* (Fig. 7). These spores resemble the prominently winged forms of other genera as *Lomagramma* and *Arthropteris* and are especially similar to those of *Stigmatopteris sancti-gabriellii*. *Bolbitis* spores are often incompletely developed with a shriveled central cell but have a well-formed wall (Fig. 8). The elaborate reticulate winged formation of spores in the Asiatic species *B. appendiculata* (Willd.) Iwats. (Figs. 9, 10) are unlike others in *Bolbitis* and resemble those of the African species *Lomariopsis Rossii* Holtt. A series of SEM micrographs of *Bolbitis* spores (Hennipman, 1977) shows the alignment in the sporangium and variation in the surface structure.

Cytology

The diploid number, $n = 41$, is most frequent among cytological records for *Bolbitis,* summarized by Hennipman (1977). Two American species, reported by Walker from Jamaica (1966), include the widespread *B. pergamentacea* with $2n = 82$, and *B. aliena,* a tetraploid with $n = 82$, from rocky sites in the mountains. These and records of paleotropical species, from a broad geographic range, especially in Asia, include 33 diploids, 4 tetraploids and 9 triploids. The latter are regarded as autotriploids by Hennipman on the basis of morphological similarities. Three hybrids are cytologically determined as diploid and several others are proposed on the intermediate morphology of the plants.

Observations

The leaves of *Bolbitis* exhibit a heteroblastic series from the juvenile to the adult plants. A detailed study of the leaf architecture

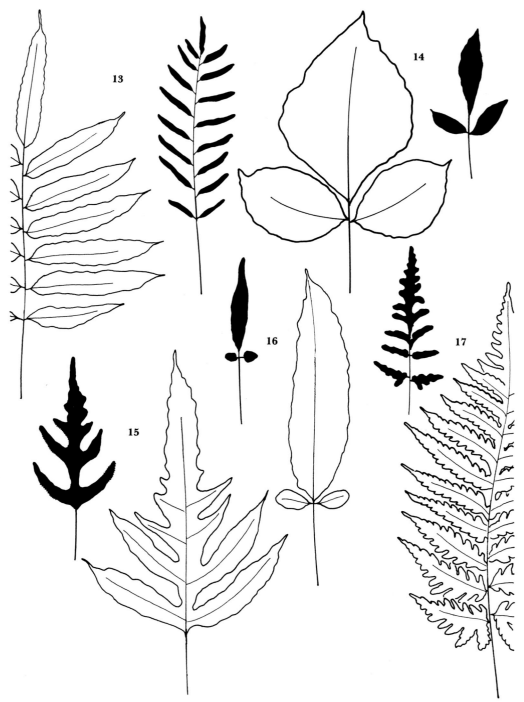

Figs. 93.13–93.17. *Bolbitis,* fertile leaves (black) and sterile leaves (outline), ×0.2. **13.** *B. serratifolia.* **14.** *B. oligarchica.* **15.** *B. hastata.* **16.** *B. pandurifolia.* **17.** *B. umbrosa.* (After Hennipman, 1977.)

and venation pattern of these series and a discussion of the evolutionary potential that is derived from the addition or the elimination of stages is presented by Hennipman (1977). The addition of stages could result in a more complex leaf architecture and venation, while the deletion of later stages could result in an adult form similar to one of the juvenile stages of the ancestral species, a process known as neoteny. This is probably a widespread phenomenon in the ferns that deserves more attention for it also appears to have occurred in other groups as *Adiantum* and *Lindsaea*.

Literature

Hennipman, E. 1977. A monograph of the fern genus *Bolbitis* (Lomariopsidaceae). Leiden Bot. Ser. 2. 331 pp. Leiden Univ. Press.
Walker, T. G. 1966. See reference under the tribe Bolbitideae.

94. *Lomariopsis*
Figs. 94.1–94.14

Lomariopsis Fée, Mém. Fam. Foug. 2 (Hist. Acrost.): 10, 66. 1845. Type: *Lomariopsis sorbifolia* (L.) Fée (*Acrostichum sorbifolium* L.).

Description

Terrestrial or rupestral, usually becoming scandent-epiphytic; stem long-creeping or scandent to ca. 15 m, slender to rather stout, bearing scales and very few to many fibrous roots; leaves moderately to strongly dimorphic (the fertile pinnae then much narrower than the sterile), ca. 20 cm to 1.5 m long, rather closely to usually widely spaced, lamina 1-pinnate with entire pinnae, imparipinnate, glabrate to slightly and minutely scaly, veins free; sporangia borne over the abaxial surface of the pinnae, no paraphyses or indusium; spores ellipsoidal, monolete, the laesura short, less than $\frac{3}{4}$ the spore length, with prominent, erose, winglike formation, the surface rugose of more or less fused strands, or prominently echinate. Chromosome number: $n = 39$; $2n = 32, 62, 78, 164$.

The scandent stem is usually densely scaly (Fig. 4) especially toward the apex. The veins of the fertile pinnae are free (Fig. 5) as in the sterile pinnae (Fig. 6). Sporangia are borne over the abaxial surface of the fertile pinna (Fig. 5). A plant of *Lomariopsis Fendleri* with strongly dimorphic leaves is shown in Fig. 3, with a juvenile plant on its stem. The small scales borne among the sporangia are not considered to be paraphyses since they are of the same type as on the abaxial surface of the sterile pinnae.

Systematics

Lomariopsis is a nearly pantropical genus of about 45 species, with perhaps 15 in America. The genus is characterized by a dorsiventral stem and stele, a 1-pinnate and imparipinnate lamina, and articulate pinnae with free veins. Underwood (1906) recognized the group as a section of *Stenochlaena*, and Holttum (1932) distinguished it as a separate genus from among others with scandent stems and acrostichoid sporangia.

Fig. 94.1. *Lomariposis Fendleri,* Río Reventazón, Costa Rica. (Photo W. H. Hodge.)

Fig. 94.2. Distribution of *Lomariopsis* in America.

Figs. 94.3–94.8. *Lomariopsis*. *3.* Plant of **L. Fendleri** with strongly dimorphic leaves, and scandent stem; a juvenile plant on stem at lower right, × 0.3. **4.** Scaly scandent stem of **L. Fendleri**, × 2.5. **5.** Abaxial surface of fertile pinna of **L. nigropaleata**, sporangia partly removed below to show the veins, × 4. **6.** Veins of sterile pinna of **L. nigropaleata**, × 2. **7.** Apex of rachis of **L. japurensis**, the stalk of the apical pinna is continuous, the lateral pinna is articulate at the base, × 4. **8.** Apex of rachis of **L. Wrightii**, the spur, left, represents the abortive apical pinna, a lateral, articulate pinna in the apical position, × 4.

Lomariopsis is related clearly to the other genera of the tribe Bolbitideae, but especially to *Elaphoglossum*.

The American species were revised by Underwood (1907) and they were summarized and some new species were described by Holttum (1940).

Tropical American Species

The American species of *Lomariopsis* are in need of a modern revision based on field studies and the more ample collections than those available to Underwood in 1907. The following arrangement of species is adapted from Holttum (1940).

Apical pinna articulate, a lateral pinna occupies the position of the more or less abortive true apical pinna (Fig. 8). In some specimens there is a relatively large remnant of the apical pinna while in others there is hardly a trace of it; in most specimens it is represented by a greenish stub or spur. A similar modification also occurs in the imparipinnate pinnae of *Cyathea conformis*. Three species are recognized in the Greater Antilles: *Lomariopsis jamaicensis* (Underw.) Holtt., *L. Underwoodii* Holtt., and *L. Wrightii* D. C. Eaton.

Apical pinna not articulate (Fig. 7), rachis narrowly alate throughout or at least in the apical half; the leaves of juvenile plants, when known, are 1-pinnate (Fig. 3). Species that belong in this group are, for example, *Lomariopsis Kunzeana* (Underw.) Holtt. and *L. sorbifolia* (L.) Fée; also *L. Fendleri* D. C. Eaton (*L. vestita* Fourn.), *L. latiuscula* (Maxon) Holtt., and *L. Maxonii* (Underw.) Holtt., which are somewhat variable in the extent of development of the alate rachis.

Apical pinna not articulate, rachis narrowly alate only toward the apex; the leaves of juvenile plants, when known, are simple. Species of this group are, for example, *Lomariopsis erythrodes* (Kze.) Fée, *L. japurensis* (Mart.) J. Sm. (*L. Prieuriana* Fée), *L. nigropaleata* Holtt. and *L. recurvata* Fée.

Ecology (Fig. 1)

Lomariopsis is primarily a genus of wet primary forests, but occasionally it grows in swamp forests or in thickets.

In America it grows in lowland rain forests and montane forests, rarely in brushy vegetation or in rocky places or on cliffs. It grows at relatively low elevations, from ca. sea level to 2200 m, mostly below 1000 m.

Plants are initially terrestrial or rupestral but may become scandent-epiphytic on trees or bushes and then retain their terrestrial connection. There is no pronounced dimorphism between terrestrial and scandent stems as in *Lomagramma*.

Geography (Fig. 2)

Lomariopsis occurs in tropical and subtropical America, in tropical Africa, in Madagascar and islands of the Indian Ocean, and in the eastern Himalayas and southern China through Malaysia to the Philippine Islands, in the Bonin Islands, New Guinea, and

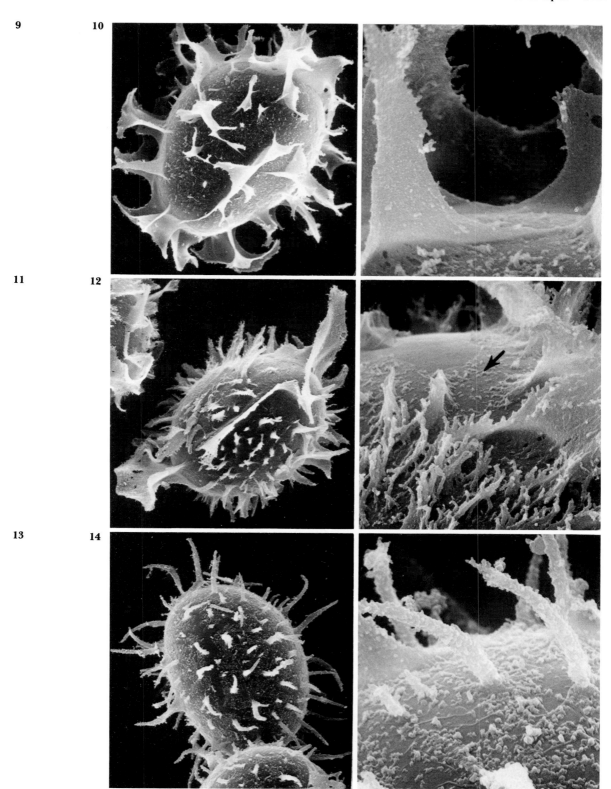

Figs. 94.9–94.14. *Lomariopsis* spores, × 1000, surface detail, × 5000. **9, 10. L. latiuscula,** Mexico, *Dressler & Jones 142.* **9.** Irregular, erose folds. **10.** Echinate surface detail of winglike folds. **11, 12. L. recurvata,** Mexico, *Matuda 3652.* **11.** Irregular more or less fused wings. **12.** Surface detail, lower rugate perispore (at arrow) between smooth exospore, above, and surface strands, below. **13, 14. L. nigropaleata,** Peru, *Klug 2082.* **13.** Diffuse echinate elements. **14.** Detail of surface deposit.

Queensland in Australia, and east through the Pacific islands to Tahiti.

In America *Lomariopsis* ranges from Veracruz in Mexico through Central America and in southern Florida in the United States, the Antilles, and in South America south to Bolivia and Santa Catarina in southeastern Brazil.

Spores

Lomariopsis spores have prominent erose, winglike folds as in *L. latiuscula* (Fig. 9). The wings are somewhat echinate with rugate or more or less fused stranded material adjacent to the wings as in the surface detail of *L. latiuscula* and *L. recurvata* (Figs. 10, 12). The major winglike projections may be more or less fused as in *L. recurvata* (Fig. 11). A thin, rugate lower perispore stratum is formed between the elaborate outer layer and the smooth exospore surface as in the abraded spore of *L. recurvata* (Fig. 12). Prominent echinate elements in spores of *L. nigropaleata* appear to form a base on which the outer spore surface is deposited, as in the surface detail (Figs. 13, 14). The elaborate winglike folds of Asiatic species figured by Holttum (1978, p. 317) are similar to those of the American species. However, some African species have quite distinctive, reticulate spores similar to those of *Bolbitis appendiculata*. The perispore in *Lomariopsis* generally resembles, but is more irregularly erose than that of *Lomagramma guianensis* and *Bolbitis*.

Cytology

Chromosome numbers for *Lomariopsis* are not reported for American tropical species but are known for a species of Malaya $2n = 164$ and several from Ghana with $2n = 32, 62,$ and 78 (Roy and Manton, 1966). The Asiatic species is correlated with the base number 41 of the family but those of west Africa are not readily aligned. Roy and Manton suggest that there may be fusion of small chromosomes that are about a third the size of larger ones in their figure. These distinctive chromosome numbers in the paleotropical species will be useful to indicate relationships when cytological records for the genus are reported from the American tropics.

Literature

Holttum, R. E. 1932. On *Stenochlaena, Lomariopsis* and *Teratophyllum* in the Malayan region. Gards. Bull. Straits Settl. 5: 245–313.

Holttum, R. E. 1940. New species of *Lomariopsis.* Kew Bull. 1939: 613–628.

Holttum, R. E. 1978. Reference under the Tribe Bolbitideae.

Roy, S. K., and I. Manton. 1966. The cytological characteristics of the fern subfamily Lomariopsidoideae *sensu* Holttum. Jour. Linn. Soc. Bot. 59: 343–347.

Underwood, L. M. 1906. The genus *Stenochlaena.* Bull. Torrey Bot. Cl. 38: 35–50.

Underwood, L. M. 1907. American ferns, VII.—The American species of *Stenochlaena.* Bull. Torrey Bot. Cl. 38: 591–603.

95. *Lomagramma*
Figs. 95.1–95.9

Lomagramma J. Sm., Jour. Bot. (Hook.) 4: 152. 1841. Type: *Loma-gramma pteroides* J. Sm.

Description

Terrestrial or sometimes rupestral, usually becoming scandent-epiphytic; stem long-creeping or scandent to ca. 20 m long, rather slender, bearing scales and few to usually many fibrous roots; leaves dimorphic (the fertile segments narrower than the sterile), ca. 25 cm to 1.5 m long, widely spaced, lamina 1-pinnate or rarely 2-pinnate, usually imparipinnate, glabrate to slightly scaly, veins anastomosing without included free veins; sporangia borne over the abaxial surface and rarely on the margin and upper surface of the segments, mixed with usually few paraphyses, no indusium; spores more or less ellipsoidal to spheroidal, monolete, the laesura about $\frac{1}{2}$ the spore length with prominent winglike folds and echinate surface, or smooth without perispore. Chromosome number: $2n = 82$.

A portion of the scaly, scandent stem of *Lomagramma guianensis* is shown in Fig. 4. The veins are fully anastomosing, except sometimes at the margin (Fig. 8). The abaxial surface of the fertile segment is covered by sporangia (Fig. 2), which usually protrude beyond the sterile margins (Fig. 3).

Systematics

Lomagramma is a tropical genus of about 20 species, with a single one, *Lomagramma guianensis* (Aubl.) Ching (Fig. 9), in America. The genus is characterized by a scandent-epiphytic stem, dorsiventral stem and stele, and the lamina 1- to 2-pinnate, usually imparipinnate, with articulate pinnae and anastomosing veins. *Lomagramma guianensis* was formerly treated as *Bolbitis guianensis* (Aubl.) Vareschi, and earlier called *Leptochilus guianensis* (Aubl.) C. Chr. It was excluded from *Lomagramma* by Holttum (1937,

Fig. 95.1. Distribution of *Lomagramma guianensis.*

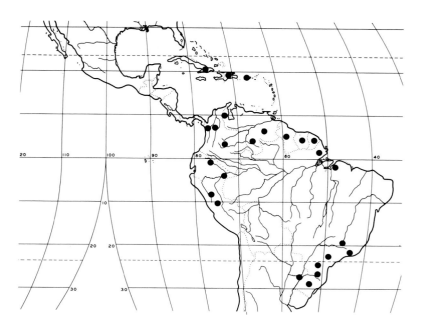

2

3

4

5

6

7

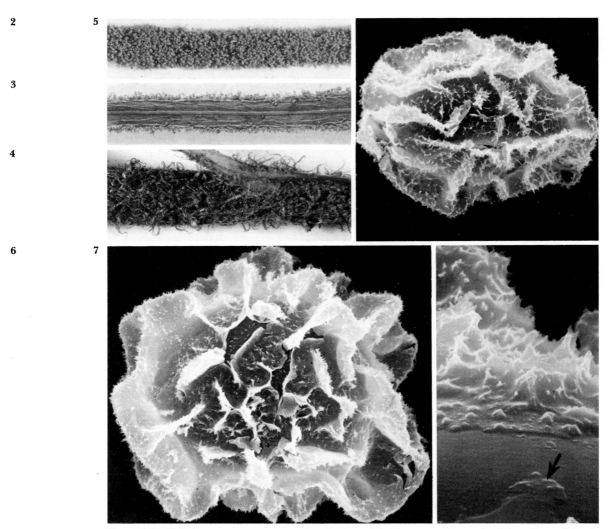

Figs. 95.2–95.7. *Lomagramma guianensis.* **2.** Acrostichoid sporangia on abaxial surface of fertile pinna, ×3. **3.** Adaxial surface of fertile pinna, ×3. **4.** Portion of scandent stem, ×2. **5–7.** Spores, ×1000. **5.** Winglike folds and echinate deposit Peru, *Schunke 1372.* **6.** Prominent winglike folds and echinate surface, portions of lower rugose stratum in abraded areas near top, Brazil, *Dusén 14305.* **7.** Detail of abraded surface with portions of lower, rugose stratum (arrow), Brazil, *Handro 2214,* ×5000.

1978) and from *Bolbitis* by Hennipman (1977). *Lomagramma guianensis* is distinguished from both genera only by the slender paraphyses, which is not sufficient grounds for treating it as a separate genus. It is included in *Lomagramma* on the basis of the articulate pinnae and presence of paraphyses. The prominently winged spores of the American plants differ from the nearly smooth spores lacking perispore in paleotropical species. The slender, gland-tipped paraphyses of *Lomagramma guianensis* differ from the large, irregularly expanded ones of paleotropical species. Holttum (1937) revised the paleotropical species, and later (1978) those of the Flora Malesiana region.

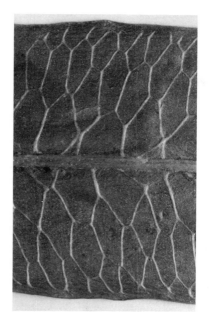

Fig. 95.8. Venation of sterile pinna of *Lomagramma guianensis,* × 5.

Ecology

Lomagramma is primarily a genus of wet, primary forests, seldom occurring in disturbed forests or secondary vegetation. In America *Lomagramma guianensis* grows at relatively low altitudes between ca. 50 and 600 m. Plants initially are terrestrial, becoming scandent on trees but retaining their terrestrial connection. The terrestrial stem is slender and bears only sterile leaves. The scandent stem is stouter and bears slightly to quite different sterile leaves as well as fertile ones.

Geography (Fig. 1)

Lomagramma grows in America, and from the eastern Himalayas and southern China, through Malaysia to the Philippine Islands, New Guinea and east in the Pacific islands to Tahiti.

Fig. 95.9. Dimorphic leaves and portion of stem. *Lomagramma guianensis,* × 0.3.

Lomagramma guianensis occurs in the Greater Antilles, except Jamaica; and in South America from Pará in Brazil west to Colombia and south to Peru; it is disjunct in southeastern Brazil and adjacent Argentina.

Spores

The spores of *Lomagramma guianensis* from Brazil and Peru are similar with prominent wings and echinate surface but clearly differ in size (Figs. 5, 6). Abraded spores show the wall is formed of a thin lower perispore below the elaborate surface wings, and overlays a fairly smooth exospore (Figs. 6, 7). These spores resemble those of several American species of *Bolbitis* rather than the erose forms in *Lomariopsis*. The smooth spores of the paleotropical species, treated in sodium hydroxide, do not have an expanded outer layer, indicating that the perispore is lacking.

Cytology

The chromosome number is not known in the American species but two from New Guinea are reported as diploid with $2n = 82$ by Roy and Manton (1966). The mitotic chromosomes figured in that work have marked constrictions and satellites, are relatively large, and of more uniform size than those of *Lomariopsis*. The chromosome number 41 is consistent with the base number in bolbitoid genera.

Literature

Hennipman, E. 1977. A monograph of the fern genus *Bolbitis* (Lomariopsidaceae). Leiden Bot. Ser. 2. 331 pp. Leiden Univ. Press.

Holttum, R. E. 1937. The genus *Lomagramma*. Gards. Bull. Straits Settl. 9: 190–221.

Holttum, R. E. 1978. Reference under the tribe Bolbitideae.

Roy, S. K. & I. Manton. 1966. The cytological characteristics of the fern subfamily Lomariopsidoideae *sensu* Holttum. Jour. Linn. Soc. Bot. 59: 343–347.

96. *Elaphoglossum*
Figs. 96.1–96.34

Elaphoglossum J. Sm., Jour. Bot. (Hook.) 4: 148. 1841, *nom. conserv.*
   Type: *Elaphoglossum conforme* (Sw.) J. Sm. (*Acrostichum conforme* Sw.).
Aconiopteris Presl, Tent. Pterid. 236. 1836. Type: *Aconiopteris subdia-*
   phana (Hook. & Grev.) Presl (*Acrostichum subdiaphanum* Hook. &
   Grev.) = *Elaphoglossum nervosum* (Bory) Christ.
Peltapteris Link, Fil. Sp. 147. 1841. Type: *Acrostichum peltatum* (Sw.) Sw.
   (*Osmunda peltata* Sw.) = *Peltapteris peltata* (Sw.) Morton = *Elaphoglos-*
   sum peltatum (Sw.) Urban.
Rhipidopteris Fée, Mém. Fam. Foug. 2 (Hist. Acrost.): 14, 78. 1845, *nom.*
   superfl. for *Peltapteris* Link and with the same type.
Hymenodium Fée, *ibidem* 2: 20, 90. 1845. Type: *Hymenodium crinitum* (L.)
   Fée (*Acrostichum crinitum* L.) = *Elaphoglossum crinitum* (L.) Christ.
Dictyoglossum J. Sm., Comp. Curtis' Bot. Mag. 72: 18. 1846, *nom. superfl.*
   for *Hymenodium* Fée and with the same type.
Microstaphyla Presl, Epim. Bot. 160. 1849. Type: *Microstaphyla bifurcata*
   (Jacq.) Presl (*Osmunda bifurcata* Jacq.) = *Elaphoglossum bifurcatum*
   (Jacq.) Mickel.

Description

Terrestrial, rupestral or epiphytic; stem very short- to long-creeping, very slender to moderately stout, including the persistent petiole bases, bearing scales and very few to many fibrous roots; leaves monomorphic or usually more or less dimorphic (the fertile often differ from the sterile in size, shape and length), to strongly dimorphic (the fertile lamina entire to somewhat lobed and the sterile pinnately or flabellately dissected), ca. 2 cm to 2 m long, borne in a cluster to widely spaced, lamina simple, usually entire with smooth or rarely crenate, broadly crenate or deeply crenate-serrate margins, or lobed at the base, or 2-lobed to flabellately dissected, or pedate, or pinnatisect to 1-pinnate with often furcate pinnae, glabrate to slightly or densely scaly, rarely with some minute gland-tipped trichomes, veins free or rarely connected at the margin, or rarely anastomosing without included free veins; sporangia densely borne over the abaxial surface, except for a usually narrow sterile margin, mixed with paraphyses or not, sometimes with scales, no indusium; spores more or less ellipsoidal, monolete, the laesura about $\frac{1}{2}$ the spore length, with low folds, these sometimes short and saccate or winglike with echinate to reticulate surface, or the folds strongly perforate or fenestrate, or sometimes without folds and echinate, reticulate or verrucate. Chromosome number: $n = 40, 41, 82$, ca. 164; $2n = 82$, ca. 164; ca. 250.

The fertile lamina is completely covered by sporangia on the abaxial side and is usually flat (Fig. 5). In a few species it is conduplicate as in *Elaphoglossum spathulatum* (Bory) Moore (Fig. 6). The leaf disarticulates above the base of the petiole in many species (Fig. 7) and the persistent base is called a phyllopodium. Some of the diversity in lamina shape is shown in Figs. 17–26. The usually free veins may be close (Fig. 12) or widely spaced (Fig. 11) and are rarely anastomosing as in *E. crinitum* (Fig. 13). Some of the diverse types of lamina scales are shown in Figs. 14–16.

Systematics   *Elaphoglossum* is a pantropic and south-temperate genus of about 500 species with perhaps 350 in America. It is characterized by an entire fertile and sterile lamina which is either scaly or bears vestiges of scales, by acrostichoid sporangia and at least usually by a dorsiventral stem and stele.

This large and complex genus has not had a modern revision; thus an infrageneric classification is not presented but rather the diversity of major taxonomic characters is reviewed. A considerable number of distinct groups have been proposed in several different infrageneric systems. The monograph of Christ (1899) recognized 53 groups within the genus in four taxonomic ranks. Brade (1961) had 44 groups in his treatment of the species of Brazil. Mickel (1980) and Mickel and Atehortúa (1980) have a new classification within nine sections and 21 subsections.

The four genera that have been proposed appear to represent minor evolutionary elements within *Elaphoglossum,* although a few species with dissected lamina are strikingly different from the

Fig. 96.2. *Elaphoglossum petiolatum* (Sw.) Urban, Dominica. (Photo W. H. Hodge.)

vast majority with the lamina entire. The character on which *Aconiopteris* has been distinguished is the free veins fused only at their tips. This condition occurs in unrelated species such as *Elaphoglossum nervosum* (Bory) Christ of St. Helena, *E. longifolium* (Jacq.) J. Sm. of America and *E. alatum* Gaud. (Anderson and Crosby, 1966) of the Hawaiian Islands.

Hymenodium, still maintained by some authors as Proctor (1977), is distinguished by anastomosing veins. On the basis of this character the genus includes species that are not closely related as *E. crinitum* of America and *E. crassifolium* (Gaud.) Anders. & Crosby of the Hawaiian Islands, and perhaps *E. pachyphyllum* (Kze.) C. Chr. of America.

Peltapteris has the sterile lamina flabellately veined or dissected but is closely connected to *Elaphoglossum* by a series of species such as *E. ovatum* (Hook. & Grev.) Moore, *E. deltoideum* (Sod.) Christ and *E. squamipes.* The chromosome number $n = 40$ in *E. peltatum* is evidently an aneuploid derivative of $n = 41$ which is characteristic for the genus.

The genus *Microstaphyla* includes species with a pinnatisect or 1-pinnate sterile lamina and usually furcate pinnae and, in *E. bifurcatum* (Jacq.) Mickel, with a dissected fertile lamina. The two elements within this group are clearly referable to *Elaphoglossum.* The type of the genus, *Elaphoglossum bifurcatum,* is referred to section *Lepidiglossa* Christ, by Mickel and Atehortúa. The other element *Elaphoglossum Moorei* (Fig. 19) and related species of the Andes are included in section *Squamipedia* Mickel & Atehortúa.

The chromosome numbers and characteristics of the spores support the position of *Elaphoglossum* in the Dryopteridaceae; however, alliances of the genus within the family are uncertain. Similarities in the entire form of the leaves suggest possible affinities to some species of *Bolbitis, Oleandra* and *Lomariopsis,* and except for the latter, the spores with folds or prominent wings and elaborate perispore are also similar.

The Brazilian species of *Elaphoglossum* have been revised by

Fig. 96.3. *Elaphoglossum peltatum,* sterile leaves, Costa Rica. (Photo W. H. Hodge.)

Alston (1958) and by Brade (1961), and the species of the French Antilles by Morton (1948).

Tropical American Species

The classification of *Elaphoglossum* species, especially those of the Andes, is in need of study for they are poorly known and application of their names needs to be verified by examination of the original specimens. Most of the diversity of the systematic characters is illustrated in American species.

The **stem** is very slender, ca. 1–2 mm in diameter, and long-creeping in *Elaphoglossum Andreanum* Christ and *E. squamipes* (Hook.) Moore, while it is rather short and stout in *E. Orbignyanum* (Fée) Moore and *E. decoratum.*

The **leaf** disarticulates near the base of the petiole in a number of species, such as *Elaphoglossum Lingua* Brack. This appears as a slight swollen zone, and the part below the swollen area, the phyllopodium, is darker colored and often scaly. There does not appear to be an abscission layer as the petiole breaks in an irregular manner (Fig. 7) but a study by Bell (1951) indicates there is some anatomical differentiation in the region. Fertile leaves may be much shorter than the sterile as in *Elaphoglossum Herminieri* (Fée) Moore (Fig. 24) or they are often longer as in *E. scalpellum* (Mart.) Moore (Fig. 20). The lamina is diverse in both size and shape. In *E. crinitum* (L.) Christ (Fig. 22) the lamina may be ca. 50 cm long and 25 cm wide. It may be very narrow as in *E. Eatonianum* (E. Britt.) C. Chr. (Fig. 17), or very small as in *E. piloselloides* (Presl) Moore (Fig. 23). Other lamina forms are shown in

Fig. 96.4. Distribution of *Elaphoglossum* in America.

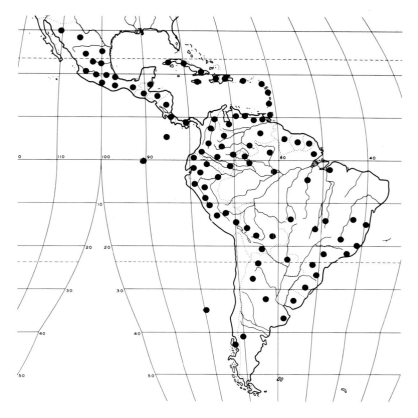

E. cardiophyllum (Hook.) Moore (Fig. 26), *E. boragineum* (Sod.) Christ (Fig. 21), and *E. apodum* (Kaulf.) J. Sm. (Fig. 25) a species with nearly sessile leaves. Dissected form of the lamina is shown in *E. peltatum* (Sw.) Urban (Fig. 18), and *E. Moorei* (E. Britt.) Christ (Fig. 19). Both the fertile and sterile lamina are pedate in *E. Cardenasii* Wagner.

The **scales** are exceptionally diverse and often abundant on stem, petiole and lamina. Long, blackish scales on the petiole of *Elaphoglossum crinitum,* as in many species, curl and twist and appear as large trichomes (Fig. 9). Petiole scales of *E. decoratum* (Kze.) Moore are exceptionally large, light brown, broad and flat (Fig. 10). Examples of variation in lamina scales are shown in *E. trichophorum* (Sod.) C. Chr. (Fig. 15), *E. Dombeyanum* (Fée) Houlst. & Moore (Fig. 16) and *E. plicatum* (Cav.) C. Chr. (Fig. 14). These are sometimes concentrated at the margin as in *E. Lindbergii* (Kuhn) Rosenst. (Fig. 8).

Ecology (Figs 1–3)

Elaphoglossum is primarily an epiphytic genus, although some species are terrestrial or rupestral. The epiphytes mostly grow on tree trunks near the ground, sometimes on branches among the forest canopy. Many species grow in wet places where they may be terrestrial, rupestral, or epiphytic. In drier sites, high epiphytes are absent and plants are more often confined to a single habitat.

In tropical America *Elaphoglossum* is primarily a genus of wet

5   6   7

8   9   10

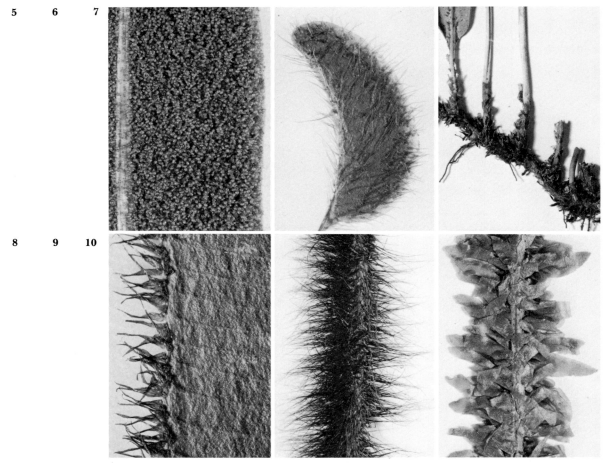

Figs. 96.5–96.10. *Elaphoglossum.* **5.** Acrostichoid sporangia on portion of fertile lamina of *E. lepto-phyllum* (Fée) Moore, ×5. **6.** Conduplicate fertile lamina of *E. spathulatum,* ×4. **7.** Stem with leaves and phyllopodia, *E. Lingua,* ×1. **8.** Scales concentrated at the margin of the lamina of *E. Lindbergii,* ×8. **9.** Petiole scales of *E. crinitum,* ×1.5. **10.** Petiole scales of *E. decoratum,* ×2.5.

mountain forests and cloud forests. It is most commonly a low epiphyte but often grows on wet rocks or in the forest soil. Species sometimes grow in crevices or on ledges of cliffs, and uncommonly in soil in brushy disturbed areas. At the high elevations in the Andes, the plants are in turf or rocky places. *Elaphoglossum* grows from near sea level to ca. 4500 m, most often at 1000–3000 m.

Plants that grow in sites that are occasionally dry may show different responses to desiccation. Some species such as *E. Calaguala* (Kl.) Moore and *E. Mathewsii* (Fée) Moore, have leaves curled lengthwise with the abaxial surface exposed, while in *E. hirtum* (Sw.) C. Chr. the leaves become spirally curled. *Elaphoglossum proliferans* Maxon & Morton has a sterile leaf with an apical bud that roots, producing a new plant. *Elaphoglossum Wageneri* (Kze.) Moore of Colombia and a few others also have sterile leaves with proliferous apices.

Ecological studies on *Elaphoglossum,* conducted at two localities, near Manaus, Brazil (Tryon and Conant, 1975), demonstrated that at a local site plants of both *E. discolor* (Kuhn) C. Chr.

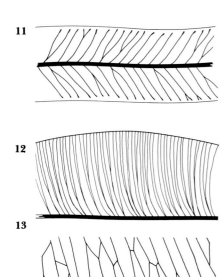

Figs. 96.11–96.13. Venation of sterile lamina of *Elaphoglossum*, ×0.5. **11.** widely spaced veins of *E. Aubertii* (Desv.) Moore. **12.** Closely spaced veins of *E. barbatum* Hieron. **13.** Anastomosing veins of *E. crinitum*.

and *E. glabellum* J. Sm. were most numerous in more humid portions of a microenvironmental gradient. Both species have coriaceous leaves and their distribution probably reflects the microenvironment suitable for the gametophyte rather than for the sporophyte. At one locality, along a densely wooded stream bank, *E. discolor* was most frequent on trees within 1 m of the stream border where 75% of the plants occurred, and they were correspondingly less frequent 2–3 m beyond the stream border. At another site, in a campina forest, plants of *Elaphoglossum glabellum* were abundant on tree trunks. The plants were rather uniformly distributed on the trunks at levels ranging from the ground up to 180–200 cm and infrequent above that level.

Geography (Fig. 4)

Elaphoglossum occurs from tropical and south-temperate America to Africa, southern India, Ceylon and Malaysia, and from the Himalayas and southern China north to Japan, also it extends to New Guinea, New Caledonia and northeastern Australia, and across the Pacific to the Hawaiian Islands and Easter Island.

In America *Elaphoglossum* occurs in northern Mexico through Central America, in the Antilles, and in South America south to Buenos Aires and Neuquén in Argentina, and also in Valdivia, Llanquihue and Chiloe in Chile; also on Cocos Island, the Galápagos Islands and the Juan Fernandez Islands. It is especially rich in species in the Andes from Colombia to Bolivia, where perhaps more than 150 species occur.

Spores

The diverse types of surface structure of *Elaphoglossum* spores generally appear to be derived from a basic folded type with a more or less echinate-reticulate surface as in *E. ovatum* (Fig. 27) and *E. conforme* (Sw.) J. Sm. (Fig. 28). The folds may be expanded into wings, as in *E. Plumieri* Moore (Fig. 29), or may be strongly perforate, as in *E. latifolium* (Sw.) J. Sm. (Fig. 34). Spores without prominent folds are reticulate, as in *E. eximium* (Mett.) Christ (Fig. 32), echinate or tuberculate. The size generally ranges between 25 and 70 μm in longest dimension. Spores with low folds generally tend to be smaller than elaborately reticulate ones. Large-leaved species, such as *E. decoratum* and *E. crinitum* (Fig. 33), have some of the smallest spores while those with small leaves, such as *E. peltatum* (Fig. 30), have large ones. Spores of the latter have a low, echinate surface with broad, raised areas, characteristic of several species including those formerly placed in *Peltapteris*. They resemble spores of species in the *Dryopteris Carthusiana* complex especially *D. intermedia* (Fig. 74.19). The perispore of *Elaphoglossum* may consist of three strata, the central one often a columnar structure between the surface and basal strata as in *E. latifolium* (Fig. 34). The perispore is less complex in *E. squamipes* with a thin, lower stratum overlayed by the outer, echinate formation (Fig. 31). Erdtman

Figs. 96.14–96.16. *Elaphoglossum* scales on abaxial surface of the lamina. **14. E. plicatum,** ×14. **15. E. trichophorum,** ×20. **16. E. Dombeyanum,** ×20.

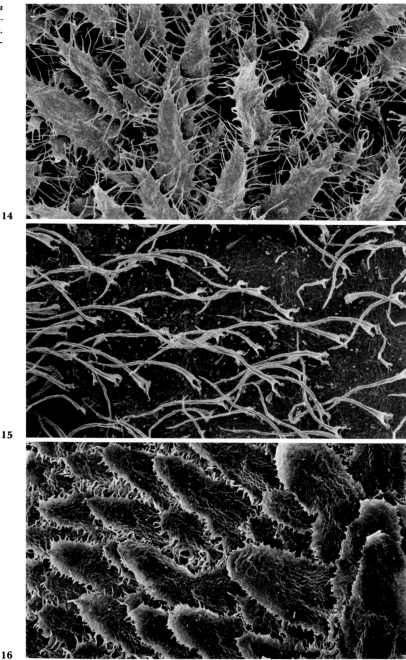

and Sorsa (1971) recognize the columnar structure and distinguish two strata in a number of paleotropical species. Mickel and Atehortúa (1980) indicate the utility of the spores in confirming relationships of species, and illustrate the strongly echinate spores of *E. crinipes* C. Chr. and the verrucate form in *E. Bellermannianum* (Kl.) Moore, in addition to the diversity included here.

Figs. 96.17–96.26. *Elaphoglossum,* sterile leaves, ×0.25 except as indicated. **17.** *E. Eatonianum.* **18.** *E. peltatum,* ×0.5. **19.** *E. Moorei,* ×0.5. **20.** *E. scalpellum,* the fertile leaf longer and narrower than the sterile, ×0.5. **21.** *E. borangineum.* **22.** *E. crinitum,* the fertile leaf smaller than the sterile. **23.** *E. piloselloides,* plant, ×0.5. **24.** *E. Herminieri,* the fertile leaf much shorter than the sterile. **25.** *E. apodum.* **26.** *E. cardiophyllum.*

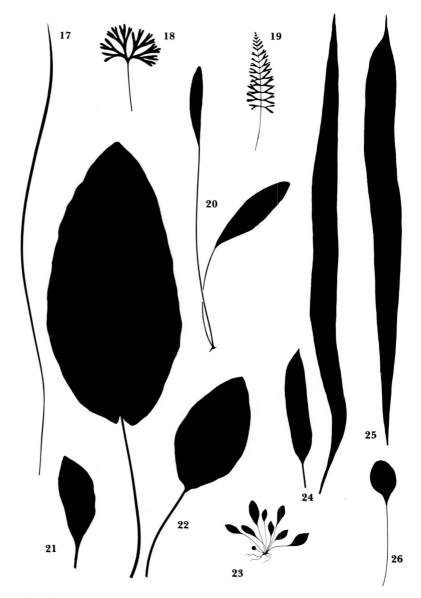

Cytology

The chromosome number of *Elaphoglossum,* mainly $n = 41$, is based on records from distant geographic regions, although the cytology of the genus is inadequately known especially from South America. Records from the American tropics are largely diploid species from the Antillean area, mainly based on the studies of Jamaican plants by Walker (1966, 1973). A few tetraploids are also known and a triploid is reported as a hybrid between *E. chartaceum* (Jenm.) C. Chr. and *E. latifolium. Elaphoglossum crinitum,* sampled from disjunct localities in Jamaica and also Puerto Rico, is consistently $n = 41$. Chromosome numbers of three species from Tristan da Cunha include diploid, tetraploid and hexaploid levels. The latter reported as $2n = $ ca. 250 represents the highest recorded for the genus (Manton and Vida,

27   28   29

30   31

32   33   34

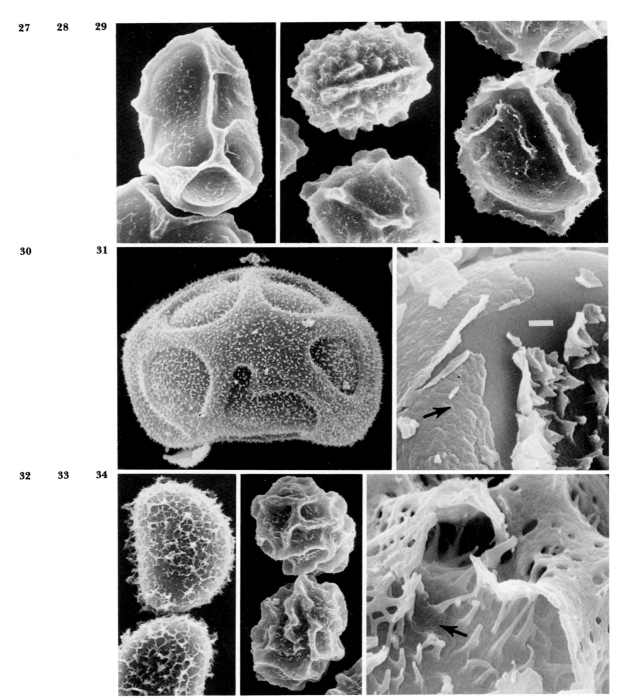

Figs. 96.27–96.34. *Elaphoglossum* spores, × 1000; wall profile or surface detail, × 5000. **27.** *E. ovatum,* low folds sparsely echinate, Ecuador, *Barclay & Juajibioy 8873.* **28.** *E. conforme,* shorter saccate folds, Mexico, *Pringle 13637.* **29.** *E. Plumieri,* perforate, winglike folds, Dominica, *Hodge 20.* **30.** *E. peltatum,* echinate surface and broad, raised areas, Guatemala, *Hatch & Wilson 88.* **31.** *E. squamipes,* echinate surface, lower right, lower perispore left (at arrow), white bar on exospore, Bolivia, *Bang 1799.* **32.** *E. eximium,* reticulate surface, Honduras, *Barkley & Hourcade 40046.* **33.** *E. crinitum,* surface folds, Jamaica, *Wilson & Webster 455.* **34.** *E. latifolium,* abraded surface, columnar structure of central perispore, at center, outer perforate surface, right and left, and lower perispore (arrow). Mexico, *Pringle 10307.*

1968). The report of $n = 40$ for *E. peltatum* from Costa Rica, formerly in *Peltapteris*, is exceptional for *Elaphoglossum* and evidently an aneuploidy derivative. Chromosome numbers reported for species of the paleotropics are mostly tetraploid or diploid and rarely hexaploid.

Literature

Alston, A. H. G. 1958. The Brazilian species of *Elaphoglossum*. Bol. Soc. Broter. 32: 1–32.

Anderson, W. R., and M. R. Crosby. 1966. A revision of the Hawaiian species of *Elaphoglossum*. Brittonia 18: 380–397.

Bell, P. R. 1951. Studies in the genus *Elaphoglossum*, III. Anatomy of the rhizome and frond. Ann. Bot. n.s. 15: 347–357.

Brade, A. C. 1961. O gênero *Elaphoglossum* (Polypodiaceae) no Brazil, I. Chaves para determinar as espécias Brasileiras. Rodriguesia 32: 21–48.

Christ, H. 1899. Monographie des genus *Elaphoglossum*. Denkschr. Schweiz Naturfors. Gesells. 36: 1–159.

Erdtman, G., and P. Sorsa. 1971. Reference under the tribe Dryopterideae.

Manton, I. and G. Vida. 1968. Cytology of the fern flora of Tristan da Cunha. Proc. Roy. Soc. Bot. B. 170: 361–379.

Mickel, J. T. 1980. Relationships of the dissected elaphoglossoid ferns. Brittonia 32: 109–117.

Mickel, J. T., and L. Atehortúa. 1980. Subdivision of the genus *Elaphoglossum*. Amer. Fern Jour. 70: 47–68.

Morton, C. V. 1948. Notes on *Elaphoglossum*, II. The species of the French West Indies. Amer. Fern Jour. 38: 202–214.

Proctor, G. R. 1977. Pteridophyta, *in*: R. A. Howard, Flora of the Lesser Antilles 2. 414 pp., Arnold Arboretum, Harvard Univ., Cambridge, Mass.

Tryon, R. M., and D. S. Conant. 1975. The ferns of Brazilian Amazonia. Acta Amazonica 5: 23–34.

Walker, G. T. 1966. Reference under the tribe Bolbitideae.

Walker, G. T. 1973. Additional cytotaxonomic notes on the pteridophytes of Jamaica. Trans. Roy. Soc. Edinburgh 69: 109–135.

19. Aspleniaceae

Description

Aspleniaceae Frank, Syn. Pflanzenkr. (Leunis), ed. 2, 3: 1465. 1877. Type: *Asplenium* L.

Stem usually erect, to long-creeping, stout to slender, sometimes branched, dictyostelic, or rarely filiform and protostelic, more or less indurated, bearing scales and rarely also trichomes; leaves ca. 1 cm to 3 m long, entire, pinnate, or rarely dichotomous, circinate in the bud, petiole without stipules; sori usually elongate, sometimes very long, borne on a vein, covered by a short to very long indusium, or rarely within a nearly marginal pouch formed by the indusium and opposed leaf-tissue; sporangia usually long-stalked, the stalk 1-rowed below its apex, the annulus interrupted by the stalk; homosporous, spores without chlorophyll. Gametophyte epigeal, with chlorophyll, more or less cordate, or somewhat elongate, with broad thin margins and slightly thickened centrally, glabrous or with 1- to few-celled trichomes, the archegonia borne on the lower surface, especially near the sinus, the antheridia 3-celled, borne on the lower surface, usually on or near the central cushion.

Comments on the Family

The Aspleniaceae are a family of seven genera, with *Ceterach* Willd., and *Diellia* Brack. confined to the Old World, *Schaffneria* and *Holodictyum* confined to America, and *Asplenium, Pleurosorus* and *Camptosorus* in both hemispheres. *Boniniella* Hayata and *Sinephropteris* Mickel are Asiatic genera of uncertain status. The former is perhaps allied to species of *Asplenium* section *Hymenasplenium* on the basis of its dorsiventral stem and short phyllopodia. *Sinephropteris* (*Phyllitis Delavayi* (Franch.) C. Chr.) is perhaps a diminutive relative of *Phyllitis schizocarpa* Copel. of New Guinea and could be treated as a section of *Asplenium.*

The Aspleniaceae are unusually homogeneous for a large group of about 700 species. Nearly all species are similar in characters of the vascular bundles of the petiole, stem scales, sori and indusia, and chromosome number. There is a single X-shaped xylem element in the vascular bundle near the base of the petiole, or two C-shaped elements in separate bundles, each facing toward a side of the petiole. The scales are clathrate, with thickened side and end walls and are often iridescent. Except in a few species, the sori and indusia are more or less elongate and borne along a vein. The chromosome number is consistently $n = 36$ or multiples of that number, except for a few reports of $n = 40$.

Current classifications accept *Asplenium* as a large genus, but with a variable number of small segregate genera. Pichi-Sermolli (1977) recognized 13 genera, that included about 40 species, in addition to *Asplenium.* Holttum (1949) recognized five segregate genera and Copeland (1947) eight. The treatment adopted here recognizes six segregate genera that have differences supporting their position as distinctive divergent lines within the family. All exindusiate species are placed in separate genera: *Ceterach* with the lamina densely scaly beneath and *Pleurosorus* with pubescent leaves. *Asplenium Dalhousiae,* sometimes associated with *Ceterach,* has strongly fenestrate spores, distinct from those of *Ceterach* with prominent folds and slightly perforate surface, which does not support a close affinity of these groups. Three genera with fully anastomosing veins are also recognized: *Camptosorus* with costal areolae, and sori irregularly disposed, *Holodictyum* with the venation as in *Camptosorus* but with sori confined to the outer arc of the costal areolae, and *Schaffneria* with an entire, rhombic-obovate lamina that lacks a costa. The small genus *Diellia,* endemic to the Hawaiian Islands, has lindsaeoid sori. It is clearly derived from the *Asplenium Trichomanes* group and probably should be recognized as a section of *Asplenium. Pleurosoriopsis,* often placed in the Aspleniaceae, belongs in the Polypodiaceae.

Phyllitis, Diplora, Loxoscaphe, and others sometimes recognized as genera are included here with *Asplenium.* The unusual scolopendrioid sori with indusia in facing pairs that characterize

Phyllitis and *Diplora* is a variable character in some species and also occurs infrequently in *Asplenium serratum*.

The sori included in marginal pouches in *Loxoscaphe* are quite distinctive in *A. gibberosum* (Forst.) Mett. of Fiji, but in *A. Mannii* Hook., *A. Beccarianum* Cesati, and *A. flaccidum* Forst. they intergrade to a nearly typical *Asplenium* sorus.

The fossil record of the Aspleniaceae is essentially unknown. *Asplenium alaskanum* Hollick from the Tertiary of Alaska may belong here, or may be an athyrioid fern.

The Aspleniaceae are a rather morphologically isolated family, but perhaps there is some alliance to the Dryopteridaceae. Species in both families have the petiole with two vascular bundles, similar spores with similar surface and lower echinate formations, and the chromosome number of $n = 40$. Some species of Dryopteridaceae have clathrate scales as in the Aspleniaceae.

Key to Genera of Aspleniaceae in America

a. Veins wholly free to fully anastomosing and then the areolae divergent toward the margin. b.
  b. Lamina entire and elliptic to elongate and with a costa, or lamina pinnatifid or more complex. c.
    c. Sori indusiate, leaf glabrous or somewhat scaly, rarely pubescent with non-glandular trichomes.   97. *Asplenium*
    c. Sori exindusiate, leaf pubescent with gland-tipped trichomes.   98. *Pleurosorus*
  b. Lamina entire, more or less rhombic-obovate, lacking a costa.   101. *Schaffneria*
a. Veins fully anastomosing, with costal areolae, the long axis of the areolae nearly parallel to the costa. d.
  d. Sori irregularly disposed on many veins.   99. *Camptosorus*
  d. Sori only on the outer arc of the costal areolae.   100. *Holodictyum*

Literature

Copeland, E. B. 1947. Genera Filicum. 247 pp. Chronica Botanica, Waltham, Mass.

Holttum, R. E. 1949. The classification of ferns. Biol. Reviews 24: 267–296.

Pichi-Sermolli, R. E. G. 1977. Tentamen Pteridophytorum genera in taxonomicum ordinem redigendi. Webbia 31: 313–512.

97. *Asplenium*
Figs. 97.1–97.48

Asplenium l., Sp. Pl. 1079. 1753; Gen. Pl., ed. 5, 485. 1754. Type: *Asplenium marinum* L.

Phyllitis Hill, Brit. Herb. 525. 1757. Type: *Phyllitis scolopendrium* (L.) Newm. = *Asplenium scolopendrium* L.

Scolopendrium Adans., Fam. Plantes 2: 20. 1763. Type: *Scolopendrium vulgare* J. E. Sm. = *Asplenium scolopendrium* L.

Caenopteris Berg., Acta Acad. Sci. Imp. Petrop. 6: 249. (1782). 1786. Type: *Caenopteris rutifolia* Berg. = *Asplenium rutifolium* (Berg.) Kze.

Darea Juss., Gen. Pl. 15. 1789. Type: none designated.

Glossopteris Raf., Anal. Nat. Tab. Univ. 205. 1815, *nom. superfl.* for *Scolopendrium* Adans. and with the same type.

Acropteris Link, Hort. Reg. Bot. Berol. 55. 1833; Handb. Erken. Gew. 3: 23. 1833. Type: *Acropteris septentrionalis* (L.) Link (*Acrostichum septentrionale* L.) = *Asplenium septentrionale* (L.) Hoffm.

Antigramma Presl, Tent. Pterid. 120. 1836. Type: *Antigramma repanda* (Presl) Presl (*Scolopendrium repandum* Presl) = *Asplenium brasiliense* Sw. (*Antigramma brasiliensis* (Sw.) Moore).

Neottopteris J. Sm., Jour. Bot. (Hook.) 3: 409. 1841. Type: *Neottopteris vulgaris* J. Sm., *nom. superfl.* for *Asplenium nidus* L. (*Neottopteris nidus* (L.) Hook.).

Amesium Newm., Hist. Brit. Ferns 10. 1844., *nom. superfl.* for *Acropteris* Link and with the same type.

Thamnopteris Presl, Epim. Bot. 68. 1851, not Brogn. 1849 (fossil Osmundaceae), *nom. superfl.* for *Neottopteris* J. Sm., and with the same type. *Asplenium* section *Thamnopteris* Presl, Tent. Pterid. 105. 1836. Type: *Asplenium nidus* L. (*Thamnopteris nidus* (L.) Presl). The name was published in generic rank only in the names of species.

Tarachia Presl, Epim. Bot. 74. 1851. Type: *Tarachia furcata* (Thunb.) Presl (*Asplenium furcatum* Thunb.) = *Asplenium aethiopicum* (Burm. f.) Bech.

Loxoscaphe Moore, Jour. Bot. Kew Misc. 5: 227. 1853. Type: *Loxoscaphe concinna* (Schrad.) Moore (*Davallia concinna* Schrad.) = *Asplenium concinnum* (Schrad.) Kuhn = *Asplenium theciferum* (HBK.) Mett.

Micropodium Mett., Ann. Mus. Lugd. Batav. 2: 232. 1866, not Saporta, 1861 (fossil Leguminosae?). Type: *Micropodium longifolium* (Presl) Mett. (*Scolopendrium longifolium* Presl, not *Asplenium longifolium* Schrad.) = *Asplenium Haenkei* Copel.

Diplora Baker, Jour. Bot. 11: 235. 1873. Type: *Diplora integrifolia* Baker (not *Asplenium integrifolium* (Blume) Mett.) = *Diplora Durvillei* (Bory) C. Chr. (*Scolopendrium Durvillei* Bory, not *Asplenium Durvillei* Kuhn) = *Asplenium scolopendropsis* F. Muell.

Asplenidictyum (Hook.) J. Sm., Hist. Fil. 333. 1875. *Asplenium* section *Asplenidictyum* Hook., Icon. Plant. *t. 937.* 1854. Type: *Asplenium Findlaysonianum* Hook. (*Asplenidictyum Findlaysonianum* (Hook.) J. Sm.)

Eremopodium Trev., Atti Ist. Veneto V, 3: 589. 1877. Type: *Eremopodium sundense* (Blume) Trev. = *Asplenium sundense* Blume.

Triphlebia Baker, Malesia (Beccari) 3:41. 1886. Type: *Triphlebia pinnata* (Kze.) Baker (*Scolopendrium pinnatum* Kze., not *Asplenium pinnatum* Copel.) = *Asplenium Haenkei* Copel.

Biropteris Kümm., Mag. Bot. Lapok. 19: 2. (1920). 1922. Type: *Biropteris antri-Jovis* Kümm. = *Asplenium Scolopendrium* L.

Hymenasplenium Hayata, Bot. Mag. (Tokyo) 41: 712. 1927. Type: *Hymenasplenium unilaterale* (Lam.) Hayata = *Asplenium unilaterale* Lam. *Asplenium* section *Hymenasplenium* (Hayata) Iwats., Acta Phytotax. Geobot. 27: 44. 1975.

Ceterachopsis (J. Sm.) Ching, Bull. Fan Mem. Instit. Biol. Bot. 10: 8. 1940. *Asplenium* section *Ceterachopsis* J. Sm., Hist. Fil. 317. 1875. Type: *Asplenium alternans* Hook. = *Asplenium Dalhousiae* Hook. (*Ceterachopsis Dalhousiae* (Hook.) Ching.

Description   Terrestrial, rupestral, or epiphytic; stem erect or decumbent and small to rather stout, or rarely long-creeping and sometimes filiform, bearing scales or rarely predominently trichomes, and many to sometimes few, usually long and fibrous roots; leaves monomorphic or rarely dimorphic (the fertile, in *A. dentatum* L., longer and more erect, or in *A. dimorphum* Kze., larger and more complex with narrower segments), ca. 1 cm to 3 m long, borne in a crown, a cluster, or spaced, lamina simple, entire to pinnatifid, to 5-pinnate, or rarely (in *A. septentrionale* (L.) Hoffm.) subdichotomous, glabrous to somewhat scaly, rarely pubescent, veins free or rarely anastomosing without included veinlets; sori elongate to very long (to ca. 8 cm), borne on a nearly unmodified

Fig. 97.1. *Asplenium serratum,* Serra Ricardo Franco, Mato Grosso, Brazil. (Photo P. G. Windisch.)

vein well back of the tip, covered by an indusium attached to the vein, or sori within a nearly marginal pouch formed by the indusium and opposed leaf-tissue; spores more or less ellipsoidal, to somewhat spheroidal, monolete, the laesura $\frac{2}{3}$ to $\frac{3}{4}$ the spore length, with prominent, often winglike, folds and the adjacent surfaces fairly smooth, echinate-reticulate or perforate, or wholly echinate or reticulate with a columelliform meshwork. Chromosome number: $n = 36,\ 39$ or $40,\ 40,\ 72,\ 76,\ 108,\ 144,\ 216,\ 288$; $2n = 72,\ 80,\ 116,\ 120,\ 144,\ 152,\ 216$; apogamous $= 108,\ 210$ or $211,$ ca. $212,\ 216,$ ca. $270,\ 277–280,\ 288,$ ca. $330–357,$ ca. $346.$

The sori of *Asplenium* vary from very short (Fig. 5) to elongate (Fig. 6) to very long (Fig. 7); they may be at a wide angle to the costa (Figs. 7, 11) or nearly parallel to it (Fig. 10). The segments are usually pinnately, but are sometimes flabellately veined (Fig. 12). Veins are rarely anastomosing as in *Asplenium Purdieanum* (Fig. 8) and *A. brasiliense* (Fig. 9). In the latter species, the indusia are facing in pairs (Fig. 9) as in *Asplenium Scolopendrium*. Some examples of lamina diversity are shown in Figs. 13–32.

Systematics

Asplenium is a nearly worldwide genus of about 650 species with about 150 in tropical America. The great majority of species have never been satisfactorily aligned in an infrageneric classification. They evidently represent a multiplicity of evolutionary lines that are not readily classified into subgenera or sections. The problems related to segregate genera are considered in the family treatment.

Maxon published a revision of the *Asplenium salicifolium* group (1908) and of the group of *Asplenium Trichomanes* (1913), and Weatherby (1931) a revision of the small group of *Asplenium fragile*. More recently, the genus has been treated for Venezuela by Morton and Lellinger (1966) and for Chile by Looser (1944). Descriptions and notes on many American species were made by Hieronymus (1918, 1919) in his critical studies of *Asplenium.*

Tropical American Species

In tropical America, *Asplenium* consists of distinctive species, small species-groups, and a few, large, poorly defined complexes. While the latter include series of related species, they may also contain elements that are not closely related. The species, especially of the Andean region, are not well known. The few critical systematic studies have seldom included an assessment of variation over a broad geographic range, or have been based on inadequate materials. Analyses of polyploid and hybrid complexes are essential for a modern treatment of the species.

The following synopsis presents the diversity of the genus in the American tropics. All species placed under one heading are not necessarily closely related. The species with a simple lamina are placed first as a matter of convenience, although they are undoubtedly derived; the primitive species are probably among those with a decompound lamina.

Synopsis of the Diversity of Tropical American *Asplenium*

a. Lamina simple, entire to pinnatifid.   b.
  b. Lamina deeply pinnatifid.   *A. Dalhousiae* Hook.
  b. Lamina entire.
     In *A. brasiliense* Sw. (Fig. 9) and *A. Douglasii* Hook. & Grev. (*Antigramma plantaginea* (Schrad.) Presl), the veins are fully anastomosing. In *A. angustum* Sw., *A. serratum* L. (Figs. 1, 7, 32), and a few other species, the veins are free and the sori are on the acroscopic side of the veins. In *A. Scolopendrium* L. (*Phyllitis Lindenii* (Hook.) Maxon), the veins are free and the sori are in facing pairs, on the acroscopic and basiscopic sides of the veins.
a. Lamina 1-pinnate or more complex   c.
  c. Stem stout, decumbent.
     In *A. Jamesonii* Hook., *A. squamosum* L. (Fig. 13), and *A. tucumanense* Hieron., the lamina is 2-pinnate to 2-pinnate-pinnatifid. In *A. pseudonitidum* Raddi and a few related species, it is ca. 3-pinnate. In *A. scandicinum* Kaulf. the lamina is 3-pinnate-pinnatifid to 5-pinnate.
  c. Stem rather small and usually erect, to rather stout and creeping.   d.
     d. Lamina 2-pinnate or more complex, the pinnae equilateral at the base and usually overlaying the rachis.
        A. cristatum Lam., *A. cuneatum* Lam. (Fig. 15), *A. Eatonii* Davenp., *A. radicans* L., *A. rutaceum* (Willd.) Mett. (Fig. 14), and about 20 other species. *Asplenium radicans* var. *cirrhatum* (Rich.) Rosenst. has a 1-pinnate lamina with a prolonged, proliferous rachis (Fig. 19).
     d. Lamina 1-pinnate to 1-pinnate-pinnatifid, or if 2-pinnate then the pinnae inequilateral at the base.   e.

e. Sorus within a nearly marginal pouch formed by the indusium and opposed leaf-tissue.

> *A. theciferum* (HBK.) Mett. (Figs. 5, 16).

e. Sorus superficial, covered by an indusium attached along the vein. f.

> f. Lamina 1-pinnate with large pinnae, usually imparipinnate or nearly so, the pinnae more or less equilateral at the base. g.
>
>> g. Veins anastomosing, at least toward the margin.
>>
>>> *A. Purdieanum* Hook. (Figs. 8, 24).
>>
>> g. Veins free. h.
>>
>>> h. Stem moderately stout, usually creeping, pinnae strongly serrate.
>>>
>>>> *A. dissectum* Sw. (*A. bissectum* Sw.) (Fig. 31), *A. serra* Langsd. & Fisch. (Fig. 10), and a few other species.
>>>
>>> h. Stem erect to decumbent, pinnae entire to crenate.
>>>
>>>> *A. falcinelllum* Maxon, *A. Feei* Fée, *A. neogranatense* Fée, *A. oligophyllum* Kaulf. (Fig. 30), and several other species.
>
> f. Lamina 1-pinnate, the pinnae small and/or inequilateral at the base, or lamina more than 1-pinnate. i.
>
>> i. Rachis broadly alate throughout.   *A. alatum* Willd.
>>
>> i. Rachis not alate, or narrowly alate mostly toward the apex. j.
>>
>>> j. Stem filiform, ca. 0.5 mm in diameter, bearing many trichomes in addition to a few scales.
>>>
>>>> *A. delicatulum* Presl (Fig. 17), *A. filicaule* Baker, *A. quitense* Hook. (Fig. 18), *A. repens* Hook., and a few other species.
>>>
>>> j. Stem a few mm in diameter or thicker, bearing only scales. k.
>>>
>>>> k. Stem creeping, dorsiventral.
>>>>
>>>>> *A. delitescens* (Maxon) Gómez (Fig. 11), *A. laetum* Sw., *A. obtusifolium* L., and three other American and five paleotropical species. This group is *Asplenium* section *Hymenasplenium* (Iwatsuki, 1975; Smith, 1976).
>>>>
>>>> k. Stem erect or short-decumbent. l.
>>>>
>>>>> l. Basal pinnae small, usually much shorter than the longest ones. m.
>>>>>
>>>>>> m. Rachis straw-colored to greenish, or if darker then greenish-alate.
>>>>>>
>>>>>>> In *A. fragile* Presl, *A. Gilliesii* Hook. (Fig. 29), *A. triphyllum* Presl, and a few other species, and the probably unrelated *A. Clutei* Gilbert, the pinnae are about as broad as long. In *A. harpeodes* Kze., *A. mucronatum* Presl (Fig. 21), *A. Raddianum* Gaud. (Fig. 26), *A. sessilifolium* Desv., and about 20 other species, the pinnae are about twice as long as broad, or longer.

m. Rachis brown to blackish, not greenish-alate.
  A. castaneum Schlect. & Cham., *A. extensum* Fée (Fig. 6), *A. formosum* Willd., *A. monanthes* L. (Figs. 2, 25), *A. resiliens* Kze. and about 15 other mostly related species.

l. Basal pinnae of moderate size, not or only slightly smaller than the longest ones.   n.

n. Leaf pubescent with whitish, multicellular trichomes.
  A. pumilum Sw. (Fig. 20) and the apparently related *A. leucothrix* Maxon.

n. Leaf glabrous to somewhat scaly.   o.

o. Pinnae auriculate, the auricle usually overlaying the rachis.
  A. auriculatum Sw. (Fig. 28), *A. hastatum* Kze., *A. salicifolium* L., and a few other species.

o. Pinnae not auriculate, or if so the auricle not overlaying the rachis.   p.

p. Basal pinnae pinnately veined or cleft.
  A. auritum Sw. (Figs. 3, 22, 23), *A. cuspidatum* Lam. (*A. fragrans* Sw.), and the unrelated *A. abscissum* Willd.

p. Basal pinnae flabellately veined or cleft.
  A. dimidiatum Sw., *A. erosum* L. (Fig. 12), *A. praemorsum* Sw. (Fig. 27) and a few related species.

Ecology (Figs. 1–3)

Asplenium is primarily a genus of moist or wet forests, where it grows in ravines, along streams, on the forest floor, among rocks or on cliffs or as an epiphyte. Boreal species are mostly rupestral and some are closely associated with a particular type of rock.

In tropical America, *Asplenium* usually grows in moist habitats, or in drier sites on cliffs, in lava beds or among rocks. Plants grow in lowland rain forests, in montane forests, cloud forests and gallery forests, in wet woods, moist ravines, on wet rocks along streams, in the spray of waterfalls, and rarely (*A. obtusifolium*) as a rheophyte. *Asplenium* species also grow in disturbed sites such as thickets, road banks, secondary forests, and on old walls and masonry. In alpine areas, plants grow on cliffs or among rocks. *Asplenium* occurs from sea level to above 3500 m, most commonly from 500 to 2500 m, but in the Andes it grows to 4700 m (*A. triphyllum*) and to 5100 m (*A. castaneum*).

Proliferous buds occur on different parts of the leaf, for example, in the axils of the pinnae in *Asplenium commutatum* Kuhn, on the petiole in *A. fragile,* on the upper surface of the pinnae in *A. bulbiferum* Forst., at the rachis tip of an otherwise normal lam-

Fig. 97.2. *Asplenium monanthes,* Volcán Cotopaxi, Ecuador. (Photo W. H. Hodge.)

ina apex in *A. stellatum* Colla and *A. potosinum* Hieron., or at the tip of a prolonged, naked rachis in *A. radicans.* The most extreme modification is in the paleotropical species *Asplenium Mannii* Hook. and *A. bipinnatifidum* Baker which have some leaves reduced to a petiole and naked rachis, the whole axis resembling a stolon, which forms new rooted plants at the tip.

In a few species, colonies are formed by long-creeping stems. In *Asplenium quitense, A. filicaule,* and *A. delicatulum* the filiform stem, at intervals, bears small, erect stems, each with a few leaves. These plants may later persist independently as the creeping stem decays.

Vegetative reproduction is notably frequent in *Asplenium* (Faden, 1973), about a third of the species having buds on the leaf. The often large colonies formed by these species may have a primary role in polyploid speciation by increasing the longev-

Fig. 97.3. *Asplenium auritum,* Tapanti, Costa Rica. (Photo W. H. Hodge.)

ity of sterile hybrids and therefore the opportunities for the initiation of apogamy or chromosome doubling. An example of this potential is the sterile triploid *Asplenium Fawcettii* Jenm. (Walker, 1966) which forms colonies near the summit of Blue Mountain Peak in Jamaica.

Geography (Fig. 4)

Asplenium is essentially a worldwide genus that occurs on all continents, most Atlantic islands, and in the Pacific eastward to the Hawaiian Islands, Pitcairn Island and Easter Island. It grows north to lat. 70°N in Europe and to lat. 61°N in Greenland and south to southernmost South America, Amsterdam Island and New Zealand.

In America *Asplenium* occurs from Greenland to Alaska, south to Magellanes in Chile and Tierra del Fuego in Argentina; also on Bermuda, Ilha Trindade, Cocos Island, the Galápagos Islands, the Juan Fernandez Islands, and the Falkland Islands.

Asplenium is most diverse in the Andes of Colombia to Bolivia where over 100 species occur. There are about 60 species in Guatemala and southern Mexico, over 40 in Hispaniola and about 50 in southeastern Brazil.

Many species, such as *Asplenium auritum, A. theciferum, A. abscissum, A. praemorsum, A. formosum, A. castaneum,* and *A. monanthes,* have extensive ranges through much of tropical America.

Fig. 97.4. Distribution of *Asplenium* in America, south of lat. 35° N.

Figs. 97.5–97.12. Venation, sori and indusia of ***Asplenium***. **5.** Sori within a pouchlike structure, ***A. theciferum***, ×10.0. **6.** Pinnae with sori nearly covered by indusia, ***A. extensum***, ×2.5. **7.** Portion of lamina with long, parallel sori, **A. serratum**, ×1.0. **8.** Portion of fertile lamina with anastomosing veins, ***A. Purdieanum***, ×2.0. **9.** Portion of lamina with anastomosing veins, the sori in facing pairs, ***A. brasiliense***, ×2.0. **10.** Portion of pinna, with free veins, and sori near the costa, ***A. serra***, ×3.0. **11.** Portion of pinna with free veins and divergent sori, ***A. delitescens***, ×2.0. **12.** Portion of fertile pinna with flabellate veins, ***A. erosum***, ×2.0.

There are strong relationships between American species of *Asplenium* and those of the Old World. Some tropical American species that also occur in the paleotropics, or have close relatives there, are *Asplenium cuneatum, A. delitescens, A. formosum, A. monanthes, A. praemorsum, A. pteropus* Kaulf. and *A. theciferum.* The boreal species have European/eastern Asian relations, while those of southern South America, such as *Asplenium trilobum* Cav., *A. obliquum* Forst., and *A. dareoides* Desv., have circumaustral relations.

Spores

The basic structural similarities of the spores, along with the uniformity of the indusium and chromosome number, indicate the homogeneous composition of the genus. Three main spore types, costate-alate, echinate and reticulate, recognized by Viane and Van Cotthem (1977) represent the main forms but the alate types are especially varied. Spores have been examined with the SEM from Kenya (Viane and Van Cotthem, 1977) and all of the Australian species and a few from Malesia by Puttock and Quinn (1980). Those of the *A. Adiantum-nigrum* L. complex examined by Shivas (1969) have a relatively uniform costate-alate surface and size tends to correspond with different ploidy levels. Several other European species shown by Brownsey and Jermy (1973) have varied perforate surfaces and less prominent wings. The following review is based on these SEM studies and a survey of a hundred other species mainly of the American tropics.

Costate-alate spores (Figs. 33–37, 44) are characteristic of most species. Those with echinate-reticulate areas adjacent to the winged folds as in *A. delitescens* (Fig. 34) and *A. Purdieanum* (Fig. 35) are more common in America than those with a nearly smooth surface as in *A. castaneum* (Fig. 33). The lower echinate structure as in *A. Feei* (Fig. 36), in profile in *A. squamosum* (Fig. 37), is usually fused with the surface. Perforate or fenestrate surfaces between folds are somewhat developed in *A. Scolopendrium* (Fig. 38) and are more extensive in *A. auritum* (Fig. 39), *A. Clutei* (Fig. 40), and *A. theciferum* (Fig. 44). The most elaborate development of the inner perispore is in the perforate spores of *A. Dalhousiae* (Fig. 48).

Echinate spores (Figs. 41–43) are of two forms in the American species. They tend to be spheroidal in *A. falcinellum* (Fig. 41) with prominent, coarse spines adjacent to reticulate surfaces in *A. neogranatense* (Fig. 42). This form represents a widely distributed type characteristic of several species in Africa and also in the Philippines. The other echinate spores have irregular processes without reticulate areas as in *A. repens* (Fig. 43). This form seems to represent a reduction in perispore deposit characteristic of a group of diminutive species of the American tropics.

Reticulate spores (Figs. 45–47) are the most complex and have the surface meshwork fused to a lower columelliform structure as in the wall profile of *A. Eatonii* (Fig. 45). There appear to be two subtly different reticulate forms. In *A. Eatonii* the

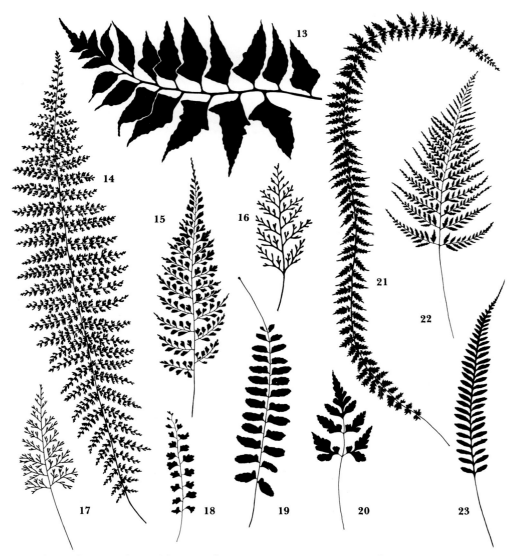

Figs. 97.13–97.23. Lamina architecture of *Asplenium*. **13.** *A. squamosum,* pinna, ×0.25. **14.** *A. rutaceum,* ×0.5. **15.** *A. cuneatum,* ×0.25. **16.** *A. theciferum,* ×0.5. **17.** *A. delicatulum,* ×0.5. **18.** *A. quitense* ×0.5. **19.** *A. radicans* var. *cirrhatum,* with elongate, proliferous rachis tip, ×0.25. **20.** *A. pumilum,* ×0.25. **21.** *A. mucronatum,* an epiphyte with pendent leaves and ascending pinnae, ×0.25. **22.** *A. auritum,* with 2-pinnate lamina, ×0.25. **23.** *A. auritum,* with 1-pinnate lamina, ×0.25.

meshwork is echinate (Fig. 46) while in *A. serra* (Fig. 47) the projecting elements are not developed and the perforations tend to be larger. Spores of the African *A. Friesiorum* C. Chr. are similar and the leaves also resemble those of *A. serra*.

Intermediates between the main forms particularly in spores with strongly perforated to reticulated surfaces suggest there are similar systems of wall formation within the genus. The general structure resembles that of spores in several genera of the Dryopteridaceae as *Tectaria, Polystichum, Oleandra* and *Diplazium*. The lower echinate perispore is also well developed in *Tectaria*, but in *Asplenium* it is strongly fused to the outer reticulum. Di-

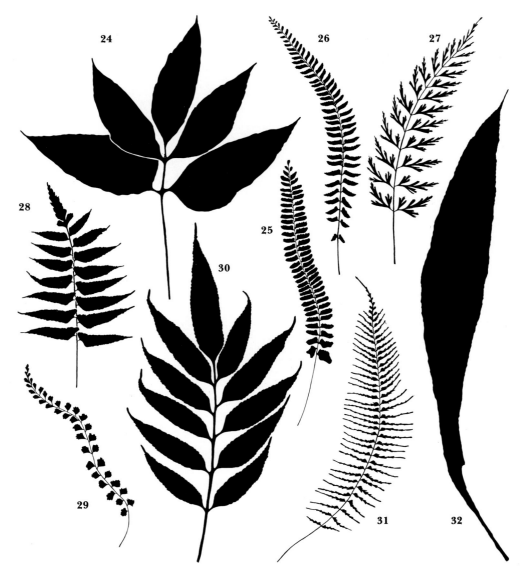

Figs. 97.24–97.32. Lamina architecture of *Asplenium*. **24.** *A. Purdieanum,* ×0.25. **25.** *A. monanthes,* ×0.25. **26.** *A. Raddianum,* ×0.25. **27.** *A. praemorsum,* ×0.25. **28.** *A. auriculatum,* ×0.25. **29.** *A. Gilliesii,* ×0.5. **30.** *A. oligophyllum,* ×0.25. **31.** *A. dissectum,* ×0.25. **32.** *A. serratum,* ×0.25.

versity of the surface structure in *Tectaria* spores (Figs. 69.23–69.37) is similar to but not as great as that in *Asplenium.*

Cytology

The cytology of *Asplenium* in the American tropics is scarcely known although the genus has been extensively studied especially in temperate America and Europe. Species are mostly diploid or tetraploid based on $n = 36$ with exceptional reports of $2n = 39$ II or 40 II in *A. repandulum* Kze. from Chiapas, Mexico (Smith and Mickel, 1977), and $n = 40$ in *A. unilaterale* Lam. of

Figs. 97.33–97.40. *Asplenium* spores, × 1000.   **33.** *A. castaneum,* low folds, these and adjacent surface nearly smooth, Mexico, *Pringle 6150.*   **34.** *A. delitescens,* alate folds and adjacent reticulate-echinate surface, British Honduras, *Gentle 2714.*   **35.** *A. Purdieanum,* alate folds and echinate reticulate-surface, Honduras, *Edwards P-690.*   **36.** *A. Feei,* abraded surface with lower, echinate perispore, Galápagos Islands, *Stewart 861,* × 5000.   **37.** *A. squamosum,* wall profile, the coarse, columnar structure fused to the outer surface, Peru, *Vargas 15645,* × 10,000.   **38.** *A. Scolopendrium,* low folds and reticulate-perforate surface, Haiti, *Ekman 7804.*   **39.** *A. auritum,* low folds and reticulate-perforate surface, Jamaica, *Proctor 19537.*   **40.** *A. Clutei,* sparse, low folds, perforate surface, Haiti, *Holdridge 1349.*

India (Bir, 1973). Higher ploidy levels to octoploid are reported in New Zealand and to 16-ploid in the neotropics.

The American tropical species are largely known from the work of Walker (1966, 1973) in Jamaica, reporting about equal numbers of diploids and tetraploids but also higher levels to 16-ploid. Tetraploid and octoploid plants are known in *A. radicans; A. erosum* is octoploid with $n = 144$ as well as 16-ploid with 288; and *A. Fawcettii* is recognized as a local triploid hybrid with $n = 108$. In Jamaica, *A. resiliens* is a sexual hexaploid $n = 108$, while it is known elsewhere in America as an apogamous triploid with 108. The widespread *A. auritum* var. *auritum* is a sexual tetraploid in Jamaica with $n = 72$ while the local var. *obtusum* Mett. (as var. *macilentum* (Kze.) Moore) is an apogamous octoploid with 288. The record of a sexual tetraploid of *A. monanthes* with $2n = 72$ II from Chiapas, Mexico, reported by Smith and Mickel is exceptional, for this widely distributed species is triploid, apogamous with 108 in Veracruz, Mexico, and Azores, Madeira, Tristan da Cunha, and the southeastern United States.

The European aspleniums include about equal numbers of diploids and tetraploids among reports of some 38 taxa. Eleven of the tetraploids have originated as genomic alloploids, six as autopolyploids and numerous hybrids are recognized between the aspleniums as well as one with *Ceterach*. The cytogenetic analysis of *Asplenium* by Lovis (1973, 1977) reviews the literature, including early study of the central European species by Meyer (1957) and the series of collaborative works by Brownsey, Lovis, Manton, Reichstein, Shivas, Sleep, and Vida. An analysis of the role of polyploidy in evolution by Vida (1976) includes a synopsis of the diverse leaf forms among the European diploid and tetraploid aspleniums. These extensive studies supply the most complete evidence on cytogenetic systems in the ferns and provide a basis for examination of the higher polyploid levels in the tropical species.

In temperate America a complex in the Appalachian region involves a core of three diploids, two species of *Asplenium* and *Camptosorus*. These have given rise to tetraploids and triploids as depicted in the morphological and cytological assessment by Wagner (1954) and confirmed by chromotographic studies of Smith and Levin (1963) and Harborne et al. (1973).

The New Zealand species are wholly polyploid according to Brownsey (1977a, 1977b), and include nine widely distributed tetraploids and eight largely endemic octoploids. Morphologically uniform populations of apogamous *Asplenium flabellifolium* Cav. are known with different aneuploid numbers. The Australian plants are reported as ca. 212, while in New Zealand there are populations with 210–211 and also 277–280.

The cytological assessment of the *Asplenium unilaterale* complex in the eastern Himalayan area by Bir (1973) explains the exceptional records of $n = 40$ and 76. Three varieties of the species are reported as $n = 40$ while the widespread sexual tetraploid of southern India with $n = 76$ is considered a dibasic hybrid. Its origin is proposed as a cross between a sexual form with

Figs. 97.41–97.48. *Asplenium* spores, × 1000. **41.** *A. falcinellum,* echinate with adjacent reticulate surfaces. Costa Rica, *Madison 733.* **42.** *A. neogranatense,* detail of coarse echinate elements and reticulum, Colombia, *Cuatrecasas 12550.* **43.** *A. repens,* irregularly echinate, Ecuador, *Spruce 5336.* **44.** *A. theciferum,* alate folds and echinate-reticulate areas, Colombia, *Cuatrecasas 19492.* **45, 46.** *A. Eatonii,* Mexico, *Pringle* in 1914. **45.** Wall profile, the smooth exospore below, and columelliform perispore fused to the reticulate surface, above, × 10,000. **46.** Reticulate surface with echinate elements. **47.** *A. serra,* reticulate surface with lower columelliform structure, Dominican Republic, *Gastony et al. 638.* **48.** *A. Dalhousiae,* wall profile with branched columelliform perispore, Nepal, *Hara et al. 794,* × 5000.

36 and one with 40, that formed a diploid with 76 single chromosomes and then doubled to become a sexual tetraploid.

Five Japanese species of *Asplenium* with $2n = 144$, and karyotypes of four with $2n = 72$ were examined by Tatuno and Kawakami (1969) and Kawakami (1970). They arrange the chromosomes into six basic types each consisting of 12 chromosomes and on the basis of these analyses propose the primary base number $b = 12$, for *Asplenium*.

Most cytological modifications of apogamous ferns are of the classical type described by Döpp (1932) and Manton (1950) which entail the formation of a restitution nucleus in the mitotic division of an 8-celled sporangium prior to meiosis. Another cytological compensatory system in apogamous ferns was recognized in the African *Asplenium aethiopicum* (Burm.) Bech., by Braithwaite (1964). In these plants restitution nuclei are formed in the 16-celled sporangia through an abbreviated meiotic sequence giving rise to spore diads with the same chromosome number as the sporophyte.

Literature

Bir, S. S. 1973. Cytology of Indian Pteridophyta. Glimpses in Plant Research 1: 28–119.

Braithwaite, A. F. 1964. A new type of apogamy in ferns. New Phytol. 63: 293–305.

Brownsey, P. J. 1977a. A taxonomic revision of the New Zealand species of *Asplenium*. New Zealand Jour. Bot. 15: 39–68.

Brownsey, P. J. 1977b. *Asplenium* hybrids in the New Zealand flora. New Zealand Jour. Bot. 15: 601–637.

Brownsey, P. J., and A. C. Jermy. 1973. A fern collecting expedition to Crete. Brit. Fern Gaz. 10: 331–348.

Döpp, W. 1932. Die Apogamie bei *Aspidium remotum* A. Br. Planta (Berl.) 17: 86–152.

Faden, R. B. 1973. Some notes on the gemmiferous species of *Asplenium* in tropical east Africa. Amer. Fern Jour. 63: 85–90.

Harborne, J. B., C. A. Williams, and D. M. Smith. 1973. Species specific kaempferol derivatives in the Appalachian *Asplenium* complex. Biochem. Syst. 1: 51–54.

Hieronymus, G. 1918. Aspleniorum species novae et non satis notae. Hedwigia 60: 210–266.

Hieronymus, G. 1919. Kleine Mitteilungen über Pteridophyten, II. Hedwigia 61: 4–39.

Iwatsuki, K. 1975. Taxonomic studies of Pteridophyta, X. Acta Phytotax. Geobot. 27: 39–55.

Kawakami, S. 1970. Karyological studies on Aspleniaceae 2. Chromosome of seven species of Aspleniaceae. Bot. Mag. (Tokyo) 83: 74–81.

Looser, G. 1944. Los *Asplenium* (Filices) de Chile. Lilloa 10: 233–264.

Lovis, J. D. 1973. Phyletic problems in Aspleniaceae, *in*: A. C. Jermy et al., The phylogeny and classification of the ferns, pp. 211–228, Bot. Jour. Linn. Soc. 67, Suppl. 1.

Lovis, J. D. 1977. Evolutionary patterns and processes in ferns. Advances Bot. Research 4: 229–440.

Manton, I. 1950. Problems of Cytology and Evolution in the Pteridophyta. 316 pp. Cambridge Univ. Press, Cambridge, England.

Maxon, W. R. 1908. *Asplenium salicifolium* and confused species. Contrib. U.S. Nat. Herb. 10: 475–481.

Maxon, W. R. 1913. *Asplenium Trichomanes* and its American allies. Contrib. U.S. Nat. Herb. 17: 133–153.

Meyer, D. E. 1957. Zur zytologie der Asplenien Mitteleuropas (I-XV). Ber. Deutsch Bot. Ges. 70: 57–66.

Morton, C. V., and D. B. Lellinger. 1966. The Polypodiaceae subfamily Asplenioideae in Venezuela. Mem. New York Bot. Gard. 15: 1–49.

Puttock, C. F., and C. J. Quinn. 1980. Perispore morphology and the taxonomy of the Australian Aspleniaceae. Aust. Jour. Bot. 28: 305–322.

Shivas, M. G. 1969. A cytotaxonomic study of the *Asplenium Adiantum-nigrum* complex. Brit. Fern Gaz. 10: 68–80.

Smith, A. R. 1976. *Diplazium delitescens* and the neotropical species of *Asplenium* sect. *Hymenasplenium.* Amer. Fern Jour. 66: 116–120.

Smith, A. R., and J. T. Mickel. 1977. Chromosome counts for Mexican ferns. Brittonia 29: 391–398.

Smith, D. M., and D. A. Levin. 1963. A chromatographic study of reticulate evolution in the Appalachian *Asplenium* complex. Amer. Jour. Bot. 50: 952–958.

Tatuno, S., and S. Kawakami. 1969. Karyological studies in Aspleniaceae 1. Karyotypes of three species in *Asplenium.* Bot. Mag. (Tokyo) 82: 436–444.

Viane, R., and W. van Cotthem. 1977. Spore morphology and stomatal characters of some Kenyan *Asplenium* species. Ber. Deutsch Bot. Ges. 90: 219–339.

Vida, G. 1976. The role of polyploidy in evolution, Evolutionary Biology, pp. 267–294. Praha.

Wagner, W. H. 1954. Reticulate evolution in the Appalachian aspleniums. Evolution 8: 103–118.

Walker, T. G. 1966. A cytotaxonomic survey of the pteridophytes of Jamaica. Trans. Roy. Soc. Edinburgh 66: 169–237.

Walker, T. G. 1973. Additional cytotaxonomic notes on pteridophytes of Jamaica. Trans. Roy. Soc. Edinburgh 69: 109–135.

Weatherby, C. A. 1931. The group of *Asplenium fragile* in South America. Contrib. Gray Herb. 95: 49–52.

98. *Pleurosorus*
Figs. 98.1–98.5

Pleurosorus Fée, Mém. Fam. Foug. 5 (Gen. Fil.): 179. 1852. Type: *Pleurosorus papaverifolius* (Kze.) Fée (*Gymnogramma papaverifolia* Kze.).

Description

Rupestral; stem erect, small, bearing scales and very many, long, fibrous roots; leaves monomorphic, ca. 3–15 cm long, borne in a cluster, lamina 1-pinnate to 2-pinnate, pubescent or glandular-pubescent, veins free; sori elongate, borne on the nearly unmodified veins, exindusiate; spores somewhat ellipsoidal to spheroidal, monolete, the laesura $\frac{3}{4}$ the spore length, with few, coarse folds sometimes with echinate margins, the surface fairly smooth. Chromosome number: $n = 36, 72$; $2n = 72$.

Comments on the Genus

Pleurosorus is a genus of four species of widely disjunct regions. *Pleurosorus papaverifolius* (Kze.) Fée (Fig. 2) occurs in central Chile to southern Argentina (Fig. 1), *P. hispanicus* (Cosson) Morton in southern Spain and Morocco, *P. rutifolius* (R. Br.) Fée in Australia, Tasmania and New Zealand, and *P. subglandulosus* (Hook. & Grev.) Tind. in Australia. The leaves are glandular-pubescent in *P. papaverifolius* (Fig. 3) and *P. subglandulosus*, pubescent in the other two species, and the sori of all species are exindusiate.

Asplenium Petrarchae (Guérin) DC. of Europe has pubescent leaves and a hybrid has been synthesized between the diploid ssp. *bivalens* (D. Meyer) Lovis & Reichst. and *Pleurosorus hispanicus* (Lovis, 1973). This suggests that the species of Europe may be derived from *Asplenium* and should be placed in that genus. The three austral species may have a similar origin but there are no *Asplenium* species presently in South America or Australia/New Zealand that appear to be closely allied. Thus *Pleurosorus* is maintained here pending further information on its origin, particularly of the species in America.

Pleurosorus papaverifolius grows in exposed or shaded crevices of dry rocks or cliffs. It occurs in the Province of Coquimbo south to Bio-Bio in central Chile and in Andean Argentina from Mendoza to Santa Cruz.

Spores of *Pleurosorus papaverifolius* (Fig. 4) and *P. hispanicus* (Fig. 5) are quite smooth with few prominent folds while the species of Australia have shorter intersecting folds. A lower colu-

Fig. 98.1. Distribution of *Pleurosorus* in America.

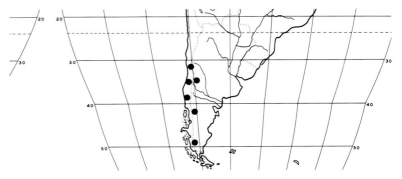

2

3   4   5

Figs. 98.2–98.5. *Pleurosorus.* **2, 3.** *P. papaverifolius.* **2.** Plant, × 1.0. **3.** Fertile lamina with gland-tipped trichomes, × 1.5. **4, 5.** Spores, × 1000. **4.** *P. papaverifolius* with few, low folds and fairly smooth surface, Argentina, *Hunnewell 15866.* **5.** *P. hispanicus,* laesura across upper spore, the lower in end view, Spain, *Font Quer* in 1925.

melliform structure is fused to the outer spore surface as in *Asplenium*.

The chromosome number is not reported for the American species but *Pleurosorus hispanicus* is diploid with $n = 36$ (Meyer, 1964) and *P. rutifolius* is tetraploid with $n = 72$ in New Zealand (Brownlie, 1958).

Literature

Brownlie, G. 1958. Chromosome numbers in New Zealand ferns. Trans. Roy. Soc. New Zealand 85: 213–216.

Lovis, J. D. 1973. Phyletic problems in Aspleniaceae, *in*: A. C. Jermy et al., The phylogeny and classification of ferns, pp. 211–228. Bot. Jour. Linn. Soc. 67, Suppl. 1.

Meyer, D. 1964. Über neue and selten Asplenien Europas. Ber. Deutsch. Bot. Ges. 77: 3–13.

99. *Camptosorus*
Figs. 99.1–99.5

Camptosorus Link, Hort. Reg. Bot. Berol. 69. 1833. Type: *Camptosorus rhizophyllus* (L.) Link (*Asplenium rhizophyllum* L.).

Description

Rupestral; stem erect, small, bearing scales and many, long, fibrous roots; leaves monomorphic, ca. 5–35 cm long, borne in a cluster, lamina entire, ca. oblong to lanceolate-attenuate to linear-elliptic-attenuate, glabrate, veins anastomosing without included veinlets; sori short to elongate, borne in various positions on the nearly unmodified veins, not paraphysate, covered by an indusium attached along the vein; spores ellipsoidal, monolete, the laesura $\frac{2}{3}$ to $\frac{3}{4}$ the length, with prominent winglike folds and reticulate-echinate adjacent areas. Chromosome number: $n = 36$; $2n = 72$.

Comments on the Genus

Camptosorus includes two species, *C. rhizophyllus* (L.) Link (Fig. 2) of eastern North America and *C. sibiricus* Rupr. of northeastern China to Japan and north to Siberia and Kamtchatka. It is clearly related to *Asplenium* but there is no intermediate species that indicates a direct affinity between the genera. The irregularly disposed sori and fully anastomosing veins with costal areolae (Figs. 3, 4) are distinctive features of *Camptosorus*.

Fig. 99.1. Distribution of *Camptosorus* in America.

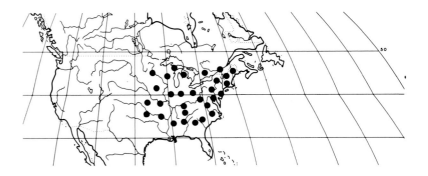

2

3   4   5

Figs. 99.2–99.5. *Camptosorus rhizophyllus.* **2.** Plant with plantlets formed on three leaf apices, ×0.5. **3.** Portion of fertile leaf with immature sori along anastomosing veins, costa left, ×4.5. **4.** Portion of leaf with old sori, some removed at upper left, ×3.0. **5.** Spores with reticulate-echinate surface and alate folds, Vermont, *Blanchard* in 1910, ×1000.

Camptosorus rhizophyllus grows from Maine and southern Quebec west to southeastern Minnesota and south to Georgia and eastern Oklahoma (Fig. 1). It usually grows on calcareous rocks in damp, shaded habitats. Large colonies are often formed in moss mats by means of the proliferous leaf tips (Fig. 2). Sometimes it grows on sandstone, less often on granite, and rarely it is terrestrial or epiphytic on the base of trees.

The spores are unusually small with prominent winglike folds and reticulate-echinate areas (Fig. 5). They are of the costate-alate type similar to those of many species of *Asplenium*.

The chromosome number $n = 36$ is documented in plants

from Virginia and Indiana. Sterile hybrids are formed between *Camptosorus rhizophyllus* and *Asplenium platyneuron* (L.) BSP. where the two species grow in proximity. At Havana Glen, Alabama, the hybrid has become tetraploid and fertile. A rare hybrid is reported between *Camptosorus rhizophyllus* and *Asplenium Ruta-muraria* L. var. *cryptolepis* (Fern.) Wherry from southern Ohio.

The Asiatic species *Camptosorus sibiricus* is reported as $n = 36$ from several localities in Japan. Somatic chromosomes, $2n = 72$ were karyotyped by Kawakami (1970) and the primary base number $b = 12$ proposed, as in *Asplenium*. The chromosome morphology, compared to that of *Asplenium varians* Hook. & Grev., is remarkably similar, although the *Camptosorus* chromosomes are about half the size of those in the *Asplenium* species.

Literature

Kawakami, S. 1970. Karyological studies on Aspleniaceae, 2. Chromosome of seven species in Aspleniaceae. Bot. Mag. (Tokyo) 83: 74–81.

100. *Holodictyum*
Figs. 100.1–100.7

Holodictyum Maxon, Contrib. U.S. Nat. Herb. 10: 481. 1908. Type: *Holodictyum Ghiesbreghtii* (Fourn.) Maxon (*Asplenium Ghiesbreghtii* Fourn.).

Fig. 100.1. *Holodictyum Ghiesbreghtii,* Rio Malila, Hidalgo, Mexico. (Photo L. D. Gómez.)

Fig. 100.2. Distribution of *Holodictyum.*

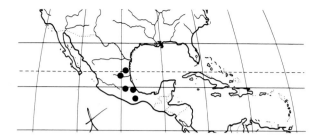

Description

Rupestral; stem erect, rather small, bearing scales and many fibrous roots; leaves monomorphic, ca. 30–50 cm long, borne in a cluster, lamina entire, linear-lanceolate, glabrous, veins anastomosing without included veinlets; sori elongate, borne on the nearly unmodified outer arc of a costal areola, not paraphysate, covered by an indusium with its margin toward the costa; spores more or less ellipsoidal, monolete, the laesura $\frac{2}{3}$ to $\frac{3}{4}$ the spore length, with few prominent folds, the surface perforate or fenestrate. Chromosome number: not reported.

The lamina is long-tapered and extends nearly to the base of the subsessile leaf (Fig. 3). The sori are only on the outer arc of the veins forming costal areolae (Fig. 4) and the indusia open toward the costa (Fig. 5).

Systematics

Holodictyum (Fig. 1) is a rare, monotypic genus of Mexico based on *H. Ghiesbreghtii* (Fourn.) Maxon. A second species *H. Finkii* (Baker) Maxon, recognized in the original description of the genus, is included under *H. Ghiesbreghtii* as it differs in only a few, minor characters. The sori, borne only on the outer arc of costal areolae, is the principal feature of *Holodictyum*. The genus does not seem to have a close affinity with others of the family.

Ecology and Geography

Holodictyum (Fig. 1) grows on or among wet rocks in the states of Tamaulipas, San Luis Potosi, Hidalgo, Veracruz and Oaxaca, Mexico (Fig. 2); it occurs at 400–1200 m.

Spores

The spores have a perforate surface with occluded lumen at the base of projecting folds (Fig. 6). A lower columelliform structure is fused to the surface meshwork (Fig. 7) as in the reticulate type of spores in *Asplenium*.

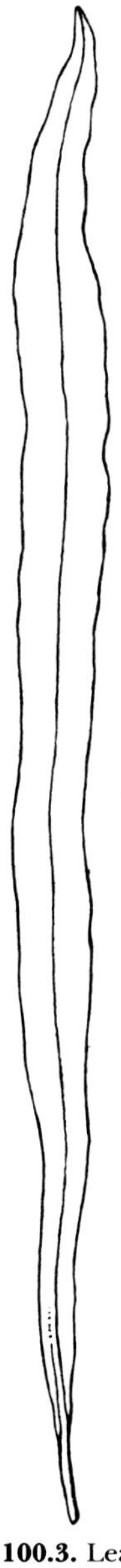

Fig. 100.3. Leaf of *Holodictyum Ghiesbreghtii,* × 0.5.

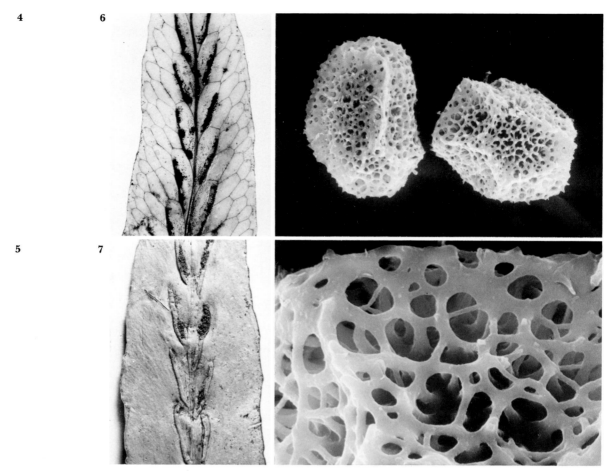

Figs. 100.4–100.7. *Holodictyum Ghiesbreghtii.*   **4.** Portion of cleared lamina, with anastomosing veins and sori on the costal areole, × 2.0.   **5.** Portion of lamina with indusia facing the costa, × 2.0.   **6, 7.** Spores, Mexico, *Palmer 336.*   **6.** Fenestrate spores with few, somewhat projecting, folds, × 1000.   **7.** Detail of perforate surface and lower columelliform structure, × 5000.

101. *Schaffneria*
Figs. 101.1–101.5

Schaffneria Fée, Mém. Fam. Foug. 7: 56. 1857. Type: *Schaffneria nigripes* Fée.

Description   Rupestral; stem erect, small, bearing scales and fibrous roots; leaves monomorphic, borne in a cluster, ca. 2–15 cm long, lamina entire, more or less rhombic-obovate, glabrate, flabellately veined, the veins anastomosing without included veinlets; sori

Fig. 101.1. Distribution of *Schaffneria*.

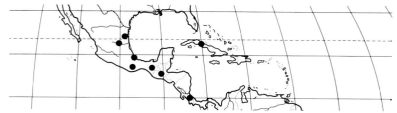

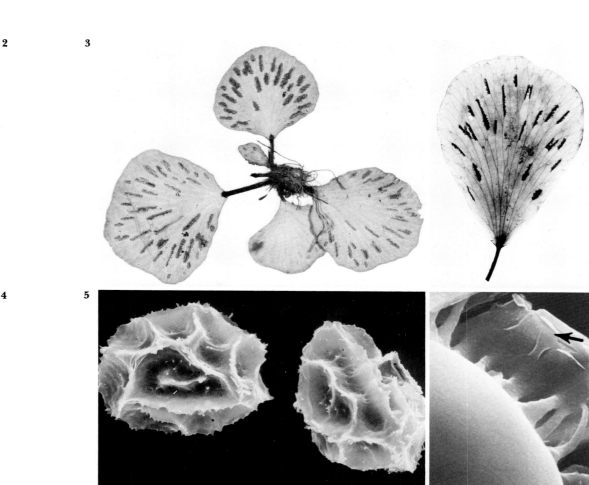

Figs. 101.2–101.5. *Schaffneria nigripes.* **2.** Plant, × 1. **3.** Lamina with flabellately anastomosing veins and no costa, × 1.5. **4, 5.** Spores. **4.** Costate-alate surface, Guatemala, *Tuerckheim II 1292*, × 1000. **5.** Wall profile, the inner columelliform structure fused to outer surface (arrow), the smooth exospore, left, Mexico, *Pringle 3366*, × 5000.

short to long, borne on nearly unmodified veins, not paraphysate, covered by an indusium; spores ellipsoidal, monolete, the laesura $\frac{3}{4}$ the spore length with few prominent folds and somewhat reticulate-echinate adjacent areas. Chromosome number: not reported.

Schaffneria nigripes (Fig. 2) has small, entire, thickened leaves that lack a central costa (Fig. 3).

Systematics

Schaffneria is a monotypic genus of tropical America based on *S. nigripes* Fée. It is the most distinctive of the genera allied to *Asplenium* in its leaf shape, venation and lack of a costa. *Schaffneria Delavayi* (Franch.) Tard. is a superficially similar, but unrelated, species of southeast Asia, that has been placed in the monotypic genus *Sinephropteris* (Mickel, 1976).

Ecology and Geography

Schaffneria nigripes grows on mossy rocks and ledges, in wet forests or canyons at ca. 1000–1500 m. It is a rare species that occurs in Tamaulipas in Mexico, south to Costa Rica and in Cuba (Fig. 1).

Spores

The surface, with prominent, intersecting folds and adjacent areas scarcely reticulate-echinate (Fig. 4), is fused with the lower, columelliform structure (Fig. 5). These spores are of the costate-alate type, as in *Asplenium,* with a compact surface and few prominent folds.

Literature

Mickel, J. T. 1976. *Sinephropteris,* a new genus of scolopendroid ferns. Brittonia 28: 326–328.

20. Davalliaceae

Description

Davalliaceae Frank, Syn. Pflanzenk. (Leunis), ed. 2. 3: 1474. 1877. Type: *Davallia* Sm.
Nephrolepidaceae Pic.-Ser., Webbia 29: 8. 1974. Type: *Nephrolepis* Schott.
Gymnogrammitidaceae Ching, Acta Phytotax. Sinica 11: 12. 1966. Type: *Gymnogrammitis* Griffith.

Stem erect, decumbent, or long-creeping, moderately stout to slender, or very rarely reduced, with a dictyostele, more or less indurated, bearing scales; leaves ca. 15 cm to 3 m long, entire to usually pinnate, circinate in the bud, petiole without stipules; sporangia borne in roundish to somewhat elongate, rarely linear, abaxial sori, at the apex of a vein, near or back of the margin, or on an intramarginal commissure, paraphysate or usually not, indusiate or rarely exindusiate, indusium small and roundish, lunate, reniform, or orbicular with a narrow sinus, or rarely linear and attached at the base, or roundish to elongate and attached also along the sides, sporangia usually long-stalked, the stalk 2-rowed below its apex, the annulus interrupted by the stalk; homosporous, spores without chlorophyll. Gametophyte epigeal, with chlorophyll, more or less cordate, or sometimes broader, or elongate, slightly thickened centrally, with thin margins, usually with 1- to few-celled trichomes, archegonia borne on the lower surface, usually on the central cushion, antheridia 3-celled, borne on the lower surface, especially toward the margins.

Comments on the Family

The Davalliaceae are a family of about 10, or perhaps fewer, genera of the Old World with *Nephrolepis* also in America. The family is predominantly in tropical Asia and includes *Araiostegia* Copel., *Leucostegia* Presl, *Davallodes* Copel., *Davallia* Sm. (*Scyphularia* Fée), *Humata* Cav. and *Gymnogrammitis* Griffith, as well as the usually recognized *Trogostolon* Copel., and *Parasorus* vAvR. These genera are mostly epiphytes with a dorsiventral stele and leaves that disarticulate near the base of the petiole. The position of *Nephrolepis* in the family is perhaps doubtful, but it is placed here, as discussed under the genus, on the basis of its spore characters.

There are no certain reports of fossil Davalliaceae.

The family Davalliaceae was earlier allied to the dicksonioiddennstaedtioid ferns (Bower, 1928; Copeland 1947), largely on the basis of the cup-shaped indusium and sori that are often nearly marginal. However, this alliance is not supported by the cytology or spore characters. It is currently considered (Sen et al. 1972) as having affinities to the Dryopteridaceae, especially the group of *Tectaria.*

Detailed information on the comparative anatomy of the Davalliaceae, and genera that have been associated with it, is contained in the extensive study of Pérez Arbeláez (1928) and in a recent paper by Kato & Mitsuta (1979).

Literature

Bower, F. O. 1928. The Ferns (Filicales), Vol. 3, 306 pp. Cambridge Univ. Press, Cambridge, England.

Copeland, E. B. 1947. Genera Filicum. 247 pp. Chronica Botanica, Waltham, Mass.

Kato, M., and S. Mitsuta. 1979. Stelar organization in davallioid ferns. Phytomorph. 29: 362–369.

Sen, T., U. Sen, and R. E. Holttum. 1972. Morphology and anatomy of the genus *Davallia, Araiostegia,* and *Davallodes,* with a discussion of their affinities. Kew Bull. 27: 217–243.

Pérez Arbeláez, E. 1928. Die natürliche Gruppe der Davalliaceen (Sm.) Kfs. Bot. Abhandl. Goebel 14: 1–96.

102. *Nephrolepis*

Figs. 102.1–102.15

Nephrolepis Schott, Gen. Fil. *t. 3.* 1834. Type: *Nephrolepis exaltata* (L.) Schott (*Polypodium exaltatum* L.).

Leptopleuria Presl, Tent. Pterid. 136. 1836. Type: *Leptopleuria abrupta* (Bory) Presl (*Dicksonia abrupta* Bory) = *Nephrolepis abrupta* (Bory) Mett.

Lepidoneuron Fée, Mém. Fam. Foug. 5 (Gen. Fil.): 301. 1852. Type: *Lepidoneuron biserratum* (Sw.) Fée (*Aspidium biserratum* Sw.) = *Nephrolepis biserrata* (Sw.) Schott.

Lindsayoides Nakai, Ord. Fam. Trib. Nov. 202. 1943. Type: *Lindsayoides acutifolia* (Desv.) Nakai (*Lindsaea acutifolia* Desv.) = *Nephrolepis acutifolia* (Desv.) Christ.

Description Terrestrial, rupestral, or epiphytic; stem erect or decumbent, sometimes reduced, bearing scales and few to many fibrous roots; leaves monomorphic or partly dimorphic (the fertile pinnae borne toward the apex of the lamina and narrower, lobed to deeply pinnatifid), ca. 15 cm to 3 m long, borne in a cluster, lamina 1-pinnate, the pinnae entire or nearly so or sometimes the fertile to deeply pinnatifid, glabrous to somewhat pubescent and/or scaly, veins free; sori round to slightly elongate, borne on a slightly raised receptacle, on the vein-ends, near or back of the margin, or (in *N. acutifolia* (Desv.) Christ) the sporangia borne on a more or less continuous intramarginal commissure connecting the vein-ends, not paraphysate, covered by a lunate to reniform or an orbicular indusium with a narrow sinus, or (in *N. acutifolia*) a linear indusium; spores ellipsoidal, monolete, the laesura usually ca. $\frac{1}{2}$ the spore length, the surface irregularly tuberculate to rugose. Chromosome number: $n = 41$, 82; $2n = 82$.

The typical forms of the sori and indusia of *Nephrolepis* are shown in Figs. 3 and 4. The stem is very poorly developed and apparently bears only living leaves in *N. occidentalis* (Fig. 7) and in *N. undulata* (Sw.) J. Sm. of Africa. It is probably annual in these species, the plants reproducing primarily by means of stolons and sometimes scaly tubers (Fig. 8). The stem is dictyostelic while the stolons, produced at the base of the petiole, are protostelic and they often bear more roots than the stem itself. The leaf sometimes has a loose bud at the apex and apparently is then indeterminate.

The linear, nearly marginal sorus and indusium of *Nephrolepis acutifolia* provides an example similar to that of *Tectaria panamensis* of the extraordinary variation of those structures within a closely related evolutionary line.

Systematics *Nephrolepis* is a pantropical and subtropical genus of perhaps 20 species, with six native to tropical America. It is a highly distinctive genus characterized by abundant stolons, disarticulate, inequilateral and usually entire pinnae, and spores with a relatively thin, tuberculate-rugose perispore. There are hydathodes on the upper surface of the pinnae that secrete a calcareous substance (Guttenberg, 1934) that often dries as a thin, whitish, circular cover.

Fig. 102.1. *Nephrolepis cordifolia* on a palm trunk, Turrialba, Costa Rica. (Photo W. H. Hodge.)

Nephrolepis is usually considered to be related either to the oleandroid or the davallioid ferns. While its alliance with these groups has not been clearly established, the staurocytic type of stomata suggests closer relations to the davallioid genera (Van Cotthem, 1970) as do the coarsely tuberculate spores.

The native American species of *Nephrolepis* were treated by Tryon (1964), and the group of *Nephrolepis cordifolia* has been revised by Pichi-Sermolli (1969).

Tropical American Species

The American species of *Nephrolepis* are characterized by a cluster of distinctive characters, but they are all variable so that precise definition and accurate identification is sometimes difficult. Nauman (1979) described a rather widespread hybrid of *Nephrolepis biserrata* and *N. exaltata* from southern Florida, and indicates that, except for the sterile spores, it intergrades between the two parents. Other species probably also hybridize and this may be the basis of some of the taxonomic problems at the species level.

Fig. 102.2. Distribution of native *Nephrolepis* in America.

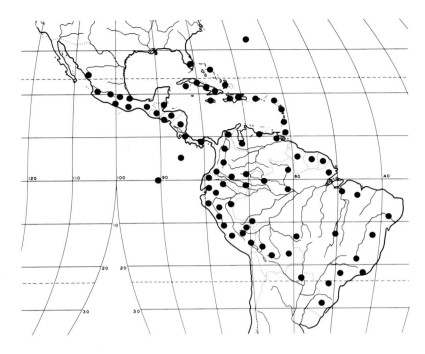

The following key includes the six native American species and *Nephrolepis multiflora* of the paleotropics which is rather widely adventive in the American tropics. Variations of several cultivated species with leaves that are forked or more than 1-pinnate occasionally become established, probably from garden refuse.

Observations on the cultivated species and their horticultural forms are made by Morton (1958).

Key to Species of *Nephrolepis* in America

a. Indusium orbicular with a narrow sinus (Fig. 3), to orbicular-reniform, at least toward the base of the pinna, mostly facing the margins.   b.
  b. Most pinnae with the base very unequal, the acroscopic side acutely to subacutely auriculate, the basiscopic side semicuneate to somewhat convex (Fig. 9).   *N. rivularis* (Vahl) Krug
  b. Most pinnae with the base equilateral (Fig. 5), cuneate to auriculate, or unequal and the basiscopic side rounded to lobed.   c.
    c. The longer pinnae rather abruptly reduced to a prolonged apex to acuminate, scales at the base of the petiole more or less spreading, indusia mostly or all orbicular with a narrow sinus.
                                          N. biserrata (Sw.) Schott
    c. The longer pinnae obtuse to shortly acute, indusia toward the pinna apex often reniform or lunate.   d.
      d. Adaxial side of the pinna-costa short-pubescent, scales at the base of the petiole usually closely appressed.
                                         N. multiflora (Roxb.) Morton
      d. Adaxial side of the pinna-costa glabrous or somewhat long-pubescent and/or scaly, scales at the base of the petiole more or less spreading.   *N. exaltata* (L.) Schott
a. Indusia reniform to lunate, mostly facing the apex of the pinna.   e.
  e. Upper pinnae, or most of them, with the base semicuneate on the basiscopic side (Fig. 6), stolons not bearing tubers.
                                    N. pectinata (Willd.) Schott
  e. Upper pinnae (but not necessarily the apical ones) with the base rounded on the basiscopic side.   f.
    f. Pinnae usually coriaceous, the veins obscure, stolons wiry, sometimes bearing tubers, stem perennial.   *N. cordifolia* (L.) Presl
    f. Pinnae thin-herbaceous, the veins evident, stolons rather soft, usually bearing tubers, evidently annual from tubers or stolons.
                                            N. occidentalis Kze.

3   4   5

6   7   8

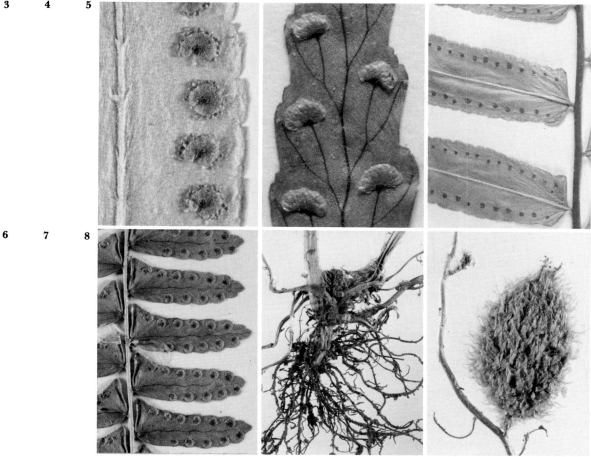

Figs. 102.3–102.8. *Nephrolepis.* **3.** Orbicular indusia with a narrow sinus, *N. exaltata,* × 8. **4.** Lunate indusia of *N. occidentalis,* × 8. **5.** Base of pinnae of *N. biserrata,* × 1. **6.** Central pinnae of *N. pectinata,* × 2.5. **7.** Small stem of *N. occidentalis* bearing two petioles at top, stolons and roots, × 2.5. **8.** Scaly tuber attached to stolon of *N. cordifolia,* × 2.5.

Ecology (Fig. 1)

Nephrolepis plants are ecologically adaptable, growing in a wide variety of habitats. They are usually in mesic and partly shaded places, but sometimes in wet sites or in rather dry habitats.

In America *Nephrolepis* grows in rain forests, elfin and cloud forests, or gallery forests; also in savannahs and grasslands, in open rocky places, and in swamps, marshes and bogs. Species are often epiphytic and especially frequent on the persistent leaf bases of palm trunks. They are also rupestral, growing on boulders, in lava beds, on old walls or on cliffs. *Nephrolepis* also grows well in more or less disturbed places such as roadsides, rough pastures, clearings, thickets and secondary forests. The freely stoloniferous plants often give rise to colonies of several meters in extent. *Nephrolepis* grows from sea level up to 3000 m in the Andes, but most commonly below 1500 m.

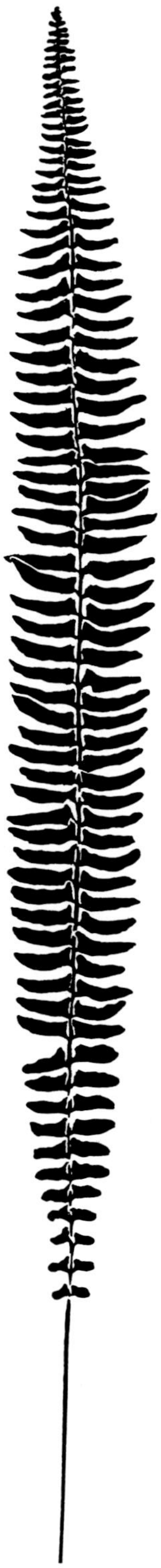

Fig. 102.9. Leaf of *Nephrolepis rivularis,* × 0.25.

Geography (Fig. 2)

Nephrolepis is widely distributed from America, Africa and Indo-Malesia to China, Japan, Queensland in Australia, New Zealand, and through the islands of the Pacific to the Hawaiian Islands and Pitcairn Island.

In America native species occur from Sinaloa in Mexico and southern Florida in the United States, southward to Bolivia, Paraguay and Rio Grande do Sul in southeastern Brazil; also in Bermuda, the Bahamas, Cocos Island, and the Galápagos Islands. All of the species are rather widely distributed and *N. cordifolia* and *N. biserrata* also grow in the Old World. Some of the cultivated species are adventive, in the southern United States and in northern Argentina, beyond the native range of the genus.

Spores

The perispore formation is not evident from the general aspect of the surface in the tuberculate spores of *Nephrolepis exaltata* (Fig. 10) and *N. rivularis* (Fig. 11), or in the irregularly rugose spores of *N. biserrata* (Fig. 13) and *N. occidentalis* (Fig. 15). The surface contours are formed by the perispore that overlays a smooth, thicker exospore as in *N. rivularis* and *N. biserrata* (Figs. 12, 14). The presence of perispore in *Nephrolepis* has previously been questioned. Nayar (1964) and Braggio (1966) describe the spores without perispore while Erdtman and Sorsa (1971) indicate that it is probably present. Spores of the paleotropical species surveyed with SEM by Liew (1977) are similar to those of the neotropics.

The irregularly tuberculate or rugose surface of the spores of *Nephrolepis* is clearly distinct from the folded or winged formation characteristic of the Dryopteridaceae. They are generally similar to spores of the davallioid genera although smaller with surface contours formed by the perispore rather than the exospore as in other genera of the Davalliaceae.

Cytology

The chromosome number of *Nephrolepis* seems to be consistently $n = 41$ or 82 aside from a few uncertain reports of ca. 40. Records of *N. exaltata* and *N. biserrata* from Jamaica are uniformly diploid (Walker, 1966, 1973) while *N. pectinata* is known as tetraploid there and reported as diploid with $2n = 41$ II in Chiapas, Mexico (Smith and Mickel, 1977). A collection of *N. cordifolia,* (reported as *N. pendula*) from the Galápagos Islands with $n = $ ca. 40 and $2n = 82$ showed irregular chromosome pairing (Jarrett et al., 1968). On the basis of the chromosome numbers the relationship of *Nephrolepis* appears to be with either Dryopteridaceae or Davalliaceae but the spores suggest an alliance closer to the Davalliaceae.

10 11 12

13 14 15

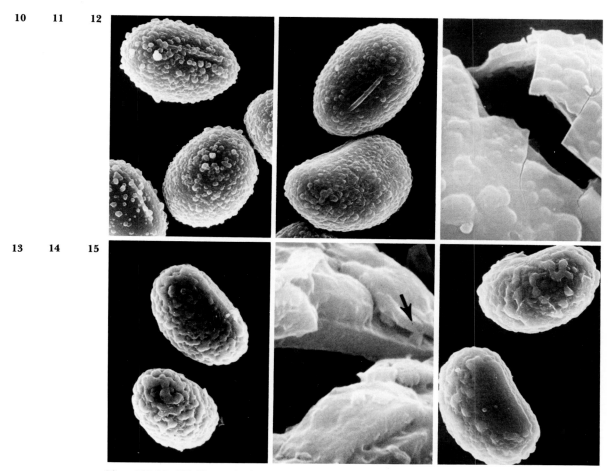

Figs. 102.10–102.15. *Nephrolepis* spores, ×1000. **10.** *N. exaltata,* tuberculate, laesura horizontal, upper spore, Dominican Republic, *Gastony et al. 132.* **11, 12.** *N. rivularis,* Puerto Rico, *Stimson 3394.* **11.** Tuberculate perispore, laesura center, upper spore. **12.** Abraded wall, tuberculate perispore, exospore, below, ×5000. **13, 14.** *N. biserrata,* Bolivia, *Steinbach 5406.* **13.** Rugosetuberculate perispore, laesura at top of upper spore. **14.** Wall section, perispore (arrow), exospore, below, ×10,000. **15.** *N. occidentalis* rugose-tuberculate perispore, laesura, right, lower spore, Brazil, *Irwin et al. 21683.*

Literature

Braggio, G. 1966. Morfologia delle spore e systematica delle davalliales. Webbia 21: 725–764.

Erdtman, G., and P. Sorsa. 1971. Pollen and Spore Morphology/Plant Taxonomy. Pteridophyta. 302 pp. Almqvist & Wiksell. Stockholm.

Guttenberg, H. von. 1934. Studien an pflanzen der Sunda-Inseln. Ann. Jard. Bot. Buitenz. 44: 1–62.

Jarrett, F. M., I. Manton, and S. K. Roy. 1968. Cytological and taxonomic notes on a small collection of living ferns from the Galápagos. Kew Bull. 22: 475–480.

Liew, F. S. 1977. Scanning electron microscope studies of the spores of Pteridophytes. XI. The family Oleandraceae (*Oleandra, Nephrolepis* and *Arthropteris*). Gardens' Bull. Singapore 30: 101–110.

Morton, C. V. 1958. Observations on cultivated ferns, V. The species and forms of *Nephrolepis*. Amer. Fern Jour. 48: 18–27.

Nauman, C. E. 1979. A new *Nephrolepis* hybrid from Florida. Amer. Fern. Jour. 69: 65–70.

Nayar, B. K. 1964. Palynology of modern pteridophytes, *in:* Advances in Palynology, ed. P. K. K. Nair. 438 pp. Lucknow. Nat. Bot. Gard.

Pichi-Sermolli, R. G. E. 1969. Taxonomical notes on *Nephrolepis cordifolia* (L.) Presl and related species. Ann. Mus. Civ. Stor. Nat. Genova 77: 270–277 (1968).

Smith, A. R., and J. T. Mickel. 1977. Chromosome counts for Mexican ferns. Brittonia 29: 391–398.

Tryon, R. 1964. The ferns of Peru (Dennstaedtieae to Oleandreae). Contrib. Gray Herb. 194: 1–253.

Von Cotthem, W. 1970. Comparative morphological study of the stomata in the Filicopsida. Bull. Jard. Bot. Nat. Belg. 40: 81–239.

Walker, T. G. 1966. A cytotaxonomic survey of the pteridophytes of Jamaica. Trans. Roy. Soc. Edinburgh 66: 169–237.

Walker, T. G. 1973. Additional cytotaxonomic notes on the pteridophytes of Jamaica. Trans. Roy. Soc. Edinburgh 69: 109–135.

21.   Blechnaceae

Blechnaceae (Presl) Copel., Gen. Fil. 155. 1947.

Blechneae Presl, Epim. Bot. 103. 1851, as Blechnaceae. Type: *Blechnum* L.

Stenochlaenaceae Ching, Acta Phytotax. Sinica 16(4): 18. 1978. Type: *Stenochlaena* J. Sm.

Description

Stem decumbent and small to stout, to erect and slender to massive, or long-creeping, or scandent, dictyostelic, hardly sclerotic to very indurated, bearing scales; leaves ca. 10 cm to 15 m long, usually pinnatisect to 1-pinnate-pinnatifid, rarely entire or 2-pinnate, circinate in the bud, petiole without stipules; sori elongate, on an outer arc of an areole, or on a continuous vascular commissure, or partly acrostichoid, covered by a short to very long indusium open toward the axis, or exindusiate; sporangia with a 2- to 3-rowed stalk, the annulus interrupted by the stalk, or rarely (in *Brainea*) not interrupted; homosporous, spores without chlorophyll or with chlorophyll (green). Gametophyte epigeal, with chlorophyll, more or less cordate to elongate, with broad thin margins and slightly thickened centrally, glabrous or usually with 1-celled trichomes, the archegonia borne on the lower surface, the antheridia 3-celled, borne on the lower, or also on the upper surface or on the margin.

Comments on the Family

The Blechnaceae are a moderately large family of nine genera and about 175 species. *Salpichlaena* is an American genus and *Blechnum* and *Woodwardia* are also in the Old World. *Diploblechnum* Hayata (invalid), *Doodia* R. Br., *Pteridoblechnum* Hennipm., *Brainea* J. Sm. and *Sadleria* Kaulf. are confined to the Old World, the last two are not basically different from *Blechnum*. *Stenochlaena* J. Sm. is also an Old World genus of distinctive morphology and distant from the other genera, but it evidently belongs to the blechnoid alliance, having the unfolding leaves tinged with red (Stone, 1981).

Woodwardia arctica (Heer) Brown from the Paleocene of Wyoming and *W. columbiana* Knowlton from the Pleistocene of Oregon are clearly referrable to the family.

The Blechnaceae are a distinctive family, without obvious affinities to other ferns. The sorus and indusium, the occurrence of the chromosome number of $n = 36$, and the structure of the spores show similarities to the Aspleniaceae, but in other characters the families are perhaps not closely allied.

The Blechnaceae usually have the unfolding leaves tinged with red, due to the presence of 3-deoxyanthocyanins (Crowden and Jarman, 1974). The spores are sometimes green, as reported for *Blechnum nudum* in a study of the gametophytes of the family by Stone (1961), and also in other species.

Key to American Genera of Blechnaceae

a. Sori borne on the outer arc of an areole. 103. *Woodwardia*
a. Sporangia borne on a vascular commissure. b.
    b. Leaf to ca. 2 m long, determinate, lamina entire to 1-pinnate.
                                        104. *Blechnum*
    b. Leaf climbing to 15 m, indeterminate, lamina 2-pinnate.
                                      105. *Salpichlaena*

Literature

Bower, F. O. 1928. The ferns, Vol. 3. Cambridge Univ. Press, Cambridge, England.

Crowden, R. K., and S. J. Jarman. 1974. 3-Deoxyanthocyanins from the fern *Blechnum procerum*. Phytochem. 13: 1947–1948.

Lloyd, R. M. 1976. Spore morphology of the Hawaiian genus *Sadleria* (Blechnaceae). Amer. Fern Jour. 66: 1–7.

Pichi-Sermolli, R. E. G. 1977. Tentamen pteridophytorum genera in taxonomicum ordinem redigendi. Webbia 31: 313–512.

Stone, B. C. 1981. Personal communication, including observations by Haji Mohamed in Malaya.

Stone, I. G. 1961. The gametophytes of the Victorian Blechnaceae. Australian Jour. Bot. 9: 20–36.

Walker, T. G. 1973. Evidence from cytology in the classification of ferns, *in:* A. C. Jermy et al., The phylogeny and classification of the ferns, pp. 91–110. Bot. Jour. Linn. Soc. 67, Suppl. 1.

103. *Woodwardia*
Figs. 103.1–103.13

Woodwardia Sm., Mém. Acad. Turin 5: 411. 1793. Type: *Woodwardia radicans* (L.) Sm. (*Blechnum radicans* L.).
Anchistea Presl, Epim. Bot. 71. 1851. Type: *Anchistea virginica* (L.) Presl (*Blechnum virginicum* L.) = *Woodwardia virginica* (L.) Sm.
Lorinseria Presl, Epim. Bot. 72. 1851. Type: *Lorinseria areolata* (L.) Presl (*Acrostichum areolatum* L.) = *Woodwardia areolata* (L.) Moore.
Chieniopteris Ching, Acta Pytotax. Sinica 9: 37. 1964. Type: *Chieniopteris Harlandii* (Hook.) Ching = *Woodwardia Harlandii* Hook.

Description

Terrestrial or rarely rupestral; stem decumbent, the apex more or less erect, short-creeping and stout, or long-creeping and moderately stout to rather slender, bearing scales and usually many, sometimes thin, fibrous roots; leaves monomorphic or (in *W. areolata*) dimorphic (the fertile longer than the sterile and with narrower segments), ca. 25 cm to 3 m long, borne in a loose

Fig. 103.1. *Woodwardia Martinezii,* north of Jalapa, Veracruz, Mexico. (Photo W. H. Hodge.)

crown to widely spaced, lamina deeply pinnatifid to 1-pinnate-bipinnatifid, usually sparsely scaly and sometimes also minutely glandular, veins partly anastomosing without included free veinlets, those toward the margin free; sori elongate, borne on the outer arc of an areole adjacent to the segment axis, sometimes sunken, not paraphysate, covered by an indusium open toward the axis; spores ellipsoidal, monolete, laesura $\frac{1}{2}$ to $\frac{2}{3}$ the spore length with folds more or less elaborated in wings. Chromosome number: $n = 34, 35, 68; 2n = 68$.

The stem and usually the petiole base of *Woodwardia* have rather large, often long, brown to reddish-brown, thin, and nearly concolorous scales (Fig. 5). Sterile segments of the leaves have one or more series of areolae and beyond these free veins (Fig. 6).

Systematics   *Woodwardia* is a circumboreal genus of about 12 species, with four of them in America. Six additional species have been described from China by Chiu (1974). The genus is characterized by anastomosing veins and indusiate sori borne on the outer arc of areolae. The three segregate genera sometimes recognized represent diversity within *Woodwardia* rather than major evolutionary divergence.

Bower (1928) and Pichi-Sermolli (1977) considered *Woodwardia* to be derived from *Blechnum* by division of the "coenosorus" into separate sori. However, it seems more likely that the

Fig. 103.2. Distribution of **Woodwardia** in America.

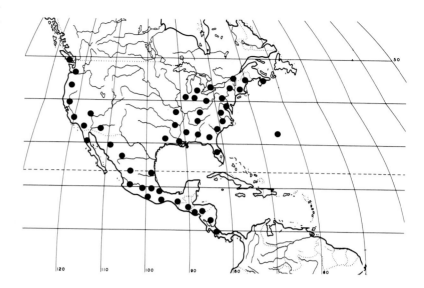

long vascular commissure of *Blechnum* is a specialization and that *Woodwardia* represents a less derived element in the family (Holttum, 1947).

American Species

The species of the western United States has usually been separated from the common one of Mexico and Central America since Maxon (1919) enumerated several differences between the two. However, they are not distinct morphologically or geographically, and are considered here as a single species, *Woodwardia fimbriata* (Fig. 8). This is most closely related to *Woodwardia orientalis* Sw. of eastern Asia and to *W. radicans* (L.) Sm. of Europe. *Woodwardia Martinezii* (Fig. 9) appears closely related to *W. Kempei* Copel. of eastern Asia, rather than to *W. fimbriata*. The temperate *Woodwardia virginica* and *W. areolata* have similar chromosome numbers and spores and seem more distantly related to the Asiatic species.

Key to American Species of *Woodwardia*

a. Leaves dimorphic.   *W. areolata* (L.) Moore
a. Leaves monomorphic.   b.
    b. Stem long-creeping, leaves distant.   *W. virginica* (L.) Sm.
    b. Stem rather short, leaves adjacent (Fig. 7).   c.
        c. Most sori long to very long (Fig. 3).   *W. Martinezii* Weath.
        c. Most sori short (Fig. 4).
            W. fimbriata Sm. (*W. Chamissoi* Brack., *W. spinulosa* Mart. & Gal.)

Ecology (Fig. 1)

Woodwardia is primarily a genus of mountainous regions where it grows usually in forests on hillslopes, along streams or in canyons. It also grows in moist meadows, in damp rocky places and at forest borders.

In Mexico and Central America, *Woodwardia* usually grows in oak woods or pine and oak woodland, less often in cloud forest,

3   4   5

6   7

Figs. 103.3–103.7. *Woodwardia*. 3. Portion of fertile lamina of ***W. Martinezii*,** the long sori mature and confluent, ×1. **4.** Portion of fertile pinna of ***W. fimbriata*,** ×5. **5.** Base of petiole with scales, ***W. fimbriata*,** ×0.5. **6.** Portion of sterile pinna of ***W. fimbriata*,** basal anastomosing veins, ×5. **7.** Longisection of stem of ***W. fimbriata*,** ×0.5.

especially along streams, in canyons, or sometimes on brushy hillsides, on moist rocks or on cliffs. *Woodwardia areolata* and *W. virginica* usually grow in acidic bogs, swamps or wet woods. In Mexico and Central America, *Woodwardia* grows from 500 to 2800 m, most often at 2000 m or higher.

Geography (Fig. 2)

Woodwardia occurs in North America, the Azores to the Canary Islands and southern Europe, and from the Himalayas to Java and New Guinea, Formosa and the Philippine Islands, and to Korea and Japan.

In America, *Woodwardia fimbriata* occurs from southern British Columbia south to Costa Rica, and *W. Martinezii* in the states of Hidalgo, Puebla and Veracruz, Mexico. *Woodwardia virginica* occurs from Nova Scotia and southern Quebec south to Florida, westward to northern Illinois, Missouri and eastern Texas, and is also on Bermuda. The range of *Woodwardia areolata* is similar but not as extensive.

The center of diversity of the genus is in China, where six or more species occur.

Figs. 103.8, 103.9. *Woodwardia* leaves. 8. *W. fimbriata*, ×0.15. 9. *W. Martinezii*, ×0.30.

Spores

Spores of *Woodwardia Martenizii* are rugulose with low folds (Fig. 10) and those of *W. fimbriata* somewhat rougher with plates and short rods (Figs. 12, 13). Those of *Woodwardia virginica* are of the same general form but have a diffuse, papillate surface (Fig. 11). Spores of the Japanese species included in the SEM survey of Blechnaceae spores (Mitui, 1979) have short rods similar to those of *W. fimbriata*. The perispore of *Woodwardia* was regarded by Mitui as a derived type developed from spores with a single perispore layer. This concept is contrasted with that proposed for *Sadleria* by Lloyd (1976) in which the multilayered perispore as in *S. cyatheoides* Kaulf. is regarded as the primitive form.

Cytology

The chromosome number $n = 34$ is reported in *Woodwardia fimbriata* (as *W. Chamissoi*) by Manton and Sledge (1954) and in several species of Asia. Cytological studies of *Woodwardia* in Japan by Mitui (1968) show the chromosomes of *W. orientalis* are more than twice as large as those of *Blechnum orientale* L. Cytotypes of *Woodwardia orientalis* are ecologically differentiated in Japan with the coastal var. *formosana* Rosenst., diploid, $n = 34$, and the inland var. *orientalis* tetraploid. The record of $n = 35$ in the American *Woodwardia virginica* is photographically documented in material from southern Ontario (Britton, 1964). The same number is reported for specimens from southern Michigan, and also for a cultivated specimen of *Woodwardia areolata* (Wagner,

10   **11**

12   **13**

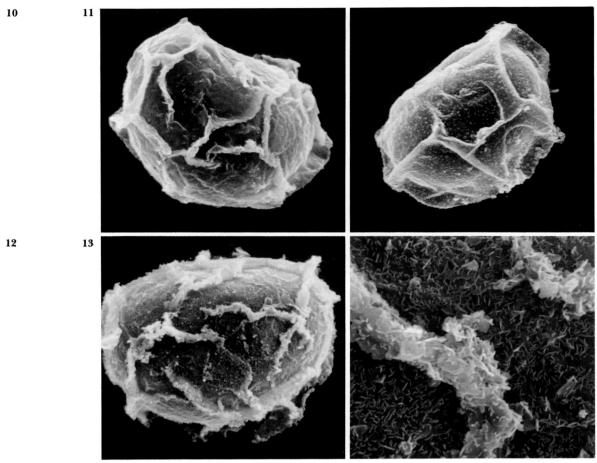

Figs. 103.10–103.13. *Woodwardia* spores, × 1000.   **10. *W. Martinezii*,** folds and rugulose surface, Mexico, *Barrington 392.*   **11. *W. virginica*,** folds and diffuse papillae, Connecticut, *Eames 8248.*   **12, 13. *W. fimbriata*,** Guatemala, *Williams et al. 41511.*   **12.** Irregular surface folds.   **13.** Detail of surface rods and plates, × 5000.

1955). Cytological study of a wider sample of *Woodwardia* species will determine whether there are consistent aneuploid differences between the species of Asia and America.

Literature

Bower, F. O. 1928. Reference under the family.

Britton, D. M. 1964. Chromosome numbers of ferns in Ontario. Canad. Jour. Bot. 64: 1149–1154.

Chiu, P. 1974. On the genus *Woodwardia* Sm. from the mainland of Asia. Acta Phytotax. Sinica 12: 237–248.

Holttum, R. E. 1947. A revised classification of leptosporangiate ferns. Jour. Linn. Soc. (Bot.) 53: 123–158.

Lloyd, R. M. 1976. Reference under the family.

Manton, I., and W. A. Sledge. 1954. Observations on the cytology and taxonomy of the Pteridophyte flora of Ceylon. Phil. Trans. Roy. Soc. London. Ser. B 238: 127–185.

Maxon, W. R. 1919. Notes on American ferns, XIV. Amer. Fern Jour. 9: 67–73.

Mitui, K. 1968. Chromosomes and speciation in ferns. Sci. Report Tokyo Kyoiku Daigaku 13: 285–333.

Mitui, K. 1979. Spore morphology of the fern genera, *Blechnum, Struthiopteris* and *Woodwardia* (Blechnaceae). Bull. Nippon Dental Univ. 8: 139–145.

Pichi-Sermolli, R. E. G. 1977. Reference under the family.

Wagner, W. H. 1955. Cytotaxonomic observations on North American ferns. Rhodora 57: 219–240.

104. *Blechnum*
Figs. 104.1–104.31

Blechnum L., Sp. Pl. 1077. 1753; Gen. Pl. ed. 5, 485. 1754. Type: *Blechnum occidentale* L., as *Blechnum orientale*.

Struthiopteris Scop., Fl. Carn. 168. 1760. Type: *Struthiopteris Spicant* (L.) Scop. (*Osmunda Spicant* L.) = *Blechnum Spicant* (L.) Roth.

Lomaria Willd., Berl. Mag. 3: 160. 1809. Type: *Lomaria nuda* (Labill.) Willd. (*Onoclea nuda* Labill.) = *Blechnum nudum* (Labill.) Luerss. *Blechnum* section *Lomaria* (Willd.) Keys., Polypod. Cyath. Herb. Bung. 16. 1873.

Stegania R. Br., Prod. Nov. Holl. 152. 1810. Type: *Stegania nuda* (Labill.) R. Br. (*Onoclea nuda* Labill.) = *Blechnum nudum* (Labill.) Luerss. *Lomaria* section *Stegania* (R. Br.) J. Sm., Hist. Fil. 304. 1875.

Parablechnum Presl, Epim. Bot. 109. 1851. Type: *Parablechnum procerum* (Forst.) Presl (*Osmunda procera* Forst.) = *Blechnum procerum* (Forst.) Sw. *Blechnum* section *Parablechnum* (Presl) Moore, Ind. Fil. xxv. 1857.

Distaxia Presl, *ibidem* 110. 1851. Type: *Distaxia fraxinea* (Willd.) Presl = *Blechnum fraxineum* Willd.

Mesothema Presl, *ibidem* 111. 1851. Type: *Mesothema australe* (L.) Presl = *Blechnum australe* L. *Blechnum* section *Mesothema* (Presl) J. Sm., Hist. Fil. 301. 1875.

Spicanta Presl, *ibidem* 114. 1851, *nom. superfl.* for *Struthiopteris* Scop. and with the same type.

Blechnopsis Presl, *ibidem* 115. 1851. Type: *Blechnopsis orientalis* (L.) Presl = *Blechnum orientale* L., as *Blechnum occidentale*. *Blechnum* section *Blechnopsis* (Presl) J. Sm., Hist. Fil. 301. 1875.

Blechnopsis subgenus *Diafnia* Presl, *ibidem* 119. 1852. Type: *Blechnopsis serrulata* (L. C. Rich.) Presl = *Blechnum serrulatum* L. C. Rich. *Blechnum* section *Diafnia* (Presl) J. Sm., Hist. Fil. 301. 1875.

Orthogramma Presl, *ibidem* 121. 1851. Type: *Orthogramma Gilliesii* (Hook. & Grev.) Presl. (*Lomaria Gilliesii* Hook. & Grev., *Blechnum Gilliesii* (Hook. & Grev.) Mett.) = *Blechnum chilense* (Kaulf.) Mett.

Lomaridium Presl, *ibidem* 154. 1851. Type: *Lomaridium Plumieri* (Desv.) Presl (*Lomaria Plumieri* Desv.) = *Blechnum binervatum* (Poir.) Mort. & Lell. *Lomaria* section *Lomaridium* (Presl) J. Sm., Hist. Fil. 303. 1875.

Lomaria subgenus *Paralomaria* Fée, Mém. Fam. Foug. 5 (Gen. Fil.): 69. 1852. Type: *Lomaria procera* (Forst.) Spreng. (*Osmunda procera* Forst.) = *Blechnum procerum* (Forst.) Sw.

Blechnidium Moore, Ind. Fil. clv. 1859. Type: *Blechnidium melanopus* (Hook.) Moore = *Blechnum melanopus* Hook.

Lomaria section *Lomariocycas* J. Sm., Hist. Fil. 305. 1875. Type: *Lomaria Boryana* (Sw.) Willd. (*Onoclea Boryana* Sw., *Blechnum Boryanum* (Sw.) Schlect.) = *Blechnum tabulare* (Thunb.) Kuhn. *Blechnum* section *Lomariocycas* (J. Sm.) Morton, Amer. Fern Jour. 49: 3. 1959.

Lomaria section *Loxochlena* J. Sm. Hist. Fil. 304. 1875. Type: *Lomaria punctulata* (Sw.) J. Sm. = *Blechnum punctulatum* Sw.

Homophyllum Merino, Ann. Soc. Hist. Nat. 1898: 108. Type: *Homophyllum blechniforme* Merino = *Blechnum Spicant* (L.) Roth.

Spicantopsis Nakai, Bot. Mag. (Tokyo) 47: 180. 1933. Type: *Spicantopsis niponica* (Kze.) Nakai (*Lomaria niponica* Kze.) = *Blechnum niponicum* (Kze.) Makino.

Description

Terrestrial, rupestral, sometimes epiphytic or scandent-epiphytic; stem erect, rather small to arborescent and massive, to ca. 3 m tall, or decumbent, or long-creeping, or scandent to 10 m or more, bearing scales and usually many long, sometimes fine, fibrous roots; leaves monomorphic or dimorphic (the fertile usually longer than the sterile and with narrower segments), ca. 10 cm to 2.5 m long, lamina entire (in *B. lanceola*) to usually deeply pinnatifid or 1-pinnate, or (in *B. microbasis* (Baker) C. Chr.) 2-pinnate, glabrous or more or less scaly, rarely pubescent or glandular-pubescent, veins free or (in *B. Heringeri* and *B. melanopus* Hook.) partly anastomosing without included free veinlets; sori borne on a usually long to sometimes short vascular commissure usually near the segment axis, sometimes the sporangia acrostichoid beyond the commissure, no paraphyses, covered by an indusium open toward the axis; spores monolete, elliptical, laesura $\frac{1}{2}$ to $\frac{3}{4}$ the spore length, slightly papillate to nearly smooth, rugose, or with folds, somewhat winglike, reticulate, or coarsely echinate. Chromosome number: $n = 28, 29, 31, 32, 33, 34, 36$, ca. $37, 56, 62, 64, 68, 99$; $2n = 66, 68, 124$.

There is considerable diversity in *Blechnum* in the type of stem and stem scales, which often are persistent at the base of the petiole. *Blechnum Schomburgkii* and allied species have an arborescent stem that is strongly indurated (Figs. 3, 5) with acicular, somewhat thickened, long scales (Fig. 6). The stem of species as *Blechnum Kunthianum* is long and scandent-epiphytic (Fig. 7) with a dense covering of rather large and soft, mostly concolorous scales (Fig. 8); that of *Blechnum loxense* is short (Fig. 9), hardly sclerotic and green and mucilaginous within and the petiole base has large, broad, thin scales (Fig. 10). The vascular commissure bears the sporangia, with the indusium attached on the outer side, (Fig. 11). A selection of the diversity of lamina architecture is in Figs. 12–20.

Systematics

Blechnum is a nearly worldwide genus of perhaps 150 species, with about 50 in America. It is characterized by sporangia borne on a vascular commissure, 1-pinnate or less complex lamina, and the indusium open on the side toward the segment axis. Most of the species have dimorphic leaves. Although *Blechnum* is often considered to be an old genus, it is relatively homogeneous for a large group. The usually poorly differentiated species groups based on relatively few characters, are not indicative of great age. *Blechnum* is not sufficiently well known to provide the information for an adequate infrageneric classification. Study of the whole genus, especially the cytology and spore characters will be necessary to provide data for establishing relationships within the group.

Fig. 104.1. *Blechnum chiriquanum,* Tapanti, Costa Rica. (Photo W. H. Hodge.)

The genus *Blechnidium,* based on a Himalayan species with partly anastomosing veins, has been accepted by Pichi-Sermolli (1977), but *Blechnum Heringeri* Brade of Brazil, similar in venation but otherwise unrelated, shows that this character is not of generic significance.

The North American species with dimorphic leaves were revised by Broadhurst (1912) and the American species with monomorphic leaves by Murillo (1968). Looser (1947) gave an extended account of the genus in Chile, and later (1958) a new key and illustrations, and De la Sota has published revisions of southern South American species (1972).

Tropical American Species

The tropical American species of *Blechnum* have mostly been described and maintained on the basis of local floristic work, with few studies of a broad geographic scope. This has led to the recognition of an excessive number of species and many of the groups need a modern taxonomic study.

Hybrids involving *Blechnum occidentale* are discussed under the section on cytology. In addition to these, two hybrids from Argentina, involving *Blechnum australe* were studied by Rolleri (1976).

Fig. 104.2. *Blechnum occidentale*, Macchu Picchu, Peru. (Photo A. F. Tryon.)

The following synopsis provides a framework of the diversity among the American species, especially in the stem and leaf characters.

Synopsis of American *Blechnum*

1. *Blechnum occidentale* Group
Stem rather small, decumbent, leaves monomorphic or nearly so.

A group of about 10 species, including *Blechnum asplenioides* Sw. (*B. polypodioides* Raddi, *B. blechnoides* (Sw.) C. Chr., *B. unilaterale* Sw.), *B. australe* L. (*B. auriculatum* Cav.), *B. fraxineum* Willd. (Fig. 16), *B. glandulosum* Link (*B. confluens* auths., not Schlect. & Cham.) (Fig. 12), *B. lanceola* Sw. (Fig. 13), and *B. occidentale* L (*B. confluens* Schlect. & Cham.) (Figs. 2, 14).

Blechnum occidentale is adventive in the Hawaiian Islands.

2. *Blechnum serrulatum* Group
Stem subterranean, long-creeping, with ascending branches bearing monomorphic leaves, lamina 1-pinnate with articulate pinnae.

Blechnum serrulatum L. C. Rich. is a very distinctive species of tropical and subtropical America and Malesia-Australia (*B. indicum* auths., not Burm f. which is an *Asplenium* (Morton, Amer. Fern Jour. 60: 122. 1970). It is undoubtedly adventive in one of the two regions.

3. *Blechnum penna-marina* Group
Small plants, the stem creeping to erect, leaves dimorphic.

A group of about eight species, including *Blechnum Spicant* (L.) Roth in North America and *B. asperum* (Kl.) Sturm in southern South America. *Blechnum andinum* (Baker) C. Chr. and *B. penna-marina* (Poir.) Kuhn occur on mountains in the tropics and to the far south; *B. stoloniferum* (Fourn.) C. Chr. is in Mexico and Guatemala (Fig. 15).

Fig. 104.3. ***Blechnum Werckleanum,*** Cerro de la Muerte, Costa Rica. (Photo W. H. Hodge.)

4. *Blechnum lineatum* Group

Usually large plants with a rather stout, decumbent, short-creeping to semiscandent stem, leaves dimorphic.

A group of 15–20 poorly known species. The croziers sometimes produce copious mucilage. Some species are *Blechnum chilense* (Kaulf.) Mett., *B. chiriquanum* (Broadh.) C. Chr., *B. costaricense* (Christ) C. Chr., *B. divergens* (Kze.) Mett. (Fig. 18), *B. jamaicense* (Broadh.) C. Chr., *B. Lehmannii* Hieron., *B. L'Herminieri* (Kze.) Mett., *B. lineatum* (Sw.) Hieron., *B. pteropus* (Kze.) Mett. (Fig. 17), *B. Sprucei* C. Chr., and *B. varians* (Fourn.) C. Chr. (Fig.1).

5. *Blechnum loxense* Group

Stem short-creeping, decumbent, hardly sclerotic, green and mucilaginous within, leaves hardly dimorphic.

The unusual stem characters of *Blechnum loxense* (HBK.) Salomon are perhaps also in some other species, which can be established by field observations.

6. *Blechnum fragile* Group

Stem scandent-epiphytic, leaves dimorphic, the sterile pinnatisect.

A group of 10–15 species, or perhaps fewer, including *Blechnum binervatum* (Poir.) Mort. & Lell. (*B. Plumieri* (Desv.) Mett.), *B.*

Fig. 104.4. Distribution of ***Blechnum*** in America. The genus is represented in the western United States and Canada by ***Blechnum Spicant***, which ranges westward to Atka Island in the Aleutian Islands.

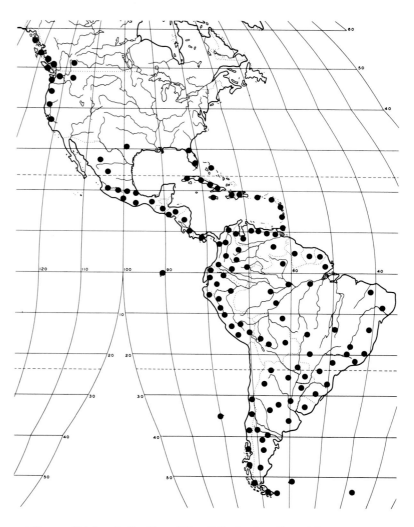

ensiforme (Liebm.) C. Chr. (Fig. 20), *B. fragile* (Liebm.) Mort. & Lell. (*B. polypodioides* (Sw.) Kuhn), and *B. Kunthianum* C. Chr. (*B. meridense* (Kl.) Mett.

7. *Blechnum Buchtienii* Group

Stem subarborescent to arborescent, usually massive, scales long, acicular, more or less thickened and often curved, leaves dimorphic.

A group of 10 or perhaps fewer species, more or less confined to tropical mountains and also in the far south, with the same or related species on the Juan Fernandez Islands, Tristan da Cunha Islands and in South Africa. *Blechnum Spannagelii* Rosenst. evidently belongs here, as does *B. brasiliense* Desv. (Fig. 19), although the latter has monomorphic leaves. Among other species often recognized are *Blechnum Buchtienii* Rosenst., *B. imperiale* (Fée) Christ, *B. insularum* Mort. & Lell., *B. magellanicum* (Desv.) Mett., *B. Schomburgkii* (Kl.) C. Chr., *B. Underwoodianum* (Broadh.) C. Chr., and *B. Werckleanum* (Christ) C. Chr. (Fig. 3).

5   6   7

8

9   10   11

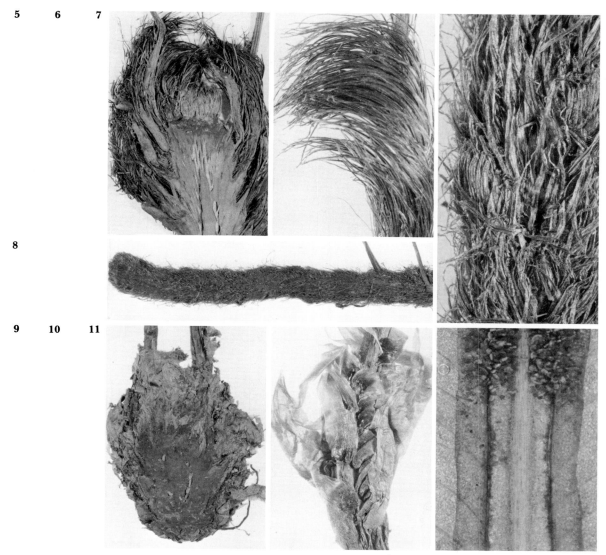

Figs. 104.5–104.11. *Blechnum.* 5. Longisection of stem apex of *B. Schomburgkii,* ×0.5. **6.** Scales at base of petiole of *B. Schomburgkii,* ×1.25. **7.** Apical portion of stem of *B. Kunthianum,* ×0.66. **8.** Scales on stem of *B. Kunthianum,* ×4. **9.** Longisection of stem of *B. loxense,* ×0.5. **10.** Scales at base of petiole of *B. loxense,* ×2. **11.** Sporangia, removed from vascular commissure, below, and indusium of *B. fraxineum,* ×10.

Ecology (Figs. 1–3)

Blechnum is primarily a genus of wet mountain forests, usually terrestrial, although some species also grow on rocks or on cliffs, and some are epiphytes. It also occurs in swamps, in thickets, at forest borders, in grasslands and sometimes in disturbed sites. *Blechnum obtusatum* (Labill.) Mett. of New Caledonia is sometimes a rheophyte. The species of the far south, such as *Blechnum penna-marina,* grow in woodland, in rocky places, on cliffs and on moorland.

In tropical America *Blechnum* grows in lowland rain forests, oak, pine and evergreen montane forests and cloud forests, on hillsides, in ravines, and along streams and rivers. It also grows in swamps and bogs, in gallery forest, savannahs and on páramos. Some species occur in disturbed thickets, in partly cleared forests, on bare clay soil of rivers, streams and road banks, along ditches or on old stone walls. Epiphytic species, particularly *Blechnum fragile,* frequently grow on the trunks of Cyatheaceae.

Some species as *Blechnum filiforme* (A. Cunn.) Ettingsh. of New Zealand have different growth forms; a terrestrial phase with slender stems and small, 1-pinnate leaves with coarsely toothed pinnae margins, and a scandent phase with a thicker stem, larger, 1-pinnate sterile leaves with the pinnae margins serrulate, and also fertile leaves. It resembles *Maxonia* in this terrestrial-epiphytic dimorphism. Several species are stoloniferous and establish large colonies, and a few have proliferous buds on the leaf. *Blechnum Sprucei* roots at the tip of the much-prolonged and somewhat modified apical portion of the lamina, as does the closely related *B. longicauda* C. Chr. of the Juan Fernandez Islands. *Blechnum asperum* of Chile and adjacent Argentina has dimorphic sterile leaves; one form has the lamina reduced to small lobes along the rachis which is elongated and bears proliferous buds.

Blechnum grows from sea level to about 3500 m, most commonly between 1000 and 3000 m.

Geography (Fig. 4)

Blechnum is a widely distributed genus, occurring in America, Europe to the Ural Mountains, Africa and eastward to Japan, New Guinea, Australia and New Zealand; it is also widely distributed through the Pacific to Rapa, Tahiti and the Marquesas Islands, but is not native in the Hawaiian Islands.

In America *Blechnum Spicant* occurs from California to Alaska and Atka Island in the Aleutian Islands. The other species are in Texas and Florida in the United States and commonly southward to Tierra del Fuego; also on the Galápagos Islands, the Juan Fernandez Islands, Staten Island, the Falkland Islands, and South Georgia.

Blechnum Spicant has a disjunct distribution, in northwestern North America and in Europe east to the Ural Mountains, and is represented in Japan by *B. niponicum* (Kze.) Makino. *Blechnum penna-marina* has a circumantarctic distribution, occurring on most southern islands from South America east to New Zealand.

Spores

Spores of American blechnums are often slightly papillate to nearly smooth and have a well-developed lower reticulate layer as *Blechnum jamaicense* (Fig. 21, 22). The lower reticulate formation is less elaborate in spores of *B. occidentale* and the surface is fairly smooth (Fig. 23). The spherical surface deposit as in *Blechnum serrulatum* (Fig. 24) is characteristic of spores of many genera especially in the Polypodiaceae, but is uncommon in the

Figs. 104.12–104.20. Lamina architecture of *Blechnum.* **12.** *B. glandulosum,* fertile leaf, × 0.25.   **13.** *B. lanceola,* fertile leaf and shorter sterile leaf, × 0.5.   **14.** *B. occidentale,* fertile leaf, × 0.125.   **15.** *B. stoloniferum,* fertile and sterile leaf, × 0.5.   **16.** *B. fraxineum,* fertile and shorter sterile leaf, × 0.25.   **17.** *B. pteropus,* fertile and sterile leaf, × 0.125.   **18.** *B. divergens,* fertile and sterile leaf, × 0.125.   **19.** *B. brasiliense,* fertile leaf, × 0.125.   **20.** *B. ensiforme,* fertile and sterile leaf, × 0.125.

Blechnaceae. The slender strands on the winged folds in *Blechnum loxense* (Fig. 25) comprise a network that is filled in to form the compact surface as in the spores of *B. costaricense* (Fig. 26) and also the reticulate perispore below (Fig. 27). A few species with small spores, as *B. Kunthianum,* have a sparsely rugulose surface (Fig. 28) and scarcely developed lower perispore. The arborescent species as *Blechnum Schomburgkii* and *B. Buchtienii* have low rugose spores (Figs. 29, 31), which are formed by a series of irregular sheaths as in the abraded spore of *B. insularum* (Fig. 30).

The development of the spore wall in *Blechnum Spicant* is known from the TEM work of Lugardon (1971) following stages from meiosis through germination. The exospore formed by a completely ensheathing layer, lacking a middle stratum, was proposed as characteristic of advanced groups of the Filicales. In general structure and diversity *Blechnum* spores resemble those of *Sadleria.* Variation shown in spores of that group by Lloyd (1976) was regarded as evidence of the polyphyletic composition of *Sadleria.* Spores of *Woodwardia,* in contrast to these, are relatively uniform.

Cytology

Chromosome numbers in *Blechnum* range between $n = 28$ and 99 with nearly consecutive series from 28 to 36. Reports of tropical American species, largely based on Walker's Jamaican studies (1966, 1973a) represent different series as *Blechnum lineatum* and *B. fragile* (as *B. polypodioides*) with $n = 29$; *B. serrulatum* (as *B. indicum*), $n = 36$; *B. occidentale* and *B. glandulosum* (as *B. unilaterale*), $n = 62$; and *B. Underwoodianum,* $n = 66$. Tetraploid hybrids of *Blechnum glandulosum* and *B. occidentale* with $n = 62$ are reported from the Galápagos Islands by Jarrett et al. (1968) and from Jamaica. The cytology of this hybrid in Jamaica, and variation in leaf form within the complex, including a triploid involving *Blechnum occidentale,* have been analyzed by Walker (1973a). The highest number for the American species reported as $2n = 198$ II, in *Blechnum Lehmannii* from Honduras, was considered to be a dodecaploid based on 33 (Smith and Mickel, 1977). The temperate species *Blechnum Spicant* appears to be uniformly $n = 34$ or $2n = 68$ in Europe, Iceland and western North America. In the southern hemisphere, *Blechnum penna-marina* is represented by two cytotypes, $n = 66$ on Tristan da Cunha and the Falkland Islands, and $2n = 68$ in New Zealand. The cytological records for *Blechnum* cover a broad geographic scope and include evidence of aneuploidy, polyploidy and hybrid complexes. The reports of $n = 29$, 62 and 66 in species of both the American tropics and those of the Old World, suggests there may be alliances between groups in these regions. The review of *Blechnum* cytology by Walker (1973b) suggests that 33 may represent the primitive number with both loss and gain of chromosomes, but evidence in support of this is not strong.

21 22 23

24 25

26 27 28

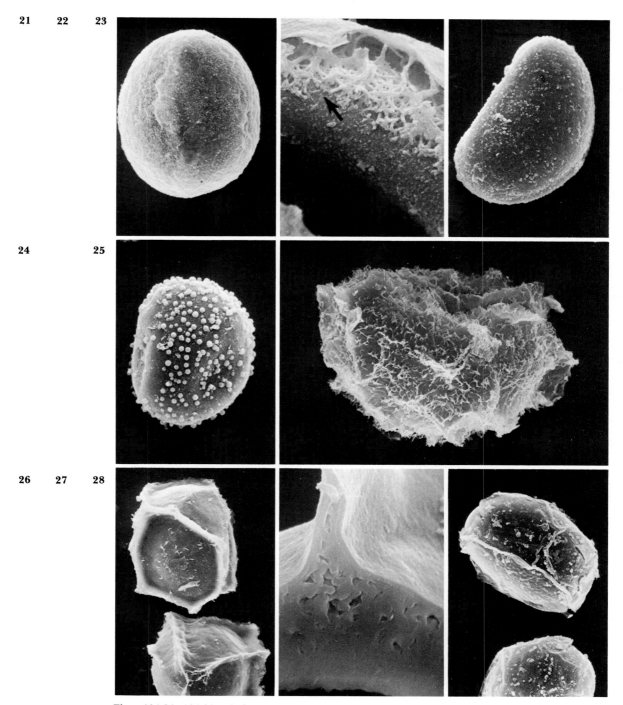

Figs. 104.21–104.28. ***Blechnum*** spores, × 1000, except 26, × 500. **21, 22.** ***B. jamaicense,*** Jamaica, *Maxon 10199.* **21.** Rugulose surface, laesura at center. **22.** Wall profile, reticulate formation below compact surface, exospore below (arrow), × 5000. **23.** ***B. occidentale,*** rugulose to nearly smooth, Peru. *Tryon & Tryon 5395.* **24.** ***B. serrulatum,*** spherical deposit on slightly foliose surface, Colombia, *Schultes & Cabrera 19649.* **25.** ***B. loxense,*** compact winged folds and fine surface strands, Ecuador, *Mexia 7533.* **26, 27.** ***B. costaricense,*** Costa Rica, *Scamman 7169.* **26.** Winglike folds, × 500. **27.** Detail of surface folds and wall profile. × 10,000. **28.** ***B. Kunthianum,*** low folds and papillate surface, Brazil, *Dusén* in 1914.

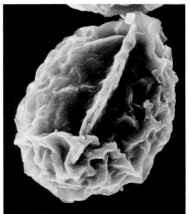

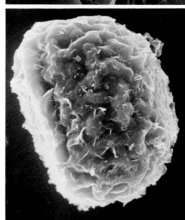

Figs. 104.29–104.31. *Blechnum* spores, × 1000. **29. *B. Schomburgkii,*** rugose folds laesura at center, Colombia, *Soejarto 1485.* **30. *B. insularum,*** abraded surface with foliose sheaths, Puerto Rico, *Shafer 3581* (US), × 5000. **31. *B. Buchtienii,*** rugose folds, Costa Rica, *Scamman 6032.*

Observations

Variations in the sorus and indusium, sometimes resembling those of *Asplenium,* have been noted by Bower (1928) and by Schelpe (1952), especially in *Blechnum punctulatum* Sw. These are probably atavistic variants, although Bower considered them to be of phyletic significance. Pinnatifid pinnae occur in some species and are also atavistic.

Studies of gametophytes, grown from spores of four populations of *Blechnum Spicant* from the Olympic Peninsula, Washington, and a population from northern Idaho (Cousens, 1979) showed each had its characteristic early ontogeny and sex expression. This work suggests that fern populations may be genetically isolated from each other rather than connected by frequent gene flow.

Literature

Bower, F. O. 1928. Reference under the family.

Broadhurst, J. 1912. The genus *Struthiopteris* and its representatives in North America, I, II. Bull. Torrey Bot. Cl. 39: 257–278; 357–385.

Cousens, M. I. 1979. Gametophyte ontogeny, sex expression and genetic load as measures of population divergence in *Blechnum Spicant.* Amer. Jour. Bot. 66: 116–132.

Jarrett, F. M., I. Manton, and S. K. Roy. 1968. Cytological and taxonomic notes on a small collection of living ferns from Galapagos. Kew Bull. 22: 475–480.

Lloyd, R. M. 1976. Reference under the family.

Looser, G. 1947. Los *Blechnum* (Filices) de Chile. Revis. Univers. Católica Chile 32 (2): 7–104.

Looser, G. 1958. Clave de los *Blechnum* (Filicales) de Chile. Revis. Univers. Católica Chile 43: 123–128.

Lugardon, B. 1971. Contribution à la connaissance de la morphogenese et de la structure des parois sporales chez les Filicinées isosporées. 257 pp. Thése, Univ. Paul Sabatier, Toulouse.

Murillo, M. T. 1968. *Blechnum* subgenero *Blechnum* en Sur America. Nova Hedwigia 16: 329–366.

Pichi-Sermolli, R. E. G. 1977. Reference under the family.

Rolleri, C. 1976. Estudio de la morfología foliar comparada de especies e híbridos interespecíficos del género *Blechnum* subgénero *Blechnum.* Bol. Soc. Argent. Bot. 17: 5–24.

Smith, A. R., and J. T. Mickel. 1977. Chromosome counts for Mexican ferns. Brittonia 29: 291–298.

Schelpe, E. A. C. L. E. 1952. A revision of the African species of *Blechnum.* Jour. Linn. Soc. Lond. 53: 487–510.

Sota, E. R. de la. 1972. Notas sobre especias austrosudamericanas del género *Blechnum* L., III–V. Bol. Soc. Argent. Bot. 14: 177–197.

Walker, T. G. 1966. A cytotaxonomic survey of the pteridophytes of Jamaica. Trans. Roy. Soc. Edinburgh 66: 169–237.

Walker, T. G. 1973a. Additional cytotaxonomic notes on the pteridophytes of Jamaica. Trans. Roy. Soc. Edinburgh 69: 109–135.

Walker, T. G. 1973b. Reference under the family.

105. *Salpichlaena*
Figs. 105.1–105.9

Salpichlaena Hook., Gen. Fil. *t. 93*, 1842, as *Salpichloena* in text, corrected in index. Type: *Salpichlaena volubilis* (Kaulf.) Hook. (*Blechnum volubile* Kaulf.)

Description

Terrestrial; stem long-creeping, rather slender to moderately stout, bearing scales and many, long fibrous roots; leaves monomorphic or partly dimorphic (the fertile pinnae with narrower segments than the sterile pinnae) to 15 m long, rather closely spaced, lamina 2-pinnate, the pinnae imparipinnate, with large, entire segments, lamina glabrous or sometimes the axes short-pubescent or the segments with a few scales, veins free, with the ends connected by a marginal vein, sori on a long vascular commissure parallel and close to the costa, sometimes the sporangia also on the inner surface of the indusium, no paraphyses, covered by an indusium open toward the costa and arching completely over the sporangia; spores ellipsoidal, monolete, the laesura $\frac{2}{3}$ to $\frac{3}{4}$ the spore length, with spherical deposit over the papillate-rugulose surface. Chromosome number: $n = 40$.

The scales of the stem are rather small, rigid, dark-brown to blackish, sometimes with a lighter border. The sterile segments and broad fertile ones have a cartilaginous border and a marginal vein (Fig. 7). The indusium is firm and close to the costa (Fig. 6). The leaves of juvenile plants are entire (Fig. 2) and become progressively more complex, with 3-foliolate (Fig. 3) and 1-pinnate (Fig. 4) leaves produced prior to those with a 2-pinnate lamina (Fig. 5).

Systematics

Salpichlaena is a monotypic genus of tropical America based on *S. volubilis* (Kaulf.) Hook. The high-climbing leaves are distinctive and have a counterpart only in the leaves of *Lygodium*. The genus is also distinctive in the marginal vein and the long, firm indusium that envelops the sporangia. *Salpichlaena* is not closely allied to other genera of the family. The sorus and the indusium resemble those of *Blechnum,* but the spores with sparse perispore, and the chromosome number of $n = 40$ are unlike the other genera.

Fig. 105.1. Distribution of *Salpichlaena.*

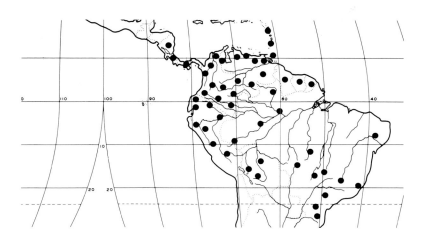

Figs. 105.2–105.5. *Salpichlaena* leaves.   **2.** Leaf of juvenile plant, ×0.25.   **3.** 3-foliate leaf of older plant, ×0.25.   **4.** 1-pinnate leaf of still older plant, ×0.25.   **5.** Sterile pinna and fertile pinna, with narrow segments, from large, climbing leaf, ×0.125.

The pinnae segments of *Salpichlaena volubilis* vary in size, shape and form of the apex and base. The fertile segments are usually broad, up to 2.5 cm in width, but they may be very narrow and then veins do not extend beyond the vascular commissure bearing the sporangia. *Lomaria volubilis* Hook. was described on the basis of this variation with narrow fertile segments, but it does not merit recognition even as a form.

Ecology and Geography (Fig. 1)

Salpichlaena grows in rain-forests, montane forests and inundated forests, also in thickets, in palm swamps and it may persist in partly cleared areas. It grows from near sea level to 3000 m, most commonly up to 1500 m. The leaf is scandent to at least 15 m into the tops of trees by means of the twining rachis and is partially supported by the distant pairs of subopposite pinnae.

Salpichlaena occurs from Nicaragua to Panama, in the Lesser Antilles, and in South America south to Bolivia and Santa Catarina in southeastern Brazil.

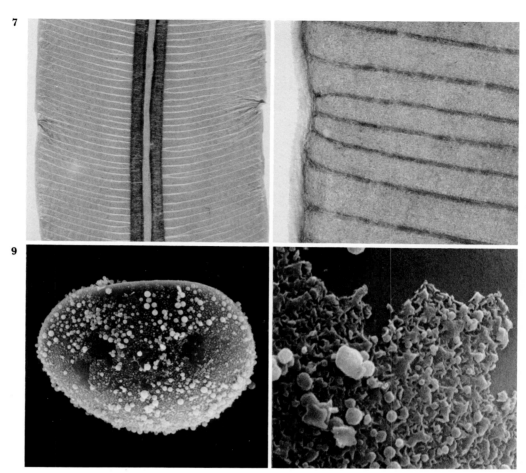

Figs. 105.6–105.9. *Salpichlaena.* **6.** Portion of fertile segment with long, firm indusium adjacent to the costa, × 2.5. **7.** Sterile segment with marginal vein connecting ends of lateral veins, and cartilaginous border, × 10. **8, 9.** Spores. **8.** Scattered spherical deposit lateral view, arrow at laesura, Colombia, *Killip 7761,* × 1000. **9.** Detail of papillate-rugulose perispore above smooth exospore, Ecuador, *Mexia 8426,* × 5000.

Spores

The scattered spherical deposit and papillate-rugulose surface (Fig. 8) differ from the several types of spores in *Blechnum.* The thin perispore formation over the relatively smooth exospore (Fig. 9) also differs from the elaborate wall structure of most *Blechnum* species.

Cytology

Salpichlaena volubilis is reported as $n = 40$ from Trinidad (Walker, 1973). This number is not included in the large series, known in *Blechnum* or in *Woodwardia,* but is characteristic of several genera of the Davalliaceae and Dryopteridaceae.

Literature

Walker, T. G. 1973. Reference under the family.

Family 22.   Polypodiaceae

Polypodiaceae Bercht. & J. S. Presl, Přirozen. Rostl. 1: 272. 1820. Type: *Polypodium* L.

Family synonyms are placed under the tribes, where they correspond more closely with the taxonomy.

Description

Stem erect to short-creeping or very long-creeping, small to rather massive, or slender to filiform, with a solenostele or dictyostele, indurated or hardly sclerotic, bearing scales; leaves ca. 3 cm to 2 m long, entire, pinnatifid or variously forked or pinnate, circinate in the bud, petioles without stipules; sori round to elongate, borne on a vein or at its tip or at the junction of veins, or the sporangia borne on a vascular commissure or on a special network of fertile veins, exindusiate; sporangia with a 1- or 2-rowed or apically 3-rowed stalk, the annulus interrupted by the stalk; homosporous, spores with chlorophyll (green) or without chlorophyll. Gametophyte epigeal, with chlorophyll, obcordate, spathulate or ligulate, or initially filamentous-branched prior to maturity, more or less thickened centrally, glabrous or often with various types of trichomes, the archegonia borne on the lower surface, usually in the central region, or near the sinus, the antheridia 3-celled, borne on the lower or sometimes on the upper surface, mostly apart from the archegonia.

Comments on the Family

The Polypodiaceae are a family of worldwide distribution with about 40 genera, 12 of them in America. It includes 1000 or more species nearly all of which are epiphytes. The sorus is exindusiate, often with paraphyses, or trichomes or glands similar to those of the lamina, it is usually round, sometimes elongate or the sporangia are in long lines or generally distributed along the veins. The sporangia may be borne on a clearly defined receptacle that extends beyond the associated veins. This differs from the acrostichoid condition in which the sporangia are borne on the surface of unmodified leaf tissue. The number of genera differs widely among current authors, depending on whether the diverse elements within the family are considered at the rank of genus, or poorly distinguished evolutionary lines are placed within a genus. A major problem in classification, is the recognition of convergent characters apart from those indicative of evolutionary affinity.

Fossils that seem to be accurately placed in the family are *Astralopteris* Tidw., Rushf. & Reveal, a drynarioid fern from the lower Cretaceous of the western United States, and *Polypodium paleoserratum* Kolak., a polypodioid fern from the Pliocene of Russia.

The Polypodiaceae are an isolated group, most often considered to have distant relationships with the Gleicheniaceae (Bower, 1928; Pichi-Sermolli, 1977). However, the analysis of several characters by Jarrett (1980) concludes that the evidence is unconvincing, and suggests that a relationship to the advanced indusiate leptosporangiate ferns is more likely.

The genera allied to *Grammitis* are often treated in a separate family from those related to *Polypodium* on the basis of differences in the petiole, lamina indument, sporangia, spores and gametophytes. However, most of the characters of one group may occur in the other. This is evident in the gametophyte studies by Stokey and Atkinson (1958). Grammitid gametophytes are usually filamentous and tardily develop a platelike form in contrast to the initially cordate type in *Polypodium*. However, the Old World genera *Crypsinus* Presl and *Selliguea* have filamentous gametophytes much as in the grammitids. It is also shown in the studies on the sporangia by Wilson (1959) who recognized the 1-row stalk in the grammitids as distinct from the 2-rowed form in *Polypodium*. The Old World genus *Dictymia,* which on other characters is clearly placed with *Polypodium,* has 1-rowed sporangia stalks as in *Grammitis*. Since there is no conclusive evidence supporting discrete evolutionary lineages, a single family is recognized.

The tribe **Polypodieae** typically has the petiole articulate at or near the base, the lamina with short, if any, trichomes, the sporangium with a 2-rowed stalk, the spores ellipsoidal-monolete, and not green, and the gametophyte cordate to strap-shaped. It is a large, diverse tribe of 35 or more genera.

The tribe **Loxogrammeae** has the petiole articulate or nonarticulate, the lamina becoming glabrous, the sporangium with a 1-rowed stalk, or 2-rowed by later cell divisions, the spores are green, and spheroidal-trilete or ellipsoidal-monolete, with both forms in some species, and the gametophyte is strap-shaped and branched. It is a small tribe of two genera, with some characters of the tribe Polypodieae and others of the Grammitideae.

The tribe **Grammitideae** typically has the petiole continuous, the lamina with long, stiff trichomes, the sporangium with a 1-rowed stalk, the spores tetrahedral-globose and green (containing chlorophyll), and the gametophyte at first filamentous and branched, later becoming elongate or cordate. The tribe is composed of the large genus *Grammitis* and perhaps a few closely related genera, and the distinctive *Pleurosoriopsis*.

Key to American Genera of Polypodiaceae

a. Sterile lamina with all, or nearly all, of the veins free, or if irregularly anastomosing then the petiole conspicuously pubescent and the spores spheroidal, trilete. b.
  b. Spores spheroidal, trilete, petiole continuous with the stem, breaking irregularly with age, leaf usually with some to many long, rather rigid trichomes.   117. *Grammitis,* p. 747
  b. Spores ellipsoidal, monolete, petiole disarticulating with a clean scar, or rarely irregularly disarticulating, leaf without trichomes or with short or somewhat tortuous trichomes. c.
    c. Lamina entire.   110. *Microgramma,* p. 715
    c. Lamina subdichotomously forked.   108. *Dicranoglossum,* p. 708
    c. Lamina pinnatifid, 1-pinnate, or more complex.
        106. *Polypodium,* p. 688
a. Sterile lamina with regularly anastomosing veins, the petiole not or rarely conspicuously pubescent, spores ellipsoidal, monolete. d.
  d. Stem bearing tubers.   114. *Solanopteris,* p. 735
  d. Stem lacking tubers. e.
    e. Sporangia borne on a long inframarginal commissure or along a series of anastomosing veins. f.

f. Sporangia borne on a long inframarginal commissure, leaves monomorphic. g.
    g. Lamina entire.   113. *Neurodium*, p. 732
    g. Lamina subdichotomously forked. 108. *Dicranoglossum*, p. 708
f. Sporangia borne on anastomosing veins, leaves dimorphic. h.
    h. Lamina entire, with small peltate scales.
        109. *Marginariopsis*, p. 712
    h. Lamina more or less dichotomously lobed or forked, with stalked stellate trichomes.   115. *Platycerium*, p. 739
e. Sporangia borne in round to elongate sori. i.
    i. Immature sori covered by stalked, peltate, roundish paraphyses.
        107. *Pleopeltis*, p. 702
    i. Immature sori with nonpeltate paraphyses or not paraphysate, sometimes covered by scales attached at the edge of the receptacle. j.
        j. Lamina pinnatifid or more complex.   106. *Polypodium*, p. 688
        j. Lamina entire. k.
            k. Stem scales elongate, not clathrate, attached well beyond the base.   110. *Microgramma*, p. 715
            k. Stem scales roundish to usually broad, clathrate, attached at or near the base or rarely near the center. l.
                l. Sori elongate, oblique to nearly parallel to the costa.
                   116. *Loxogramme*, p. 743
              l. Sori roundish. m.
                m. Sori at the vein-tip, lamina glabrous or with a few scales.   106. *Polypodium*, p. 688
                m. Sori at the junction of the veins, or below a vein-tip, or rarely some at the tip and then leaf pubescent. n.
                  n. Sori borne at the junction of veins.
                   112. *Niphidium*, p. 727
                  n. Sori borne on a single vein, below the tip or rarely some at the tip or a few at the junction of veins.
                   111. *Campyloneurum*, p. 722

Literature

Bower, F. O. 1928. The Ferns, Vol. 3, pp. 1–306. Cambridge Univ. Press, Cambridge, England.

Christensen, C. 1929. Taxonomic fern-studies. I. Revision of the polypodioid genera with longitudinal coenosori (Cochlidiinae and "Drymoglossinae"); with a discussion of their phylogeny. Dansk Bot. Ark. 6 (3): 1-93.

Jarrett, F. 1980. Studies in the classification of the leptosporangiate ferns: I. The affinities of the Polypodiaceae *sensu stricto* and the Grammitidaceae. Kew Bull. 34: 825–833.

Pichi-Sermolli, R. E. G. 1977. Tentamen pteridophytorum genera in taxonomicum ordinem redigendi. Webbia 31: 313–512.

Stokey, A. G., and L. R. Atkinson. 1975. The gametophyte of the Grammitidaceae. Phytomorph. 8: 391–403.

Wagner, F. S. 1980. New basic chromosome numbers for genera of neotropical ferns. Amer. Jour. Bot. 67: 733–738.

Walker, T. G. 1966. A cytological survey of the pteridophytes of Jamaica. Trans. Roy. Soc. Edinburgh 66: 169–231.

Wilson, K. A. 1959. Sporangia of the ferns genera allied with *Polypodium* and *Vittaria*. Contrib. Gray Herb. 185: 97–127.

22a. Tribe Polypodieae

Platyceriaceae (Nayar) Ching, Acta Phytotax. Sinica 16 (3): 18. 1978.
Polypodiaceae subfamily *Platycerioideae* Nayar, Taxon 19: 223. 1970.
Type: *Platycerium* Desv.
Drynariaceae Ching, Acta Phytotax. Sinica 16 (4): 19. 1978. Type: *Drynaria* (Bory) J. Sm.

A large and complex tribe with 10 genera in America and three of these also in the Old World. With few exceptions, the American genera seem to have their closest relations to others of America, although their exact affinities are not always clear.

Polypodium Bradeorum Rosenst. (*P. colysoides* Copel.) is a highly variable, rare fern of Veracruz, Mexico, British Honduras, Nicaragua and Costa Rica. It belongs in the tribe but its relations are uncertain. It has been studied by Evans and Mickel (1969), and by Gómez (1977) who recognized it as the genus, *Pseudocolysis* Gómez. *Polypodium Bradeorum* certainly does not belong in any of the American genera of the Polypodiaceae. The species is extremely variable and appears to have the attributes of a hybrid between widely different parents as in the case of *Pleuroderris Michleriana*. The fertile leaves vary from entire to pinnatifid and the sori from roundish to long and oblique. It is difficult to suggest putative parents without a knowledge of associated species, however, it occurs within the range of *Loxogramme mexicana*, which also has elongate, oblique sori.

At least 25 genera are confined to the Old World where the tribe shows much greater diversity than in America. These may be grouped into six alliances: (1) *Drynaria* (Bory) J. Sm. and allies, including *Aglaomorpha* Schott, *Photinopteris* J. Sm. and *Thayeria* Copel; (2) *Microsorium* Link and allies, including *Lecanopteris* Reinw. and perhaps a few other genera; (3) *Pleopeltis* and allies, including *Belvisia Mirbel, Lemmaphyllum Presl* and *Lepisorus* (J. Sm.) Ching; (4) *Polypodium, Goniophlebium* (Bl.) Presl and *Thylacopteris* J. Sm.; (5) *Pyrrosia* Mirbel and *Platycerium;* and (6) *Selliguea* Bory and allies, including *Arthromeris* (Moore) J. Sm., *Christiopteris* Copel, and *Oleandropsis* Copel. The position of some genera such as *Dictymia* J. Sm. and *Polypodiopteris* Reed needs clarification.

Literature

Evans, A. M., and J. T. Mickel, 1969. A re-evaluation of *Polypodium Bradeorum* and *P. colysoides*. Brittonia 21: 255–260.

Gómez, L. D. 1977. Contribuciones a la pteridología centroamericana, II. Novitates. Brenesia 10/11: 115–119.

106. *Polypodium*
Figs. 106.1–106.42

Polypodium L., Sp. Pl. 1082. 1753; Gen. Pl. ed. 5, 485. 1754. Type: *Polypodium vulgare* L.

Marginaria Bory, Dict. Class. d'Hist. Nat. 6: 587. 1824. Species listed in 10: 176. 1826. Type: *Marginaria polypodioides* (L.) Tidestr. (*Acrostichum polypodioides* L.) = *Polypodium polypodioides* (L.) Watt.

Synammia Presl., Tent. Pterid. 212. 1836. Type *Synammia triloba* (Cav.) Presl (*Polypodium trilobum* Cav., 1802, not Houtt., 1783) = *Synammia Feuillei* (Bert.) Copel. = *Polypodium Feuillei* Bert.

Phlebodium (R. Br.) J. Sm., Jour. Bot. (Hook.) 4: 58. July, 1841. *Polypodium* section *Phlebodium* R. Br., Pl. Jav. Rar. (Bennett & Brown; Horsfield) 4. 1838. Type: *Polypodium aureum* L. (*Phlebodium aureum* (L.) J. Sm.).

Chrysopteris Link, Fil. Sp. 120. Sept., 1841, *nom. superfl.* for *Phlebodium* and with the same type.

Lepicystis (J. Sm.) J. Sm., Lond. Jour. Bot. 1: 195. 1842. *Goniophlebium* section *Lepicystis* J. Sm., Jour. Bot. (Hook.) 4: 56. 1841. Type: *Goniophlebium incanum* J. Sm. (*Polypodium incanum* Sw., *nom. superfl.*) (*Lepicystis incana* (J. Sm.) J. Sm.) = *Polypodium polypodioides* (L.) Watt.

Description Terrestrial, rupestral or epiphytic; stem short- to long-creeping (to 5 m or more), moderately stout to slender, bearing scales and few to many, usually long, fibrous roots; leaves monomorphic to somewhat dimorphic (the fertile longer than the sterile), ca. 4 cm to 2 m long, borne in a cluster to widely spaced, lamina pinnatifid, pinnatisect, or 1-pinnate, rarely to 2-pinnate-pinnatifid, or (in *P. glaucophyllum*) entire, glabrous, pubescent or sparsely to densely scaly, veins free to anastomosing with or without free included veinlets; sori round to elongate, borne at the tip of a vein or at the junction of veins, on a usually slightly to moderately raised or rarely sunken receptacle, paraphysate or not, exindusiate; spores ellipsoidal, monolete, the laesura $\frac{1}{2}$ to $\frac{2}{3}$ the length, often with a low, verrucate, tuberculate or papillate surface, sometimes with more or less winglike folds. Chromosome number: $n = 37, 74, 111$; apogamous $= 111$.

The leaves of *Polypodium* disarticulate, leaving a clean scar above the base of the petiole, which is called a phyllopodium. The lamina is ofen glabrous (Figs. 5–8), sometimes pubescent, or with a few scales on the axes. In many species there are small (Fig. 9) to rather large scales (Fig. 10) on the lamina that sometimes provide a dense cover. The veins may be free (Fig. 5) or variously anastomosing (Figs. 6–8) and the sori are borne at the tip of a veinlet (Figs. 5, 8) or at the junction of veins (Fig. 8).

Paraphyses are not conspicuous in *Polypodium*. They are usually filamentous, sometimes branched, or they may be expanded at the apex, or as in *Polypodium vulgare* (*sens. lat.*) they may resemble sporangia (Martens and Pirard, 1943; Peterson and Kott, 1974). Some species of the *Polypodium pycnocarpum* and *P. lepidopteris* groups have the immature sori covered by scales. These are not considered to be paraphyses since they are attached at the periphery of the receptacle.

Fig. 106.1. *Polypodium dissimile,* Río Palenque, Ecuador. (Photo W. H. Hodge.)

Systematics

Polypodium is a genus of about 150 species in America and mostly in extratropical regions of the Old World. It is characterized by a pinnatifid, or more complex, lamina with nonarticulate pinnae, sori borne at the tip or at the junction of veins, and paraphyses that, when present, do not cover the immature sorus.

American species are sometimes included in *Goniophlebium* (Bl.) Presl on the basis of the areolae with a free, excurrent veinlet. Also it has been incorrectly typified (Pichi-Sermolli, 1973) by an American species. However, *Goniophlebium* is an Old World genus with articulate pinnae and probably not closely related to American elements.

Among the American genera of the Polypodiaceae, *Polypodium* appears to be related to *Pleopeltis* through species of the *P. pycnocarpum* group. There are undoubtedly affinities with other genera but these are not clear. One of the basic problems in the family is whether entire or more complex leaves are primitive. The plants often have entire, juvenile leaves and later produce a series of increasingly complex ones, as in *Polypodium aureum* (Fig.

Fig. 106.2. *Polypodium aureum,* Pedregal Las Vigas above Jalapa, Veracruz, Mexico. (Photo W. H. Hodge.)

19). Although this suggests that neotony and precocious fertility may operate in the evolution of entire leaves, evidence is lacking for the evolutionary direction of leaf complexity.

The scaly-leaved species of *Polypodium* with free veins have been revised by Maxon (1916a) and also those with anastomosing veins (1916b). Weatherby revised the *Polypodium polypodioides* group (1939) and the group of *Polypodium lepidopteris* in Brazil (1947). A revision of the *Polypodium pectinatum* group was done by Evans (1969). Species of Argentina and southern Brazil have been treated by de la Sota (1960, 1965, 1966) who also has revised the *Polypodium lepidopteris* (*P. squamatum*) group (1968).

Tropical American Species

There have been taxonomic revisions of a few groups of American *Polypodium,* but most alliances have received little or no attention. Revisionary work, employing new characters, is needed not only for a modern assessment of the species but also for evidence on the relations of species groups.

Hybrids among the American tropical species have been proposed between *Polypodium Friedrichsthalianum* and *P. thyssanolepis* (Gómez, 1975), and also with *P. furfuraceum* (Gómez, 1976; Wagner et al., 1977). The irregularly lobed or pinnatifid leaves

Fig. 106.3. *Polypodium guttatum,* Pedregal Las Vigas above Jalapa, Veracruz, Mexico, growing with *Pleopeltis macrocarpa,* right. (Photo W. H. Hodge.)

of *Polypodium semipinnatifidum* (Fée) Mett. suggest that this is a hybrid, undoubtedly involving *Polypodium glaucophyllum,* since it has a similar long-creeping stem with rounded scales, but the other parent is unknown.

The following synopsis of the tropical American species is presented as a guide for further study. The groups are based on a combination of characters of the stem scales, lamina indument, venation, soral arrangement and spores. The extent to which the characters may be convergent is not clear but a final classification cannot be expected until the relations of species groups and the evolutionary development of characters are better known.

The complex of *Polypodium vulgare,* including *P. Scouleri* Hook. & Grev. and *P. vulgare* L. (*sens. lat.*), occurs rarely in the mountains of Mexico and Guatemala. It is primarily a circumboreal group of terrestrial plants with free veins. Christensen (1928) discussed its alliances with tropical and subtropical elements but it is still uncertain whether the closest relations are with American or Asiatic species. Among tropical American species *Polypodium vulgare* may be allied to *Polypodium fissidens* and *P. puberulum* in the *Polypodium plesiosorum* group.

Among the few species of Chile, and the Juan Fernandez Islands, *Polypodium masafuerae* Phil. clearly belongs in the *Polypodium pycnocarpum* group. The other species, *Polypodium Feuillei* Bert., *P. Espinosae* Weath., and *P. intermedium* Colla, are certainly of American rather than circum-austral affinity. They may be rather distantly allied to species such as *Polypodium Catharinae* and *P. lasiopus* of the *Polypodium loriceum* group.

Fig. 106.4. Distribution of ***Polypodium*** in America, south of lat. 35° N.

1. *Polypodium plesiosorum* Group

Lamina without scales or with a few confined to the rachis and costae, stem scales (Fig. 11) brown, not or slightly clathrate, sometimes short-pubescent (Fig. 12), veins free or anastomosing, sori in one row between the costa and margin, spores prominently verrucate, coarsely tuberculate, or somewhat folded.

A group of about 20 species, for example: *Polypodium aureum* L. (Figs. 2, 19), *P. fissidens* Maxon, *P. Lowei* C. Chr., *P. macrodon* Hook. (not Baker, Syn. Fil. 318), *P. Martensii* Mett., *P. plectolepis* Hook., *P. plesiosorum* Kze., *P. puberulum* Schlect. & Cham., *P. sororium* Willd. (*P. dissimile* auths, not L.) and *P. subpetiolatum* Hook.

2. *Polypodium loriceum* Group

Lamina without scales or with a few confined to the rachis and costae, stem scales (Fig. 14) clathrate, not pubescent, veins free or usually anastomosing, sori in one or more rows between the costa and margin, spores coarsely and prominently tuberculate, coarsely verrucate, or sometimes with more or less winglike folds.

A group of 35–40 species, within which the following intergrading alliances may be recognized.

Species with two or more rows of sori between the costa and margin and a pinnatisect to usually 1-pinnate lamina are, for example: *Polypodium adnatum* Kl., *P. brasiliense* Poir., *P. Caceresii* Sod. (Fig. 20), *P. decumanum* Willd., *P. decurrens* Raddi, *P. Fendleri* D. C. Eaton, *P. fraxinifolium* Jacq., and *P. triseriale* Sw.

5 6 7

8 9 10

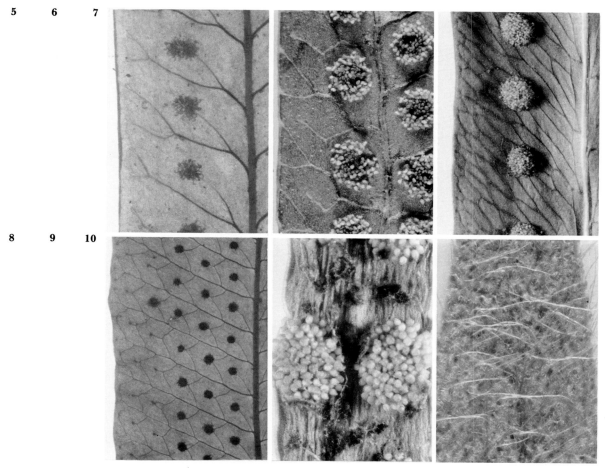

Figs. 106.5–106.10. Portions of pinna-segments of *Polypodium*. **5.** Sori and free veins of *P. sororium*, ×6. **6.** Sori and partially anastomosing veins of *P. loriceum*, ×8. **7.** Sori and anastomosing veins of *P. aureum*, ×4. **8.** Sori at vein-tips and at the junction of veins and anastomosing veins of *P. adnatum*, ×2. **9.** Sori and small scales on segment of *P. Buchtienii*, ×10. **10.** Long, acicular scales on segment of *P. pyrrholepis*, ×10.

Species with the lamina 1-pinnate and large, thin, finely clathrate (Fig. 13) and iridescent stem scales are: *Polypodium Kunzeanum* C. Chr. (Fig. 21), (*P. cordatum* Kze., not Desv.) and *P. sessilifolium* Desv.

Species with the lamina pinnatifid to pinnatisect, and attenuate stem scales are, for example: *Polypodium Catharinae* Langsd. & Fisch., *P. dissimile* L. (*P. chnoodes* Spreng.), *P. Gilliesii* C. Chr., *P. lasiopus* Kl. and *P. latipes* Langsd. & Fisch.

Species with the stems very long, green, and bearing sparse, roundish, dark scales (Fig. 14) similar to those of *P. fraxinifolium* in the first alliance are: *Polypodium dasypleuron* Kze., *P. Funkii* Mett., *P. loriceum* L., *P. maritimum* Hieron., and *P. subandinum* Sod. (Fig. 22).

Polypodium glaucophyllum Kl. has the stem and scales similar to others in the *P. loriceum* alliance. It is unique in the genus in its entire lamina, and sori usually between prominent lateral veins in one to several rows between the costa and margin.

11   12   13

14 15 16 17 18

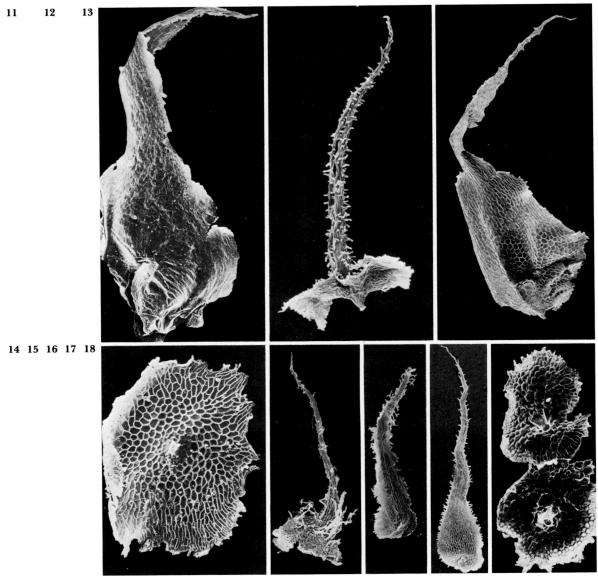

Figs. 106.11–106.18. Stem scales of *Polypodium.* **11.** Nonclathrate scale of *P. Lowei,* ×10. **12.** Short-pubescent scale of *P. macrodon,* ×15. **13.** Clathrate scale of *P. sessilifolium,* ×10. **14.** Clathrate scale of *P. maritimum,* ×35. **15.** Scale of *P. pectinatum* with long trichomes at the base, ×8. **16.** Scale of *P. polypodioides* with sclerotic central stripe, ×15. **17.** Elongate scale of *P. fimbriatum,* ×8. **18.** Round-ish scales of *P. Collinsii,* ×50.

3.   *Polypodium pectinatum* Group

Lamina usually with few scales confined to the rachis and costae, stem scales reddish-brown, not or somewhat clathrate, often pubescent at the base, veins usually free or sometimes anastomosing, sori in one row between the costa and margin, spores verrucate.

A group of about 25 species with a pectinately pinnatisect lamina (Fig. 23). The systematic revision by Evans (1969) has clarified the species; however, the relationship of the group to others in *Polypodium* is not resolved.

Some of the species are: *Polypodium atrum* Evans, *P. bolivianum*

Fig. 106.19. Leaves from juvenile to nearly adult plants of ***Polypodium aureum,*** the size and complexity of the lamina corresponds to the size of the stem, × 0.5.

Rosenst., *P. curvans* Mett., *P. dispersum* Evans, *P. eurybasis* C. Chr., *P. filicula* Kaulf., *P. pectinatum* L., and *P. pumula* Willd.

4. *Polypodium pycnocarpum* Group

Lamina usually conspicuously although sometimes sparsely scaly, stem scales (Fig. 16) with a dark, sclerotic central stripe and paler, narrow to broad borders, veins free to anastomosing, sori in one row between the costa and margin, spores shallowly verrucate or slightly papillate, to nearly smooth, with dense spherical deposit.

A group of about 25 species, most of them with a pinnatisect lamina, for example: *Polypodium Buchtienii* Rosenst. (Fig. 24), *P. guttatum* Maxon, *P. macrolepis* Maxon, *P. murorum* Hook., *P. plebejum* Schlect. & Cham., *P. pleopeltifolium* Raddi, *P. polypodioides* (L.) Watt., *P. pycnocarpum* C. Chr., *P. thyssanolepis* Kl., and *P. tridens* Kze.

Species with the lamina 2-pinnate-pinnatifid are *Polypodium macrosorum* Fée and *P. monosorum* Desv. (Fig. 25).

5. *Polypodium lepidopteris* Group

Lamina more or less densely scaly, stem scales long, narrow, bright reddish-brown and concolorous or, dark, roundish, and closely imbricate, veins anastomosing, sori in one row between the costa and margin, spores papillate with dense spherical deposit.

Nectaries in the form of excavated areas borne on a small auricle near the base of the pinna-segments on the acroscopic side are present in all species.

A group of about 15 species that form an especially distinctive alliance on the basis of the nectaries and papillate spores.

Species with long, narrow stem scales (Fig. 17) are, for example: *Polypodium fimbriatum* Maxon (Fig. 26), *P. lepidopteris* (Langsd. & Fisch.) Kze., *P. lepidotrichum* (Fée) Maxon, *P. minarum* Weath., *P. monoides* Weath., *P. pyrrholepis* (Fée) Maxon, and *P. squamatum* L.

Species with roundish, closely imbricate stem scales (Fig. 18) are: *Polypodium Collinsii* Maxon, *P. Mickelii* Sota, *P. myriolepis* Christ, and *P. sancta-rosae* (Maxon) C. Chr.

6. Scaly-Leaved Species of Uncertain Position

The following four highly distinctive species are not clearly allied with those of the previous groups. Some may perhaps represent species of hybrid origin between morphologically diverse parents.

Polypodium Friedrichsthalianum Kze. has stem scales light brown, concolorous, clathrate, the lamina is 2-pinnate-pinnatifid, and the spores are usually not well filled or irregular. Plants in culture have been reported as $n =$ ca. 37, and $n = 35$ which was regarded as anomalous by Smith and Mickel (1977).

Polypodium furfuraceum Schlect. & Cham. has stem scales light brown to whitish, nearly concolorous, not clathrate, the lamina is narrow and pectinately pinnatisect.

Polypodium Munchii Christ has stem scales brownish, not clathrate, the lamina is pinnatifid with few large segments, with large sori.

Polypodium fallax Schlect. & Cham. (*P. margaritiferum* Christ) has a very long, slender stem with small, dark, pubescent scales.

Ecology (Figs. 1–3)

Polypodium grows on trees usually as low epiphytes, or on rocks, soil, or rotted wood. It is most often terrestrial or rupestral in the northern and southern parts of its range and at its altitudinal limits, while in the wet tropics it is most often epiphytic.

In tropical America, *Polypodium* usually grows in lowland rain forests, in montane forests or cloud forests, also in pine and oak woods and sometimes in seasonally dry forests. It also occurs in gallery forests, in shrubby savannahs, in rough pastures, where it grows on relict trees and living fence posts, in partially cutover land, and as an epiphyte in oil palm, cacao and coffee plantations. It grows in a variety of niches such as cliffs, mossy boulders, moist sites in lava beds, on rocky banks, on old stone walls, on masonry and on tile roofs. *Polypodium* grows from sea level to 4400 m, most commonly up to ca. 2500 m.

Polypodium filicula and *P. funiculum* Fée have long, very slender stems with few scales and roots that gives rise at intervals to short, more or less erect, scaly stems with abundant roots, and leaves. This habit of growth is like that of *Asplenium filicaule* and allied species, and as in those species, the short stems with leaves eventually are separated from the creeping stem and become independent plants. De la Sota (1960) reports another case of vegetative reproduction in *Polypodium Singeri* Sota, in which especially long roots may give rise to new plants.

Geography (Fig. 4)

Polypodium is widely distributed in America, and across Eurasia (*P. vulgare, sens. lat.*) with extensions southward to the Himalayas and Taiwan, and is also in southern Africa (*P. Ecklonii* Kze.). The *P. vulgare* complex is on widely distant islands as Iceland, the Azores, Madeira, the Canary Islands and Kerguelen Island, and *P. pellucidum* Kaulf. is on the Hawaiian Islands. *Polypodium*

Figs. 106.20–106.26. Leaves of *Polypodium* **20.** *P. Caceresii,* ×0.25. **21.** *P. Kunzeanum,* ×0.25. **22.** *P. subandinum,* ×0.5. **23.** *P. pectinatum,* ×0.5. **24.** *P. Buchtienii,* ×0.5. **25.** *P. monosorum,* ×0.25. **26.** *P. fimbriatum,* ×0.25.

has been erroneously reported as occurring in New Zealand. *Polypodium viride* Gilbert, described from New Zealand, was shown by Morton (1958), to be based on a specimen from the United States. *Polypodium vulgare* var. *auritum* Gilbert is another case, not considered by Morton, involving an erroneous New Zealand label, for the specimen evidently is *Polypodium pellucidum.*

In America, *Polypodium* has a broad distribution in the United States and ranges northeastward to southern Greenland and northwestward to Alaska. It is widely distributed from Mexico and the Greater Antilles southward to southern Chile and Argentina; also on Guadaloupe Island, Cocos Island, the Galápagos Islands, the Juan Fernandez Islands and on Ilha Trindade.

The species of *Polypodium* are concentrated in the Greater Antilles, southern Mexico to Colombia and Venezuela, south to Peru and in southeastern Brazil. There are relatively few species

Figs. 106.27–106.34. *Polypodium* spores, ×1000. **27–31.** *P. plesiosorum* group. **27.** *P. Martensii,* verrucate proximal face, Mexico, *Bourgeau 1039.* **28.** *P. puberulum,* verrucate, part of distal face below, Mexico, *Hale & Soderstrom 20919.* **29.** *P. fissidens,* verrucate, the perispore with low folds, Guatemala, *Molina 15938.* **30, 31.** *P. aureum,* Mexico, *Pringle 2582.* **30.** Coarsely tuberculate proximal face. **31.** Detail of partly abraded surface (arrow) edge of perispore. **32–34.** *P. loriceum* group. **32.** *P. adnatum,* tuberculate, Ecuador, *Harling et al. 9134.* **33.** *P. fraxinifolium,* coarsely tuberculate, Brazil, *Mexia 4929.* **34.** *P. sessilifolium,* low, verrucate exospore under winglike folds, Ecuador, *Camp E-4230.*

in other parts of America, perhaps ten in each of the following regions: the United States, northern Mexico from the states of Coahuila to Baja California, the Lesser Antilles, Amazonian Brazil, and Argentina, excluding the provinces of Missiones and Corrientes where several Brazilian species reach their southern limit. Three species occur in Chile.

The genus is poorly represented on isolated islands and insular endemism usually is not strong. One species occurs on Guadaloupe Island, two on Cocos Island, seven on the Galápagos Islands including the endemics *P. insularum* (Morton) Sota and *P. tridens,* two on the Juan Fernandez Islands, one *P. intermedium* endemic, and two species, both endemic, on Ilha Trindade, *P. insulare* Brade and *P. trindadense* Brade.

Spores

The spore type is consistent within species alliances, except for the groups of *Polypodium plesiosorum* and *P. loriceum.* In each of these there appears to be a predominant type of spore and other forms that are common to both groups. Mature spores of all species of *Polypodium* have a thin perispore formed over and often following the contours of the verrucate, tuberculate or rugulose exospore (Figs. 31, 39).

Species in the *Polypodium plesiosorum* group commonly have verrucate spores as in *P. Martensii* and *P. puberulum* (Figs. 27, 28). A few species have low folds that overlay the verrucate exospore as in *P. Eatonii* Baker *P. subpetiolatum,* and *P. fissidens* (Fig. 29). These are similar but less prominent than the winglike forms in the *P. loriceum* group. Coarsely tuberculate spores as in *P. aureum* (Figs. 30, 31) represent another type common to this and the next group.

The coarsely tuberculate spores, frequent among species of the *Polypodium loriceum* group, are often small as in *P. adnatum* and *P. fraxinifolium* (Figs. 32, 33). Spores of several species with winglike folds that overlay the verrucate exospore are unusual in the genus. This form, recognized by Lloyd (1969, 1981) in *P. dissimile* (as *P. chnoodes*), also occurs in *P. Kunzeanum, P. mindense* Sod. and *P. sessilifolium* (Fig. 34).

Species of the *Polypodium pectinatum* group uniformly have rather low, verrucate spores as in *P. atrum* (Fig. 35). The larger size and prominent laesura in *P. dispersum* (Fig. 36) represents an apogamous form with 32 rather than the usual 64 spores per sporangium.

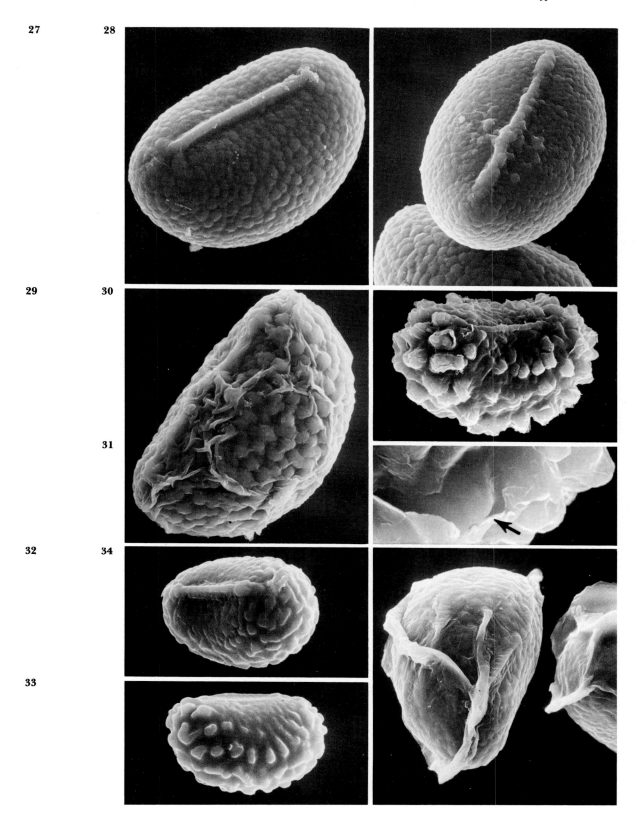

35   **36**

37   **38**

39

40   **41**

42

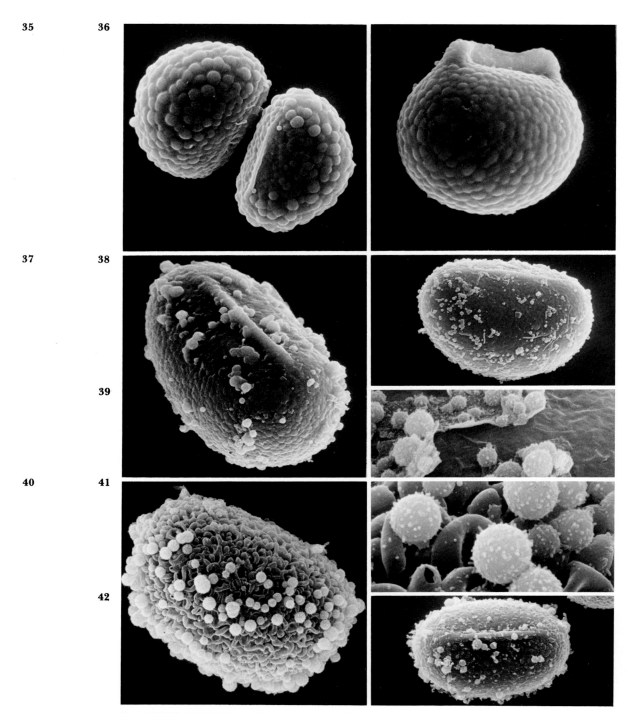

Figs. 106.35–106.42. *Polypodium* spores, × 1000, detail, × 5000.   **35, 36. *P. pectinatum group.*   35. *P. atrum,*** verrucate, the proximal face tilted, right, Mexico, *Seaton 424.*   **36. *P. dispersum,*** verrucate, the laesura at top, Colombia, *Cuatrecasas 12135.*   **37–39. *P. pycnocarpum*** group.   **37. *P. monosorum,*** coarse spherical deposit, Colombia, *Cuatrecasas 19061.*   **38. *P. polypodioides,*** coarse spherical deposit, Peru, *Tryon & Tryon 5190.*   **39. *P. tridens,*** abraded surface of thin perispore with spherical deposit and rugulose stratum below, Galápagos Islands, *Wiggins 18349,* × 5000.   **40–42. *P. lepidopteris*** group.   **40, 41. *P. minarum,*** Brazil, *Williams 5495.*   **40.** Prominent papillate surface with spherical deposit.   **41.** Detail of deflated papillae and spherical deposit, × 5000.   **42. *P. lepidotrichum,*** sparsely papillate proximal face, Mexico, *Bourgeau 2883.*

Species of the *Polypodium pycnocarpum* group have spores with coarse, spherical deposit which overlays a relatively smooth to slightly papillate surface as in *P. monosorum* and *P. polypodioides* (Figs. 37, 38). The exospore below forms a relatively smooth rugulose surface in *P. tridens* (Fig. 39).

Species in the group of *Polypodium lepidopteris* uniformly have papillate spores with a large spherical deposit, as in *P. minarum* and *P. lepidotrichum* (Figs. 40, 42). The spherical material, similar to that of the previous group, often densely covers the papillae which may be collapsed in dry specimens (Fig. 41).

Cytology

The chromosome numbers of *Polypodium* are relatively stable at either the diploid $n = 37$ and tetraploid, $n = 74$ levels. The tropical American species, reported by Smith and Mickel (1977), largely from Mexico, are mostly diploids with a few tetraploids. Both levels are known in a few wide ranging species as *Polypodium polypodioides* and *P. thyssanolepis*. In the *Polypodium pectinatum* group the 14 taxa reported by Evans (1969) include 9 diploids, 4 tetraploids and an apogamous triploid with 111 chromosomes.

In European *Polypodium vulgare,* diploid, tetraploid and hexaploid plants were reported by Manton (1950) and diploids, triploids and tetraploids have been found in *P. virginianum* in eastern North America (Shivas, 1961). This species and a few others are diploid in Japan. The *Polypodium vulgare* complex in western North America (Lang, 1969, 1971) includes two diploids, two triploids and a tetraploid that are morphologically distinct and show differences in geographic ranges. The diploid *Polypodium glycyrrhiza* D. C. Eaton is coastal, the diploid *P. amorphum* Suskd. (*P. montense* F. A. Lang) is more limited, mostly to the mountains of the northwest, while the tetraploid, *P. hesperium* Maxon is widely distributed in the west and extends inland to North Dakota. Similar cytogeographic analyses need to be made on species groups in tropical America.

Literature

Christensen, C. 1928. On the systematic position of *Polypodium vulgare.* Dansk Bot. Ark. 5 (22): 1–10.

Evans, A. M. 1969. Interspecific relationships in the *Polypodium pectinatum-plumula* complex. Ann. Mo. Bot. Gard. 55: 193–293.

Gómez, L. D. 1975. Contribuciones a la pteridología Costarricense, VIII. Brenesia 6: 49–67.

Gómez, L. D. 1976. The identity of *Polypodium furfuraceum* f. *pinnatisectum.* Amer. Fern Jour. 66: 28.

Lang, F. A. 1969. A new name for the species of *Polypodium* from northwestern North America. Madroño 20: 53–60.

Lang, F. A. 1971. The *Polypodium vulgare* complex in the Pacific northwest. Madroño 21: 235–253.

Lloyd, R. M. 1969. Occurrence of a winged perispore on a new world species of *Polypodium.* Brittonia 21: 80–82.

Lloyd, R. M. 1981. The perispore in *Polypodium* and related genera (Polypodiaceae). Canad. Jour. Bot. 59: 175–189.

Manton, I. 1950. Problems of the cytology and evolution in the Pteridophyta. pp. 316. Cambridge Univ. Press, Cambridge, England.

Martens, P., and N. Pirard. 1943. Les organes glanduleux de *Polypodium virginianum* L. La Cellule 49: 385–406.

Maxon, W. R. 1916a. *Polypodium furfuraceum* and related species. Contrib. U.S. Nat. Herb. 17: 557–579.

Maxon, W. R. 1916b. *Polypodium squamatum* and its allies. Contrib. U.S. Nat. Herb. 17: 579–596.

Morton, C. V. 1958. The identity of *Polypodium viride* Gilbert. Amer. Fern Jour. 48: 75–77.

Peterson, R. L., and L. S. Kott. 1974. The sorus of *Polypodium virginianum:* some aspects of the development and structure of paraphyses and sporangia. Canad. Jour. Bot. 52: 2283–2288.

Pichi-Sermolli, R. E. G. 1973. Fragmenta pteridologiae, IV. Webbia 28: 445–477.

Shivas, M. G. 1961. Contributions to the cytology and taxonomy of the species of *Polypodium* in Europe and America 1, Cytology. Jour. Linn. Soc. 58: 13–25.

Smith, A. R., and J. T. Mickel. 1977. Chromosome counts for Mexican ferns. Brittonia 29: 291–298.

Sota, E. R. de la. 1960. Polypodiaceae y Grammitidaceae Argentinas. Opera Lilloana 5: 5–229.

Sota, E. R. de la. 1965. Las especies escamosas del género *"Polypodium"* L. (*s.str.*) en Brasil. Revis. Mus. La Plata n.s. Sec. Bot. 9: 243–271.

Sota, E. R. de la. 1966, Revisión de las especies Americanas del grupo *"Polypodium squamatum"* L. Rev. Museo La Plata, n.s. Sec. Bot. 10: 69–186.

Sota, E. R. de la. 1968. Acerca del género *"Synammia"* Presl. Rev. Museo La Plata n.s. Sec. Bot. 11: 129–132.

Wagner, W. H., Jr., F. S. Wagner, and L. D. Gómez P. 1977. An enigmatic polypody fern from Cartago, Costa Rica. Brenesia 12/13: 81–103.

Weatherby, C. A. 1939. The group of *Polypodium polypodioides.* Contrib. Gray Herb. 124: 22–35.

Weatherby, C. A. 1947. *Polypodium lepidopteris* and its relatives in Brazil. Contrib. Gray Herb. 165: 76–82.

107. *Pleopeltis*
Figs. 107.1–107.13

Pleopeltis Willd., Sp. Pl. 5: 211. 1810. Type: *Pleopeltis angusta* Willd.

Description Epiphytic or sometimes rupestral; stem long-creeping, slender, bearing scales and usually many fibrous roots; leaves monomorphic to slightly dimorphic (the fertile longer and narrower than the sterile), ca. 5–35 cm long, usually borne at intervals, lamina entire or subdichotomously forked into a few segments, sparsely and minutely to densely and prominently scaly, especially beneath, veins anastomosing with or without included free veinlets; sori round to somewhat elongate, usually borne at the junction of two or more veins, on a slightly to moderately raised receptacle, paraphysate, exindusiate; spores ellipsoidal, monolete, the laesura $\frac{2}{3}$ to $\frac{3}{4}$ the spore length, the surface shallow to prominently verrucate. Chromosome number: $n = 34, 35, 37$, ca. 70, 74; $2n = $ ca. 210.

Fig. 107.1. *Pleopeltis macrocarpa* near Jalapa, Veracruz, Mexico. (Photo W. H. Hodge.)

Fig. 107.1. *Pleopeltis macrocarpa* near Jalapa, Veracruz, Mexico. (Photo W. H. Hodge.)

The stem of *Pleopeltis* is more or less flattened and the sori are borne in a single row on each side of the costa (Fig. 2). The stem scales are clathrate, or subclathrate in *Pleopeltis percussa,* and in most species have a dark central portion and thinner, lighter borders; they often bear trichomes at the base. The anastomosing veins of the sterile lamina are shown in Fig. 5 and of the fertile lamina, with the sori at the junction of veins, in Figs. 4 and 12. *Pleopeltis percussa* has abundant, persistent, filamentous, branched paraphyses (Fig. 11) in the mature sorus. The lamina is entire except in *Pleopeltis angusta* which is subdichotomously forked (Fig. 6).

Systematics

Pleopeltis is a genus of about 10 species, all American with one of them also in Africa to India and Ceylon. The genus is characterized by the roundish, stalked, peltate paraphyses covering the immature sorus (Fig. 7) which are absent in the mature sorus (Fig. 8). *Marginariopsis,* the only other American genus with paraphyses of this type, is probably a derived genus, as may be *Dicranoglossum. Pleopeltis* seems to be closely related to some of the scaly-leaved groups of American *Polypodium,* which may also have scales covering the immature sorus, although in these they are mostly or wholly lamina scales attached at the edge of the receptacle.

Pleopeltis percussa and *P. fuscopunctata* are included in the genus primarily on the basis of their stalked, peltate paraphyses that cover the immature sorus.

The Old World *Lepisorus* is not included in *Pleopeltis,* although it is similar in a number of characters, because of its probable Asiatic origin. Further studies of the two groups are needed to

determine if the similarities are due to convergence or to evolutionary affinity.

The group of *Pleopeltis macrocarpa* in America has been revised by Weatherby (1922), under the name of *Polypodium lanceolatum*.

Tropical American Species

Most species of *Pleopeltis* have stem scales with a dark, sclerotic center, and a more or less densely and prominently scaly, entire, coriaceous lamina. These form a closely related species-group comprised of *Pleopeltis macrocarpa* (Willd.) Kaulf. (*Polypodium lanceolatum* L., not *Pleopeltis lanceolata* Kaulf.), *P. revoluta* (Willd.) A. R. Sm. (*Polypodium astrolepis* Liebm.), *P. polylepis* (Kze.) Fourn. (*Polypodium peltatum* Cav., not *Pleopeltis peltata* vAvR.), *P. erythrolepis* (Weath.) Pic.-Ser., *P. Conzattii* (Weath.) R. & A. Tryon, *P. panamensis* (Weath.) Pic.-Ser., and *P. fructuosa* (Weath.) Lell.

Pleopeltis angusta Willd. has stem scales similar to the above group but a more or less subdichotomously forked lamina. *Pleopeltis percussa* (Cav.) Hook. & Grev. has abundant filamentous, branched paraphyses in the mature sorus, and stem scales that are similar to those of *Microgramma. Pleopeltis fuscopunctata* (Hook.) R. & A. Tryon has a papyraceous lamina and broad, concolorous, thin stem scales.

Hybrids involving species of *Pleopeltis* and *Polypodium* have been detected by irregularities in leaf morphology. *Polypodium Bartlettii* Weath. was suggested as a hybrid of *Pleopeltis polylepis* and *Polypodium polypodioides* (Weatherby, 1935). An analysis of *Polypodium leucosporum* Kl., a rather widely distributed tropical fern, indicated this was derived as a hybrid of *Pleopeltis macro-*

Fig. 107.3. Distribution of *Pleopeltis* in America.

carpa and *Polypodium thyssanolepis* and perhaps also involves other *Polypodium* species (Wagner and Wagner, 1975).

Polypodium sordidulum Weath., with rather irregularly lobed leaves, is related to *Pleopeltis revoluta* and may be a hybrid involving that species.

Ecology (Figs. 1, 2)

Pleopeltis is usually epiphytic in rain forests, montane forests, and secondary forests and in oak and pine woods, where it grows on the trunks and large branches of trees, and occasionally on fallen trunks and branches. It also grows in secondary forests, in cutover land on small trees and on living fence posts; also in plantations of cacao, coffee, orange or mango. Less often it grows on boulders, on ledges or in crevices of rocks, and rarely on old stone walls. *Pleopeltis* occurs from sea level to 3400 m, most often below 2000 m.

Geography (Fig. 3)

Pleopeltis occurs in America, Africa, Madagascar and the Mascarenes to India and Ceylon. Only *Pleopeltis macrocarpa* (*sens. lat.*) occurs in the Old World.

In America, *Pleopeltis* grows from southwestern Texas in the United States, northern Mexico and the Greater Antilles, southward to Buenos Aires in Argentina, and is disjunct in Con-

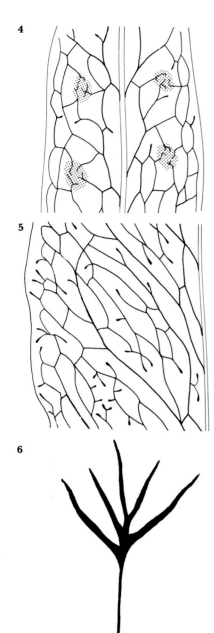

Figs. 107.4–107.6. *Pleopeltis* **4.** Portion of a fertile leaf of *P. macrocarpa*, showing venation, the receptacles (shaded), and the vascular supply to the sorus, ×3.5. **5.** Portion of a sterile lamina of *P. macrocarpa*, venation, ×3.5. **6.** Lamina of *P. angusta*, ×0.33.

cepción and Valdivia in Chile. *Pleopeltis revoluta* occurs on Cocos Island and *P. macrocarpa* on the Galápagos Islands and the Juan Fernandez Islands. All species, except *Pleopeltis fuscopunctata*, occur in Mexico or Central America, while only two species occur in the Greater Antilles, one in the Lesser Antilles and four in South America.

Spores

Spores of *Pleopeltis* are usually verrucate, quite prominently in *P. macrocarpa* and *P. percussa* (Figs. 10, 13), or rather shallowly in *P. revoluta* (Fig. 9). They are often large, nearly 70 μm as in specimens of *P. macrocarpa* from Peru (Fig. 10) and up to 80 μm in collections from Ethiopa. The verrucate formation is similar to that of spores in other genera of the family especially those of *Dicranoglossum, Campyloneurum,* and *Marginariopsis.*

Cytology

Chromosome records of *Pleopeltis* are mostly diploid, $n = 37$, although there are a few divergent numbers. *Pleopeltis percussa* is diploid in both Costa Rica (Smith and Mickel, 1977) and Peru (Evans, 1963). The cytology of *P. macrocarpa* is more complex. Study of the species in Jamaica (as *Polypodium lanceolata*) by Walker (1966, 1973a) showed the presence of a tetraploid $n = 74$, also a pentaploid hybrid with 74 pairs and 37 single chromosomes, and two tetraploid hybrids with 148 chromosomes and irregular pairing. These strongly suggest the existence of other fertile cytotypes that may be difficult to detect in this group because of similarities in leaf morphology. Plants of *P. macrocarpa* from the Galápagos Islands are reported by Jarrett et al. (1968) as hexaploid, $2n = 210$. They note that African plants from Cameroon Mountain are $n = $ ca. 70, and a diploid $n = 35$ is reported from India. A cultivated plant of *Pleopeltis angusta*, originating from Guatemala, was reported as consistently $n = 34$ II by Smith and Mickel (1977) although the same plant was previously reported as $n = 37$ (Sorsa, 1966).

The Old World *Lepisorus* is often segregated from *Pleopeltis* although the relationships of the two are uncertain. The cytological records of *Lepisorus* are the most diverse in the Polypodiaceae and according to Mitui (1971) and Walker (1973b) include $n = 22, 23, 25, 26, 35, 36, 37, 47,$ ca. $50, 51, 52, 70, 75; 2n = 50,$ 75, ca. 95, ca. 140. These were regarded by Mitui as representing two phyletic elements, one based on $x = 25$ or 26 and the other on $x = 35$ which correlate with two distinct groups based on differences in rhizome scales and paraphyses.

Literature

Evans, A. M. 1963. New Chromosome observations in the Polypodiaceae and Grammitidaceae. Caryologia 16: 671–677.

Jarrett, F. M., I. Manton, and S. K. Roy. 1968. Cytological and taxonomic notes on a small collection of living ferns from Galapagos. Kew Bull. 22: 475–480.

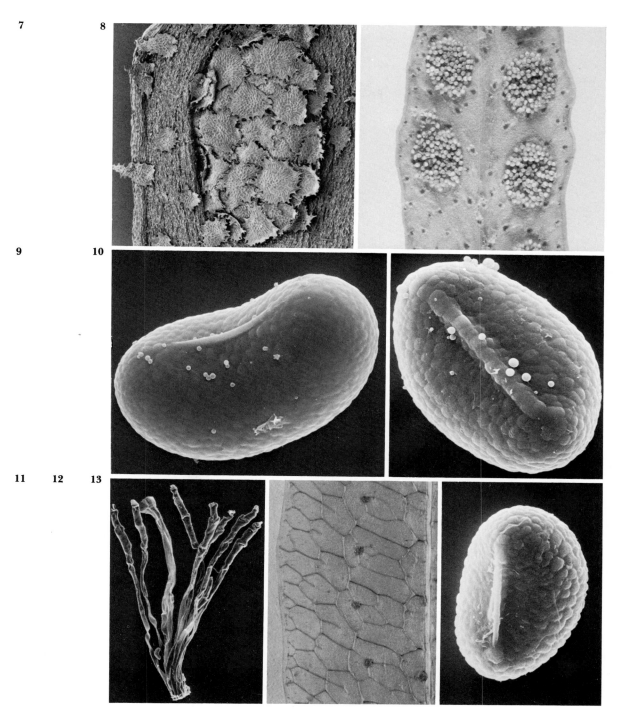

Figs. 107.7–107.13. *Pleopeltis.* **7.** Immature sorus of ***P. macrocarpa*** covered by peltate paraphyses, ×20. **8.** Mature sori of ***P. angusta***, ×8. **9, 10.** Spores, ×1000. **9. *P. revoluta***, shallowly verrucate, Costa Rica, *Weber 6004*. **10. *P. macrocarpa***, verrucate, Peru, *Wurdack 633*. **11.** Branched filamentous paraphyses from mature sorus of ***P. percussa***, ×90. **12.** Portion of lamina of ***P. fuscopunctata***, showing venation and sori, ×3. **13.** Spore of ***P. percussa***, verrucate, Colombia, *Plowman et al. 2322*, ×1000.

Mitui, K. 1971. Correlation between the chromosome numbers and morphological characters in the genus *Lepisorus*. Jour. Jap. Bot. 46: 83–96.

Smith, A. R., and J. T. Mickel. 1977. Chromosome counts for Mexican ferns. Brittonia 29: 391–398.

Sorsa, V. 1966. Chromosome studies in the Polypodiaceae. Amer. Fern Jour. 56: 113–119.

Wagner, W. H., and F. S. Wagner. 1975. A hybrid polypody from the New World tropics. Fern Gaz. 11: 125–135.

Walker, T. G. 1966. Reference under the family.

Walker, T. G. 1973a. Additional cytotaxonomic notes on the pteridophytes of Jamaica. Trans. Roy. Soc. Edinburgh 69: 109–135.

Walker, T. G. 1973b. Evidence from cytology in the classification of ferns, 91–110. *in*: A. C. Jermy, J. A. Crabbe, and B. A. Thomas (Eds.), The phylogeny and classification of ferns. Bot. Jour. Linn. Soc. 67, Suppl. 1.

Weatherby, C. A. 1922. The group of *Polypodium lanceolatum* in North America. Contrib. Gray Herb. 65: 3–14.

Weatherby, C. A. 1935. On certain Mexican and Central American ferns. Amer. Fern Jour. 25: 52–59.

108. *Dicranoglossum*

Figs. 108.1–108.9

Dicranoglossum J. Sm., Bot. Voy. Herald (Seemann) 232. 1854, *nom. nov.* for *Cuspidaria* Fée and with the same type.

Eschatogramme Trev., Atti Ist. Veneto II, 2: 168. 1851, *nom. nud.*

Cuspidaria Fée, Mém. Soc. Mus. Hist. Nat. Strasbourg 4: 201. 1850, not DC. 1838 (Bignoniaceae). Type: *Cuspidaria furcata* (L.) Fée (*Pteris furcata* L.) = *Dicranoglossum furcatum* (L.) J. Sm.

Description

Epiphytic; stem small, short-creeping, bearing scales and rather fibrous roots; leaves monomorphic, borne in a loose cluster, ca. 10–40 cm long, lamina more or less subdichotomously forked into two to several segments, slightly scaly, veins free or anastomosing without included free veinlets; sori round, terminal on the vein-tips, or elongate, or in a long inframarginal line, receptacle slightly raised, not paraphysate, exindusiate; spores ellipsoidal, monolete, the laesura $\frac{1}{2}$ to $\frac{2}{3}$ the spore length, with a shallow verrucate surface. Chromosome number: $n = 36$.

The lamina of *Dicranoglossum* bears small scales (Fig. 7). The sporangia are typically borne on a long, narrow receptacle served by a vascular commissure (Fig. 5), and form a broad band at maturity (Fig. 7).

Systematics

Dicranoglossum is an American genus of perhaps three species. While it may be allied to the Old World *Lepisorus* by its chromosome number of $n = 36$, it is more likely an American derivative

Fig. 108.1. *Dicroanoglossum polypodioides.* Río Palenque, Ecuador. (Photo W. H. Hodge.)

of *Pleopeltis,* with relations to *Marginariopsis,* both of which have similar spores. *Dicranoglossum* is readily distinguished from related genera by its forked leaf (Fig. 3) which is sessile or subsessile and often irregularly disarticulate. *Pleopeltis angusta* which often has a similar lamina has a petiolate leaf.

The genus was revised by Christensen (1929), under the formerly widely used name *Eschatogramme.*

Tropical American Species

The species of *Dicranoglossum* are poorly distinguished, primarily on quantitative characters such as the width of the segments, the size and abundance of the lamina scales, and the free to rather fully anastomosing venation. Christensen (1929) stated "I find it very natural to consider the genus monotypic," although in practice he recognized four species. One of these, *Dicranoglossum panamense,* with long lines of sporangia (Fig. 5), cannot be maintained as distinct from *D. polypodioides* (Fig. 4), with separate, round sori, since there are all gradations between the two extreme soral conditions. The variation in other species needs modern assessment to confirm the taxonomy.

Fig. 108.2 Distribution of *Dicranoglossum*.

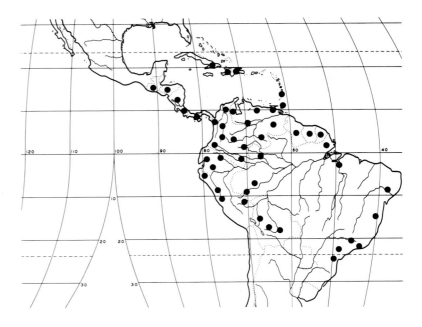

Figs. 108.3–108.6. *Dicranoglossum.* **3.** Subdichotomously forked leaf of *D. polypodioides,* ×0.25. **4.** Portion of fertile segment of *D. polypodioides,* with discrete sori, the receptacles in black at the vein-tips, ×2.5. **5.** Portion of fertile segment of *D. polypodioides,* with vascular commissure connecting the vein-ends, ×2.5 **6.** Portion of sterile segment of *D. Desvauxii,* ×2.5.

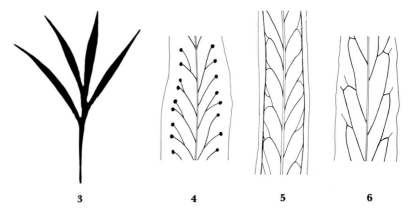

3   4   5   6

Key to Species of *Dicranoglossum*

a. Sterile portion of segments with free veins, or occasionally with some areolae beyond the costa, or rarely with a few costal areolae.   b.
   b. Scales on the lamina beneath ca. 0.2 mm long, circular to short-cusped.
            D. polypodioides (Hook.) Lell. (*D. panamense* (C. Chr.) Gómez)
   b. Scales on the lamina beneath ca. 0.5 mm long, mostly more or less ovate-lanceolate.   *D. furcatum* (L.) J. Sm.
a. Sterile portion of segments predominantly with costal areolae (Fig. 6), scales on the lamina beneath ca. 0.2 mm long, circular to short-cusped.
            D. Desvauxii (Kl.) Proctor

Ecology (Fig. 1)

Dicranoglossum usually grows in wet forests, frequently along streams, or rarely in thickets. It commonly grows in moss mats at the base of trees or on their lower trunks and branches. It also grows on small trees and on shrubs. *Dicranoglossum* occurs most often from sea level to ca. 700 m, rarely up to 1500 m.

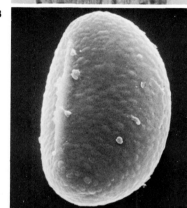

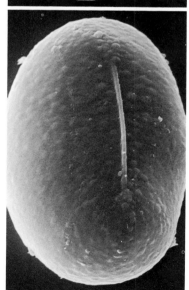

Geography (Fig. 2)

Dicranoglossum is distributed from Guatemala through Central America, in Cuba, Hispaniola, St. Vincent and Trinidad, and in South America south to Bolivia and to Paraná in southeastern Brazil.

Spores

Dicranoglossum spores have shallow verrucate surfaces and a relatively short laesura. The smaller spores of *D. polypodioides* (Fig. 8) corresponds to the diploid record for the species, and contrast with the larger ones of *D. furcatum* (Fig. 9). They are reported as chlorophyll-bearing in *D. Desvauxii* (Atkinson, 1974) but the question is raised whether this might represent early stages in germination before the wall has cracked rather than a characteristic of the spores. The relatively low, verrucate surface of *Dicranoglossum* spores resembles that of several genera of the Polypodiaceae especially *Pleopeltis* and *Marginariopsis*.

Cytology

The record of $n = 36$ for *Dicranoglossum polypodioides* (as *D. panamense*) from Costa Rica (Wagner, 1980) is an uncommon number in the Polypodiaceae. This, as well as the records of $n = 35$ and 34 in a few genera, suggests that loss or fusion of chromosomes is a derived condition as compared to the 37 pairs in most genera.

Literature

Atkinson, L. R. 1974. Gametophyte of *Dicranoglossum Desvauxii*. Phytomorph. 24: 49–56.
Christensen, C. 1929. Reference under the family.
Wagner, F. S. 1980. Reference under the family.

Figs. 108.7–108.9. *Dicranoglossum.* **7.** Portion of fertile segment of ***D. Desvauxii,*** with small scales and broad bands of mature sporangia, × 8. **8, 9.** Shallowly verrucate spores. × 1000. **8. *D. polypodioides,*** laesura at left, Peru, *Schunke 5190*. **9. *D. furcatum,*** laesura at center, Haiti, *Ekman 4710*.

109. *Marginariopsis*
Figs. 109.1–109.6

Marginariopsis C. Chr., Dansk Bot. Ark. 6 (3): 42. 1929. Type: *Marginariopsis Wiesbaurii* (Sod.) C. Chr. (*Drymoglossum Wiesbaurii* Sod.).

Description   Epiphytic; stem long-creeping, slender, bearing scales and rather few and short fibrous roots; leaves dimorphic (the fertile much narrower than the sterile and sometimes petiolate rather than sessile), ca. 8–15 cm long, usually borne at intervals, lamina entire, slightly scaly, veins anastomosing with included free veinlets; sori borne on a broad receptacle between the costa and margin, extending in continuous or sometimes broken lines for most of the length of the fertile lamina; spores ellipsoidal, monolete,

Fig. 109.1. *Marginariopsis Wiesbaurii,* Balsa, Costa Rica. (Photo L. D. Gómez.)

Fig. 109.2. Distribution of *Marginariopsis*.

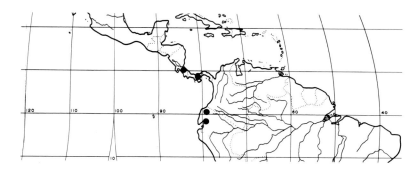

the laesura $\frac{2}{3}$ the spore length, shallowly verrucate with sparse, spherical deposit. Chromosome number: $n = 35$.

The stem of *Marginariopsis* is somewhat flattened and bears rather broad, thin and finely clathrate scales. The leaves disarticulate more or less cleanly and the sterile lamina has many excurrent and recurrent free veinlets (Fig. 4). The anatomy of *Marginariopsis* has been studied by Chandra (1978).

Systematics   *Marginariopsis* is a monotypic American genus based on *M. Wiesbaurii* (Sod.) C. Chr. It is distinctive in its dimorphic leaves, stalked, peltate paraphyses that cover the immature sporangia and long receptacle (Fig. 3) that is furnished by a separate vascular supply from that of the lamina. It is evidently most closely related to *Pleopeltis* which has similar paraphyses.

Marginariopsis Wiesbaurii was one of three species placed by Maxon (1912) in the Old World genus he called *Pteropsis* Desv. (then equivalent to *Drymoglossum* and now called *Pyrrosia*). The Costa Rican element, *Pteropsis Underwoodiana* Maxon and *Pteropsis martinicensis* (Christ) Maxon were also included. Christensen (1929) considered the material from Costa Rica to be conspecific with that in Ecuador and also pointed out that the third species was based on a mixture of Asiatic polypodioid ferns, presumably cultivated in Martinique.

3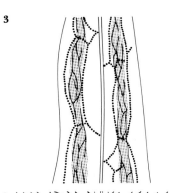

4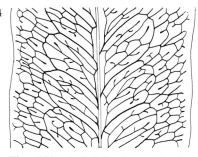

Figs. 109.3, 109.4. Portions of lamina of *Marginariopsis*. **3.** Fertile lamina, the principal vein system is shown by coarse dotted lines, the largely independent veins that serve the receptacle by fine lines, and the receptacle is shaded, ×5.0. **4.** Sterile lamina, venation, ×1.5.

Ecology and Geography (Figs. 1, 2)

The following notes on the ecology of *Marginariopsis* in Costa Rica (Fig. 1) have been provided by Luis Diego Gómez. "*Marginariopsis* is a frequent high-canopy epiphyte of the tropical and lower tropical forests, from sea level to 2000 m. It thrives under good light conditions and high relative humidity. It may form massive populations on pollards and live fence posts (it prefers *Cupressus, Erythrina* and *Acnistus*). Fertile fronds are sparse but present throughout the year in areas with evenly spread rainfall, while in areas with well defined dry and rainy seasons, fertile fronds flush in late May to June and then in November and December."

The genus is known from Costa Rica, Panama and Ecuador (Fig. 2) but it is probably more widely distributed because it may be overlooked due to its general resemblance to several other polypodioid genera.

5   6

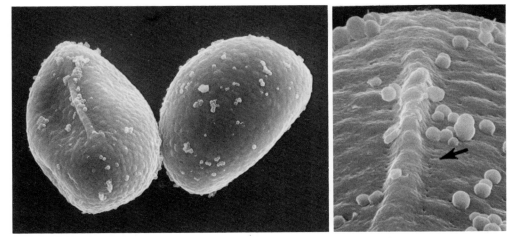

Figs. 109.5, 109.6. Spores of *Marginariopsis Wiesbaurii,* Costa Rica, *J. D. Smith 6941.* **5.** Shallowly verrucate surface, left, proximal face laesura at center, right, distal face, × 1000. **6.** Surface detail with spherical deposit, perforate at arrow, × 5000.

Spores

Marginariopsis spores are shallowly verrucate with scattered spherical deposit (Fig. 5) and perforations especially near the laesura (Fig. 6). They are reported as green with few chloroplasts that break down almost immediately after collecting (Lloyd and Klekowski, 1970). The shallow verrucate surface is similar to that of spores of several other genera as *Pleopeltis, Dicranoglossum* and *Campyloneurum.*

Cytology

The report of $n = 35$ in *Marginariopsis* (Wagner, 1980) distinguishes the genus from others which commonly have $n = 37$ in the Polypodiaceae, however reports of 34 and 35 are known in *Pleopeltis.* The less common lower numbers may be the result of loss or fusion of chromosomes and represent a derived condition in the family.

Literature

Chandra, S. 1978. Studies on the morphology of the monotypic fern genus *Marginariopsis* (Polypodiaceae). Brenesia 14–15: 337–348.

Christensen, C. 1929. Reference under the family.

Lloyd, R. M., and E. J. Klekowski. 1970. Spore germination and viability in Pteridophyta: Evolutionary significance of chlorophyllous spores. Biotropica 2: 129–137.

Maxon, W. R. 1912. The American species of *Pteropsis.* Contrib. U.S. Nat. Herb. 16: 51–52.

Wagner, F. S. 1980. Reference under the family.

110.   *Microgramma*
Figs. 110.1–110.27

Microgramma Presl, Tent. Pterid. 213. 1836. Type: *Microgramma persi-cariifolia* (Schrad.) Presl (*Polypodium persicariifolium* Schrad.).

Craspedaria Link, Fil. Sp. 117. 1841. Type: *Craspedaria vacciniifolia* (Langsd. & Fisch.) Link (*Polypodium vacciniifolium* Langsd. & Fisch.) = *Microgramma vacciniifolia* (Langsd. & Fisch.) Copel.

Mecosorus Kl., Linnaea 20: 404. 1847, *nom. superfl.* for *Microgramma* and with the same type.

Anapeltis J. Sm., Cat. Cult. Ferns 5. 1857. Type: *Anapeltis lycopodioides* (L.) J. Sm. (*Polypodium lycopodioides* L.) = *Microgramma polypodioides* (L.) Copel.

Lopholepis (J. Sm.) J. Sm., Lond. Jour. Bot. 1: 195. 1842, not Decaine, 1839 (Gramineae). *Goniophlebium* section *Lopholepis* J. Sm., Jour. Bot. (Hook.) 4: 56. 1841. Type: *Goniophlebium piloselloides* (L.) J. Sm. (*Polypodium piloselloides* L.), *Lopholepis piloselloides* (L.) J. Sm. = *Microgramma piloselloides* (L.) Copel. *Microgramma* subgenus *Lopholepis* (J. Sm.) Lell., Amer. Fern Jour. 67: 59. 1977.

Description

Epiphytic or rupestral; stem long-creeping, often slender, bearing scales and few to many short to long, fibrous roots; leaves monomorphic to dimorphic (the fertile narrower and longer than the sterile), ca. 1–35 cm long, lamina entire, slightly pubescent, slightly to moderately scaly, or glabrous, veins free to anastomosing usually with included free veinlets; sori round to elongate, borne in one row on each side of the costa on, or usually at the tip of, a single vein or at the junction of veins, or (in *M. persicariifolium*) associated with but not always following the veins, on a hardly to somewhat raised receptacle, paraphysate or not, exindusiate; spores ellipsoidal, monolete, the laesura $\frac{1}{2}$ to $\frac{3}{4}$ the length, verrucate, coarsely tuberculate, or rugose, often with granulate deposit. Chromosome number: $n = 37, 74$.

In *Microgramma,* the venation of the sterile lamina varies from wholly free in *M. chrysolepis* (Fig. 8) to anastomosing with the free veinlets excurrent (directed toward the margin, Fig. 11), or more complex with some free veinlets recurrent (directed toward the costa, Figs. 6, 9, 10, 19). Some species with strongly dimorphic leaves have the veins free in the fertile lamina as in *M. reptans* (Fig. 12) and anastomosing in the sterile lamina (Fig. 13).

In *Microgramma megalophylla,* the sori are often very large (Fig. 14), sometimes 6 mm in diameter. The sori have distinctive paraphyses in some species as the filamentous type in *M. megalophylla* (Fig. 17) or broader, scalelike forms in *M. nitida* (Fig. 18). However, in *M. piloselloides* the paraphyses may be quite similar to the lamina scales although somewhat narrower. In *M. chrysolepis* the young sorus is covered by scales that are mostly attached at the perimeter of the sorus rather than on the receptacle.

Systematics

Microgramma is an American genus of about 13 species, with one of them in Africa to the Mascarenes. The elongate, nonclathrate stem scales, attached well above their base (Fig. 7) are a distinctive feature of the genus and all species have a somewhat to strongly flattened stem. *Microgramma* appears to be most closely related to *Campyloneurum* and perhaps also to *Pleopeltis.*

Fig. 110.1. *Microgramma lycopodioides*, Sylvania, Dominica. (Photo W. H. Hodge.)

Fig. 110.2. Distribution of *Microgramma* in America.

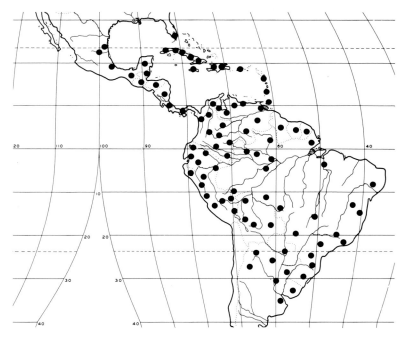

Within the genus diversity is especially evident in the glabrous to scaly and monomorphic to dimorphic leaves (Figs. 3, 4, 5), recurrent or only excurrent free veinlets, and in sori that have filamentous or scalelike paraphyses or are non-paraphysate. These characters are not sufficiently correlated in the species to justify the recognition of infrageneric taxa.

Tropical American Species

The species of *Microgramma* need a critical revision to assess the variation of species, especially on a geographic basis, and to determine the relationships of the species. The following key includes the more distinctive species and their common synonyms. A few poorly distinguished species allied to *M. piloselloides* are *M. reptans* (Cav.) A. R. Sm. (*M. ciliata* (Willd.) Alston), *M. tecta* (Kaulf.) Alston, *M. cordata* (Desv.) Crabbe, and *M. acatallela* Alston. Those allied to *Microgramma geminata* are *Polypodium Thurnii* Baker, *P. loretense* Maxon, and *P. recreense* Hieron.

A hybrid between *Microgramma lycopodioides* and *Polypodium plesiosorum* Kze. was described by Gómez (1975).

Key to American Species of *Microgramma*

a. Stem very slender, ca. 0.5–1.0 mm wide, recurrent free veinlets absent or few in the sterile lamina.   b.
  b. Sterile lamina with free veins.   *M. chrysolepis* (Hook.) Crabbe
  b. Sterile lamina with anastomosing veins.   c.
    c. Paraphyses more or less contorted, lamina glabrous.
          M. heterophylla (L.) Wherry (*Polypodium Swartzii* Baker)
    c. Paraphyses stiffly projecting beyond the sorus, lamina more or less scaly.   *M. piloselloides* (L.) Copel.
a. Stem slender to broadly flattened, ca. 2.0–30 mm wide, if slender then recurrent free veinlets of regular occurrence, at least in the sterile lamina.   d.
  d. Stem very strongly flattened, ca. 10–30 mm broad, sori large to 6 mm in diameter.   *M. megalophylla* (Desv.) Sota
  d. Stem slightly to moderately flattened, ca. 2.0–5.0 mm broad.   e.
    e. Leaves monomorphic.   f.
      f. Fertile and sterile leaves linear.
          M. rosmarinifolia (Kunth) R. & A. Tryon
      f. Fertile and sterile leaves narrowly elliptic or broader.   g.
        g. Sori elongate, oblique to the costa.
           M. persicariifolia (Schrad.) Presl
        g. Sori roundish.   h.
          h. Leaves sessile to subsessile, membranous.
             M. Lindbergii (Kuhn) Sota
          h. Leaves short-petiolate to rarely subsessile, papyraceous to coriaceous.   *M. geminata* (Schrad.) R. & A. Tryon
    e. Leaves dimorphic or sometimes subdimorphic.   i.
      i. Stem scales spreading.   *M. crispata* (Fée) R. & A. Tryon
                        (*M. Galatheae* (C. Chr.) Crabbe)
      i. Stem scales appressed.   j.
        j. Stem scales not or hardly toothed or ciliate.
          M. nitida (J. Sm.) A. R. Sm. (*Polypodium Palmeri* Maxon)
        j. Stem scales clearly toothed or ciliate, especially at the stem apex.   k.
          k. Sterile lamina glabrous beneath or with a few dissected brownish scales on the costa, leaves subdimorphic.
           M. lycopodioides (L.) Copel. (*M. Baldwinii* Brade, *Polypodium prominulum* Maxon, *P. surinamense* Jacq.)
          k. Sterile lamina with a few to many usually dissected whitish scales beneath, leaves usually dimorphic.   l.
            l. Sori mostly on single vein-tips.
              M. vacciniifolia (Langsd. & Fisch.) Copel.
            l. Sori mostly served by 2 or 3 veins.
                M. squamulosa (Kaulf.) Sota

Ecology (Fig. 1)

Microgramma is an epiphyte in swamp forests, in rain forests and cloud forests, often along rivers and streams. It also occurs in secondary forests, in brushy forest clearings and in thickets, as

3   4

5   6   7

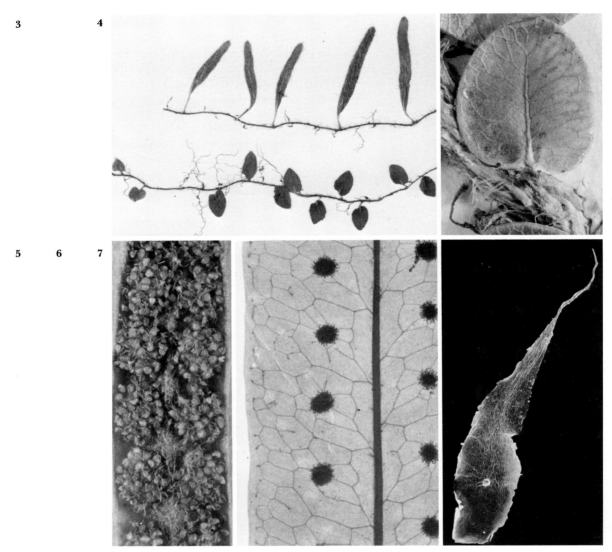

Figs. 110.3–110.7. *Microgramma.*   3. Stems of *M. tecta* with fertile leaves, above, and with sterile leaves, below, × 1.   **4.** Sterile leaf attached to stem, *M. vacciniifolia,* × 5.   **5.** Portion of fertile leaf of *M. vacciniifolia,* × 10.   **6.** Portion of fertile leaf of **M. Lindbergii,** with venation and sori, × 3.   **7.** Nonclathrate stem scale of **M. nitida,** attached at a point above the base, × 15.

well as in cacao and coffee plantations, in hedge rows and on living fence posts. Plants can persist for some time on branches that have fallen to the ground. Species grow less often on wet or moist rocks and rarely on old stone walls. As an epiphyte, *Microgramma* occurs at the base of trees, on the trunk where plants may climb to 10 m or more, on tree branches, and sometimes the stem twines on the slender branches and twigs of shrubs.

Microgramma is mostly a genus of low altitudes, from sea level to ca. 500 m; less frequently it grows to 1000 or 2000 m, and *M. chrysolepis* occurs up to 3200 m.

Figs. 110.8–110.13. *Microgramma* leaves, ×2.5, the black, roundish areas indicate the receptacles. **8.** Sterile leaf of *M. chrysolepis*. **9.** Portion of fertile leaf of *M. lycopodioides*. **10.** Portion of sterile leaf of *M. lycopodioides*. **11.** Portion of sterile leaf of *M. piloselloides*. **12.** Portion of fertile leaf of *M. reptans*. **13.** Portion of sterile leaf of *M. reptans*.

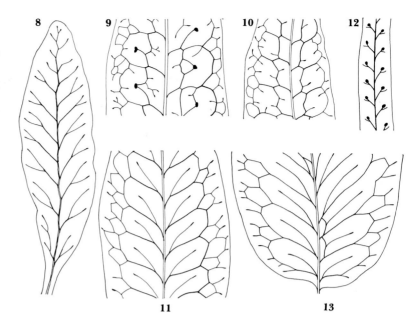

Geography (Fig. 2)

Microgramma is widely distributed in the American tropics where it occurs from Tamaulipas in Mexico and southern Florida, generally southward to Catamarca and Buenos Aires in Argentina. *Microgramma mauritiana* (Willd.) Tard. (*M. owariensis* (Desv.) Alston) occurs in Africa, Madagascar and the Mascarene Islands and is close to if not the same as *M. lycopodioides*.

The genus centers on the eastern slopes of the Andes and adjacent Amazon basin, where about 10 species occur.

Spores

Microgramma spores are tuberculate as in *M. persicariifolia* (Fig. 20) and *M. piloselloides* (Figs. 21, 22) and the tubercules are usually less prominent on the distal face as in *M. vacciniifolia* (Fig. 24) and *M. reptans* (Figs. 25, 26). The laesura is often short as in *M. lycopodioides* (Fig. 23). The rugose spores of *M. megalophylla* (Fig. 27), nearly twice as large as others, possibly represent a higher ploidy level.

Cytology

Reports for *Microgramma lycopodioides* are uniformly diploid, $n = 37$, from Jamaica, Cuba, Puerto Rico and Costa Rica and this or the closely allied species of Africa is also diploid. *Microgramma pilloselloides* is tetraploid with $n = 74$ in Jamaica (Walker, 1966). De la Sota and Cassa de Pazos (1980) report two chromosome counts obtained from gametophytes. Their record of $n = 37$ for *Microgramma vacciniifolia* is consistent with other counts for the genus but differs from $n = 36$ reported by Evans (1963). They also report *Microgramma Mortoniana* Sota as an allotetraploid hybrid with $n = 74$, derived from *M. squamulosa* and *M. vacciniifolia*.

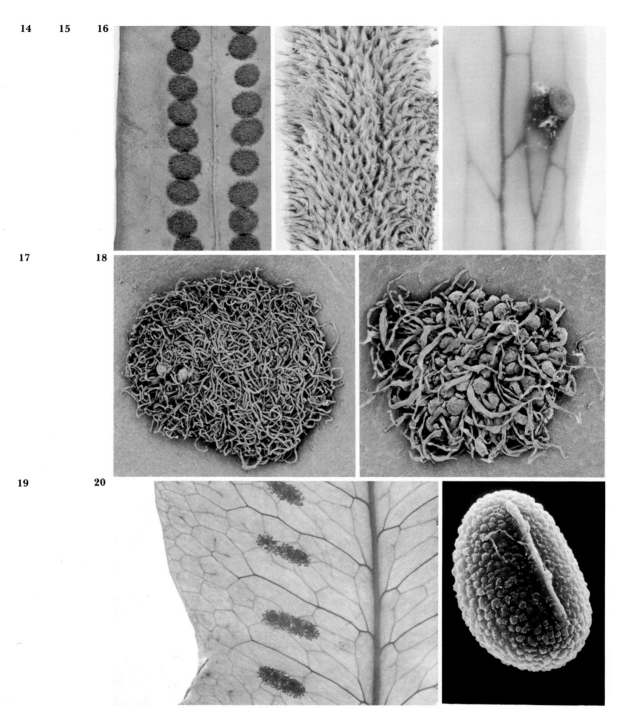

Figs. 110.14–110.20. *Microgramma.*   **14.** Portion of fertile leaf of *M. megalophylla,* sori ca. 6 mm in diameter, × 1.3.   **15.** Upper side of densely scaly, flattened stem of *M. megalophylla,* × 2.   **16.** Cleared stem of *M. megalophylla,* 25 mm wide and 2 mm thick, with a larger central meristele and smaller lateral ones, clean scar of a petiole, right, × 2.   **17.** Young sorus of *M. megalophylla* with filamentous para-physes, × 12.   **18.** Young sorus of *M. nitida* with scalelike paraphyses, × 24.   **19.** Portion of fertile leaf of *M. persicariifolia,* with venation and sori, × 4.   **20.** Spore of *M. persicariifolia,* tuberculate proximal face and long laesura, Brazil, *Irwin et al. 17869,* × 1000.

21   22   23

24   25

26   27

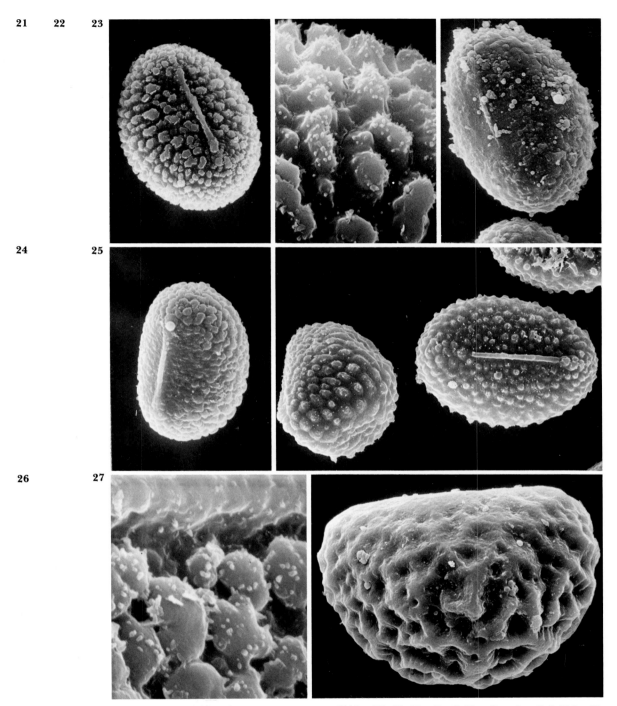

Figs. 110.21–110.27. *Microgramma* spores, × 1000. **21, 22.** *M. piloselloides,* Ecuador, *Bell 206.* **21.** Tuberculate proximal face with laesura. **22.** Detail of tubercles with echinate deposit, × 5000. **23.** *M. lycopodioides,* proximal face, short laesura, left, Jamaica, *Orcutt 3052.* **24.** *M. vacciniifolia,* proximal face, shallowly tuberculate, laesura, left, Brazil, *Dusén 14159.* **25, 26.** *M. reptans.* **25.** Proximal face, right, and end view, left, with coarse tubercles, Costa Rica, *Burger & Stolze 5591.* **26.** Detail of tubercles, Colombia, *Klug 1725,* × 10,000. **27.** *M. megalophylla,* rugose distal face, Brazil, *Krukoff 9031.*

Observations

The species of *Microgramma* have a more or less flattened stem that reaches an extreme development in *M. megalophylla* in which it may be up to 30 mm broad and only 2 mm thick. It is densely scaly above (Fig. 15) and the scales below are somewhat shorter and broader. Roots are borne mostly in narrow zones near the edges and the central part of the stem may be raised above the substrate. The strongly dorsiventral stem has a nearly straight central, primary meristele and three or more smaller lateral ones that branch and rejoin (Fig. 16). De la Sota (1963) discusses the anatomy of this unusual species.

Literature

Evans, A. M. 1963. New chromosome observations in the Polypodiaceae and Grammitidaceae. Caryologia 16: 671–677.

Gómez, L. D. 1975. Contribuciones a la pteridología Costarricense, VIII. La hibridación en al trópico: *Microgramma* × *Polypodium* y *P. aspidiolepis* Baker. Brenesia 6: 49–57.

Sota, E. R. de la. 1963. "*Microgramma megalophylla* (Polypodiaceae *s. str.*)" una interesante especie de Amazonas. Bol. Soc. Argent. Bot. 10: 158–165.

Sota, E. R. de la, and L. Cassa de Pazos. 1980. Recuentos cromosómicos en protalos de dos especies de "*Microgramma*" Presl (Polypodiaceae *s. str.* Pteridophyta.) Bol. Soc. Argent. Bot. 19: 69–73.

Walker, T. G. 1966. Reference under the family.

111. *Campyloneurum*
Figs. 111.1–111.20

Campyloneurum Presl, Tent. Pterid. 189. 1836. Type: *Campyloneurum repens* (Aubl.) Presl (*Polypodium repens* Aubl.).

Cyrtophlebium (R. Br.) J. Sm., Jour. Bot. (Hook.) 4: 58. 1841, *nom. superfl.* for *Campyloneurum* and with the same type. *Polypodium* subgenus *Cyrtophlebium* R. Br., Pl. Jav. Rar. (Bennett & Brown; Horsfield) 4. 1838, no species included; typified by J. Sm., as genus.

Hyalotricha Copel., Amer. Fern Jour. 43: 12. 1953, not Dennis, 1949 (Discomycetes). Type: *Hyalotricha anetioides* (Christ) Copel. (*Polypodium anetioides* Christ) = *Campyloneurum anetioides* (Christ) R. & A. Tryon.

Hyalotrichopteris Wagner, Taxon 27: 548. 1978, *nom. nov.* for *Hyalotricha* Copel. and with the same type. (*Hyalotrichopteris anetioides* (Christ) Wagner).

Description Terrestrial, rupestral or usually epiphytic; stem short- to rather long-creeping, small to slightly stout or slender, bearing scales and a few to many, long, fibrous roots; leaves monomorphic, ca. 5 cm to 2 m long, borne in a loose cluster to widely spaced, lamina entire, narrowly linear to broadly elliptical, glabrous, with a few scales, or finely pubescent, veins anastomosing with included free veinlets, sori round or roundish, borne on a free vein or rarely at the junction of two veins, below the tip, rarely subterminal or at the tip, on a slightly to somewhat raised receptacle, not paraphysate or rarely minutely paraphysate, exindusiate; spores ellipsoidal, monolete, the laesura $\frac{2}{3}$ to $\frac{3}{4}$ the spore

Fig. 111.1. *Campyloneurum angustifolium,* Tapanti, Costa Rica. (Photo W. H. Hodge.)

length, the surface shallowly to prominently verrucate with scattered, spherical deposit. Chromosome number: $n = 37, 74$.

The stem scales of *Campyloneurum* are more or less clathrate (Fig. 5) as in most genera of the Polypodiaceae. Most species have prominent, lateral veins and the sori are borne in pairs between them (Fig. 3). Species with very narrow leaves have a single row of sori on each side of the costa (Fig. 12) while those with broad leaves may have five or more rows (Figs. 3, 13). In the very large leaf of *C. pascoense,* one or two additional sori may be interposed between the pairs (Fig. 14).

The sori are nearly always borne below the tip of a free veinlet (Figs. 12–14), although often in *Campyloneurum anetioides* (Fig. 11) and sometimes in *C. trichiatum* and *C. occultum* they may be near or at the tip. Paraphyses of *Campyloneurum anetioides* are small glandular trichomes or sporangiasters as shown by Wagner and Farrar (1976). The leaves of all species of *Campyloneurum* are entire, but diverse in size and shape (Figs. 6–10).

Systematics   *Campyloneurum* is an American genus of about 20–25 species. It forms a relatively homogeneous group of species characterized by sori borne below the free vein tip without conspicuous paraphyses and by clathrate stem scales and monomorphic leaves. The genus appears to be most closely related to *Niphidium* and *Microgramma.*

Fig. 111.2 Distribution of *Campyloneurum.*

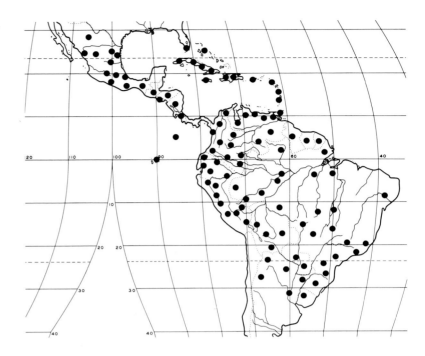

Polypodium decurrens and *P. Fendleri.* (*Campyloneurum magnificum* Moore), with 1-pinnate leaves, have generally been accepted as species of *Campyloneurum* because a single pinna closely resembles the lamina of *C. repens* (Aubl.) Presl and similar species in shape, texture and venation. However, the terminal position of the sori on the veins indicates their alliance with species of *Polypodium* rather than with *Campyloneurum.*

The monotypic *Hyalotrichopteris* is included in *Campyloneurum* because the unusual trichomes with a protruding basal gland cell and the inconspicuous paraphyses also occur in other species such as *C. trichiatum* and *C. occultum.* The basal gland cell of the trichomes suggest that they may be reduced from scales. The

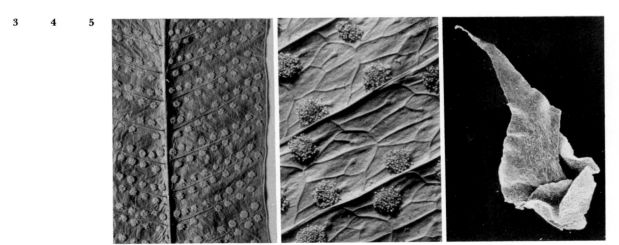

Figs. 111.3–111.5. *Campyloneurum.* **3.** Portion of the fertile lamina of *C. repens*, ×1.4. **4.** Sori and prominent veins of *C. pascoense*, ×3.0. **5.** Stem scale of *C. leucorhizon,* ×8.

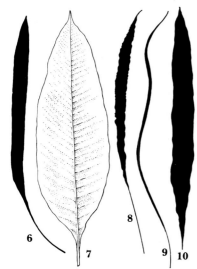

Figs. 111.6–111.10. Leaves of *Campyloneurum.* **6.** *C. Jamesonii,* ×0.5. **7.** *C. sphenodes,* ×0.25. **8.** *C. lapathifolium,* ×0.25. **9.** *C. angustifolium,* ×0.125. **10.** *C. pascoense,* ×0.04.

petioles of *C. anetioides* are sometimes articulate and some sori are born below the vein-tip as in other *Campyloneurum* species. Its simple venation is identical with that of leaves of juvenile plants of species as *C. phyllitidis* (L.) Presl. The small leaves of *Campyloneurum anetioides* may have been derived through neotony from species with larger leaves and more complex venation.

Tropical American Species

The species of *Campyloneurum* need a modern revision to critically evaluate the characters and assess their variation. They are often separated on technical characters, although leaf size and shape are generally useful for recognition of a number of species. *Campyloneurum angustifolium* (Sw.) Fée, one of the most variable and widely distributed species, usually has narrow, pendent leaves (Figs. 1, 9); *C. aglaolepis* Alston has the narrowest lamina and simple venation (Fig. 12); *C. pascoense* R. & A. Tryon has the largest leaves, up to 1.9 m long and 18 cm broad with a coriaceous texture and prominent veins (Fig. 4); *C. tucumanense* (Hieron.) Ching also has large but membranous-papyraceous leaves up to 1 m long and 14 cm broad; *C. anetioides* (Christ) R. & A. Tryon has the smallest leaves ca. 3–10 cm long with a basal gland on the lamina trichomes, and the same type of trichome is on the larger-leaved *C. trichiatum* (Rosenst.) Ching and *C. occultum* (Christ) Gómez; *C. brevifolium* (Link) Link and *C. coarctatum* (Kze.) Fée have a broad lamina more or less contracted at the base; *C. sphenodes* (Kl.) Fée has the lamina elliptic to broadly so with an attenuate tip (Fig. 7) and the leaves are distant on a long, slender stem; and *C. lapathifolium* (Poir.) Ching has the lamina margins usually slightly sinuate.

Ecology (Fig. 1)

Campyloneuron is largely a genus of epiphytes growing in humid forests; often high climbing on tree trunks, or on branches overhanging streams or on banks in damp soil; also in clearings or disturbed forests; at higher altitudes it occurs in shaded elfin forests and extends up to the páramo where it grows among rocks, on cliff faces and soil banks; less often in open sun in pastures or on walls. It has a wide altitudinal range above 100 m, it

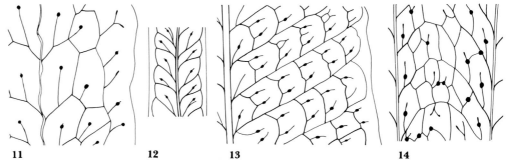

Figs. 111.11–111.14. Fertile leaves of *Campyloneurum,* the black roundish areas indicate the receptacles. **11.** *C. anetioides,* ×2.5. **12.** *C. aglaolepis,* ×2.5. **13.** *C. repens,* ×2.5. **14.** *C. pascoense,* ×1.25.

15   16

17   18

19   20

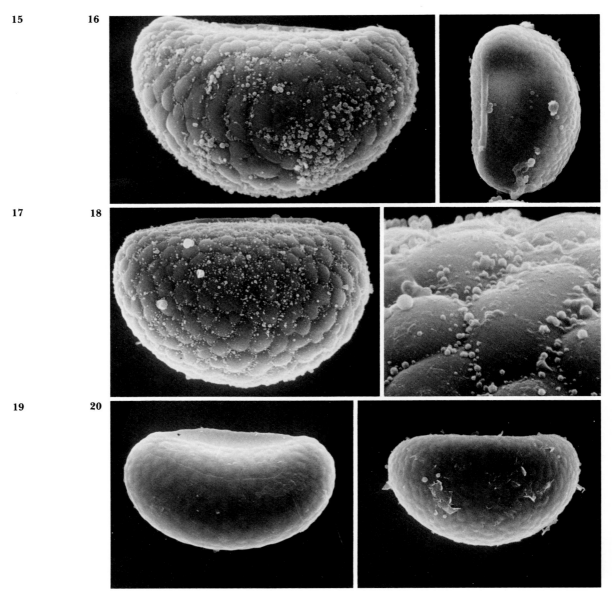

Figs. 111.15–111.20. *Campyloneurum* spores, × 1000.   **15.** *C. angustifolium,* verrucate with spherical deposit, laesura at top, Colombia, *Smith & Idrobo 1319.*   **16.** *C. pascoense,* spore from indehiscent sporangium, laesura left. Peru, *Soukup 3352.*   **17, 18.** *C. leucorhizon,* Peru, *Soukup 3142.* **17.** Verrucate with spherical deposit, laesura at top.   **18.** Surface detail, × 5000.   **19.** *C. trichiatum,* shallowly verrucate, laesura at top, Peru, *Killip & Smith 23781.*   **20.** *C. phyllitidis,* verrucate distal surface, Mexico, *Seaton 422.*

is most common up to 1000 m, but sometimes grows to 3000 m or more. In the Andes a few species, including *Camplyoneurum Jamesonii* Fée may occur to 4000 m.

Geography (Fig. 2)

Campyloneurum is distributed from Chihuahua in Mexico, southern Florida and the Bahamas, south to Catamarca and Entre Ríos in Argentina, and Uruguay; also on Cocos Island and the

Galápagos Islands. At least 20 species occur in the Andes of Colombia to Bolivia and fewer elsewhere, with only four species, with few records, in the Amazon basin of Brazil.

Spores

Campyloneurum spores are more or less verrucate with a spherical deposit as in *C. angustifolium* (Fig. 15) and *C. leucorhizon* (Kl.) Fée (Figs. 17, 18), or smoother as in the less mature spores of *C. pascoense* (Fig. 16) and in *C. trichiatum* (Fig. 19). The verrucate surface is similar to that of *Pleopeltis* spores and distinct from the tuberculate-granulate spores of *Microgramma*. Small spores of *Campyloneurum phyllitidis* (Fig. 20) may represent a diploid plant, while the large spores of *C. angustifolium* may represent a tetraploid.

Collections of *Campyloneurum anetioides* from Costa Rica (*Skutch 2753*) and from Panama (*Seibert 310*) have 64 spores per sporangium, differing from the report by Wagner and Farrar (1976) of regularly 32 spores per capsule. The spores are especially large up to 100 μm and may reflect the tetraploid level of the species.

Cytology

The cytological records for *Campyloneurum* are consistently diploid $n = 37$ or tetraploid $n = 74$. Both levels are known in several species, such as *Campyloneurum xalapense* Fée, *C. latum* Moore, *C. angustifolium* and *C. phyllitidis*. In the last two species the tetraploid appears to be more widely distributed. Figures of the tetraploid *Campyloneurum anetioides* (as *Hyalotricha anetioides*) in Wagner and Farrar (1976) show the chromosomes in well-paired bivalents.

Literature

Wagner, W. H., and D. R. Farrar. 1976. The central American fern genus *Hyalotricha* and its family relationships. Syst. Bot. 1: 348–362.

112. *Niphidium*
Figs. 112.1–112.10

Niphidium J. Sm., Hist. Fil. 99. 1875. Type: *Niphidium americanum* (Hook) J. Sm. (*Polypodium americanum* Hook.) = *Niphidium longifolium* (Cav.) Mort. & Lell.

Anaxetum Schott, Gen. Fil. *t.l.* 1834, not *Anaxeton* Gaertn. 1791 (Compositae). Type: *Anaxetum crassifolium* (L.) Schott (*Polypodium crassifolium* L.) = *Niphidium crassifolium* (L.) Lell.

Pleuridium (Presl) Fée, Mém. Fam. Foug. 5 (Gen. Fil.): 273. 1852, not Bridel, Mant. Musc. 10. 1819 (Musci). *Phymatodes* section *Pleuridium* Presl, Tent. Pterid. 196. 1836. Type: *Phymatodes crassifolia* (L.) Presl (*Polypodium crassifolium* L.) = *Niphidium crassifolium* (L.) Lell.

Fig. 112.1. *Niphidium nidulare,* near Turrialba, Costa Rica. (Photo W. H. Hodge.)

Pessopteris Maxon Contrib. U.S. Nat. Herb. 10: 485. 1908, *nom. nov.* for *Anaxetum* Schott, and with the same type (*Pessopteris crassifolia* (L.) Maxon).

Polypodium subgenus *Anaxetum* C. Chr., Ind. Fil. Suppl. 3: 12. 1934. Type: *Polypodium crassifolium* L. = *Niphidium crassifolium* (L.) Lell.

Description Terrestrial, rupestral or epiphytic; stem short- to long-creeping, rather slender to moderately stout, bearing scales and usually a dense mat of rather soft roots (epiphytic) or many long fibrous roots (terrestrial); leaves monomorphic, ca. 25 cm to 1.25 m long, borne in a cluster or at intervals, lamina entire, densely to slightly scaly or glabrous, veins anastomosing with included free veinlets; sori round to sometimes oblong, borne at the junction of veins, in a single series between the main lateral veins and in two or more rows on each side of the costa, on a slightly to somewhat raised receptacle, paraphysate or not, exindusiate; spores ellipsoidal, monolete, the laesura $\frac{1}{2}$ to $\frac{2}{3}$ the length, the surface smooth with scattered spherical deposit. Chromosome number: $n = 74$.

The stem scales are clathrate, usually dark and with lighter margins. In some species the sporangia are often setose (Fig. 7). Paraphyses are absent in many species but are present as sporangiasters in some, especially in *Niphidium crassifolium.*

Fig. 112.2. *Niphidium crassifolium,* Prov. Pinchincha, Ecuador. (Photo W. H. Hodge.)

Systematics

Niphidium is an American genus of 10 or fewer species. It is distinguished especially by the sori that are borne in a single series between the main lateral veins (Fig. 6) and by the receptacle that is served by a ring of veins (Fig. 4).

Niphidium and *Pessopteris* were placed in separate genera until recently united by Lellinger. This alliance is based on several similarities, although *Niphidium longifolium* is distinctive in its densely scaly lamina (Fig. 5) and its cyclocytic stomata (Sen and Hennipman, 1981). The species of *Pessopteris* have few scales on the lamina and its polocytic or copolocytic types of stomates are common in the Polypodiaceae.

Niphidium is evidently most closely related to *Campyloneurum,* from which it differs in the pattern of the sori and their position on the veins.

Lellinger (1972) revised *Niphidium* in the inclusive sense adopted here.

Tropical American Species

Niphidium longifolium (Cav.) Mort. & Lell. (*N. americanum* (Hook.) J. Sm.) of Ecuador is the most distinctive species with a dense mat of long-ciliate scales that cover the lower surface of the

Fig. 112.3. Distribution of **Niphidium.**

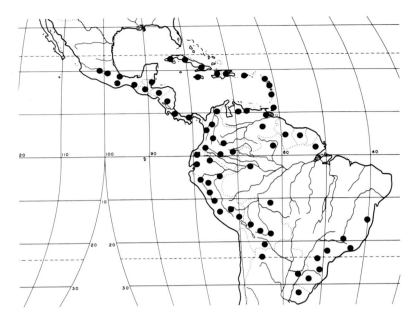

leaves. *Niphidium nidulare* (Rosenst.) Lell. of Costa Rica has black-ish stem scales and *N. carinatum* Lell. of the Andes from Colombia to Bolivia, has scales involute beyond the base. Other species segregated from *Niphidium crassifolium* (L.) Lell., primarily on the basis of stem scale characters seem to be less distinctive and some perhaps may represent subspecies or varieties. Additional work on this complex, especially field studies to assess local variation and cytology, are needed to determine the status of the *Niphidium crassifolium* segregates.

Ecology (Figs. 1, 2)

Niphidium is usually rupestral or epiphytic, sometimes terrestrial. It grows in a variety of habitats and has an extensive altitudinal range. Species may grow in dense, wet rain forests, in cloud forests, through a range of habitats to dry rocky slopes, open scrubby places or páramo. It grows on the ground, on dead stumps, decaying logs, or fallen branches, on living fence posts, in tree branches and on tree trunks, as well as on earth or among rocks. It grows from sea level to 3900 m, most commonly below 2000 m.

Geography (Fig. 3)

Niphidium is distributed from Querétero in central Mexico and the West Indies, southward to Jujuy in Argentina and disjunctly in Uruguay north to Bahia in Brazil. It is absent from most of central and northeastern Brazil.

Spores

The spores of *Niphidium* are quite large and relatively smooth with short laesura. Those of *N. longifolium* (Fig. 10) are about 60–70 μm and in *N. crassifolium* they may be 80 μm in longest dimension. The somewhat abraded spore of the latter in Fig. 8

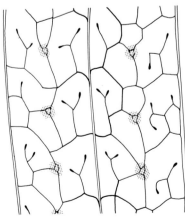

Fig. 112.4. Venation in a portion of a fertile leaf of **Niphidium crassifolium,** the shaded areas indicate the receptacles, ×4.0.

5 6

7 8

9 10

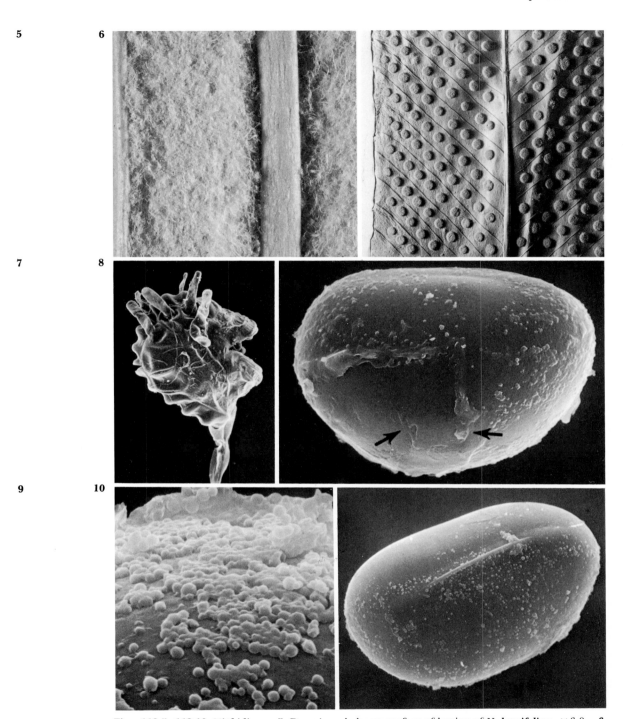

Figs. 112.5–112.10. *Niphidium.* **5.** Densely scaly lower surface of lamina of *N. longifolium,* ×6.0. **6.** Portion of lamina with single rows of sori between veins, *N. crassifolium,* ×1.0. **7.** Setose sporangium of *N. crassifolium* with spores in stomium at right, ×100. **8, 9.** *N. crassifolium* spores, Mexico, *Riba et al. 313.* **8.** The surface somewhat abraded at arrows, ×1000. **9.** Detail of fused spherical deposit, ×5000. **10.** *N. longifolium,* laesura at center, Ecuador, *Hitchcock 20817,* ×1000.

shows the thin strata forming the surface. The spherical deposition may be more or less fused and form compact areas on the surface (Fig. 9). *Niphidium* spores are unusual in the family in their large size, short laesura and relatively smooth surface although *Campyloneurum* and some groups of *Polypodium* also have large spores. They most closely resemble *Platycerium,* but spores of that genus have an exceptionally short laesura.

Cytology

The chromosome number of *Niphidium crassifolium* (as *Polypodium crassifolium*) in Jamaica is $n = 74$ (Walker, 1966), a tetraploid level that is common in many genera of the Polypodiaceae. The large size of the spores of a collection from Veracruz, Mexico, suggests this may also represent the tetraploid.

Literature

Lellinger, D. B. 1972. A revision of the fern genus *Niphidium.* Amer. Fern Jour. 62: 101–120.
Sen, U., and E. Hennipman. 1981. Structure and ontogeny of stomata in Polypodiaceae. Blumea 27: 175–201.
Walker, T. G. 1966. Reference under the family.

113.   *Neurodium*
Figs. 113.1–113.9

Neurodium Fée, Mém. Mus. Hist. Nat. Strasbourg 4: 201. 1850. Type: *Neurodium lanceolatum* (L.) Fée (*Pteris lanceolata* L.).
Paltomium Presl, Epim. Bot. 156. 1851, *nom. superfl.* for *Neurodium* and with the same type.
Heteropteris Diels, Nat. Pflanz. 1 (4): 305. 1899, *nom. superfl.* for *Neurodium* and with the same type.

Description   Epiphytic or rarely rupestral; stem rather small, short-creeping, bearing scales and many long, fibrous roots; leaves monomorphic, ca. 10–40 cm long, loosely clustered or somewhat spaced, lamina entire, glabrous, veins anastomosing with included free veinlets; sori borne on an intramarginal commissure in short to long lines on each side of the lamina toward its apex, not paraphysate or with filamentous paraphyses, exindusiate; spores ellipsoidal, monolete, the laesura $\frac{1}{2}$ to $\frac{3}{4}$ the length, verrucate to irregularly tuberculate with dense spherical deposit. Chromosome number: $n = 37$.

Systematics   *Neurodium* is a distinctive, monotypic American genus based on *N. lanceolatum* (L.) Fée. The narrowly elliptical lamina is fertile toward the apex where it is often somewhat contracted. The sporangia are borne on an inframarginal commissure (Fig. 3) parallel to the margin, usually continuous, but sometimes more or less interrupted. The stem scales are small, dark and usually bear several trichomes (Fig. 4). The relationship of *Neurodium* to

Fig. 113.1. *Neurodium lanceolatum,* Tela, Honduras. (Photo L. D. Gómez.)

Fig. 113.2. Distribution of *Neurodium.*

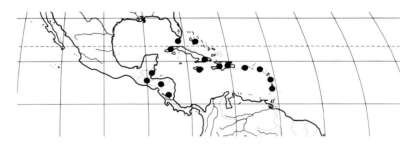

other polypodioid ferns is not clear, but the tuberculate spores suggest a possible relationship with *Microgramma.*

Neurodium sinensis Christ, an unrelated Asiatic species is correctly treated as *Lepisorus sinensis* (Christ) Ching.

Heteropteris Fée, X Session, Congr. Sci. France, Strasbourg, 1842, 1: 178. 1843 was cited by Fée as a synonym of *Neurodium* in his "Genera Filicum" and this was evidently the reason that Diels adopted the name. However, in the first paper, which apparently is seldom consulted, Fée listed only "*Heteropteris formosa* Fée" without description or reference to previous literature.

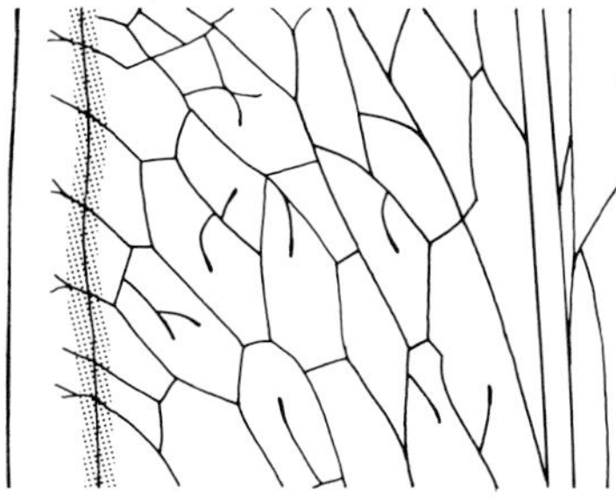

Ecology and Geography (Figs. 1, 2)

Neurodium grows in wet woods, in coppices, sometimes in wooded swamps or in coffee plantations. It is usually epiphytic (Fig. 1) but rarely grows on rocky banks. It occurs from near sea level to ca. 600 m.

Neurodium is distributed (Fig. 2) from the Florida Keys and the Bahamas, in all of the Greater Antilles and many of the Lesser Antilles south to the Grenadines, and in Central America from

Fig. 113.3. Portion of fertile lamina of *Neurodium lanceolatum,* receptacle indicated by shaded area, × 7.0.

4   5   6

7   8   9

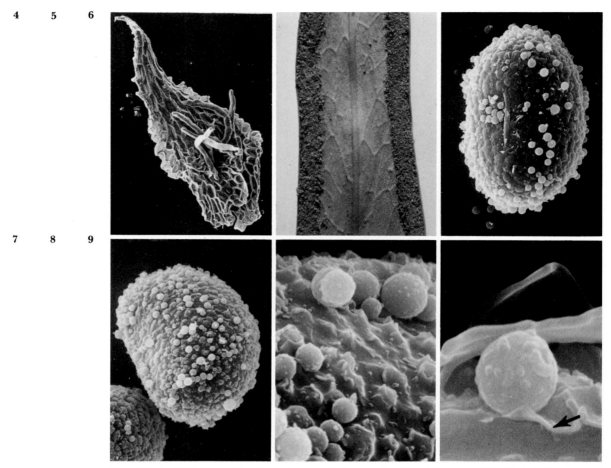

Figs. 113.4–113.9. *Neurodium lanceolatum.*   **4.** Stem scale with trichomes, × 50.   **5.** Apical portion of fertile lamina with sori near and parallel to margins, × 3.   **6–9.** Spores, × 1000.   **6.** Proximal face shallowly tuberculate, laesura central, British Honduras, *Schipp S-780.*   **7–9.** Jamaica, *Clute 288.*   **7.** Lateral view, laesura left.   **8.** Surface detail with spherical deposit, × 5000.   **9.** Profile, outer perispore stratum at arrow, exospore below, × 10,000.

Guatemala and British Honduras south to Nicaragua. The reports of Posthumus (1928) and Proctor (1977) of *Neurodium* in French Guiana require confirmation.

Spores

Neurodium spores are tuberculate with dense spherical deposit (Fig. 8). A profile of the wall shows the thin perispore with a sphere fused to it, and, below, the compact exospore (Fig. 9). Specimens from Jamaica (Fig. 7) are similar but somewhat more prominently tuberculate than those from British Honduras (Fig. 6). *Neurodium* spores resemble those of *Microgramma lycopodioides* in their relatively shallow tuberculate surface and short laesura.

Cytology

The photograph of a meiotic cell of *Neurodium lanceolatum* from Jamaica in Walker (1966) clearly shows 37 bivalents.

Literature

Posthumus, O. 1928. The ferns of Surinam and of French and British Guiana. 196 pp. Malang, Java.

Proctor, G. R. 1977. Pteridophyta, *in*: R. A. Howard, Flora of the Lesser Antilles, Vol. 2. pp. 414. Arnold Arboretum of Harvard University, Cambridge, Mass.

Walker, T. G. 1966. Reference under the family.

114. *Solanopteris*
Figs. 114.1–114.9

Solanopteris Copel., Amer. Fern Jour. 41: 75, 128. 1951. Type: *Solanopteris bifrons* (Hook.) Copel. (*Polypodium bifrons* Hook.).

Microgramma subgenus *Solanopteris* (Copel.) Lell. Amer. Fern Jour. 67: 59. 1977.

Description Epiphytic; stem long-creeping, slender, usually with highly modified hollow tubers borne on short lateral branches, bearing scales and rather few roots; leaves nearly monomorphic to dimorphic (the fertile longer and narrower than the sterile), ca. 3–25 cm long, borne at intervals on the creeping stem, or sometimes on the tubers, lamina entire to pinnately lobed, glabrous or slightly and minutely scaly, veins anastomosing with included free veinlets; sori round to elongate, or in a long, broad line, receptacle more or less raised, served by few to many veins, paraphysate, exindusiate; spores ellipsoidal, monolete, the laesura about $\frac{1}{2}$ the spore length, echinate with granulate deposit on the spines and adjacent surfaces. Chromosome number: $n = 37$.

The paraphyses of *Solanopteris* are usually abundant and filamentous with a series of short, inflated cells at the apex. Recurrent, free veins are frequent in the sterile lamina (Fig. 6) and in the fertile lamina the receptacle is supplied by several veins (Fig. 5).

Systematics *Solanopteris* is a tropical American genus of three or four species. It is highly distinctive in its echinate spores and stem tubers, and its relationship to other less specialized genera in the family is uncertain. Some species of *Polypodium* as *P. Funkii* and its allies, have similar long-creeping, pruniose (whitish), somewhat flattened stems with roundish peltate scales but these resemblances may be superficial.

The generic name was first published as *Solenopteris* which would suggest that it might be derived from *Solenopsis,* a genus of ants, although Copeland did not indicate the source of the name. It was changed to *Solanopteris* in the index and errata in the same publication; thus it appears that the name is based on *Solanum,* the genus of the potato.

Solanopteris tuberosa (Maxon) Rauh appears to be an illegitimate combination since the reference to the original publication is not provided in the text. However, this information is clearly legible in the photograph of the holotype.

Fig. 114.1. *Solanopteris bifrons,* Río Bobonaza, Pastaza, Ecuador. (Photo C. R. Sperling.)

Fig. 114.2 Distribution of *Solanopteris.*

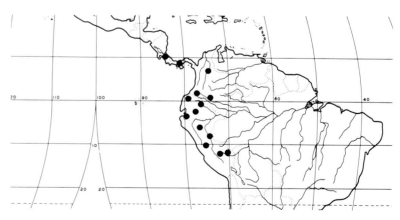

3

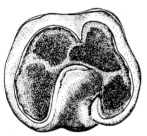

4

Figs. 114.3, 114.4. *Solanopteris*. 3. Part of plant of ***Solanopteris bifrons*** with sterile and fertile leaves, ×0.5. **4.** Tuber of ***Solanopteris Brunei,*** longitudinal section, the opening leading to the internal cavities, below, ×1.

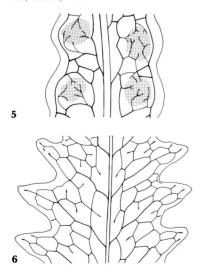

5

6

Tropical American Species

Solanopteris bifrons (Hook.) Copel. is most distinctive, with pruinose tubers and usually pruinose stems, roundish sori and pinnately lobed sterile leaves (Fig. 3). *Solanopteris Brunei* (Christ) Wagner is characterized by a tuber that is beset with long, narrow scales, roundish sori and entire to repand sterile leaves. The status of *S. Bismarkii* Rauh and *S. tuberosa* (Rauh, 1973) is uncertain because of limited collections. In both species the sporangia are borne on a long receptacle.

Ecology (Fig. 1)

Solanopteris is an epiphyte, usually in the canopy of trees in wet forests, less often it grows on small trees or shrubs. It occurs at altitudes of ca. 200–1800 m, most commonly at 300–1000 m.

According to Gómez (1974), there is evidence that plants grow best when the tubers are inhabited by ants, and that *Solanopteris* often grows on trees that provide them with food from floral nectaries.

Geography (Fig. 2)

Solanopteris is distributed from Costa Rica south to Cuzco and Madre de Dios in southern Peru. *Solanopteris Brunei* grows from Costa Rica to Nariño in southern Colombia; the other species are confined to South America.

Since the genus is often a high epiphyte, obscure in the canopy, it undoubtedly is more common than the relatively few collections might indicate.

Spores

The echinate spores of *Solanopteris* (Fig. 7) are unique among the American genera of the Polypodiaceae although spinulose spores are known in a few paleotropical genera as *Colysis* and *Crypsinus*. The spines of *Solanopteris* are solid, formed of material similar to that covering the adjacent surfaces and are readily detached from the smooth stratum below (Fig. 8). The dense, granulate surface deposit (Fig. 9) resembles that on *Microgramma* spores rather than the spherical deposition in many genera of the family. There appear to be fewer spines in *Solanopteris bifrons* than *S. Bismarckii* as shown in the work of Rauh (1973). The echinate formation possibly may be an adaptation for adhering to the ants which may serve in the local distribution of the spores. However, observations on the transport by ants have not been reported.

Figs. 114.5, 114.6. Veins and sori of ***Solanopteris bifrons,*** ×3. **5.** Portion of fertile lamina, receptacles shaded. **6.** Portion of sterile lamina.

7

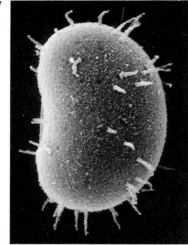

8

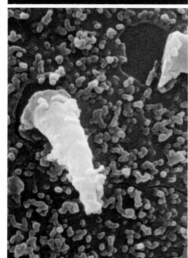

9

Figs. 114.7–114.9. Spores, *Solanopteris bifrons,* Colombia, *Cuatrecásas 11209.* **7.** Echinate spores, laesura, left, × 1000. **8.** Detail of granulate surface and broken bases of two spines, × 10,000. **9.** Detail of granulate deposit on spines, × 10,000.

Cytology

A diploid $n = 37$ has been reported for *Solanopteris Brunei* from Costa Rica (Wagner, 1976). The number supports the relationship of the genus within the main group of genera of the Polypodiaceae.

Observations

Solanopteris is the only American fern with modified stems that are often inhabited by ants. Stems of the paleotropical genus *Lecanopteris* also support ant colonies. The adaptive relation of the tubers (Fig. 4) has been considered especially in respect to their possible function in water storage, the debris collected within as a source of nutrition, and as a residence for ants that may control adjacent competitive plants. The spiny spores are unusual and could have an adaptative function for local dispersal by ants as noted above. The anatomy and morphology of the tubers has been discussed by Senn (1910), Hagemann (1969), and Rauh (1955). The insect relationships and ecology of *Solanopteris Brunei* were studied by Gómez (1974), and a detailed review of the species is provided by Wagner (1972). The most complete account, especially of the morphology of the species, is presented by Rauh (1973).

Literature

Gómez, L. D. 1974. Biology of the potato-fern *Solanopteris Brunei.* Brenesia 4: 37–61.

Hagemann, W. 1969. Zur Morphologie der Knolle von *Polypodium bifrons* Hook. und *P. Brunei* Wercklè. Mém. Soc. Bot. France 1969: 17–27.

Rauh, W. 1955. Botanische Mitteilungen aus den Anden, I. Morphologische und anatomische Beobachtungen an *Polypodium bifrons* Hook. Akad. Wiss. Lit. Mainz, Abhandl. Math.-Natur. Klasse 1955: 45–57.

Rauh, W. 1973. *Solanopteris Bismarckii* Rauh. Akad. Wiss. Lit. Mainz, Math.-Natur. Klasse 1973: 223–256; and Trop. Subtrop. Pflanzenwelt 5: 5–38.

Senn, G. 1910. Die Knollen von *Polypodium Brunei* Wercklè. Verhand. Naturfor. Ges. Basel 21: 115–125.

Wagner, F. S. 1976. The chromosome number of *Solanopteris Brunei* (Wercklè ex Christ) Wagner. Brenesia 9: 81–82.

Wagner, W. H., 1972. *Solantoperis Brunei,* a little-known fern epiphyte with dimorphic stems. Amer. Fern Jour. 62: 33–43.

115. *Platycerium*
Figs. 115.1–115.10

Platycerium Desv., Mém. Linn. Soc. Paris 6: 213. 1827. Type: *Platycerium alcicorne* Desv., *nom. nov.* for *Acrostichum alcicorne* Sw. 1801, not Willemet, 1796.

Alcicornium Gaud., Freyc. Voy. Uranie, Bot. 48. 1825, *nom. provis.;* Underw. Mem. Torrey Bot. Cl. 6: 275. 1899; Bull. Torrey Bot. Cl. 30: 673. 1903; *ibidem* 32: 587. 1905, *nom. nud.*

Neuroplatyceros Fée, Mém. Fam. Foug. 2: 25. 1845, *nom. superfl.* for *Platycerium* and with the same type.

Description   Epiphytic; stem moderately stout, ascending, bearing scales and many fibrous roots; leaves nearly monomorphic (in *P. elephantotis*) or dimorphic (the nest leaves at least basally, appressed, more or less erect, with a continuous petiole, ca. 25 cm to 1 m long, the fertile erect-spreading to usually pendent, with a disarticulate petiole, ca. 50 cm to 3 m long), borne in a cluster, lamina

Fig. 115.1. *Platycerium andinum,* near Tarapoto, San Martín, Peru. (Photo W. H. Hodge.)

Figs. 115.2, 115.3. *Platycerium andinum,* near Tarapoto, San Martín, Peru. (Photo W. H. Hodge.) **2.** Venation of nest leaves. **3.** Fertile areas on pendent leaves.

2

3

entire to strongly lobed to dichotomously branched, pubescent, veins anastomosing with or without included free veinlets; sporangia borne along the nearly unmodified veins in certain areas of the lamina, paraphysate or not, exindusiate; spores ellipsoidal, monolete, the laesura $\frac{1}{4}$ to $\frac{1}{2}$ the spore length, the surface nearly smooth with scattered spherical deposit. Chromosome number: $n = 37$; $2n = 74$.

The sporangia are confined to particular areas on the fertile leaf, where the fertile veins form an additional network close to

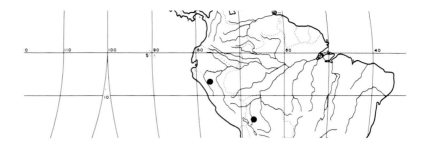

Fig. 115.4. Distribution of *Platycerium* in America.

the abaxial surface, connected with the primary central vein system (Figs. 5, 6). When fully mature, the sporangia appear to cover the fertile surface, however, they are usually borne only on the veins, although in *Platycerium coronarium* (Mull.) Desv. and *P. Ridleyi* Christ they are achrostichoid.

The stellate trichomes that occur in the fertile area may be essentially like those on the sterile portions of the lamina (Fig. 7), while in other species they are clearly modified and are considered to be paraphyses.

Systematics

Platycerium is a genus of 15 species, with one, *P. andinum* Baker (Figs. 1–3) in America. This is most closely related to *P. quadridichotomum* (Bonap.) Tard. and *P. elephantotis* Schweinf. of Africa. It is a highly specialized epiphytic genus with the nest leaves modified for humus collecting. The characteristic stellate trichomes on the leaf of *Platycerium* indicate a close relationship with *Pyrrosia,* as do the codesmocytic stomata (Sen and Hennipman, 1981).

Platycerium is one of the few fern genera in which the modern taxonomy is based on living plants. The genus has long been a greenhouse or tropical garden speciality and recently all species have been brought into cultivation. The morphology and phylogeny of *Platycerium* is treated by Hoshizaki (1972). Critical nomenclatural notes have been published by de Joncheere (1967, 1974), and the genus is being monographed by Hennipman and Roos (in prep.). We are indebted to E. Hennipman who has reviewed the treatment of this genus.

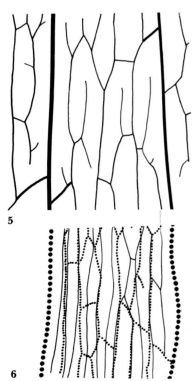

Figs. 115.5, 115.6. *Platycerium andinum,* near Tarapoto, San Martín, of lamina, ×3.5. **6.** Venation of fertile portion of a lamina. The fertile vein system near the lower surface shown by solid lines, central vein system, dotted lines, and two large veins by large dots, ×3.5.

Ecology (Fig. 1)

Platycerium is an epiphytic genus of rain forests and in southeast Asia of monsoon forests and gallery forests. In primitive forests it may be a very high epiphyte, it also grows on low branches in other places, and is typical of low to mid-elevations. It often is on *Elaeis* palms.

In America *Platycerium* grows from 100 to 1200 m on the eastern slopes of the Andes. A note on a R. Spruce collection from Peru indicates that it "Sometimes forms masses so large that branches break down from their weight" and R. S. Williams notes that in Bolivia it grows "in clumps 6 ft. or more across."

7

8

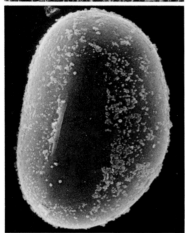

9

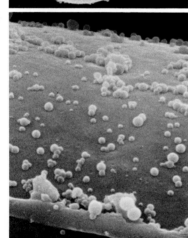

10

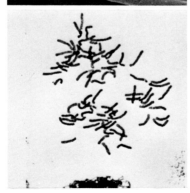

Figs. 115.7–115.10. *Platycerium andinum,* Peru, *Schunke 2165.*   **7.** Stellate trichomes on upper surface of lamina, × 8.0.   **8, 9.** Spores.   **8.** Distal face, short laesura center, left, × 1000.   **9.** Detail of surface deposit, laesura at base, × 5000.   **10.** Mitotic chromosomes from root, $2n = 74$.

Geography (Fig. 4)

Platycerium grows in Andean South America, in tropical Africa, Madagascar and islands of the Indian Ocean, and in Burma and southern China to the Philippine Islands, Java and east to New Guinea, and in eastern Australia south to New South Wales; also on Lord Howe Island.

Platycerium andinum is known from several collections in the Department of San Martín, Peru, and a few from the province of La Paz, Bolivia.

Spores

Platycerium spores are usually tan or brown but are green in *P. Wallichii* Hook., and in a few species the spore number may be reduced to eight per sporangium. The surface is relatively smooth, the laesura short (Fig. 8) and the spherical deposition quite dense (Fig. 9). Spores of *Platycerium* are generally smoother and lack the verrucate surface structure characteristic of many genera of the Polypodiaceae. Except for the short laesura they most closely resemble spores of *Niphidium.*

Cytology

The chromosome numbers reported for *Platycerium* from Africa, India, Malesia and Australia are uniformly $n = 37$. The report of *P. Hillii* Moore from Queensland by Fabbri (1957) clearly shows meiotic cells with $n = 37$. The 74 mitotic chromosomes in the American *P. andinum* (Fig. 10) show a diverse morphology and size that should allow them to be readily karyotyped.

Observations

The unique forked leaves of *Platycerium* and its epiphytic habit adapted to growth in baskets or as hanging plants have made the species popular in cultivation. *Platycerium bifurcatum* (Cav.) C. Chr. is one of the most widely cultivated greenhouse species and, in frost-free regions, *P. Willinckii* Moore and *P. alcicorne* Desv. are often grown. The species of *Platycerium,* their cultivated forms, and cultural requirements are treated by Joe (1964).

Literature

Fabbri, F. 1957. Sondaggi citogenetici nelle Polypodiaceae sensu stricto. Caryologia 10: 402–407.

Hennipman, E., and M. C. Roos. (in prep.). A monograph of the fern genus *Platycerium.*

Hoshizaki, B. J. 1972. Morphology and phylogeny of *Platycerium* species. Biotropica 4: 93–117.

Joe, B. 1964. A review of the species of *Platycerium* (Polypodiaceae). Baileya 12: 69–126.

Joncheere, G. J. de. 1967. Notes on *Platycerium* Desv., I. Blumea 15: 441–451.

Joncheere, G. J. de. 1974. Nomenclatural notes on *Platycerium* (Filices). Blumea 22: 53–55.

Sen, U., and Hennipman, E. 1981. Structure and ontogeny of stomata in Polypodiaceae. Blumea 27: 175–201.

22b. Tribe Loxogrammeae

Loxogrammeae (Pic.-Ser) R. & A. Tryon, Rhodora, 84: 129. 1982.
Loxogrammaceae Pic.-Ser., Webbia 29: 11. 1974. Type: *Loxogramme* (Bl.) Presl.

A tribe of two genera, *Loxogramme* and the closely related *Anarthropteris* Copel., represented by a single species, *A. lanceolata* (J. D. Hook.) Pic.-Ser. (*Anarthropteris Dictyopteris* (Mett.) Copel.) of New Zealand and the New Hebrides.

116. *Loxogramme*
Figs. 116.1–116.8

Loxogramme (Bl.) Presl, Tent. Pterid. 214. 1836. *Antrophyum* section *Loxogramme* Bl., Fl. Jav. Filices 73. 1829. Type: *Antrophyum lanceolatum* (Sw.) Bl. (*Grammitis lanceolata* Sw.) = *Loxogramme lanceolata* (Sw.) Presl.

Description Rupestral or usually epiphytic; stem short- to rather long-creeping, slender, bearing scales and numerous, rather soft roots; leaves monomorphic, subdimorphic or (in *L. dimorpha* Copel.) strongly dimorphic (the fertile much narrower than the sterile ca. 3–60 cm long), borne in a loose cluster to rather widely spaced, entire or rarely irregularly lobed, glabrous when mature, veins anastomosing, with or without included free veinlets; sori elongate, nearly parallel to the costa or usually oblique, on a scarcely to slightly raised receptacle associated with narrow areolae, or in a long line on and near a single vein or series of veins and narrow areolae parallel to the costa, paraphysate or not, exindusiate; spores spheroidal, trilete with laesurae $\frac{1}{2}$ to $\frac{3}{4}$ the radius, or often ellipsoidal, monolete with the laesura $\frac{1}{4}$ to $\frac{2}{3}$ the length, shallowly to somewhat prominently papillate. Chromosome number: $n = 35, 36, 70; 2n = 72$.

Systematics *Loxogramme* is a nearly pan-tropical genus with subtemperate extensions, of about 30 species with only one, *L. mexicana* (Fée) C. Chr., in America (Figs. 2–4).

The genus is characterized by anastomosing veins, clathrate stem scales, elongate sori, glabrescent leaves, an entire lamina, and green spores.

Fig. 116.1. Distribution of *Loxogramme*.

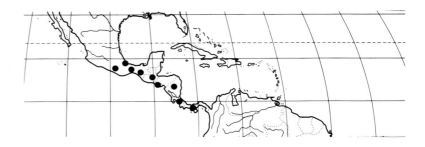

The American *Loxogramme mexicana* is closest to the African *L. lanceolata* (Sw.) Presl, a similar geographic relation to that of *Platycerium*, which also has a single American species with affinities in Africa.

Loxogramme, with *Anarthropteris*, has been placed with the polypodioid or with the grammitid ferns, or as a group more or less isolated from either of these (Pichi-Sermolli, 1974). It shares characters of the gametophyte (Nayar, 1968), petiole (Konta, 1974), sporangium stalk, and spores with both polypodioids and grammitids. *Loxogramme* may have been derived from an ancestor of the other groups, or possibly is a derivative of one of them with characters convergent toward the other.

The treatment of *Loxogramme* has been reviewed by Michael G. Price, who is currently preparing a revision of the genus.

Ecology

Loxogramme is usually a low epiphyte on mossy tree trunks in wet, dark forests. It also grows in rocky places, especially along streams, on boulders, wet cliffs and or on logs in the forest. Rarely it grows in more open habitats.

Loxogramme mexicana grows on wet rocks and on tree trunks at elevations mostly between 1000 and 2000 m.

Geography (Fig. 1)

Loxogramme occurs in America, Africa to India and Ceylon and northeastward to Japan and eastward to New Guinea, and through the Pacific to Tahiti.

Loxogramme mexicana occurs in Jalisco and Veracruz, Mexico, southward to Panama.

Spores

Loxogramme mexicana spores are exceptionally large and Fig. 5, showing the short laesura and low papillate surface, is at a lower magnification than Figs. 6 and 7. The marked size differences illustrated in species studied by Chang Yu-Lung (1963) suggest that polyploid changes, as indicated in the cytology, have occurred in a number of species. A thin perispore overlays and conforms to a shallow exospore as in the detail of *L. involuta* (D. Don) Presl (Fig. 8). The papillae are somewhat coarser and prominent in *Loxogramme salicifolia* (Mak.) Mak. (Fig. 6) and in the subtrilete spore of *L. chinensis* Ching (Fig. 7) with a shortened

Fig. 116.2. Plant of *Loxogramme mexicana*, ×0.5.

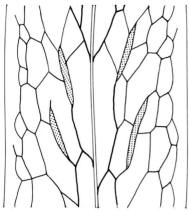

Fig. 116.3. Venation, with receptacle stippled, ***Loxogramme mexicana,*** × 3.

laesura. The strata generally resemble some spores of *Polypodium* as has been indicated by Mitui (1977) in a survey of spore wall structure of species in the Polypodiaceae. Spores of *Loxogramme* are exceptional in their diverse forms ranging from trilete-spheroidal to monolete-ellipsoidal within the same species. The variant forms are described in *Loxogramme "lanceolata"* (Nayar, 1963) and reviewed in 20 species by Chang Yu-Lung (1963). Developmental studies of these variant forms will be of special interest as they apply to the two classes, trilete and monolete, designated for pteridophyte spores.

Cytology

The chromosome number is not reported for the American species but the large spore size of *Loxogramme mexicana* from Veracruz (Fig. 5) suggests that it may be polyploid. Among the relatively few cytological records of Old World species of *Loxo-*

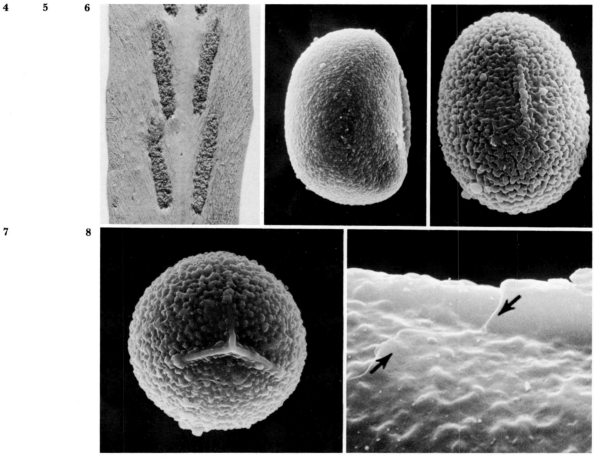

Fig. 116.4–116.8. *Loxogramme.* **4.** Portion of fertile leaf, elongate sori, ***Loxogramme mexicana,*** × 2.0. **5–8.** Spores. **5.** ***L. mexicana,*** low, papillate laesura at right, Mexico, *Copeland 156,* × 600. **6.** ***L. salicifolia,*** prominently papillate, laesura center, Japan, *Furuse in 1961,* × 1000. **7.** ***L. chinensis,*** with one laesura arm shorter, China, *Tsai 62620,* × 1000. **8.** ***L. involuta,*** detail of abraded spore, perispore at arrows, India, *Hara 319,* × 5000.

gramme there are twice as many diploids, mostly $n = 35$, than tetraploids. *Loxogramme involuta* is reported as tetraploid, $n = 70$ from eastern India (Panigrahi and Patnaik, 1961) while there are records of diploid, $n = 36$ and $2n = 72$, from southern India (Roy et al., 1971). These numbers suggest that aneuploid changes may have occurred in addition to the increase in ploidy level in this Indian species.

Literature

Chang, Yu-Lung, 1963. Studies in the spore morphology of *Loxogramme* Presl. Acta Bot. Sinica 11: 26–43.

Konta, F. 1974. Frond articulation in the fern genus *Loxogramme*. Acta Phytotax. Geobot. 26: 119–126.

Mitui, K. 1977. Spore wall structure of some Japanese species in Polypodiaceae, *s. st.* Bull. Nippon Dental Univ., Gen. Ed. 6: 119–124.

Nayar, B. K. 1963. Spore morphology of *Loxogramme*. Grana Palynol. 4: 388–392.

Nayar, B. K. 1968. The gametophytes and juvenile leaves of *Loxogramme*. Amer. Fern Jour. 58: 19–29.

Panigrahi, G., and S. N. Patnaik. 1961. Cytology of some Polypodiaceae in eastern India. Nature 191: 1207, 1208.

Pichi-Sermolli, R. E. G. 1974. Fragmenta Pteridologiae V. Webbia 29: 1–16.

Roy, R. P., B. M. Sinha, and A. R. Sakya. 1971. Cytology of some ferns of the Kathmandu valley. Brit. Fern Gaz. 10: 193–199.

22c. Tribe Grammitideae

Grammitideae Presl, Tent. Pterid. 205. 1836, as Grammitaceae. Type: *Grammitis* Sw.

Grammitidaceae (Presl) Ching, Sunyatsenia 5: 264. 1940, as Grammitaceae.

Pleurosoriopsidaceae Ching, Acta Phytotax. Sinica 16(4): 17. 1978. Type: *Pleurosoriopsis* Fomin.

A tribe of two, or perhaps a few more, genera with *Pleurosoriopsis* the only element distinct from the large genus *Grammitis*. *Pleurosoriopsis* Fomin, represented by the single species *P. Makinoi* (Makino) Fomin, has been studied by Mashuyama (1975), Kurita and Ikebe (1977), and Nayar (1977). It has affinities to *Grammitis* in the long, stiff trichomes on the petiole, the germination of the spores within the spore wall and within the sporangium, the granulate spore surface, and the gametophytes with gemmae. It is unusual, among other characters, in its very slender stem with a simple type of stele, the 2-pinnate to 2-pinnate-pinnatifid lamina, the more or less elongate sori and the chromosome number of $2n = 144$.

Other genera of the Old World often currently accepted are *Acrosorus* Copel., *Adenophorus* Gaud., *Calymmodon* Presl, *Nematopteris* vAvR., *Oreogrammitis* Copel., *Prosaptia* Presl, and *Scleroglossum* vAvR. These probably belong in an infrageneric classification of *Grammitis*.

Literature

Kurita, S., and C. Ikebe. 1977. On the systematic position of *Pleurosoriopsis Makinoi* (Maxim.) Fomin. Jour. Jap. Bot. 52: 39–48.

Masuyama, S. 1975. The gametophyte of *Pleurosoriopsis Makinoi* (Maxim.) Fomin. Jour. Jap. Bot. 50: 105–113.

Nayar, B. K. 1977. On the gametophytes of *Pleurosoriopsis Makinoi.* Jour. Jap. Bot. 52: 107–109.

117. *Grammitis*
Figs. 117.1–117.34

Grammitis Sw., Jour. Bot. (Schrad.) 1800 (2): 17. 1802. Type: *Grammitis marginella* (Sw.) Sw. (*Polypodium marginellum* Sw.).

Xiphopteris Kaulf., Berl. Jahrb. Pharm. 21: 35. 1820. Type: *Grammitis myosuroides* (Sw.) Sw. (*Polypodium myosuroides* Sw., *Xiphopteris myosuroides* (Sw.) Kaulf.). *Grammitis* section *Xiphopteris* (Kaulf.) Presl, Tent. Pterid. 208. 1836.

Cochlidium Kaulf., Berl. Jahrb. Pharm. 21: 36. 1820. Type: *Grammitis graminoides* (Sw.) Sw. (*Acrostichum graminoides* Sw., *Cochlidium graminoides* (Sw.) Kaulf.). [Section *Pleurogramme*].

Pleurogramme (Bl.) Presl, Tent. Pterid. 223. 1836. *Antrophyum* section *Pleurogramme* Bl., Fl. Jav. Fil. 69. 1828. Type: *Taenitis linearis* Kaulf. (*Pleurogramme linearis* (Kaulf.) Presl) = *Grammitis seminuda* (Willd.) Willd. *Grammitis* section *Pleurogramme* (Bl.) R. & A. Tryon, Rhodora 84: 128. 1982.

Ctenopteris Kze., Bot. Zeit. 4: 425. 1846. Type: *Ctenopteris venulosa* (Bl.) Kze. (*Polypodium venulosum* Bl. = *Grammitis venulosa* (Bl.) R. & A. Tryon. [Section *Cryptosorus*].

Cryptosorus Fée, Mém. Fam. Foug. 5 (Gen. Fil.): 231. 1852. Type: *Cryptosorus Blumei* Fée, *nom. superfl.* for *Polypodium obliquatum* Bl. = *Grammitis obliquata* (Bl.) Hassk. *Polypodium* section *Cryptosorus* (Fée) Fourn., Ann. Sci. Nat. V, 18: 282. 1873. *Grammitis* section *Cryptosorus* (Fée) R. & A. Tryon, Rhodora 84: 128. 1982 (Morton, Contrib. U. S. Nat. Herb. 38: 90. 1967, invalid).

Micropteris Desv., Mém Linn. Soc. Paris 6: 217. 1827. Type: *Micropteris serrulata* (Sw.) Desv. = *Grammitis serrulata* (Sw.) Sw. [Section *Xiphopteris*].

Polypodium section *Grammitastrum* Fourn., Ann. Sci. Nat. V, 18: 282. 1873. Type: *Polypodium pseudaustrale* (Fourn.) Fourn. = *Grammitis pseudaustralis* Fourn. *Grammitis* section *Grammitastrum* (Fourn.) Morton, Contrib. U.S. Nat. Herb. 38: 89. 1967, incorrectly attributed to Fournier.

Lomaphlebia J. Sm., Hist. Fil. 182. 1875. Type: *Lomaphlebia linearis* J. Sm. = *Grammitis graminea* (Sw.) Ching. [Section *Grammitastrum*].

Glyphotaenium (J. Sm.) J. Sm., Hist. Fil. 187. 1875. *Ctenopteris* subgenus *Glyphotaenium* J. Sm., Bot. Voy. Herald (Seemann) 227. 1854. Type: *Ctenopteris crispata* J. Sm. (*Glyphotaenium crispatum* (J. Sm.) J. Sm.) = *Grammitis crispata* (J. Sm.) Morton *Grammitis* section *Glyphotaenium* (J. Sm.) R. & A. Tryon, Rhodora 84: 128. 1982.

Enterosora Baker, Timehri 5: 218. 1886. Type: *Enterosora Campbellii* Baker = *Grammitis trifurcata* (L.) Copel. [Section *Glyphotaenium*].

Grammitis subgenus Melanoloma Copel., Phil. Jour. Sci. 80: 253. 1951. Type: *Grammitis marginella* (Sw.) Sw. (*Polypodium marginellum* Sw.). [Section *Grammitis*].

Nanopteris Vareschi, Fl. Venez. 1: 881. 1969, invalid.

Fig. 117.1. *Grammitis asplenifolia,* Dominica. (Photo W. H. Hodge.)

Description

Terrestrial, usually epiphytic or rupestral; stem usually small, short-creeping (to erect to 50 cm in *G. erecta*), bearing scales or rarely these absent and numerous to rather sparse, slender, fibrous roots; leaves monomorphic or rarely the apical fertile portion modified, ca. 2 cm to 1 m long, borne in a cluster, lamina entire to 3-pinnate, with trichomes and sometimes glands, rarely densely pubescent, veins usually free, sometimes anastomosing with or without included free veinlets; sori round to elongate, borne on the vein or at the tip, or the sporangia in short to long lines on each side of the costa and associated with free or anastomosing veins, paraphyses present or not, exindusiate; spores tetrahedral-globose, trilete, the laesurae $\frac{2}{3}$ to $\frac{3}{4}$ the radius, usually prominently papillate with more or less dense spherical deposit. Chromosome number: $n = 32, 33, 36, 37, 74, 132–138$ (I); $2n = 72, 74$.

The stem scales are commonly concolorous, nonclathrate and glabrous, but in some species they are strongly clathrate and in others prominently pubescent; some small species may lack scales. Rarely, the lamina has a dark, sclerotic border (Fig. 7) that persists after the lamina tissue deteriorates. The leaves usually bear long trichomes, especially on the petiole, but they

Fig. 117.2. *Grammitis moniliformis,* páramo on Cotopaxi, Ecuador. (Photo W. H. Hodge).

may also be abundant on the segments (Fig. 10). The sori are usually roundish (Figs. 12, 13) or elongate (Figs. 11, 14) and may be confluent at maturity (Fig. 11). Paraphyses with large globose cells (Fig. 8) occur in some species, and in some the sporangia bear long trichomes (Fig. 9). Some of the diversity of lamina architecture in American species of *Grammitis* is shown in Figs. 15–24.

Systematics *Grammitis* is a pantropical and austral genus of about 400 species, with about 175 in America. The nomenclature and description of the genus is restricted to the elements that pertain to groups in America, and do not include the various segregate genera of the Old World that might be placed here. In America, *Grammitis* is characterized by continuous, nonarticulate petioles, long, usually rigid trichomes on the leaf, free venation, and spherical, trilete, green spores.

Fig. 117.3. *Grammitis percrassa,* Volcán Barba, Costa Rica. (Photo W. H. Hodge.)

The several segregate genera recognized by Copeland (1947) and by Pichi-Sermolli (1977) that include American species appear to be intergrading evolutionary lines within *Grammitis*. These are recognized as six sections in the following synopsis that will serve to present a framework for the diversity of the genus. In each section there clearly are closely related species, but some may have closer affinities to other sections. In the four pantropic sections, it is doubtful that there are many close relations between American and Old World elements since the potential for convergence of characters seems unusually great in these mostly small plants of simple form.

Relationships of *Grammitis* are uncertain except for a general affinity to the tribes Polypodieae and Loxogrammeae, with which it shares several common features, including prevalent chromosome numbers of $n = 37$ or 36.

The earliest typification of *Xiphopteris* was evidently by Desvaux in 1827, when he removed *Grammitis serrulata,* one of the two original species, to his genus *Micropteris*. The type of *Lomaphlebia* is *L. linearis* J. Sm., a new species, rather than a transfer of *Grammitis linearis* Sw., which is a superfluous name and a synonym of *Grammitis magellanica*. Nomenclatural corrections in sectional names are necessary in Morton's work (1967), for names attributed to Fournier under *Grammitis* which were actually published under *Polypodium*. *Grammitis* section *Cryptosorus* is invalid because the basionym *Cryptosorus* Fée is not cited, and *Grammitis* section *Grammitastrum* is a new combination, incorrectly attributed to Fournier.

The basic revisions of American *Grammitis* are those of Maxon (1914, 1915, 1916), and of Christensen (1929) on section *Pleurogramme* (*Cochlidium*). The treatments of Copeland (1952a, 1952b) generally follow the work of Maxon, while his treatment of section *Cryptosorus* (*Ctenopteris*) in America (1956) is incomplete and inadequate. Bishop (1977, 1978) has reworked small groups covered by both Maxon and Copeland. De la Sota (1960) has published on the species of Argentina, Morton (1967) on those of Ecuador, and Rodriguez (1974) has treated the species of Chile. The two Chilean species are also included in the treatments of Parris (1975) on Australia and Parris and Given (1976) on New Zealand. Brade (1966) treated a portion of the genus in Brazil.

Tropical American Species

Studies of the species of sections *Xiphopteris,* *Grammitastrum,* *Grammitis* and *Pleurogramme* have provided revisions for about one-third of the American species. However, knowledge of variation in many of the small species that grow sequestered among mosses is limited for collections are still inadequate. Other species, especially those of section *Cryptosorus* are in much need of critical study.

Among the few cytological records of American *Grammitis,* some unusual numbers as $n = 32$ and 33 suggest that the chromosome numbers may eventually provide needed information on species relationships.

Fig. 117.4. *Grammitis serrulata,* Dominica. (Photo W. H. Hodge.)

Fig. 117.5. *Grammitis rostrata,* Cerro Zurqui, Costa Rica. (Photo W. H. Hodge.)

The principal features and characteristic species of the sections are included in the following synopsis.

Section *Cryptosorus*

Lamina pinnatisect to 2-pinnate, pinnae or pinna-segments pinnately veined, with two or more sori.

About 100, or perhaps more, species in America.

This is the largest and most diverse section, including some rather well-defined species groups. The group of *Grammitis moniliformis* (Sw.) Proctor (Fig. 2) has a creeping, often rather long, stem with clathrate, usually thin and irridescent scales. Among other species are *Grammitis erecta* Morton, *G. firma* (J. Sm.) Morton, and *G. flabelliformis* (Poir.) Morton (*Polypodium rigescens* Willd.) (Fig. 16) and *G. peruviana* (Desv.) Morton. The group of *Grammitis curvata* (Sw.) Ching has the pinnatisect lamina reduced to a sinuate base and often many, large, whitish glands and similar but longer paraphyses. Some of the other species of this group are *Grammitis amylacea* (Copel.) Morton (Fig. 12), and *G. subsessilis* (Baker) Morton (Fig. 22); also *G. podocarpa* (Maxon) Seymour and *G. pozuzoensis* (Baker) R. & A. Tryon (Figs. 13, 15) which may have the pendent leaves elaborately or irregularly branched. A few species with articulate petioles as *Grammitis Mathewsii* (Mett.) Morton, *G. Kalbreyeri* (Baker) Morton, and *G. semihirsuta* (Kl.) Morton (Fig. 20) are not closely related. Some other species of the section are *Grammitis achilleifolia* (Kaulf.) R. & A. Tryon (Fig. 19), *G. apiculata* (Kl.) Seymour (Fig. 17), *G. asplenifolia* (L.) Proctor (Fig. 1), *G. cultrata* (Willd.) Proctor, *G. gradata* (Baker) R. & A. Tryon (Fig. 21), *G. heteromorpha* (Hook. & Grev.) Morton (Fig. 18), *G. taxifolia* (L.) Proctor, and *G. variabilis* (Kuhn) Morton (Fig. 10).

Section *Glyphotaenium*

Lamina slightly sinuate to rather deeply lobed, mesophyll spongy, veins free or sometimes anastomosing, sori round to elongate, usually irregularly disposed.

Fig. 117.6. Distribution of *Grammitis* in America.

A small American group that could be placed in section *Cryptosorus. Grammitis uluguruensis* (Reimers) Copel. of Africa is similar, but its characters may be convergent. *Grammitis curvata,* above, has spongy mesophyll but in other characters seems unrelated. Among the species are *Grammitis crispata* (J. Sm.) Morton, *G. Harrisii* (Jenm.) A. R. Sm., *G. percrassa* (Baker) Seymour (Fig. 3), and *G. trifurcata* (L.) Copel. (Fig. 14).

Section *Xiphopteris*

Lamina lobed to pinnatisect, segments with a simple or 1-forked vein and one sorus.

Perhaps 30 species in America. The apical fertile portion of the lamina is nearly entire in *Grammitis Jamesonii* (Hook.) Morton (Fig. 11), *G. myosuroides* (Sw.) Sw., *G. serrulata* (Sw.) Sw. (Fig. 4), and a few other species. It is not modified in other species as *Grammitis delitescens* (Maxon) Proctor, *G. Hartii* (Jenm.) Proctor, *G. Knowltoniorum* (Hodge) Proctor, *G. taenifolia* (Jenm.) Proctor, and *G. trichomanioides* (Sw.) Ching.

Section *Grammitastrum*

Lamina entire or sometimes forked, the margin unmodified, sori round to elongate.

About 15 species in America that could be placed in section *Grammitis* which differs only by the modified border of the lamina. The vein-tips are connected by an intramarginal vein in *Grammitis graminea* (Sw.) Ching and *G. turquina* (Maxon) Copel.

7 8 9

10 11

12 13 14

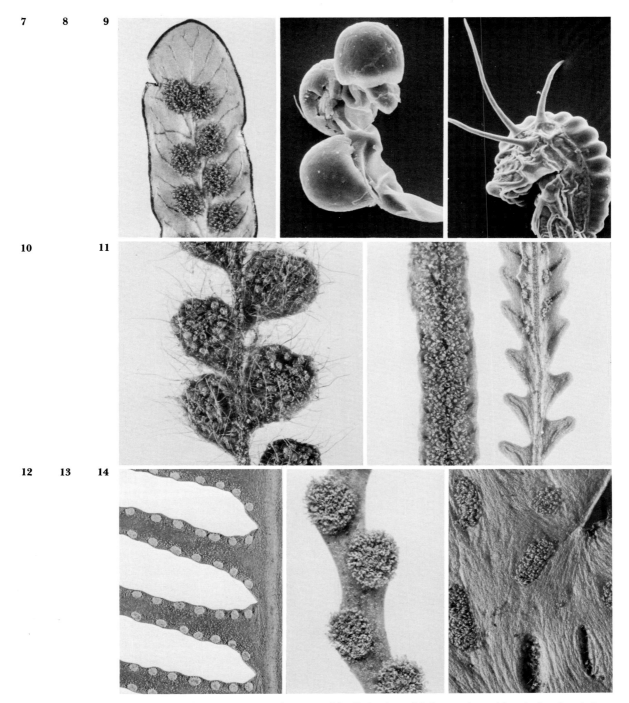

Figs. 117.7–117.14. *Grammitis.* **7.** Apex of fertile lamina of *G. leptopodon,* with a dark, sclerotic border, × 5. **8.** Glandular paraphysis of *G. jungermannioides,* × 500. **9.** Sporangium of *G. asplenifolia* with trichomes, × 160. **10.** Portion of fertile pinna of *G. variabilis,* × 8. **11.** Portions of fertile leaves of *G. Jamesonii,* confluent mature sori, left, and immature sori, right, × 8. **12.** Portion of fertile lamina of *G. amylacea,* × 2. **13.** Portion of fertile segment of *G. pozuzoensis,* × 8. **14.** Portion of a fertile lamina of *G. trifurcata,* sori removed, below, showing sunken condition in dry tissue, × 6.

In other species they are free as in *Grammitis Connellii* (Wright) A. C. Sm., *G. furcata* Hook. & Grev. (Fig. 23), *G. jungermanioides* (Kl.) Ching, *G. magellanica* Desv., *G. parietina* (Kl.) Fée (Fig. 24), *G. Poeppigiana* (Mett.) Schelpe (*G. Armstrongii* Tindale), and *G. tepuiensis* (A. C. Sm.) Vareschi.

Section *Grammitis*

Lamina entire, the margin dark, sclerotic.

A small group of about 15 species with a few in America, for example, *Grammitis fluminensis* Fée, *G. leptopodon* (Wright) Copel. (Fig. 7), *G. limbata* Fée, and *G. marginella* (Sw.) Sw.

Section *Pleurogramme*

Lamina entire or forked, sporangia borne toward the apex in short to long, usually continuous lines on each side of the costa.

A small American group including *Grammitis graminoides* (Sw.) Sw. (*Cochlidium graminoides* (Sw.) Kaulf.), *G. punctata* Raddi (*G. paucinervata* Fée, *Cochlidium punctatum* (Raddi) Bishop), *G. rostrata* (Hook.) R. & A. Tryon (Fig. 5) (*Cochlidium rostratum* (Hook.) C. Chr.), and *G. seminuda* (Willd.) Willd. (*Cochlidium seminudum* (Willd.) Maxon).

Ecology (Figs. 1–5)

Grammitis is predominantly a genus of wet forests, especially on mountains of the tropics where it grows on mossy trees or less often on mossy rocks along streams.

In America, *Grammitis* usually grows in cloud forests and elfin forests on mossy trunks and branches of trees and shrubs and may persist in cutover cloud forests and on trees in subalpine meadows. Less often *Grammitis* grows in swamp forests and lowland rain forests. Although usually epiphytic, it also grows on trunks and branches of fallen trees, on mossy logs and on decaying wood, or in caves and caverns of rocky places and lava beds, and in rocky soil or on rocks along streams. In the páramo, species such as *Grammitis moniliformis* grow on soil banks and in exposed, rocky ledges and cliffs. *Grammitis tepuiensis* may grow in the relatively dry campina forest of Amazonian Brazil. The leaves of large species of the cloud forest are flaccid and pendent, while small species of open rocky habitats have erect leaves. *Grammitis* grows from sea level to ca. 4500 m, most commonly from 2000 to 3000 m.

Geography (Fig. 6)

Grammitis occurs in America, on the Atlantic Islands of St. Helena, Tristan da Cunha and Gough, in Africa, Madagascar and the Mascarenes, and eastward to southern Japan, New Guinea and eastern Australia and New Zealand; it is widely distributed in the Pacific east to the Hawaiian Islands, Tahiti and Rapa.

In America *Grammitis* occurs from central Mexico and the Greater Antilles, southward to South America and to Tierra del Fuego in Argentina; also on Cocos Island, the Juan Fernandez Islands, Staten Island, the Falkland Islands and South Georgia.

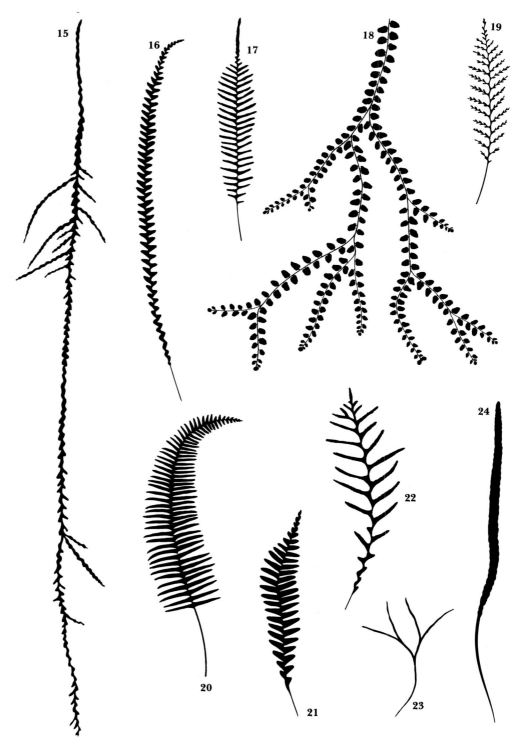

Figs. 117.15–117.24. Lamina architecture of *Grammitis*. **15.** Pendent leaf of *G. pozuzoensis*, ×0.5. **16.** *G. flabelliformis*, ×0.5. **17.** *G. apiculata*, ×0.5. **18.** Portion of pendent leaf of *G. hetero-morpha*, ×0.5. **19.** *G. achilleifolia*, ×0.5. **20.** *G. semihirsuta*, ×0.25. **21.** *G. gradata*, ×0.5. **22.** *G. subsessilis*, ×0.5. **23.** *G. furcata*, ×0.5 **24.** *G. parietina*, ×0.5.

Figs. 117.25–117.34. *Grammitis* spores, × 1000.   **25.** *G. Knowltoniorum*, proximal face, papillate with spherical deposit, Dominica, *Hodge 54*.   **26.** *G. jungermannioides*, prominently papillate distal face, Costa Rica, *Scamman & Holdridge 8077*.   **27.** *G. peruviana*, prominently papillate distal face, Peru, *Ferreyra 16753*.   **28.** *G. Knowltoniorum*, detail of papillate surface with spherical deposit, Dominica, *Hodge 54*, × 5000.   **29.** *G. moniliformis*, proximal face depressed between laesurae, Costa Rica, *Weber 6232*.   **30.** *G. punctata*, irregularly papillate distal surface. Brazil, *Tryon & Tryon 6884*.   **31, 32.** *G. trifurcata*, Costa Rica, *Standley & Valerio 51633*.   **31.** Abraded surface at laesura, perispore raised above exospore, × 10,000   **32.** Papillate spores, proximal face above, distal below.   **33.** *G. delitescens*, broad laesurae, distal face. Costa Rica, *Scamman & Holdridge*, in April 1956.   **34.** *G. asplenifolia*, short laesura, arm at arrow. Mexico, *Dressler 1602*.

The two species of southern Chile and Argentina have a circumantarctic distribution. *Grammitis Poeppigiana* occurs on the Falkland Islands, South Georgia, Tristan da Cunha, Gough, in South Africa, on Marion, Crozet and Kerguelen Islands, in southeastern Australia, Tasmania and New Zealand, and on the Auckland Islands. *Grammitis magellanica* is not as widely distributed but also occurs on Amsterdam Island and Macquarie Island.

The center of diversity of *Grammitis* in America is in the Andes from Colombia to Peru, where there are about 80 species. Other centers are in Costa Rica and Jamaica, with about 50 species each.

Spores

Grammitis spores are more or less papillate with scattered spherical deposit as in *G. Knowltoniorum* (Figs. 25, 28) and the walls are often more or less collapsed when dry (Fig. 29). The abraded spore with raised perispore (Fig. 31) suggests this is thicker than in spores of *Polypodium* or *Loxogramme*. The laesurae are usually broad and often nearly equal to the radius in length (Figs. 25, 32, 33). In *Grammitis asplenifolia* the laesurae length varies (Fig. 34) as in some species of *Loxogramme*. The marked disparity in spore size of *Grammitis jungermannioides*, *G. peruviana* and *G. punctata* (Figs. 26, 27, 30) appears to exceed that often reflecting different ploidy levels. These size differences are also evident in the figures of *Grammitis* spores of Colombia (Murillo and Bless, 1974), southern Argentina (Morbelli, 1980) and some Asiatic species (Chang et al., 1971). Stokey and Atkinson (1958) report *Grammitis* spores regularly bear chlorophyll and germinate within the spore wall. The spheroidal form and papillate surface with spherical deposit of *Grammitis* spores generally resemble the trilete type in *Loxogramme*.

Cytology

Cytological reports for *Grammitis* are predominately $n = 37$ with few lower at $n = 36, 33$, and 32. The record of $n = 36$ in *Grammitis attenuata* Kze. from southern India is clearly documented by photographs of meiotic and mitotic cells (Kuriachan, 1967), but ca. 36 in *G. trichomanoides* from Brazil is uncertain (Araujo, 1976). The lowest numbers in *Grammitis* and in American species

25 26 27

28 29

30 31

32 33 34

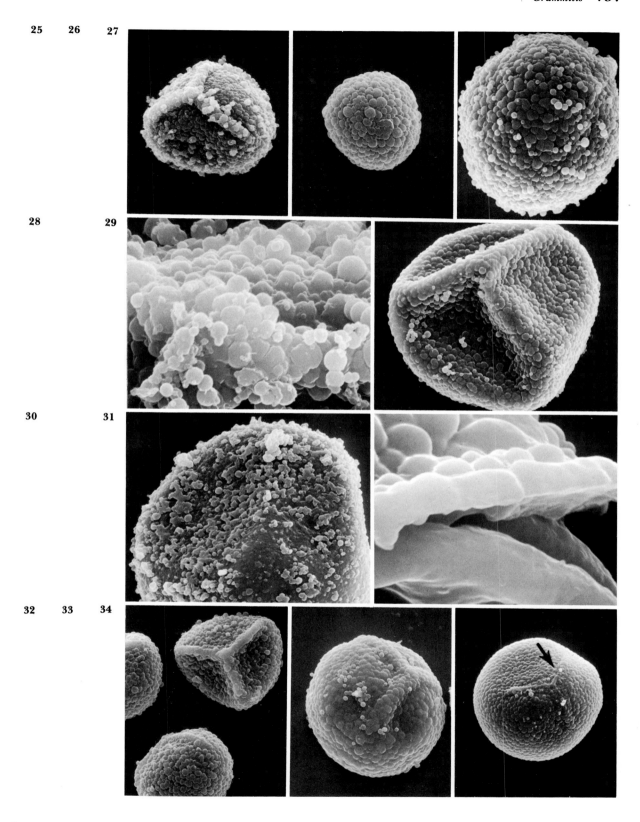

of the Polypodiaceae are $n = 33$ in *G. Hartii* from Jamaica (Walker, 1966) and *G. rostrata* (as *Cochlidium*), and $n = 32$ in *G. limbata* (as *Xiphopteris*), the latter two based on material from Costa Rica (Wagner, 1980). *Grammitis moniliformis* with $n =$ ca. 74 in Jamaica is one of the uncommon tetraploids. There are relatively few reports of chromosome numbers for *Grammitis*, but on the basis of these there appears to be greater cytological diversity among the American species than those of the Old World.

Literature

Araujo, I. da Silva. 1976. Report of chromosome numbers, *in*: IOBP chromosome reports 53: ed. A. Löve. Taxon 25: 483.

Bishop, L. E. 1977. The American species of *Grammitis* section *Grammitis*. Amer. Fern Jour. 67: 101–106.

Bishop, L. E. 1978. Revision of the genus *Cochlidium* (Grammitidaceae). Amer. Fern Jour. 68: 76–94.

Brade, A. C. 1966. Os gêneros *Xiphopteris* e *Grammitis* no Brasil. Sellowia 18: 73–85.

Chang, Y. L., I. C. Hsi, C. T. Chang, K. C. Kao, N. C. Tu, H. C. Sun, and C. C. Kung. 1971. Sporae Pteridophytorum Sinicorum. 451 pp. Academia Sinica, Science Press. Peking.

Christensen, C. 1929. Reference under the family.

Copeland, E. B. 1947. Genera Filicum. 247 pp. Chronica Botanica. Waltham, Mass.

Copeland, E. B. 1952a. *Grammitis*. Phil. Jour. Sci. 80: 93–271.

Copeland, E. B. 1952b. The American species of *Xiphopteris*. Amer. Fern Jour. 42: 41–52, 93–110.

Copeland, E. B. 1956. *Ctenopteris* in America. Phil. Jour. Sci. 84: 381–473.

Kuriachan, P. I. 1967. Cytological observations on some south Indian ferns. Cytologia 32: 500–506.

Maxon, W. R. 1914. Notes upon *Polypodium duale* and its allies. Contrib. U.S. Nat. Herb. 17: 398–406.

Maxon, W. R. 1915. *Polypodium marginellum* and its immediate allies. Bull. Torrey Bot. Cl. 42: 219–225.

Maxon, W. R. 1916. *Polypodium trichomanoides* and its American allies. Contrib. U.S. Nat. Herb. 17: 542–557.

Morbelli, M. A. 1980. Morfología de las esporas de Pteridophyta presentes en la región Fuego-Patagónica, República Argentina. Opera Lilloana 28: 1–138.

Morton, C. V. 1967. The genus *Grammitis* in Ecuador. Contrib. U.S. Nat. Herb. 38: 85–123.

Murillo, M. T., and M. J. M. Bless. 1974. Spores of recent Colombian pteridophytes, 1. Trilete spores. Rev. Palaeobot. Palynol. 18: 223–269.

Parris, B. S. 1975. A revision of the genus *Grammitis* Sw. (Filicales: Grammitidaceae) in Australia. Bot. Jour. Linn. Soc. 70: 21–43.

Parris, B. S., and D. R. Given. 1976. A taxonomic revision of the genus *Grammitis* Sw. (Grammitidaceae: Filicales) in New Zealand. New Zealand Jour. Bot. 14: 85–111.

Pichi-Sermolli, R. E. G. 1977. Reference under the family.

Rodriguez, R. 1974. Revisión del género *Grammitis* (Filices) en Chile. Bol. Soc. Biol. Concepción 47: 159–170.

Sota, de la, E. R. 1960. Polypodiaceae y Grammitidaceae Argentinas. Opera Lilloana 5: 1–229.

Stokey, A. G., and L. R. Atkinson. 1958. The gametophyte of the Grammitidaceae. Phytomorph. 8: 391–403.

Wagner, F. S. 1980. Reference under the family.

Walker, T. G. 1966. Reference under the family.

Family 23.   Marsileaceae

Marsileaceae Mirbel, Hist. Nat. Veg. (Lam. & Mirb.) 5: 126. 1802. Type: *Marsilea* L.

Pilulariaceae Wettst., Handb. Syst. Bot. 2: 81. 1903. Type: *Pilularia* L.

Description

Stem short- to long-creeping, slender, often branched, solenostelic, hardly indurated, bearing trichomes; leaves ca. 1–40 cm long, with 4, 2 or no leaflets at the apex of the petiole, circinate in the bud, petiole without stipules; sori borne within sporocarps, indusiate, with indehiscent, short- to long-stalked, not annulate, megasporangia and microsporangia; heterosporous, spores without chlorophyll. Gametophytes minute, the megagametophyte at the apex and partly protruding from the ruptured megaspore wall, the microgametophyte within the microspore, which bursts to release the sperm.

Comments on the Family

The Marsileaceae are a family of three genera, all in tropical America. The genera are clearly distinct and not very closely allied.

There are numerous detailed studies on the Marsileaceae centering on the form and development of the elaborate reproductive structures and simplified leaves. Physiological and experimental work on factors influencing leaf form have also utilized these plants. The selection of papers cited under the genera will provide access to these studies. However, some of the basic problems on the nature of the leaflets and the sporocarp are not resolved, and the phyletic relationships of the three genera are not wholly certain.

The Marsileaceae are usually aligned near the Schizaeaceae. Many of the characters that have been used to support relationships of the two families, as the solenostele, dichotomous venation, stem trichomes, the basal position of the sporangia on the leaf, and vestigial apical annulus of the microsporangium of *Pilularia*, are characteristics of primitive leptosporangiate ferns rather than those especially of the Schizaeaceae. The family is exceptional in its highly specialized reproductive structures along with relatively simple vegetative structures. The phyletic assessment of these is difficult, for adaptation to the aquatic habitat undoubtedly has influenced the morphology of the plants.

Fossil *Marsilea* has been reported from the Cretaceous, and later periods.

Key to Genera of Marsileaceae

a. Leaflets at the apex of the petiole.
    b. Four leaflets.   118. *Marsilea*
    b. Two leaflets.   119. *Regnellidium*
a. Petiole without leaflets.   120. *Pilularia*

Literature

Bierhorst, D. W. 1971. Morphology of Vascular Plants. 560 pp. Macmillan, New York.

Lugardon, B. 1971. Contribution a la connaissance de la morphogénese et de la structure des parois sporales chez les Filicinées isosporées. 257 pp. Thèse, Univ. Paul Sabatier, Toulouse.

118. *Marsilea*
Figs. 118.1–118.9

Marsilea L., Sp. Pl. 1099. 1753; Gen. Pl. ed. 5, 485. 1754. Type: *Marsilea quadrifolia* L.

Spheroidea Dulac, Pl. Dept. Haut-Pyrénées. 39. 1867, *nom. superfl.* for *Marsilea* L. and with the same type.

Palustral or aquatic; stem usually long-creeping, frequently branched, slender, bearing trichomes and, at the often distant, internodes, usually long roots; leaves ca. 1–40 cm long, the petiole terminated by two adjacent pairs of narrowly cuneate to broadly flabellate leaflets, glabrous or pubescent, veins more or less anastomosing, usually connected at the margin; sori borne within one to several stalked, indurated sporocarps attached to the petiole or at its base, enclosed by a diaphanous indusium, with microsporangia and megasporangia; megaspores somewhat ellipsoidal with an apical papilla-like laesura, the surface papillate, microspores spheroidal, trilete, the surface slightly rugulose. Chromosome number: $n = 20$; $2n = 40, 60$.

Marsilea is a nearly worldwide genus of perhaps 50 species, with about 12 in America. The tropical American species are poorly known for there are relatively few collections and these often lack sporocarps. There is need for a taxonomic revision based on more adequate materials, on field observations and experiments to determine the variation of characters under different environmental conditions. Species are based on characters such as the indument and stomata, type of margin and shape of the leaflets, and characters of the sporocarps such as the number on a petiole, the position of the stalk, the number of sori in a sporocarp and the number of megasporangia and of microsporangia in a sorus.

O. Kuntze (Rev. Gen. Pl. 2: 823. 1891) adopts the name *Zalusianskya* Necker (Acta Theod. Palat. Phys. 3: 303. 1775) for *Marsilea* L. The status of Necker's name has not been determined, but Kuntze's use of it is superfluous.

Gupta (1957) studied the epidermal and sporocarp characters of several American species, and (1962) reviewed the systematics and morphology of the genus, with special reference to the species of India.

Tropical American Species

Some of the species of Mexico are *Marsilea Fournieri* C. Chr., *M. macropoda* A. Br. (Fig. 3), *M. mexicana* A. Br., *M. mollis* Robins. & Fern., and *M. vestita* Hook. & Grev.; *Marsilea Nashii* Underw. is in the Bahamas, and *M. ancylopoda* A. Br. and *M. punae* Sota in South America. *Marsilea polycarpa* Hook. & Grev. (Fig. 4) is the most widely distributed species, in Central America, the West Indies and northern South America.

Fig. 118.1. *Marsilea mexicana*, growing with *Azolla*, Guatemala. (Photo W. H. Hodge.)

Ecology (Fig. 1)

Marsilea is primarily a genus of seasonally wet habitats, growing in shallow water and at the edge of ponds, lakes or rivers; also in ditches, rice paddies, marshes and river flats.

In tropical America it grows in ephemeral pools, along irrigation canals and ditches, on creek banks, seasonally flooded river bottoms and creeks, lake edges, wet grasslands, sometimes in partly saline habitats. It is found mostly at low elevations, below 500 m but rarely up to 3600 m.

Plants may be submerged except for the floating leaflets, or partly emergent, or in wet mud. Species as *Marsilea Fournieri* and *M. vestita* that commonly grow in seasonally dry habitats, produce sporocarps as the soil dries. Other species as *Marsilea polycarpa* grow in permanent water, and produce sporocarps mostly on emergent leaves. Spores contained within the sporocarp remain viable for nearly 70 years (Allsopp, 1952); thus species are adapted to survive in arid regions that receive infrequent rain.

Geography (Fig. 2)

Marsilea is distributed from America, Africa and Europe and adjacent Russia to the Himalayas, China, and to Japan, also to Malaya, the Philippine Islands, New Guinea, Australia and New Caledonia; in the Pacific it occurs on the Fiji, Society, and Hawaiian Islands. *Marsilea polycarpa* is probably introduced in the Society Islands from tropical America.

In America, *Marsilea* grows from British Columbia, Alberta and Saskatchewan in Canada, southward and eastward to Florida, in the United States, and from northern Mexico, the Greater Antilles and the Bahamas, southward to Buenos Aires

Fig. 118.2. Distribution of *Marsilea* in America, south of lat. 35° N.

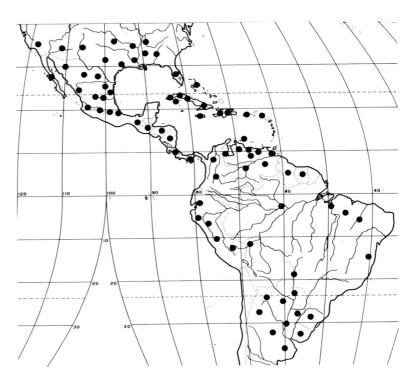

in Argentina. Although the distribution is extensive, there are relatively few collections from the tropics.

It is long known that waterfowl feed on *Marsilea* sporocarps but Malone and Proctor (1965) established that undamaged sporocarps will pass through the digestive tract of several kinds of waterfowl. The work confirmed the supposition that the species could be dispersed by birds.

Spores

The single megaspore that develops within the megasporangium is somewhat longer than broad, ellipsoidal or ovate with the apical, papilla-like laesura sometimes depressed (Fig. 5). The microspores are trilete and the proximal face smoother than the distal one (Figs. 6, 7). The shallow papillate surface of both megaspore (Fig. 8) and microspore (Fig. 9) does not reveal the complex stratification shown in TEM studies of the spore wall. The studies of Pettitt (1971, 1979) on the megaspore wall in *Marsilea Drummondii* A. Br. and *M. quadrifolia* L. show the elaborate perispore consists of a lower reticulate layer and an outer prismatic formation of radially aligned chambers. Developmental studies of the single perispore layer in the microspore show this layer is formed by deposition of tapetal material. Similarities of *Marsilea* and *Pilularia* microspores and resemblance of the exospore structure of these to the stranded formation in *Blechnum* spores is noted by Lugardon (1971).

Cytology

Reports of $n = 20$ for *Marsilea* are mainly based on Indian species (Mehra and Loyal, 1959). *Marsilea minuta* L. consists of a

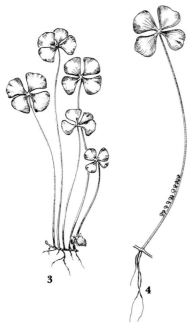

Figs. 118.3, 118.4. *Marsilea,* portion of plants, ×0.5 **3.** *M. macropoda.* **4.** *M. polycarpa.*

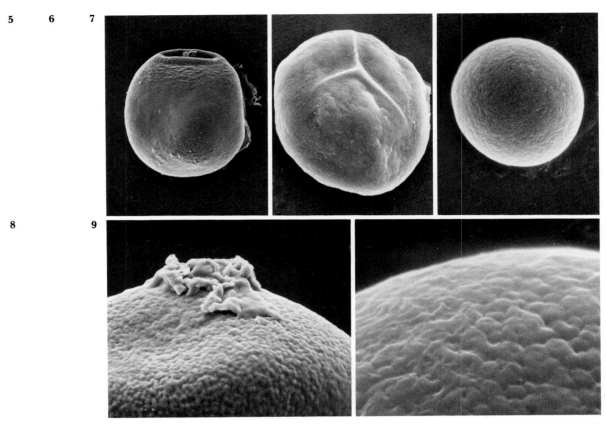

Figs. 118.5–118.9. *Marsilea macropoda* spores, Mexico, *Pringle 1975.* **5.** Megaspore, partly coated with gelatinous sporocarp material, papilla-like laesura depressed, at top, × 80. **6.** Microspore, trilete, proximal face, × 800. **7.** Microspore, slightly rugulose distal face, × 800. **8.** Detail of megaspore apex, × 3000. **9.** Detail, distal face of rugulose microscope, × 5000.

complex of reproductively polymorphic strains including a diploid $n = 20$ and a sterile autotriploid, $2n = 60$, with both cytotypes widespread in India. Reproductive changes in the species through decline or elimination of sexual systems and development of extensive vegetative systems have added to the complexity. Chromosome number of American species are not reported, and several early counts, prior to use of the squash technique, are not reliable, although indicative of high ploidy levels in some species.

Observations

Studies on *Marsilea* have concentrated on morphology, anatomy, reproductive structures, and growth and development. A series of papers, "Etudes sur les Hydropteridales," from the laboratory of M. Martens, Institut Carnoy, Louvain (Boterberg, 1965; Demalsy-Feller, 1956; Tournay, 1951; Feller, 1953) have thoroughly investigated the anatomy and histology of the reproductive and vegetative structures. Several detailed studies on the morphological nature of the leaf and sporocarp of *Marsilea* have resulted in different concepts of these structures. Puri and Garg (1935) interpret the sporocarp in the usual way as a highly modified, fertile pinna. However, they suggest that the four leaflets are a single pinna rather than, as more commonly interpreted, two pairs of pinnae. Since the nature of the leaves is not clear, the term leaflet is applied here to circumvent the misapplication of the term pinna.

Although the sporocarps of *Marsilea* are often interpreted as highly modified pinnae, Bierhorst (1971) points out that the sporocarp initial has three cutting faces, unlike fern pinnae, but is similar to the sporangial stalk initial of the Filicales (suborder Polypodineae).

Experimental hybridization attempted by Buchholz and Selett (1941), between species of *Marsilea* yielded short-lived sporophytes, but crosses with *Pilularia* were unsuccessful.

The work of Allsopp (1963) examined the influence of nutritional levels on the heteroblastic sequence of leaves of juvenile plants, and the variation in characters between terrestrial and aquatic forms of a species. White (1961) demonstrated the presence of vessels in the roots of *Marsilea* and (1968) determined that the size and complexity of a leaf was directly correlated with the size of the apical cell of the stem.

Literature

Allsopp, A. 1952. Longevity of *Marsilea* sporocarps. Nature 169: 79–80.

Allsopp, A. 1955. Investigations on *Marsilea* 5. Cultural conditions and morphogenesis, with special reference to the origin of land and water forms. Ann. Bot. n.s. 19: 247–264.

Bierhorst, D. W. 1971. Reference under the family.

Boterberg, A. 1956. Genèse et différenciation des parois sporales chez *Marsilea diffusa* Lepr. La Cellule 58: 81–106.

Buchholz, J. T., and J. W. Selett. 1941. The hybridization of water ferns - *Marsilea* and *Pilularia*. Amer. Nat. 75:90–93.

Demalsy-Feller, M.-J. 1957. Gamétophytes et gamétogenèse dans le genre *Marsilea*. Le Cellule 58: 171–207.

Feller, M. J. 1953. Sporocarpe et sporogenèse chez *Marsilea hirsuta* R. Br. La Cellule 55: 307–377.

Gupta, K. M. 1957. Some American species of *Marsilea* with special reference to their epidermal and soral characters. Madroño 14: 113–127.

Gupta, K. M. 1962. *Marsilea*. Bot. Monogr. 2, pp. 1–113. Council Sci. & Industr. Research. New Delhi.

Lugardon, B. 1971. Reference under the family.

Malone, C. R., and V. W. Proctor. 1965. Dispersal of *Marsilea mucronata* by water birds. Amer. Fern Jour. 55: 167–170.

Mehra, P. N., and D. S. Loyal. 1959. Cytological studies in *Marsilea* with particular reference to *Marsilea minuta* L. Res. Bull. Punjab Univ. 10: 357–374.

Pettitt, J. M. 1971. Some ultrastructural aspects of sporoderm formation in pteridophytes, *in:* G. Erdtman and P. Sorsa, Pollen and Spore Morphology. IV, pp. 227–251 Almqvist & Wiksell, Stockholm.

Pettitt, J. M. 1979. Ultrastructure and cytochemistry of spore wall morphogenesis. *in:* A. F. Dyer, The Experimental Biology of Ferns. pp. 213–252. Academic Press, London.

Puri, V., and Garg, M. L. 1953. A contribution to the anatomy of the sporocarp of *Marsilea minuta* L. with a discussion of the nature of the sporocarp in the Marsileaceae. Phytomorph. 3: 190–209.

Tournay, R. 1951. Le sporophyte de *Marsilea,* recherches sur *M. hirsuta* R. Br. La Cellule 54: 164–218.

White, R. A. 1961. Vessels in roots of *Marsilea*. Science 133: 1073–1074.

White, R. A. 1968. A correlation between the apical cell and the heteroblastic leaf series in *Marsilea*. Amer. Jour. Bot. 55: 485–493.

119. *Regnellidium*
Figs. 119.1–119.7

Regnellidium Lindm., Ark. för Bot. 3 (6): 2. 1904. Type: *Regnellidum diphyllum* Lindm.

Description

Aquatic or palustral; stem usually long-creeping, slender, bearing trichomes and long roots at and between internodes; leaves ca. 5–20 cm long, borne at short to distant intervals, with a pair of broadly cuneate to subreniform leaflets at the apex of the petiole, pubescent to glabrate, veins free or fused at the margin; sori born within a single, stalked, indurated sporocarp attached near the petiole base, enclosed by a diaphanous indusium, with megasporangia and microsporangia; megaspores spheroidal with an apical papilla-like laesura, the microspores spheroidal, with a papilla-like laesura, the surface with prominent, coarse folds. Chromosome number: $n = 19, 20$; $2n = 38, 40$.

Systematics

Regnellidium is a monotypic genus based on *R. diphyllum* Lindm. (Fig. 1) of southern Brazil and adjacent Argentina.

Latex-bearing ducts were described in the petiole and leaflets by Laboureau (1952). These and a usually lower chromosome number and elaborate perispore formation of the spores indicate more derived features than in *Marsilea* and *Pilularia*.

Ecology and Geography

Regnellidium grows among aquatic vegetation along lake shores and in stagnant ponds, in Rio Grande do Sul and southern Santa Catarina, Brazil, and in Corrientes, Argentina (Fig. 2).

Spores

The spheroidal megasporangium (Fig. 3) closely envelopes the megaspore which is partly exposed in Fig. 6. The papillate megaspore surface (Fig. 4) is more complex than in megaspores of *Marsilea* and *Pilularia*. Sections of the wall shown in the work of Chrysler and Johnson (1939) have the papillate formation overlaying the prismatic layer in *Regnellidium* megaspores. The strongly folded surface and papillate laesura of the microspore (Figs. 5, 7) are also more elaborate than in the microspores of *Marsilea* or *Pilularia*.

Cytology

The chromosome number of *Regnellidium diphyllum* is reported as $2n = 38$ (Jain and Raghuvanshi, 1973) based on cultivated plants, but they indicate that there were 40 chromosomes in 20% of the cells analyzed. Their account of B or accessory chromosomes in these plants is the first record of these for the pteridophytes. The occurrence of the accessories may relate to fragmentation of, and account for, fewer large chromosomes. Deviation in chromosome numbers in *Regnellidium* may be indicative of a mechanism for aneuploid change through loss of chromosomes.

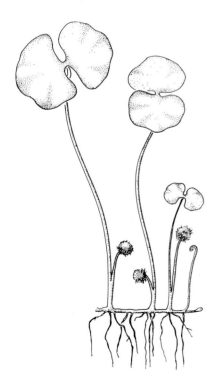

Fig. 119.1. Portion of a plant of ***Regnellidium,*** × 0.5.

Fig. 119.2 Distribution of *Regnellidium.*

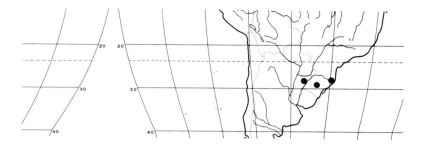

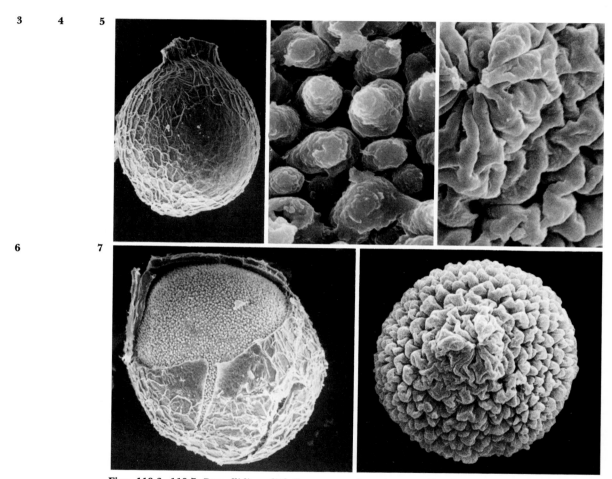

Figs. 119.3–119.7. *Regnellidium diphyllum* sporangia and spores, Brazil, *Rau in 1937.* **3.** Spheroidal megasporangium, ×70. **4.** Detail of papillate megaspore surface, ×2000. **5.** Detail of folded microspore surface, the papilla-like laesura at upper left, ×2800. **6.** Megaspore within abraded sporangium, ×100. **7.** Microspore, compact surface folds and 3-lobed laesura, ×1000.

Observations

The morphology, anatomy and reproductive structures of *Regnellidium* are included in the original description of Lindman. Several comprehensive studies have been made on the general morphology and anatomy (Johnson and Chrysler, 1938), sporogenesis (Chrysler and Johnson, 1939) and on the gametophytes and embryology (Higinbotham, 1941).

Literature

Chrysler, M. A., and D. S. Johnson. 1939. Spore production in *Regnellidium*. Bull. Torrey Bot. Cl. 66: 263–279.

Higinbotham, N. 1941. Development of the gametophytes and embryo of *Regnellidium diphyllum*. Amer. Jour. Bot. 28: 282–300.

Jain, S. J., and S. A. Raghuvanshi. 1973. Cytogenetic study of *Regnellidium diphyllum* with special reference to supernumeraries. Caryologia 26: 469–481.

Johnson, D. S., and M. A. Chrysler. 1938. Structure and development of *Regnellidium diphyllum*. Amer. Jour. Bot. 25: 141–156.

120.   *Pilularia*

Figs. 120.1–120.8

Pilularia L., Sp. Pl. 1100. 1753; Gen. Pl. ed. 5, 486. 1754. Type: *Pilularia globulifera* L.

Calamistrum O. Ktze., Rev. Gen. Pl. 2: 822. 1891, *nom. superfl.* for *Pilularia* L. and with the same type.

Description

Aquatic or palustral; stem usually long-creeping, slender, bearing trichomes and, at the usually long internodes, many short roots; leaves ca. 1–10 cm long, filiform, pubescent or glabrate, with a single vascular strand; sori born within the single, short-stalked, indurated sporocarp attached at the petiole base, enclosed by a diaphanous indusium, with megasporangia and microsporangia; megaspores somewhat ovate with an apical papilla-like laesura, the surface compact rugulose, microspores spheroidal, trilete, the laesurae short, about $\frac{1}{8}$ to $\frac{1}{4}$ the spore radius, the surface with compact folds. Chromosome number: $n = 10$; $2n = 20$.

Systematics

Pilularia is a widely distributed genus of about five species. It is a very distinctive genus but not readily distinguished in the field from small grasses and sedges, except by the sporocarps or circinate leaves.

Ecology

Pilularia grows in low depressions that are seasonally dry, such as shallow rock pools, clay depressions, ditch margins, small ponds and also in marshes and on lake margins where the water level fluctuates. In the United States it has recently spread into artificial lakes especially those in which the water level is periodically lowered to control the growth of rooted aquatics. *Pilularia* may grow submerged, but evidently produces sporocarps only on emergent plants. The development of colonies in recently formed lakes in Texas and Tennessee (Dennis and Webb, 1981) suggest that *Pilularia* is disseminated by waterfowl.

Fig. 120.1. Portion of a plant of *Pilularia americana* with sporocarps, × 2.

Geography (Fig. 2)

Pilularia is a genus of perhaps five species of wide but disjunct distribution. *Pilularia americana* A. Br. (Fig. 1) has a scattered distribution in the United States and occurs in Chile; also in Venezuela, Colombia, Bolivia, southern Brazil and Argentina where it is usually named *P. Mandonii* A. Br. *Pilularia globulifera* L. is in Europe and the Ural region, and *P. minuta* A. Br. is Mediterranean. The other species are *Pilularia nova-zealandiae* Kirk of New Zealand and *P. nova-hollandiae* A. Br. of Australia and Tasmania.

Spores

Both mega- and microsporangia are produced within the densely pubescent sporocarp (Fig. 3). A single megaspore develops within and is closely enveloped by the megasporangium, (Fig. 4). The megaspore, free from the sporangium, has a rugulose surface and prominent apical papilla-like laesura (Fig. 5). The elongate microsporangium (Fig. 6) encloses many spheroidal, trilete microspores. The somewhat rugose surface of these is formed by the compact, folded perispore which overlays a smoother exospore (Figs. 7, 8). Sections of *Pilularia globulifera* microspores, examined by Lugardon (1971), show the prominent perispore folds and a complex lower exospore consisting of a compact central area between outer and inner reticulate zones. The structure of these is generally similar to *Regnellidium* but there are differences in shape and especially in the type of surface.

Fig. 120.2. Distribution of *Pilularia* in America.

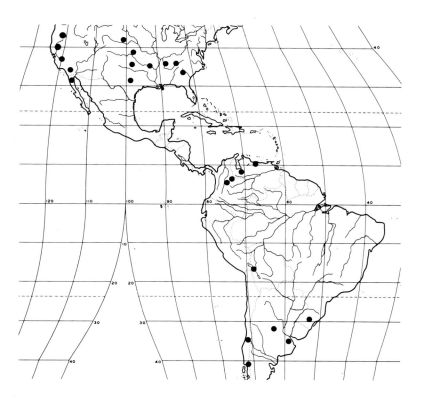

3 4 5

6 7 8

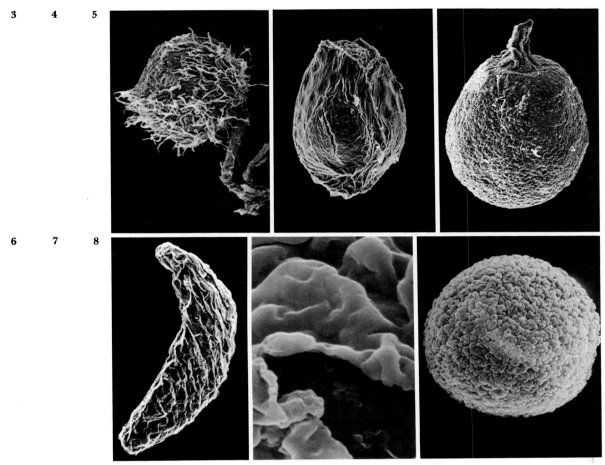

Figs. 120.3–120.8 *Pilularia americana*, California, *Pollard* in 1955. **3.** Sporocarp, ×12. **4.** Megasporangium, megaspore partly exposed at top, ×80. **5.** Megaspore, rugulose, papilla-like laesura at top, ×90. **6.** Microsporangium, ×120. **7.** Microspore, detail of folded perispore surface, exospore below, at lower right, ×10,000. **8.** Spheroidal microspore, compact surface folds, ×1000.

Cytology

The chromosome number $2n = 20$, for *Pilularia americana* (as "*P. caroliniana* A. Br.," an invalid name) from Arkansas, is reported by Chopra (1960). The mitotic chromosomes from roottips were of uniform size and meiotic cells had 10 condensed bivalents. The numbers correlate with $2n = 40$ in *Marsilea* and *Regnellidium* and support the base number 10 proposed for the family. Some divergent numbers reported, prior to the use of the squash technique, need to be verified.

Observations

Experimental studies in *Pilularia americana* (McMillan et al., 1968) show that the occurrence of the species in Texas may be restricted to the growing conditions characteristic of semipermanent pools on granitic substrates. Plants tolerated various levels of submergence, but greatest sporophyte development occurred under conditions of slow drying. Plants that were completely desiccated died, unlike the frequent associate *Isoetes litho-phila* Pfeiff. in which leaf regeneration occurred following desiccation.

The development and structure of the leaf, sporocarp and sporangia of *Pilularia* were studied in detail by Johnson (1898).

Literature

Chopra, N. 1960. Chromosome number of *Pilularia caroliniana* A. Br. Castanea 25: 116–119.

Dennis, W. M., and D. H. Webb. 1981. The distribution of *Pilularia americana* A. Br. (Marsileaceae) in North America, north of Mexico. Sida 9: 19–24.

Johnson, D. S. 1898. On the leaf and sporocarp of *Pilularia*. Bot. Gaz. 26: 1–24.

Lugardon, B. 1971. Reference under the family.

McMillan, C., and twelve others. 1968. Factors influencing the narrow restriction of *Pilularia americana* in Texas. Southwest. Nat. 13: 117–127.

Family 24. Salviniaceae

Salviniaceae Reichenb., Bot. Damen Kunst. Freunde Pflanzenw. 255. 1828. Type: *Salvinia* Séguier.

Azollaceae Wettst., Handb. Syst. Bot. 2: 77. 1903. Type: *Azolla* Lam.

Description

Stem creeping, small and slender, often branched, nearly pro-tostelic or solenostelic, not indurated, usually bearing trichomes; floating leaves ca. 0.5–2.5 cm long, not circinate in the bud, pet-iolate leaves without stipules; sori borne on the lobe of a leaf or on a branched leaf, indusiate, with the sporangia stalked, not an-nulate, the sori either megasporangiate or microsporangiate; heterosporous, spores without chlorophyll. Megagametophyte and microgametophyte minute, partly protruding from the spore wall.

The structure bearing the sporangia is considered to be a sorus, surrounded by an indusium, and accordingly the special term sporocarp is not used.

Comments on the Family

The living Salviniaceae are represented by two genera, *Salvinia* and *Azolla*, floating aquatics of wide distribution in the tropics. Although *Azolla* is sometimes included in a separate family, it has many basic similarities with *Salvinia* and the fossil record indicates that they are both elements of an old group adapted to aquatic life (Hall, 1974).

The relationships of the Salviniaceae to other ferns is unknown. It has been allied with the Hymenophyllaceae, in a general way, based on its simple vegetative features, hydrophilic nature, and indusiate sori, but none of these characters provide a direct connection to that family.

Salvinia and *Azolla* species are both known from the Upper Cretaceous. The fossil record of the family is an unusually rich one with many species and several genera (Hall, 1968; Hall, 1974; Hall and Swanson, 1968; Hills and Topal, 1967).

Key to Genera of *Salviniaceae*

a. Leaves in whorls of three, two floating, ca. 5–25 mm long, with anastomosing veins and not lobed, one submerged, highly branched.   121. *Salvinia*
a. Leaves alternate, ca. 0.5–1.5 mm long, without veins, bilobed.   122. *Azolla*

Literature

Hall, J. W. 1968. A new genus of the Salviniaceae and a new species of *Azolla* from the late Cretaceous. Amer. Fern Jour. 58: 77–88.
Hall, J. W. 1974. Cretaceous Salviniaceae. Ann. Mo. Bot. Gard. 61: 354–367.
Hall, J. W., and N. P. Swanson. 1968. Studies on fossil *Azolla: Azolla montana*, a Cretaceous megaspore with many small floats. Amer. Jour. Bot. 55: 1055–1061.
Hills, L. V., and B. Gopal. 1967. *Azolla primaeva* and its phylogenetic significance. Canad. Jour. Bot. 45: 1179–1191.

121.   *Salvinia*
Figs. 121.1–121.12

Salvinia Séguier, Fl. Veron. 3: 52. 1754. Type: *Salvinia natans* (L.) All. (*Marsilea natans* L.).

Description

Floating aquatics; stem short, creeping, slender, bearing trichomes and no roots; leaves borne in whorls of three, dimorphic (two green, floating, entire, oblong to suborbicular, ca. 0.5–2.5 cm long, often pubescent, usually papillate on the upper surface, veins anastomosing, one submerged, pendent in the water, highly branched, bearing many trichomes); sporangia borne in stalked sori on the submerged leaf, enclosed by the indusium, with either megasporangia or microsporangia; spores trilete, enclosed within a vacuolate tapetal formation, the megaspore surface relatively plane, perforate, the microspore surface somewhat rugulose. Chromosome number: $n = 9$. $2n = 18$, 54, chromosomes 45, 63.

The large floating leaves of *Salvinia auriculata* are shown in Fig. 3 and also the strongly dissected submerged leaves, which become matted out of water.

Systematics

Salvinia is a widely distributed genus of about 10 species, with seven of them in America.

Shaparenko (1956) recognized four sections, and these were adopted by de la Sota (1962); they are not recognized here since the genus is a small one and especially since the differentiating characters are not fundamental, as in the infrageneric classification of *Azolla*.

Fig. 121.1. *Salvinia auriculata,* San José, Costa Rica. (Photo W. H. Hodge.)

Fig. 121.2. Native distribution of *Salvinia* in America.

De la Sota has provided detailed studies of most of the American species (1962, 1963, 1964) and Mitchell and Thomas (1972) supply additional information, illustrations and a key. A modern anatomical study of *Salvinia* was done by Bonnet (1955).

Tropical American Species

The studies of de la Sota and Mitchell and Thomas have clarified the species of American *Salvinia.* However, their work was primarily in southern Brazil and adjacent Argentina, and equiv-

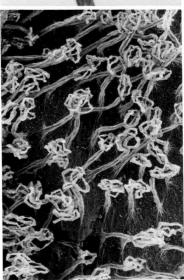

Figs. 121.3, 121.4. *Salvinia auriculata.* **3.** Plant with broad floating leaves and below, the matted dissected submerged leaves, × 1.0. **4.** Papillae on upper surface of floating leaves with apical trichomes joined at the tip to form a cagelike structure, × 10.

alent study of the species in other regions needs to be done, especially to assess the species in the *Salvinia auriculata* complex.

The following key is adapted from Mitchell and Thomas (1972). It does not account for the sterile, open water form of *Salvinia Herzogii* described by those authors, which will key to *S. molesta*.

Key to American Species of *Salvinia*

a. Papillae confined to the margin of the floating leaves.   *S. Sprucei* Kuhn
a. Numerous papillae on the upper surface of the floating leaves.   b.
    b. Floating leaves oblong, ca. three times as long as broad.
          S. oblongifolia Mart.
    b. Floating leaves ovate to suborbicular.   c.
      c. Trichomes at the apex of the papillae free, spreading.
          S. minima Baker (*S. rotundifolia* auths., not Willd.)
      c. Trichomes at the end of the papillae fused at their apex (Fig. 4) (*S. auriculata* complex).   d.
        d. Sporangia mostly empty or with abortive spores.
            S. molesta Mitchell
        d. Sporangia with normal spores.   e.
          e. Submerged leaves sessile, sori on spreading branches.
            S. auriculata Aubl. (*S. rotundifolia* Willd.)
         e. Submerged leaves stalked, sori on an elongate branch system.   f.
          f. Sori mostly subsessile.   *S. Herzogii* Sota
          f. Sori with long stalks.   *S. biloba* Raddi

Ecology (Fig. 1)

Salvinia grows in lakes, rivers, ditches and ponds; also in marshes and in rice fields; very rarely it is terrestrial in continuously wet sites. It grows in stagnant water and sometimes in slightly brackish water, often among aquatic vegetation, but sometimes forming large floating mats.

Some species of *Salvinia* grow very vigorously under suitable conditions and become a serious pest, impeding navigation in waterways and reducing waterflow in drainage ditches. The aggressive growth of *Salvinia molesta* in the artificial Lake Kariba in Zambia and Rhodesia has received considerable attention (Mitchell, 1969).

In America *Salvinia* grows at low elevations, usually below 100 m, rarely up to 1200 m.

Geography (Fig. 2)

Salvinia occurs naturally in America, Africa and Madagascar, and Europe, eastward to India and China, southward to Sumatra and southwestern Australia and northward to Manchuria and Japan.

In America *Salvinia* grows naturally in Florida, in the United States, in Cuba and from Mexico southward in South America to Buenos Aires in Argentina.

The natural range of *Salvinia* is difficult to determine as it is frequently grown in aquaria, in greenhouses and in tropical botanical gardens, and may be readily introduced from these or other sources. The wide distribution of the sterile *Salvinia molesta* in Africa and Ceylon is evidence of how the native range of the genus may be altered.

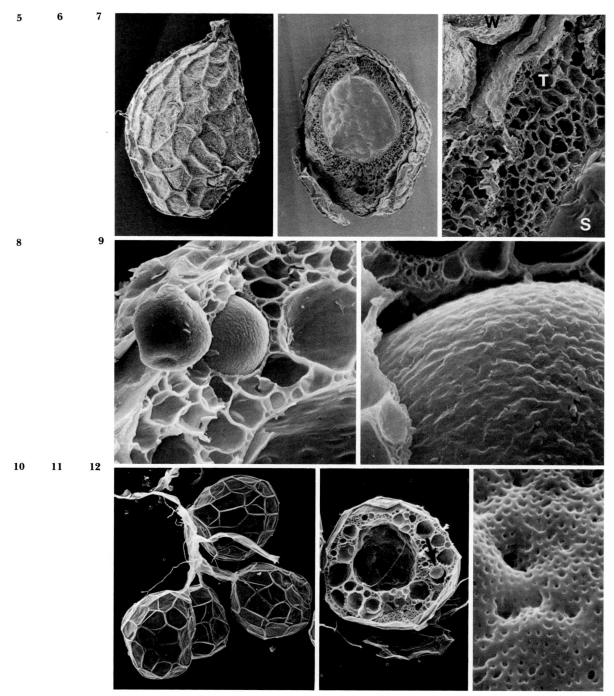

Figs. 121.5–121.12. *Salvinia* spores and associated structures. **5–7.** *S. natans*, Russia, *Margittai* in 1931. **5.** Megasporangium, ×80. **6.** Section of megasporangium and associated structures; sporangium wall, vacuolate tapetal formation surrounding the central megaspore. **7.** Detail of megasporangium section: wall (W), tapetal formation (T), megaspore (S), ×500. **8–11.** *S. auriculata*, Brazil, *Gardner 1217*. **8.** Microspores within vacuolate tapetal formation, ×1000. **9.** Rugulose microspore wall and portions of surrounding tapetum, ×5000. **10.** Microsporangia, ×100. **11.** Section of microsporangium, a spore (arrow) within large cavity of vacuolar tapetal structure, ×120. **12.** Detail of perforate wall of megaspore, ×5000.

Salvinia is considered to be introduced in Bermuda, also Georgia and sporadically in other parts of the eastern United States.

Spores

The megasporangia are somewhat pear-shaped in *Salvinia natans* (L.) All. (Fig. 5). As the single megaspore develops, a thick vacuolate formation, originating from the tapetum, envelops the spore (Figs. 6, 7). The outer layer of the megaspore is relatively smooth with a perforate surface (Figs. 7, 12). The microspores also develop within a vacuolate matrix (Fig. 11), surrounded by the microsporangium (Fig. 10). The microspores are formed within large vacuoles and have a somewhat rugulose surface (Figs. 8, 9). The vacuolar material is sometimes termed perispore in accounts of sporogenesis of *Salvinia*. This term, however, does not seem appropriate because the material seems to be a special feature of the Salviniaceae, and is not equivalent to perispore as recognized in other ferns.

Cytology

The chromosome numbers reported for *Salvinia* represent a polyploid series, the diploid, $2n = 18$ in *S. natans* is known from Japan, Italy and Kashmir. In the American tropics, *Salvinia auriculata* is reported as hexaploid, $2n = 54$, from Trinidad, with 32 microsporocytes and normal spores. *Salvinia Herzogii,* from Brazil, is a heptaploid, with 63 irregular pairing chromosomes and abnormal spores. On the basis of this cytological evidence Schneller (1980, 1981), considered *Salvinia Herzogii* to be a hybrid, and probably not one of the parents involved in *S. molesta,* the aggressive weed in Africa. A base number of nine was determined by karyological analyses of *Salvinia natans* and of an allopentaploid, $2n = 45$, *S. cucullata* Bory (Tatuno and Takei, 1969, 1970). They also used differences in chromosome morphology in *Salvinia natans* for analyses of reciprocal translocations in two populations in Japan, and correlated the origin of new chromosomal types with the decline of microspore fertility. The genus is of cytological interest for its large, morphologically diverse chromosomes, and low chromosome number.

Literature

Bonnet, A-L.-M. 1955. Contribution a l'étude des Hydroptéridées: Reserches sur *Salvinia auriculata* Aubl. Ann. Sci. Nat. Bot. XI, 16: 529–600.

Mitchell, D. S. 1969. The ecology of vascular hydrophytes on Lake Kariba. Hydrobiologia 34: 448–464.

Mitchell, D. S., and P. A. Thomas. 1972. Ecology of waterweeds in the neotropics, an ecological survey of the aquatic weeds *Eichhornia crassipedes* and *Salvinia* species and their natural enemies in the neotropics. Technical papers in Hydrology 12: 1–50. UNESCO, Paris.

Schneller, J. J. 1980. Cytotaxonomic investigations of *Salvinia Herzogii* de la Sota, Aquatic Bot. 9: 279–283.

Schneller, J. J. 1981. Chromosomes numbers and spores of *Salvinia auriculata* Aublet *s. str.* Aquatic Bot. 10: 81–84.

Shaparenko, K. K. 1956. History of the Salvinias. (In Russian). Act. Inst. Bot. Komarov, Ser. VIII, Palaeobotanica 2: 1–44.

Sota, E. R. de la. 1962. Contribución al conocimiento de las "Salvinia-ceae" neotropicales. I–III. Darwiniana 12: 465–520.

Sota, E. R. de la. 1963. Contribución al conocimiento de las "Salvinia-ceae" neotropicales, IV. Darwiniana 12: 612–623.

Sota, E. R. de la. 1964. Contribución al conocimiento de las "Salvinia-ceae" neotropicales, V. Darwiniana 13: 529–536.

Tatuno, S., and M. Takei. 1969. Cytological studies of Salviniaceae I. Karyotype of two species in the genus *Salvinia*. Bot. Mag. (Tokyo) 82: 403–408.

Tatuno, S., and M. Takei. 1970. Cytological studies of Salviniaceae II. Bot. Mag. (Tokyo) 83: 67–73.

122. *Azolla*
Figs. 122.1–122.13

Azolla Lam., Encycl. 1: 343. 1783. Type: *Azolla filiculoides* Lam.
Rhizosperma Meyen, Reise 1: 337. 1834. Type: *Azolla pinnata* R. Br.
    Azolla section *Rhizosperma* (Meyen) Mett., Linnaea 20: 273. 1847.

Description

Floating aquatics; stem short, creeping, slender, sometimes bearing trichomes, and usually short roots; leaves ca. 0.5–1.5 mm long, bilobed, the upper lobe usually minutely to strongly papillate, without veins; sporangia borne in short-stalked sori on the under-lobe, enclosed by an indusium, with either one megasporangium or several microsporangia; spores trilete, the megaspores coarsely and irregularly tuberculate with a perforate surface, the microspores smooth, embedded in massulae. Chromosome number: $n = 22$; $2n = 44$.

The megaspore has accessory structures usually referred to as a columella, bearing floats, which in the living species are either three or nine in number. The microsporangia contain a few to several massulae, within which the microspores are imbedded. Projecting structures called glochidia are often on the outer surface of the massulae.

Systematics

Azolla is a widely distributed genus of six species, with four of them in America. It is notable in having the most complex reproductive structures of any plant. Among the living species two sections may be recognized.

Carpanthus Raf. is sometimes cited as a synonym of *Azolla*, but Merrill (1943) showed that it was a Scrophulariaceae.

The basic work on the American species is that of Svenson (1944).

Synopsis of *Azolla*

Section *Rhizosperma*
Megaspore with nine floats, massulae without glochidia or with a few unbarbed ones.

This section is composed of the two Old World species, *Azolla nilotica* Mett. and *A. pinnata* R. Br.

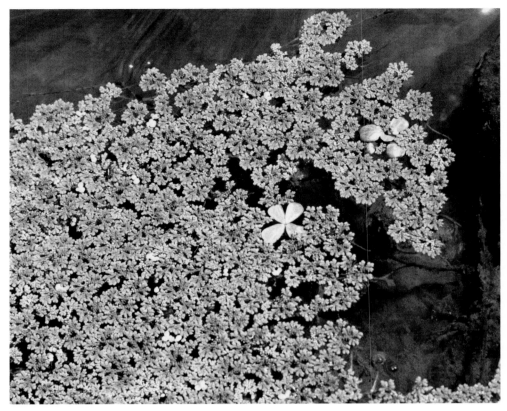

Fig. 122.1. *Azolla mexicana,* Guatemala. (Photo W. H. Hodge.)

Section *Azolla*

Megaspore with three floats, massulae with glochidia barbed at the tip.

Four American species constitute this section, which is thought to be advanced over section *Rhizosperma* because of the fewer floats.

Tropical American Species

Identification of *Azolla* species is difficult because plants are not often fertile and especially the megaspores are rarely produced. Sterile plants of some species have distinctive characters although usually fertile material is needed for certain identification. The following key is adapted from Svenson (1944).

Key to American Species of *Azolla*

a. Leaves ca. 0.5 mm long, stem subdichotomously branched, glochidia with few or no septae.   *A. caroliniana* Willd.
a. Leaves ca. 1 mm long. b.
  b. Stem subdichotomously branched, glochidia with several septae.
                                A. mexicana Presl
  b. Stem pinnately branched. c.
    c. Leaves ovate, plants sometimes 3 cm or to 6 cm long, glochidia with few or no septae.   *A. filiculoides* Lam.
    c. Leaves suborbicular, plants 1–2 cm long, glochidia with several septae.   *A. microphylla* Kaulf.

Fig. 122.2. *Azolla filiculoides,* Iquitos, Peru. (Photo Alice F. Tryon.)

Ecology (Fig. 1)

Azolla floats on the surface of ponds, ditches, in small pools, often in stagnant water, in quiet water of streams, lakes and rivers, less often in wet marshes. It grows among other aquatic vegetation or often forms a velvety covering over the water. *Azolla* grows from near sea level to ca. 3600 m in the Andes.

Geography (Fig. 2)

Azolla is widely distributed in America, also in Africa and Madagascar and eastward to New Guinea and New Caledonia and northward to Japan. It is said to be introduced in Europe, the Hawaiian Islands, New Zealand and southern Africa.

Azolla is generally, although usually locally, distributed in the United States, and occurs in British Columbia, and it has been reported from Alaska. Stations may readily become extinct through severe winters or habitat alteration in the northern part of its range, where previous records may not reflect its current distribution. To the south it is generally distributed from Mexico and the Greater Antilles, southward to southern Chile; also on the Galápagos Islands and the Falkland Islands.

Spores

The megaspore of *Azolla microphylla* is surrounded by an indusium (Fig. 4) and the delicate megasporangium wall disintegrates as the spore enlarges. The apical portion of the indusium persists as a cap partly covering the "swimming apparatus" of three or more floats and a central columella (Fig. 5). The thick spore wall obscures the triradiate laesura (Fig. 7) and consists of three main strata, an inner stranded exospore, a central reticulate layer and a perforate surface (Fig. 8). The perforate surface becomes invaginated and covered with numerous long strands that appear to be affixed to the inner part of the perforate wall (Figs. 6, 8). The megaspore surface of *A. filiculoides* from Brazil

Fig. 122.3. Distribution of *Azolla* in America, south of lat. 35° N.

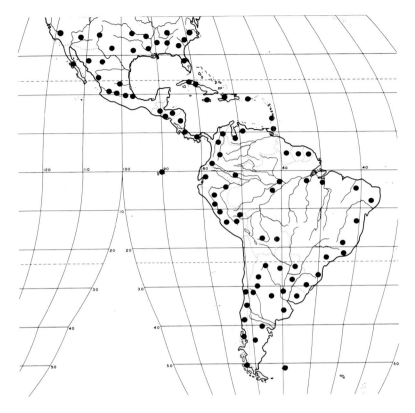

(Fig. 6) is similar to that shown in a collection from Georgia (Bates and Brown, 1981). The microsporangia of *Azolla filiculoides* (Fig. 12) are enclosed within a spheroidal indusium (Fig. 11). The microspores are embedded in massulae (Fig. 9) that in American species have glochidia radiating from the surface. The glochidia are composed of single cells (Fig. 10) or have a few to several septae, and barbed apices.

Thin sections of the megaspore wall of *Azolla pinnata* and of several fossil species from the Tertiary of western Europe have been examined by Kempf (1969). A longisection of the megaspore and associated structure of the fossil *Azolla nana* Dorofeev is remarkably well preserved and is formed of strata similar to those in spores of living species.

Cytology

The chromosome number $n = 22$ in *Azolla pinnata* at diakinesis is reported from Punjab, India, by Loyal (1958). The chromosomes are noted as smaller than those of *Salvinia* and perhaps the smallest among ferns. However, the chromosomes of *Selaginella* are unusually small and probably represent the smallest among pteridophytes. Chromosome numbers are not reported for tropical American species of *Azolla*. Reports of a larger number $n = 48$ in *A. caroliniana*, based on cultivated material, or sections made prior to 1950, need to be verified.

Figs. 122.4–122.13. *Azolla* spores. **4, 5, 7, 8. A. microphylla,** Galápagos Islands, *Fosberg 44814.* **4.** Indusium surrounding megaspore, × 100. **5.** Apical indusial cap lifted showing two floats, the megaspore below, × 80. **6. A. filiculoides** coarsely tuberculate surface of the megaspore with long strands, Brazil, *Ule* in 1889, × 500. **7.** Trilete megaspore, the wall and adjacent structures removed from the apex, × 100. **8.** Section of megaspore wall with the perforate outer layer lifted, the stranded exospore (X) below, central reticulate layer (R), and inner part of perforate layer (F) above, × 1000. **9–12. A. filiculoides,** Uruguay, *Rosengurth Herb. B 781.* **9.** Four massulae with radiating glochidia more or less deflated, × 120. **10.** Detail of nonseptate glochidia with apical barbs, × 900. **11.** Indusium surrounding the microsporangia, × 18. **12.** Microsporangia, a portion of the stalk below, × 120. **13.** Megaspore, stranded exospore surface, *Fosberg 44814,* × 5000.

Observations

Plants of *Azolla* have colonies of the blue-green alga, *Anabaena Azollae* in cavities of the upper leaf lobe. Rarely have plants been found without the alga association. Under suitable growing conditions, *Azolla* can multiply very rapidly, and the *Anabaena* fixes considerable amounts of atmospheric nitrogen. The nitrogen fixing capacity may be equivalent to that of legumes. For these reasons, *Azolla* has been cultivated for many years as a green manure in rice fields of Vietnam, and recently its use has been expanded along with active agricultural studies of *Azolla*-rice cultivation (Lumpkin and Plucknett, 1980).

Details of the anatomy and morphology of *Azolla* have been investigated by Demalsy (1953, 1958).

Literature

Bates, V. M., and E. T. Browne. 1981. *Azolla filiculoides* new to the southeastern United States. Amer. Fern Jour. 71: 33–34.

Demalsy, P. 1953. Le sporophyte d'*Azolla nilotica* Decaisne. La Cellule 56: 7–60.

Demalsy, P. 1958. Nouvelles recherches sur le sporophyte d'*Azolla.* La Cellule 59: 235–268.

Kempf, E. K. 1969. Elektronmikroskopie der Sporodermis von känozoischen Megasporen der Wasserfarn-Gattung *Azolla.* Paläont. Z. 43: 95–108.

Loyal, D. 1958. Cytology of two species of Salviniaceae. Curr. Sci. 27: 357–358.

Lumpkin, T. A., and D. L. Plucknett. 1980. *Azolla:* Botany, physiology, and use as a green manure. Econ. Bot. 34: 111–153.

Merrill, E. D. 1943. New names for ferns and fern allies proposed by C. S. Rafinesque, 1806–1838. Amer. Fern Jour. 33: 41–56.

Svenson, H. K. 1944. The New World species of *Azolla.* Amer. Fern Jour. 34: 69–84.

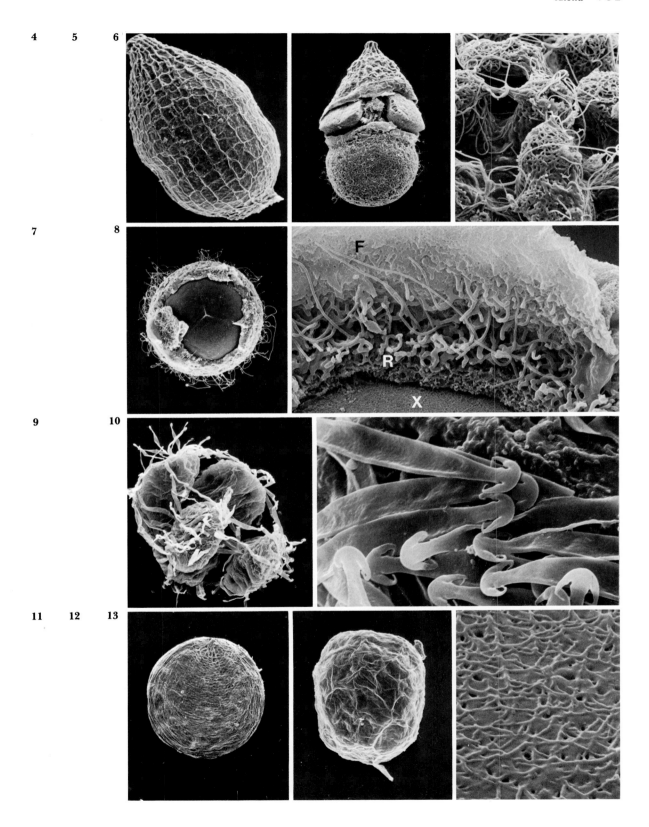

Family 25.   Psilotaceae

Psilotaceae Kanitz, Növényrends. Áttek. 43. 1887. Type: *Psilotum* Sw.
Tmesipteridaceae Nakai, Ord. Fam. Trib. Nov. 206. 1943. Type: *Tme-sipteris* Bernh.

Description

Stem protostelic or solenostelic, flaccid to somewhat indurated, lacking indument and roots; leaves small to minute, borne alternately, with a single vein or none; two or three sporangia joined in a sessile, thick-walled synangium; homosporous, spores without chlorophyll. Gametophyte subterranean, mycorrhizic, fleshy, without chlorophyll, elongate, branched, with the archegonia and antheridia rather uniformly distributed or in definite zones.

Comments on the Family

The Psilotaceae are a family of two genera, *Tmesipteris* Bernh. of western Malesia, Australia and the Pacific and *Psilotum* pantropical in distribution.

The family, although without a fossil record, was long considered to represent the most primitive of living vascular plants, related to *Psilophyton* Dawson and *Rhynia* Kidst. & Lang, ancient fossils of simple structure. The review of early vascular plants by Gensel (in White et al., 1977), however, shows no relations to the Psilotaceae.

After an extensive series of studies, D. W. Bierhorst concluded that the Psilotaceae were primitive elements in the Filicopsida. This interpretation of the position of the family provoked further interest in its phylogeny and resulted in a symposium on the taxonomic and morphological relationships of the Psilotaceae (White et al., 1977). The relationship of *Stromatopteris* to *Psilotum* and *Tmesipteris,* particularly stressed by Bierhorst, was reviewed by Kaplan, among others, who considered that the leafy shoot of the latter two genera is not homologous to the leaf of *Stromatopteris.* In a later paper, Cooper-Driver (1977) showed that the Psilotaceae were characterized by the presence of biflavonyls, which were absent in the Stromatopteridaceae, Gleicheniaceae and Schizaeaceae. In addition, the Psilotaceae lacked the flavonols and proanthocyanidins that were present in the three fern families.

The Psilotaceae have a number of characters that occur in the Filicopsida as the details of early ontogeny of the leaf, the subterranean, cylindrical gametophytes, multiflagellate sperm, and special aspects of the rhizoids and gametangia. These and the unequivocal evidence from the structure and wall formation of *Psilotum* spores are indicative of relationships with the ferns; thus Psilotaceae are placed in the class Filicopsida, as a separate subclass Psilotidae.

Literature

Cooper-Driver, G. 1977. Chemical evidence for separating the Psilotaceae from the Filicales. Science 198: 1260–1261.

White, R. A., et al. 1977. Taxonomic and morphological relationships of the Psilotaceae. Brittonia 29: 1–68, including the following individual papers:

Bierhorst, D. W., The systematic position of *Psilotum* and *Tmesipteris,* pp. 3–13.

Gensel, P. G., Morphologic and taxonomic relationships of the Psilotaceae relative to evolutionary lines in early land vascular plants, pp. 14–29.

Kaplan, D. R., Morphological status of the shoot systems of Psilotaceae, pp. 30–53.

Wagner, W. H. Jr., Systematic implications of the Psilotaceae, pp. 54–63.

White, R. A., Introduction to the symposium, pp. 1–2.

123. *Psilotum*

Figs. 123.1–123.10

Psilotum Sw., Jour. Bot. (Schrad.) 1800(2): 8, 109. 1802, *nom. nov.* for *Hoffmannia* Willd., not Sw. and with the same type. (*Psilotum triquetrum* Sw., *nom. superfl.* for *Lycopodium nudum* L.) = *Psilotum nudum* (L.) Beauv.

Hoffmannia Willd., Mag. Bot. (Roemer & Usteri) 2 (6): 15. 1789, not Sw. 1788 (genus of Rubiaceae). Type: *Hoffmannia aphylla* Willd., *nom. superfl.* for *Lycopodium nudum* L. = *Psilotum nudum* (L.) Beauv.

Bernhardia Willd., Akad. Wiss. Erfurt 2: 11. 1802, *nom. nov.* for *Hoffmannia* Willd., not Sw. and with the same type. (*Bernhardia dichotoma* Willd., *nom. superfl.* for *Lycopodium nudum* L.) = *Psilotum nudum* (L.) Beauv.

Tristeca Mirbel, Hist. Nat. Veg. (Lam. & Mirb.) 3: 478. 1802. Type: *Lycopodium nudum* L. (*Tristeca nuda* (L.) Jaume St. Hil.) = *Psilotum nudum* (L.) Beauv.

Description Terrestrial, rupestral or usually epiphytic; stem compactly branched in the substrate, the aerial stems green, erect to pendent, with several dichotomous divisions, ca. 20 cm to 1 m long, without roots, leaves minute, alternate, leafy stem glabrous, large synangia borne singly in the axil of forked fertile leaves; spores elongate-ellipsoidal monolete, the laesurae $\frac{1}{2}$ to $\frac{2}{3}$ the spore length, the surface coarsely rugose to shallowly and compactly verrucate. Chromosome number: $n = 52, 104$, ca. 210; $2n = 104, 208$.

Systematics *Psilotum* is a highly distinctive genus of wide distribution in the tropics and with some extensions to subtemperate regions. Notable characters of the genus are the dichotomously branched stem (Fig. 4), the bifurcate sporophylls (Fig. 5), and the large synangium (Fig. 6). Two pantropical species are usually recognized: *Psilotum nudum* (L.) Beauv., with 3-angled branches to ca. 1.2 mm wide and *P. complanatum* Sw. with flattened branches ca. 2 mm or more wide. *Psilotum* has a simple morphology, probably through reduction from more complex ancestors. Its position in the plant kingdom, along with *Tmesipteris,* is discussed under the family.

Among the three names published for this genus in 1802, *Psilotum* has been generally adopted.

Fig. 123.1. *Psilotum complanatum,* epiphyte in cloud forest, with *Polypodium,* Veracruz, Mexico. (Photo W. H. Hodge.)

Ecology (Figs. 1, 2)

Psilotum is terrestrial, rupestral or epiphytic. It grows most commonly as an epiphyte on tree trunks or branches, or in the crevices of rocks.

In America, *Psilotum* grows in the soil of wet forests, on fallen branches and trunks of trees, at the base of trees, in swamp forests, montane forests, sometimes mixed subtropical forests, and also in cloud forests. It is often an epiphyte on tree ferns and palms. It grows, less frequently, in rocky places, in crevices of rocks on cliffs in seepage areas, or on wet rocks. *Psilotum* occurs from sea level to about 1400 m, usually below 500 m.

Geography (Fig. 3)

Psilotum is widely distributed in America, Africa and Madagascar, and eastward to Ceylon, India and China and northward to Japan and Korea, and eastward to Australia and New Zealand;

Fig. 123.2. *Psilotum complanatum*, on trunk of tree fern, Veracruz, Mexico. (Photo W. H. Hodge.)

Fig. 123.3. Distribution of *Psilotum* in America.

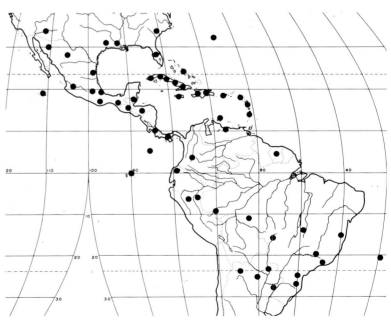

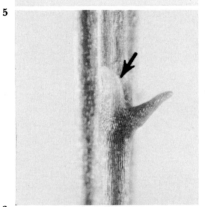

Figs. 123.4–123.6 *Psilotum nudum.*
4. Portion of aerial stem, with synangia, × 0.4. **5.** Bifurcate sporophyll, arrow at one lobe, × 10.0. **6.** Synangium with subtending sporophyll, × 10.0.

and through the Pacific to the Hawaiian Islands, the Society Islands, Pitcairn Island and Easter Island.

In America it occurs in the southern United States southward through Mexico, Central America and the West Indies, and in South America to northern Argentina; also on Bermuda, Ilha Trinidade, the Revillagigedo Islands, Cocos Island and the Galápagos Islands. *Psilotum complanatum* is less widely distributed than *P. nudum*, especially in South America, where it appears to be infrequent.

Spores

The rugose surface formed by the perispore is not completely developed in the young spore of *Psilotum nudum* (Figs. 7, 8). The lower part of the mature, rugose spore (Fig. 9) is partially coated with the lipid that is copious within the cell. The spore surface of *Psilotum complanatum* is generally similar but with somewhat more discrete verrucate elements (Fig. 10). Detailed studies of the development and wall structure of spores of *Psilotum nudum* (as *P. triquetrum*) by Lugardon (1979) supplies new data on the wall formation as well as evidence relevant to the systematic relationship of the genus. Except for a special "pseudo-endospore" formed between the exospore and cell, the structure is characteristic of the Filicopsida and is clearly distinct from that of other pteridophytes. The surface, formed by a relatively thin perispore, resembles that of monolete spores of *Gleichenia*.

Cytology

The chromosome numbers for native plants of *Psilotum nudum* reported over a wide range, Australia, New Zealand, Malaya, Ceylon and Japan, are largely tetraploid, $n = 104$. Ten populations from India and Ceylon were all tetraploid (Ninan, 1956). The record of tetraploids at two stations in Jamaica (Walker, 1966) suggests these may also be widely distributed in the American tropics. An octoploid, $n = $ ca. 210, from New Caledonia, is the highest level known for the genus. The relatively few diploid records $n = 52$, are from Ceylon, (Manton, 1942), the Rykuku and Okinawa Islands and Shikoku, Japan (Mitui, 1975, 1976). In addition to the high numbers, the chromosomes of *Psilotum* are exceptionally large. They range from 4.5–18.0 μm in length, the largest among some 20 genera compared by Abraham et al. (1962), although this did not include *Salvinia* or the Hymenophyllaceae which also have large chromosomes.

Reports of the chromosome number for the related genus *Tmesipteris* of the Old World $n = 104$ and 208 are within the same polyploid series as *Psilotum*. These high numbers have evolved by replication of whole chromosomes sets from 52 which itself evidently represents duplication of a complement possibly based on $x = 13$.

7   8

9   10

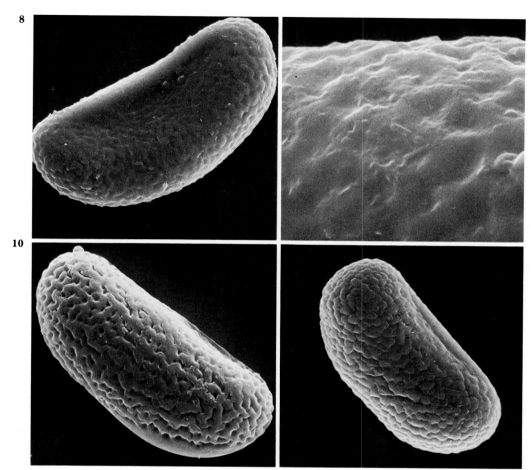

Figs. 123.7–123.10. *Psilotum* spores, × 1000.   **7, 8. *P. nudum*,** Grenada, *Hunnewell 19414*.   **7.** Young spore, shallowly rugose.   **8.** Surface detail, × 5000.   **9. *P. nudum*** mature spore, laesura at top, lipid partially covering surface at base of spore, Bahamas, *Small & Carter 8674*.   **10. *P. complanatum*,** shallowly verrucate, Solomon Islands, *Kajewski 1878*.

Literature

Abraham, A., C. A. Ninan, and P. M. Mathew. 1962. Studies on the cytology and phylogeny of pteridophytes. 7. Observations on one hundred species of south Indian ferns. Jour. Ind. Bot. Soc. 41: 339–421.

Lugardon, B. 1979. Sur la formation du sporoderme chez *Psilotum triquetrum* Sw. (Psilotaceae). Grana 18: 145–165.

Manton, I. 1942. A note on the cytology of *Psilotum* with special reference to vascular prothalli from Rangitoto Island. Ann. Bot. n.s. 6: 283–293.

Mitui, K. 1975. Chromosome numbers of Japanese pteridophytes. Bull. Nippon Dental Coll., Gen. Ed. 4: 221–271.

Mitui, K. 1976. Chromosome numbers of some ferns of Rykuku Islands. Jour. Jap. Bot. 51: 33–41.

Ninan, C. A. 1956. Cytology of *Psilotum nudum* (L.) Beauv. La Cellule 57: 307–318.

Walker, T. G. 1966. A cytotaxonomic survey of the pteridophytes of Jamaica. Trans. Roy. Soc. Edinburgh 66: 169–237.

Family 26.   Equisetaceae

Equisetaceae A. P. DC., Fl. Franc. (Lam. & DC.) ed. 3, 2: 580. "1815", 1805. Type: *Equisetum* L.

Description

Stem jointed, more or less indurated, the aerial usually green, the subterranean with numerous wiry roots; leaves each with a single vein, borne in whorls, joined in a sheath; sporangia thin walled, borne on stalked, peltate sporangiophores that form a strobilus; homosporous, spores with chlorophyll. Gametophyte green, epigeal, rather irregularly thallose strap-shaped, branched, the archegonia often mostly on the thickened portions, the antheridia on the lobes, sometimes the gametophytes unisexual or functionally so.

Comments on the Family

Equisetum has an extensive fossil record extending from the Jurassic, or if the similar *Equisetites* Sternb. is included, it extends back to the Permian.

The gametophytes of *Equisetum* have been the subject of many detailed studies. The following three papers provide access to the extensive literature in addition to the data presented. The gametophyte morphology in several species examined by Duckett (1979a) show that two types of gametophytes correspond with the sporophyte characters in distinguishing two subgenera. Crossing experiments (Duckett, 1979b) also emphasize these differences, for six hybrids were synthesized in each of the two subgenera but hybrids were not formed from intersubgeneric crosses. Study of wild gametophytes from six sites in Britain (Duckett and Duckett, 1980) indicate there are significantly different sex ratios between populations demonstrating the labile nature of the sex-determining mechanism in *Equisetum*.

Literature

Duckett, J. G. 1979a. Comparative morphology of the gametophytes of *Equisetum* subgenus *Hippochaete*. Bot. Jour. Linn. Soc. 79: 179–203.

Duckett, J. G. 1979b. An experimental study of the reproductive biology and hybridization in the European and North American species of *Equisetum*. Bot. Jour. Linn. Soc. 79: 205–229.

Duckett, J. G., and A. R. Duckett. 1980. Reproductive biology and population dynamics of wild gametophytes of *Equisetum*. Bot. Jour. Linn. Soc. 80: 1–40.

124. *Equisetum*
Figs. 124.1–124.12

Equisetum L., Sp. Pl. 1061. 1753; Gen. Pl. ed. 5, 484. 1754. Type: *Equisetum fluviatile* L.

Hippochaete Milde, Bot. Zeit. 23: 297. 1865. Type: *Hippochaete hyemalis* (L.) C. Börn. = *Equisetum hyemale* L. *Equisetum* subgenus *Hippochaete* (Milde) Baker, Handb. Fern-Allies. 4. 1887.

Presla Dulac, Fl. Haute-Pyrénées. 25. 1867, *nom. superfl.* for *Equisetum* L. and with the same type.

Allostelites C. Börn., Fl. Deutsche Volk 59, 283. 1912. Type: *Allostelites arvense* (L.) C. Börn. = *Equisetum arvense* L. [Subgenus *Equisetum*.]

Description Terrestrial, palustral or aquatic; stem subterranean, short- to long-creeping, freely branched, bearing erect, aerial stems which are jointed, longitudinally ridged and usually hollow, ca. 10 cm to 8 m long, often with whorls of branches; leaves small, in whorls at nodes, the lower portion laterally fused in sheaths, the upper portion in more or less prolonged teeth; sporangia large, several borne on each stalked, peltate, apically flattened

Fig. 124.1. *Equisetum giganteum,* Pastaza, Ecuador. (Photo B. Øllgaard.)

Fig. 124.2. *Equisetum giganteum,* Santo Domingo de los Colorados, Ecuador. (Photo W. H. Hodge.)

sporangiophore, in a condensed terminal strobilus; spores spheroidal with circular aperture, and four paddle-shaped elators, the surface with small granulate and larger spherical deposit. Chromosome number: $n = 108$; $2n = 216$.

The *Equisetum* strobilus (Fig. 5) is composed of sporangiophores that are sometimes called sporophylls. A detailed study of these structures by Page (1972a) show they are partly modified from a leaf and partly a cauline appendage, thus the term sporangiophore is more appropriate. The small leaves that surround each node are laterally fused below and form coarse teeth above (Figs. 6, 7).

Systematics

Equisetum is a widely distributed genus of about 15 species, with 13 in America and five of these in the tropics. The species of *Equisetum* were revised by Hauke, those of subgenus *Hippochaete* (1961, 1963) and of subgenus *Equisetum* (1978). The infrageneric classification is adopted from these monographs although only the subgenera are presented here; the several sections may be found in the revisions. Both *Equisetum* and *Hippochaete* were

Fig. 124.3. *Equisetum bogotense,* Volcán Cotopaxi, Ecuador. (Photo W. H. Hodge).

recognized as genera by Holub (1972), with several subgenera and sections.

Equisetum arvense L. was chosen as the type of *Equisetum* by Pichi-Sermolli (1971) but since that species was removed to a separate genus by Börner, the earlier lectotype, *Equisetum fluviatile* must be restored.

Synopsis of *Equisetum*

Subgenus *Hippochaete*

Aerial stems mostly evergreen, stomata in regular, long rows, sunken (Fig. 9), strobili usually apiculate.

A subgenus of two sections and seven species, all except *Equisetum ramosissimum* Desf. native in America, and that species has recently been introduced (Hauke, 1979). All of the tropical American species, except *Equisetum bogotense* belong in this subgenus.

Subgenus *Equisetum*

Aerial stems deciduous, stomata irregularly disposed in broad bands, superficial (Fig. 10), strobili blunt.

A subgenus of three sections, one with two subsections, including eight species, all in America except for *Equisetum diffusum* Don. In the American tropics, *Equisetum bogotense* is the only representative of this subgenus.

Tropical American Species

The few *Equisetum* species of the American tropics are ecologically and geographically varied but can be determined by clusters of characters. The prominent deposits of silica and other surface features of the stem have traditionally been used in the species taxonomy. Recent SEM studies (Page, 1972b) of the silica formation especially on the ridges, furrows, and on the cells associated with the stomata provide precise, systematically pertinent details.

Several hybrids known in *Equisetum* include *E. hyemale* × *myriochaetum* known in Guatemala and Mexico (Hauke, 1963).

The following key has been adapted from the treatments of Hauke which include additional characters of the species.

Key to Species of *Equisetum* in Tropical America

a. Aerial stem unbranched, or irregularly branched.   b.
  b. Sheaths about as long as broad.   *E. hyemale* L.
  b. Sheaths definitely longer than broad.   c.
    c. Teeth promptly deciduous or mostly so.   *E. laevigatum* A. Br.
    c. Teeth persistent.   *E. bogotense* HBK.
a. Aerial stems with regular whorls of branches.   d.
    d. Teeth mostly persistent, stem ridges, especially of the branches, flat or bluntly tuberculate (Fig. 8) in profile.   *E. giganteum* L.
    d. Teeth promptly deciduous, or mostly so, stem ridges, especially of the branches, ascending serrate in profile.
                                     E. myriochaetum Schlect. & Cham.

Ecology (Figs. 1–3)

Equisetum grows in a diversity of habitats and often forms dense colonies from the freely branching underground stems. It occurs in open sterile soils, thickets, bogs, shores of ponds or lakes, and in dense forests.

In the American tropics *Equisetum* grows along stream and river banks, along the margins of lakes, in ditches and marshes, and also in drier sites, such as road banks, in rock slides and in alpine turf. *Equisetum* grows from sea level to 4200 m, most frequently from 500 to 2500 m, and *Equisetum bogotense* occurs at the highest altitudes, above 4000 m.

Geography (Fig. 4)

Equisetum occurs in America, Eurasia and Africa and east to northeastern Asia, the Philippine Islands and New Guinea. In the Pacific it is found on New Caledonia and the Fiji Islands.

In America, *Equisetum* grows from northern Alaska and Greenland, south to Mexico, Central America and in the

Fig. 124.4. Distribution of *Equisetum* in America, south of lat. 35° N.

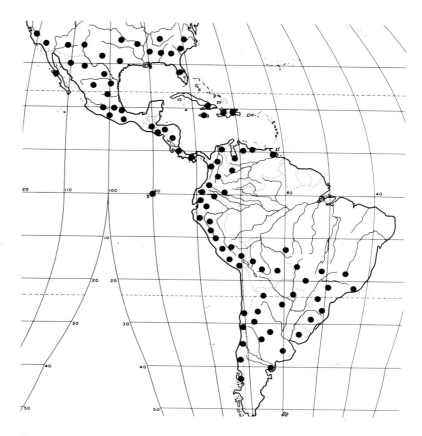

Greater Antilles, except Puerto Rico. It is widespread in South America from Venezuela to Colombia and Bolivia and to southeastern Brazil and southern Chile and Argentina. It appears to be absent from the Amazon basin and northeastern Brazil.

Equisetum bogotense occurs from Costa Rica south to Argentina and Chile and the range of *E. giganteum* is slightly larger, including Guatemala and the Greater Antilles. *Equisetum myriochaetum* is distributed from Mexico to Peru, while *E. hyemale* and *E. laevigatum* are largely temperate, with limited tropical ranges. The former grows south to Mexico and Guatemala, and the latter south to northwestern Mexico.

Spores

The chlorophyll-bearing spores of *Equisetum*, with a circular aperture and four paddle-shaped elators, are unique in the pteridophytes. The deposit of fine granules and coarser spherical material that covers the spores is also formed on the elators as in *Equisetum bogotense* (Fig. 11). The two pairs of elators are fused to the perispore; only one of the pairs is included in Fig. 12. The elators are bound to the perispore around the sides of the circular aperture near the point where the exospore and perispore join. *Equisetum* spores sectioned by Lugardon (1969) show the wall consists largely of two exospore strata which form a special pluglike obturator below the circular aperture. A relatively thick endospore layer is laid down adjacent to the cell membrane and the spore surface is formed by a thin perispore. Studies of *Equisetum* chloroplasts and cell organelles by Gullvåg

5   6   7

8   9

10   11   12

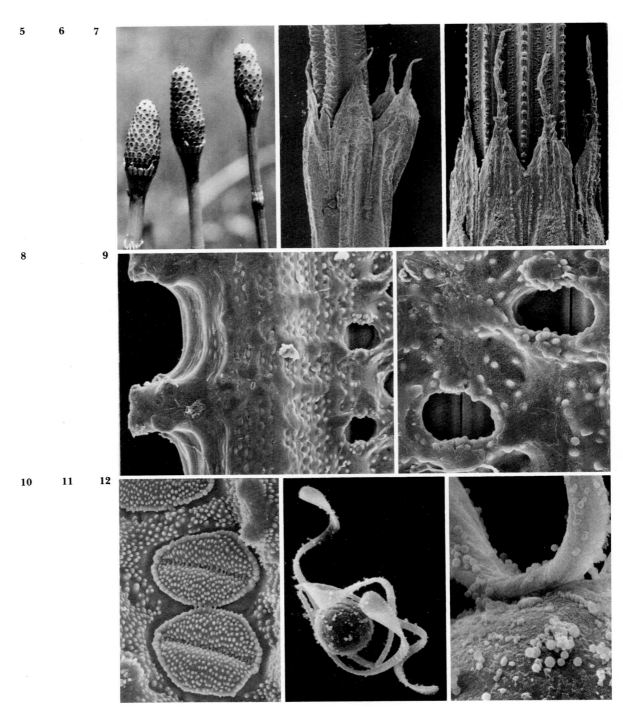

Figs. 124.5–124.12. *Equisetum*.   5. Strobili of *E. giganteum,* ×2.   **6.** Whorled leaves with apical teeth, *E. bogotense,* ×14.   **7.** Whorled leaves with apical teeth and stem ridges with tubercles, *E. giganteum,* ×20.   **8, 9.** *Equisetum giganteum.*   **8.** Profile of tubercules on stem ridge, silica papillae on stem surface and stomatal pores, right, ×280.   **9.** Detail of stomatal pores and silica deposition on stem, ×600.   **10.** *E. bogotense,* dense silica papillae on stomatal cells and adjacent parts of stem, ×600   **11.** *E. bogotense,* spore with four paddle-shaped elators, Colombia, *Killip & Smith 18818,* ×400.   **12.** *E. giganteum,* detail of spore apex and base of a pair of elators, Argentina, *Meyer 12342,* ×2000.

(1971) indicate moderate amounts of starch are present in immature spores and mature ones have a special proteinaceous storage unlike that of other pteridophytes. *Equisetum* spores also differ from those of most other Pteridophyta in their exceptionally high water content as well as short viability.

Cytology

The chromosome number $n = 108$ in *Equisetum* is remarkable in the constancy of the number, although there are few documented records from America and none from the American tropics. A résumé of chromosome numbers of *Equisetum* (Bir, 1960) covers a wide geographic range including all of the British species and many from Europe and India. An account of *E. arvense* from Japan (Kurita, 1977) notes the length of chromosomes 1.5–4.5 μm at prophase and the occurrence of one large and two or three small nucleolei, as characteristic of several species. The chromosomes of species in subgenus *Hippochaete* are larger than those in subgenus *Equisetum* and support other morphological differences distinguishing these groups. Cytological analyses of hybrids, as *E.* × *trachyodon* A. Br. (*E. hyemale* × *variegatum* Weber & Mohr), with $n = 216$ univalents and aborted spores, provide data for interpreting its status (Bir, 1960). The uniform records of $n = 108$ for the genus suggest these represent a highly derived level, and extinction of lower chromosome numbers.

Literature

Bir., S. S. 1960. Chromosome numbers of some *Equisetum* species of the Netherlands. Acta Bot. Neerland 9: 224–334.

Gullvåg, B. M. 1971. The fine structure of some Pteridophyte spores before and during germination, *in:* G. Erdtman and P. Sorsa. Pollen and Spore Morphology/Plant Taxonomy, pp. 252–295. Almqvist & Wiksell, Stockholm.

Hauke, R. L. 1961. A resume of the taxonomic reorganization of *Equisetum* subgenus *Hippochaete* Amer. Fern Jour. 51: 131–137, continued in 52: 29–35, 57–63, and 123–130.

Hauke, R. L. 1963. A taxonomic monograph of the genus *Equisetum* subgenus *Hippochaete*. Beih. Nova Hedwigia 8: 1–123.

Hauke, R. L. 1978. A taxonomic monograph of *Equisetum* subgenus *Equisetum*. Nova Hedwigia 30: 385–455.

Hauke, R. L. 1979. *Equisetum ramosissimum* in North America. Amer. Fern Jour. 69: 1–5.

Holub, J. 1972. Bemerkungen zu den tschechoslowakischen taxa der familie Equisetaceae. Preslia 44: 112–130.

Kurita, S. 1977. Miscellaneous notes on the cytology of pteridophytes 1. Chromosome number of *Equisetum arvense* L. Chromosome Inform. Serv. 23: 4–6.

Lugardon, B. 1969. Sur la structure fine des parois sporales d'*Equisetum maximum* Lam. Pollen et Spores 11: 449–474.

Page, C. N. 1972a. An interpretation of the morphology and evolution of the cone and shoot of *Equisetum*. Bot. Jour. Linn. Soc. 65: 359–397.

Page, C. N. 1972b. An assessment of inter-specific relationships in *Equisetum* subgenus *Equisetum*. New Phytol. 71: 355–369.

Pichi-Sermolli, R. E. G. 1971. Names and types of the genera of fern-allies. Webbia 26: 129–194.

Family 27.   Lycopodiaceae

Lycopodiaceae Mirbel, Hist. Nat. Veg. (Lam. & Mirb.) 4: 293. 1802.
Type: *Lycopodium* L.
Phylloglossaceae Kze., Bot. Zeit. 1: 722. 1843. Type: *Phylloglossum* Kze.
Urostachyaceae Rothm., Fedde Repert. 54: 58. 1944, *illegit.*, based on
Urostachys Hert., *illegit.*
Huperziaceae Rothm., Fedde Repert. 66: 236. 1962, *nom. nov.* for Uro-
stachyaceae.

Description

Stem protostelic, or (in *Phylloglossum*) sometimes partly solenostelic, indurated or not, sometimes brittle, usually lacking indument, sometimes short-pubescent, bearing usually few, fleshy,
dichotomously branched roots; leaves simple, ca. 2 mm to
20 mm long, with one vein; sporangia sessile to short-stalked,
single on the adaxial surface of a leaf usually near the base, or on
the stem near the axil; homosporous, spores without chlorophyll. Gametophyte superficial, basally tuberous and mycorrhizic, with more or less erect lobes bearing chlorophyll and non-
mycorrhizic; or subterranean, without chlorophyll, mycorrhizic
and either erect and radially symmetrical or prostrate-horizontal
and radially or bilaterally symmetrical, the archegonia more or
less sunken, the antheridia superficial, mostly borne apart from
the archegonia.

**Comments
on the
Family**

The Lycopodiaceae are a family of two genera, *Lycopodium* with
about 250 species and *Phylloglossum* Kze. of Australia, Tasmania
and New Zealand, with the single species, *P. Drummondii* Kze.

The fossil record of *Lycopodium* is obscure due to a lack of sufficiently well preserved materials that can with certainty be
placed in the genus. *Lycopodites* Lindl. & Hutt. is known from as
early as the Devonian and extends to the Upper Cretaceous.
Some of the many species assigned to it may be allied to *Selaginella* or to other fossil lycopodiaceous genera. *Lycopodites creta-
ceum* Berry of the Cretaceous is perhaps the most certain representative of *Lycopodium* in the fossil record.

The Lycopodiaceae have only distant affinities to the Selaginellaceae and Isoetaceae.

The types of gametophytes in *Lycopodium* (Bierhorst, 1971;
Boivin 1950; Bruce, 1979a, 1979b) correlate with the three subgenera. In subgenus *Cernuistachys* the spores germinate quickly
and produce a gametophyte that is superficial, partly green, and
only partly mycorrhizic. In the subgenera *Lycopodium* and *Selago*,
the spores germinate after a few years and produce subterranean gametophytes or ones that are buried in humus when
epiphytic, without chlorophyll and wholly mycorrhizic. Gametophytes of subgenus *Lycopodium* are erect and radially symmetrical, while in subgenus *Selago* they are more or less horizontal in
the substrate, either radially or bilaterally symmetrical. There
are other characters and variants, of the three basic types of gametophytes, in addition to those noted above.

Gametophytes have been described in so few species that the
diversity of forms may not be known and the taxonomic value of

their characters cannot be assessed. Thus the gametophytes are discussed here, rather than under the subgenera of *Lycopodium,* as it may be premature to imply their systematic importance.

Literature

Bierhorst, D. W. 1971. Morphology of Vascular Plants. 560 pp. Macmillan, New York.

Boivin, B. 1950. The problem of generic segregates in the form-genus *Lycopodium.* Amer. Fern Jour. 40: 32–41.

Bruce, J. G. 1979a. Gametophyte of *Lycopodium digittatum.* Amer. Jour. Bot. 66: 1138–1150.

Bruce, J. G. 1979b. Gametophyte and young sporophyte of *Lycopodium carolinianum.* Amer. Jour. Bot. 66: 1156–1163.

125. *Lycopodium*
Figs. 125.1–125.27

Lycopodium L., Sp. Pl. 1100. 1753; Gen. Pl., ed. 5, 486. 1754. Type: *Lycopodium clavatum* L.

Selago Hill, Brit. Herb. 533. 1757, and Boehm., Defin. Gen. Pl. (Ludwig) Ed. Boehmer 484. 1760, and Schur, Enum. Pl. Transsilv. 825. 1866, are all latter homonyms of *Selago* L. (Scrophulariaceae) and all based on *Lycopodium Selago* L. [Subgenus *Selago*].

Huperzia Bernh., Jour. Bot. (Schrad.) 1800(2): 126. 1802. Type: *Huperzia Selago* (L.) Schrank. & Mart. = *Lycopodium Selago* L. [Subgenus *Selago*].

Plananthus Mirbel, Hist. Nat. Veg. (Lam. & Mirb.) 3: 476. 1802. Type: *Plananthus Selago* (L.) Beauv. = *Lycopodium Selago* L. [Subgenus *Selago*].

Lepidotis Mirbel, Hist. Nat. Veg. (Lam. & Mirb.) 3:477. 1802. *nom. superfl.* for *Lycopodium* L. and with the same type.

Copodium Raf., Amer. Monthly Mag. & Crit. Rev. 2: 44. 1817, *nom. superfl.* for *Lycopodium* L. and with the same type.

Diphasium Presl, Abhandl. Böhm. Gesell. (Bot. Bemerk.) V, 3: 883. 1845. Type: *Diphasium Jussiaei* (Poir.) Presl = *Lycopodium Jussiaei* Poir. [Subgenus *Lycopodium*].

Lycopodium subgenus *Selago* Baker, Handb. Fern-Allies. 8. 1887. Type: *Lycopodium Selago* L.

Lycopodium subgenus *Subselago* Baker, Handb. Fern-Allies. 8. 1887. Type: *Lycopodium squarrosum* Forst. [Subgenus *Selago*].

Lycopodium subgenus *Lepidotis* Baker, Handb. Fern-Allies. 8. 1887, *nom. superfl.* for subgenus *Lycopodium* L. and with the same type.

Lycopodium subgenus *Urostachya* Pritzel, Nat. Pflanz. 1 (4): 592. 1901, *nom. superfl.* for *Lycopodium* subgenus *Selago* Baker and with the same type.

Lycopodium subgenus *Rhopalostachya* Pritzel, Nat. Pflanz. 1 (4): 601. 1901, *nom. superfl.* for subgenus *Lycopodium* and with the same type.

Lycopodium subgenus *Cernuistachys* Hert., Engl. Bot. Jahrb. 43 (Beibl. 98): 29. 1909. Type chosen here: *Lycopodium cernuum* L.

Lycopodium subgenus *Inundatistachys* Hert., Engl. Bot. Jahrb. 43 (Beibl. 98): 29. 1909. Type chosen here: *Lycopodium inundatum* L. [Subgenus *Cernuistachys*]. In Rev. Sudam. Bot. 8: 94–98. 1950, Herter used the name *Cernuistachys* but not *Inundatistachys*.

Urostachys Hert., Beih. Bot. Centralbl. 39 (2): 249. 1922, *nom. superfl.* for *Huperzia* Bernh. and with the same type.

Phlegmariurus Holub, Preslia 36: 21. 1964. Type: *Lycopodium phlegmaria* L. *Lycopodium* section *Phlegmariurus* Hert., Engl. Bot. Jahrb. 43 (Beibl. 98): 30, 37. 1909, *nom. superfl.* for *Lycopodium* section *Phlegmaria* Pritzel, Nat. Pflanz. 1 (4): 599. Type: *Lycopodium phlegmaria* L. [Subgenus *Selago*].

Lycopodiella Holub, Preslia 36: 22. 1964. Type: *Lycopodiella inundata* (L.) Holub = *Lycopodium inundatum* L. [Subgenus *Cernuistachys*].

Palhinhaea Vascon. & Franco, Bull. Soc. Broter. II, 41: 24. 1967. Type: *Palhinhaea cernua* (L.) Vascon. & Franco = *Lycopodium cernuum* L. [Subgenus *Cernuistachys*].

Diphasiastrum Holub, Preslia 47: 104. 1975. Type: *Diphasiastrum complanatum* (L.) Holub = *Lycopodium complanatum* L. [Subgenus *Lycopodium*].

Lycopodium subgenus *Lycopodiella* (Holub) Øllgaard, Amer. Fern Jour. 69: 49. 1979. [Subgenus *Cernuistachys*].

Only infrageneric names pertinent to the classification adopted are included in the above list.

Description

Terrestrial, rupestral or epiphytic; the stem decumbent with erect branches, or pendent, or short to long, prostrate-creeping with more or less erect branches, or scrambling or (in *L. casuarinoides* Spring and *L. volubile* Forst.) climbing to 10 m or more, rather slender, sparingly to profusely branched; leaves ca. 2–20 mm long, borne spirally around the stem, or subverticillate, sometimes in ranks, or apparently opposite; sporangia large, single on the adaxial side of the leaf or on the stem near the axil, sessile or short-stalked, associated with unmodified leaves or with well-differentiated sporophylls borne in a strobilus; spores trilete, somewhat spheroidal, the proximal face often compressed to somewhat depressed, often with an equatorial ridge, the laesura $\frac{2}{3}$ to $\frac{3}{4}$ the radius, the surface rugulate, rugate, reticulate, foveolate-fossulate, clavate, or scabrate. Chromosome number: $n = 23, 34, 35, 48, 70, 78, 104, 110,$ ca. $128,$ ca. $132, 136, 156,$ ca. $165, 208, 264$; $2n = 46, 48, 68, 264, 272,$ ca. $340, 528.$

Systematics

Lycopodium is an essentially worldwide genus of about 250 species with about half of them in America. Five segregate genera have been recognized by Pichi-Sermolli (1977) but these are not based on a modern review of the species. As in the cases of *Selaginella* and *Equisetum,* there are many characteristics of *Lycopodium* by which the genus is readily recognized and that unify the genus. Several subgeneric groups within *Lycopodium* are recognized in the works of Nessel (1939) and Herter (1950) but none clearly represents major divergent evolutionary lines. At least 20 species-alliances can be recognized but these are presently not all well known or clearly defined. In this account, *Lycopodium* is treated as a single genus with three groups as subgenera. It is expected that alliances may be recognized with certainty as data from new collections and field studies, as well as information on structure, growth and development of the plants are obtained.

About 20 species of *Lycopodium* occur in the United States and Canada, most of them not represented in the tropics. The more common ones include *Lycopodium inundatum* L. of subgenus *Cernuostachys; L. annotinum* L., *L. obscurum* L., and *L. tristachyum*

Fig. 125.1. *Lycopodium cernuum,* Tapanti, Costa Rica. (Photo W. H. Hodge.)

Pursh of subgenus *Lycopodium;* and *L. lucidulum* Michx. and *L. Selago* L. of subgenus *Selago.*

Among six species in Chile and southern Argentina, five belong to subgenus *Lycopodium: Lycopodium confertum* Willd., *L. magellanicum* (Beauv.) Sw., *L. chonoticum* Phil., and *L. paniculatum* Poir. belonging to one species-alliance with affinities to *L. fastigiatum* R. Br. of Australia and New Zealand and *L. spurium* of tropical South America; and *Lycopodium Gayanum* Fée & Remy belonging to another alliance, with affinities to *L. scariosum* Forst., also of Australia and New Zealand and to *L. Jussiaei* of tropical America. The sixth species is *Lycopodium fuegianum* Roiv. of subgenus *Selago.*

An early account of the tropical American species by Underwood and Lloyd (1906) recognized only 42 species. The entire genus was revised by Nessel (1939) in a large and somewhat unwieldy treatment. A more useful account of the Brazilian species was provided by Nessel (1927). The species of Chile and southern Argentina were covered by Looser (1961), and Rolleri (1980) has revised the species of Herter's *Lycopodium* section *Crassistachys,* recognizing 56 species in this large group of subgenus *Selago.*

Studies by Øllgaard on characters of the sporangia (1975) and on the branching patterns (1979a) and by Bruce (1976) on mucilage canals are pertinent to the infrageneric classification.

Fig. 125.2 *Lycopodium crassum,* Páramo de Tufino, Carchi, Ecuador. (Photo B. Øllgaard.)

Fig. 125.3 *Lycopodium subulatum,* Sevilla de Oro, Azuay, Ecuador, (Photo B. Øllgaard.)

Fig. 125.4. *Lycopodium reflexum,* north of Jalapa, Veracruz, Mexico. (Photo W. H. Hodge.)

The treatment of *Lycopodium* has been prepared in collaboration with R. James Hickey. Systematic and ecological information was provided by B. Øllgaard, who is currently preparing a treatment of *Lycopodium* for the "Flora of Ecuador."

Synopsis of *Lycopodium*

Subgenus *Cernuistachys*

Terrestrial; roots emerge directly from the stem; main stem with lateral branches and with upright ones originating from the upper side of the stem; with mucilage canals; sporophylls somewhat to quite distinct from sterile leaves; sides of the sporangium with the cell walls straight, thin, unlignified, except for nodular or semi-annular thickenings.

A subgenus of 25 or more species, with about 15 in tropical America.

Subgenus *Lycopodium*

Terrestrial; roots emerge directly from the stem; main stem with lateral branches and with upright ones originating from the sides of the stem; with or without mucilage canals; sporophylls

Fig. 125.5. *Lycopodium Trencilla,* Volcán Cayambe, Ecuador. (Photo C. R. Sperling.)

quite distinct from sterile leaves; sides of the sporangium with the cell walls sinuate, thin, lignified.

A subgenus of about 30 species, with 6 or more in tropical America.

Subgenus *Selago*

Terrestrial, rupestral, or epiphytic; roots usually emerge after descending through the cortex of the stem; stem dichotomously branched, or rarely with a main stem; without mucilage canals; sporophylls similar to the sterile leaves below the strobilus; sides of the sporangium with the cell walls sinuate, thick, lignified.

A subgenus of about 200 species, with about 100 in tropical America.

Tropical American Species

The estimate of about 125 species of *Lycopodium* in the American tropics is based on the premise that there are fewer than now currently recognized. Most studies have been directed toward distinguishing different kinds of *Lycopodium* and few have considered the probability that local and regional variation may occur within the species.

Landslides and other disturbed, open places in the mountains of Ecuador are suggested by Øllgaard (1979b) as suitable habitats for the development of a flora rich in terrestrial lycopods. These sites include variants not found in more mature vegetation, and the increased variability of species is regarded as an additional source of diversity for speciation. This concept merits

Fig. 125.6. Distribution of *Lycopodium* in America, south of Canada.

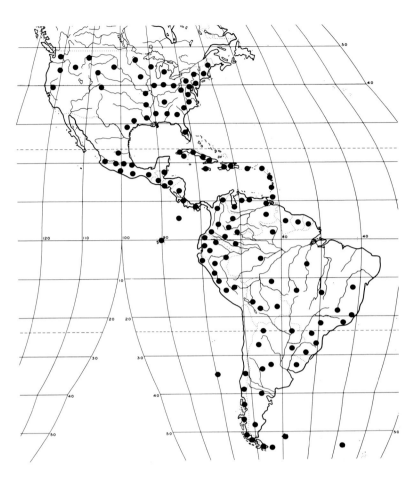

further consideration and perhaps could be extended to the epiphytic species.

Several hybrids have been proposed in *Lycopodium* from temperate and boreal regions. One of the best established among these is *Lycopodium lucidulum* × *Selago* (*L.* × *Buttersii* Abbe) which is quite intermediate between these two distinctive species. The occurrence of hybrids, along with the high chromosome number, suggest that alloploidy may have a role in the evolution of *Lycopodium*.

Some of the diversity of tropical American species is presented for each of the subgenera.

Subgenus *Cernuistachys: Lycopodium cernuum* L. (Fig. 1) has erect, much branched fertile stems to 1 m or more in height, bearing many rather small strobili, the fertile stems arise near the base of long, arching stems that root at the tip, the sporophylls (Fig. 8) are well differentiated from the sterile leaves (Fig. 7). The related *Lycopodium Lehmannii* Heiron. often has strongly incurved leaves and it and *L. glaucescens* Presl, with larger strobili, have a stem that may scramble among shrubs to several meters.

Lycopodium alopecuroides L. has an erect, simple or once branched, fertile stem with rooting branches more or less arching, a large strobilus with poorly differentiated sporophylls and small sporangia.

Lycopodium carolinianum L. has erect, simple fertile stems and a prostrate, sparsely branched rooting stem, strongly differentiated sporophylls and large sporangia. *Lycopodium contextum* Mart. is similar but with taller branches and smaller sporangia.

These species grow in open, moist habitats, especially in sandy or peaty soil in savannahs.

Subgenus *Lycopodium: Lycopodium complanatum* L. has a very long-creeping, superficial, often scrambling stem bearing erect branches with flattened branchlets and much reduced 4-ranked leaves with the lateral ones larger; the strobili are borne on a long peduncle that is usually branched at the apex. *Lycopodium Fawcetii* Lloyd & Underw. is similar, if not the same species. *Lycopodium Jussiaei* Poir. is similar to *L. complanatum* but with 5-ranked leaves, the upper two ranks larger (Fig. 12), and mostly with unbranched peduncles.

Lycopodium clavatum L. has the same habit, but with moderately large leaves borne around the stem and long, usually branched peduncles. The closely related *Lycopodium contiguum* Kl. has the strobilus sessile or nearly so. Another relative, *L. vestitum* Poir., has the apical portion of the leaf whitish-scarious (Fig. 13).

Lycopodium spurium Willd. differs in having a shallowly subterranean stem, with rather short aerial branches, and the strobili short-pendunculate.

These species usually grow in shrubby, partly open habitats in cloud forests or páramo.

Subgenus *Selago: Lycopodium rufescens* Hook. has short, freely branched stems with usually rigidly patent, broad, short, thick, stiff leaves, with a large interior cavity at the base, and sporophylls similar to the sterile leaves. *Lycopodium crassum* Willd. (Fig. 2) and *L. saururus* Lam. are larger, with narrow imbricate leaves. These three species represent a rather large group of páramo species that usually grow in open, boggy places, or among shrubs or occasionally in drier sites. *Lycopodium Trencilla* Baker (Fig. 5) has the largest leafy stem in the genus, and grows to nearly a meter long and ca. 3 cm broad. It has been collected only on Volcán Cayambe and the Llanganti Mountains, Ecuador. Many species are partly to entirely red in color, especially those that grow above 4000 m where frosts are frequent.

Lycopodium dichotomum Jacq. has rather long, thin, narrow, spreading sterile leaves and similar sporophylls. The leaves of *Lycopodium taxifolium* L. are broader and more ascending, while those of *L. linifolium* L. are narrow, rather widely spaced and have the base twisted. These species are mostly pendent epiphytes.

Lycopodium subulatum Poir. (Fig. 3) has the basal branches with rather large, spreading leaves, other branches bear small, clasping leaves which are sterile near the base of the branch and bear sporangia beyond (Fig. 9). *Lycopodium callitrichifolium* A. Br. and

7　8　9　10

11　12　13

Figs. 125.7–125.13. *Lycopodium.* **7.** Portion of sterile branch of *L. cernuum,* ×6. **8.** Portion of strobilus of *L. cernuum,* ×8. **9.** Portion of strobilus of *L. subulatum,* ×8. **10.** Portion of fertile branch of *L. molongense,* ×6. **11.** Portion of fertile branch of *L. compactum,* ×6. **12.** Portion of sterile branches of *L. Jussiaei,* ×3. **13.** Portion of sterile branch of *L. vestitum,* ×6.

L. myrsinites Lam. (*L. Skutchii* Maxon) are similar. These species are usually long-pendent, much branched epiphytes of the cloud forest.

Lycopodium molongense Hert. has long slender branches, the ones beyond the base bearing folded leaves with the margins nearly meeting. The sporophylls (Fig. 10) are also folded above the attachment of the sporangium.

Other species of the subgenus include *Lycopodium compactum* Hook. (Fig. 11), *L. reflexum* Lam. (Fig. 4), *L. serratum* Thunb., and *L. Lindavianum* Hert.

Ecology (Figs. 1–5)

Lycopodium is largely a genus of wet forests in tropical mountains. However, its ecology is diverse, as some species grow in the far north or in tropical alpine regions where frost is frequent.

A few species grow in seasonally dry habitats, as *Lycopodium carolinianum* with a variety in Africa that produces perennating tubers (Ballard, 1950). The tubers, which are dormant during the dry season, can survive light fires (Kornas, 1975).

In extratropical regions *Lycopodium* is terrestrial or rupestral, and grows on rocks and cliffs, in sandy or gravely sterile soil, in open habitats or among shrubby vegetation, or on the forest floor.

In the American tropics many species are epiphytic on trunks and branches of trees in cloud forests or rain forests where the pendent stems may reach 2 m in length. The terrestrial species grow in wet sandy or peaty depressions in savannahs, or similar local habitats in forested areas, or sometimes on bare clay soils. They also occur in rocky places or among shrubs, especially in or near the cloud forest zone, but are rarely terrestrial in the forest. The most striking habitat, where *Lycopodium* may be abundant, is the high páramo or alpine moorland with environments for which relatively few vascular plants are adapted. The terrestrial species are also often found on road cuts and land fills, and especially among regenerated vegetation of landslides. Species grow from sea level to 4400 m, mostly above 2500 m and several are above 4000 m.

Some *Lycopodium* species, especially *L. clavatum* and *L. complanatum* have very long-creeping stems that may form colonies of unusual dimension and age. Oinonen (1967) reports a clone of *Lycopodium complanatum* in southern Finland somewhat over 200 m in extent and 800 years in age.

Geography (Fig. 6)

Lycopodium is the most widely distributed genus of Pteridophyta, having an extensive range on all continents, where it occurs from north of the arctic circle south to southernmost South America and Africa and Tasmania. It is present on all of the mid-Atlantic islands and on the southern islands of South Georgia, Marion, Kerguelen, Amsterdam, Auckland and the Antipodes. In the Pacific it ranges from New Zealand and the Ryukyu Islands eastward to the Hawaiian Islands, the Marquesas, Tahiti and Rapa.

In the northern portion of its range in America, *Lycopodium* occurs from the north coast of Alaska east to Greenland and southward to northern California, Idaho, Colorado, Minnesota to eastern Texas and to Florida. It is apparently absent from central California, southern Idaho, New Mexico and North Dakota, south to west-central Texas and northern Mexico, except for an occurrence in the Black Hills of South Dakota. It also occurs in central Mexico and Central America, the West Indies, and generally in South America south to Tierra del Fuego in Argentina; also on Cocos Island, the Galápagos Islands, the Juan Fernandez Islands, Falkland Islands, and South Georgia.

The species of American *Lycopodium* are not sufficiently known to determine centers of species diversity and endemism. However, the Andes of Colombia to Peru has a very rich *Lycopodium* flora.

Spores

Three of the five basic spore types in *Lycopodium* recognized by Wilce (1972) occur in tropical American species. The *rugate (rugulate) type* characteristic of subgenus *Cernuistachys,* is prominently ridged on both faces as in *Lycopodium contextum* (Fig. 14) or coarsely tuberculate on the proximal face as in *L. alopecuroides* (Figs. 16, 17). The surface is less prominently rugate in *L. cernuum* spores of Bolivia (Fig. 15), but specimens from South Africa have prominent ridges in the equatorial area similar to those of spores noted in species above.

The *reticulate type,* characteristic of subgenus *Lycopodium,* includes several different forms. A group with larger areoles and relatively smooth proximal face, as *L. Jussiaei* (Fig. 18) is distinguished from those of *L. clavatum* with both faces reticulate (Fig. 19). Detail of the reticulate surface shows the fused papillate perispore (Figs. 20, 21). The mature wall of *L. clavatum* (Fig. 20) is mainly formed by a laminated exospore structure cemented together with sporopollenin.

The *foveolate-fossulate type,* in subgenus *Selago,* is the most common form of spore among the tropical American species of *Lycopodium.* The low ridges and foveolate surfaces are similar in spores of *Lycopodium saururus* (Fig. 26) and *L. rufescens* (Fig. 27) and characteristic of many species in the subgenus. Slight variants as in the more prominent ridges in *Lycopodium Lindavianum* vel aff. (Fig. 23), somewhat parallel orientation of ridges, concealed pits, and larger spore size in *L. molongense* (Figs. 24, 25) do not appear to characterize distinct groups as in the species with reticulate spores. The spherical spore tetrads, especially characteristic of subgenus *Selago,* as in the three abutting members of the tetrad of *L. linifolium* (Fig. 22) are formed by compression of the proximal faces of the spores.

The wall of *Lycopodium* spores consists largely of a laminated exospore above a granulate inner exospore, and a thin surface layer of perispore. The perispore was not included in the figures of *Lycopodium* spores sectioned by Afzelius et al. (1954), and Erdtman and Sorsa (1971) indicate that *Lycopodium* spores are without perine. However, spores of *L. annotinum* sectioned by Lugardon (1978) show the perispore forms an outer limited component of the wall. The small papillate structures in Figs. 20 and 21 form part of the perispore surface.

The ontogeny and structure of *Lycopodium* spores differ from other pteridophytes in the centripetal development of the exospore, from the outside toward the inner part of the spore. This is unlike the developmental sequence of spore walls in most pteridophytes and resembles the succession in the formation of Angiosperm pollen walls. Sections of *Lycopodium annotinum* spores, for comparative studies of wall development in pteriodophytes (Gullvåg, 1971), show cell organelles involved in the formation of the exospore. The exine is considered to be formed on unit membranes in a manner similar to the formation of nexine 2 in pollen walls.

Because of availability of *Lycopodium* spores in large quantity,

Figs. 125.14–125.21. *Lycopodium* spores, × 1000. **14.** *L. contextum,* proximal face with equatorial ridge, left, rugate distal face, right, Colombia, *Schultes 6906.* **15.** *L. cernuum,* nearly smooth proximal face above, rugate distal face, below, Bolivia, *Krukoff 10348.* **16, 17.** *L. alopecuroides,* Brazil, *Irwin et al. 13500.* **16.** Tuberculate to rugate proximal face with equatorial ridge. **17.** Detail of central part of proximal face, × 5000. **18.** *L. Jussiaei,* reticulate distal face, left, relatively smooth proximal face, right, Peru, *Stork & Horton 10143.* **19.** *L. clavatum,* reticulate, proximal face, left, distal right, Colombia, *Barkley & Juajibioy 6791.* **20.** *L. clavatum,* detail of reticulate structure and papillate perispore deposit, Galápagos Islands, *Stewart 1014,* × 10,000. **21.** *L. contiguum,* detail of ridges, and of papillate perispore deposit, Colombia, *Mexia 84,* × 10,000.

they often have been used in experimental work, beginning with that of Zetzsche and Huggler (1928) who coined the term sporopollenin. This was identified as a component of both pollen and spore walls which forms the most inert outer wall. Experiments of Southworth (1974) testing the solubility of pollen and spore walls with a series of solvents showed different effects on the sporopollenin of *Lycopodium* and that of pollen. Most pollen exines disintegrated in organic bases related to 2-aminoethanol while the *Lycopodium* spores were not affected by this or any other solvents. The remarkable stability of the *Lycopodium* spores suggest that the sporopollenin of these differs from that of pollen.

Cytology

The extensive range in chromosome numbers of *Lycopodium* between $2n = 46$ to 528, and especially the many high numbers, suggests complex relationships in the group. Deletion of some early inaccurate counts helps to clarify the records and verification of doubtful numbers will allow a more precise cytological assessment. Reports from tropical America (Walker, 1966; Sorsa in Fabbri, 1965) are relatively few but depict cytological complexities of the three subgenera.

Differences in chromosome size and morphology, noted in several accounts on the cytology of *Lycopodium* suggest that karyotype analyses may be useful in clarifying relationships. The two especially large chromosomes in *Lycopodium complanatum* may be counted as multiples and a possible source of error in the reports on that complex. The high chromosome numbers have likely been derived by an increase in ploidy levels although few multiple series are evident. Aneuploidy, as detected in the *L. cernuum* complex, further complicates the cytology and appears to have obscured the polyploid relationships.

Subgenus *Cernuistachys: Lycopodium cernuum,* reported as $n =$ ca. 165 from Jamaica and Trinidad, is not aligned with the series of aneuploid numbers, $n = 104, 110, 136$, reported for the species from India (Bir, 1973). Temperate species of the subgenus, *L. carolinianum* with $n = 35, 70$ from the southern United States, and *L. inundatum* with $n = 78$ from Europe and eastern North America are likewise uncorrelated.

Subgenus *Lycopodium:* The record of $n = 34$ for *Lycopodium clavatum* in Jamaica is consistent with reports for the species over

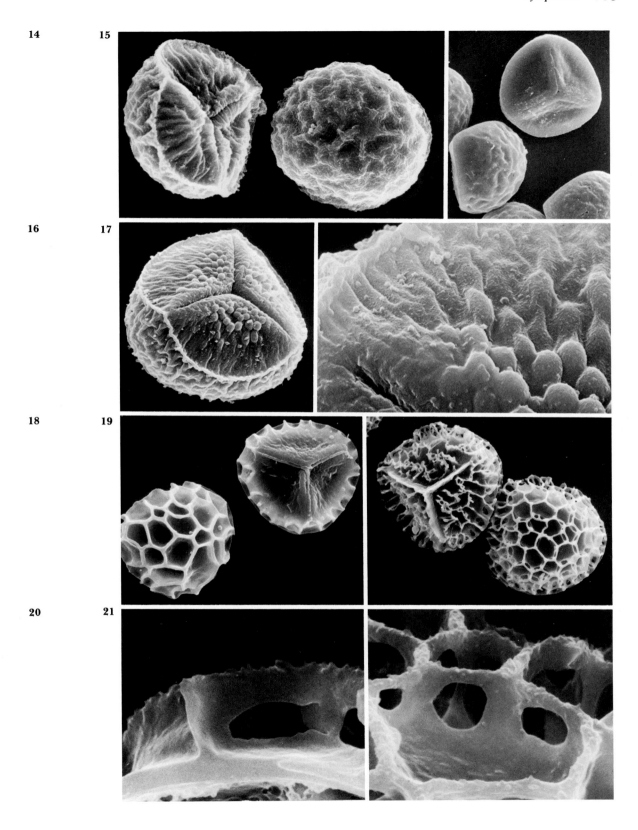

a wide geographic range in Britain, Finland, eastern Canada, and India. The same number is known for the temperate species *L. obscurum.* The record of $n = 23$ in *Lycopodium Fawcettii,* from Jamaica, also known in other species of the *L. complanatum* complex, indicates there are two cytological lineages in the subgenus.

Subgenus *Selago:* The records of $n = $ ca. 128 in *Lycopodium reflexum* from Jamaica, $n = $ ca. 132 in *L. dichotomum,* and $n = 130-140$ in *L. linifolium,* from Puerto Rico are characteristic of the high chromosome levels of the subgenus. The highest number encountered in the genus, $n = 264$, in *L. serratum,* from India is represented in this group.

Observations

The few chemical analyses of *Lycopodium* suggest that further studies of broad scope may provide data useful in the systematics of the genus. The study of phenolics (Towers and Maas, 1965) found that syringic acid was present in seven species of subgenus *Lycopodium* and absent in eight species of the other two subgenera. Investigation of flavonoid glycosides (Markham and Moore, 1980) found that *Lycopodium scariosum* and *L. cernuum* were chemically quite distinct. A survey of seven species of *Lycopodium* in northeastern North America by Marion and Manske (1946 and references therein), isolated a total of 26 alkaloids, including nicotine which had previously been known only in flowering plants. The species contained between four and nine alkaloids, including some that were unique in each species. In nine species (Braekman et al., 1974) the major alkaloid was usually lycopodine and minor components were unique to some species or occurred in different combinations.

Literature

Afzelius, M., G. Erdtman, and F. Sjöstrand, 1954. On the fine structure of the outer part of the spore wall of *Lycopodium clavatum* as revealed by the electron microscope. Svensk Bot. Tidskr. 48: 155–161.

Ballard, F. 1950. *Lycopodium carolinianum* in tropical Africa. Amer. Fern Jour. 40: 74–83.

Bir, S. S. 1973. Cytology of Indian Pteridophytes. Glimpses in Plant Research 1: 28–119.

Braekman, J. C., L. Nyembo, P. Bourdoux, K. Kahindo, and C. Hootele. 1974. Distribution des alcaloides dans le genre *Lycopodium.* Phytochem. 13: 2519–2528.

Bruce, J. G. 1976. Development and distribution of mucilage canals in *Lycopodium.* Amer. Jour. Bot. 63: 481–491.

Erdtman, G., and P. Sorsa, 1971. Pollen and Spore Morphology/Plant Taxonomy. IV. Pteridophyta. 302 pp. Almqvist & Wiksell, Stockholm.

Fabbri, F. 1965. Secondo supplemento alle tavole cromosomiche delle pteridophyta di Alberto Chiarugi. Caryologia 18: 675–731.

Gullväg, B. M. 1971. The fine structure of some pteridophyte spores before and during germination, *in:* G. Erdtman and P. Sorsa. Pollen and Spore Morphology/Plant Taxonomy. IV. Pteridophyta. pp. 252–295. Almqvist & Wiksell, Stockholm.

Herter, G. 1950. Systema Lycopodiorum. pp. 67–86 from Rev. Sudam Bot. 8: 67–86. 1949; pp. 91–116 from Rev. Sudam. Bot. 8: 93–116. 1950. Montevideo, Uruguay.

22 23
24 25
26 27

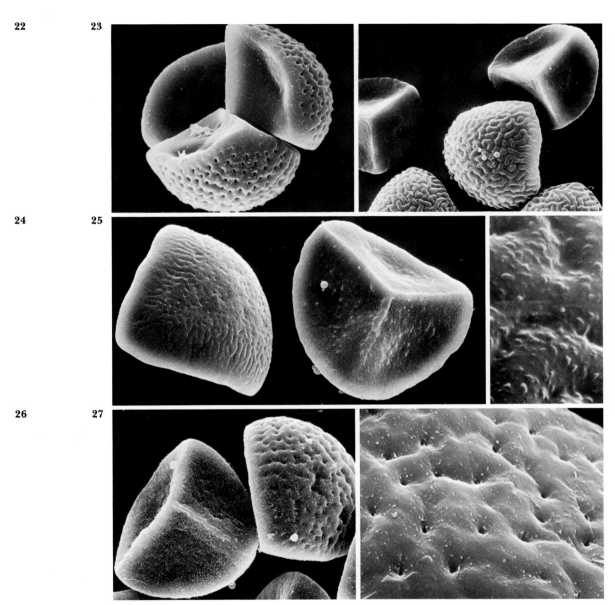

Figs. 125.22–125.27. *Lycopodium* spores, × 1000. **22.** *L. linifolium,* three members of a tetrad with abutting, smooth, proximal faces, Mexico, *Nevling & Gómez-Pompa 1636.* **23.** *L. Lindavianum* vel aff. rugate distal face below, proximal face above, Ecuador, *Holm-Nielsen et al. 6091.* **24, 25.** *L. molongense,* Colombia, *Ewan 15621.* **24.** Parallel ridges, distal face, left, proximal face, right. **25.** Detail of ridges, the foveolae obscure, × 10,000. **26.** *L. saururus,* proximal face, left, rugate distal face, right, Costa Rica, *Cronquist & Muños 8854.* **27.** *L. rufescens,* detail of foveolate-fossulate surface, Colombia, *Killip & Smith 17967,* × 5000.

Kornas, J. 1975. Tuber production and fire resistance in *Lycopodium carolinianum* L. in Zambia. Acta Soc. Bot. Polon. 44: 653–663.

Looser, G. 1961. Los pteridófitos o helechos de Chile, I. Revis. Univers. Católica Chile 46: 213–262.

Lugardon, B. 1978. Isospore and microspore walls of living pteridophytes: Identification possibilities with different observation instruments. Palyn. Conf. Lucknow 1: 152–163.

Marion, L., and R. H. F. Manske. 1946. The alkaloids of *Lycopodium* species. VIII. *Lycopodium sabinaefolium* Willd. Canad. Jour. Res. B, 24: 63–65.

Markham, K. R., and N. A. Moore. 1980. Comparative flavonoid glycoside biochemistry as a chemotaxonomic tool in the subdivision of the classical 'genus' *Lycopodium*. Biochem. Syst. Ecol. 8: 17–20.

Nessel, H. 1927. As Lycopodiáceas do Brasil. Arch. bot. estad. São Paulo 1: 358–533.

Nessel, H. 1939. Die Bärlappgewächse (Lycopodiaceae). 404 pp. Gustav Fischer, Jena.

Oinonen, E. 1967. Soral regeneration of ground pine (*Lycopodium complanatum* L.) in southern Finland in the light of the dimensions and age of its clones. Acta Forest. Fennica 83 (3): 1–85.

Øllgaard, B. 1975. Studies in the Lycopodiaceae, I. Observations of the structure of the sporangium wall. Amer. Fern Jour. 65: 19–27.

Øllgaard, B. 1979a. Studies in the Lycopodiaceae, II. The branching patterns and infrageneric groups of *Lycopodium sensu lato*. Amer. Fern Jour. 69: 49–61.

Øllgaard, B. 1979b. *Lycopodium* in Ecuador—habits and habitats, *in:* K. Larsen and P. Holm-Nielsen, Eds., Tropical Botany, pp. 381–395. Academic Press, London, New York.

Pichi-Sermolli, R. E. G. 1977. Tentamen pteridophytorum in taxonomicum ordinem redigendi. Webbia 31: 315–512.

Rolleri, C. H. 1980. Sinopsis de las especies de *Lycopodium* L. (Lycopodiaceae, Pteridophyta) de la seccion *Crassistachys* Herter. Rev. Museo. La Plata, n.s. Sec. Bot. 13: 61–114.

Southworth, D. 1974. Solubility of pollen exines, Amer. Jour. Bot. 61: 36–44.

Towers, G. H. N., and W. S. G. Maas. 1965. Phenolic acids and lignins in the Lycopodiales. Phytochem. 4: 57–66.

Underwood, L. M., and F. E. Lloyd. 1906. The species of *Lycopodium* of the American tropics. Bull. Torrey Bot. Cl. 33: 101–124.

Walker, T. G. 1966. A cytotaxonomic survey of the pteridophytes of Jamaica. Trans. Roy. Soc. Edinburgh 66: 169–237.

Wilce, J. H. 1972. Lycopod spores, 1. General spore patterns and the generic segregates of *Lycopodium*. Amer. Fern Jour. 52: 65–79.

Zetzsche, F., and K. Huggler. 1928. Untersuchungen über die Membran der Sporen und Pollen, 1. *Lycopodium clavatum* L. Liebigs Ann. 461: 89–108.

Family 28.   Selaginellaceae

Selaginellaceae Milde, Hoher. Sporenpfl. Deutschl. Schweiz 136. 1865, as Selaginelleae. Type: *Selaginella* Beauv.

Description

Stem mono- or polyprotostelic or siphonostelic, indurated or not, branched, bearing rather few, often long roots usually at a branch of the stem; leaves simple, ca. 0.5–10 mm long, with one vein; sporangia short-stalked, single, near the axil of a leaf; heterosporous, spores without chlorophyll. Megagametophytes minute, partly protruding from the megaspore wall, microgametophytes developed wholly within the microspore, the wall rupturing to release the spermatozoids.

Comments on the Family

The Selaginellaceae are a distinctive family including the single genus *Selaginella*, only distantly related to others as the Lycopodiaceae and Isoetaceae. Heterospory, and the presence of vessels in some species of subgenus *Selaginella*, indicate the family is specialized. The vascular system may consist of a single proto-stele or sometimes two or three steles, and in *S. exaltata* there may be several steles (Mickel and Hellwig, 1969).

Selaginellites Zeller, a genus of the Carboniferous and later periods, includes some older species that have been placed in *Selaginella*. Some species of *Lycopodites* Lindl. & Hutt., of the Cretaceous and earlier, may belong to *Selaginella* or possibly an allied extinct genus.

The name Selaginellaceae Milde is illegitimate being a latter homonym of the illegitimate Selaginellaceae Willk. 1861, but is retained pending its conservation.

Literature

Mickel, J. T., and R. L. Hellwig. 1969. Actino-plectostely, a complex new stelar pattern in *Selaginella*. Amer. Fern Jour. 59: 123–134.

126. *Selaginella*
Figs. 126.1–126.27

Selaginella Beauv., Magasin Encycl. 5: 478. 1804; and Prod. Fam. Aethéog. 101. 1805, *nom. conserv.* Type: *Selaginella spinosa* Beauv., *nom. nov.* for *Lycopodium selaginoides* L. = *Selaginella selaginoides* (L.) Link.

Selaginoides Séguier, Pl. Veron. 3: 51. 1754. Type: *Lycopodium selaginoides* L. = *Selaginella selaginoides* (L.) Link.

Selago Browne, Nat. Hist. Jam. 82. 1756, not. L. 1753. Type: *Lycopodium serpens* Poir. = *Selaginella serpens* (Poir.) Spring.

Lycopodioides Boehm., Defin. Gen. Pl. (Ludwig) Ed. Boehmer. 485. 1760. Type: *Lycopodium denticulatum* L. = *Selaginella denticulata* (L.) Spring. [Subgenus *Stachygynandrum*].

Mirmau Adans., Fam. Plantes 2: 491. 1763, *nom, superfl.* for *Selaginoides* Séguier and with the same type.

Trispermium Hill, Gen. Hist. Nat. 2 (Hist. Pl.): 112. 1773, *nom. superfl.* for *Lycopodioides* Boehm. and with the same type.

Polycocca Hill, *ibidem* 116. 1773., *nom. superfl.* for *Selaginoides* Séguier and with the same type.

Stachygynandrum Mirbel, Hist. Nat. Veg. (Lam. & Mirbel) 3: 477. 1802. Type: *Lycopodium flabellatum* L. (*Stachygynandrum flabellatum* (L.) Beauv.) = *Selaginella flabellata* (L.) Spring. *Selaginella* subgenus *Stachygynandrum* (Beauv.) Baker, Jour. Bot. 21: 3. 1883.

Didiclis Mirbel, *ibidem* 477. 1802. Type: *Lycopodium ornithopodioides* L. (*Didiclis ornithopodioides* (L.) Jaume St. Hil. = *Selaginella ornithopodioides* (L.) Spring. [Subgenus *Stachygynandrum*].

Gymnogynum Beauv., Prod. Fam. Aethéog. 103. 1805. Type: *Gymnogynum domingense* Beauv. = *Selaginella plumosa* (L.) Presl (*Lycopodium plumosum* L.). [Subgenus *Stachygynandrum*].

Diplostachyum Beauv., *ibidem* 104. 1805, as *Diplostachium*. Type: *Diplostachyum helveticum* (L.) Beauv. (*Lycopodium helveticum* L.) = *Selaginella helvetica* (L.) Spring. [Subgenus *Stachygynandrum*].

Selaginella subgenus *Homostachys* Baker, Jour. Bot. 21: 4. 1883. Type: not selected. [Subgenus *Stachygynandrum*].

Selaginella subgenus *Heterostachys* Baker, *ibidem* 4. 1883. Type: not selected. [Subgenus *Stachygynandrum*].

Fig. 126.1. *Selanginella hispida,* Mt. Albion, St. Ann, Jamaica. (Photo W. H. Hodge.)

Heterophyllum C. Börn., Fl. Deutsche Volk 110, 285. 1912. Type: *Heterophyllum helveticum* (L.) C. Börn. (*Lycopodium helveticum* L.) = *Selaginella helvetica* (L.) Spring. [Subgenus *Stachygynandrum*].

Hypopterygiopsis Sakurai, Bot. Mag. (Tokyo) 57: 255. 1943. Type: *Hypopterygiopsis reptans* Sakurai = *Selaginella Sakuraii* Miller, *nom. nov.*, not *Selaginella reptans* Sod. [Subgenus *Stachygynandrum*].

Description   Terrestrial, rupestral or rarely epiphytic; stem slender, branched, sometimes dichotomously, prostrate-creeping or with ascending branches, or pendent-epiphytic, or erect from a usually stoloniferous base; leaves ca. 0.5–10 mm long, with one vein, borne in a close spiral or alternately in four ranks; sporangia large, borne in or near the axil of a well-differentiated sporophyll, in a quadrangular, or (in *S. deflexa* and *S. selaginoides*) a cylindrical strobilus, megasporangia commonly basal in the strobilus, usually with four megaspores, larger and different in size and color than the microsporangia that are usually borne above them and have many microspores; megaspores tetrahedral-globose, trilete, the laesurae $\frac{3}{4}$ to usually equal to the radius, often with a more or less prominent equatorial ridge, rugose-reticulate, rugose, papillate, tuberculate, granulate, rarely nearly smooth on the proximal face, microspores tetrahedral-globose, trilete, often compressed or the proximal face more or less depressed, the laesurae $\frac{1}{2}$ to equal to the radius, usually finely to coarsely echinate, rugose, papillate, perforate-cristate, or granulate. Chromosome number: $n = 8, 9, 10, 12, 18, 36; 2n = 14, 16, 18, 20, 24, 27, 36, 48–50, 60$.

Fig. 126.2. *Selaginella flabellata,* Dominica. (Photo W. H. Hodge.)

Plants of *Selaginella* vary greatly in size. Some small species have stems about 3 cm long, while larger ones have stems 50 cm to ca. 1 m long, or exceptionally to 5 m long. The roots of *Selaginella* have traditionally been called rhizophores; however, Webster and Steeves (1964a) conclusively established that they are roots. Leaves may be borne in a closely imbricate spiral around the stem (Fig. 5) or in most species, in four ranks with two large lateral leaves and two smaller median leaves (Figs. 7, 10, 11). At the leaf base, there is a small, adaxial ligule which is quite inconspicuous on the mature leaf. The strobilus is quadrangular (Fig. 6) in all but two species.

Systematics

Selaginella is a nearly worldwide genus of about 700 species, with about 270 of them in America. The two types of sporangia and associated heterospory are distinctive features of the genus. However, sterile materials may sometimes be confused with some flat-branched lycopodiums and very small sterile species with liverworts. The new bryophyte genus *Hypopterygiopsis* was based on a small species of *Selaginella.*

There have been several infrageneric classifications of *Selaginella,* as that of Hieronymus (1901) with two subgenera, 12 other

Fig. 126.3. *Selaginella pallescens,* Pedregal Esquilón, near Jalapa, Veracruz, Mexico, stems curled inward in desiccated state. (Photo W. H. Hodge.)

Fig. 126.4. Distribution of *Selaginella* in America, south of lat. 35° N.

infrageneric taxa, in addition to 44 species-groups. Walton and Alston (1938) recognized four subgenera and seven series and four species-groups. We recognize two subgenera; in subgenus *Selaginella,* the infrageneric classification is probably adequate (Tryon, 1955), while in the much larger subgenus *Stachygynandrum* a modern classification based on more fundamental characters than those used previously is needed.

The segregation of *Selaginella* into three genera by Rothmaler (1944) is based on an inadequate study of the genus.

There have been several special studies of the tropical American species. Tryon (1955) reviewed the subgenus *Selaginella* and revised the species of the *S. rupestris* group. Alston (1936) revised the species of Brazil and (1939) those of Argentina, Uruguay and Paraguay, and (1952) those of the West Indies, and (1955) the North American species of subgenus *Stachygynandrum*. Alston et al. (1981) treat about 130 species of tropical South America. Bautista et al. (1975) presented an illustrated treatment of the Amazon species of Brazil, and Gregory and Riba (1979) the species of Veracruz, Mexico.

Synopsis of *Selaginella*

Subgenus *Selaginella*

Leaves similar (Fig. 5), sporophylls similar.

A subgenus of about 50 species, with 35 in America. In addition to the 44 species of the *Selaginella rupestris* (L.) Spring group, which are predominently American, the other species are: the circumboreal *S. selaginoides* (L.) Link and the related *S. deflexa* Brack. of the Hawaiian Islands, *S. pygmaea* (Kaulf.) Alston of southern Africa, and *S. gracillima* (Kze.) Alston and *S. uliginosa* (Labill.) Spring of Australia. The latter two are the only annual species in the genus.

Subgenus *Stachygynandrum*

Leaves, at least on the branches, dimorphic (Fig. 10), sporophylls similar or dimorphic.

A subgenus of about 625 species, with about 220 in America. The extratropical American species are: *Selaginella Douglasii* (Hook. & Grev.) Spring of the northwestern United States and British Columbia, *S. ludoviciana* A. Br. of the southeastern United States, and *S. apoda* (L.) Morren of the eastern United States and adjacent Canada south to Mexico.

Tropical American Species

The species of the American tropics are rather well known through the studies of Alston, but additional work on them is needed to incorporate new materials, including new species.

Many of the characters used to group or distinguish species of *Selaginella* relate to the following features. The roots of species with prostrate stems may arise from the upper side, that is, on the side of the stem corresponding to the upper surface of the leaves, or they may arise on the lower side. The stem of some species is enlarged and jointed (articulate) at the site of a branch (Fig. 9). Species with erect leafy stems may have the apex much prolonged into a rooting tip (flagelliform). The single leaf often borne where the stem branches, the axillary leaf (Fig. 9), may have special characters. The arrangement of the megasporangia and the microsporangia in the strobilus (Horner and Arnott, 1963) is sometimes a useful species character.

Some of the diversity in the tropical American species of *Selaginella* is presented in the following account.

In subgenus *Selaginella,* the only tropical species are those of the *Selaginella rupestris* group. These usually form loose or densely intricate mats (Fig. 8), as in *S. peruviana* (Milde) Hieron., *S. Sartorii* Hieron., *S. Sellowii* Hieron., and *S. Wrightii* Hieron. In *Sellaginella rupincola* Underw. the short stems are erect and in *S. extensa* Underw. the stems are long and often epiphytic-pendent.

In subgenus *Stachygynandrum*, several groups may be recognized on the basis of habit, although these are not well defined due to species intermediate between the main types.

1. Main stem and leafy branches prostrate or partly ascending, rooted throughout or at least well beyond the base, main stems with median and lateral leaves.

Small plants with lateral leaves ca. 1 mm long, the branches mostly 2 cm long. *Selaginella cladorrhizans* A. Br., *S. hispida* (Willd.) Urban (Fig. 1), *S. microphylla* (HBK.) Spring, and *S. porphyrospora* A. Br.

Larger plants, lateral leaves 2–4 mm long, branches ca. 4–6 cm long. *Selaginella delicatissima* A. Br., *S. denudata* (Willd.) Spring, *S. eurynota* A. Br., *S. Kunzeana* A. Br., and *S. marginata* Willd.) Spring.

2. Main stem erect, usually from a stoloniferous base, rooted at or near the base, most of the leaves of the main stem, below the branches, more or less isomorphic, borne around the stem.

Primary branches simple or flabellately branched, ultimate branches rather few, lateral leaves 5–10 mm long. *Selaginella articulata* (Kze.) Spring, *S. bombycina* Spring, *S. sericea* A. Br. and *S. speciosa* A. Br.

Primary branches freely and usually pinnately branched, ultimate branchlets many, lateral leaves ca. 2 mm long. *Selaginella anceps* (Presl) Presl, *S. flabellata* (L.) Spring (Fig. 2), *S. Galeottii* Spring, *S. oaxacana* Spring and *S. pedata* Kl. In *Selaginella erythropus* (Mart.) Spring and *S. haematodes* (Kze.) Spring, the main stem is red. *Selaginella exaltata* (Kze.) Spring is the largest American species, with the scambling stem to 5 m long.

3. Leafy braches closely clustered, often forming a rosette. In *Selaginella lepidophylla* (Hook. & Grev.) Spring and *S. pilifera* A. Br., the main stem has median and lateral leaves below the branches, while in *S. convoluta* (Arn.) Spring and *S. pallescens* (Presl) Spring, the main stem has the leaves more or less isomorphic, at least toward the base.

Ecology (Figs. 1–3)

Selaginella species usually grow in mesic, shaded forests. They are mostly terrestrial but also grow on damp rocks, by streams or waterfalls, or on shaded, wet cliffs. Other species grow in arctic-alpine turf, in rocky deserts or in xeric scrubland. Others grow in thickets, rather dry woods, in marshes and in secondary forests. *Selaginella angustiramea* Muell. & Bak. of New Guinea is sometimes an epiphyte growing on branches at a height of 25 m.

In tropical America, *Selaginella* grows in wet primary rain forests, in secondary rain forests, in oak and pine woods, and on road banks and in rough pastures. It often occurs on wet cliffs, on rocks by streams and near waterfalls, and also on mesic lava

Figs. 126.5–126.11. Leafy stems and strobili of *Selaginella*. **5.** Portion of leafy stem, *S. Sellowii,* ×10. **6.** Apical portion of strobilus of *S. anceps,* ×10. **7.** Lower parts of strobili on branches with dimorphic leaves, *S. denudata,* ×8. **8.** Portion of mat of stems, *S. Underwoodii* Hieron., ×1. **9.** Lower side of leafy stem of *S. arthritica* Alston with joint just below the branches and axillary leaf between them, ×10. **10.** Upper side of leafy branch of *S. exaltata,* with large lateral leaves and smaller, ascending median leaves, ×10. **11.** Part of leafy branches of *S. lepidophylla,* ×10.

beds. *Selaginalla extensa* is sometimes a low epiphyte with long-pendent stems. Species of the *Selaginella rupestris* group mostly grow in semixeric habitats that are periodically dry, although some as *S. eremophila* Maxon grow in dry, rocky deserts where the rainfall is low and sporadic. Species of seasonally dry habitats have the capacity to survive for long periods in the dry condition. A study of *Selaginella densa* Rydb. by Webster and Steeves (1964b) found that mats would resume growth when watered after 2 years and 9 months of desiccation.

Species confined to calcareous rocks include *Selaginella lepidophylla* and *S. pilifera*. These and a few other "resurrection plants" curl their spreading rosette of leafy branches inward under desiccation to form a tight ball. *Selaginella pallescens* (Fig. 3) often is similar in its response to dry conditions.

In tropical America, *Selaginella* grows from near sea level to 4000 m. Species of subgenus *Stachygynandrum* grow to ca. 2200 m, but mostly below 1500 m, while some species of subgenus *Selaginella,* grow higher, *S. Sellowii* to 2700 m and *S. peruviana* to 4000 m.

Geography (Fig. 4)

Selaginella is widely distributed in America, Africa and Europe, east to the Bering Straits, to Kamchatka, Japan, and to New Guinea and Australia also in the Pacific east to the Hawaiian Islands, the Marquesas, Tahiti and Rapa.

In America, *Selaginella* occurs from northern Alaska east to Greenland, at lat. 64° N, and south to Mendoza and Buenos Aires in Argentina; it is also on Cocos Island. It is certainly better represented in the Amazon basin than shown in Fig. 4, for 31 species are known from that region. There are apparently no native species in Chile. Early reports of six species (Sturm, 1858), including *Selaginella chilensis* (Willd.) Spring, were all based on erroneous labels (Alston, 1939).

The two subgenera have somewhat equivalent ranges in tropical America, except that subgenus *Selaginella* is absent from Nicaragua to Panama, Cocos Island, the Lesser Antilles and most of the Greater Antilles, and from most of Brazil.

The center of species diversity and of endemism in the *Selaginella rupestris* group is in the southwestern United States, from Texas to southern California, and Mexico. Twenty-three of the 44 species occur in this region and 15 are endemic to it. There are few tropical species of the group except in Mexico. *Selaginella Sellowii* is in eastern Cuba, *S. Steyermarkii* Alston in Guatemala and there are four species in South America.

Subgenus *Stachygynandrum* centers on the eastern slopes of the Andes from Colombia to Bolivia, where perhaps as many as 100 species occur.

Selaginella species can migrate readily on continental areas (Tryon, 1971), but probably due to heterospory, are poorly represented on isolated islands. A single species occurs on Cocos Island, only seven in the Lesser Antilles, and few on Pacific islands. *Lycopodium,* by comparison, has more species, usually about twice as many, on the islands where *Selaginella* occurs.

12   13   14

15   16

17   18   19

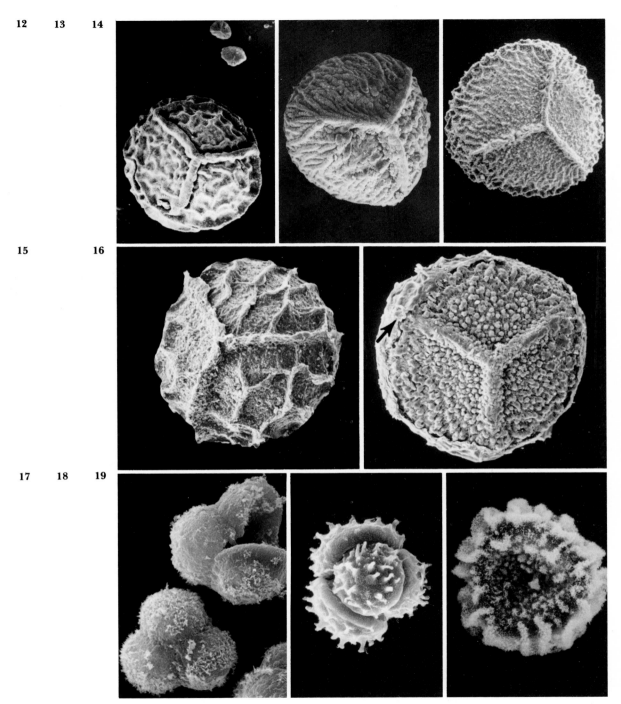

Figs. 126.12–126.19. *Selaginella* spores. **12.** *S. Sartorii,* megaspore rugose-reticulate, and microspores, right above, Mexico, *Kenoyer 1566,* ×150. **13.** *S. minima,* rugose megaspore, proximal face, Panama, *Johnston 327,* ×200. **14.** *S. porphyrospora,* fine, compact reticulate, megaspore, proximal face, Mexico, *Pringle,* Nov. 1902, ×150. **15.** *S. lepidophylla,* coarsely reticulate megaspore, Mexico, *Pringle 11275,* ×150. **16.** *S. delicatissima,* tuberculate megaspore, proximal face, surrounded by megasporangium wall, at arrow, Mexico, *Botteri 91,* ×150. **17.** *S. estrellensis,* microspore tetrads partly separated at top, Costa Rica, *Skutch 3371,* ×1000. **18.** *S. tarapotensis* tetrad, echinate microspores, Mexico, *Mexia 9270a,* ×1000. **19.** *S. delicatissima* compact tetrad, spores with perforate, cristate ridges, Mexico, *Botteri 91,* ×1000.

Spores

Selaginella microspores are more diverse than the megaspores as evident in the micrographs of some of the tropical American species.

Relative size differences of the two spore types are evident in some species (Figs. 12, 20, 24). The megaspores of *S. Sartorii* are rather small while the microspores are relatively large (Fig. 12) in comparison to those of *S. Galeottii* (Fig. 20) and *S. marginata* (Fig. 24). *Selaginella Sartorii* megaspores have a low, rugose-reticulate formation characteristic of subgenus *Selaginella*. The laesurae, as in most *Selaginella* megaspores, are equal to the radius and frequently a ridge is more or less developed in the equatorial region. The rugae may be compact and form more or less parallel striae as in *S. minima* Spring (Fig. 13). A low, compact reticulum, as in *S. porphyrospora* (Fig. 14) is characteristic of megaspores of many species in subgenus *Stachygynandrum*. The meshwork may be somewhat coarser as in *S. stenophylla* A. Br. (Fig. 27) or very coarse as in *Selaginella lepidophylla* (Fig. 15), *S. Galeottii* (Fig. 20) and *S. marginata* (Fig. 24). The proximal and distal faces may be uniform as in *S. Galeottii* or more often differ as in *S. microdendron* Baker (Fig. 25). In that species, as in *S. delicatissima* (Fig. 16), the proximal face is papillate to tuberculate and the distal face is reticulate.

Selaginella microspores may be dispersed in tetrads enclosed within a special wall as in *S. estrellensis* Hieron. and *S. delicatissima* (Figs. 17, 19), or dispersed singly as *S. microdendron* (Fig. 26, below) or cling together as in *S. tarapotensis* Baker (Fig. 18) and *S. Galeottii* (Fig. 22). They often adhere to the megaspores as in *S. Galeottii* and *S. marginata* (Figs. 20, 24). The fine echinate processes of the megaspore surface as in *S. microdendron* (Fig. 26, above), as well as the usually echinate microspores appear to be adaptations for dispersal of the two types of spores together.

The complex stratification of the megaspore wall evident in the section of *S. Galeottii* (Fig. 23) shows long stranded material adjacent to the protoplast, an open irregular matrix of short strands, and above this the outer part of the wall composed of a regular gridlike structure (Fig. 21). Scanning electron and TEM micrographs show the gridlike formation is organized in hexagonal units and there are heavy deposits of silica in the outer parts of the wall (Tryon and Lugardon, 1978). This gridlike material is also characteristic of other species as *S. marginata* and *S. myosurus* (Sw.) Alston. The megaspore wall of other species may consist of a less complex, labyrinth-like formation in *A. delicatissima*, *S. microdendron,* and in *S. Martensii* Spring which has been examined with TEM, SEM and X-ray spectography (Tryon and Lugardon, 1978). Sections of *Selaginella* megaspore wall figured by Morbelli (1977) include similar forms and others with more compact, coarse particles.

Study of wall formation in *Selaginella Broadwayii* Hieron. along with TEM micrographs of other species by Pettitt (1971) provide an insight to the wall development in *Selaginella*. A section of a young tetrad prior to initiation of the spore walls shows there are two mucopolysacchride layers, one around each of the

20 21 22

23 24

25 26 27

26

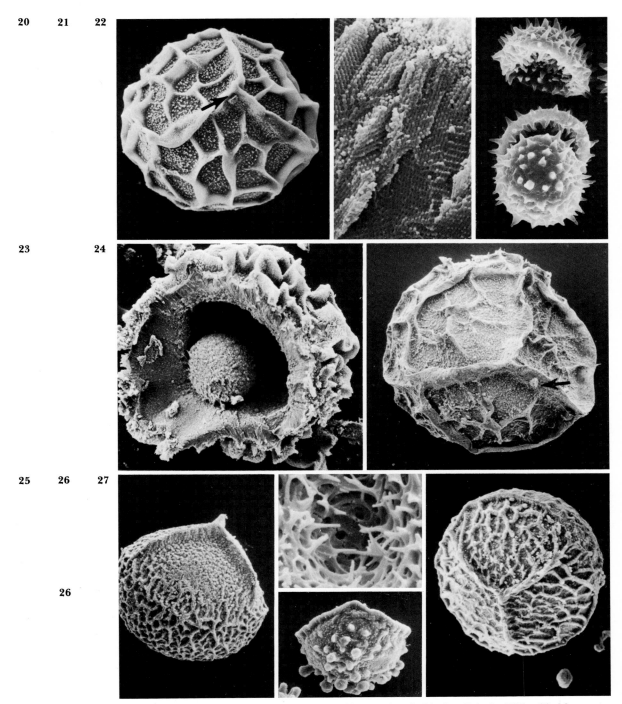

Figs. 126.20–126.27. *Selaginella* spores. **20–23.** *S. Galeottii,* Mexico, *Pringle 6170.* **20.** Megaspore with microspore on laesura, at arrow, × 100. **21.** Gridlike upper echinate formation of megaspore wall, × 5000. **22.** Echinate microspores, × 1000. **23.** Section of megaspore wall, the shrunken spheroidal protoplast at center covered with and attached to wall by inner stranded exospore, the center an open mesh of short exospore strands, the gridlike outer part below the coarse surface reticulum, × 190. **24.** *S. marginata* reticulate megaspore with microspore on laesura, at arrow. Mexico, *Langlassé 371,* × 100. **25, 26.** *S. microdendron,* British Honduras, *Schipp 99.* **25.** Megaspore, lateral view, the coarsely papillate proximal face above, reticulate distal face below, × 150. **26.** Above, detail of echinate megaspore surface, × 5000; below, clavate microspore, × 1000. **27.** *S. stenophylla* megaspore, reticulate proximal face and tetrad of microspores below, Mexico, *Purpus 2389,* × 150.

spores and another layer enclosing the tetrad. The manner of incorporation of sporopollenin particles into the exospore, which forms the major surface contours of the microspores, was also examined. *Selaginella* microspore walls sectioned by Lugardon (1972) show the outermost perispore layer and the special para-exospore layers which are characteristic of *Selaginella* and *Isoetes,* but quite distinct from spores of most other pteridophytes.

Cytology

Chromosome numbers are low in *Selaginella,* between $n = 7$ ($2n = 14$) and 36, although there are few records relative to the large size of the genus, they cover a broad geographic range. The neotropical species are mostly $2n = 20$ as *Selaginella Broadwayii, S. cladorrhizans, S. mnioides* A. Br., *S. trifurcata* Baker, *S. viticulosa* Kl., all from Trinidad, and *S. Krugii* Hieron. from Puerto Rico, and *S. sulcata* (Poir.) Spring from Brazil. A few with $2n = 18$ in the American tropics include *S. Hartii* Hieron. and *S. platybasis* Baker from Trinidad and *S. oaxacana* from Costa Rica. The temperate species are uniformly $2n = 18$ from western North America, Europe and Australia. This number appears to be most widely distributed for in addition to the records from temperate areas, $2n = 18$ is reported for species of Borneo, India and Malaya. Reports of $n = 12$ and $2n = 24$ for species of India suggest the possibility of six as a base number upon which 18 and higher multiples could be derived.

Reports of chromosome numbers for some 60 species of *Selaginella,* including 10 from the American tropics (Jermy et al., 1967), and eight from south India (Kuriachan, 1963, and Fabbri, 1965) are the major accounts of tropical species of *Selaginella.* Reports based on low numbers as 6, 7 and 8 are infrequent and records largely appear to be based on either 9 or 10. Some general correlations are drawn between the base numbers, the growth patterns and the ecology of the species by Jermy et al. (1967). Those with chromosome numbers based on 10 have determinate lateral branch systems and simultaneous cone production, and grow in dense forests. In contrast, species with chromosome numbers based on 9 have more or less indeterminate branch systems and sporadic cone production, and are characteristic of more open habitats in temperate regions. They propose that aneuploidy, accompanied by changes in coning habit, has had a major role in the evolution and spread of *Selaginella* from tropical to temperate regions.

Selaginella chromosomes are the smallest among pteridophytes. Those of a species of India range between 1.0 and 1.5 μm (Abraham et al., 1962). Marked differences in chromosome morphology and size are evident in comparisons of eight species by Jermy et al. (1967). Size appears to be inversely correlated with chromosome number, since species with low numbers as $2n = 14$ in *S. Schlechteri* Hieron. of New Guinea have the largest chromosomes.

Literature

Abraham, A., C. A. Ninan, and P. M. Mathew. 1962. Studies in the cytology and phylogeny of the pteridophytes VII. Observations on one hundred species of South Indian ferns. Jour. Ind. Bot. Soc. 41: 339–342.

Alston, A. H. G. 1936. The Brazilian species of *Selaginella*. Fedde Repert. 40: 303–319.

Alston, A. H. G. 1939. The "Selaginellae" of Argentina, Uruguay, and Paraguay. Physis 15: 251–257.

Alston, A. H. G. 1952. A revision of the West Indian species of *Selaginella*. Bull. Brit. Mus. (Nat. Hist.) Bot. 1: 27–47.

Alston, A. H. G. 1955. The heterophyllous Selaginellae of continental North America. Bull. Brit. Mus. (Nat. Hist.) Bot. 1: 221–274.

Alston, A. H. G., A. C. Jermy and J. M. Rankin. 1981. The genus *Selaginella* in tropical South America. Bull. Brit. Mus. (Nat. Hist.), Bot. 9 (4): 233–330.

Bautista, H. P., M. E. van den Berg, and P. B. Cavalcante. 1975. Flora Amazónica, 1. Pteridófitas. Bol. Mus. Goeldi, n.s. Bot. 48: 1–48.

Fabbri, F. 1965. Secondo supplemento alle tavole cromosomiche della pteridophyta di Alberto Chiargui. Caryologia 18: 675–731.

Gregory, D., and R. Riba. 1979. Flora de Veracruz, Selaginellaceae, fasc. 6: 1–36.

Hieronymus, G. 1901. Selaginellaceae, in Engler & Prantl, Nät. Pflanz. 1 (4): 621–715.

Horner, H. T., and H. J. Arnott. 1963. Sporangial arrangement in North American species of *Selaginella*. Bot. Gaz. 124: 371–383.

Jermy, A. C., K. Jones, and C. Colden. 1967. Cytomorphological variation in *Selaginella*. Jour. Linn. Soc. Bot. 60: 147–158.

Kuriachan, P. I. 1963. Cytology of the genus *Selaginella*. Cytologia 28: 376–380.

Lugardon, B. 1972. Sur la structure fine et la nomenclature des parois microsporales chez *Selaginella denticulata* (L.) Link et *S. selaginoides* (L.) Link. C. R. Acad. Sci. Paris 274: 1656–1659.

Morbelli, M. A. 1977. Esporas de las especies Argentina de *Selaginella* (Selaginellaceae-Pteridophyta) Obra Cent. Mus. La Plata 3: 121–150.

Pettitt, J. M. 1971. Some ultrastructural aspects of sporoderm formation in pteridophytes, *in:* G. Erdtman & P. Sorsa. Pollen and Spore morphology/Plant taxonomy, pp. 227–251, Almqvist & Wiksell, Stockholm.

Rothmaler, W. 1944. Pteridophyten-Studien, 1. Fedde Repert. 54: 55–82.

Sturm, J. W. 1858. Enumeratio plantarum vascularium cryptogamicarum Chilensium. Abhandl. Natur. Ges. Nurenburg 1858: 151–202. Reprint 1–52. 1858.

Tryon, A. F., and B. Lugardon. 1978. Wall structure and mineral content in *Selaginella* spores. Pollen et Spores 20: 315–339.

Tryon, R. 1955. *Selaginella rupestris* and its allies. Ann. Mo. Bot. Gard. 42: 1–99.

Tryon, R. 1971. The process of evolutionary migration in species of *Selaginella*. Brittonia 23: 89–100.

Walton, J., and A. H. G. Alston. 1938. Lycopodiinae, 500–506, *in:* Verdoorn (Ed.), Manual of pteridology. 640 pp. M. Nijoff, The Hague.

Webster, T. R., and T. A. Steeves. 1964a. Developmental morphology of the root of *Selaginella Kraussiana* A. Br. and *Selaginella Wallacei* Hieron. Canad. Jour. Bot. 42: 1665–1676.

Webster, T. R., and T. A. Steeves. 1964b. Observations on drought resistance in *Selaginella densa* Rydb. Amer. Fern Jour. 54: 189–196.

Family 29.   Isoetaceae

Isoetaceae Reichenb., Bot. Damen Kunst. Freunde Pflanzenw. 309. 1828. Type: *Isoetes* L.

Description

Stem protostelic, the stele anchor-shaped or cylindrical, not indurated, bearing many, long, firm, usually dichotomously branched roots; leaves borne in a close spiral or distichous, more or less acicular, with 1 vein; sporangia very large, single, sessile, thin-walled, and more or less embedded near the base of the leaf on the adaxial side; heterosporous, spores without chlorophyll. Megagametophytes and microgametophytes minute, partly protruding from the megaspore or microspore wall.

Comments on the Family

The Isoetaceae are a family of one genus, *Isoetes*. It is a very distinctive family without close relationship to other pteridophytes, although with more affinity to the Selaginellaceae than to other families.

Fossils of *Isoetes*, considered by some as *Isoetites* Münster, are known as early as the Upper Triassic. Brown (1939) provides morphological details of some of these early fossils.

Literature

Brown, R. W. 1939. New occurrence of the fossil quillworts called *Isoetites*. Jour. Wash. Acad. Sci. 48: 358–361.

127.   *Isoetes*
Figs. 127.1–127.23

Isoetes L., Sp. Pl. 1100. 1753; Gen. Pl. ed. 5, 486. 1854. Type: *Isoetes lacustris* L.

Calamaria Boehm., Defin. Gen. Pl. (Ludwig) Ed. Boehm. 500. 1760. *nom. superfl.* for *Isoetes* and with the same type.

Cephaloceraton Gennari, Comment. Soc. Crittogam. Ital. 1: 111. 1862. Type: *Cephaloceraton histrix* (Bory) Genn. = *Isoetes histrix* Bory.

Isoetella Gennari, *ibidem* 1: 114. 1862. Type: *Isoetella Durieui* (Bory) Genn. = *Isoetes Durieui* Bory.

Stylites Amstutz, Ann. Mo. Bot. Gard. 44: 121. 1957. Type: *Stylites andicola* Amstutz = *Isoetes andicola* (Amstutz) Gómez. *Isoetes* subgenus *Stylites* (Amstutz) Gómez, Brenesia 18: 4. 1980.

Description

Aquatic, palustral or terrestrial; stem erect or (in *I. tegetiformans*) shortly horizontally extended, rather small or to 6 cm in diameter, rarely branched; leaves ca. 2 cm to 1 m long, borne in a close spiral cluster, or distichous, narrowly ligulate to filiform, with one large megasporangium or microsporangium borne near the base on the adaxial side; megaspores spheroidal, trilete, the laesurae $\frac{3}{4}$ to equal the radius, the surface smooth, tuberculate, echinate, cristate, or reticulate, with an equatorial ridge, microspores ellipsoidal, monolete, the laesura $\frac{1}{4}$ to nearly equal the spore length, the surface spinose, papillate, denticulate-muricate, echinate, cristate, tuberculate, or smooth. Chromosome number: n = 10, 11, ca. 29, ca. 48, 54–56, 55 + 1; $2n$ = ca. 20, 22, 24–26, 44, 44 + 1, ca. 48, ca. 50, ca. 58, 66, ca. 100, 110; apogamous = 22 + 1, 33 + 1.

Fig. 127.1. *Isoetes Storkii,* in lake in Volcán Poas, Costa Rica. (Photo W. H. Hodge.)

Isoetes has many unusual features among the Pteridophyta. Among these are the dichotomously branched roots, the anchor-shaped stele, the presence of a cambium, and the largest sporangium known in vascular plants, which may be up to 15 mm long. The more or less triangular ligule (Fig. 6) is above the sporangium (Figs. 6, 7) and the labium, borne at the base of the ligule, may partly cover it. The sporangium is usually partly covered by the velum, a membranous outgrowth of the leaf.

Systematics

Isoetes is a widely distributed genus of about 150 or more species, some 50–60 of them American and of these about 35 occur south of the United States.

Previous attempts to provide an infrageneric classification on the basis of the ecology of the plants, the number of lobes on the stem, or on megaspore characters are clearly unsuccessful and, at the present time, there is no suitable basis for subdivisions of the genus. The development of an infrageneric classification based on evolutionary lines is one of the major problems in *Isoetes* taxonomy.

The study of Kubitzki and Borchert (1964) on *Isoetes triquetra* raised questions concerning the distinctiveness of the leaves and roots of *Stylites*. Diversity in the form of the stem, shown in *Isoetes tegetiformans* Rury (Rury, 1978), and observations of R. J. Hickey

Fig. 127.2. Distribution of *Isoetes* in America, south of lat. 35° N.

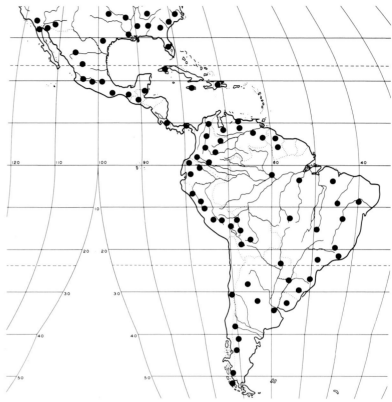

on recent collections from Bolivia enforce the conclusion that the characters of *Stylites* are within the range of variability of *Isoetes.*

The basic taxonomic revision of *Isoetes* by Pfeiffer (1922) is supplemented by studies of the South American species by Weber (1922, 1934) and by Pastore (1936) of the species of Argentina. Other works of special interest are the morphological studies of *Stylites* by Rauh and Falk (1959a, 1959b) and the anatomical study of the genus by Paolillo (1963).

The treatment of this genus has been prepared in close collaboration with R. James Hickey.

Tropical American Species

The species of the American tropics are poorly known because of inadequate collections which do not provide a basis for an understanding of the local or geographic variation. Recent intensive collecting and study, especially of Andean *Isoetes,* indicates that there are considerably more species than currently recognized.

Some of the species are: *Isoetes andicola* (Amstutz) Gómez, *I. andina* Hook., *I. cubana* Baker, *I. Ekmanii* Weber, *I. Killipii* Mort., *I. Lechleri* Mett., *I. novogranadensis* Fuchs, *I. ovata* Pfeiff., *I. Storkii* Palmer, and *I. triquetra* A. Br. The name *Isoetes andina* Hook. (Brit. Ferns *t. 56,* note. 1861) is adopted for the species of the northern Andes usually called *Isoetes triquetra,* a different species of southern Peru. Although Hooker said it "is probably distinct" (from *I. lacustris*), other comments indicate he accepted it as a new species.

3   4

5   6   7

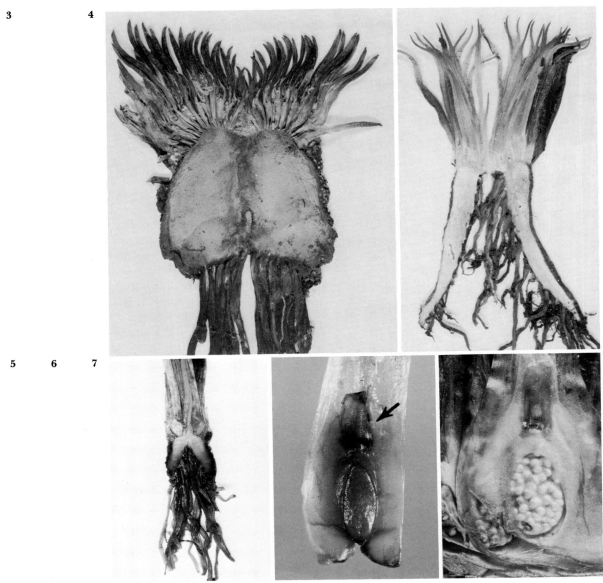

Figs. 127.3–127.7. *Isoetes.* **3.** Plant of *I. andina* in longisection, ×1. **4.** Plant of *I. andicola* in longisection, ×1. **5.** Longisection of stem and base of leaves of *I. Lechleri,* ×1. **6.** Microsporangium of *I. novogranadensis,* with the ligule above, at arrow, ×4. **7.** Megasporangium of *I. novogranadensis,* ×4.

Ecology (Fig. 1)

Isoetes grows in a wide variety of habitats where the soil is saturated with water for at least a portion of the year. It most frequently grows in lakes, ponds and rivers or in ephemeral pools.

In the American tropics species grow in lakes, usually in rather shallow water, in pools or in streams, in grassy turf at the border of lakes, in wet open sandy soil, and in alpine bogs, while one species, *Isoetes Ekmanii,* grows in saline tidal flats on the Isla Martín García, Argentina. *Isoetes* grows from sea level to 4200 m, most frequently above 2000 m.

Figs. 127.8–127.16. *Isoetes* megaspores, × 100; microspores, × 1000. **8–10.** *I. andina,* Colombia, *Cuatrecasas & Idrobo 27038.* **8.** Megaspore, lateral view, laesurae at top, proximal face coarsely papillate, distal face somewhat cristate. **9.** Surface detail of stranded silica deposition, × 500. **10.** Echinate microspore. **11–13.** *I. Lechleri,* Bolivia, *Hickey 758.* **11.** Megaspore, smooth proximal face and few microspores. **12.** Microspore, detail of fused outer reticulate surface, × 5000. **13.** Microspores with inner echinate formation at left and center, the outer reticulate formation, right. **14–16.** *Isoetes andicola,* Peru, *Saunders 1154.* **14.** Megaspore, proximal face, plane. **15.** Detail of finely granulate megaspore surface, partly removed on laesura, × 500. **16.** Granulate microspore.

Geography (Fig. 2)

Isoetes is a widely distributed genus, occurring in America, Europe and eastward through Asia to Kamchatka, in Africa and Madagascar, and India and China northward to Japan and southeastward to Sumatra, New Guinea, Australia and Tasmania and New Zealand. Except for its presence on New Zealand, it is absent from oceanic islands.

In America, *Isoetes* occurs from southern Greenland and Labrador, west to Northwest Territories, Canada, coastal Alaska and the Aleutian Islands, and southward to Mexico and Central America, the Greater Antilles, and in most of South America south to Magellanes in Chile.

Isoetes species often have broad to moderately broad ranges, but in some regions, for example, in New Guinea (Croft, 1980) and in southern Peru and Bolivia there is considerable local endemism.

American centers of species diversity are the southeastern United States, with nine species, the Andes of western Venezuela to Ecuador, with six or seven species, the Andes of Peru and Bolivia, with eight to ten species, and in Guanabara (Rio de Janeiro), Brazil, and adjacent states, with seven species.

Spores

The irregular surface of *Isoetes andina* megaspores (Fig. 8) show the morphological gradation between the proximal and distal faces. The fine rods and strands that form the outer wall (Fig. 9) apparently consist of silica which is heavily deposited on the outer part of the spore (Robert et al., 1973). The surface structure of *I. andina* is similar to that of several other American species as *I. Killipii* in which the surface elements are fused into reticulate forms. Microspores are more or less granulate (Fig. 16) to echinate (Fig. 10), in most species with little resemblance to the megaspore surface. The megaspores of *Isoetes andicola* are quite smooth with a granulate surface (Figs. 14, 15) and the microspores are also granulate (Fig. 16). *Isoetes Lechleri* megaspores are smooth but smaller (Fig. 11) than those of *I. andicola.* The microspores have an outer more compact surface formed over an echinate structure (Figs. 12, 13). The megaspores of *Isoetes cubana* are tuberculate with slender strands connecting the tubercules that are more or less covered with a granulate deposit (Fig. 18), and the microspores are coarsely echinate (Fig. 19). The clavate surface of *Isoetes ovata* megaspores is covered with

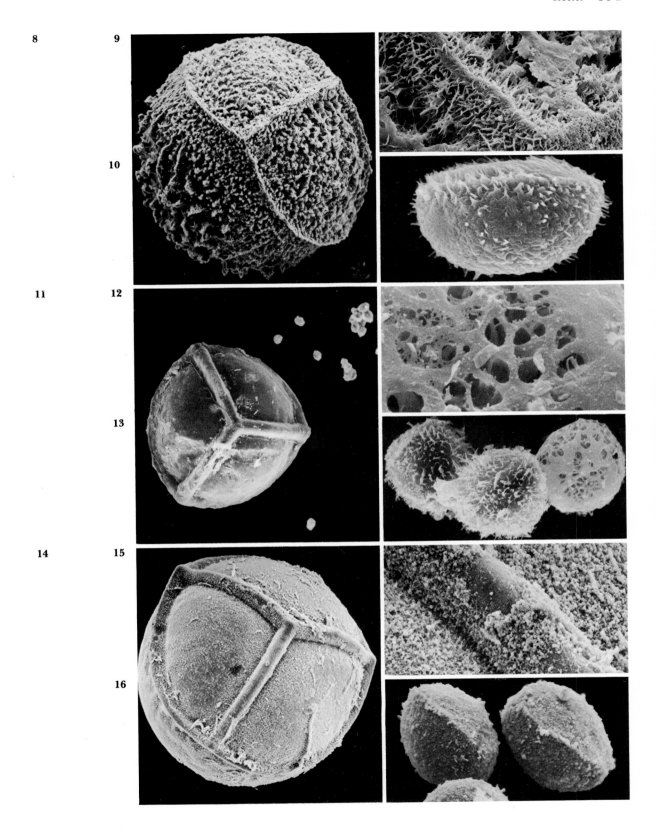

short echinate projections (Figs. 21, 23). A section of the wall (Fig. 21) shows this consists largely of compact granulate material above coarser laminated strata. The microspores of this species and *I. cubana* have coarse, echinate projections formed of compact, granulate particles (Figs. 19, 20).

Silica in the form of silica gel constitutes the outer part of the megaspore wall of *Isoetes setacea* Lam. and is deeply embedded in the exospore as shown in the SEM and infrared spectra analyses of Robert et al. (1973). The microspore walls of several species of *Isoetes* sectioned by Lugardon (1973, 1979) show that a special para-exospore layer of anastomosing plates is formed below the perispore. The abundant silica deposition and para-exospore formations are special features of *Isoetes* that are also characteristic of *Selaginella* spores. The tetrad associations of *Isoetes* spores are peculiar for the microspores are monolete and aligned in square tetrads (Berthet and Lecocq, 1977) in contrast to the trilete megaspores that are in a tetrahedron arrangement more characteristic of pteridophyte spores.

Cytology

The lowest chromosome number for *Isoetes* is $n = 10$ in *I. histrix* Bory reported for plants from Britain and Morocco (Manton, 1950), and the highest $2n = 110$ in *I. lacustris* L., from several localities in Europe, and in *I. macrospora* Dur., from several stations in northeastern North America. These last two species are considered to be dodecaploids. There are three diploids, $2n = 22$, in northeastern North America, *I. Engelmannii* A. Br., *I. echinospora* Dur., and *I. Eatonii* Dodge, and three tetraploids, $2n = 44$, *I. acadiensis* Kott, *I. riparia* A. Br., and *I. Tuckermanii* Engelm. (Kott and Britton, 1980).

In tropical America, *Isoetes andicola* var. *gemmifera* (Rauh) Gómez (*Stylites andicola* Rauh) has been reported as $2n = $ ca. 50 (48–52) by Rauh and Falk (1959a). Other photographically documented records by R. J. Hickey (personal communication) include, two diploids with $2n = 22$, *Isoetes alcalophila* Halloy from Argentina and *I. Storkii* from Costa Rica, and three tetraploids from Bolivia, with $2n = 44$, *I. glacialis* Aspl., *I. Lechleri*, and an undescribed species.

Although a wide range of chromosome numbers are reported from Europe, India, Africa and Japan, the recent American records are mostly $2n = 22$ or 44. *Isoetes* chromosomes are readily fragmented by pressure in squash preparations, and several of the deviating numbers may actually be based on $x = 11$.

Studies of *Isoetes cormandelina* L. f. in India by Ninan (1958) report the occurrence of an extra chromosome in both diploid apogamous plants with $22 + 1$ and apogamous triploids with $33 + 1$. The origin of apogamy was considered to be the result of accumulated structural hybridity. The karyotype analyses of *Isoetes asiatica* Makino by Tatuno (1963) showed marked differences in chromosome structure due to negative heteropycnotic regions. It was shown that the complement of 22 chromosomes

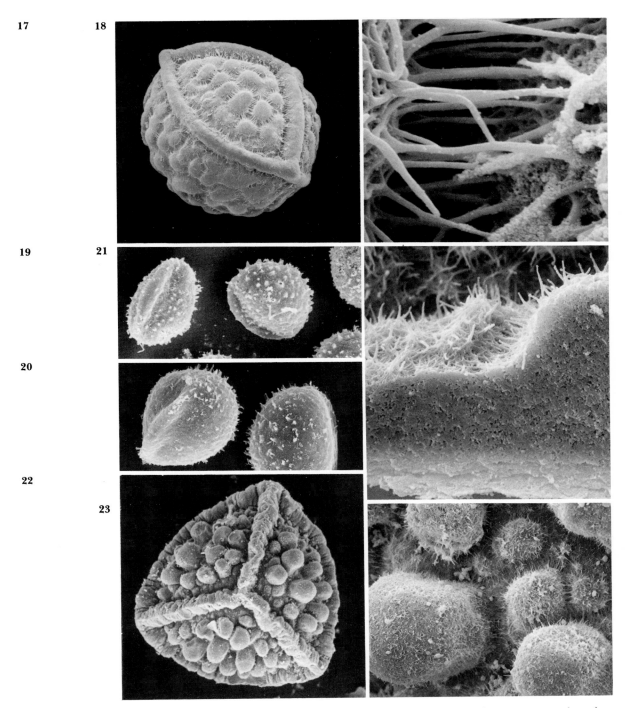

Figs. 127.17–127.23. *Isoetes* spores. **17–19.** *L. cubana, Wright 3912.* **17.** Megaspore, tuberculate, proximal face, above, × 150. **18.** Detail of silica strands and granulate deposit, × 5000. **19.** Microspores, echinate, × 1000. **20–23.** *I. ovata,* British Guiana, *Maguire & Fanshaw 23569.* **20.** Microspores, echinate, laesura below on left spore, × 1000. **21.** Section of mature megaspore wall and surface detail, × 2000. **22.** Clavate megaspore, proximal face, × 100. **23.** Detail of clavate projections and silica surface strands, × 500.

in the species is derived from an initial set of six, consisting of two sets of three kinds of chromosomes. This involves loss of a small pair of chromosomes after doubling of the initial set.

Literature

Berthet, R., and M. Lecocq. 1977. Morphologie sporale des especes Francaise du genre *Isoetes* L. Pollen et Spores 19: 341–359.

Croft, J. R. 1980. A taxonomic revision of *Isoetes* L. (Isoetaceae) in Papuasia. Blumea 26: 177–190.

Kott, L. and D. M. Britton. 1980. Chromosome numbers for *Isoetes* in northeastern North America. Canad. Jour. Bot. 58: 980–984.

Kubitzki, K., and R. Borchert. 1964. Morphologische Studien an *Isoetes triquetra* A. Braun und Bemerkungen über das Verhältnis der Gattung *Stylites* E. Amstutz zur Gattung *Isoetes*. Ber. Deutsch. Bot. Ges. 77: 227–235.

Lugardon, B. 1973. Nomenclature et structure fine des parois aceto-resistantes des microspores d'*Isoetes*. C. R. Acad. Sci. Paris, 276: 3017–3020.

Lugardon, B. 1979. Sur la structure fine de l'exospore dans les divers groupes de Ptéridophytes actuelles (microspores et megaspores), *in*: I. K. Ferguson and J. Muller, The evolutionary significance of the exine, pp. 231–250 Linn. Soc. Sympos. Ser. 1.

Manton, I. 1950. Problems of Cytology and Evolution in the Pteridophyta. 316 pp. Cambridge Univ. Press, Cambridge, England.

Ninan, C. A. 1958. Studies on the cytology and phylogeny of the Pteridophytes. Jour. Ind. Bot. Soc. 37: 93–103.

Paolillo, D. J., Jr. 1963. The developmental anatomy of *Isoetes*. Illinois Biol. Monog. 31: 1–130.

Pastore, A. I. 1936. Las Isoetaceas Argentinas. Rev. Museo La Plata. n.s. Sec. Bot. 1: 3–30.

Pfeiffer, N. E. 1922. Monograph of the Isoetaceae. Ann. Mo. Bot. Gard. 9: 79–232.

Rauh, W., and H. Falk. 1959a. *Stylites* E. Amstutz, eine neue Isoetacee aus den Hochanden Perus, 1. Heidelberg. Akad. Wissens. Mathnatur. Kl. 1959 (1): 3–83.

Rauh, W., and H. Falk. 1959b. *Stylites* E. Amstutz, eine neue Isoetacee aus den Hochanden Perus, 2. Heidelberg. Akad. Wissens. Math-Natur. Kl. 1959 (2): 87–160.

Robert, D., F. Roland-Heydacker, J. Denizot, J. Laroche, P. Fougeroux, and L. Davignon. 1973. La paroi megasporale de l'*Isoetes setacea* Bosc. ex Delile, Etudes en microscopies photonique et electroniques. Localization et nature de la silice entrant dans sa constitution. Adansonia 13: 313–332.

Rury, P. M. 1978. A new and unique, mat-forming Merlin's-grass (*Isoetes*) from Georgia. Amer. Fern Jour. 68: 99–108.

Tatuno, S. 1963. Zytologische untersuchungen der Pteridophyten 1. Chromosomen von *Isoetes asiatica* Makino. Cytologia 28: 293–304.

Weber, U. 1922. Zur Anatomie and Systematik der Gattung *Isoetes* L. Hedwigia 63: 219–262.

Weber, U. 1934. Neue südamerikanische *Isoetes*-Arten. Ber. Deutsch. Bot. Ges. 52: 121–125.

Index to Names

The page number indicates the primary or sole place where the name is mentioned. Synonyms are in italic.